# Handbook of Precision Agriculture

## *Principles and Applications*

# *FOOD PRODUCTS PRESS®*
## *Crop Science*
## Amarjit S. Basra, PhD
## Editor in Chief

*The Lowland Maya Area: Three Millennia at the Human-Wildland Interface* edited by A. Gómez-Pompa, M. F. Allen, S. Fedick, and J. J. Jiménez-Osornio

*Biodiversity and Pest Management in Agroecosystems, Second Edition* by Miguel A. Altieri and Clara I. Nicholls

*Plant-Derived Antimycotics: Current Trends and Future Prospects* edited by Mahendra Rai and Donatella Mares

*Concise Encyclopedia of Temperate Tree Fruit* edited by Tara Auxt Baugher and Suman Singha

*Landscape Agroecology* by Paul A. Wojtkowski

*Concise Encyclopedia of Plant Pathology* by P. Vidhyasekaran

*Molecular Genetics and Breeding of Forest Trees* edited by Sandeep Kumar and Matthias Fladung

*Testing of Genetically Modified Organisms in Foods* edited by Farid E. Ahmed

*Fungal Disease Resistance in Plants: Biochemistry, Molecular Biology, and Genetic Engineering* edited by Zamir K. Punja

*Plant Functional Genomics* edited by Dario Leister

*Immunology in Plant Health and Its Impact on Food Safety* by P. Narayanasamy

*Abiotic Stresses: Plant Resistance Through Breeding and Molecular Approaches* edited by M. Ashraf and P. J. C. Harris

*Teaching in the Sciences: Learner-Centered Approaches* edited by Catherine McLoughlin and Acram Taji

*Handbook of Industrial Crops* edited by V. L. Chopra and K. V. Peter

*Durum Wheat Breeding: Current Approaches and Future Strategies* edited by Conxita Royo, Miloudi M. Nachit, Natale Di Fonzo, José Luis Araus, Wolfgang H. Pfeiffer, and Gustavo A. Slafer

*Handbook of Statistics for Teaching and Research in Plant and Crop Science* by Usha Rani Palaniswamy and Kodiveri Muniyappa Palaniswamy

*Handbook of Microbial Fertilizers* edited by M. K. Rai

*Eating and Healing: Traditional Food As Medicine* edited by Andrea Pieroni and Lisa Leimar Price

*Physiology of Crop Production* by N. K. Fageria, V. C. Baligar, and R. B. Clark

*Plant Conservation Genetics* edited by Robert J. Henry

*Introduction to Fruit Crops* by Mark Rieger

*Generations Gardening Together: Sourcebook for Intergenerational Therapeutic Horticulture* by Jean M. Larson and Mary Hockenberry Meyer

*Agriculture Sustainability: Principles, Processes, and Prospects* by Saroja Raman

*Introduction to Agroecology: Principles and Practice* by Paul A. Wojtkowski

*Handbook of Molecular Technologies in Crop Disease Management* by P. Vidhyasekaran

*Handbook of Precision Agriculture: Principles and Applications* edited by Ancha Srinivasan

*Dictionary of Plant Tissue Culture* by Alan C. Cassells and Peter B. Gahan

*Handbook of Potato Production, Improvement, and Postharvest Management* edited by Jai Gopal and S. M. Paul Khurana

*Carbon Sequestration in Soils of Latin America* edited by Rattan Lal, Carlos C. Cerri, Martial Bernoux, Jorge Etchevers, and Eduardo Cerri

# Handbook of Precision Agriculture
## *Principles and Applications*

Ancha Srinivasan, PhD
Editor

CRC Press
Taylor & Francis Group
Boca Raton London New York

CRC Press is an imprint of the
Taylor & Francis Group, an **informa** business

First published 2006 by Food Products Press.

Published 2022 by CRC Press
Taylor & Francis Group
6000 Broken Sound Parkway NW, Suite 300
Boca Raton, FL 33487-2742



ISBN-13: 978-1-56022-954-4 (hbk)
ISBN-13: 978-1-56022-955-1 (pbk)



**Visit the Taylor & Francis Web site at**
**http://www.taylorandfrancis.com**

**and the CRC Press Web site at**
**http://www.crcpress.com**

**Library of Congress Cataloging-in-Publication Data**

Handbook of precision agriculture : principles and applications / Ancha Srinivasan, editor.
p. cm.
Includes bibliographical references and index.
ISBN-13: 978-1-56022-954-4 (hard : alk. paper)
ISBN-10: 1-56022-954-3 (hard : alk. paper)
ISBN-13: 978-1-56022-955-1 (soft : alk. paper)
ISBN-10: 1-56022-955-1 (soft : alk. paper)
1. Precision farming. I. Srinivasan, Ancha.
S494.5.P73H36 2006
631—dc22

2005032077

Cover design by Marylouise E. Doyle.

# Handbook of Precision Agriculture
## *Principles and Applications*

Ancha Srinivasan, PhD
Editor

CRC Press
Taylor & Francis Group
Boca Raton London New York

CRC Press is an imprint of the
Taylor & Francis Group, an **informa** business

First published 2006 by Food Products Press.

Published 2022 by CRC Press
Taylor & Francis Group
6000 Broken Sound Parkway NW, Suite 300
Boca Raton, FL 33487-2742



ISBN-13: 978-1-56022-954-4 (hbk)
ISBN-13: 978-1-56022-955-1 (pbk)



**Visit the Taylor & Francis Web site at**
**http://www.taylorandfrancis.com**

**and the CRC Press Web site at**
**http://www.crcpress.com**

**Library of Congress Cataloging-in-Publication Data**

Handbook of precision agriculture : principles and applications / Ancha Srinivasan, editor.
p. cm.
Includes bibliographical references and index.
ISBN-13: 978-1-56022-954-4 (hard : alk. paper)
ISBN-10: 1-56022-954-3 (hard : alk. paper)
ISBN-13: 978-1-56022-955-1 (soft : alk. paper)
ISBN-10: 1-56022-955-1 (soft : alk. paper)
1. Precision farming. I. Srinivasan, Ancha.
S494.5.P73H36 2006
631—dc22

2005032077

Cover design by Marylouise E. Doyle.

## *CONTENTS*

## ABOUT THE EDITOR

**Ancha Srinivasan, PhD,** is a Principal Researcher and Manager at the Institute for Global Environmental Strategies (IGES) in Hayama, Japan. Prior to joining IGES, he worked at premier institutions such as the International Crops Research Institute for the Semi-Arid Topics, the Hokkaido National Agricultural Experiment Station, and the Japan International Research Center for Agricultural Sciences (JIRCAS). Dr. Srinivasan served as a Principal Investigator for several funded projects in sustainable agriculture, information systems, and the *environment*, and recently developed a special interest in policy issues of adaptation to climate change in agriculture. His work has been widely published in national and international journals, and he serves on the editorial board of the *Journal of Crop Improvement* (Haworth).

*Handbook of Precision Agriculture*

doi:10.1300/5627_a

# Contributors

**Matthew L. Adams,** CSBP Futurefarm, Subiaco, Western Australia, Australia.

**Richard E. Baird,** Entomology and Plant Pathology Department, Mississippi State University, Mississippi State, Mississippi.

**P. M. Basson,** AIDS Centre, Kovsie Health, University of the Free State, Bloemfontein, South Africa.

**Karsten D. Bjerre,** Department of Plant Biology, Royal Veterinary and Agricultural University, 40 Thorvaldsensvej, Frederiksberg, Denmark.

**Simon Blackmore,** Unibots Ltd. Bedfordshire, United Kingdom.

**Rodolfo Bongiovanni,** National Institute for Agricultural Technology (INTA), Córdoba, Argentina.

**Judith M. Bradow,** U.S. Department of Agriculture, Agricultural Research Service, Southern Regional Research Center, New Orleans, Louisiana.

**Robert G. V. Bramley,** CSIRO Land and Water and Cooperative Research Centre for Viticulture, Glen Osmond, Southern Australia, Australia.

**Carl R. Camp,** U.S. Department of Agriculture, Agricultural Research Service, Florence, South Carolina.

**Greg Carlson,** professor of soil physics, South Dakota State University, Brookings, South Dakota.

**Svend Christensen,** Danish Institute of Agricultural Sciences, Research Centre Bygholm, Horsens, Denmark.

**David E. Clay,** South Dakota State University, Brookings, South Dakota.

**Sharon A. Clay,** South Dakota State University, Brookings, South Dakota.

**Thomas S. Colvin,** Oxford Farms, Cambridge, Iowa.

*Handbook of Precision Agriculture*

doi:10.1300/5627_b

**Simon E. Cook,** Centro Internacional de Agricultura Tropical, Cali, Colombia.

**Josse De Baerdemaeker,** Department of Agro-Engineering and Economics, Leuven, Heverlee, Belgium.

**Robert A. Dunn,** Entomology and Nematology Department, University of Florida, Gainesville, Florida.

**Robert G. Evans,** U.S. Department of Agriculture, Agricultural Research Service, Northern Plains Agricultural Research Laboratory, Sidney, Montana.

**Daniele Focht,** consultant, Insolo Solucoes Agricolas, Parana, Brazil.

**Spyros Fountas,** University of Thessaly, Magnissias, Greece.

**David W. Franzen,** North Dakota State University, Fargo, North Dakota.

**Roland Gerhards,** University of Hohenheim, Stuttgart, Germany.

**Maarten Geypens,** Catholic University of Leuven, Heverlee, Belgium.

**Hans W. Griepentrog,** Royal Veterinary and Agricultural University, Taastrup, Denmark.

**Silvia Haneklaus,** Institute of Plant Nutrition and Soil Science (FAL), Bundesallee, Braunschweig, Germany.

**Lee F. Johnson,** Institute of Earth System Science and Policy, California State University Monterey Bay and, Earth Science Division, NASA/Ames Research Center, Moffett Field, California.

**Richard M. Johnson,** U.S. Department of Agriculture, Agricultural Research Service, Sugarcane Research Unit, Houma, Louisiana.

**Lise N. Jørgensen,** Danish Institute of Agricultural Sciences, Department of Crop Protection, Research Centre Flakkebjerg, Slagelse, Denmark.

**Jess Lowenberg-DeBoer,** Department of Agricultural Economics, Purdue University, West Lafayette, Indiana.

**Ntsikane Maine,** Department of Agricultural Economics, University of the Free State, Bloemfontein, South Africa.

The late **Colin McKenzie,** formerly with Alberta Agriculture Food and Rural Development, Crop Diversification Center South, Brooks, Alberta, Canada.

**W. T. (Wimpie) Nell,** Department of Agricultural Economics, University of the Free State, Bloemfontein, South Africa.

**Ramakrishna R. Nemani,** School of Forestry, University of Montana, Missoula, Montana.

**Jørgen E. Olesen,** Danish Institute of Agricultural Sciences, Tjele, Denmark.

**Søren M. Pedersen,** Danish Research Institute of Food Economics, Frederiksberg, Denmark.

**G. Stuart Pettygrove,** Department of Land, Air, and Water Resources, University of California, Davis, California.

**Richard E. Plant,** Departments of Agronomy and Range Science and Biological and Agricultural Engineering, University of California, Davis, California.

**Piet Reyns,** LVH Rilland Centre of Expertise, Rilland, The Netherlands.

**Jimmy R. Rich,** University of Florida, Quincy, Florida.

**Alvaro Roel,** Instituto Nacional de Investigaciones Agropecuarias, Treinta y Tres, Uruguay.

**Glaucio Roloff,** Department de Solos e Engenharia Agricola, Universidade Federal do Parana, Curitibas Parana, Brazil.

**E. John Sadler,** U.S. Department of Agriculture, Agricultural Research Service, Coastal Plains Soil, Water, and Plant Research Center, Florence, South Carolina.

**Ewald Schnug,** Institute of Plant Nutrition and Soil Science (FAL), Bundesallee, Braunschweig, Germany.

**Sakae Shibusawa,** Faculty of Agriculture, Tokyo University of Agriculture and Technology, Fuchu, Tokyo, Japan.

**John V. Stafford,** Silsoe Solutions, Ampthill, United Kingdom.

**Ludo Vanongeval,** AMINAL Directoraat-Generaal, Brussels, Belgium.

**Brett M. Whelan,** Australian Centre for Precision Agriculture, McMillan Building, the University of Sydney, New South Wales, Australia.

**Michael A. White,** Department of Geography and Earth Resources, Utah State University, Logan, Utah.

**Shelley A. Woods,** Alberta Agriculture Food and Rural Development, Agriculture Center, Lethbridge, Alberta, Canada.

**Anne F. Wrona,** National Cotton Council, Memphis, Tennessee.

# Foreword

I am happy that a book on precision agriculture edited by Dr. Ancha Srinivasan is being published by The Haworth Press after the FAO World Food Summit +5 meeting in Rome and the World Summit on Sustainable Development (WSSD) organized by the United Nations at Johannesburg. WSSD identified agricultural progress as the best safety net against hunger and deprivation in most developing countries. WSSD placed sustainable agricultural progress and poverty eradication on the top of global political agenda.

The world is witnessing two cultures of agriculture. On the one hand, farming in industrialized countries is becoming essentially agri-business, supported by heavy inputs of technology, capital, and subsidy. Farming in most developing countries remains a way of life and the primary means to livelihood. "Mass production" technologies have the advantage of scale, in contrast to the "production by masses" situation prevailing in developing countries.

The smaller the farm, the greater is the need for marketable surplus. The ultimate goal of any movement for the technological upgrading of agriculture should be happy farming families and sustainable farming systems. The green revolution of the 1960s in crops such as wheat and rice helped to enhance productivity per unit of land. It had some potential ecological problems, which I highlighted in the following remarks I made at the Indian Science Congress Session held at Varanasi in January 1968 (i.e., before the term *green revolution* was coined by Dr. William Gaud of the United States in September 1968).

> Exploitative agriculture offers great dangers if carried out with only an immediate profit or production motive. The emerging exploitative farming community in India should become aware of this. Intensive cultivation of land without conservation of soil fertility and soil struc-

*Handbook of Precision Agriculture*

doi:10.1300/5627_c

> ture would lead, ultimately, to the springing up of deserts. Irrigation without arrangements for drainage would result in soils getting alkaline or saline. Indiscriminate use of pesticides, fungicides, and herbicides could cause adverse changes in biological balance as well as lead to an increase in the incidence of cancer and other diseases, through the toxic residues present in the grains or other edible parts. Unscientific tapping of underground water will lead to the rapid exhaustion of this wonderful capital resource left to us through ages of natural farming. The rapid replacement of numerous locally adapted varieties with one or two high-yielding strains in large contiguous areas would result in the spread of serious diseases capable of wiping out entire crops, as happened prior to the Irish potato famine of 1854 and the Bengal rice famine in 1942. Therefore the initiation of exploitative agriculture without a proper understanding of the various consequences of every one of the changes introduced into traditional agriculture, and without first building up a proper scientific and training base to sustain it, may only lead us, in the long run, into an era of agricultural disaster rather than one of agricultural prosperity.

In order to overcome the ecological and social problems associated with green revolution technologies, I recommended a paradigm shift from a commodity-centered to an integrated natural resources management-centered production methodology. I referred to the latter as an "ever-green revolution," or advances in productivity in perpetuity. In a recent book, Dr. E. O. Wilson referred to this concept of an ever-green revolution in the following words.

> The problem before us is how to feed billions of new mouths over the next several decades and save the rest of life at the same time, without being trapped in a Faustian Bargain that threatens freedom and security. No one knows the exact solution to this dilemma. *The benefit must come from an evergreen revolution.* The aim of this new thrust is to lift food production well above the level obtained by the green revolution of the 1960s, using technology and regulatory policy more advanced and even safer than those now in existence.

Reducing the cost of production without reducing yield is the need of the hour. Precision agriculture helps to harness the best in science and technology for achieving cost reduction coupled with productivity enhancement. The various components of precision agriculture are dealt with by eminent scientists in this book in an authoritative manner.

Using the knowledge and information contained in this book, agricultural scientists can design along with farming families, a precision farming methodology for each cropping system in a location specific manner. If this is done, precision agriculture will help to improve the productivity, profitability and sustainability of the major farming systems in the world.

I hope this book will not only be widely read but also used for stimulating a precision farming revolution in the world.

*M. S. Swaminathan, Chairman*
*M.S. Swaminathan Research Foundation*
*Center for Research on Sustainable*
*and Rural Development*
*Chennai, India*

# Preface

Since the early 1990s, the world has witnessed the growth of a body of knowledge on a new concept of farming known as precision agriculture, which could be attributed largely due to market and societal pressures such as global trade, environmental concerns, and changing consumer preferences. While farming itself can never be as precise as it should be, the goal is largely to symbolize human efforts toward perfectionism in the world's oldest profession. Precision agriculture builds upon the use of several technologies and is backed by the latest advances not only in traditional areas of agronomic science and technology but also in fields other than agriculture. Precision agriculture is a comprehensive and innovative approach to solve problems of the green-revolution agriculture, where input application does not consider the concept of spatial and temporal variability within the field adequately. When applied correctly, precision agriculture has the potential to fundamentally alter decision making in farming and to simultaneously achieve the multiple objectives of enhancing input use efficiency, reducing environmental pollution, and increasing farm profits and product quality. Research in the past 15 years has shown that it holds promise for enhanced agricultural profitability. An indication of the growing interest in precision agriculture can be ascertained from the fact that seven international, four European and one Asian conference were held on this topic by July 2005.

*Handbook of Precision Agriculture: Principles and Applications* is an attempt to synthesize the latest knowledge in this area of scientific endeavor. The emphasis here is more on principles and practice rather than the detailed accounts of progress in its adoption in each country or region. Several international experts contributed their most up-to-date research in accessible format, particularly for graduate students, researchers, and members of the interested general public.

It is a pleasure and honor to acknowledge all those colleagues, who offered help and constructive criticism throughout the evolution of this book.

*Handbook of Precision Agriculture*

doi:10.1300/5627_d

I am especially grateful to Dr. Amarjit Singh Basra, the managing editor of Food Products Press, for inviting me to edit this book and for encouraging me to proceed with its completion whenever I faced setbacks. Without his instrumental guidance and unwavering support, the book would never have materialized. I wish to thank all 51 contributors for their excellent contributions and patience. As many as 125 reviewers (listed separately) from various parts of the world provided helpful comments in revising various chapters. For the final oversight of the manuscript editing and the proofs, I acknowledge the kind assistance provided by various Haworth Press staff including Peg Marr and others.

It is my hope that the book will provide the readers with a useful insight into theory and practice of precision agriculture, promote exchange of knowledge and increase the quality of research and applications in precision agriculture worldwide.

# Acknowledgments

The editor would like to extend thanks to all individuals listed here, who have offered to review various chapters of this book. They are an essential part of publication process of this book and I deeply appreciate their help and gracious efforts:

Chris O. Andrew, Lanny Ashlock, H Auernhammer, Richard Baird, Vethaiya Balasubramanian, Jos Balendonck, Leonard L. Bashford, Graeme Batten, Stuart Birrell, Simon Blackmore, Greg Blumhoff, Jim Bordovsky, Johan Bouma, Mario Bragachini, Rob Bramley, Kevin F. Bronson, Roger Brook, David Bruce, Don Bullock, Christopher Choi, Rex Clark, Graeme Cox, Charles Crissman, Bill Day, Markus Demmel, Richard Earl, Steve Ebelhar, Ken Evans, Eli Feinerman, Johann Frahm, Aida Frese, Lance Gibson, Joseph Giles, Tom Goddard, Dick Godwin, Daan Goense, Carroll E. Goering, Mike Gooding, Cary J. Green, Mark D. Greenspan, Ardell Halvorson, Shufeng Han, Harry Harris, Anton Haverkort, Dale Heermann, Christo Helm, Georges Hofman, Teresa Hopping, Daniel Humburg, John Hummel, Liz Humphreys, Keith Jaggard, Dan Jaynes, Chris J. Johannsen, Rob Kelly, Newell Kitchen, James Kotcon, Len Kryzanowski, J.K. Ladha, David Lamb, Robert Lascano, Johann Laubscher, Joe Lauer, Guido Lemoine, Charvat Lesprojekt, Holger Lilienthal, Dan Long, Roger Longhorn, James Lowenberg-DeBoer, Drew J. Lyon, Monika MacDevette, Nik Nasrudin Mahmud, Robin Matthews, Alex McBratney, Jonathan Medway, Haddish Melakeberhan, Graciela Metternicht, David Molden, Mark Morgan, David J. Mulla, Dan Munk, Lloyd Murdock, Richard Nelson, Chris Nussey, Ivan O'Halloran, Calvin Perry, Lars Pierce, Fran Pierce, Richard E. Plant, Aziz Rahman, Andres Ravelo, Alejandro Reca, Harold Reetz, Jim Reeves, Sandra Richardson, John Sadler, John K. Saichuk, Sally Schneider, Ewald Schnug, Mark Schrock, Stephen W. Searcy, Eduardo Segarra, Sakae Shibusawa, Clinton Shock, Nathan Slaton, Steve Sonka, Azhar Stapa, Timothy S. Stombaugh, Joe Street, Kenneth A. Sudduth, Randal K. Taylor, Lars Thylén, Allen Torbert, Mikio Umeda, Bernard Vigier, Paul Voroney, Regis Voss, James Walcott, Jeremy Wallace, Robert Wample, Dwayne Westfall, Brett Whelan, Keith Whigham, Brian Wienhold, and Chris Williams.

*Handbook of Precision Agriculture*

doi:10.1300/5627_e

# *PART I: PRINCIPLES, TECHNOLOGIES, AND MANAGEMENT ISSUES*

## Chapter 1

# Precision Agriculture: An Overview

Ancha Srinivasan

### *INTRODUCTION*

Agriculture dominates the world's land use decisions. The urgent need for doubling farm production over the next 25 years on less land with less water through further intensification would inevitably involve substantial social, economic, and environmental costs. Identification of tools to minimize such costs through enhanced productivity and economic profits while simultaneously conserving the environment is, therefore, crucial. Precision agriculture (PA) is one of such tools catching worldwide attention since the early 1990s. Research interest in PA grew so rapidly that by July 2005 as many as seven international conferences in the United States and five European conferences in the United Kingdom, Denmark, France, Germany, and Sweden respectively, and one Asian conference in Malaysia were held. Further, an international journal *Precision Agriculture* is published by Springer Netherlands (formerly by Kluwer Academic Publishers).

PA is defined and termed in many ways, but the underlying concept remains the same. It is known as precision farming, information-intensive agriculture, prescription farming, target farming, site specific crop management, variable rate management, variable rate technology (VRT), farming by soil, grid soil sampling agriculture, grid farming, Global Positioning Systems (GPS) agriculture, farming by the inch, farming by the foot, and so on. In simple terms, PA can be defined as a holistic and environmentally friendly strategy in which farmers can vary input use and cultivation methods—including application of seeds, fertilizers, pesticides, and water, variety selection, planting, tillage, harvesting—to match varying soil and crop conditions across a field. A few other definitions of PA are given as follows.

- Careful tailoring of soil and crop management to fit the different conditions found in each field (Johannsen, 1996).

*Handbook of Precision Agriculture*

doi:10.1300/5627_01

- Information gathering, management planning, and field operations that improve the understanding and management of soil and landscape resources so the cropping inputs of management practices are utilized more efficiently than with conventional "one-fits-all" strategies (Kitchen et al., 1996).
- Farmers use GPS technology involving satellites and sensors on the ground and intensive information management tools to understand variations in resource conditions within fields. They use this information to more precisely apply fertilizers and other inputs and to more accurately predict crop yields (The U.S. House Committee on Agriculture, 2000).
- Application of technologies and principles to manage spatial and temporal variability associated with all aspects of agricultural production (Pierce and Nowak, 1999). In other words, it is nothing but "doing the right thing, at the right time, in the right place, in the right way." It applies to virtually every aspect of agriculture, from planting to harvest.

These definitions suggest that there are at least three elements critical to the success of PA: information, technology, and management.

## *BASICS OF PRECISION AGRICULTURE*

It has been long recognized that crops and soils within a field and/or region are both spatially and temporally variable. Growers tried to manage such variability to a limited extent mainly by intuition. As early as the 1620s, colonists observed site-specific practices of Indian farmers, who placed fish directly at the roots of each crop plant (National Research Council, 1997). Developments in geo-spatial information and communication technologies especially in the late twentieth century have made it possible, however, to manage such variability much more precisely than before.

PA, therefore, differs from conventional farming as it involves determining variation more precisely and linking spatial relationships to management actions, thereby allowing farmers to look at their farms, crops and practices from an entirely new perspective, thereby leading to

1. reduction in costs,
2. optimization of yields and quality in relation to the productive capacity of each site,
3. better management of the resource base, and

4. protection of the environment. PA also provides a framework of information with which farmers can make rational management decisions (Srinivasan, 2001a,b). In the future, PA may even enable us to trace farm products to their genetics and environmental conditions thereby providing a significant degree of control over food quality and safety.

Field adoption of PA may be represented as a five-step cyclical process including data collection, diagnostics, data analysis, precision field operations, and evaluation (Figure 1.1). Tools and techniques to accomplish them are discussed later. The evaluation of economic profitability, safety, and environmental impacts of precision field operations becomes a part of the data collection process for the next season. The PA system can be considered, therefore, the agricultural system of the twenty-first century, as it symbolizes a better balance between reliance on traditional knowledge and information- and management-intensive technologies. While the approach of PA is information-intensive and encompasses several technologies, application of its principles per se requires neither large scale operation nor heavy equipment. Because of its focus on technology use, some people consider PA as nothing but using information technologies to turn data into decisions. It is worth noting, however, that there are several opportunities and ways in which PA concepts and principles can be applied in diverse agronomic settings. Thus PA must be viewed more of a management approach rather than a prescriptive technology.

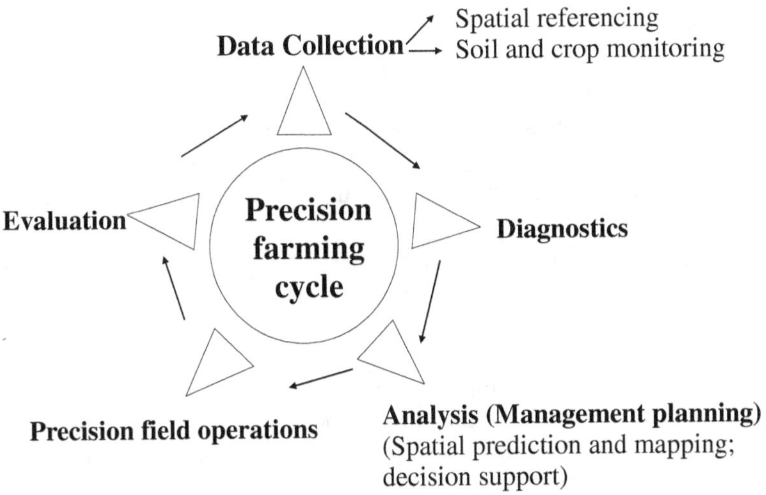


FIGURE 1.1. Precision farming cycle and its components.

***Precision Agriculture and Sustainability***

Precision agriculture promises to revolutionize farming as it offers a variety of benefits in profitability, productivity, sustainability, crop quality, environmental protection, on-farm quality of life, food safety, and rural economic development (National Research Council, 1997). In the short term, the diagnostic and database-building benefits are considerable. Growers can predict and correct problems such as water and nutrient stresses, diseases, and pests more efficiently. Database-building benefits will be in the form of accurate farm record keeping of inputs, property, machinery and labor, and efficient monitoring of environmental quality through recording the amounts and location of input applications. In the long term, farmers can optimize yields through finding locations that produce maximum profit margins, and through avoiding introduction of inappropriate cropping systems. Knowledge on crop response enables more precise targeting of external inputs, information on spatial and temporal response of varieties may help in selecting right genotypes and observations of consistent poor performance of a cropping system may encourage the withdrawal of such systems.

On a regional level too, spatial and temporal characterization of farms can help identify emerging trends in factors that impact sustainability, either in geographically adjoining regions or in specific agro-ecological zones, and to target possible solutions. Building geo-referenced databases for each field in a region will help identify causes and factors that underlie productivity features in each cropping system, and provide a synthesis within and among fields of the extent and severity of factors that constrain or enhance sustainability. For instance, data gathered in PA systems can help in identifying areas with high pollution potential. Adoption of PA on a large scale increases opportunities for skilled employment in farming, and provides new tools for evaluating multi-functional character of agriculture.

***What Is Critical for Precision Agriculture?***

At least three criteria are required for implementation of PA:

1. Clear evidence of significant spatial and temporal variability in soil and crop conditions within a field and in fields within a region.
2. Ability to identify and quantify such variability. It is possible to identify variation even with conventional techniques (e.g., based on grower intuition) but positioning and information technologies are of great help in quantifying such variation. Indeed successful adoption of PA

indeed depends upon blending the best of farmer's knowledge on field variability with information and communication technologies.

3. Ability to reallocate inputs and adjust management practices to improve productivity and profitability while minimizing environmental degradation. For instance, if data on precise weed distribution is available, spot- or patch- spraying of herbicides can be done instead of whole-field treatments.

The development of cost-effective techniques to discern such information, to develop site-specific or field-specific recommendations and data analysis tools, and to determine the appropriate scales of analysis and measurement are, therefore, prerequisites for PA.

It is important to note that PA principles and technologies are applicable in a variety of conditions irrespective of cropping system (mono- and multiple cropping), farm size (small and large), farming method (organic and chemical farming), crop type (field crops, horticultural crops, genetically engineered [GE] and non-GE crops), method of water management (rainfed and irrigated), and input levels in a system. The principles are applicable in livestock industry and forestry too. The driving forces for adoption of PA include economic pressures, environmental concerns and social/demographic changes, and vary with each context hence the method of implementation too varies. The overall profitability from adopting PA, however, varies. For example, although small farms are conducive to intensive agriculture and PA, small farmers are severely handicapped in using modern technologies because of restrictions on capital, market, and human skills. Indeed in many small farms all over the world, farmers have been applying fairly simple examples of PA for many years, such as adding manure or straw to hilltops or saline areas. Blackmore (1999) listed some examples to show how PA principles can be applied in a developing country context (e.g., tea in Tanzania and Sri Lanka; dates in Iran and Saudi Arabia) without incurring excessive costs.

Applying PA in cropping systems of one area often can have positive impact on other areas. For example, simple field- or site-specific measures (e.g., reduced tillage, input reallocation in fields) aimed at minimizing erosion and nutrient runoff of upland watersheds can have dramatic effects on sustainability of systems in lowlands.

It is important to note PA is not necessarily appropriate for every situation. For example, grain yield maps and soil fertility maps can help us determine and then vary optimal rate of fertilizers for each management unit. However, many pest and disease problems are more dependent on short-term weather patterns and do not follow consistent patterns across a field. Indeed for some pests, it may be best to always spray the entire field because even a few survivors could rapidly multiply, spread to the rest of the field and continue to cause problems. In such cases, precision spraying of pesticides is of little use. Insect and disease patterns across a field also can have a large temporal variation because of their temperature-dependence. In such cases, timing of pesticide application is more important than variable application. In contrast, perennial weeds such as Canada thistle, velvet leaf, and quackgrass grow in the same patches year after year. Application of precision spraying techniques in such cases would be more cost-effective than blanket applications.

## *TOOLS FOR IMPLEMENTATION OF PRECISION AGRICULTURE*

The role of various technologies in the evolution of PA is discussed in detail in Chapter 2, hence only an outline is given here. To achieve the ultimate goal of optimizing productivity and profitability in each unit of land, three basic requirements must be met:

1. Ability to identify each field location,
2. Ability to collect, interpret, and analyze data at an appropriate scale and frequency,
3. Ability to adjust input use and farm practices to maximize benefits from each location.

Global Positioning Systems (GPS), geographic information systems (GIS), remote and proximal sensing, variable rate technology (VRT) and decision support systems (DSS) are employed to meet these needs (Figure 1.2). In addition, new devices (e.g., GIS-enabled personal digital assistants, GPS-equipped yield and quality monitors), communication (e.g., Internet) and data compression technologies are useful for efficient collection and delivery of data, services, and products.

### *GPS*

GPS is a worldwide radio-navigation system formed from a constellation of 24 satellites and their ground stations, which provides geospatial accuracy to farm practices by enabling farmers to identify each field site. With the decision by the U.S. government to turn off selective availability (an artificial error introduced into the satellite data to reduce the positional accuracy to 100 meters) effective May 1, 2000, GPS accuracy has considerably

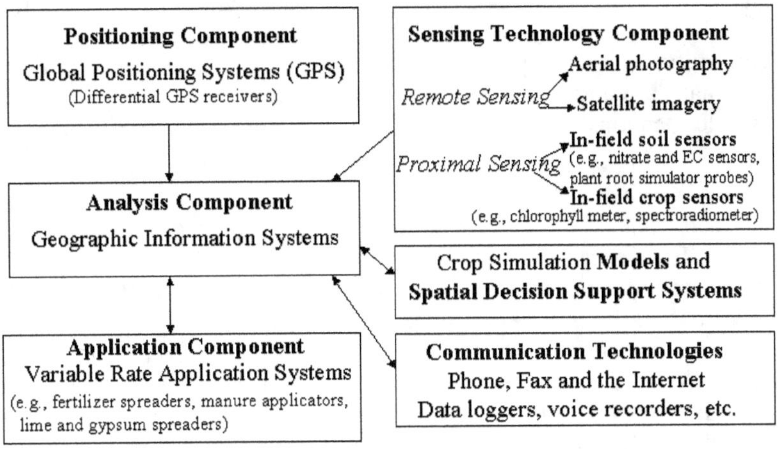


FIGURE 1.2. Tools for implementation of precision agriculture.

improved to about 20 meters. However, differential GPS (GPS receiver used along with a ground reference station) is necessary to get accuracy of 1-3 m critical for yield mapping, crop scouting, and variable input applications. GPS can also be used to create field boundary and topography maps. A series of maps with each field identified with the crop grown in each season can help in tracking crop rotations in a region from year to year. Recently, GPS auto-guidance (automatic steering for application equipment) is attracting attention in the United States, Australia, Brazil, and South Africa as a measure to save labor.

## *GIS*

GIS is a software application for computerized data storage, retrieval, and transformation and is used to manage and analyze spatial data relating crop productivity and agronomic factors. Data may be derived for different fields from various sources including existing digital maps, data digitized from maps and photographs, topographic surveys, soil or crop sampling, and sensor data with location information derived by using GPS. GIS can display analyzed data in maps that allow (1) better understanding of interactions among yield, fertility, pests, weeds, and other factors, and (2) decision making based on such spatial relationships. A GIS for PA contains base maps such as land ownership, crop cover, soil type, topography, N, P, K, and other nutrient levels, soil moisture, pH, etc. Data on rotations, tillage, nutrient, and pesticide applications, yields, etc., is also stored. GIS is used to create fertility, weed and pest intensity maps, and for making prescription maps that show recommended application rates of farm chemicals at various field locations.

***Remote Sensing***

Low-flying aircraft and satellites have become a major source of spatial data due to reduction in both cost and time of image acquisition and delivery. Although the use of remote sensing in agriculture is several decades old, improvements in spatial, spectral, and temporal resolutions of recent satellites (e.g., IKONOS) and systems such as airborne videography, and recent developments in new hyper-spectral sensors are increasingly becoming useful to determine size, location, and cause of variation. Imagery can show all fields in a region and spot anomalies related to crop vigor and biomass much earlier than ground inspections, thereby improving the efficiency of crop scouting in large fields and allowing prompt remedial treatments. Temporal changes in vigor, as determined from NDVI [Normalized Difference Vegetation Index] analysis of images acquired at critical times, can be used to characterize spatial and temporal dynamics of crop performance and predict crop yields. Upon integration of imagery with other data layers in a GIS, maps can be prepared. GPS receivers can then be used to locate weak spots and apply corrective measures. Likewise, images of different crops planted in rotation can reveal distinct crop vigor responses due to agronomic strategies adopted in previous crops. Imagery can also be used to monitor land use changes at regional level. In several tropical countries, cloud cover persists for most part of the cropping season and in such cases acquisition of radar imagery may be useful.

***Proximal Sensing***

Technologies employing electronic in-field crop and soil sensors to quantify variability in crop and soil conditions are also in development. Soil sensors for testing pH, EC [electrical conductivity], N content and organic matter are currently in use in North America. The introduction of Veris soil pH sensor in the fall of 2003 indeed signaled a new stage in PA as it automates the soil sampling process (Lowenberg-Deboer, 2004). Likewise crop sensors for determining N status (chlorophyll meter), water stress, disease or weed problems, grain moisture levels (useful in optimizing harvest operations and storage and/or drying requirements), and protein are also in use. The

Hydro-N sensor to measure plant N, which was developed in Germany in the mid-1990s, is now in use in several parts of Europe. Similarly, GreenSeeker technology, which was licensed from Oklahoma State University in the United States and marketed in the winter wheat region is increasingly becoming popular in several regions (Lowenberg-Deboer, 2004). Research on plant disease sensors and other sensors is in progress. Such sensors provide quick and cost-effective ways to identify spatial variability, as they can collect data at a level of intensity not economically feasible with physical sampling and laboratory analysis. Yield monitors are another group of proximal sensors. Indeed, yield monitoring and mapping are the most widely used PA components today. Yield monitors for corn, wheat, potatoes, peanuts, and soybeans have been already developed while the monitors for cotton, sugar beets, sugar cane, onions, etc., are in advanced stages of development.

### *VRT*

VRT provides "on-the-fly" delivery of field inputs. A GPS receiver is mounted on a truck so that a field location can be recognized. An in-vehicle computer with the input recommendation maps controls the distribution valves to provide a suitable input mix by comparing to the positional information received from the GPS receiver. VRT systems are either map-based or sensor-based. VRT has been largely used for applying N, P, K, lime, pesticides and herbicides. For example, the instruments SOILECTION and FALCON, which were developed by AgChem, USA, are now used for spatially variable application of fertilizer and pesticides. Variable rate seeders and variable rate irrigation systems are currently being studied. Systems such as manure applicators, which enable precise application of animal wastes, are also in development.

### *Communication Technologies*

Technologies such as fax, cable TV, and the Internet (World Wide Web) offer capability to send, receive, and aggregate information at much lower costs than ever before. For example, transmitting a remotely sensed image of a field with crop stress zones, or digital photos of problem areas via the Internat can enable the grower to take prompt in-season remedial measures. Likewise, posting to the Web sites of crop yield maps in different rotations over two to three years enables crop consultants to examine and evaluate them in preparing appropriate recommendations. In this context, image compression technologies will be increasingly useful.

### *New Devices*

Several devices integrating geospatial and communication technologies are in development. For instance, a differential GPS-enabled voice recognition system permits data collection with simple, easy-to remember voice commands regarding weed or insect problems, crop lodging, and other farm operations. Likewise, systems employing GIS-based wireless application protocols (e.g., MapInfo's MapXtend) and GIS-linked Infielder Crop Records System are useful for mapping various field operations and stress zones.

### *Crop Models and DSS*

These provide an integrated framework for assessing the degree of sustainability of a farming system. They can be used for understanding yield variability in both space and time so as to optimize production, indicate future trends, and prescribe suitable actions to minimize environmental impact. Process-oriented crop simulation models, such as CERES and CROPGRO, integrate the effects of temporal and multiple stress interactions on crop growth processes under different environmental and management conditions. Basso et al. (2001) reported that a combination of crop model (e.g., CROPGRO) and remote sensing could be used to identify management zones and causes for yield variability. Matthews et al. (2000) reviewed the utility of crop models and reported that the use of models as DSS had major impacts in irrigation scheduling (e.g., LORA in France) and pest management (e.g., EPIPRE and SIRATAC). Use of models with GIS permits mapping of adaptation zones for individual crops and rotation systems and help in targeting sustainable practices to defined regions.

Recent advances in Web-based GIS can provide access to spatial analysis of cropping systems through the Web in a scalable, threaded format. Jensen et al. (2000), through an example of Pl@nteInfo, showed that it is possible to build a Web-based DSS, in which personalized advice on farm conditions can be given in real-time to farmers. By creating an information and communication system called CropMD in Pennsylvania, which enables recording, tracking, aggregating, and assessing field-specific agronomic data, Pruss (2000) reported that it could be used in rapidly identifying fields with nutrient levels in excess of crop needs. Efforts to create Web-based DSS for integrated crop management in legumes are underway in Hokkaido (Srinivasan, 1997, 1998).

In several developed countries, customized services in PA such as grid soil sampling, yield mapping, variable rate application of fertilizers and pesticides are now offered through the private sector. Custom services can decrease the cost and increase the efficiency by distributing capital costs for specialized equipment over more land and by using the skills of PA specialists more effectively.

## *CURRENT STATUS, UNCERTAINTIES, AND FUTURE TRENDS*

The current status of adoption of PA in various crops and cropping systems is discussed in detail in Part II of this book. PA technologies are being currently applied in cereals such as rice, wheat, corn, and barley, but also in soybeans, potatoes, sugar beets, sugar cane, etc. Some components of PA technologies are also being tested in vegetable crops such as onion, tomato, and fruit crops such as citrus, grapes, apple, pear, etc.

Among all PA technologies, yield monitoring and mapping seems to be the most used (Exhibit 1.1) followed by grid soil sampling and variable rate fertilization. Lowenberg-DeBoer (2003) provided an overview of status of adoption of PA technologies in the United States and other countries. For example, in 2001 there were about 30,000 headers in the United States equipped with yield monitors (compared to only 800 in Australia, 560 in Argentina, 400 each in the United Kingdom and Denmark, 150 in Germany, 50 in France, 6 in the Netherlands and 5 in Belgium) and were used to monitor yields in as much as 34 percent of acreage in corn, 25 percent in soybeans, 10 percent in wheat, 8 percent in potatoes, and about 1 percent in sugar crops. Yield mapping with GPS, however, was done in only about 11 percent for corn, 8 percent for soybeans and 2 percent for wheat. It was also reported that by 2002, 50 percent farm retail dealers in the United States offered grid soil sampling using GPS. However, variable rate fertilization was limited. In 2000, VRT was used in 11 percent of corn areas, 7 percent of soybean areas, and 3 percent of wheat areas, 10 percent of potato areas, and 9 percent of sugar beet areas. Variable rate seed or pesticide application was on about 1 to 3 percent of grain acreage (Lowenberg-DeBoer, 2003). As per a recent Croplife/Purdue Survey, which is the most complete guage of the status of precision technology in U.S. agriculture, GPS guidance was the most widely used, while field mapping, remote sensing, and VRT did not see much growth (Erickson, 2005). However, the higher energy and fertilizer prices in recent years are considered to drive a greater adoption of PA technologies in the future (Lowenberg-BeBoer, 2006).

In terms of geographic scope, the adoption of PF in major continents and countries is discussed in detail in Part III of this book. Adoption of PA so far is mostly restricted to the United States, Canada, northern Europe, and Australia. Even in the United States, where modern PA has more or less originated, less than 5 percent of all farms use precision technologies. However, in the "heartland" region, dominated by corn and soybean production, 11 percent of the farms adopted PA in 2000 (Daberkow and McBride, 2000). PA first found widespread acceptance as a management practice in the Corn Belt of the Midwest, helping to increase corn, wheat, and soybean yields and reduce production costs (Rains and Thomas, 2000). Indeed as much as 60 percent of the U.S. farms using PA are in the Corn Belt. In general, grain farms have the highest PA adoption rate (14 percent), especially those with annual gross sales over $500,000 (18 percent). Among the most likely to try PA are farmers who are younger, better educated, have low debt to asset ratios, and large farm acreage (Grusy, 2003). In the United States, average prices for selected PA services are estimated to be around US$1-10 per acre depending on service (e.g., yield monitor analysis: $1.33; Soil sampling $3.10; controller driven application: $10.10) but these costs are likely to come down over time (Parcell, 2000). In both the United States and Canada, some aspects such as grain yield monitors and VRT of fertilizer on some higher value crops are moving toward becoming standard practice but variable application of other inputs is still limited. Several countries in Europe (UK, France, Germany, Denmark, and others) and Australia adopted PA techniques to some extent, especially in cereal farming. Among the developing countries, Argentina, Brazil, South Africa, China, India, Malaysia, and others have begun to adopt some PA components, especially on research farms, but the adoption is still very limited.

An overall assessment of progress in PA adoption worldwide indicates that it is slow and still very limited compared to other innovations introduced about the same time (e.g., glyphosate resistant soybeans, Bt maize). However, many researchers and agribusinesses feel that its full potential is yet to be realized on commercial farms. Currently, three types of uncertainty seem to contribute to slow progress in PA adoption. First, the tools available to assist with PA are somewhat difficult to understand and use (steep learning curve) besides their high cost and obsolescence over time. Many people feel that PA technologies came to the market in an undeveloped form, and that determining the appropriateness of specific technologies is still complex. Further research is necessary to determine such appropriateness in each country, region, and farm. In Australia, for example, high soil testing costs discouraged the U.S.-style variable-rate fertilizer and growers are searching for alternative ways to develop variable-rate fertilizer recommendation maps. Because of severe soil compaction problems in that country, GPS guidance for controlled traffic is considered to be the best starting point for precision farming. Second, we still have limited understanding on how crop inputs should be varied even after quantifying variation in soil and crop conditions. Third, persistent questions about profitability have constrained its rapid adoption. Indeed, documentation of the economic and environmental benefits of PA is incomplete. Some studies indicate that PA is profitable only on farms bigger than 100 ha. Gandonou et al. (2003), for example, reported that the break-even acreage for buying the PA equipment versus custom hiring is as much as 511 acres (204 ha) for variable rate application of two fertilizers and lime. Analysis of latest developments indicated that the progress in development of PA hardware and software is impressive but progress in management issues is still limited. There is thus a huge gap to be closed between successful data collection to successful data analysis. The decision support systems specifically tailored to the PA approach are, therefore, crucial. Further, the time investment in PA is perceived as happening exactly when farmers have a time deficit, e.g., at planting or harvest times.

In light of the above uncertainties, prediction of future trends in PA adoption is difficult. Further, PA is not a single technology but a suite of management strategies, technologies, and practices that can be introduced in many different combinations of products and services. Despite slow progress in adoption so far, infrastructure necessary for PA is gradually building up in various countries. In developed countries, a growing number of PA service providers in the private sector are offering advice to farmers in adopting various PA components (Exhibit 1.1).

In developing countries too, the diffusion of PA techniques is improving as large estate farms, corporate farms, and farms with high value crops are gradually enhancing their management capability to handle new information tools. Future progress, therefore, depends on further collaborations between policy makers, research communities, private sector, and the farmers.

## *EPILOGUE*

Management of field variability to improve profitability and minimize adverse environmental impacts is crucial for sustainable agriculture whether one uses traditional methods or modern technologies. PA is an emerging agronomic management systems approach that has the potential to realize sustainability by increasing profits and optimizing crop yields while minimizing the adverse environmental impacts through location-specific management. In a modern sense, PA may be referred to as the "digital face" of agriculture. Information, technology, and management are three key ingredients in PA. As all these ingredients are important to any farm in any country, PA is expected to be practiced at various levels. The future adoption of PA depends largely on how well researchers can develop cost-effective technologies and knowledge necessary to guide the new technologies (especially in terms of developing simple and practical applications) and

**EXHIBIT 1.1. List of Precision Agriculture Services**

Georeferenced field boundary mapping
Georeferenced soil sampling (i.e., grid, map linked or location-specific)
Georeferenced field scouting (for weeds/insects/diseases)
GPS-aided yield mapping
Remote sensing (aerial photographs and satellite imagery)
Variable tillage
Variable rate seeding
Variable planting of genotypes
Variable rate application of fertilizers (N, P, and K)
Variable rate lime application (for pH adjustment)
Variable rate herbicide/pesticide application
Variable water management
Yield monitoring
GPS-aided yield mapping
Site-specific grain quality monitoring and mapping

how quickly the farmers take up such technologies and applications in improved farm decision making. The authors of later chapters in this book examine such issues in depth.

## REFERENCES

Basso, B., J.T. Ritchie, F.J. Pierce, R.P. Braga, and J.W. Jones. 2001. Spatial validation of crop models for precision agriculture. *Agricultural Systems* 68(2): 97-112.

Blackmore, S. 1999. Developing the principles of precision farming. In *Proceedings of Agrotech 99*. Barretos, Brazil, November 15-19, 1999, Barretos Institute of Technology.

Daberkow, S.G. and W.D. McBride. 2000. Adoption of precision agriculture technologies by U.S. farmers. In *Proceedings of the Fifth International Conference on Precision Agriculture*. Madison, WI: ASA-CSSA-SSA.

Erickson, B. 2005. Ten years of precision survey finds adoption follows returns. *Site-Specific Management Center Newsletter,* Purdue University, October 2005.

Gandonou, J-M., C. Dillon, S. Shearer, and T. Stombaugh. 2003. Is my operation big enough to buy precision agriculture equipment? Site-Specific Issues, Harvest 2003. University of Kentucky Cooperative Extension Service Newsletter. Available online at: http://www.bae.uky.edu/~precag/PrecisionAg/030910 percent 20PA percent20newsletter.pdf.

Grusy, R. 2003. Precision farming in Central Kentucky: Evaluating public and private sector influence on the adoption decision. *University of Kentucky Cooperative Extension Service Soil Science News and Views,* 23(2): 6p.

Jensen, A.L., Boll, P.S., Thysen, I., and Pathak, B.K. 2000. Pl@nteInfo—A Web-based system for personalized decision support in crop management. *Computers and Electronics in Agriculture* 25:271-293.

Johannsen, C.J. 1996. Overview of precision farming. In *Proceedings of Information Ag Conference*. Champaign, IL, July, 1996, pp. 53-54.

Kitchen, N.R., K.A. Sudduth, S.J. Birrell, and S.C. Borgelt. 1996. Missouri precision agriculture research and education. In *Proceedings of the 3rd International Conference on Precision Agriculture* (pp. 1091-1100). Madison, WI: ASA. CSAA. SSSA.

Lowenberg-DeBoer, J. 2003. Precision farming or convenience agriculture? In *Proceedings of the 11th Australian Agronomy Conference*. The Regional Institute Ltd. Available online at http://www.regional.org.au/au/asa/2003/i/6/lowenberg .htm.

Lowenberg-DeBoer, J. 2004. Precision agriculture creates global opportunities. *Farm Chemicals International* June: 28-30.

Lowenberg-DeBoer, J. 2006. Effect of higher energy and fertilizer prices on precision agriculture adoption. *Site-Specific Management Center Newsletter,* Purdue University, February 2006.

Matthews, R., Stephens, W., Hess, T., Mason, T., and Graves, A. 2000. Application of crop/soil simulation models in developing countries. Institute of Water and Environment, Cranfield University, Silsoe, UK.

National Research Council, Board on Agriculture, Committee on Assessing Crop Yield: Site-Specific Farming, Information Systems, and Research Opportunities. Sonka, S.T., M.E. Bauer, E.T. Cherry, J.W. Colburn Jr., R.E. Hemlich, D.A. Joseph, J.B. Leboeuf, E. Lichtenberg, D.A. Mortensen, S.W. Searcy, S.L. Ustin, and S.J. Ventura. 1997. *Precision agriculture in the 21st century: Geospatial and information technologies in crop management.* Washington, DC: National Academy Press.

Parcell, J. 2000. Use and perceptions of precision agriculture technologies by professional chemical services. *University of Missouri Farm Management Newsletter* FM 2000-3.

Pierce, F.J. and Nowak, P. 1999. Aspects of precision agriculture. *Advances in Agronomy* 67:1-85.

Prusss, J. 2000. Crop management information systems initiatives. In Summary report of the Forum on the Application of Information Technologies for Small Growers, May 9, 2000—Second International Conference on Geospatial Information in Agriculture and Forestry, Lake Buena Vista, FL. Veridian ERIM International, Inc., Ann Arbor, MI.

Rains, G.C. and D.L. Thomas. 2000. Precision farming: An introduction. The University of Georgia Cooperative Extension Service Bulletin 1186 Available at http://www.ces.uga.edu/pubcd/81186.htm.

Srinivasan, A. 1997. Crop information systems for food legumes in Hokkaido, Japan. In *GIS in agricultural research: Awareness package.* (Denisov, N., Heberlein, C., Czaran, L. and Simonett, O. Eds.). UNEP/DEIA/TR.97-9. Case Study No. 19. United Nations Environment Program, Global Resources Information Database, Arendal, Norway.

Srinivasan, A. 1998. Crop information systems for soybean and azuki bean in Hokkaido, Japan. In *Proceedings of the 18th Environmental System Research Institute User Conference,* San Diego, July 27-31, 1998. CD-ROM.

Srinivasan, A. 2001a. Relevance of precision farming technologies to sustainable agriculture in Asia and the Pacific. In S. Suthipradit, C. Kuntha, S. Lorlowhakarn, and J. Rakngan (eds.), *Sustainable agriculture: Possibility and direction* (pp. 325-338). NSTDA, Bangkok, Thailand.

Srinivasan, A. 2001b. Site-specific management for selected cropping systems in Asia: Progress and prospects. In *Promoting global innovation of agricultural science and technology and sustainable agriculture development session 6: Information technology of agriculture* (Organizing Committee, ICAST Eds.) Proceedings of the International Conference on Agricultural Science and Technology, Beijing, China. pp. 180-190.

The U.S. House Committee on Agriculture. 2000. Agriculture glossary. Available online at http://agriculture.house.gov/info/glossary/p.htm.

Chapter 2

# The Role of Technology in the Emergence and Current Status of Precision Agriculture

John V. Stafford

## *THE BEGINNINGS OF PRECISION AGRICULTURE*

For more than ten years, precision agriculture has gained an increasing profile in the agricultural community. Although many farmers are yet to practice any part of this new concept, considerable research effort has been expended, advisors are eager to learn, and agricultural and other engineering companies are producing equipment of increasing sophistication to enable the practice of precision agriculture. The basis of precision agriculture—the spatial and temporal variability in soil and crop factors within a field—has been appreciated for centuries.[1] In past centuries, the much smaller size of fields and their delineation by natural boundaries, such as water courses and change of soil type, may have enabled farmers to vary treatment manually. However, with the enlargement of fields, intensive production and mechanization in the latter half of the past century, it was not possible to take account of within-field spatial variability without significant developments in technology. Figure 2.1 shows a 60 ha area in Denmark which was farmed as 22 fields in 1775, 5 fields in 1875, and 1 field in 1995.

The first seeds of precision agriculture were set early in the twentieth century by researchers such as Mercer and Hall (1911), Eden and Maskell (1928) (the first yield maps), and Linsley and Bauer (1929) (advocating spatially variable application of lime). However, widespread interest in precision agriculture was arguably initiated by Johnson et al. (1983) who developed the concept of *custom prescribed tillage.* They were visionaries in terms of how automation, sensing systems, location systems, and "information technology" would transform agricultural crop production as the technology came on-stream. They stated,

*Handbook of Precision Agriculture*

doi:10.1300/5627_02

> Future machinery used in production agriculture will be automatically controlled to prescribe cultural practices, based on soil, crop and climate. Some soil and crop information may be sensed on-the-go and stored in a computer on board the prime mover or field machine. This computer, in turn, could be programmed to make real time decisions based on this information to control cultural practices such as fertilizer, herbicide, and pesticide application. Important to this concept is a general spatial position-sensing system that can pinpoint the position of a machine in the field at any time. (p. 117)

At about the same time, Matthews (1983) also foresaw the need for greater precision in crop management and that that would require "continuous variation of the quantity of inputs to match the needs of soil or crop in a localized region."

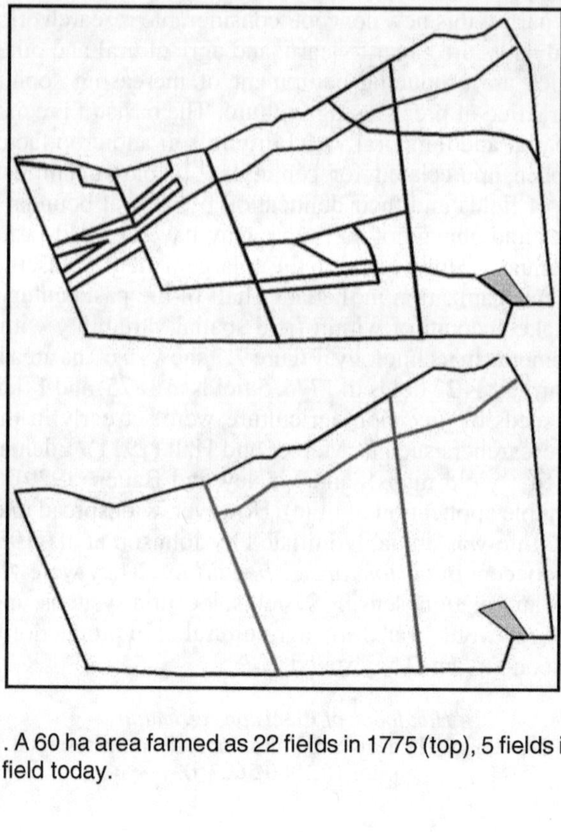

FIGURE 2.1. A 60 ha area farmed as 22 fields in 1775 (top), 5 fields in 1875 (bottom), and 1 field today.

Probably the first real application of precision agriculture was the "on-the-go" fertilizer blending and distribution system developed by Soil Teq in the United States (Fairchild, 1988) which used information from aerial photography and grid soil samples to generate a fertilizer application map. Positioning of the field vehicle, however, was by dead-reckoning as GPS was not sufficiently developed for civilian use.

Essential to the application of precision agriculture is the availability of a location positioning system of sufficient accuracy and reliability. This was increasingly available from the early 1990s in the form of the U.S. Global Positioning System, a military satellite-based positioning system first developed in the 1970s. Without GPS as a reliable, reasonable accuracy, reasonable cost, always available location system, it is unlikely that precision agriculture would have gained its current high profile exposure.

Although the term *precision agriculture* is widely interpreted as applying only to within-field spatial variability, a more complete definition would be "applying the right treatment in the right place at the right time." This definition includes the need to choose the correct treatment, and recognizes that timing is critical for many agricultural operations. For example, timing is more important than dose, or even the choice of chemical, for many pesticides. The effect on crop growth and productivity of temporal variability is, in many cases, greater than spatial variability, due primarily to seasonal weather variation (Pierce and Nowak, 1999).

Four drivers may be identified that provide the motivation for widespread uptake of precision agriculture: environmental legislation, traceability, and public concern regarding farming practices; the fourth, and negative, driver may be public aversion to genetic modification of crops. Precision agriculture provides an alternative and realistic means to reduce and optimize the use of agrochemicals.

## *THE BASIS FOR PRECISION AGRICULTURE: INFORMATION TECHNOLOGY*

Precision agriculture as a crop management concept is rooted in information generation, storage, manipulation, and utilization. Without the "information revolution" particularly with regard to computer processing power and data storage, it is unlikely that precision agriculture would have developed to its current status. GPS, of course, is just one information source. Data must be translated into useful information and there is increasing research into the interpretation of spatially variable soil and crop data, crop management on a localized basis, and the decision support systems needed to ensure rationally based spatially variable applications. The connection between technology

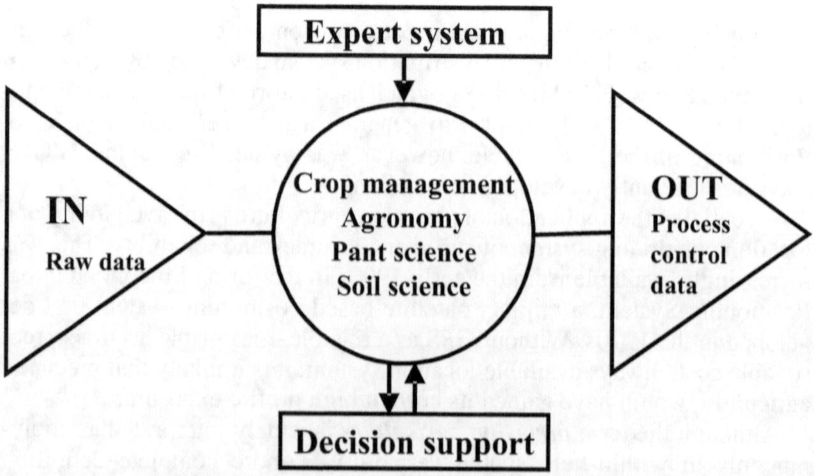


FIGURE 2.2. Information flow in precision agriculture.

functions and management functions may be illustrated by the information flow diagram shown in Figure 2.2. Crop management, decision support, and the relevant sciences are central to precision agriculture and must use real data on spatial variability. Similarly, they are without value if the output process control data cannot be utilized in appropriate control, data transfer, and precision application systems. Thus the essential technologies for precision agriculture, as illustrated in Figure 2.3, are sensing and monitoring systems, and control, data transfer and precision application systems.

## *SPATIAL LOCATION*

In-field positioning is required in order to map the sensed soil and crop factors and for the control of application equipment. The position resolution required covers a wide range, depending on the field operation, with ballpark figures as follows:

| | |
|---|---|
| Variable fertilizer application | 30 m |
| Yield mapping | 10 m |
| Variable application of herbicide | 1 m |
| Spray overlap avoidance | 10 cm |
| Row crop planting | 10 cm |
| Seedbed structure | 5 cm |
| Plant scale/leaf treatment | 1 cm |

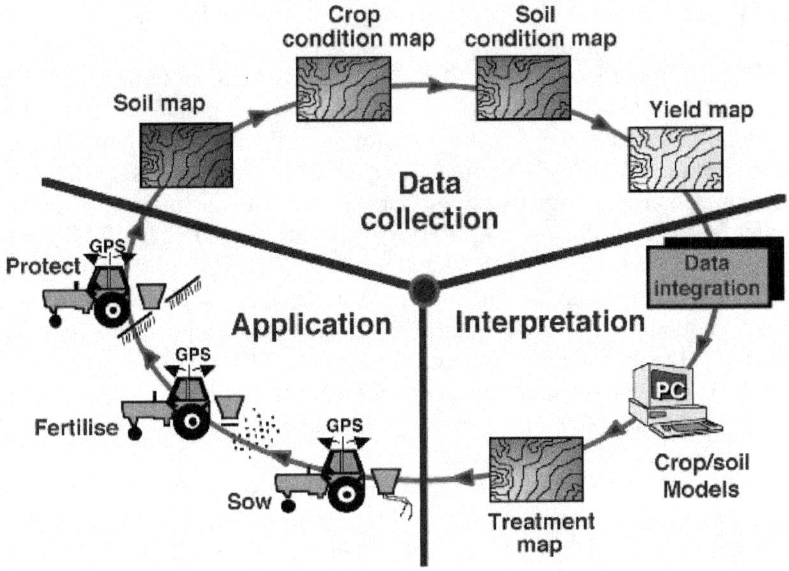


FIGURE 2.3. The precision agriculture cycle. (See corresponding PLATE 2.3 in central gallery.)

These figures are open to considerable debate, but they illustrate the large range of position resolution that is required.

The Global Positioning System is the de facto location system for precision agriculture. Other positioning technologies such as radio systems, microwave systems, and laser have been advocated in the past. Some of these systems depend on locally installed reference points. The attraction of the Global Positioning System is the establishment of global reference points (the constellation of satellites) with very accurately tracked positions with positioning signals available around the globe 24 hours a day.

## *BASICS OF GPS*

GPS is based on a constellation of satellites under the control of the U.S. Interagency GPS Executive Board (IGEB). Each satellite transmits a composite signal in the microwave L-Band which may be picked up by a microwave receiver. Provided the receiver has line-of-sight to sufficient satellites, the position of the receiver can be determined, either by measuring the time delay between transmitting and receiving the signal, or by phase-shift measurement of the signal carrier.

The constellation consists of 24 satellites which orbit in 6 different planes at a height of 20,200 km in order to provide global coverage. Depending on latitude, 6 to 12 satellites are more than 10° above the horizon throughout the day. Each satellite transmits two carrier waves, each modulated by two or three signals using a spread-spectrum technique. The carrier center frequencies are 1575.42 MHz (the L1 carrier) and 1227.60 MHz (the L2 carrier). One signal carries navigation information (clock correction, almanac, ephemeris data). The other two signals consist of pseudo-random binary sequences; one is designated the C/A (coarse/acquisition) code and is designed to provide positioning for civil users; the other, designated the P code, is encrypted and provides much higher position resolution for military use. The provision of two modulated carriers allows for compensation of ionospheric errors in P code receivers.

Each satellite has an atomic clock to precisely time the signal codes. A GPS receiver generates a duplicate of the code sequence and time shifts it to correlate with the received code sequence. Thus the delay between transmission and reception can be determined and so the range from each satellite can be calculated by reference to the velocity of propagation of electro-magnetic (EM) radiation. There are four unknowns in determining the position of a receiver. As well as the x, y, z coordinates, the receiver clock introduces a common bias error to the determination of range. Because of this, the computed range is usually called the "pseudo-range." Thus reception of at least four satellite signals is required to determine the three-dimensional position of a receiver.

Much higher position resolution to the millimeter level can be achieved by measuring carrier phase shift between transmission and reception; a technique referred to as kinematic GPS. The wavelength of the carrier is about 19 cm and so phase shift measurement to 1 percent would yield a range measurement of 2 mm (Goad, 1996). The major disadvantage of kinematic GPS is uncertainty over carrier cycle count; when so-called "cycle-slip" occurs, then position is lost. However, techniques exist for rapidly recovering position information after cycle slip, which may be due, for instance, to momentary loss of satellite signal. Phase shift smoothing is used in current pseudo-range receivers to improve position resolution.

The satellites' almanac/ephemeris data are based on the WGS84 geodetic datum, and so receivers compute position in latitude/longitude. The ephemeris is the daily log of the satellites' positions in space. Most receivers incorporate software transforms to local geodetic datums and to planar coordinate systems such as UTM (Universal Transverse Mercator).

### *Accuracy and Errors*

As with any instrumentation system, GPS is subject to a number of error sources. The main cause of low position resolution in the past was the deliberate downgrading of the satellite signals by the U.S. military to limit enemy utilization of the system. This downgrading, which took the form of dithering of the time signal and inaccurate ephemeris data, was known as Selective Availability (SA). The effects could be largely overcome by using differential correction whereby a GPS receiver received correction signals from a ground reference station to correct the computed position.

Five to ten years ago, GPS pseudo-range position computation with differential correction attained a position resolution of the order of 5 m. The significant error components that contributed to this resolution are shown in Table 2.1.

In the early days of precision agriculture, GPS was unreliable for dynamic positioning within the field. Typical position resolution with differential operation was 5m (2 × rms figure) with a skewed Gaussian error distribution extending out to tens of meters (Stafford and Ambler, 1994). The incomplete constellation of satellites accentuated the problem of signal obscuration by trees and buildings (Lachapelle and Henriksen, 1995; Brock and Karakurt, 1999) and multipath reflections were a significant cause of poor positioning. GPS receivers were bulky and expensive. In recent years, the general perception is that GPS is a mature technology with considerable, sophisticated processing power built into commercial receivers that have reduced in price and size very significantly.[2] The satellite constellation is

TABLE 2.1. GPS error sources.

| Source | GPS, m | DGPS, m |
|---|---|---|
| Satellite clock | 15 | 0.1 |
| Ephemeris | 40 | 1.0 |
| Orbit | 5 | 0.1 |
| Ionosphere | 12 | 1.0 |
| Troposphere | 3 | 0.5 |
| Multipath | 2 | 2.8 |
| Receiver noise | 0.5 | 0.7 |
| Total rms | 44.8 | 3.3 |

complete and receivers typically receive signals from 8 to 12 satellites above the horizon. Most receivers used in precision agriculture are 12 channel and use phase smoothed pseudo-range positioning with claimed submeter accuracies. Thus, GPS might be perceived as an available tool needing no further development.

### *Current Developments*

Important developments in positioning will benefit precision agriculture. Some aspects of precision agriculture, such as avoidance of spray overlap and application control near sensitive areas such as field margins, will be practiced at a smaller and smaller scale requiring higher resolution positioning. Targeting of field inputs to some crops will be at the plant scale, and even down to leaf scale, requiring much enhanced reliability and accuracy of dynamic positioning. Traceability will require recording precise application information tagged with very accurate position data.

Differential GPS was developed to counter the effects of SA. However, depending on the accuracy of reference station survey, DGPS also compensates for errors inherent in GPS such as those due to refraction in the ionosphere and troposphere. A development of DGPS that has been implemented worldwide is to gather information from an array of ground reference stations and generate correction signals that are transmitted as pseudo-GPS signals from geo-stationary communication satellites. The U.S. system is known as WAAS (wide area augmentation system) and uses 25 ground reference stations sending data to two uploading stations on the East and West Coasts. A similar system, EGNOS (Euro Geostationary Navigation Overlay Service) has been implemented in Europe.

A European global navigation satellite system (GNSS), Galileo, has been under discussion for some time (Spiller et al., 1998) and was subject to a "definition study" funded by the European Union and the European Space Agency in 2000. The system is being developed by a public-private partnership and will be operational by 2008 (Divis, 1999). Such a system should provide improved positioning accuracy and reliability compared to GPS. The system, based on 30 satellites, will provide three levels of service: mass market (free), commercial (subscription), and high security (subscription). Figure 2.4 is a general schematic of the system.

Possible technical enhancements in GNSS must be set in the context of the advanced technology and processing power already incorporated into GPS receivers. The accuracy of the Standard Positioning Service based on pseudo-range calculation of the Coarse/Acquisition signal from the satellites is limited by the repetition rate (1.023 MHz) of the pseudo-random

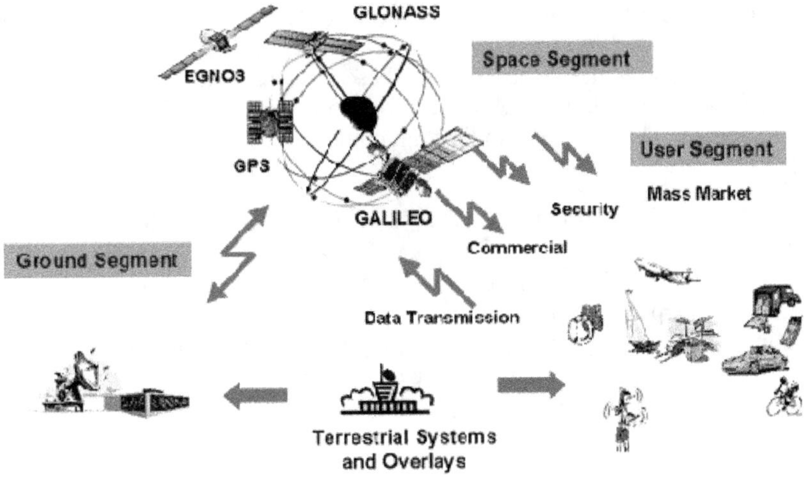


FIGURE 2.4. The planned European Galileo GNSS. (See corresponding PLATE 2.4 in central gallery.)

binary code transmitted on one satellite carrier (the L1 signal). If civilian access were allowed to the Precise Positioning Service where the precise (P) code is transmitted on the L2 signal at ten times the rate (10.23 MHz), then a significant improvement in positioning accuracy could be achieved. Developments of the U.S. GPS system are being put in place to update the architecture of the whole system. A civilian signal will be included with the L2 (renamed L2C) carrier and an additional microwave L band (L5) carrier will also be transmitted to enhance position, timing and velocity.

Kinematic GPS, where position is determined by measuring the phase shift in the satellite carrier signal between transmission and reception, gives the potential for centimeter accuracy. However, the use of kinematic GPS in precision agriculture has been limited because of cycle ambiguity and cycle slip with momentary loss of satellite signal. Phase shift smoothing is, however, being used in current pseudo-range receivers to improve position resolution. By using a double-differencing technique (Spilker and Parkinson, 1996; Langley, 1998), real time kinematic (RTK) GPS is possible to provide high resolution dynamic positioning. The robustness of RTK GPS will certainly be improved and increasingly feature in precision agriculture.

Similarly, integration of navigation systems (Gordon, 1998) will be enhanced to provide reliable and robust positioning—if one system fails to reach its positioning specification momentarily then there is at least one other system to take over. Such systems have been developed using GPS and other field machinery navigation data such as velocity and heading (e.g., Stafford and Bolam, 1996; van Bergeijk et al., 1998). Future systems are likely to integrate visual (camera-based) navigation (such as those described by Southall et al., 1998 and Morimoto et al., 2005) with GPS for precision application at the individual plant level.

## *INFORMATION ACQUISITION: SENSORS*

Precision agriculture is information-intense; a lot of position tagged, sensed data, i.e. mapped data, are required to generate treatment maps. The ideal to attain is real-time, robust, low-cost mapping systems for soil, crop, and environment variables. However, the only commercial systems available to date are yield mapping, crop reflectance, and soil conductivity mapping systems. Satellite remote-sensed images are available but, in general, are not processed into a usable form, are too expensive, or are not timely.

Soil and crop information, such as soil moisture and acidity, crop vigor and disease presence can be obtained—but only by laborious and costly manual sampling and analysis or by visual surveying. Cost-effective mapping systems will be developed within a decade that will enable collection of essential information for implementation of precision agriculture practices.

### *Soil Sensors*

The complex interaction between soil and crop growth means that it is difficult to infer spatially variable treatment from mapped soil factors. However, sensing and mapping of soil factors provide decision support information to the crop manager in identifying factors limiting to growth and yield in various parts of the field. For instance, a map of soil nutrient availability may alert the manager to areas of deficiency but cannot be used directly for variable application of fertilizer. On the other hand, a compaction map or soil pH map may provide direct information for varying subsoiling or liming treatment.

### *Soil Electrical Conductivity*

Research studies (e.g., Lund et al., 1999; Sudduth et al., 1995; McBratney et al., 2005) have shown how mapping soil electrical conductivity can be a good surrogate measurement for spatially variable factors that are not easy to sense and map such as soil type and moisture content (Figure 2.5). The technique is available through commercial services using direct conductivity measurement (e.g., the Veris System[3] which uses soil-cutting discs) and

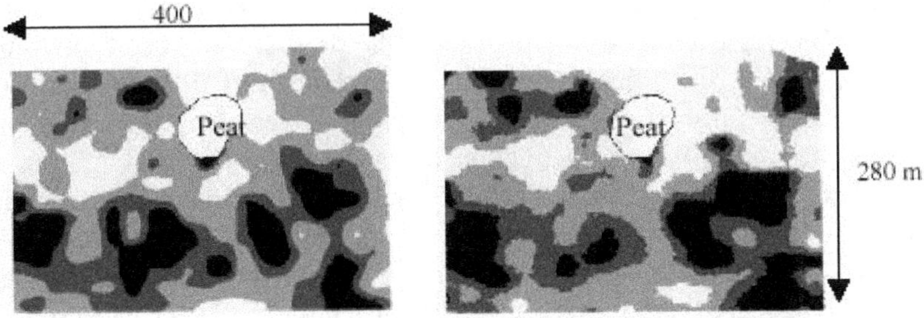


FIGURE 2.5. Estimating clay content (left) from EMC mapping (right). Darker = higher clay content. *Source:* Adapted from Nehmdahl and Greve, 2001.

indirectly using a noncontact electromagnetic induction probe such as the Geonics EM38.[4] Many research studies have been undertaken to develop different approaches to mapping conductivity and to seek correlation with other soil properties. Although electromagnetic induction probes are usually trailed along the soil surface to take measurements, a variation of the direct measurement technique has been demonstrated to map variation in soil electrical conductivity at three depths (Dabas et al., 2001). As well as showing correlation with soil properties such as moisture content, management zones, and soil boundaries can be delineated using conductivity mapping (King et al., 2001; Nehmdahl and Greve, 2001; Fridgen et al., 2000; Tarr et al., 2005).

The concept of surrogate sensing and mapping is being taken up increasingly where factors such as specific soilborne pests (e.g., potato cyst nematode, Evans et al., 2003) are difficult or impossible to sense directly. Another example is the use of a sequence of yield maps in an analysis procedure described by Lark and Stafford (1997) to divide a field into a limited number of subregions which, in many cases, delineate soil boundaries.

### *Soil Mechanical Properties*

The precision agriculture concept will be increasingly applied to soil manipulation operations such as seedbed preparation to optimize seedbed structure and subsoiling to only those areas of a field requiring it. Sensors need to be developed to characterize and map structure. There have been some research studies using video cameras and ultrasonic probes to characterize seedbeds (Scarlett et al., 1997; Stafford and Ambler 1990) but commercial equipment is not available. A microwave radar sensor has been used to characterize aggregate size distribution in a seedbed (Rouveure et al., 2001). The variation of soil strength and the presence and depth of compacted layers across a field provide the potential to vary tillage treatments and subsoil to only those areas with compacted layers. The savings in energy input and reduction in traction damage to soil may be significant (Mouazen and Ramon, 2005). Such information may also help in the interpretation of yield variability. Variations on the principle of measuring soil strength with penetrometers and the incorporation of force sensing elements in the soil engaging elements of tillage equipment are possible routes for monitoring and mapping soil strength (Stafford and Hendrick, 1988; Drummond et al., 2000; Castrignanò, 2001; Adamchuk and Christenson, 2005). The measurement and mapping of tillage draught force may also provide a surrogate for soil properties and thus provide another information layer for precision agriculture (McLaughlin and Burtt, 2000; Hayhoe et al., 2000; van Bergeijk, Goense, and Speelman, 2001; Hanquet et al., 2004).

### *Soil Moisture Content*

Available soil moisture has been shown in many research studies to be the main driver of yield and so mapping of soil moisture variation could be a very important information source for precision agriculture. However, despite many research studies, there has been little progress in developing reliable, dynamic techniques for moisture measurement. Stafford (1988) reviewed techniques for soil moisture measurement demonstrating that there are many options. Some, such as microwave attenuation (Whalley and Stafford, 1992), time domain reflectometry (Wraith and Das, 1998), soil conductivity (Morgan et al., 2000), and capacitance (Bobert et al., 2001), have been used to map soil moisture distribution. However, suitable sensing systems are not available commercially.

Soil color, as determined spectroscopically, can be related to soil moisture distribution but also to organic matter content. A fiber optic spectrophotometric device was developed by Sudduth and Hummel (1993) for mapping soil organic matter and moisture content. With calibration for specific soil types, further data analysis has shown that such techniques may be used to determine soil moisture and organic matter content of both surface and subsurface soils (Hummel et al., 2001). More recently, a multiwavelength spectrophotometric device has been mounted on a cultivator tine for deducing several soil parameters at depth (Shibusawa et al., 2005).

### *Soil pH*

Site specific application of lime to correct pH has been shown to be cost-effective both in terms of reduced lime use and increase in crop yield (Mulla et al., 2000; Viscarra Rossel and McBratney, 2001; Lowenberg-DeBoer and Hallman, 2000). Mapping of pH within the field by manual sampling on an intensive grid can be expensive but bare soil reflectometry in the visible and NIR spectra (via aerial or satellite imaging) has been shown to correlate with pH variation (Mulla et al., 2000). Mid-IR reflectometry has also been shown to be an indicator of pH (Viscarra Rossel et al., 2001). There is potential for the use of ion selective electrodes, both in their traditional form and as ion selective field effect transistors (ISFETs), to be used in rapid, on-the-go sensing systems for pH (Viscarra Rossel and McBratney, 1997; Walvoort et al., 2000). However, much development is required with regard to sensor reliability and ruggedness and to automated soil sampling systems (Viscarra Rossel et al., 2005). Adamchuk et al. (1999) described an experimental automated soil sampling system based on a soil cutting shank and a commercial pH sensor controlled by a PC in the tractor cab. A schematic is shown in Figure 2.6.

### *Available Soil Nitrogen*

Variable application of fertilizer, particularly nitrogen, has figured high in discussions on precision agriculture. Crop fertilizer requirement may be determined by reference to soil and/or crop properties. Rapid, on-the-go measurement of soil nitrate availability has been investigated in a number of research studies. It could, for instance, be determined indirectly by measuring nitrous oxide fluxes as described by Ambus and Robertson (1998) using a trace gas analyzer. Such a system might be field vehicle mounted so that rapid sampling and mapping can be undertaken. However, most research has concentrated on direct measurement that requires preparation of a soil solution so that reliable and repeatable contact with the sensing device can be achieved. Automated soil sample acquisition and preparation present numerous problems (e.g., Adsett and Zoerb, 1991; Birrell and Hummel, 1997). Once a soil sample solution has been extracted, there are several methods for determining soil nitrate. One that is directly analogous to analysis lab techniques is the use of colorometric test strips (Scholefield and Titchen, 1995; Hartz, 1994). Roth et al. (1991) used a handheld reflectometer to determine N concentration from nitrate-sensitive strips.

Ion selective electrodes (ISEs) are routinely used for monitoring specific chemical ions in water environments. Nitrate sensitive ISEs are thus poten-

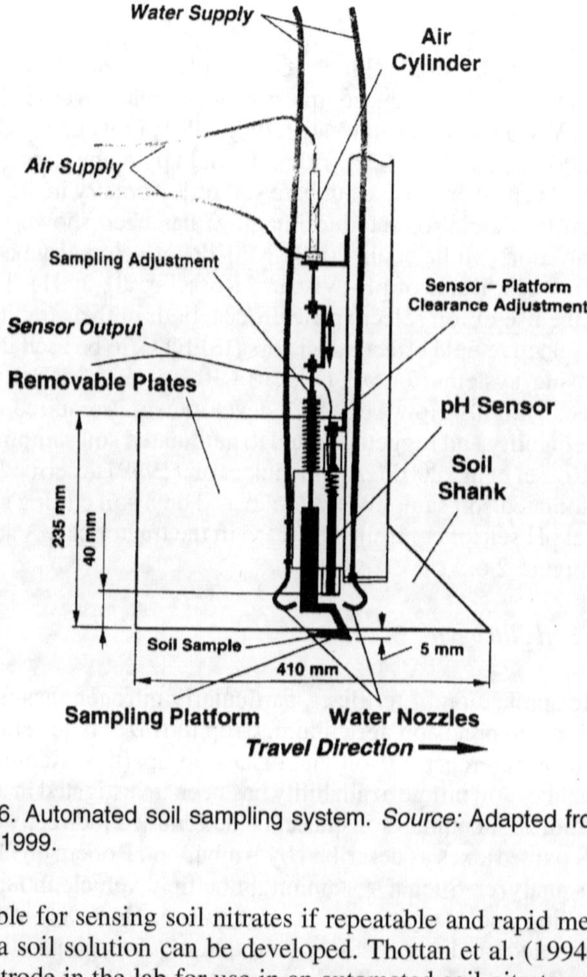


FIGURE 2.6. Automated soil sampling system. *Source:* Adapted from Adamchuk et al., 1999.

tially suitable for sensing soil nitrates if repeatable and rapid methods for extracting a soil solution can be developed. Thottan et al. (1994) tested a nitrate electrode in the lab for use in an automated soil nitrate monitoring system and achieved a cycle time of 4 s. A quicker responding and more stable nitrate-sensitive device is produced by applying a nitrate-sensitive membrane to the gate of a field effect transistor (FET) and monitoring the FET current. Two types of ISFET device were used for nitrate measurement in water-based systems in studies reported by Campanella et al. (1995). Birrell and Hummel (2001) used flow injection analysis (FIA) with ISFETs as the sensing element to investigate real-time soil nitrate sensing (Figure 2.7). The system was successful in measuring soil nitrates in manually extracted soil solutions with a cycle time of 1.25 s. As with pH sensors,

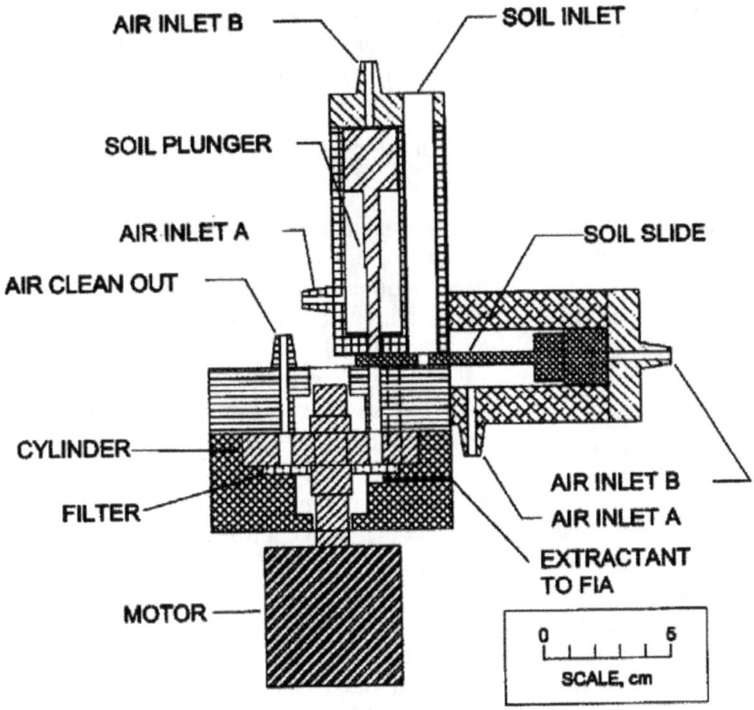


FIGURE 2.7. Automated soil extraction system for flow injection analysis. *Source:* Adapted from Birrell and Hummel, 2001.

the main restrictions to implementation of these sensor systems are reliability, ruggedness, and automated soil sampling.

### *Crop Sensors*

If the "crop is the best sensor of its own environment" (Legg and Stafford, 1998) then sensing systems that can tap into what the crop is "saying" may provide information on crop condition necessary to direct spatially variable inputs. Many studies and developments, mainly based on spectral reflection characteristics of the crop, have potential to enable inputs to be varied to optimize use and reduce environmental damage. Yield mapping, the first crop mapping system to be commercialized, provides valuable information to the crop manager to use in the following season.

However, the goal is to develop sensing systems which provide timely information on crop condition at critical growth stages so that inputs can be varied spatially to optimize crop performance.

## *CROP CONDITION*

Information on crop condition can be derived primarily from the spectral reflectance characteristics of the crop and be monitored by radiometric and hyperspectral sensing devices mounted on platforms ranging from field vehicle to satellite. A scanning radiometer system, mounted on a tractor, has been described by Stafford and Bolam (1998) and its use for crop density and canopy cover by Scotford and Miller (2005). A commercial, tractor-mounted radiometric system has been introduced by Hydro-Precise which can map spatial variability in the crop although they are marketed as a real-time variable nitrogenous fertilizer application system.

Satellite remote sensing has held out much promise for within-field monitoring but has failed to deliver in practice. Problems include timeliness of images because of cloud cover, cost, poor spatial resolution, and lack of processing to produce image data of use to the crop manager (Steven, 1993). Over recent years, a number of projects have sought to extract useful information for the farmer from remote sensed images. Some have led to precommercial trials and then commercial services sold to farmers on a per hectare basis (Coquil and Bordes, 2005). In recent years high spatial resolution, frequent revisit time satellites such as IKONOS have been launched. The imaging scanners on IKONOS have a ground resolution of 1 m panchromatic (450-900 nm) and 4 m multispectral (3 visible and 1 near infrared band). The revisit frequency is 1 to 3 days. Such satellites therefore have the potential to provide the data required for precision agriculture. Another potentially useful source of remote sensed data are SAR (synthetic aperture radar), images from satellites such as the European "Terra SAR," which have the advantage of penetrating clouds (Holmes et al., 2004). The reflected image is sensitive to surface structure and moisture content but still requires interpretation as with the visible and near infrared images. An example of a flooded area in the Po Valley, Italy, is shown in Figure 2.8. There is potential for combining SAR and optical data to improve information for crop management (Moran et al., 1997).

Processing and interpretation of the image data is essential and the problems of identifying causes for the observed effects on crop reflectance, observed by Steven (1993), remain. However, integrating image data with other ground-derived spatial data to determine underlying causes holds out promise for a solution in the current decade. The use of hyperspectral data,

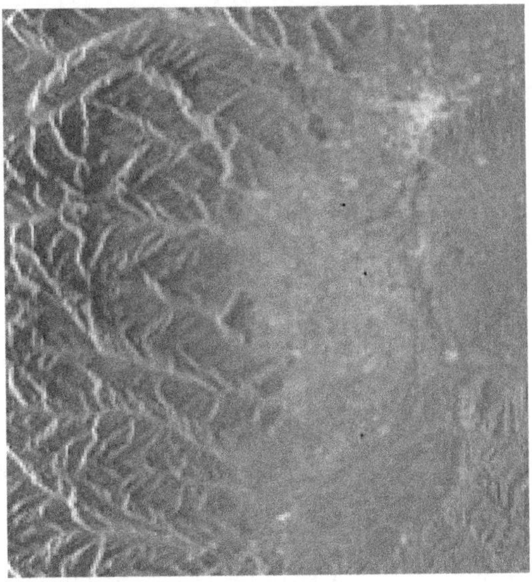

FIGURE 2.8. Synthetic aperture radar image of flooded area.

such as provided by the Hyperion hyperspectral sensor onboard the NASA EO-1 satellite, has the potential to provide information on parameters such as chlorophyll content, leaf area index, and canopy water content (McNairn et al., 2001).

Mapping of the spatial variability in plant stress may enable variable treatment for stress factors such as disease, nutrient deficiency, or insufficient moisture. However, as mentioned above, discrimination between factors using spectral reflectance is difficult because all induce leaf chlorosis with few distinguishing spectral characteristics. Developments in gene manipulation of crop plants, which continue to be the subject of research,[5] may enable differentiation of stress factors because gene expression depends on the stressing factor. An optical sensor with high specificity may then be used to identify crop areas that are, for instance, suffering a specific disease attack. Leaf fluorescence may be one such mechanism that may be sensed (Schächtl et al., 2005).

Although genetic manipulation to enable sensing of spatially variable crop factors may become technically feasible in the next decade, public antipathy toward the concept of genetic manipulation may inhibit its practical implementation. Another application for "smart plants" may be in weed/crop discrimination.

## *WEED DETECTION*

The very similar spectral characteristics of crop and weed species makes discrimination technically difficult as shown by the relatively small success of research over the last ten years to develop weed detection techniques (Stafford and Benlloch, 1997). Genetic manipulation of crop plants, if acceptable to the public, might be used to change spectral reflectance characteristics sufficiently to enable robust discrimination from weeds. Variable herbicide application strategies (Christensen et al., 1999) will, however, require species as well as presence/absence information on weed distributions in crops. Robust weed mapping approaches will be developed based on sensing multiple weed attributes such as spectral characteristics, morphological information, crop/weed geometries and distribution patterns within crops (Robbins, 1998). High classification accuracy between weeds and crops has been obtained using multi-spectral image analysis (Chapron et al., 1999), canopy reflectance (Vrindts et al., 2002) and neural network classifiers based on spectral reflectance data (Moshou et al., 2002; Bossu et al., 2005). Discrimination of weed from crop is much easier in wide-spaced row crops (e.g., Marchant et al., 1998; Marchant et al., 2001; Lee et al., 1999; Hague et al., 2006) but there are prospects for real time detection of weeds in cereal crops using multi-spectral image acquisition and image analysis techniques (Gerhards and Sokefeld, 2001; Nordmeyer et al., 2005).

## *GRAIN YIELD*

The first commercial sensing system for mapping within field variability from a field vehicle was yield mapping for small grain cereals. The essentials of yield mapping systems are a grain flow sensor and a location system. Various sensing technologies have been implemented for grain flow measurement. Some have determined grain flow into the combine by indirect methods such as measurement of engine speed (Schueller and Bae, 1987) or the torque on the grain tank filling auger. For direct measurement of grain flow, Vansichen and De Baerdemaeker (1991) used a strain-gauged change-of-momentum plate, Howard et al. (1993) used weigh cells on an extended clean grain elevator and Stafford et al. (1996) used electrical capacitance. Low energy X-rays have also been investigated for measuring grain flow (Arslan et al., 2000). Commercial systems include strain-gauged change-of-momentum plates, gamma ray absorption, a tipping bucket arrangement, and a light interruption technique in the clean grain elevator. The techniques can be divided into those which measure volumetric flow and those which measure mass flow of grain. Because of the variation in grain bulk density and moisture content, a mass flow sensor is preferable. Techniques have been described in detail by Arslan and Colvin (2002).

The design of combine harvesters is such that grain flow cannot be measured at the point of inflow of material (the header) but must be measured at a point where clean grain flows, such as the output auger to the grain holding tank. The tortuous path from the combine header to this point means that grain flow variation with time at the header is not truly reflected by the sensor output. Although commercial yield mapping systems assume a simple time lag for simplicity, the transfer function may be quite complex involving convolution of the input grain flow. Modeling of the transfer function (Lark and Stafford, 1997; Whelan and McBratney, 2000) has shown how spatial resolution of yield variation is reduced. Other errors in yield mapping due, for instance, to location and cut width (Stafford et al., 1997), may be quite significant and have been analyzed by Blackmore and Moore (1999) and Arslan and Colvin (2002).

## *GRAIN QUALITY*

Cereal quality is becoming more important than yield especially as the price of cereals reduces on world markets. There is well-substantiated evidence that quality as well as yields are spatially variable within fields (Reyns et al., 2000; Stafford, 1999) and systems will be developed to exploit such variation to add value to the harvested crop. Long et al. (1998) have shown that nitrogen fertilizer can be targeted according to a previous season grain quality (protein content) map to improve the efficiency of uptake and reduce spatial variation in quality in the current season. Thylen et al. (1999) and Stafford (1999) have shown that sorting grain from a field into different quality bands may be economically worthwhile. The technical developments required are of on-combine sorting systems and online quality sensors. The latter will probably be based on near-infrared absorption techniques (Engel et al., 1997; Long and Rosenthal, 2005).

## *ENVIRONMENT*

There is increasing appreciation of how subtle changes in field topography (slope, aspect, depressions) can affect crop development (Bishop and McBratney, 2002), final yield (Yang et al., 1998), and yield quality (Long et al., 1998). The spatial variability in topography is captured in a digital elevation model (DEM) which can be laboriously generated by conventional survey methods. With the wider availability and lower cost of high resolution GPS (such as kinematic GPS), it is possible to generate DEMs at reasonable cost. Phase-smoothed differential GPS can generate DEMs to a height resolution of 50-100 mm (Russell et al., 1999) or 120-140mm (Yao and Clark, 2000). With DEM information generated from GPS and aerial photogrammetry, it should be possible to determine variation in crop height across a field to provide another information layer in decision support.

## ASSEMBLING AND INTERPRETING INFORMATION

Precision agriculture, as mentioned previously, is "information-intense." As the amount of data generated on-farm from both mobile and static sources has increased, the need for standardized file and data formats has been recognized. On mobile equipment, the need for rationalizing the means for data flow and exchange has led to proposals for standardized data buses (Stafford and Ambler, 1988, 1993; Auernhammer et al., 1993). File and data formats have been standardized in ISO 11787, referred to as ADIS (agricultural data interchange standard) and are in use by some manufacturers. Data bus standardization for tractors and attached implements, based on the Controller Area Network (CAN) protocol, has been developed by ISO, although a German standard (DIN9684) has been available for some time (Spangler et al., 2001). The resulting standard (ISO 11783) is in 13 parts, most of which have already been published. Both of these standards are of considerable importance in the development of precision agriculture so that information can be exchanged smoothly and efficiently between sensors, processors, controllers, and software packages from different manufacturers (Hofstee and Goense, 1999).

Apart from the more general geographic information systems (GIS) such as Arc/Info and ERDAS, there has been considerable development of commercial GIS packages specifically for precision agriculture with a number of commercial systems available (such as LORIS from Kemira, AMAIS from Farmade, Agro-map from Claas, SSToolbox, and AGIS). Some of these can be described as GIS shells that enable import/export of data in various formats together with manipulation and display of data layers. Because most users do not have the necessary training and understanding to screen, filter, and integrate data and extract information that treatment maps may be based on, some precision agriculture equipment suppliers provide GIS services with increasing use of the Internet to communicate. Lütticken (2000) has proposed an open software platform, Tele-Farm, using the Internet and mobile phones for communication for the two-way exchange and processing of data between field machine, farm office, and a wider agricultural network.

It is necessary, however, to develop and incorporate models/expert systems to interpret and integrate multisource data into useful information. An important set of tools for rationalizing and summarizing spatial data is geostatistics. Commercial GIS and mapping packages tend to incorporate a standard set of geostatistical techniques (such as kriging interpolation) which cannot be readily changed by the user who, in the case of commercial producers, does not have the knowledge to correctly use such techniques. More flexible geostatistical packages, such as VESPER (Whelan et al., 2001) and GS+ for Windows (www.gammadesign.com) are available to the research community. GIS packages with analytical functions and some modeling capability include FIS (Zhang et al., 1999; Zhang and Taylor, 2000) and IDRISI (www.clarklabs.org).

## *UTILIZING INFORMATION: APPLICATION AND CONTROL*

Implementation of precision agriculture to date has mainly utilized existing field machinery and added controllers and GPS to enable spatially variable application. Thus, conventional spray booms have been used for patch spraying (Miller and Paice, 1998), and spinning disc applicators for variable fertilizer application. The potential of precision agriculture has led, however, to the development of novel, precise application techniques to achieve the precision and reliability in material placement that follow from the precision and accuracy achievable by positioning and sensing systems. The conventional spinning disc fertilizer spreader, for instance, can hardly be described as "precise"! Conventional spray nozzles typically deposit liquid over a half meter diameter circle. The variability in the physical characteristics of granular and powder materials may mean that development of novel techniques are concentrated on liquid delivery systems. Such requirements are likely to be appropriate to high-value vegetable and specialist crops while arable crops such as cereals will continue to be treated on a "management unit" basis.

## *AGROCHEMICALS*

In an era of increasing concern regarding the environmental impact of farming operations and appreciation of the economic need to reduce use of inputs, the uniform application of agrochemicals is no longer acceptable. Considerable spatial variability exists in the distribution of weeds across a field both in terms of plant density and weed species. This is due to a number of factors such as seed dispersal mechanisms, soil type, and moisture content and surface aspect. Thus, uniform application of herbicide ignores the true situation within the field and inevitably results in excess herbicide being applied.

The technologies developed within precision agriculture provide scope for targeting herbicide more accurately, thus achieving a significant degree of optimization of the use of herbicide. Weed control is essential for the successful production of arable crops with, typically in the United Kingdom, two or three applications being made to winter-grown small grain cereal crops. These weed control measures range in cost from \$20 (£13) to \$110 (£75) per hectare. Thus, if targeted application only reduces herbicide input by 50 percent, a significant saving in input costs could be made while reducing the environmental impact significantly in terms of residues on crops and leached agrochemical.

## *PATCH SPRAYING: PHILOSOPHY OF APPROACH*

Two distinctly different approaches to targeted application of herbicide can be distinguished. The first approach may be described as "real-time." A weed detection system mounted on the field vehicle detects individual weeds and passes that information to a control system that controls the spraying system mounted on the field vehicle (for example, Gerhards and Sokefeld, 2001). In the second approach, field maps of weed patches, generated by weed-sensing systems at some time prior to the spraying operation are used as the basis for sprayer control. As noted earlier, only limited success was achieved in developing sensing systems that can discriminate weed from crop, and even less in identifying specific weed species within the field. Useful developments have occurred in machine assistance to manual field surveying of weeds using, for instance, voice recognition systems (Perry et al., 2001). An example of the mapping of thistle patches using a voice recognition system and an ATV is shown in Figure 2.9. The field weed map provides data on the patchiness of weed infestation which, together with information on the reliability of weed data and a herbicide application strategy, may be used to generate a treatment map. This treatment map provides essentially a two-dimensional control signal to the sprayer controller. An essential sensing system is a means to provide dynamic position information such that the sprayer position may be continually related to the field treatment map. A control algorithm can then direct the spatial application of herbicide.

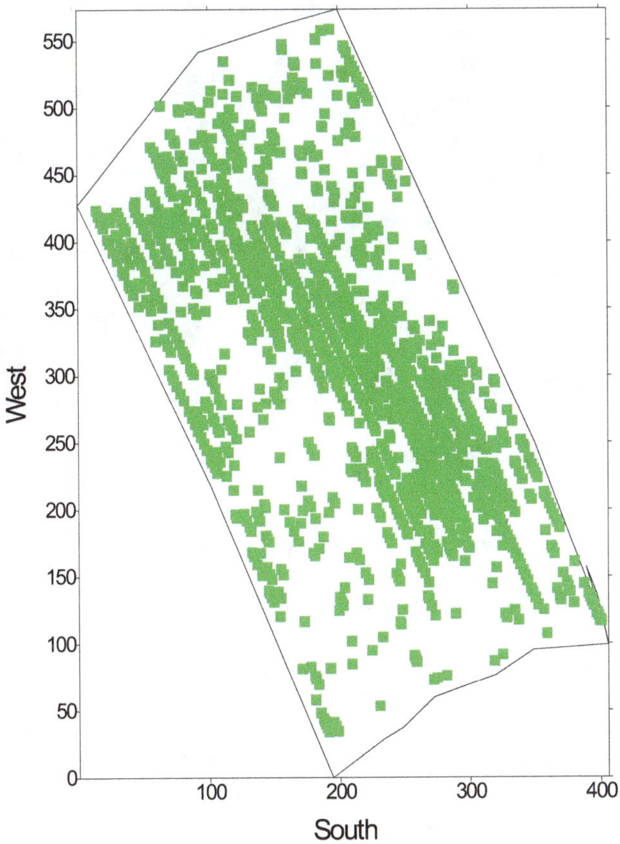


FIGURE 2.9. Mapped thistle patches from assisted manual surveying.

The second approach is realistic because it is observed that many weed species grow in patches (e.g., Wilson and Brain, 1991; Gerhards et al., 1997) that are relatively stable in location and shape from year to year. Thus, a weed map (an example is shown in Figure 2.10) generated at one time in the season, or even in the previous season, is valid at a later spraying time.

A patch spraying system requires four components: weed patch detection, field-mapping software, spatial location, and precision application. A strategy is also required for transforming field weed maps into treatment maps.

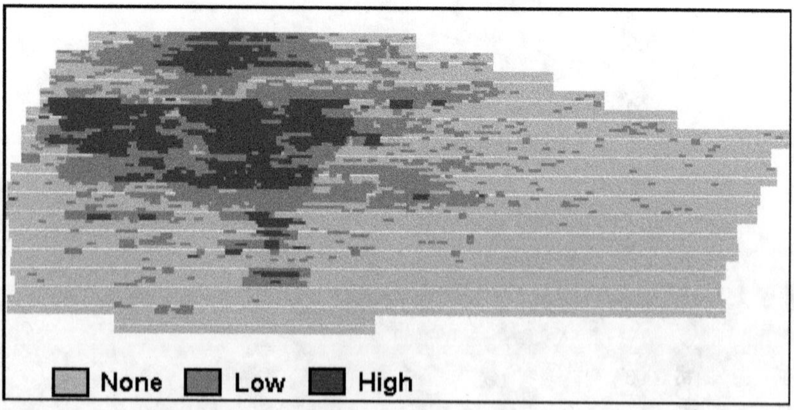


FIGURE 2.10. Distribution of blackgrass in a field of barley. The straight lines are the "tramlines" used for field vehicle guidance in NW Europe.

An experimental sprayer developed at Silsoe Research Institute in the United Kingdom was based on a 12 m boom with two parallel spray lines each with nozzles in groups of four at 0.5 m spacing. A direct injection system was used in which water from the spray tank was mixed with herbicide concentrate metered from a cylinder according to demand. The output from each boom section was separately controlled from a central control system via solenoid valves. With the two lines, four levels of control (including off) were available depending on the spray strategy implemented. For example, a base concentration could be supplied on one line with a different concentration on the second line. Thus, three levels of concentration could be applied. Alternatively, two pesticide formulations could be used and the proportion applied to different areas of the field varied.

The requirements for a spatially variable spraying system in terms of accuracy of dose, spatial resolution, dose range, and dose resolution have been reviewed by Paice et al. (1996). They considered the options for controlling sprayer output and concluded that direct injection provided advantages in terms of control, dose rate turn-down ratio, and environmental benefits. The latter include metering of concentrate only on demand, no dilute solution in the spray tank, and minimal system washing and disposal of dilute concentrate.

In a collaboration between Silsoe Research Institute and Micron Sprayers Ltd (Miller et al., 1997), a patch-spraying system was developed that was compatible with existing sprayer designs and could be retro-fitted to existing machines. The system was based on a commercial unit that provided both in-field location via a GPS system and the management and storage of a treatment map in a tractor-mounted console (the AGCO Massey Ferguson "Field Star" system). Supporting software enabled treatment maps to be generated on a farm office computer and downloaded to the tractor console via a data card.

The control principle was to switch nozzle orifice sizes and vary the pressure at the nozzle such that output could be varied almost continuously. The tractor-based control system sent a message at a frequency of 1 Hz to the sprayer containing the required application rate and the forward speed. The control system on the sprayer selected the nozzle(s) to be used and the spraying pressure to give an appropriate flow rate at a given spray quality.

The switching of nozzles and the control of spray pressure was implemented using a compressed air control. Each nozzle on the boom was fitted with a pneumatic on/off valve activated by small air lines mounted along the boom. Spraying pressure was also controlled pneumatically. Three nozzles at each spraying station on the boom gave a continuously variable output over a wide range of flow rates (Figure 2.11). In practice, good performance was obtained with two nozzles at each position along the boom. The prototype patch sprayer was compatible with existing sprayer and tractor control systems and has been used successfully to apply herbicides, fungi-

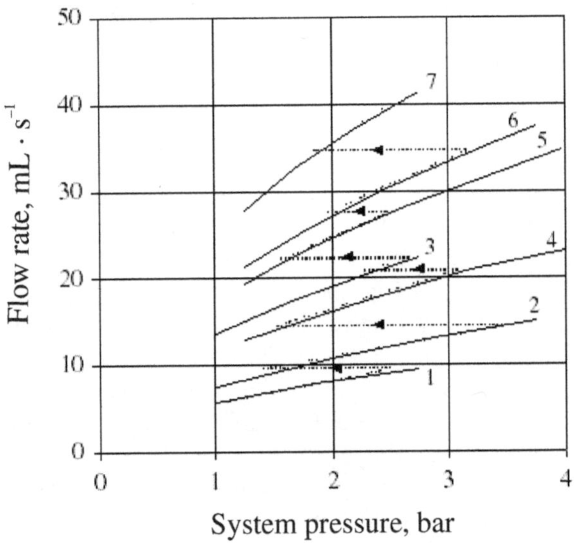


FIGURE 2.11. A typical path of increasing demand flow rate through the seven ranges of a three-nozzle system.

cides, and liquid fertilizers in a series of evaluation trials involving spatially variable application.

## FERTILIZERS

The recommendation procedures that agronomists use for determining uniform application rates for N fertilizers across a whole field are generally based on such factors as cropping history, expected yield, and soil type. Additional precision in N recommendations can be obtained by use of measured soil mineral N, especially in high residue situations. Such procedures applied in production agriculture inevitably must use average values of factors across a field or, more practically, values obtained from a few bulked soil samples or spot measurements in the crop. On a whole-field basis, such procedures and the associated average or spot values of factors are probably acceptable and generate application rates that provide the manager with a degree of insurance cover.

The precision agriculture approach is to determine N requirement on a localized basis and apply it precisely, both spatially and in applied rate. However, if this precise approach is taken, then the approximate whole-field recommendation procedure based on field averages is not sufficiently rigorous (Hergert et al., 1997). In the whole field approach, the ballpark type of recommendation is justified by the large variance on the average values of measured (or assumed) factors such as soil type and previous yield. For spatially variable N application, there must be an understanding of the soil-plant system in terms of (1) crop requirement and available sources of N and (2) how the plant takes up N and what limits that uptake.

Granular fertilizers are not particularly amenable to accurate application rate or precise spatial application. Such materials are usually metered from a bulk hopper to a spreading mechanism. The physical properties may vary considerably from batch to batch and their flow properties are influenced by ambient air humidity (Hofstee, 1992). Thus, feedback control of flow rate is required for precise control of application rate. A dynamic weighing system, accurate to 20N in a range of 6-20 kN, for a fertilizer delivery system has been described by van Bergeijk et al. (1997, 2001).

The common spinning disc application method for granular fertilizers does not equate with precise spatial positioning of the granules. Pneumatic boom spreaders should, in theory, be more precise but a more attractive approach may be to use liquid fertilizers through a modified patch sprayer (Stafford and Miller, 1997).

## *THE ROLE OF PRECISION AGRICULTURE IN THE FUTURE OF AGRICULTURE-TECHNOLOGICAL DEVELOPMENTS*

The concept of "plant scale husbandry" in which treatment is targeted according to the specific requirements of each plant has already been developed (Hague et al., 1997). The system of Hague and colleagues achieved a spatial resolution of 50 mm using image-based and inertial guidance and was equipped with precision spray nozzles that could target individual plants. Methods have been developed for dynamic discrimination of individual crop plants from weed, soil and other surface material (Marchant et al., 1998), an example being shown in Figure 2.12. Dijksterhuis et al. (1998) have shown that a tenfold increase in field vehicle guidance resolution (5 mm) is possible using real-time kinematic GPS. It may not be long, therefore, before commercial take up of plant scale precision agriculture but also development of "leaf-scale" in which differential treatments are applied to different parts of the plant.

Further developments in sensor resolution and specificity are, of course, required for this scenario and the approach may not be viable in economic terms. However, increasing consumer and supermarket demands for record keeping and traceability and environmental legislation will follow the greater precision made possible by technological development. There will then be the requirement for very precise application of inputs, which provides a challenge to innovative design and development.

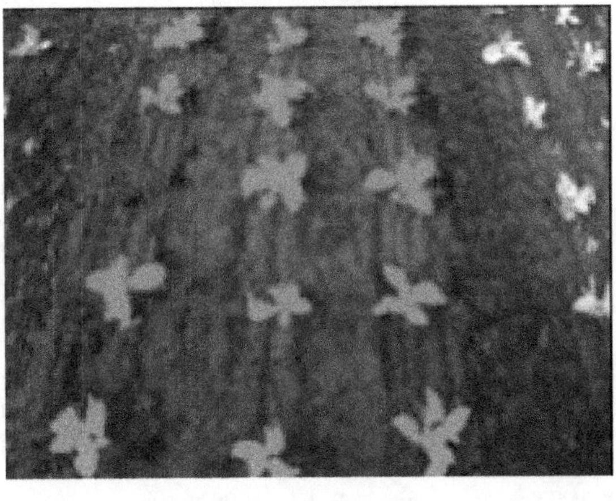

FIGURE 2.12. Discriminating individual crop plants. *Source:* Adapted from Marchant et al., 1998. (See corresponding PLATE 2.12 in the central gallery.)

Precision agriculture is a crop management concept. It should be implemented as such on a farmwide basis and practiced for all field operations concerned with the growing of crops. Environmental legislation with regard to the minimization and optimal use of inputs and market pressures for traceability and audit trails are forcing producers to take up precision agriculture. Alongside these pressures will be that of optimizing use of technologically sophisticated equipment. Thus, machinery and equipment replacement policies will be on the basis of utilizing equipment as widely as possible in all field operations. Procedures for the integration and interpretation of the masses of spatially-related information must be established both for generating rational treatment maps and as decision support to the manager.

The economic and environmental benefits of taking account of within-field spatial variability appear obvious although they have yet to be generally proven in field trials and experiments. Much of the technology is in place but further important developments will occur, particularly in the area of sensing and mapping variability. It is essential that complementary research and development in agronomy, crop, and soil science are successful in defining, explaining, and making recommendations for variable application of inputs at the very localized level.

## NOTES

1. The reader is referred to the writings of the early Jewish fathers (Midrash, Genesis Rabbah XIII, 13) and the parable of the sower in the Bible (Matthew 13:8).
2. Handheld, 12-channel GPS receivers are now available for less than £100 (U.S. $150).
3. Veris Technologies, 601 N. Broadway, Salina, Kansas.
4. Geonics Ltd, Mississauga, Ontario, Canada.
5. See, for example, *The Sunday Times,* May 17, 1998, reporting early research at IAC-Rothamsted, United Kingdom, and The Institute of Cell and Molecular Biology, Edinburgh University, United Kingdom.

## REFERENCES

Adamchuk, V.I. and P.T. Christenson. 2005. An integrated system for mapping soil physical properties on-the-go: The mechanical sensing component. In *Proceedings of the 5th European Conference on Precision Agriculture* (pp. 449-456), J.V. Stafford (ed.). Wageningen, The Netherlands: Wageningen Academic Press.

Adamchuk, V.I., M.T. Morgan, and D.R. Ess, 1999. An automated sampling system for measuring soil pH. *Transactions of American Society of Agricultural Engineers* 42(4): 885-891.

Adsett, J. F. and G.C. Zoerb. 1991. Automated field monitoring of soil nitrate levels. In: *Proceedings of the Symposium for Automated Agriculture for the 21st Century* (pp. 326-335). St. Joseph, MI: American Society of Agricultural Engineers.

Ambus, P. and G.P. Robertson. 1998. Automated near-continuous measurement of carbon dioxide and nitrous oxide fluxes from soil. *Soil Science Society of America Journal* 62(2):394-400.

Arslan, S. and T.S. Colvin. 2002. Grain yield mapping: Yield sensing, yield reconstruction and errors. *Precision Agriculture* 3(2):135-154.

Arslan, S., F. Inanc, J.N. Gray, and T.S. Colvin. 2000. Grain flow measurements with X-ray techniques. *Computers and Electronics in Agriculture* 26(1):65-80.

Auernhammer, H., M. Demmel, K. Muhr, J. Rottmeier, and K. Wild. 1993. *Yield measurement on combine harvesters.* ASAE Paper No. 931506. St. Joseph, MI: American Society of Agricultural Engineers.

Birrell, S.J., and J.W. Hummel. 1997. Multi-sensor ISFET system for soil analysis. In *Proceedings of the 1st European Conference on Precision Agriculture* (pp. 459-468). J.V. Stafford, (ed.). Oxford, UK: BIOS Scientific Publishers Ltd.,

Birrell, S.J. and J.W. Hummel. 2001. Real-time multi ISFET soil analysis system with automatic sample extraction. *Computers and Electronics in Agriculture* 32(1):45-67.

Bishop, T.F.A. and A.B. McBratney. 2002. Creating field extent digital elevation models for precision agriculture. *Precision Agriculture* 3(1):37-46.

Blackmore, S. and M. Moore. 1999. Remedial correction of yield map data. *Precision Agriculture* 1(1):53-66.

Bobert, J., F. Schmidt, R. Gebbers, T. Selige, and U. Schmidhalter. 2001. Estimating soil moisture distribution for crop management practices with capacitance probes, EM-38 And Digital Terrain Analysis. In *Proceedings of the 3rd European Conference on Precision Agriculture* (pp. 349-354), S. Blackmore, G. Gilbert (eds.)., AGRO Montpellier, France.

Bossu, J., Ch. Gée, J.P. Guillemin, and F. Truchetet. 2005. Feasibility of a real-time weed detection system using spectral reflectance. In *Proceedings of the 5th European Conference on Precision Agriculture* (pp. 123-130), J. V. Stafford (ed.). Wageningen, The Netherlands: Wageningen Academic Press.

Brock, R.H. and E. Karakurt. 1999. GPS precision and efficiency under variable forest types and crown density. American Society of Agricultural Engineers Paper No. 99-5035, St. Joseph, MI: ASAE.

Campanella, L., C. Colapicchioni, G. Crescentini, M.P. Sammartino, Y. Su, and M. Tomassetti. 1995. Sensitive membrane ISFETs for nitrate analysis in waters. *Sensors and Actuators: B-chemical* 27(1-3):329- 335.

Castrignanò, A., M. Maiorana, F. Fornaro, and N. Lopez. (2001). Mapping spatial and temporal variability of cone resistance data in an Italian field cropped to durum wheat. In *Proceedings of the 3rd European Conference on Precision Agriculture* (pp. 355-359), S. Blackmore, G. Gilbert (eds.). AGRO Montpellier, France.

Chapron, M., M. Requena-Esteso, P. Boissard, and L. Assemat. 1999. A method for recognizing vegetal species from multi-spectral images. In *Precision '99: Pro-*

*ceedings of the 2nd European Conference on Precision Agriculture* (1:239-248), J.V. Stafford (ed.). Sheffield, UK: Sheffield Academic Press.

Christensen, S., A.M. Walter, and T. Heisel. 1999. The patch treatment of weeds in cereals. In *Proceedings of the 1999 Brighton Crop Protection Conference* (2:591-600). Farnham, UK: British Crop Protection Council.

Coquil, B. and J.P. Bordes. 2005. FARMSTAR: An efficient decision support tool for near real-time crop management from satellite images. In *Proceedings of the 5th European Conference on Precision Agriculture* (pp. 873-880), J.V. Stafford (ed.). Wageningen, The Netherlands: Wageningen Academic Press.

Dabas, M., J. Tabbagh, and D. Boisgontier. 2001. Multi-depth continuous electrical profiling (mucep) for characterization of in-field variability. In *Proceedings of the 3rd European Conference on Precision Agriculture* (pp. 361-366), S. Blackmore, G. Gilbert (eds.). AGRO Montpellier, France.

Dijksterhuis, H.L., L.G. Van Willigenburg, and R.P. Van Zuydam. 1998. Centimeter-precision guidance of moving implements in the open field: A simulation based on GPS measurements. *Computers and Electronics in Agriculture* 20(3): 185-198.

Divis, D.A. 1999. Galileo—enthusiasm and money propel Europe's GNSS. *GPS World* 10(11):12-16.

Drummond, P.E., C.D. Christy, and E.D. Lund. 2000. Using an automated penetrometer and soil EC probe to characterize the rooting zone. In *Proceedings of the 5th International Conference on Precision Agriculture*, P.C. Robert, R.H. Rust, W.E. Larsen (eds.), Madison, WI: ASA-CSSA-SSSA, On CD.

Eden, T. and E.J. Maskell. 1928. The influence of soil heterogeneity on the growth and yield of successive crops. *Journal of Agricultural Science* 18:163-185.

Engel, R.E., D.S. Long, and G.R. Carlson, 1997. On-the-go grain protein sensing is near. *Better Crops with Plant Food* 81(4):20-23, Potash and Phosphate Institute, Norcross, Georgia.

Evans, K., R. Webster, A. Barker, P. Halford, M. Russell, J.V. Stafford, and S. Griffin. 2003. Mapping infestations of potato cyst nematode and the potential for spatially varying application of nematicides. *Precision Agriculture,* 4(2):149-162.

Fairchild, D.S. 1988. Soil information system for farming by kind of soil. In *Proceedings, International Interactive Workshop on Soil Resources: Their Inventory, Analysis and Interpretations for Use in the 1990s* (pp.159-164). St. Paul: University of Minnesota.

Fridgen, J.J., N.R. Kitchen, and K.A. Sudduth. 2000. Variability of soil and landscape attributes within sub-field management zones. In *Proceedings of the 5th International Conference on Precision Agriculture,* P C. Robert, R.H. Rust, W.E. Larsen (eds.), Madison, WI: ASA-CSSA-SSSA. On CD.

Gerhards, R., Sokefeld, M. 2001. Sensor systems for automatic weed detection. In *Proceedings of the 2001 Brighton Crop Protection Conference* (pp. 827-834), Farnham, UK: The British Crop Protection Council.

Gerhards, R., D.Y. Wyse-Pester, D. Mortensen, and G.A. Johnson.1997. Characterizing stability of weed populations using interpolated maps. *Weed Science* 45:249-269.

Goad, C. 1996. Surveying with the global positioning system. In *Global Positioning System: Theory and Applications* (pp. 501-518), B.W. Parkinson, J.J. Spilker (eds.). Washington, DC: American Institute of Aeronautics and Astronautics, Inc.

Gordon, G.S. 1998. Navigation systems integration. *The GEC Journal of Technology* 15(2):80-90.

Hague, T., J.A. Marchant, and N.D Tillett. 1997. A system for plant scale husbandry. In *Proceedings of the 1st European Conference on Precision Agriculture* (2:635-642), J.V. Stafford (ed.). BIOS Scientific Publishers, Oxford, UK.

Hague, T., N.D. Tillett, and H. Wheeler. 2006. Automated crop and weed monitoring in widely spaced cereals. *Precision Agriculture* 7(1): 21-32.

Hanquet, B., D. Sirjacobs, M-F. Destain, M. Frankinet, and J.C. Verbrugge. 2004. Analysis of soil variability measured with a soil strength sensor. *Precision Agriculture* 5(3): 227-246.

Hartz, T.K. 1994. A quick test procedure for soil nitrate-nitrogen. *Communications in Soil Science and Plant Analysis* 25(5-6):511-515.

Hayhoe, H.N., N.B. Mclaughlin, G.C. Topp, D.R. Lapen, Curnoe, W.E. 2000. Spectral analysis and filtering of measurements of moldboard plow draft. In *Proceedings of the 5th International Conference on Precision Agriculture,* P.C. Robert, R.H. Rust, W.E. Larsen (eds.). Madison, WI: ASA-CSSA-SSSA. On CD.

Hergert, G.W., W.L. Pan, D. R. Huggins, J.H. Grove and T.R. Peck. 1997. The adequacy of current fertilizer recommendations for site-specific management. In *Proceedings of 1st European Conference on Precision Agriculture* (pp. 371-378), J.V. Stafford (ed.). Oxford, UK: BIOS Scientific Publishers.

Hofstee, J.W. 1992. Handling and spreading of fertilizers: Part 2, physical properties of fertilizer, measuring methods and data. *Journal of Agricultural Engineering Research* 53(2):141-162.

Hofstee, J.W. and Goense, D. (1999). Simulation of a controller area network-based tractor—implement data bus according to ISO 11783. *Journal of Agricultural Engineering Research* 73(4):383-394.

Holmes, G., J. Bodley, P.J. Meadows, and H. Poilve. 2004. Integrating synthetic aperture radar with optical information for crop management applications. In *Proceedings of the 7th International Conference on Precision Agriculture*, D. Mulla (ed.). The Precision Agriculture Center, University of Minnesota, St Paul, MN. On CD.

Howard, K., J.L. Pringle, M.D. Schrock, D.K. Kuhlman, and D. Oard. 1993. An elevator-based combine grain flow sensor. American Society of Agricultural Engineers, Paper No. 93-1504. St. Joseph, MI: ASAE.

Hummel, J.W., K.A. Sudduth, and S.E. Hollinger. 2001. Soil moisture and organic matter prediction of surface and subsurface soils using an NIR soil sensor. *Computers and Electronics in Agriculture* 32(2):149-165.

Johnson, C.E., R.L. Schafer, and S.C. Young. 1983. Controlling agricultural machinery intelligently. In *Agricultural Electronics—1983 and Beyond. Proceedings of the National Conference on Agricultural Electronics Applications* (pp. 114-119). St. Joseph, MI: American Society of Agricultural Engineers.

King J.A., P.M.R. Dampney, M. Lark, T.R. Mayr, and R.I. Bradley. 2001. Sensing soil spatial variability by electro-magnetic induction (EMI): Its potential in precision farming. In *Proceedings of the 3rd European Conference on Precision Agriculture* (pp. 419-424), S. Blackmore, G. Gilbert (eds.). AGRO Montpellier, France.

Lachapelle, G. and J. Henriksen. 1995. GPS under cover: The effect of foliage on vehicular navigation. *GPS World* 6(3):26-35.

Langley, R.B. 1998. RTK GPS. *GPS World* 9(9):70-72.

Lark, R.M. and J.V. Stafford. 1997. Classification as a first step in the interpretation of temporal and spatial variability of crop yield. *Annals of Applied Biology,* 130:111-121.

Lee, W.S., D.C. Slaughter, and D.K. Giles. 1999. Robotic weed control system for tomatoes. *Precision Agriculture* 1:95-113.

Legg, B.J. and J.V. Stafford. 1998. Precision agriculture—new technologies. In *Proceedings of the Brighton Crop Protection Conference—Pests and Diseases.* (pp. 1143-1150). Farnham, UK: British Crop Protection Council.

Linsley, C.M. and F.C. Bauer, 1929. Illinois Agricultural Experiment Station, Circular 346.

Long, D.S., R.E. Engel, and G.R. Carlson, 1998. Grain protein mapping for precision management of dryland wheat. In *Proceedings of the 4th International Conference on Precision Agriculture* (pp. 787-796), P.C. Robert, R.H. Rust, W.E. Larson (eds.). Madison, WI: ASA-CSAA-SSSA.

Long, D. S., and T. Rosenthal. 2005. Evaluation of an on-combine wheat protein analyser on Montana hard red spring wheat. In *Proceedings of the 5th European Conference on Precision Agriculture* (pp. 385-392), J.V. Stafford (ed.). Wageningen, The Netherlands: Wageningen Academic Press.

Lowenberg-DeBoer, J. and A. Hallman. 2000. Value of pH soil sensor information. In *Proceedings of the 5th International Conference on Precision Agriculture,* P.C. Robert, R.H. Rust, W.E. Larsen (eds.), Madison, WI: ASA-CSSA-SSSA. On CD.

Lund, E.D., C.D. Christy, and P.E. Drummond, 1999. Practical applications of soil electrical conductivity mapping. In *Precision '99: Proceedings of the 2nd European Conference on Precision Agriculture* (2:771-780), J.V. Stafford (ed.). Sheffield, UK: Sheffield Academic Press.

Lütticken, R.E. 2000. Development of an Internet-based communication and information network to progress the implementation of precision agriculture. In *Proceedings of the 5th International Conference on Precision Agriculture.* P.C. Robert, R.H. Rust, W.E. Larsen (eds.), Madison, WI: ASA-CSSA-SSSA. On CD.

Marchant, J.M., H.J. Andersen, and C.M. Onyango. 2001. Evaluation of an imaging sensor for detecting vegetation using different waveband combinations. *Computers and Electronics in Agriculture* 32(2):101-117.

Marchant, J.A., R.D. Tillett, and R. Brivot, 1998. Real time segmentation of plants and weeds. *Real Time Imaging* 4(4):243-253.

Matthews, J. 1983. Some challenges for engineers. *Journal of the Royal Agricultural Society of England* 144:146-158.

McBratney, A.B., B. Minasny, and B.M. Whelan. 2005. Obtaining "useful" high-resolution soil data from proximally-sensed electrical conductivity/resistivity (PSEC/R) surveys. In *Proceedings of the 5th European Conference on Precision*

*Agriculture* (pp. 503-510), J.V. Stafford (ed.). Wageningen, The Netherlands: Wageningen Academic Press.

McLaughlin, N.B. and S.D. Burtt, 2000. Spatial mapping of tillage energy. In *Proceedings of the 5th International Conference on Precision Agriculture.* P.C. Robert, R.H. Rust, W.E. Larsen (eds.). Madison, WI: ASA-CSSA-SSSA. On CD.

McNairn, H., J.C. Deguise, J. Secker, and J. Shang. 2001. Development of remote sensing image products for use in precision farming. In *Proceedings of the 3rd European Conference on Precision Agriculture* (pp. 259-263). S. Blackmore, G. Gilbert (eds.). AGRO Montpellier, France.

Mercer, W.B. and A.D. Hall. 1911. The experimental error of field trials. *Journal of Agricultural Science, Cambridge* 4:107-132.

Miller, P.C.H. and M.E.R. Paice. 1998. Patch spraying approaches to optimize the use of herbicides applied to arable crops. *Journal of the Royal Agricultural Society of England* 159:70-81.

Miller, P.C.H., M.E.R. Paice, and A.D. Ganderton. 1997. Methods of controlling sprayer output for spatially variable herbicide application. In *Proceedings of the 1997 Brighton Crop Protection Conference—Weeds* (2:641-644). Farnham, UK: The British Crop Protection Council.

Moran, M. S., A. Vidal, D. Toufleau, J. Qi, T.R. Clarke, P.J. Pinter, T.A. Mitchell, Y. Inoue, and C. M. U. Neale. 1997. Combining multi-frequency microwave and optical data for crop management. *Remote Sensing of Environment* 61(1):96-109.

Morgan, C.L.S., J.M. Norman, R.P. Wolkowski, B. Lowery, G.D. Morgan, and R. Schuler. 2000. Two approaches to mapping plant available water: EM-38 measurements and inverse yield modeling. In *Proceedings of the 5th International Conference on Precision Agriculture,* P.C. Robert, R.H. Rust, W.E. Larsen (eds.). Madison, WI: ASA-CSSA-SSSA. On CD.

Morimoto, E., M. Suguri, and M. Umeda. 2005. Vision-based navigation system for autonomous transportation vehicle. *Precision Agriculture* 6(3): 239-254.

Moshou, D., H. Ramon, J. De Baerdemaeker. 2002. A weed species spectral detector based on neural networks. *Precision Agriculture* 3(3):209-224.

Mouazen, A.M., and H. Ramon. 2005. Bulk density maps as affected by implementation of a depth control system during on-line measurement of soil compaction. In *Proceedings of the 5th European Conference on Precision Agriculture* (pp. 487-494), J.V. Stafford (ed.), Wageningen, The Netherlands: Wageningen Academic Press.

Mulla, D. J., A.C. Sekely, and M. Beatty, 2000. Evaluation of remote sensing and targeted soil sampling for variable rate application of lime. In *Proceedings of the 5th International Conference on Precision Agriculture,* P.C. Robert, R.H. Rust, W.E. Larsen (eds.), Madison, WI: ASA-CSSA-SSSA. On CD.

Nehmdahl H. and M. H. Greve. 2001. Using soil electrical conductivity measurements for delineating management zones on highly variable soils in Denmark. In *Proceedings of the 3rd European Conference on Precision Agriculture* (pp. 461-466), S. Blackmore, G. Gilbert (eds.). AGRO Montpellier, France.

Nordmeyer, H., S. Aulich, and A. Kluge. 2005. Weed identification with chlorophyll fluorescence image analysis. In *Proceedings of the 5th European*

*Conference on Precision Agriculture* (pp. 147-153), J.V. Stafford (ed.), Wageningen, The Netherlands: Wageningen Academic Press.

Paice, M.E.R., P.C.H. Miller, and J., Bodle. 1996. An experimental machine for evaluating spatially selective herbicide application. *Journal of Agricultural Engineering Research* 60:107-116.

Perry, N.H., P.W. Lutman, P.C.H. Miller, and H.C. Wheeler. 2001. A map-based system for patch spraying weeds—weed mapping. In *Proceedings of the 2001 Brighton Crop Protection Conference* (pp. 841-846). Farnham, UK: The British Crop Protection Council.

Pierce, F.J., and P. Nowak. 1999. Aspects of precision agriculture. *Advances in Agronomy*. 67:1-85.

Reyns, P., P. Spaepen, and J. De Baerdemaeker. 2000. Site-specific relationship between grain quality and yield. *Precision Agriculture* 2(3):229-244.

Robbins, B. 1998. Real-time weed detection and classification via computer vision. In *Precision Weed Management in Crops and Pastures* (pp. 119-122), R.W. Medd, J.E. Pratley (eds.), Adelaide, Australia: Co-Operative Research Centre for Weed Management Systems.

Roth, G.W., D.B. Beegle, R.H. Fox, J.D. Toth, and W.P. Piekielek. 1991. Development of a quicktest kit method to measure soil nitrate. *Communications in Soil Science and Plant Analysis* 22(3-4):191-200.

Rouveure R., M. Chanet, P. Faure, and M.-O. Monod. 2001. Contactless measurement of roughness and soil moisture. In *Proceedings of the 3rd European Conference on Precision Agriculture* (pp. 473-478), S. Blackmore, G. Gilbert (eds.). AGRO Montpellier, France.

Russell, D.C., A.H. Dodson, M.J. Smith, M.D. Steven, and J.V. Stafford. 1999. Elevation modeling using GPS for use in precision agriculture. In *Proceedings of GIS Brasil '99*, Paper No. 885. Salvador, Brazil.

Scarlett, A.J., J.C. Lowe, and D.A. Semple. 1997. Precision tillage: In-field, real-time control of seedbed quality. In *Proceedings of the 1st European Conference on Precision Agriculture* (2:503-510). J.V. Stafford (ed.). Oxford, UK: BIOS Scientific Publishers.

Schächtl, J., G. Huber, F-X. Maidl, E. Sticksel, J. Schulz, and P. Haschberger. 2005. Laser-induced chlorophyll fluorescence measurements for detecting the nitrogen status of wheat. *Precision Agriculture* 6(2): 143-156.

Scholefield, D. and N.M. Titchen. 1995. Development of a rapid field test for soil mineral nitrogen and its application to grazed grassland. *Soil Use Management* 11:33-43.

Schueller, J.K. and Y.H. Bae. 1987. Spatially attributed automatic combine data acquisition. *Computers and Electronics in Agriculture* 2:119-127.

Scotford, I.M. and P.C.H. Miller. 2005. Vehicle mounted sensors for estimating tiller density and leaf area index (LAI) of winter wheat. In *Proceedings of the 5th European Conference on Precision Agriculture* (pp. 201-208), J.V. Stafford (ed.). Wageningen, The Netherlands: Wageningen Academic Press.

Shibusawa S., K. Ehara, T. Okayama, H. Umeda, and S. Hirako. 2005. A real-time multispectral soil sensor: Predictability of soil moisture and organic matter content in a small field. In *Proceedings of the 5th European Conference on*

*Precision Agriculture* (pp. 495-502), J.V. Stafford (ed.). Wageningen, The Netherlands: Wageningen Academic Press.

Skjemstad, J.O. 2001. Proximal sensing of soil pH and lime requirement by mid infrared diffuse reflectance spectroscopy. In *Proceedings of the 3rd European Conference on Precision Agriculture* (pp. 497-502), S. Blackmore, G. Gilbert (eds.). AGRO Montpellier, France.

Southall, B., J.A. Marchant, T. Hague, and B.F. Buxton. 1998. Model based tracking for navigation and segmentation. In *Proceedings 5th European Conference on Computer Vision* (pp. 797-811), H. Burkhardt, B. Neumann (eds.). Freiburg, Germany.

Spangler, A., H. Auernhammer, and M. Demmel, 2001. Stimulating use of open communication standards in agriculture (DIN9684 and ISO11783) with capable Open Source Program Library as possible reference implementation. In *Proceedings of the 3rd European Conference on Precision Agriculture* (pp. 719-724), S. Blackmore, G. Gilbert (eds.). AGRO Montpellier, France.

Spiller, J., A. Tapsell, and R. Peckham. 1998. Planning of future satellite navigation systems. *Journal of Navigation* 52(1):47-59.

Spilker, J.J. and B.W. Parkinson. 1996. Overview of GPS operation and design. In *Global Positioning System: Theory and Applications* (pp. 29-56). B.W. Parkinson, J.J. Spilker (eds.). Washington, DC: American Institute of Aeronautics and Astronautics, Inc.

Stafford, J.V. 1988. Remote, non-contact and in-situ measurement of soil moisture content: A review. *Journal of Agricultural Engineering Research* 41(3):151-172.

Stafford, J.V. 1999. An investigation into the within-field spatial variability of grain quality. In *Precision '99: Proceedings of the 2nd European Conference on Precision Agriculture* (1:353-361), J.V. Stafford (eds.). Sheffield, UK: Sheffield Academic Press.

Stafford, J.V. and B. Ambler. 1988. *A proposed standard for data communication on tractors.* American Society of Agricultural Engineers. Paper, No. 88-1535. St. Joseph, MI: ASAE.

Stafford, J.V. and B. Ambler. 1990. Computer vision as a sensing system for soil cultivator control. *Proceedings of the Institution of Mechanical Engineers.* C419:123-129. Presented at IME Conference on Mechatronics, Cambridge, UK, September 1990.

Stafford, J.V. and B. Ambler. 1993. A CAN data bus application on a patch sprayer. American Society of Agricultural Engineers. Paper 93-1534. St. Joseph. MI: ASAE.

Stafford, J.V. and B. Ambler. 1994. In-field location using GPS for spatially variable field operations. *Computers and Electronics in Agriculture* 11:23-36.

Stafford, J.V., B. Ambler, and H.C. Bolam. 1997. Cut width sensors to improve the accuracy of yield mapping systems. In *Proceedings of 1st European Conference on Precision Agriculture* (2:519-528), J.V. Stafford (ed.). Oxford, UK: BIOS Scientific Publishers.

Stafford, J.V., B. Ambler, R.M. Lark, and J. Catt. 1996. Mapping and interpreting the yield variation in cereal crops. *Computers and Electronics in Agriculture* 14 (2/3):235-248.

Stafford, J.V. and J.V. Benlloch. 1997. Machine-assisted detection of weeds and weed patches. In *Proceedings of the 1st European Conference on Precision Agriculture* (2:511-518), J.V. Stafford (eds.). Oxford, UK: BIOS Scientific Publishers.

Stafford, J.V. and H.C. Bolam. 1996. Improving reliability of position resolution using GPS for precision agriculture. *International Conference on Agricultural Engineering*. Paper No. 96G-008. AgEng96. November 23-26, Madrid, Spain.

Stafford, J.V. and H.C. Bolam. 1998. Near-ground and aerial radiometry imaging for assessing spatial variability in crop condition. In *Proceedings of the 4th International Conference on Precision Agriculture* (pp. 291-302), P.C. Robert, R.H. Rust, W.E. Larson (eds.). Madison, WI: ASA-CSSA-SSSA.

Stafford, J.V. and P.C.H. Miller. 1997. Spatially variable application of fertilizer—The precision agriculture approach. In *Proceedings International Dahlia Greidinger Symposium, Fertilization and the Environment* (pp. 200-209). J.J. Mortvedt (ed.). Haifa, Israel: Technion-Israel Institute of Technology.

Stafford, J.V. and J.G. Hendrick. 1988. Dynamic sensing of soil hard pans. *Transactions of the American Society of Agricultural Engineers* 31(1):9-13.

Steven, M.D. 1993. Satellite remote sensing for agricultural management: opportunities and logistic constraints. *Journal of Photogrammetry and Remote Sensing* 48(4):29-34.

Sudduth, K.A. and J.W. Hummel. 1993. Soil organic matter, CEC, and moisture sensing with a portable NIR spectrophotometer. *Transactions of the American Society of Agricultural Engineers* 36(6):1571-1582.

Sudduth, K.A., N.R. Kitchen, D.F. Hughes, and S.T. Drummond, 1995. Electromagnetic induction sensing as an indicator of productivity on claypan soils. In *Site-Specific Management for Agricultural Systems* (pp. 671-681), P.C. Robert, R.C. Rust, W.E. Larson (eds.), Madison, WI: American Society of Agronomy.

Tarr, A.B., K.J. Moore, C.L. Burras, D.G. Bullock, and P.M. Dixon. 2005. Improving map accuracy of soil variables using soil electrical conductivity as a covariate. *Precision Agriculture* 6(3): 255-270.

Thottan, J., J. F. Adsett, K.J. Sibley, and C.M. Macleod. 1994. Laboratory evaluation of the ion-selective electrode for use in an automated soil nitrate monitoring-system. *Communications in Soil Science and Plant Analysis* 25(17-18): 3025-3034.

Thylen, L., P.A. Algerbo, and C.G. Petterson. 1999. Grain quality variations within fields of malting barley. In *Precision '99: Proceedings of the 2nd European Conference on Precision Agriculture* (1:287-296), J.V. Stafford (ed.). Sheffield, UK: Sheffield Academic Press.

van Bergeijk, J., D. Goense, K. J. Keesman, and L. Speelman. 1998. Digital filters to integrate global positioning system and dead reckoning. *Journal of Agricultural Engineering Research* 70(2):135-143.

van Bergeijk, J., D. Goense, and L. Speelman. 2001. Soil tillage resistance as a tool to map soil type differences. *Journal of Agricultural Engineering Research* 79(4):371-387.

van Bergeijk, J., D. Goense, L.G. van Willigenburg, and L. Speelman. 1997. Dynamic weighing for accurate fertilizer application. In *Proceedings of the 1st*

*European Conference on Precision Agriculture* (1:263-272), J.V. Stafford (ed.): Oxford, UK: BIOS Scientific Publishers.

van Bergeijk, J., D. Goense, L.G. van Willigenburg, and L. Speelman. 2001. Dynamic weighing for accurate fertilizer application and monitoring. *Journal of Agricultural Engineering Research* 80(1):25-35.

Vansichen, R. and J. De Baerdemaeker. 1991. Continuous wheat yield measurement on a combine. In *Proceedings of Symposium on Automated Agriculture for the 21st Century* (pp. 346-355). St. Joseph, MI: American Society of Agricultural Engineers.

Viscarra Rossel, R.A., M. Gilbertson, L.Thylen, O. Hansen, S. McVey, and A.B. McBratney. 2005. Field measurements of soil pH and lime requirement using an on-the-go soil pH and lime requirement measurement system. In *Proceedings of the 5th European Conference on Precision Agriculture* (pp. 511-520), J.V. Stafford (ed.), Wageningen, The Netherlands: Wageningen Academic Press.

Viscarra Rossel, R.A. and A.B. McBratney. 1997. Preliminary experiments towards the evaluation of a suitable soil sensor for continuous, "on-the-go" field pH measurements. In *Proceedings of the 1st European Conference on Precision Agriculture* (2:493-501), J.V. Stafford (ed.). Oxford, UK: BIOS Scientific Publishers.

Viscarra Rossel, R.A. and A.B. McBratney, 2001. A two-factor deterministic response surface calibration model for site-specific predictions of lime requirement. *Precision Agriculture* 2(2):163-178.

Viscarra Rossel, R.A., D.J.J. Walvoort, A.B. McBratney, L.J. Janik, and J.O. Skjemstad. 2001. Proximal sensing of soil pH and lime requirement by mid-infrared diffuse reflectance spectroscopy. In *Proceedings of the 3rd European Conference on Precision Agriculture* (pp. 497-502), S. Blackmore, G. Gilbert (eds.). Montpellier, France: Agro-Montpellier.

Vrindts, E., J. De Baerdemaeker, and H. Ramon. 2002. Weed detection using canopy reflection. *Precision Agriculture* 3(1):63-80.

Walvoort, D.J.J., J. Bouma, P.D. Peters, and J.J. de Gruijter. 2000. Using ISFETs for proximal sensing in precision agriculture. In *Proceedings of the 5th International Conference on Precision Agriculture,* P.C. Robert, R.H. Rust, W.E. Larsen (eds.). Madison, WI: ASA-CSSA-SSSA. On CD.

Whalley, W.R. and J.V. Stafford. 1992. Real time sensing of soil water content from mobile machinery: Options for sensor design. *Computers and Electronics in Agriculture* 7:269-284.

Whelan, B.M. and A. McBratney, 2000. An approach to deconvoluting grain flow within a conventional combine harvester using a parametric transfer function. *Precision Agriculture* 2(4):373-382.

Whelan B.M., A.B. McBratney, and B. Minasny. 2001. VESPER—Spatial prediction software for precision agriculture. In *Proceedings of the 3rd European Conference on Precision Agriculture* (pp. 139-144), S. Blackmore, G. Gilbert (eds.). AGRO Montpellier, France.

Wilson, B.J. and P. Brain. 1991. Long-term stability of distribution of *Alopecurus Mysuroides* within cereal fields. *Weed Research* 31:367-373.

Wraith, J.M. and B.S. Das. 1998. Monitoring soil water and ionic solute distributions using time-domain reflectometry. *Soil and Tillage Research* 47(1-2):145-150.

Yang, C., C.L. Peterson, and G.J. Shropshire. Otawa, T. 1998. Spatial variability of field topography and wheat yield in the Palouse region of the Pacific Northwest. *Transactions of the American Society of Agricultural Engineers* 41(1):17-27.

Yao, H. and R.L. Clark. 2000. Evaluation of sub-meter and 2 to 5 meter accuracy GPS receivers to develop digital elevation models. *Precision Agriculture* 2(2): 189-200.

Zhang, N., E. Runquist, M. Schrock, J. Havlin, G. Kluitenburg, and C. Redulla. 1999. Making GIS a versatile analytical tool for research in precision farming. *Computers and Electronics in Agriculture* 22(2-3):221-231.

Zhang, N. and R. Taylor. 2000. Applications of a field-level geographic information System (FIS) in precision agriculture. In *Proceedings of the 5th International Conference on Precision Agriculture.* P.C. Robert, R.H. Rust, W.E. Larsen (eds.). Madison, WI: ASA-CSSA-SSSA. On CD.

# Chapter 3

# Soil Sensors for Precision Farming

Sakae Shibusawa

## *INTRODUCTION*

Site-specific management must be conducted based on a clear understanding of within/between-field variability of soil characteristics, which is encouraging the development of soil sensing/scanning technologies and mapping strategies (Figure 3.1). There are three stages with different strategies: uniform management over the field in Stage 1, within-field variable management in Stage 2, and variable management of hierarchical variability in Stage 3. Hierarchical variability implies here the variability in different scales, such as scales of microscopic soil structure, individual growing plants, land for cropping or cultivation systems, and geographical patterns. Technological innovation drives the shift of stages, and today precision farming is stepping into Stage 2, providing new devices/sensors and methodology. Moreover the shift of stages brings a large amount of data sampled/scanned from the field, which enriches information and knowledge for agronomic implication.

The grid-sampling-and-mapping approach has been the most adopted practice, while real-time sensors for soil characteristic detection are required for the progress in precision agriculture (Robert, 1999). Automated soil sampling and analysis methods are used for describing within-field variability with the help of interpolation such as kriging techniques with a limited number of soil samples, since the soil sampling approach is tedious and time-consuming. On-the-go soil sensing with Global Positioning Systems (GPS) is an attractive approach to obtain high-resolution of maps as well as to reduce the cost and time.

McBratney and Pringle (1997) referred to some soil sensors currently developed to detect moisture, nitrogen, clay content, organic matter, cation exchange capacity (CEC), salinity, and strength of soil and potassium. Soil sensing/scanning techniques are either invasive or noninvasive, in which

*Handbook of Precision Agriculture*

doi:10.1300/5627_03

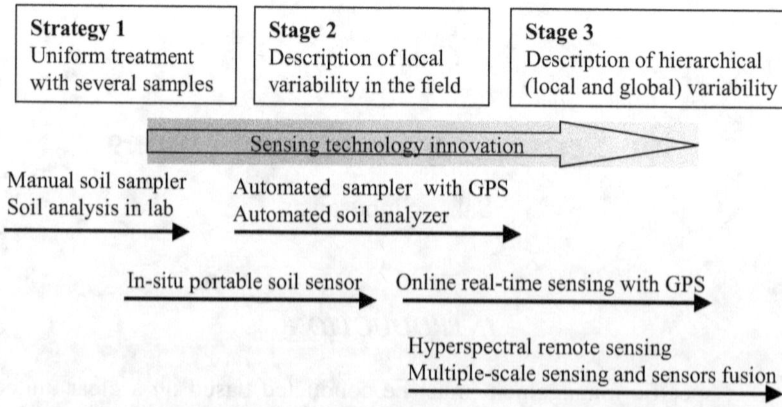


FIGURE 3.1. Three stages in development of soil sensing/scanning methodology. The stages shift from uniform treatments to locally variable treatment, followed by treatment of hierarchical variability, encouraged by sensing technology innovation such as automated soil sampling, automated in-situ soil sensing and sensors fusion.

"invasive" sensors work within the soil volume, while "noninvasive" sensors remain outside the soil. One of the most attractive invasive methods is near-infrared (NIR) with wavelengths in the range of 700-2500 nm to detect moisture, nitrogen, clay content, and CEC. Noninvasive methods involve ground penetrating radar (GPR) or time-domain reflectometry (TDR) for moisture detection, electromagnetic induction (EM) for salinity, and gamma radiometry for potassium. An EM sensor can be used for on-the-go soil sensing, and there are still many on-going trials of real-time sensing.

On the other hand, one might wonder what kinds of soil parameters must be measured or how often soil samples should be taken from the field. A general answer might be, "It depends on cropping and farming systems as well as the properties of field and soil, and the available technologies." Some recommendations have already been reported (for example, McBratney and Pringle, 1997). A farmer in Denmark said that precision farming makes one familiar with the "language of plants" (Langkilde, 1999). Precision farming practice follows spiral-development, that is, some level of knowledge is required to drive an appropriate variable-rate operation, which leads to a deeper understanding of the crops and fields. The enrichment of knowledge requires more accurate soil maps based on more precise grid design or sometimes based on a grid design locally precise and locally coarse instead of uniformly precise. That is, farm management strategy develops from uniform treatment to variable treatment and from uniform precise to variable precise toward "understanding the language of the plant."

This chapter introduces some topics on the development of on-the-go soil sensing/scanning methodology. One topic is soil sampling and mapping as a fundamental and feasible method. The second is on-the-go soil conductivity sensors providing values for several useful soil parameters simultaneously. The third is ion-selective sensors to directly measure ion activities of soil solutions. The fourth is soil reflectance sensors, followed by soil strength sensors. Finally the future development and use of soil sensors are discussed.

## *CURRENT DEVELOPMENTS AND USE OF SOIL SENSORS*

### *Soil Sampling and Mapping Approaches*

The grid-sampling-and-mapping approach has been widely used to describe the variability of soil characteristics of the field, which comprises field and laboratory work (Figure 3.2), but it is a tedious and time-consuming approach. The more detailed maps required collection of more samples even with the availability of better interpolation techniques. To improve the efficiency of the work, automation techniques are applied to tasks such as soil sampling, positioning, and soil analysis.

An all-terrain vehicle (ATV) equipped with GPS and soil sampler has been commercialized, which makes it possible not only to conduct grid sampling but also "directed sampling," referring to a design based on the spatial patterns of a field (Pockness and Boydell, 1997). Automated navigation systems are useful for controlled positioning or field scouting to produce correct maps as well as variable-rate application, and many trials on agricultural autonomous vehicles have been made using these systems

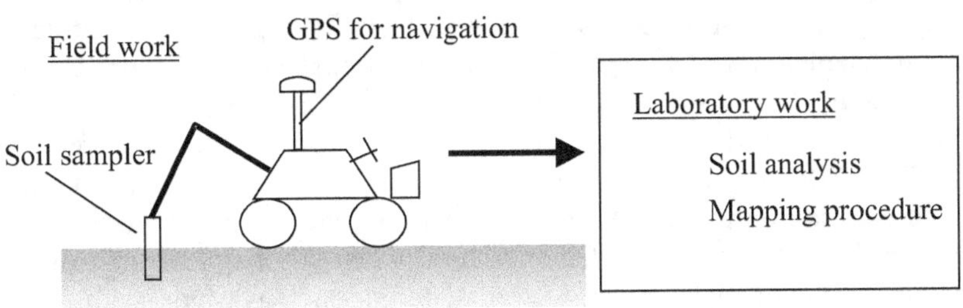


FIGURE 3.2. A typical soil sampling and mapping approach.

(e.g., Noguchi et al., 1997; Yukumoto et al., 1998). Soil-sampling devices are also important for keeping the integrity of the soil sample when density or depth information is obtained. A new design for mobile soil sampler was developed with hollow tube probes which give better results than auger types (Wright, 1996).

Hitachi Ltd. (1998, Tokyo, Japan) commercialized a new automated soil-diagnosis system in the mid-1990s. The system is composed of three components: pretreatment devices, soil analysis equipment, and diagnosis software. The pretreatment devices carried out metering of soil mass, mixing with distilled water or chemical liquids and stirring, and automated solution extraction. The soil analysis equipment includes electrodes for pH and EC measurement, chemical analysis instruments for measuring nitrogen, potassium, phosphorus, lime, and magnesium as well as an atomic absorption analysis instrument for manganese, copper, zinc, iron, lead, nickel, and cadmium. In all, 15 soil parameters are automatically measured. Finally the system produces a recommendation or suggestion on how to manage the soil. This system is very expensive but some organizations such as farmers associations can utilize it to improve their agricultural information service.

The new tools described prviously make it easy to adopt soil sampling and mapping for precision farming in the initial stage. Better methods are on-the-go sensing techniques, but soil sampling and analysis will play an important role for at least calibration purposes.

### *Soil Conductivity Sensors*

Here, soil conductivity implies the electric conductivity (EC) of bulk soil, which has been one of the most important soil parameters for crop management. Many reports and observations indicate that EC values are related to other soil properties such as clay content and are also related to yield, hence real-time EC sensing and mapping has been an attractive approach in precision farming. However, it should be noted that the usage of only the EC sensor's readings often brings about fewer requirements for the punctual utility without processing against such factors as yield or soil type.

Two different approaches have been developed and commercialized. One is based on the magnetostatic induction phenomena (Figure 3.3), in which the induction current is measured when the magnetostatic field is varied due to changes in conductive materials such as water, salinity, and clay. The other method uses electrodes inserted into the soil to measure the electric potential between the electrodes and consequently the electric re-

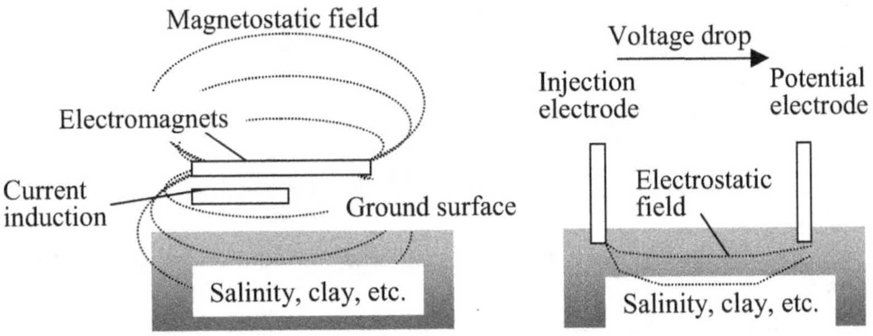


(a) Electromagnetic induction sensing (b) Resistance sensing

FIGURE 3.3. Two types of sensors for electric conductivity of bulk soil.

sistance due to the presence of conductive materials, using the magnetostatic induction approach.

Variations of soil moisture content were estimated by Kachanoski et al. (1988).

$$E_A = a\,W^2 E_W + bWE_W + E_C \tag{3.1}$$

where $E_A$ (S/m) is the bulk soil electrical conductivity, $W$ (m$^3$/m$^3$) is the volumetric soil water content, $E_W$ is the electrical conductivity of the soil solution phase, which is the same as pore water electrical conductivity in the work by Hilhorst and Balendonck (1999), $E_C$ (S/m) is the electrical conductivity of the solid phase surface, and $a$ and $b$ are regression constants. $E_A$ and $W$ were measured in-situ with EM38 and EM31 terrain meters (Geonics, Ltd., Ontario, Canada), and a time domain reflectometry (TDR) hand probe, respectively. $E_W$ was evaluated with solutions of soil samples and $E_C$ was assumed to be a function of clay content and soil texture. A graph of $W$ versus $E_A$ indicates a curvelinear relationship between the variables, which is consistent with Equation 3.1. With 52 samples of different soil textures, bulk soil electrical conductivity explained 96 percent of the spatial variation of soil moisture content using the second-order regression model below.

$$W = 0.065 + 14.6E_A - 183.0\,E_A^2 \tag{3.2}$$

Using an EM 34/3 terrain conductivity meter (Geonics, Ltd., Ontario, Canada, Williams and Hoey, 1987) measured apparent electrical conductivity ($E_A$ mS/m) in a 250 ha field; and evaluated the total soluble salts and the total clay to a depth of 15 m based on Equations 3.3 and 3.4.

$$\text{Average EC } (\mu\text{S/cm}) = -256 + 8.733\, E_A \; (R^2 = 0.777) \tag{3.3}$$

$$\text{Average clay (percent)} = 22.8 + 0.133\, E_A \; (R^2 = 0.727) \tag{3.4}$$

Average EC is the electrical conductivity determined on 1:5 (soil:water) extracts as integrated salinity.

Time domain reflectometry (TDR) is an accurate instrument for measuring water content with soil bulk electrical conductivity. Using a TDR technique, Persson (1997) measured soil solution electrical conductivity under transient conditions with varying water content, resulting in the following empirical fourth-order polynominal equation.

$$E_A = E_W (a + bW + cW^2 + dW^3 + eW^4) \tag{3.5}$$

where *a, b, c, d, e* are empirical constants, and the prediction errors are below 10 percent for $E_A$ >5.7 mS/m. Equation 3.5 calculates the soil solution electrical conductivity when the water content and bulk soil electrical conductivity are known.

Hilhorst and Balendonck (1999) developed a needle-type dielectric probe sensor, available from Delta T Devices, United Kingdom, to the measurement of pore water conductivity in the soil matrix. Calibration results showed that the soil pore water conductivity can be determined from a simultaneous measurement of the real part of permittivity and the electrical conductivity of the bulk soil independent of water content. Consequently, when the conductivity of pore water is known, the conductivity can be used to measure water content, that is, the pore water conductivity sensor can be used to calibrate soil salinity tools like EM83 or Veris 3100 (Veris Technologies, Kansas).

Using the EM38 electromagnetic induction sensor, the relationship between EM readings and depths to claypan was investigated on 30 large plots of 18 m by 190 m at the Missouri Management System Evaluation Area (Sudduth et al., 1995). Five farming systems were assigned to the plots: a minimum-till corn-soybean rotation, minimum-till corn-soybean-wheat rotation, ridge-till corn-soybean rotation, no-till corn-soybean rotation, and continuous grass-legume. The results showed that EM data were highly predictive ($R^2 = 0.81$) of depth to claypan, and that EM data were correlated with the particular crop and farming system present on each plot, and that an excessively moist soil profile was not suitable for predicting the claypan depth.

Interpretation of EM sensors' outputs often required additional information such as yield or soil properties obtained by respective ways. Sensors fusion, simultaneous measurement of different parameters relating to a target property using multiple sensors, is an efficient way to interpret EM records. An EC-sensing penetrometer, the Veris 3000 model, has been commercialized by Veris Technologies (Kansas) in order to measure both soil conductivity and soil strength (Drummond et al., 2000). This sensor is a combined penetrometer and soil EC probe of a dielectric ring. The penetrometer can measure the soil strength that gives an indicator of hardpan depth or soil texture change, and the EC probe detects the electric conductivity of bulk soil.

Sudduth et al. (2000) compared a Veris 3000 EC-sensing penetrometer with ASAE standard penetrometers (Veris Technologies, Kansas) in terms of cone index (CI) data because the dimensions of the cones were all different. The EC-sensing tip showed larger mean CI than the ASAE standard small and large cones. The increment in mean CI for the EC-sensing tip over the ASAE standard small tip was about 12 percent for an Illinois site and 24 percent for a Missouri site. The EC-sensing tip also had a lower replication error than the two ASAE standard tips, which made it possible to develop CI-dependent transformation equations to convert data collected with this tip to correspond with data from another tip. Moreover, they reconfirmed the relationships between CI, electric conductivity of bulk soil and moisture content, that is, low electric conductivity was related to high CI and to low moisture content. Penetrometer EC data exhibited trends similar to those seen with mobile EC data, which implies that the EC-sensing penetrometer can be useful for calibration or for understanding the response of mobile, profile-weighted EC measurements.

The other approach is to use electrodes inserted into the soil for measuring electrical resistance, which results in electrical conductivity as the inverse of resistance. A Veris 3100 Soil Mapping System with GPS has been commercialized as an electrode insertion type EC meter (http://www.veristech.com/about.htm). The sensor consists of six coulter-electrodes formed from disk-type blades cutting into the soil, two of which introduce an electrical potential into the soil. The remaining four coulters are spaced to measure soil electric conductivity by the voltage drop over two approximate depths: 0-30.5 cm and 0-91.5 cm. Using a Veris 3100 soil electrical conductivity sensor, Hartsock et al. (2000) investigated how many soil parameters were related to EC meter records. Results showed that the soil electric conductivity was positively related to soil Ca ($R^2 = 0.59$), Mg ($R^2 = 0.56$) and soil moisture ($R^2 = 0.72$), and inversely related to the depth to a clay increase ($R^2 = 0.38$-$0.66$, depending on the field), depth to bedrock ($R^2 = 0.33$), and depth to fragipan ($R^2 = 0.80$). Clark et al. (2000) also conducted tests to confirm soil hardpan depth with a Veris 3100 mapping system accompanied with the ASAE standard large cone penetrometer (ASAE Standard, 1999a,b). Their results indicated that the inverse of shallow electrical conductivity can predict the depth to a traffic hardpan with a correlation coefficient of about 0.6. The previous results indicate that many factors affect EC meter records and that further knowledge or information based on science and technology should be required when analyzing the data for farming operations.

Electrical resistance is the inverse of electrical conductivity. Dabas et al. (2000) have developed Multidepth Continuous Electrical Profiling to measure the electrical resistivity (ER) in the field. The sensor system has four rolling electrodes 1 m apart, the front electrodes for current injection and the rear electrodes for potential detection, connected to an electric resistance meter with a high rate of data scanning. Four more electrodes can be installed for simultaneous measurement at different depths of 0.5, 1 and 2 m. A differential global positioning system (DGPS) and a Doppler radar system were mounted on the system for precise positioning. Test runs provided clear ER maps in submeter resolution, and a high cross-correlation between ER and yield.

### *Portable Ion-Selective Sensors*

In-situ determination of ion activity in soil solutions, such as pH or nitrate-N, has been of keen interest for farmers and agricultural engineers in order to control plant growth as well as to manage and conserve both the soil fertility and the environment. Handy-type sensors using ion-selective electrodes have already been commercialized, such as the Horiba C-141 pH meter (Horiba, Ltd., Kyoto, Japan) and the Culturyzer-Mini multi-ion meter (Techno Medica Ltd, Tokyo, Japan, a kind of ion-selective electrodes) for pH, $K^+$, $Ca^{2+}$, $Mg^{2+}$, $H_2PO_4^-$, $NH_4^+$, $NO_3^-$ and EC in culture solution, but they still require laborious tasks for soil preparation, etc.

In general, on-the-go ion-selective sensors need fundamental configurations such as those shown in Figure 3.4, a soil-sampling mechanism, a soil-extraction mechanism, an extract injection/drain control system, a washout device, and a main body of ion-selective devices. All units need to be arranged for compactness and they need to perform the entire procedure within an acceptable time. One of the key points for the sensors is to get a fast response time under best ion-selectivity with low interference by other ions, which leads us to think about attractive candidates for the sensor de-

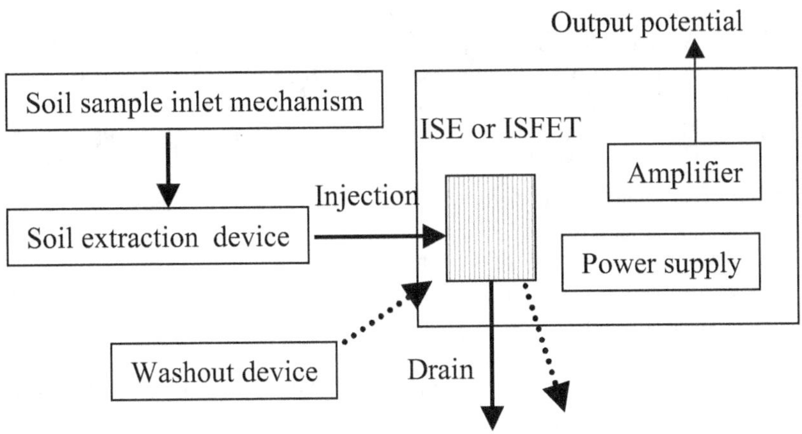


FIGURE 3.4. The composition of a real-time ion sensor using an ion-selective electrode (ISE) or an ion-selective field effect transistor (ISFET).

vices, such as ion-selective electrode (ISE) and ion-selective field effect transistor (ISFET).

Viscarra Rossel and McBratney (1997) conducted a preliminary experiment on the speed of response with four potentiometric pH sensors. Test sensors were a glass electrode, a glass micro-electrode, a metallic (Sb) electrode and a pH ISFET. An ISFET is expected as the second generation ion-selective electrode. Test solutions were standard buffered solutions at pH 3.0, 4.0, and 5.0, and soil sample extracts in 1:5 soil-0.01M $CaCl_2$, 1:5 soil-$H_2O$, and a buffer proposed by Shoemaker, McLean, and Pratt (SMP buffer) (Shoemaker et al., 1961; Tran and van Liertop, 1982). Results showed that the ISFET sensor had the highest rate of response with a mean response time of 9.1 s in the standard buffer, 7.2 s in water, 6.1 s in 0.01M $CaCl_2$, and 5.4 s in SMP buffer. Moreover, faster stirring rates resulted in an increase of response time for the ISFET sensor, and a 1:1 soil to solution ratio produced a faster response time than a 1:5 soil to solution ratio.

A trial using electrodes for an on-the-go soil pH sensor was conducted. In this kind of soil sensor, naturally moist soil samples should be dealt with in situ, which sometimes makes it difficult to keep measuring conditions constant compared with laboratory tests.

Adamchuck et al. (1998) discussed how to develop an on-the-go soil pH sensor compared with classical procedure. For example, the classical procedure follows five steps.

1. Calibrate the pH meter over an appropriate range.
2. Use a standard soil mass scale to measure a soil sample into a cup.
3. Dilute the sample with distilled water to a defined mass ratio.
4. Stir for 5-10 seconds and let stand for 10-30 min.
5. Finally, place the electrodes in the slurry and read the results.

On-the-go sensing requires

1. liminating the pre-treatment such as drying,
2. selecting electrodes in terms of handling, response time, and variation of water content,
3. soil sampling mechanism, and
4. digital components for system control, data recording as well as positioning with GPS.

Using a flat-surface combination pH electrode (Sensorex Model 450C, available from Sensorex, California) with a flat glass membrane, Adamchuck et al. (1998, 1999) confirmed that the response time was less than 5 s for 20 g soil samples mixed with 5 ml double-deionized (DDI) water. They also found that an average increase in pH of naturally moist soil (0.15 to 0.30 g/g gravimetric water content) was about 0.2 against the measurements of standard test at 1 g/g gravimetric water content. If a correlation equation between moist soil mV output and reference pH is attained, this device could be used for real-time pH sensing.

Adamchuck et al. (1998) developed a soil-sampling mechanism with a soil-sampling platform, which maintains sensor-soil contact for more than 5 seconds with low pressure applied to the sensitive glass membrane and also enables the user to adjust the sampling depth from 10-20 cm. It takes a soil sample at a defined location and depth and places the soil in contact with the electrode. The whole automated pH sensor system consists of a sampling mechanism, air compression and rinsing water circuits, a pH sensor and meter, and a GPS unit. The on-the-go soil pH sensor can run at traveling speeds of up to 7 km/h (Adamchuck and Morgan, 1999) and gives pH maps of the fields. Further validation experiments are now expected.

Use of ISFETs to sense soil soluble parameters such as nitrate-N are an attractive approach, and Birrell and Hummel (1997, 2000) have produced useful results with this approach. Based on the theory of ion-selective membranes, different types of nitrate ion-selective membranes were evaluated (Birrell and Hummel, 2000). Membranes tested were two ligands of tetradodecylammonium nitrate (TDDA) and methyltridodecylammoium nitrate (MTDA), four plasticizers of tri- (2-ethylhexyl) trimellitate (TOTM), nitrophenyloctylether (NPOE), dibutylphthalate (DBP) and tri- (2-ethylhexyl) phosphite (TEHP), and eight combinations of the two ligands and the four plasticizers. Experiments on the sensitivity and selectivity proved that all membrane compositions could be good candidates for the development of ion-selective nitrate sensors, while MTDA-ligand-based membranes resulted in slightly higher sensitivity to nitrate than TDDA-ligand-based membranes. All membranes were more sensitive to $I^-$, $SCN^-$, and $ClO_4^-$ ions than to the nitrate ion, while membranes formed with the TDDA ligand were less sensitive to these interfering ions than were the membranes formulated with the MTDA ligand. They concluded that the TDDA ligand membrane was better from a selective point of view though interference problems still remained.

Birrell and Hummel (1997) also investigated an automated soil extraction prototype system combined with ISFETs and a flow injection analysis (FIA) system. The flow injection system consisted of a variable-speed, multichannel peristaltic pump to propel the carrier and sample solution to the 2-piston Valco Model EF60 injection valve with a Valco C1-417-2460 electric actuator (Valco Instruments Co., Ltd., Texas) controlled by a relay output or TTL logic. The injection system worked to send the carrier stream through the flow cell during the washout period, and switched to direct sample solution to the flow cell during the injection period. A multi-ISFETs chip was attached to the flow cell to direct the flow stream over the ISFET membrane. The automated soil extraction system utilized a slide mechanism and soil plunger to move a constant volume of soil from the soil inlet to one of eight chambers in a rotating extraction cylinder. The system developed was able to measure soil nitrate in manually extracted soil solution ($R^2 > 0.9$), and analyze soil samples within 1.25 s. This performance was satisfactory for real-time soil sensing. On the whole, the proposed ion-selective sensors still need more investigation and improvement for commercial use.

## *Soil Reflectance Sensor*

### *Soil Reflectance Spectroscopy*

Elementary soil reflectance sensing is based on the recognition of soil color changes due to variations in moisture content, organic matter content, and so on. Sudduth and Hummel (1991), for example, investigated the possibility of using color coordinates to predict soil organic matter content and concluded that color coordinates and visible reflectance data provided considerably poorer predictions of soil organic matter content.

More attractive methods involve spectroscopic approaches using spectral reflectance in the near infrared (NIR) range. The following describes some fundamental studies to clarify the possibility of soil parameter evaluation with soil reflectance. Early efforts with spectrophotometers (Beckman DK-2A and Beckman DU, available from Beckman Instruments Inc., California, renamed Beckman Coulter Inc. in 1997) over the wavelength ranges of 185 to 3,500 nm and 210 to 1,000 nm confirmed that moisture content, organic matter, and particle size strongly influenced the reflectance and absorbance of radiation energy (Bowers and Hanks, 1965). The reflectance decreased as moisture increased, and was much more evident at 1,900 nm, a moisture-sensitive wavelength, while the oxidation of soil organic matter increased the reflectance. With both kaolinite and bentonite clays, the reflectance increased exponentially as particle size decreased.

To determine the carbonate concentration in 90 soils formed on sedimentary rocks in Israel, Ben-dor and Banin (1990) used a spectrophotometer (Mattson Alpha Centauri, wavelengths of 1,000 to 2,500 nm, 1.9 $cm^{-1}$ resolution). From the original reflectance $R_f$, absorption log $(1/R_f)$, the first derivative and the second derivative were calculated, and then multivariate linear regression analysis was conducted. Reflectance from pre-heated (600°C) and nonheated soils was compared to confirm the ability to remove spectral interference caused by hydroxyl groups. Heating to 600°C did not show a significant improvement in statistics for the calibration. Instead, optimal prediction was obtained with a calibration set containing more than 20 soils with a wide range of carbonate content and with a variation of soil properties. Finally they concluded that carbonate in soils in the concentration range of 10 to 75 percent can be satisfactorily estimated.

Based on NIR reflectance spectroscopy an attractive trial to determine many soil parameters was conducted (Reeves and Van Kessel, 1999). Target parameters were (1) various constituents (total N, total C, active N, biomass and mineralizable N, and pH), (2) field parameters (soil source, depth from which sample was obtained, type of tillage use), and (3) rate of application of nitrate ammonium $NH_4NO_3$ fertilizer. A total of 180 soil samples were collected from one well-drained and one poorly-drained silt loam field in Maryland, where long-term field experiments with no-till (NT) and plow tillage (PT) have been conducted for more than 20 years. Five depths (0-2.5, 2.5-5.0, 5.0-7.5, 7.5-12.5, 12.5-20.0 cm) and three different rates of $NH_4NO_3$ fertilization were also included in the experimental design. Soil preparation followed oven drying (105°C for 24 h) and grinding to a fine powder, and then spectral reflectance was collected in the wavelength range of 1,100 to 2,498 nm at every 2 nm with 10 nm resolution.

Using a variety of data subsets, the authors performed point averaging, derivatives, and other pretreatments (mean centering, variance scaling, multiplicative scatter correction and baseline correction), followed by Partial Least Square (PLS) Regression and the Residual Error Sum of Square (PRESS) *F*-statistic with one-cut cross-validation. Calibrations based on all samples indicated that locations were spectrally quite different while differentiation on the basis of tillage, N-rate, and depth was completely unsatisfactory. Prediction of pH, total C, total N, active N, and biomass N was more successful while more errors appeared at higher values for active N and biomass N. Mineralizable N was not determinable at all. Moreover, cross-calibration-determination analysis showed that the determination of NT samples by the PT calibration was much more satisfactory than the reverse, though a wider range in values existed in the NT samples. This implies that the method for selecting a data set for calibration is very sensitive to the validation/determination performance.

Reeves et al. (1999) conducted an experiment for evaluating manure N and carbon mineralization for more than three months with NIR spectroscopic analysis, which resulted in calibrations for nitrate, ammonia, total N, and evolved carbon dioxide. The NIR spectroscopic approach could be used to follow the temporal variations of the manure-amended soils.

Ehsani et al. (1999) also found, through simulation studies, that the NIR technique was very sensitive to calibration, and showed through experimental studies that proper calibration could help the NIR technique in detecting soil nitrate. However, it failed if the calibration equation did not contain all the constituents the actual sample contained. They inferred that there is a need for site-specific calibration. Details are as follows. In their tests two types of soil—Yolo loam and Capy clay—were used, and soil samples were enriched with three mineral-N (ammonium sulfate, calcium nitrate and ammonium nitrate) from 10 to 100 ppm, respectively. Six hundreds and sixty samples were analyzed using NIR reflectance over the 1,100 to 2,500 nm wavelengths in increments of 2 nm, with 250 scans per sample averaged for each spectrum using an NIRSystem (Silver Spring, Maryland) spectrophotometer (model 6500). Other soil samples were collected from Yolo loam and Capy clay fields after 10 application levels of calcium nitrate fertilizer and were mixed together to provide 1,000 subsamples for the respective soil types.

In the tests, a Partial Least Square (PLS) regression with 20 wavelengths resulted in high values of the coefficient of multiple determination, 0.965 for ammonium nitrate, 0.996 for ammonium sulfate and 0.919 for calcium nitrate with small standard errors below 6 ppm, respectively. Using samples pooled across soil types resulted in a good fit for ammonium nitrate ($R^2$ = 0.989, standard error = 2.8 ppm), but the results were not as good for calcium nitrate ($R^2$ = 0.82, standard error = 8.9 ppm). The other experiment proved that a calibration equation for soil samples containing 8 chemicals failed to evaluate nitrate concentration for the samples with two additional chemicals added to the original. With samples from the Yolo loam field, the results were reasonable but not as good as the laboratory test results ($R^2$ = 0.794, standard error = 7.7 ppm). The results reconfirmed again that the NIR technique is applicable only for a good calibration and that site-specific calibration will be necessary.

Ehsani et al. (2000) have further investigated the possibility of using mid-infrared reflectance spectroscopy for the determination of nitrate in soil, and reported that a strong nitrate absorption peak appeared at 7,194 nm. The wavelet denoising technique provided a higher coefficient of determination of 0.856 for nitrate prediction, and a continuous wavelet transformation could remove specular reflection effects and band overlaps.

Chang et al. (2001) presented one of the most comprehensive evaluations of NIR spectroscopy for many soil properties using the principal component regression analysis. They collected a diverse group of 802 soil samples from 448 sites in the Major Land Resource Areas, and obtained reflectance spectra in the range of 400-2,500 nm wavelengths at 2 nm intervals using a Perstrop NIR Systems 6500 scanning monochromator (Foss NIRSystems, Silver Spring, Maryland). Soil samples used were all air-dried and 8-mm sieved. Thirty-three soil parameters of chemical, physical, and biochemical properties were measured in the methods of soil parameter analysis for respective samples passing a 2-mm sieve. Averaging the spectral reflectance (optical density = 1/reflectance) data in the range of 1,300 to 2,500 nm at every five adjacent spectral points yielded 120 new spectral data, and then the first derivatives were calculated. A small subset of 30 samples used in the calibration was selected from all of those data based on the shorter square Euclidean distance to the derivative reflectance spectra. Up to 20 principal components were used in the regression calculations to found the largest $R^2$ value and smallest root mean standard error on cross validation. Soil parameters evaluated successfully were total C, total N, moisture, CEC, 1.5 MPa water, basal respiration rate, sand, silt, and Mehlich III extractable Ca ($R^2 > 0.8$). Parameters evaluated with less accuracy ($R^2$ = 0.5 to 0.8) were Mehlich III extractable metals (Fe, K, Mg, Mn), exchangeable cations (Ca, Mg, K), sum of exchangeable bases, exchangeable acidity, clay, potentially mineralizable N, total respiration rate, biomass C, and pH.

### *On-the-Go Soil Reflectance Sensors*

An on-the-go real-time soil spectrophotometer needs several units or devices to be integrated as shown in Figure 3.5. First, a soil opener to provide the bare soil surface should be designed and installed. Second a photomet-

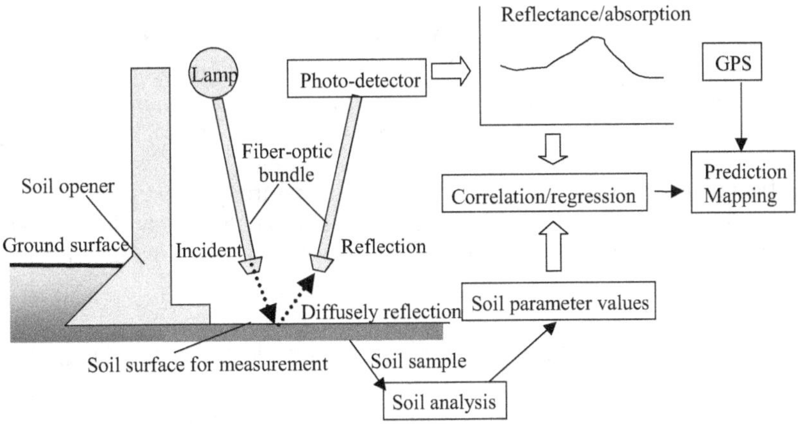


FIGURE 3.5. An image of an on-the-go real-time soil spectrophotometer.

ric system composed of an illumination lamp, fiber-optic bundles to transport the light energy, and a photo-detector with data processing system to calculate the spectral reflectance is needed. Third, a parameter prediction system involving calibration/validation equations is needed. Finally, a map-generating system combined with positioning data from GPS is needed. The design of each unit or device should be specialized to fit the natural soil conditions, and the measuring stability and high-speed measurement also requires system compactness.

Using 30 soils in seven different soil types from sand to clay, an NIR spectroscopic investigation was conducted to determine air-dried soil organic matter under 0.033 and 1.5 MPa moisture tension (Sudduth and Hummel, 1993b). Two 985-point spectra were obtained for each sample: visible/NIR data from 400.8 to 1116.8 nm at a 0.8 nm bandpass, and NIR data from 802.0 to 2,592.0 nm at a 2.0 nm bandpass. The NIR dataset (800-2,580 nm on a 20 nm spacing) was analyzed with PLS using raw reflectance ($R_f$) data, optical density transformed data defined by $\log_{10}(1/R_f)$ and Kubelka-Munk transformed data. Using the 0.033 and 1.5 MPa visible/NIR dataset with 20 different sets of initial conditions, prediction wavelengths were selected by the iterative scanning program with correlations ranging from $R^2 > 0.67$. The prediction equation used was the four-wavelength model shown below,

$$OMC = k_1 + k_2(OD_1 - OD_2)/(OD_3 - OD_4), \tag{3.6}$$

where $OMC$ = soil organic matter content ( percent), $OD_n$ = optical density at wavelength $n$, $k_n$ = multiple regression coefficient. One of the best combinations of wavelengths is listed in Table 3.1. The numerator wavelengths were chosen near two ends of the visible range, and the denominator wavelengths were chosen in the yellow-green range from 530 to 580 nm.

Li et al. (2000) have reported on a systematic investigation on NIR reflectance spectroscopy using 95 samples of naturally moist soils collected from the upland (sandy loam, 40 samples) and paddy fields (clay, 45 samples). The soil samples were provided to determine moisture content and to

TABLE 3.1. Results of soil parameter determination with soil reflectance spectra.

| Parameter | Soil | Treatment | R2cal | R2val | SEP | Wavelength (nm) | References |
|---|---|---|---|---|---|---|---|
| Moisture (percent) | Sandy loam | 2nd deriv.[a] Linear model | 0.969 | 0.968 | 0.53 | 1450 | Li et al., 2000 |
| | Silty loam | Reflectance SMLR[b] | 0.896 | — | 1.97 | 2160, 2108, 2094, 2047, 1883, 1870, 1679, 1626, 1607 | Ahmad et al., 2000 |
| SOM (percent) | Sandy loam | 1st deriv.[c] Linear model | 0.763 | 0.897 | 0.79 | 1170 | Li et al., 2000 |
| | 30 soils of clay to sand | OD[d] 4-wavelength model | 0.81 | — | 0.32 | 424.0, 808.0, 575.2, 548.8 | Sudduth and Hummel, 1991 |
| $NO_3$-N ($mg \cdot kg^{-1}$) | Sandy loam | 1st deriv. Exponential model | 0.949 | 0.953 | 9.90 | 824, 1280, 1768 | Li et al., 2000 |
| pH | Sandy loam | 1st deriv. Linear model | 0.825 | 0.906 | 0.10 | 1272 | Li et al., 2000 |
| EC ($mS \cdot m^{-1}$) | Sandy loam | 1st deriv. Linear model | 0.732 | 0.715 | 1.13 | 744, 856, 1698 | Li et al., 2000 |

*Note:* $R^2_{cal}$ and $R^2_{val}$ are the coefficients of determination for calibration and validation, respectively, and *SEP* is the standard error of prediction.

[a]2nd deriv.: Second derivative of soil reflectance spectra.

[b]SMLR: Stepwise multiple linear regression.

[c]1st deriv: First derivative of soil reflectance spectra.

[d]OD: Optical density defined as $\log_{10}$ (1/reflectance).

collect NIR reflectance just after collecting from the fields. Soil organic matter (SOM) was determined by the muffle-oven dry method, and nitrate-nitrogen ($NO_3$-N) content, pH, and EC were measured with Horiba portable sensors (Kyoto, Japan) C-141, B-173 and B-212, respectively. Variation in the parameters of the upland soil samples was 2.9-108 percent in moisture content, 31-136 mg·kg$^{-1}$ in $NO_3$-N content, 18.6-27.1 percent in SOM content, 7.2-28.0 mS·m$^{-1}$ in EC and 6.0 to 7.4 in pH. The paddy soil samples showed variations of 17.5-31.7 percent for moisture content, 88-170 mg·kg$^{-1}$ for $NO_3$-N content, 3.2-7.0 percent for SOM content, 4.4-10.5 mS·m$^{-1}$ for EC and 6.5-8.0 for pH, respectively. Using the original reflectance spectra in the 250 to 2,500 nm wavelength range on a 2 nm spacing obtained by a desktop spectrophotometer (Shimadzu, Kyoto, Japan, UV-3100PC, 20 nm peak width at half height for each spectrum), Kubelka-Munk transform and the first and second derivatives of reflectance were calculated. Selected wavelengths for the calibration by correlation analysis, multiple linear regression and validation analysis were performed. Some of the results are shown in Table 3.2.

The results showed strong moisture absorption peaks at wavelengths of 1,450, 1,840, and 2,200 nm. Treatments at 1,450 nm wavelength gave the best moisture prediction model (a type of $y = a + bx$). Coefficients of determination were 0.969 (standard error 0.979 percent) for calibration and 0.988 (standard error 0.529 percent) for validation with the upland soil using the second derivative data, and 0.709 (standard error 2.17 percent) in calibration and 0.737 (standard error 2.40 percent) in validation for the paddy soil using the Kubelka-Munk transform data.

SOM-sensitive wavelengths were 710, 920, 1,170, 1,700, 1,920 and 2,320 nm and those wavelengths (except for 1,920 nm) were used in the SOM prediction analysis because a moisture absorption band includes 1,920 nm wavelength. Using the first derivatives of reflectance at those wavelengths, a single-wavelength linear regression model predicted SOM contents with coefficients of determination greater than 0.7.

The other three parameters, $NO_3$-N, pH, and EC, are important indices of ion activity in soil solution, although there are no clearly sensitive bands in the wavelength range of 250 to 2,500 nm. Correlation analysis was performed with all treatments over the full wavelength range and the ten wavelengths with the highest scores of correlation selected after confirming the independency among the treatments. Using the ten wavelengths, regression analysis provided good prediction models, a single-wavelength model for EC, a three-wavelength model for pH, and a three-wavelength exponential model for $NO_3$-N. The results obtained led to develop the real-time spectrophotometer discussed later.

TABLE 3.2. Results of calibration and validation for soil parameter prediction using soil reflectance obtained by the real-time soil spectrophotometer.

| Parameters | Treatment | Calibration ($n$ = 15) | | Validation ($n$ = 10) | | Wavelength (nm) |
|---|---|---|---|---|---|---|
| | | $R^2_{cal}$ | *SEC* | $R^2_{val}$ | *SEP* | |
| Moisture (percent db) | 2nd deriv.[a] | 0.908 | 1.893 | 0.655 | 3.11 | 606, 1329, 1499 |
| SOM (percent) | 2nd deriv. | 0.948 | 0.259 | 0.647 | 0.56 | 606, 1311, 1238 |
| $NO_3$-N (mg·kg$^{-1}$) | 2nd deriv. | 0.803 | 3.699 | 0.539 | 47.4 | 589, 1014 |
| pH | 1st deriv.[b] | 0.714 | 0.101 | 0.541 | 0.13 | 959, 1214 |
| EC (mμS·m$^{-1}$) | 2nd deriv. | 0.735 | 23.975 | 0.650 | 4.17 | 456, 984, 1014 |

*Source:* Adapted from I made Anom, Shibusawa, Sasao, and Hirako, 2001.

*Note:* $R^2_{cal}$ and $R^2_{val}$ are the coefficients of determination for calibration and validation, respectively. SEP and SEC are the standard error of prediction and calibration, respectively.

[a]2nd deriv.: Second derivative of optical absorbance in Kubelka-Munk transform.

[b]1st deriv.: first derivative of optical absorbance in Kubelka-Munk transform. Kubelka-Munk transform = (1 – reflectance)$^2$/(2 × reflectance).

Shonk et al. (1991) developed a prototype of real-time tractor-mounted soil reflectance sensor to measure soil organic matter (SOM) content. The sensor consisted of compact transmitter and receiver modules that utilize light reflectance to measure SOM content. Eight red-light-emitting diodes (LED) (peak at 660 nm with 40 nm bandwidth) were positioned in the sensor housing and the photodiode detected the reflected radiant energy from the soil. Power to both the LEDs and photodiode was supplied by a 12-V battery and regulated to maintain a constant voltage. The sensor determined the SOM content ($R^2 > 0.893$) well under the conditions of low SOM content less than 6 percent at depths of 7.5 cm and 10 cm at 4.8-6.5 km·h$^{-1}$ traveling speeds.

Sudduth and Hummel (1993a) designed an on-the-go NIR spectrophotometer to sense soil organic matter content. Specialized features included a circular filter monochromator, a fiber optic coupling for sensing remote samples, and a software algorithm for zero baseline correction. The spectrophotometer could collect soil reflectance spectra in the range of 1,650 to 2,650 nm wavelengths under 55 nm bandwidth every 200 ms.

A 50-W, 12-V halogen lamp supplied the illumination source to circular variable filters (CVF) positioned behind a spherical bi-convex lens to focus the lamp image through the input slit (2 mm width, 10 mm radial length) onto the CVF surface. The CVF is a wavelength selection device for producing monochromatic light using circular variable interference filters having spectral characteristics that can be varied by means of physical rotation, available with a wavelength range of 1,600 to 2,900 nm on a bandwidth 55 nm (Optical Coating Laboratory, Inc., Santa Rosa, California). The CVF disc was arranged so that a quarter would cover the CVF area and three-quarters would cover the light block area. A servo-controlled DC motor-generator allowed adjustment of the filter disk rotation rate. The monochromatic light from the CVF was directed to the soil surface through a 610 mm silica fiber optic bundle with an available transmission range of 350 to 2,400 nm. Diffusely reflected light energy from the soil surface was detected by a PbS photodetector with good sensitivity to wavelengths in the range of 1,000 to 3,500 nm. Using an excitation and preamplifier circuit the spectral reflectance collected through the PbS detector was provided as signal outputs in voltage.

A feasibility study with the portable NIR spectrophotometer was also conducted to determine soil organic matter content with 30 representative Illinois soils at 1.5 MPa and 0.033 MPa moisture tension levels (Sudduth and Hummel, 1993b). Laboratory tests resulted in accurate calibration equations for organic carbon below 3 percent ($R^2 = 0.89$, $SEP = 2.3$ percent) using a PLS technique when the data were smoothed to a 40 nm data point spacing over the wavelength range from 1,640 to 2,640 nm. Variations in sample-to-sensor height from 5 to 25 mm slightly reduced the predictive capability of organic carbon when the wavelength data were normalized before calibration. The cation exchange capacity (CEC) and moisture content of the test set of 30 Illinois mineral soils were also estimated successfully in the laboratory, where the CEC levels tested were 10 to 60 $cmol \cdot kg^{-1}$ and the moisture levels were below 30 percent. But soil-moving tests showed considerable degradation of organic carbon predictive capability in the laboratory. In-furrow soil-sensing tests at slow traveling speed did not provide accurate estimates of soil organic matter content. Movement of the sample during reflectance scanning could have been a major contributor to the lack of predictive capability.

It is clear that the NIR reflectance spectroscopic approaches can be used for real-time sensing of several soil parameters using naturally moist soils, but a serious problem has emerged, that is, how to handle the fluctuations in the measuring reflectance caused by a moving-soil or moving-sensor condition.

*Advanced Real-Time Soil Spectrophotometer*

A real-time soil spectrophotometer with a GPS has been developed to collect underground soil reflectance, with which maps of several soil parameters can be produced successfully (Shibusawa et al., 2000, 2001).

An outline of this soil spectrophotometer is shown in Figure 3.6. The soil spectrophotometer was designed to collect the soil reflectance at depths of 150 to 400 mm at submeter spacings. The sensor system is composed of three main units: the sensor unit's housing, the soil penetrator and probes housing, and the external sensing/monitoring devices.

Figure 3.7 shows the soil-penetrator with the housing for micro optical devices. The penetrator tip with flat plane edge ensures uniform soil cuts, and the soil flattener behind finishes to produce a uniform surface. In the housing, seven micro optical devices were arranged. Two optical fiber probes, passing light energy with wavelengths of 400 to 2,400 nm, were used to provide an illuminated area of about 50 mm in diameter on the soil surface. Two additional optical fiber probes were used for collecting soil reflectance in the visible and NIR ranges. One fiber bundle passed reflected energy in the 400 to 900 nm wavelength range, while the other optical probe carried reflected energy in the 900 to 2,400 nm wavelength range. A micro

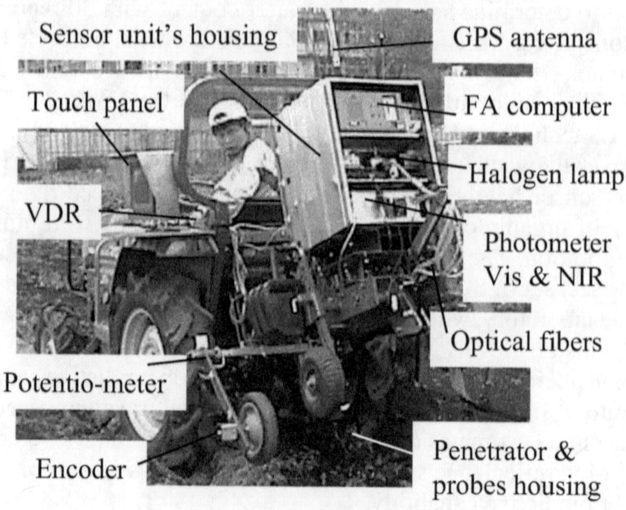


FIGURE 3.6. A tractor-mounted real-time soil spectrophotometer developed to collect underground soil reflectance in-situ. *Source:* Adapted from Shibusawa et al., 2000.

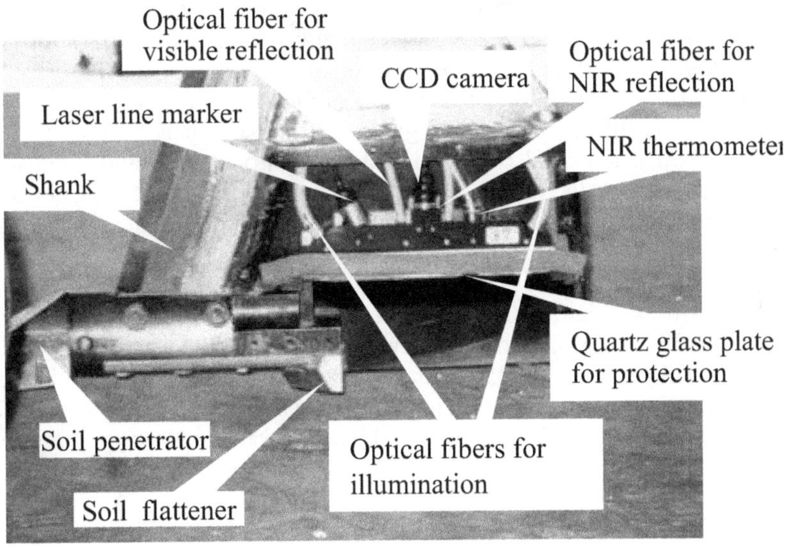


FIGURE 3.7. Soil penetrator and sensor probes arrangement.

CCD camera was adjusted to monitor a 75 mm focus point on the soil surface.

The sensor unit's housing included core devices of the system, such as a 150-W halogen lamp, a spectrophotometer (Carl Zeiss Ltd.), a personal computer (IBM, PC/AT, Pentium MMX), an RTK-GPS (Trimble MS740) receiver, etc. The spectrophotometer had a 256-channel linear photodiode array to quantify the reflected energy in the wavelength range of 400 to 900 nm. A 128-channel linear photodiode array was used to quantify the reflected energy in the wavelength range of 900 to 1,700 nm. Data scanning time was more than 4 ms. Scanned data were integrated at a single scanning treatment to get averages at respective spectra.

A video data recorder/monitor on the tractor immediately displayed the soil surface images during the experiment. The displayed images were used to watch for emergencies, such as blockage with obstacles, and the images gave information to eliminate data in data analysis. The liquid crystal monitor served as a touch control panel, and a mouse and keyboard were also available to access the personal computer.

A preliminary experiment was done to confirm the effect of soil surface disturbance on the spectral reflectance obtained with the real-time spectrophotometer in the laboratory (Shibusawa et al., 2000; Sato 2001). Ten disturbed surface conditions, for example, stripes, hollows, ditching, convex and concave soil surfaces, were provided with reference to observations of measuring soil surface in field experiments. Measurements of original absorbance shifted due to disturbed soil surface while the first derivatives lay on a single curve as shown in Figure 3.8. Another experiment of varying probe-to-soil distance also proved that no significant deviation was observed in the first derivatives within the displacement of ±15 mm in distance. Consequently, derivative operation can reduce fluctuations in the spectral reflectance scanned.

Another experiment was conducted in a 0.5 ha paddy field at the Experimental Farm of Kyoto University in December 1999. The soil has 47 percent sand, 30 percent silt, and 23 percent clay. The working speed was about 1 km·h$^{-1}$, under 20 ms scanning time and a 3 s scanning period, which resulted in a roughly 1 m scanning interval along a segment. The spectral collections were performed along the field with about 5 m spacing between segments, which resulted in more than 800 locations for soil reflectance data. The working depth was 200-250 mm.

For calibration purposes, 25 soil samples were collected at the same position and depth as the scanning points and analyzed in the laboratory for moisture, SOM, $NO_3$-N, pH, and EC. Fifteen samples were used for calibration and the remaining ten samples were used for validation. The moisture content was obtained with 110°C 24-hour oven-drying. SOM content was

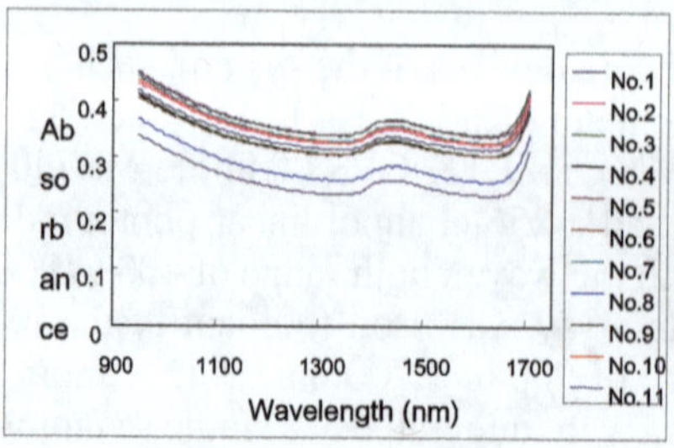


(a) Original soil absorbance

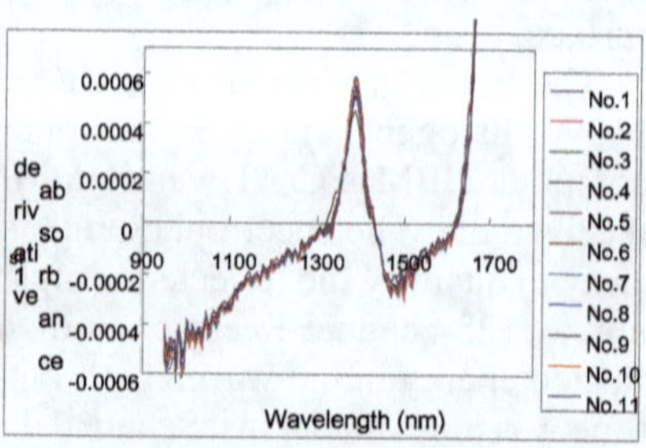


(b) The 1st derivative of the soil absorbance

FIGURE 3.8. Effect of soil surface roughness on light energy absorbance in the NIR range. The surface conditions are flat smooth (No. 1), stripes of 3 mm depth at 5 mm spacing (No. 2 and No. 3), scattered hollows with 5 mm diameter and 3 mm depth (No. 4, 10, 11), ditching of 5 mm depth at 5 mm spacing (No. 5, 6, 8), convex of 5 mm elevation (No. 7), concave of 5 mm depth (No. 9). Test soil was Kanto loam (25 percent sand, 34 percent silt, 20 percent clay) in a constant moisture content of 25 percent dry basis. *Source:* Adapted from Shibusawa et al., 2000. (See corresponding PLATE 3.8 in the central gallery.)

evaluated as ignition loss after a 4-hour burn at 800°C in a muffle oven. $NO_3$-N, pH, and EC were analyzed with the clear layer at the top of the muddy water using portable ion meters HORIBA C-141, B-212, and B-141, respectively. The test muddy water was provided by diluting 5 g dried soil with 25 g distilled water, stirring for 30 minutes, and then incubating under room conditions for 24 hours.

With the spectral reflectance, four steps were followed. The first step was linearization with a Kubelka-Munk transform, and the second step was to eliminate the optical interference in the spectral data with multiplicative scatter correction. The third step was to reduce the number of wavelengths used for calibration with correlation analysis, including derivative operation. The final step was calibration using the stepwise multiple linear regression analysis with the S-Plus Data Analysis Software (Mathematical Systems, Inc., Tokyo, Japan). The calibration model was quantified using the standard error for calibration (SEC), standard error for prediction (SEP) and coefficient of determination ($R^2$). Semivariance analysis was performed using the GS+ Geostatistics Software and soil maps were obtained by the block kriging method.

Results of calibration and validation analysis produced high $R^2$ values and fewer errors for respective soil parameters. The second derivatives of light absorption tended to provide best-fit prediction models (I Made Anom, Shibusawa, Sasao, and Hirako, 2001).

With prediction models, values for soil parameters were evaluated at 860 locations in the field. The means and standard deviations were 48.4 and 6.5 percent for moisture content, 9.51 and 1.06 percent for SOM content, 421 and 110 $mg{\cdot}kg^{-1}$ for $NO_3$-N content, 6.83 and 0.39 for pH, and 17.3 and 7.0 $mS{\cdot}m^{-1}$ for EC, respectively. Using the evaluated values as observation values, the semivariance analysis was performed. The fit of data using the spherical model for experimental semi-variograms resulted in $R^2$ values of 0.95, 0.93, 0.94, 0.99, and 0.93 for moisture content, SOM content, $NO_3$-N content, pH, and EC, respectively. Within the experimental field, the SOM content had the shortest range of spatial correlation (29.20 m), followed by $NO_3$-N content (34.50 m), moisture content (38.60 m), pH (40.40 m), and EC (46.60 m).

From the results of the semivariance analysis, the soil parameter maps were then developed (Figure 3.9). The maps were interpolated by block kriging with ten-neighborhood interpolation. Errors of kriging to observed values were estimated over 40 grids of 10 m square in illustration. Error means and standard deviations were –0.35 and 2.25 for moisture content (percent), –0.01 and 0.46 for SOM content (percent), 0.8 and 27.2 for $NO_3$-N content ($mg{\cdot}kg^{-1}$), –0.04 and 0.22 for pH, and 0.3 and 1.2 for EC ($mS{\cdot}m^{-1}$), respectively (I Made Anom Shibusawa, Sasao, and Hirako, 2001).

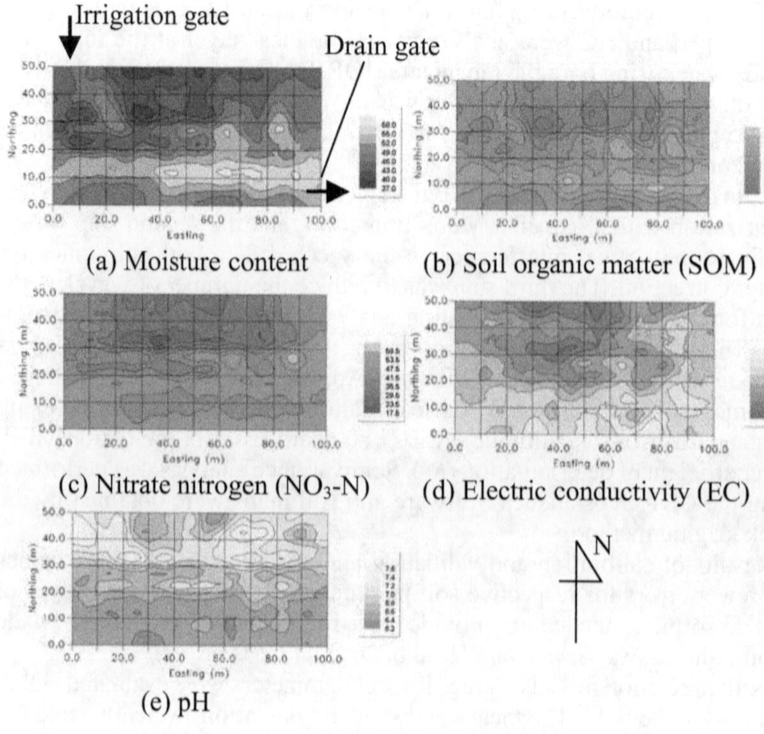


FIGURE 3.9. Kriged maps of a 0.5 ha paddy field of 50 m width and 100 m length using 860 data of soil reflectance collected by the real-time soil spectrophotometer. The working depth was about 200 mm, and spacing was 1 m on north-south segments and 5 m on east-west segments. (See corresponding PLATE 3.9 in the central gallery.)

The variability distribution of the soil parameters shows some stripes in the west-east direction, such as a high moisture belt in the east part as well as high SOM, high $NO_3$-N, high EC, and low pH belts. The irrigation gate was located at the northwest and the drain gate was at the southeast, which could have induced the stripelike maps since water flowed from north to south.

### *Soil Strength and Sensor Fusion*

Soil strength is an important index describing mechanical impedance that affects crop growth and yield, and many studies have been made to evaluate and reduce the stress to crop growth due to soil compaction (Bowen, 1981). Mechanical properties of soils have also been investigated in soil dynamics aspects, during which useful tools to evaluate and analyze soil strength were introduced (Upadhyaya et al., 1994). These aspects and information have indicated that soil strength is a function of soil moisture, density (a parameter of soil compaction), and texture (clay content as representative), as shown in Figure 3.10. When a reference tool is used to determine soil strength, the mechanical resistance of the tool comes out as a parameter of soil strength. Unfortunately, such data on mechanical resistance are temporally changeable due to moisture variation if the soil type and all other parameters are held constant. This makes it difficult to understand the variation in the field in terms of soil compaction or soil strength using a single map of mechanical resistance.

If a variation map of standardized soil strength is produced that takes into account the effects of soil moisture, density, and texture, then site-specific soil management is definitely improved. That is why sensor fusion needs to use those three parameters; moisture, density, and texture, simultaneously when the mechanical resistance is measured.

Cone penetrometer resistance, as a means of measuring soil mechanical properties, has been limited by the significant effect of soil moisture on the data collected. A technique using a portable NIR soil sensor (Sudduth and Hummel, 1993a,b) in conjunction with a bifurcated fiber optic bundle, was considered as a possible addition to a penetrometer cone to provide simultaneous soil moisture and penetration resistance data (Newman and Hummel, 1999). The modified penetrometer with NIR sensor is called an NIR penetrometer (Ahmad et al., 2000).

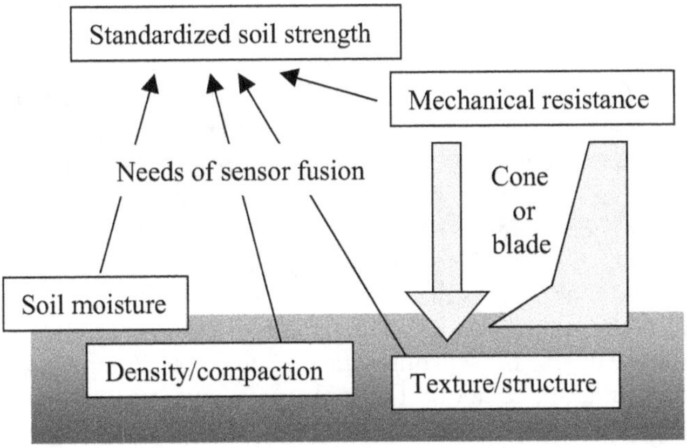


FIGURE 3.10. A concept of soil strength sensor requiring sensor fusion.

In the NIR penetrometer instrument, a silica fiber bifurcated optical cable was positioned and aligned to receive light from the illumination source, a 50-W, 12-V quartz-halogen automotive-type lamp. The CVF was retained in the instrument as the wavelength selection mechanism. The fiber optic bundle was encased in the penetrometer shaft. A 90° turning mirror was installed within the penetrometer cone to direct the light through the air gap and sapphire aperture lens on the soil surrounding the cone. Diffusely reflected energy from the soil surface reentered the cone and was transmitted through the other bifurcated cable to the exit end, where the photodetector was attached.

Sudduth and Hummel (1993b) conducted two kinds of laboratory experiments, a continuous probe movement test and a step-wise probe insertion test, with three soils to evaluate the performance of the modified standard penetrometer combined with the NIR soil sensor. The former test followed an ASAE standard test procedure (ASAE Standards, 1999a) with a rate of insertion at 2.5 $mms^{-1}$. In the latter test, the probe was inserted at the same rate and stopped at 2 cm increments to collect the soil reflectance. The moisture was determined by the gravimetric method and was evaluated with spectral reflectance using partial least square regression (PLSR) and multiple linear regression (MLR). Twenty-four wavelengths were used for the PLSR model and 4 wavelengths, 2,473, 2,309, 2,282, and 1,742 nm were used for the MLR model. They introduced standardized equations to predict cone index values at different moisture content using independent parameters, penetration resistance, dry density, clay content, and moisture content determined by the gravimetric, PLSR, and MLR methods, respectively. The force prediction equation for the data over the entire range of soil type/moisture content for the three methods yielded $R^2 = 0.895$ or higher, while the moisture prediction accuracy was not good (±2.29 percent).

Ahmad et al. (2000) conducted further investigation with the NIR penetrometer and proved that stepwise multiple linear regression (SMLR) yielded $R^2 = 0.896$ for moisture content prediction under 1.97 percent of standard error of calibration (SEC) and 2.38 percent of standard error of prediction (SEP). Nine moisture-related wavelengths were 2,160, 2,108, 2,094, 2,048, 1,883, 1,870, 1,679, 1,626, and 1,607 nm. Ahmad et al. also concluded that soil moisture, density, and clay content were significant variables in predicting the soil cone index at a lower moisture tension level.

Based on the assumption that the amount of pore space in soil will be a function of the degree of soil compaction, Clement & Stombaugh (2000) investigated a unique technique for measuring air permeability. The basic idea of the sensor was to measure the pressure created by a constant flow rate of air into the soil. The system was composed of a mass-flow control valve, a pressure transducer, a data acquisition unit, a GPS, and a tine or shank with a sensor-to-soil interface. Air entered the soil through two small orifices in a small vertical plate mounted on the bottom of a tillage shank. The thickness of the vertical plate (1.27 cm) was chosen to help the sensor slice through the soil with minimal disruption while maintaining structural durability. The sensor design prevented leakage of air through the use of the vertical plane for sensor-to-soil interface with a horizontal plate above to seal air flow vertically. The results of field experiments showed that the sensor indicated higher pressure on average in areas suspected of having highly compacted soils.

With a fusion of soil mechanical resistance sensor and moisture sensor, a unique tool was developed for evaluating the soil texture and compaction (Lui et al., 1996). The idea of the sensor fusion is based on the relationship that the draft of a tillage tool depends on its geometry, soil properties, its operating depth, and speed, as follows.

$$D = \mathrm{f}\,(w, G_i, B_D, M_C, T_C, d, S) \tag{3.7}$$

Where $D$ = draft of the tillage tool, $w$ = operation depth, $G_i$ = all other geometric parameters such as length, curvature, cutting angle, etc., $B_D$ = soil bulk density, $M_C$ = moisture content, $T_C$ = texture, $d$ = operating depth, and $S$ = travel speed. When a reference tillage tool is selected with standard geometry and when the speed and depth are kept constant, Equation 3.7 becomes:

$$D = g\,(B_D, M_C, T_C) \tag{3.8}$$

If soil moisture is independently measured, then Equation 3.8 is reduced to Equation 3.9.

$$D = g_1\,(B_D, T_C) \times g_2\,(M_C) \tag{3.9}$$

The function $g_2(M_C)$ tends to be exponential in nature. The texture/soil compaction index $TCI = g_1(B_D, T_C)$ can be determined with measurements of draft $D$ and moisture $M_C$.

The new sensor developed was composed of a load cell detecting mechanical resistance to a chisel-type tine, a dielectric-based moisture sensor, a radar gun to monitor the speed, a linear potentiometer to measure the depth, and a GPS. The device was operated at a constant speed and depth (305 mm) in the field of Yolo loam under different moisture conditions while keeping dry bulk density constant. The experiment showed that *TCI* was 49.4 kN for the untilled Yolo loam and 7.85 kN for the tilled Yolo loam in the range of moisture content 13 to 17 percent. However it was difficult for this sensor to distinguish the effects from texture such as clay content and soil compaction or bulk density independently.

One of the apparent features of soil texture, which implies here the distribution of soil particles, is the distribution of soil aggregates. There are few studies on the relationship between soil strength and soil aggregate distribution but an interesting approach to determine soil aggregates was tried.

Zou et al. (2000) applied a fiber-optic displacement sensor (Model 88nl, Series 88, Philtec Inc., Arnold, Maryland) to scan the surface of bulk samples of soil aggregates. The sensor uses a mixed bundle of glass fibers to transmit light to and receive reflected light from a target object. Over the useful operating range, the analog voltage output signal from the sensor varies proportionately with the distance from the target and the reflectivity of the target surface. Using dried and sieved test soils with fractions of <0.18, 0.18-0.25, 0.25-0.5, 0.5-1.0, 1.0-2.0, 2.0-4.0, and 4.0-8.0 mm, a laboratory experiment was conducted at 5.08 mm/s by a universal test machine. Data acquisition rates were 10, 100, 200 Hz. The mean of sensor output voltage decreased significantly as aggregate size increased, and the relationship between the number of peaks in the output and the mean aggregate diameter appeared to be of an exponential form. Sampling frequencies of 100 Hz (20 readings/mm) or greater provided more precise discrimination of soil aggregate sizes. This kind of approach has the potential to determine soil texture or soil structure.

## *FUTURE DEVELOPMENT AND PROSPECTS*

Commercialized real-time soil sensors such as the EC meters of EM series and Veris 3000 series exist now. Basically EC is an indicator of soil salinity and clay content; however, many investigations have shown that EC is well related to moisture content, soil organic matter content, and other fertility parameters. For a specific soil or field, EC meters would have high potential to determine the soil properties in question.

A sensor fusion of cone penetrometer to an NIR or EC instrument might be available for practical uses to determine standardized soil strength, since cone penetrometers are very popular for evaluating soil strength. But it is difficult to get high speed measurement if vertical penetration is adopted. Using a tillage tool for detecting mechanical resistance will be convenient for high speed on-the-go sensing.

An attractive approach involves a tractor-mounted real-time soil spectrophotometer. It could scan soil reflectance at as many points as one is interested in, providing highly accurate maps of several soil parameters simultaneously. It is expected to be commercialized in a few years. Using ion-selective electrodes is one of the most fascinating approaches for directly measuring the ion activities in soil solutions, but this requires further trials for on-the-go sensing. However, convenient portable ion electrode meters are available today.

During the past decade many attractive prototypes for on-the-go soil sensors have been developed and some of them are already on the market. One cannot predict what type of soil sensors will be developed and commercialized in the coming decade. The progress in technology development for soil sensors will be much faster than can be imagined.

The development of on-the-go soil sensors enables users to collect soil data at any spacing from the field, which could give solutions to the question: How narrow should the grid spacing be for correctly describing the field variability?

Using the real-time soil spectrophotometer, for instance, I Made Anom, Shibusawa, Sasao, Sakai, et al. (2001) evaluated kriged errors of five soil parameters (moisture, SOM, $NO_3$-N, pH and EC) due to change in sample size as well as grid spacing. More than 800 data units were collected from 0.5 ha and 1.2 ha paddy fields at 1 m by 5 m spacings and 1 m by 10 m spacings, respectively. The results showed that the kriged standard errors became minimum when the sample size was 180 (5 m × 5 m spacing) for all parameters in the 0.5 field, and 104 (10 m × 10 m spacing for pH) to 243 (4 m × 10 m for SOM) in the 1.2 ha field. Birrell et al. (1996) reported a similar result with soil-sampling approach in a 28 ha field. Soil samples were collected at a 25 m grid with additional samples taken at a 100 m grid. Two additional samples were taken at a 100 grid (total 3 samples at the 100 m grid points). At each 100 m grid point three other short-range samples were obtained at 2 m, 5 m, 10 m distance from the grid point on a straight line in a random direction. The number of whole samples was more than 700. Soil parameters discussed were potassium, phosphorous, and pH. The results showed that the 100-m-grid map including the short-range samples, which had about 120-sample data, had the same trends of the whole-data map, and that 100-m-grid maps of co-incident 28-sample data missed the considerable local variety. These kinds of efforts need to clarify soil sampling strategies for site-specific management, and real-time on-the-go soil sensors encourage the efforts.

## *CONCLUSIONS*

Entering the next stage of precision farming requires on-the-go soil sensors that enable the production of more accurate soil maps with less cost and time. An electric conductivity (EC) soil sensor that is available for practical use right now has been providing useful information, such as EC records, relating to other soil parameters. A cone penetrometer has been developed to measure standardized soil strength as a sensor-fusion instrument, in combination with an NIR sensor or EC sensor. A more fascinating method involves the soil reflectance spectroscopy method, which enables several soil parameters to be measured simultaneously, and a real-time soil spectrophotometer will be commercialized in the near future. Development of ion-selective soil sensors has been of keen interest since it can provide direct measurement of ion activities in soil solutions.

Someone once said that "data rich, application poor" is a big issue to be solved for promoting precision farming. The development of on-the-go soil sensors will provide a new platform of wise practices for precision farming.

## REFERENCES

Adamchuck, V.L., M.T. Morgan, and D.R. Ess. 1998. *Rapid determination of soil pH for precision farming.* ASAE Paper 983094, ASAE, Orlando, Florida.

Adamchuck, V.L. and M.T. Morgan. 1999. *Evaluation of automated soil pH mapping.* ASAE Paper 991100, ASAE, Toronto, Canada.

Ahmad, I.S., J.W. Hummel, K.A. Sudduth, and S.T. Drummond. 2000. Soil characterization using a near-infrared penetrometer. *Proceedings of the 5th International Conference on Precision Agriculture,* P.C. Robert, R.H. Rust, W.E. Larson (eds.). Madison, Wisconsin: ASA/CSSA/SSSA. Compiled in CD-ROM.

ASAE Standards. 1999a. ASAE EP542: Procedure for using and reporting data with the soil cone penetrometer. Approved February, 1999, St. Joseph, MI: ASAE.

ASAE Standards. 1999b. ASAE S 313. 3: Soil cone penetrometer. Approved February, 1999, St. Joseph, MI: ASAE.

Ben-dor, E. and A. Banin. 1990. Near-infrared reflectance analysis of carbonate concentration in soils. *Applied Spectroscopy,* 44(6):1064-1069.

Birrell, S.J., K.A. Sudduth and J.W. Hummel. 1996. Nutrient mapping implications of short-range variability. *Proceedings of the 3rd International Conference on Precision Agriculture* (pp. 207-216), P.C. Robert, R.H. Rust, W.E. Larson (eds.). Madison, Wisconsin, ASA/CSSA/SSSA.

Birrell, S.J. and J.W. Hummel. 1997. Multi-sensor ISFET system for soil analysis. *Precision Agriculture '97, Proceedings of the 1st European Conference on Precision Agriculture* (pp. 459-468), J.V. Stafford (ed.). Oxford, UK: BIOS Scientific Publisher Ltd.

Birrell, S.J. and J.W. Hummel. 2000. Membrane selection and ISFET configuration evaluation for soil nitrate sensing. *Transaction of the ASAE,* 43(2):197-206.

Bowen, H.D. 1981. Alleviating mechanical impedance. In *Modifying the root environment to reduce crop stress* (pp. 21-60), ASAE Monograph No. 4, G.F. Arkin, H.M. Taylor (eds.), St. Joseph, Michigan: ASAE.

Bowers, S.A. and R.J. Hanks. 1965. Reflection of radiant energy from soils. *Soil Science* 100(2):130-138.

Chang, C.W., D.A. Laird, M.J. Mausbach, and C.R. Hurburgh. 2001. Near-infrared reflectance spectroscopy—Principal components regression analyses of soil properties. *Soil Science Society of America Journal,* 65:480-490.

Clark, R.L., D.E. Kissel, F. Chen, and W. Adkins. 2000. *Mapping soil hardpans with the penetrometer and soil electrical conductivity.* ASAE Paper 001042. Milwaukee, Wisconsin: ASAE.

Clement, B.R., and T.S. Stombaugh. 2000. *Continuously-measuring soil compaction sensor development.* ASAE Paper 001041. Milwaukee, Wisconsin: ASAE.

Dabas, M., A. Brisard, J. Tabbagh, and D. Boigontier. 2000. Use of a new sub-meter multi-depth soil imaging system (MuCEP). *Proceedings of the 5th International Conference on Precision Agriculture,* P.C. Robert, R.H. Rust, W.E. Larson (eds.). Madison, Wisconsin: ASA/CSSA/SSSA. Compiled in CD-ROM.

Drummond, P.E., C.D. Christy, and E.D. Lund. 2000. Using an automated penetrometer and soil EC probe to characterize the rooting zone. *Proceedings of the 5th International Conference on Precision Agriculture,* P.C. Robert, R.H. Rust, W.E. Larson (eds.). Madison, Wisconsin: ASA/CSSA/SSSA. Compiled in CD-ROM.

Ehsani, M.R., S.K. Upadhyaya, D. Slaughter, S. Shafii, and M. Pelletier. 1999. A NIR technique for rapid determination of soil mineral nitrogen. *Precision Agriculture,* 1: 217-234.

Ehsani, M.R., S.K. Upadhyaya, D. Slaughter, L.V. Protsailo, and W.R. Fawcett. 2000. *Quantitative measurement of soil nitrate content using mid-infrared diffuse reflectance spectroscopy.* ASAE Paper 001046, Milwaukee, Wisconsin.

Hartsock, N.J., T.G. Mueller, G.W. Thomas, R.I. Barnhisel, K.L. Wells, and S.A. Shearer. 2000. Soil electrical conductivity variability. *Proceedings of the 5th International Conference on Precision Agriculture,* P.C. Robert, R.H. Rust, W.E. Larson (eds.). Madison, Wisconsin: ASA/CSSA/SSSA. Compiled in CD-ROM.

Hilhorst, M.A. and J. Balendonck. 1999. A pore water conductivity sensor to facilitate non-invasive soil water content measurements. *Precision Agriculture '99, Proceedings of the 2nd European Conference on Precision Agriculture* (pp. 211-220), J.V. Stafford, (ed.). BIOS Scientific Publisher Ltd, Oxford.

Hitachi Ltd. 1998. Welcome to HITACHI Agri World. Advertising CD-ROM, Hitachi, Ltd. (in Japanese).

I Made Anom, S.W., S. Shibusawa, A. Sasao, K. Sakai, C. Hache, and S. Hirako. 2001. Sampling strategy in soil mapping with real-time soil spectrophotometer. *Proceedings of the 2nd IFAC/CIGR Workshop on Intelligent Control for Agricultural Application,* at Bali, Indonesia, August, 2001, H.K. Purwadaria, R.J. Widodo, K.B. Seminar, H.A. Tjokronegoro, Suroso (eds.). International Federation of Automatic Control (IFAC). 40-4329.

I Made Anom, S.W., S. Shibusawa, A. Sasao, and S. Hirako. 2001. Soil parameters maps in paddy field using the real-time soil spectrophotometer. *Journal of Japanese Society of Agricultural Machinery,* 63(3):51-58.

Kachanoski, R.G., E.G. Gregorich, and I.J. van Wesenbeeck. 1988. Estimating spatial variations of soil water content using noncontacting electromagnetic inductive methods. *Canadian Journal of Soil Science,* 68:715-722.

Langkilde, N. 1999. Practical experiences with precision agriculture. *Precision Agriculture '99, Proceedings of the 2nd European Conference on Precision Agriculture* (pp. 35-46), JV. Stafford (ed.). Oxford UK: BIOS Scientific Publisher Ltd.

Li, M.Z., A. Sasao, S. Shibusawa, and K. Sakai. 2000. Soil parameters estimation with NIR spectroscopy. *J. Japanese Society of Agricultural Machinery,* 62(3): 111-120 (in Japanese).

Lui, W., S.K. Upadhyaya, T. Kataoka, and S. Shibusawa. 1996. Development of a texture/soil compaction sensor. *Proceedings of the 3rd International Conference on Precision Agriculture* (pp. 617-630), P.C. Robert, R.H. Rust, W.E. Larson (eds.). Madison, Wisconsin: ASA/CSSA/SSSA.

McBratney, A.B. and M.J. Pringle. 1997. Spatial variability in soil-implications for precision agriculture. *Precision Agriculture '97. Proceedings of the 1st European Conference on Precision Agriculture* (pp. 3-31), J.V. Stafford (ed.). Oxford UK: BIOS Scientific Publisher Ltd.

Newman, S.C. and J.W. Hummel. 1999. *Soil penetration resistance with moisture correction.* ASAE Paper 993028, Toronto, Canada.

Noguchi, N., K. Ishii, and H. Terao. 1997. Development of agricultural mobile robot using a geomagnetic direction sensor and image sensors. *Journal of Agricultural Engineering Research,* 67:1-15.

Persson, M. 1997. Soil solution electrical conductivity measurements under transient conditions using time domain reflectometry. *Soil Science Society of America Journal,* 61:997-1003.

Pockness, S., and B. Boydell. 1997. Soil sampling & precision farming. *Agri-Alternatives,* Media Discoveries, Inc., May-June 1997. 11.

Reeves, J.B. III, G.W. McCarty, and J.J. Meisinger. 1999. Near infrared reflectance spectroscopy for the analysis of agricultural soils. *Journal of Near Infrared Spectroscopy,* 7:179-193.

Reeves, J.B. III and J.S. Van Kessel. 1999. Investigations into near infrared analysis as an alternative to traditional procedures in manure nitrogen and carbon mineralization. *Journal of Near Infrared Spectroscopy,* 7:195-212.

Robert, P.C. 1999. Precision Agriculture: Research needs and status in the USA. *Precision Agriculture '99, Proceedings of the 2nd European Conference on Precision Agriculture* (pp. 19-33), J.V. Stafford (ed.). Oxford, UK: BIOS Scientific Publisher Ltd.

Sato, N. 2001. Real-time soil parameter prediction model with spectral reflectance. Master's thesis, Graduate School of Agricultural Science, Tokyo University of Agriculture and Technology (in Japanese).

Shibusawa, S. 2000. Precision farming approaches for model and control. *Proceedings of the XIV Memorial CIGR World Congress* (pp. 133-141), November 28-December 1, Tsukuba, Japan.

Shibusawa, S., S.W. I Made Anom, S. Sato, A. Sasao, and S. Hirako. 2001. Soil mapping using the real-time soil spectrophotometer. *Precision Agriculture '01,*

*Proceedings of the 3rd European Conference on Precision Agriculture,* J.V. Stafford (ed.). Oxford, UK: BIOS Scientific Publisher Ltd.

Shibusawa, S., H. Sato, S. Hirako, A. Otomo, and A. Sasao. 2000. A revised soil spectrophotometer. *Proceedings of the 2nd IFAC/CIGR International Workshop on BIO-ROBOTICS II* (pp. 225-230), S. Shibusawa, M. Monta, H. Murase (eds.). Nov. 25-26, Osaka, Japan.

Shoemaker, H.E., E.O. McLean, and P.F. Pratt. 1961. Buffer methods for determining lime requirement of soils with appreciable amounts of extractable aluminum. *Soil Science Society Proceedings,* 25:274-277.

Shonk, J.L., L.D. Gaultney, D.G. Shulze, and G.E. Van Scoyoc. 1991. Spectroscopic sensing of soil organic matter content. *Transactions of the ASAE,* 34(5): 1978-1984.

Sudduth, K A. and J.W. Hummel. 1991. Evaluation of reflectance methods for soil organic matter sensing. *Transaction of the ASAE,* 34(4):1900-1909.

Sudduth, K.A. and J.W. Hummel. 1993a. Portable near-infrared spectrophotometer for rapid soil analysis. *Transaction of the ASAE,* 36(1):185-193.

Sudduth, K.A. and J.W. Hummel. 1993b. Soil organic matter, CEC, and moisture sensing with a portable NIR spectrophotometer. *Transaction of the ASAE,* 36(6):1571-1582.

Sudduth, K.A., N.R. Kitchen, D.F. Hughes, and S.T. Drummond. 1995. Electromagnetic induction sensing as an indicator of productivity on claypan soil. *Proceedings of the 2nd International Conference on Site-Specific Management for Agricultural Systems* (pp. 671-681), P.C. Robert, R.H. Rust, W.E. Larson (eds.). Madison, Wisconsin: ASA/CSSA/SSSA.

Sudduth, K.V., J.W. Hummel, N.R. Kitchen, and S.T. Drummond, 2000. *Evaluation of a soil conductivity sensing penetrometer.* ASAE Paper 001043, ASAE, Milwaukee, Wisconsin.

Tran, T.S. and W. van Liertop. 1982. Lime requirements for attaining pH 5.5 and 6.0 of coarse texture soil using buffer pH methods. *Soil Science Society of America Journal,* 49: 1008-1014.

Upadhyaya, S.K., W.J. Chancellor, J.V. Perumpral, R.L. Schafer, W.R. Gill, and G.E. VandenBerg (eds.). 1994. *Advances in Soil Dynamics.* ASAE Monograph No.12, St. Joseph, Michigan: ASAE.

Viscarra Rossel, R.A., and A.B. McBratney. 1997. Preliminary experiments towards the evaluation of a suitable soil sensor for continuous, "on-the-go" field pH measurements. *Precision Agriculture '97. Proceedings of the 1st European Conference on Precision Agriculture* (pp. 493-501), J.V. Stafford (ed.). Oxford, UK: BIOS Scientific Publisher Ltd.

Williams, B.G., and D. Hoey. 1987. The use of electromagnetic induction to detect the spatial variability of the salt and clay contents of soils. *Australian J. Soil Research,* 25: 21-27.

Wright, N.A. 1996. New mobile soil sampler compared to hand probes and augers for fertility evaluations. *Proceedings of the 3rd International Conference on Precision Agriculture* (pp. 631-639), P.C. Robert, R.H. Rust, and W.E. Larson (eds.). Madison, Wisconsin: ASA/CSSA/SSSA.

Yukumoto, O., Y. Matsuo, N. Noguchi, and M. Suzuki. 1998. Development of tilling robot (Part1). *J. Japanese Society of Agricultural Machinery,* 60(3):37-44.

Zou, Y., D.C. Erbach, and S.J. Marley. 2000. Soil structure evaluation by using a fiber-optic sensor. *Transaction of the ASAE,* 43(6):1317-1322.

Chapter 4

# Site-Specific Nutrient Management: Objectives, Current Status, and Future Research Needs

Silvia Haneklaus
Ewald Schnug

## *INTRODUCTION*

Mineral elements were identified as essential factors for plant growth by German chemist Justus von Liebig as long ago as 1840, whereas human awareness that the fertility of land is variable in space is most likely as old as farming itself (since ca. 10000 BC) with references to this problem mentioned in the Bible (e.g. Luke 8:5-6). Soils are neither static nor homogenous in space and time, and this directly influences the concentration of plant-available nutrients. One of the most impressive examples of the spatial variation of plant-available nutrients is the phenology of the sulfur (S) supply of oilseed rape (Figure 4.1). Under humid conditions the variability of physical and hydrological soil properties within the soil profile results in a high spatio-temporal fluctuation of soil sulfate contents that becomes visible in different degrees of S deficiency symptoms (Schnug and Haneklaus, 1998). Spectral signatures from remotely sensed data were successfully used to follow up S deficiency on a large scale (Lilienthal and Schnug, 2005).

Mineral nutrients from different sources are used in agriculture to improve the quality of land. The fertilizer rate is usually calculated from the balance between the natural supply by soil and environment, the nutrient demand of the crop and inevitable losses to the environment, for example

The authors wish to express their most sincere gratitude to Dr. Kerr C. Walker (SAC, Aberdeen) for the linguistic revision of this chapter.

*Handbook of Precision Agriculture*

doi:10.1300/5627_04

FIGURE 4.1. Spatial variability of macroscopic sulfur deficiency symptoms visible in different colored patches within a flowering oilseed rape field in northern Germany. (See corresponding PLATE 4.1 in the central gallery.)

through leaching into water bodies or volatilization into the atmosphere (Isherwood, 1998; Schumann et al., 1997; Vermeulen et al., 1998). The agricultural production system fulfils the criteria of sustainability only if losses in this balance can be minimized while maximizing crop productivity and economic profits. Resources will then be saved, environmental burdens will be reduced, and a positive contribution to the income of the farmer will be made (Jackson et al., 1984).

The two major obstacles to achieving this ideal case of adjusting nutrient balances are, first, the accuracy of the methods employed to quantify the contribution from the soil, the physiological requirement of the plant, and the losses to the environment, and, second, the spatial variability of soil fertility parameters. The first problem is a subject of "agricultural chemistry," which is one of the oldest disciplines in natural sciences and was established by Liebig (1840). Aspects of agricultural chemistry, now commonly known as "plant nutrition" are not further exploited in this chapter. The extension of this discipline into space, however, which is a fairly new scientific line, is the central subject of this chapter.

Even if it would be possible to calculate precisely fertilizer rates for any combination of soil fertility, environmental losses, and yield expectations,

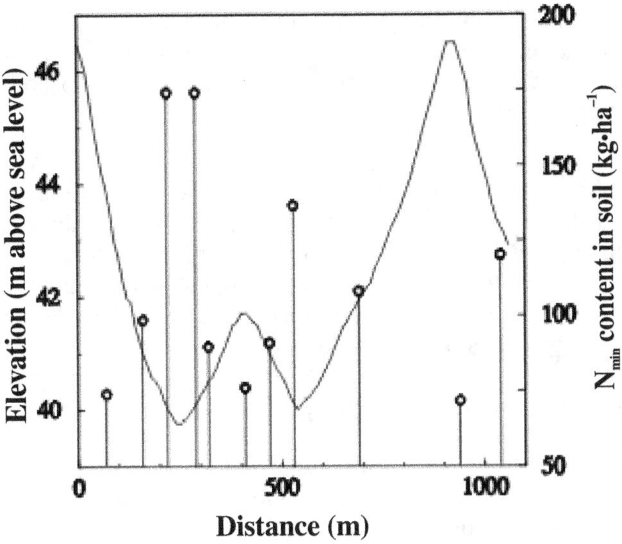


FIGURE 4.2a. Plant available nitrogen along a transect on a fluvial marsh soil in northern Germany.

the variation of soil parameters in space and the common practice of applying fertilizers uniformly on fields will cause deficiency and surplus. Figure 4.2 shows the variation of the plant-available nitrogen (N) content along a transect in northern Germany (Figure 4.2a) and patterns of surplus and deficiency in the phosphorous (P) balance on a cereal field in the same landscape (Figure 4.2b).

### *Glossary*

Within any new field of research new terms and definitions arise, which may complicate a correct assignment. A few relevant terms are, therefore, defined here:

**Local resource management (LRM):** The conceptual idea of matching agricultural inputs and operations with the spatial distribution of qualitative and quantitative features of natural resources (Schnug, Murphy, Evans, and Haneklaus, 1993).

**Precision agriculture (PA):** The techniques and measures used to apply LRM on agricultural land.

**Site-specific nutrient management (SSNM):** The application of LRM and PA to nutrients in the sense of transferring the site-specific plant nutrient demand into variable rates in dependence on the spatial variability of soil and crop parameters.

At the beginning of the 1990s the combination of modern forms of positioning, computerized processing of maps, and variable-rate fertilizer application was called computer-aided farming (CAF; Lamp and Schnug, 1990), thus adding agriculture to numerous other "computer-aided" applications. In contrast to CAF, which points out the technologies, the term SSNM focuses on the main objectives, which are to improve the sustainability of agriculture by a higher nutrient efficiency, to increase the economic profit, to protect natural resources, and to reduce losses of critical nutrients to the environment.

SSNM usually operates with digital agro-resource maps (DARMs), which are processed in geographical information systems (GIS) and are specifically designed for use in agriculture, such as with the Local Resource Information System (LORIS) (Schroeder et al., 1997). The term site in SSNM may be chosen for different fields (Dobermann and White, 1999) or independent management zones, which are areas with similar characteris-

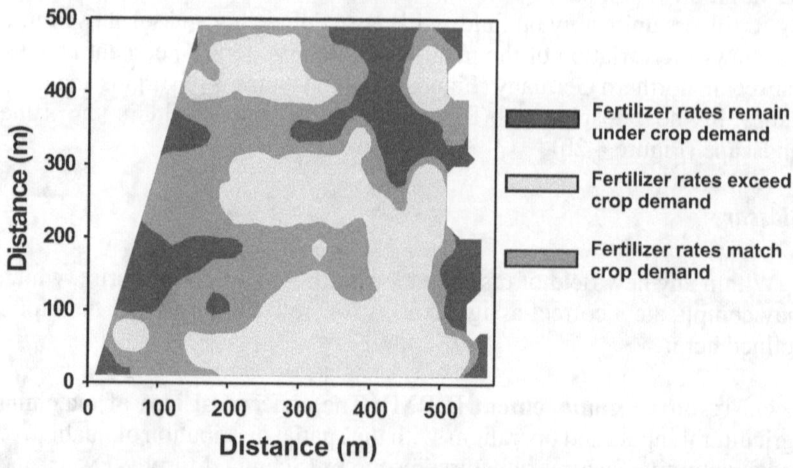


FIGURE 4.2b. Spatial variation of surplus and deficiency in the phosphorus balance on a fluvial marsh soil in northern Germany, if common methods for the estimation of the fertilizer demand together with a uniform distribution are applied.

tics derived, for example, from auto-correlation of soil and crop data, application of fuzzy algorithms, or sensor techniques (Boydell and McBratney, 1999; Lark, 2001; Mulla and Bhatti, 1997; Shatar and McBratney, 2001). A management zone was defined by Doerge (1998) as "a portion of a field that expresses a homogenous combination of yield limiting factors for which a single rate of a specific crop input is appropriate."

## *ORIGINS OF SSNM*

The problem of soil variability was addressed from many points of view and the phenomenon was related to problems for an efficient soil use long before the techniques of SSNM were available (Bouma, 1997; Burroughs, 1993). It was in the twentieth century that the breakthrough of SSNM as a strategy to address soil variability in practical agriculture succeeded because of three major developments: geostatistics, powerful computing and data storage systems, and satellite positioning and navigation (Lamp and Schnug, 1987; Schnug et al., 1985).

The quantitative assessment of spatial variability of geological and pedological features is called geostatistics, a discipline that found its way into science with the work of Matheron (1971), who created the theory of regionalized variables. This theory describes statistically the relationship between the variability of a variable and the distance between sampling points. Geostatistics is the basic science for producing maps for spatial operations such as SSNM. The generation of maps from point observations (e.g., sample locations in a field) requires an interpolation process to calculate estimates between sampling points. Methods such as triangulation or weighting using inverse distances are purely stochastic and do not take into account the geological and pedological characteristics of individual parameters. Krige (1951) realized this deficit and developed and gave name to an approach of interpolating (kriging) spatially distributed data by using weights derived from semi-variograms.

The second important development for SSNM was the provision of large data storage and fast processing systems at the end of the second millennium. Storing and processing of vast data sets in real time on personal computers is cheap and easy, and no longer a problem for SSNM. On small farms a farmer may keep spatial differences in his mind and reconsider them when delivering inputs. But with an increase of field and farm size, higher speed of field operations, a great variety of factors to be considered, and several years of observation to be reminded, the extent of information for decision making in SSNM easily outruns the storage capacity of the human brain.

Probably the most significant development for the practical implementation of SSNM in agriculture was the development of satellite-aided positioning and navigation (Bauer, 1994). Up to this time geographically correct positioning was the skill of specialists and required elaborate and long-lasting measurements and calculations. The Global Positioning System (GPS) was developed for two main purposes, i.e., to assist military forces with accurate time and positioning. The United States called its system "Navstar"; the former Soviet Union "Glonas." The principle of both systems was identical: measuring the travel time of radio signals from a set of satellites circling around the globe in order to determine the position of a mobile receiver on the surface of the globe. GPS was available for the nonmilitary public in the West from the beginning of the 1990s. The Alcatel/SEL GPS receiver was first used in agriculture in 1989 and was later donated to a German clock museum (FAL, 2002).

Important milestones in the new era of precision agriculture in the 1990s are listed here:

*Until 1988:* First application of SSNM in northern Denmark, Germany, and the United States employing variable-rate spreaders: AgriMatic (DK); CAF joint venture (D); SoilTeq (United States).

*June 30, 1988:* First demonstration of GPS-assisted SSNM at the annual exhibition of the DLG (German Agricultural Society) in Schwarzenraben, Germany, by the CAF joint venture.

*1989:* First experimental GPS-aided yield-mapping system consisting of a "yieldometer" by Jeff Claydon (1989) running in Osterhof, northern Germany, on a Deutz combine (Hansen, 1990).

*June 28, 1989:* First GPS-aided variable-rate application of a herbicide (phenmedipham) in dependence on the soil organic matter content in Birkenmoor, northern Germany, by the CAF joint venture (Bauernblatt/Landpost, 1989; Paulsen, 1989; Preusse, 1989).

*January 25, 1991:* First international workshop on computer-aided farming with SSNM as the main feature in Kiel, Germany (Kloepfer et al., 1994).

*July 1991:* First practical operation of a commercial GPS-aided yield-mapping system manufactured by Dronningborg A/S and T&O A/S, Randers (Denmark) employing a gamma ray flowmeter on a Case combine (Modern Farming, 1991).

*April 14, 1992:* First workshop on topics of PA (at this time still named soil-specific crop management) in Bloomington, Minnesota (Robert et al., 1993), with the first scientific presentation of GPS-assisted SSNM (Schnug, Murphy, Evans, Haneklaus, and Lamp, 1993).

*November:* First department for "Local Resource Management" implemented in governmental agricultural research at FAL in Braunschweig, Germany.

*1995:* Foundation of Precision Agriculture Centers at the Universities of Minnesota and Sydney (Australia); commercial release of LORIS, the first software dedicated to SSNM, a joint development of Kemira Europe, Wavre, Belgium, and FAL, Braunschweig (Schroeder et al., 1997).

*1997:* Development of "Surf-eye," later "Lassie" (Low Altitude Stationary Surveillance Instrumental Equipment) as a new data source for SSNM (Schnug et al., 1998; Lilienthal, 2003).

Since the first international workshop in St. Paul, Minnesota (Robert et al., 1993), the interest in PA and in particular its application to SSNM has rapidly developed, with a sixfold increase of contributions, oral prentations, and papers to the event (Figure 4.3).

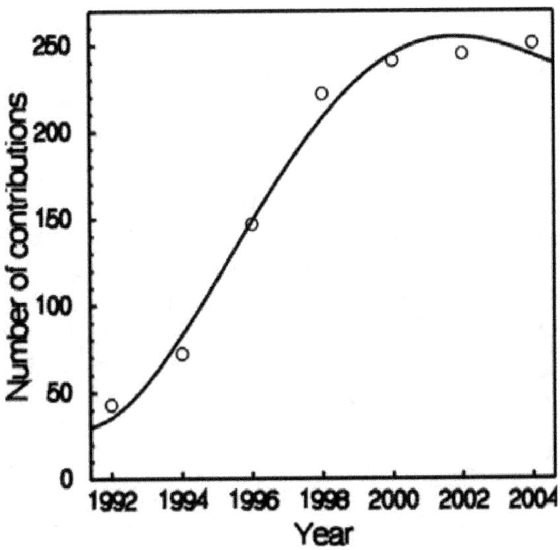


FIGURE 4.3. Number of contributions (oral presentations and posters) to the international conferences on precision agriculture, which are held every two years in St. Paul/Minneapolis, MN. *Source:* Robert et al., 1993-2001.

### *Objectives*

The main objectives of this contribution are to outline the current status of SSNM and align challenges for future research and development. SSNM is information technology applied to soil use in agriculture. Without valid information there is no SSNM and therefore potential data sources for SSNM are first discussed.

Agro-ecosystems are multifunctional and highly complex. Nutrients are a crucial part of agro-ecosystems and their pools and fluxes are influenced by many factors. SSNM aims to improve the utilization of nutrients from different sources for a better supply of plants and to reduce losses to neighboring ecosystems (Doberman and White, 1999). This implies that decision making in SSNM will have similar levels of complexity, particularly, if dynamic strategies need to be elaborated. Different approaches for decision making in SSNM are presented and discussed later.

Plant nutrients originate from very different sources, for example single nutrients as a pure chemical, such as elemental S, or fertilizers processed from waste materials combining a larger number of elements and compounds. SSNM is one part in the production process and influences not only directly the nutrient supply of the crop, but also indirectly other parameters. In SSNM different information layers and targeted field operations need to be matched spatially. The accuracy of these matches strongly affects success and costs of SNNM. Thus all components involved in the decision-making process have to be evaluated for an optimum level of accuracy, and quality standards have to be defined to achieve the best success at lowest costs.

SSNM aims at improving the economy and ecology of nutrient use in agriculture (Haneklaus and Schnug, 2002). Therefore a summary of possible effects in selected case studies is also given. Any new technology has an impact on social structures and a prognosis is made on possible social impacts of PA and SSNM. Progress in the concept of SSNM and in the related sectors needs additional research and it is addressed in the final section of this chapter.

## *DATA SOURCES FOR SSNM*

Information technology applied to fertilization is SSNM and it is pointless without spatial information. Access to valid data is meanwhile considered as a major obstacle for the further development of PA (NRC, 1997) and SSNM as well.

Information suitable for SSNM originates from different sources and quality levels. The applicability of different data sources for SSNM mainly depends on the scale of farming. Big farms usually operate on a distinctly higher level of technology than small farms, which depend much more on human resources. Thus on big farms the most efficient access to digital data covering large areas is provided for instance by ordnance maps or recording of data, employing advanced technologies such as on-the-go sensors and remote sensing. In comparison, on small farms strategies for data collection such as grid soil sampling, self-surveying, and local expertise efficiently provide georeferenced soil and crop data. Ultimately, SSNM should aim at organizing a "data warehouse" (NRC, 1997; Bauer and Guenzel, 2001) for the farm, whose products are the basis for the decision-making processes of how to improve fertilizer efficiency.

### *Map Data*

Classical soil maps provide qualitative information about permanent features such as geology, pedology, and geomorphology of an area (Hake, 1982). Traditionally these maps are hard copies, regularly referenced to regional coordinate systems. Particularly quantitative information of morphometric maps supplies relevant data for modeling nutrient fluxes (Haneklaus and Schnug, 1999b). For SSNM only maps with a scale of > 1:5000, preferably in digital form, provide suitable information (Capelle, 1999).

#### *Ordinance Survey Maps and Terrain Models*

Examples for the large-scale, overall, and consistent mapping of soil texture, stage of development, and pedogenesis are the so-called "Reichsbodenschaetzung" in Germany (Mueckenhausen et al., 1988) and the site evaluations carried out for the Brazilian agrarian reform (Sparovek et al., 2002), respectively. The "Reichsbodenschaetzung" dates back to 1935 when agricultural soils were systematically categorized in a 50 by 50 meter grid, up to a soil depth of 1 meter. Right from the start it was planned to use this information not only for tax purposes but also for agricultural and scientific questions. Today this data is partly available in digital maps in a scale of 1:5000.

Nutrients are transported not only vertically, but also laterally through landscapes. The reasons are erosion processes in the case of phosphate (Gassner et al., 2002; Gassner, 2003) and soil water fluxes in the case of nitrate and sulfate (Bloem, 1998). Therefore, particularly in ondulated areas terrain information is a very important source for decision making in SSNM

(Catt et al., 1997; Syers and Walker, 1969). Digital terrain models (DTMs) may be processed from elevation data that is available in the form of ordnance survey maps, or can be assessed directly on the farm.

*Local Knowledge*

"Local knowledge" (LK) is introduced as a new term for information that is available on a farm, either in the form of amateur maps, field files, or any other materialized form of store. However, LK also comprises knowledge from personal education, experience, or inheritance from previous generations. Making efficient use of LK is a great challenge to operate SSNM on all scales. The farmer becomes the biological interface between his or her field and crops and hard- and software components, respectively. The actual way of sampling and processing of information is in its simplicity still far away from the complex system in which growth and environmental factors are interacting. In comparison, for a well-educated farmer with ample experience and knowledge, the decision-making process in the human brain with its fuzzy neural structure interprets interactions professionally by addressing much different, often diffuse information to the target variables.

At first, LK is a valuable source of information to assign management zones (Fleming et al., 2000) and to decide on fertilizer rates. The simplest way of employing LK in SSNM would be the manual adjustment of fertilizer rates (Fixen and Reetz, 1995). This procedure is certainly limited to an upper farm size; then PA technologies need to be implemented, but LK will remain an efficient source of information and decision making. Another relevant LK for SSNM on small farms is that of geomorphology. The assessment of basic terrain information may be used to set up surprisingly efficient strategies for SSNM.

A central problem in making use of LK in SSNM is the georeferencing and parameterization of observations. But even if the accuracy of positioning may not correspond with the standards of professional maps, the value of such information for SSNM is still enormous. The second problem is more difficult to overcome, as it requires a profound understanding of natural processes. Therefore this area of expertise is a great challenge for future education and advisory systems.

*Georeferenced Field Sampling and Laboratory Analysis*

In SSNM field sampling serves the purpose to determine spatial and temporal changes of soil and crop properties. The spatial variability of soil features within a single field can be easily as high as the variability of the same parameter within the surrounding landscape (Figure 4.4).

Experiences from grid soil sampling reveal that it is superior to random sampling due to the spatial correlation of values (Peck and Melsted, 1967; Sabbe and Marx, 1987). Distances between samples taken on grassland need to be shorter than on arable land because of the absence of homogenizing soil tillage operations (Shi et al., 2000). The distinct advantage of plant sampling is its closer relationship to growth processes and wider ranges in semi-variograms of at least 100 m compared to minimum ranges of 50 m for soil parameters (Haneklaus, Ruehling, et al., 1997). The major handicap, however, is that it cannot be performed in advance for a new crop so that the period for sampling, chemical analysis, and fertilization is very short.

The aspect of temporal variability is most important when selecting an appropriate method to assess the in-field variation of extremely short-term variable soil nutrients such as N and S. So far these can only be recorded directly by online sensor techniques or derived indirectly using maps based

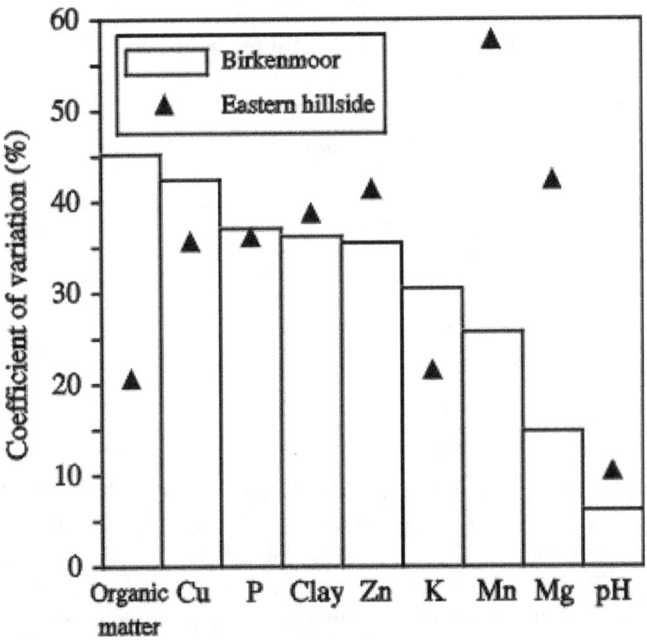


FIGURE 4.4. Coefficients of variation for different soil fertility parameters in a single field at Birkenmoor compared to those observed in the surrounding landscape, the eastern hillside of Schleswig-Holstein, northern Germany ($n$ = 169).

on crop growth models, remote sensing, and/or key parameters with strong impact on the nutrient supply as for instance topography (Franko et al. 1997; Franzen et al., 1998; Lilienthal et al., 2000; Reusch, 1997; Sogbedji et al., 2001). Dynamic modeling, taking the spatial change in which cells influence each other into account, is of particular interest for rapidly changing growth factors such as soil temperature, soil water regime, plant available N, and risk assessment of infection with soilborne fungal diseases. Russell (1996) and Parker et al. (1996) comprehensively discuss the challenges of modeling to precision agriculture.

Soil data of less mobile nutrients such as P and potassium (K) has a validity of three to five years and pH of five to ten years (Hergert, 1997). Organic matter content, soil texture, and cation exchange capacity are rated as permanent soil features (Hergert, 1997) and therefore need to be mapped only once.

Relevant to SSNM is that optimum sampling distances vary in different landscapes, depending on the scale of variation of field characteristics as a result of parent material and pedogenesis (Bouma, 1999; Burrough, 1993; McBratney and Pringle, 1997). However, spatial patterns caused by application techniques and their specific failures also affect sampling scales (Mallarino, 1996). In Table 4.1 maximum sampling distances based on range values from semi-variogram analysis are given for different parameters in different landscapes. With respect to sampling in the field it is important that samples are georeferenced and sample cores are taken within an area of not more than 10 $m^2$. The traditional mixing of 10 to 20 sample cores per 10 to 30 hectares interferes with the variability of soil features and is therefore unsuitable (Hergert, 1997). Detailed guidelines for collecting, handling, and analyzing plant and soil material are given by Reuter et al. (1997) and Collins and Budden (1998).

### *Georeferenced Field Tests*

Grid soil sampling may be profitable if the return from SSNM is high because of a reduced input of high-priced fertilizers and/or increased yield or crop quality. In general, however, grid soil sampling is far too costly to be economic. Critical also are the costs for analyzing a large number of soil samples. Although for some parameters cost-effective laboratory methods were developed such as the indirect determination of organic matter and clay content (Schnug and Haneklaus, 1997a) and the "Coca Cola" method for plant-available micronutrients (Schnug et al., 1997), the major expenditure is still the sampling procedure itself (Schnug et al., 1998).

TABLE 4.1. Semi-variogram range values for agronomically relevant soil and crop parameters in different landscapes.

| Parameter | Range (m) from | Range (m) to | Country | Reference |
|---|---|---|---|---|
| Organic matter | | 97 | Australia | McBratney and Pringle (1997) |
| | 45 | 99 | Denmark | Kristensen et al. (1995) |
| | 59 | 171 | Germany | Fromm et al. (1993); Haneklaus, Ruehling, et al. (1997) |
| | 50 | 114 | United States | Cambardella et al. (1994); Miller et al. (1988); Mulla (1993) |
| Clay | | 65 | Australia | McBratney and Pringle (1997) |
| | 37 | 450 | Denmark | Kristensen et al. (1995) |
| | 58 | 102 | Germany | Haneklaus, Ruehling, et al. (1997) |
| | | 36 | Tunis | Vauclin et al. (1983) |
| | | 200 | United Kingdom | Webster and de la Cuanaolo (1975) |
| | 66 | 93 | United States | Burden and Selim (1989); Miller et al. (1988); Mulla (1988) |
| Plant-available P | | 193 | Australia | McBratney and Pringle (1997) |
| | 60 | 148 | Denmark | Kristensen et al. (1995) |
| | 60 | 364 | Germany | Haneklaus, Ruehling, et al. (1997) |
| | | 241 | United Kingdom | Webster and McBratney (1987) |
| | 50 | 900 | United States | Cahn et al. (1994); Cambardella et al. (1994); Han et al. (1997); Mulla (1993); Pierce et al. (1995) |
| Plant-available K | 25 | 129 | Denmark | Kristensen et al. (1995) |
| | 60 | 232 | Germany | Gassner et al. (2002); Haneklaus, Ruehling, et al. (1997) |
| | | 428 | United Kingdom | Webster and McBratney (1987) |

TABLE 4.1 *(continued)*

| Parameter | Range (m) from | to | Country | Reference |
|---|---|---|---|---|
| | 40 | 174 | United States | Cahn et al. (1994); Cambardella et al. (1994); Han et al. (1997); Pierce et al. (1995) |
| Plant-available Mg | 74 | 171 | Germany | Haneklaus, Ruehling, et al. (1997) |
| | | >125 | United States | Cambardella et al. (1994) |
| Plant-available N | 23 | 54 | Denmark | Kristensen et al. (1995) |
| | <30 | 121 | Germany | Haneklaus et al. (1999) |
| | 79 | 900 | United States | Everett and Pierce (1997); Han et al. (1997) |
| pH | | 70 | Australia | McBratney and Pringle (1997) |
| | 17 | 19 | Denmark | Kristensen et al. (1995) |
| | 61 | 197 | Germany | Haneklaus, Ruehling, et al. (1997) |
| | 105 | 190 | United States | Cambardella et al. (1994); Mulla (1993); Pierce et al. (1995) |
| Yield | 67 | 105 | Germany | Haneklaus, Ruehling, et al. (1997) |
| | | 143 | United Kingdom | Schnug, Murphy, Evans, and Haneklaus (1993) |
| | 70 | 80 | United States | Miller et al. (1988); Mulla (1993) |
| | 34 | 61 | Denmark | Kristensen et al. (1995) |
| Macronutrients in plants | 105 | 167 | Germany | Haneklaus, Ruehling, et al. (1997) |

Self-surveying is a sampling strategy by which basic spatial information of, for instance, soil texture, organic matter content, and geomorphology are gathered using surplus labor on farms and human sensory capabilities (Haneklaus et al., 1998). The self-surveyor takes georeferenced soil samples in a 30-50 meter grid in the field and attributes directly to each sample the corresponding geomorphological feature and position. Such simple evaluation of morphometric features (summit, depression, convex and concave slope) when walking the land for sampling provides excellent results for decision making in SSNM (Haneklaus, Schroeder, and Schnug, 1997b). Alternatively, digital elevation data that has been recorded by laser scanning systems together with high precision differential GPS may be used to process digital terrain models using suitable software (MacMillan et al., 2000).

Then, after collection of all samples of the investigated area individual scoring schemes are set up for organic matter and soil texture, whereby the number of classes depends on the range of variation in the area. In general, darker soils are associated with higher organic matter contents and vice versa. Soil samples are classified according to the scale of colors into different categories, which is easy to conduct for untrained persons, too. The necessary conversion of such scores into "real" data would be done by a process called "local calibration," in which a set of samples with scores covering the whole range of observation is analyzed by means of common laboratory methods and from the regression between scores and "real" data the concentration of the parameter is calculated for the entire data set. The accuracy of the described method for determining the spatial variation of organic matter in soils has shown to be as precise as the differentiation between humus classes (Haneklaus et al., 1998).

In a similar procedure differences in soil texture can be assessed rapidly. Different categories of sand, silt, and clay in a soil sample can be distinguished by means of the "finger test" and converted into analytical data using a local calibration.

### *Remote Sensing*

The use of remote-sensed images in agriculture never really had a breakthrough because it contains primarily thematic information, but agricultural operations such as SSNM require quantitative values of biological, physical, and chemical properties and the difference between nutrient demand and actual nutrient supply cannot be determined in time. The major problem is cloud coverage at phenologically relevant growth stages, particularly in central Europe, and the restricted revisit time and physical availability of airplanes, respectively. Grenzdoerffer (2001) for instance revealed that during the vegetative growth period, when SSNM needs to be applied, sufficient ground visibility can be expected only on about 14 days for 140 survey hours in northern Germany.

Moran (2000) summarized the agronomic demands toward remotely sensed images by preference: image delivery within 24 hours; georeference with an accuracy of one pixel; information accuracy of the measured crop or soil conditions of 70 to 75 percent; repeat coverage ranging from twice per week to biweekly; spatial resolution of 10 to 20 m; maps providing quantitative information of measured features; fair product prices. A breakthrough with view to an instant image delivery could be the development of a digital, low-cost, remote sensing system, which provides ad-hoc georeferenced images by aerotriangulation using high accuracy GPS receivers (Grenzdoerffer, 2001).

Information on spatially variable soil and plant features that can be gathered from remotely sensed images without ground truth and used for SSNM are management zones, revealing areas with similar spectral signature (Panten, 2002; Schepers et al., 2000; Stewart and McBratney, 2001) or areas with different organic matter content (Walters et al., 1997).

Bare soil images are preferable for the investigation of key soil parameters such as organic matter content, soil texture, and topography as vegetation changes the reflectance and emission characteristics of the image. An accompanying ground campaign is necessary for a quantitative interpretation of images, because of the superimposition of the reflectance characteristics by soil texture, organic matter, water content, and ferric oxides (Barnes et al., 1997). The use of remotely sensed data for the determination of the spatio-temporal variation of short and medium stable features such as the soil nutrient status is not feasible, because no consistent relationship between spectral signature and soil nutrient status was found (Panten, 2002).

Images of the vegetated soil reveal different patterns in the field but causal factors are unknown, so that a ground truth is inevitable. Among all nutrients N has the strongest effect on the chlorophyll content of plants, and reflectance spectra are used to determine normalized difference vegetation index and red edge (Guyot and Baret, 1988; Heute, 1998), which are set in relation to the chlorophyll content or N status (Engel et al., 2000; Kokaly and Clark, 1999). However, there are much more environmental factors affecting the chlorophyll content of plant tissue than the N supply, such as drought, temperature, diseases, and nutrient disorders (Clevers, 1994; Schnug and Haneklaus, 1998). Thus, changes of the spectral signature can only be assumed to be caused by a variation of the N supply in the soil under ceteris paribus conditions, which in fact never occur under field conditions.

Important differential diagnostic criteria for identifying the true cause of visible symptoms of plants are spatial patterns and their coincidences with soil characteristics, which require the real-time surveillance of crops. This is yet not feasible because of technical restrictions with view to image acquisition and time for image processing (Panten et al., 1998). New challenges for remote sensing in SSNM come from the development of LASSIE, an innovative concept for the continuous recording of real-time images of crop and soil surfaces, which are automatically rectified and georeferenced in a GIS (Schnug et al., 2000, Lilienthal, 2003; Lilienthal and Schnug, 2004). LASSIE describes a stationary digital camera or scanner system with remote-controlled variable, horizontal, and vertical positioning on elevated places in the landscape (Figure 4.5).

Changes in crop performance become easily identifiable with LASSIE and can be addressed either directly or indirectly by an accompanying directed ground truth. The results of soil/crop analysis can then be more or less instantly transferred into variable-rate management on the fields. LASSIE offers the distinct advantages of having no temporal constraints, being independent of weather conditions, using standardized processing routines, and, above all, yielding only low maintenance costs.

FIGURE 4.5. Prototype of a LASSIE (1) installed on a grain tower 20.7 meters above ground at the experimental site of FAL (E 10.45172; N 52.29954), and original (2), rectified (3), and mosaiced (4) images taken by LASSIE. Adapted from Schnug et al., 2000.

### *On-the-Go Sensing*

On-the-go sensing is a great fiction in PA. Sensors shall supply dense information of soil and crop data and, combined with sufficient computing power, enable real-time SSNM. In a way, a farmer using his or her inherent local knowledge for SSNM could be understood as an on-the-go sensor too.

The idea of ground-based sensors for nutrients in soils and plants is still visionary (NRC, 1997). Major efforts in this field concentrate on N, the motor of plant growth and the supposedly most profitable nutrient. Intrusive methods, which combine sophisticated soil sampling, for example, with a modified circular saw, and fast on-the-go application of traditional laboratory methods (Adsett and Zoerb, 1991) exist in order to assess the spatial variation of soil water, nitrate, organic matter, pH, electrical conductivity, and soil texture (Birrell and Hummel, 1993; Lui et al., 1997; Sudduth and Borgelt; 1993; Viscarra Rossel, 2001).

On-the-go sensing of nutrients in soils and plants employing nonintrusive techniques can not be expected in the near future because even the newest developments in instrumental element analysis, such as alpha-proton X-ray spectrometry (APXS), which was employed on the *Mars Sojourner* rover, require direct contact between sample and detector. In addition, soil analysis for an assessment of the nutritional status of crops requires information about plant-available soil concentrations and thus needs at least a suspension of the soil in an extractant (Davenport and Jabro, 2001).

Nutrient sensors for plants have one very serious problem in common: the characteristics of the spectral signature of the reflection of leaves and canopies is not related to one single influencing factor (see previous discussion), an assumption on which, however, numerous investigations with N sensors and remote sensing are based (Blackmer and White, 1997; Blackmer et al., 1995; Sérélé et al., 2001; Wollring and Reusch, 1999).

One of the most promising developments in generating data for SSNM by on-the-go sensing are yield sensors, because the level of nutrient removal from a field is important information for making decisions in SSNM. Yield sensors are probably most widely used to determine the productivity of combinable crops in the world and the characteristics of the different systems are discussed comprehensively by Griepentrog (1999), Murphy et al. (1995), and Reyns et al. (2002).

## *DECISION MAKING FOR SSNM*

### *General Decisions in SSNM*

The general decisions to be made in SSNM underlie the same principles as in any other nutrient management system. Crop response to fertilization can only be expected if there is a physiological requirement of a nutrient and if other essential plant nutrients and growth factors are in an optimum state. One of the first and most crucial steps in SSNM is a ranking of the nutritional minimum factors of a site. A suitable approach to identify yield-limiting factors in the mineral nutrition of a particular site is the directed sampling in high- and low-yielding areas. If this concept is extended to a general assessment of the site, that means if information of physical soil parameters such as soil texture and hydrology is gathered, it may help to identify other yield-limiting factors that can not be managed site specifically (Schnug et al., 1994; Stewart et al., 2002).

A bias-free method to evaluate the nutritional minimum factors of a site is plant analysis, because the whole spectrum of essential mineral elements is scanned with only one method. To such data sets evaluation systems should be applied that are based on response curves such as PIPPA (Professional Interpretation Program for Plant Analysis; Schnug and Haneklaus, 1992) rather than employing static critical values.

The evaluation of the relative significance of individual plant nutrients for crop performance on a particular site is more difficult. Soil analysis is the most common method to determine the fertilizer demand of a crop. In a countless number of studies the preference of different soil tests to determine the plant-available content of a soil nutrient was explored worldwide. Soil tests differ mainly in the extraction force of the extractant, the soil: extractant ratio, and the extraction time. Thus official standard procedures for soil analysis differ nationwide, sometimes even regional. Recommended rates may be based on three alternative ways of interpreting soil test values: first, the build-up and maintenance strategy, second, the fertilizing of the crop concept, and third, the basic cation saturation approach (Voss, 1997). The first aims at increasing the available nutrient content up to a level above which no further yield increase can be expected, and maintaining this status by applying fertilizer rates equivalent to the off-take of the crop. The second and most frequently used concept matches the nutritional level (low, medium, high) by corresponding fertilizer rates, whereas the third aims at an optimum Ca:Mg:K ratio in the soil (Voss, 1997).

### *Decisions Based on Existing Recommendation Schemes*

Common rules for fertilizer application have been proposed for decision making in SSNM and were directly transferred into fertilizer rates for lime, P, K, and magnesium (Mg) (Dampney et al., 1997; Dawson, 1996; Dawson and Johnston 1997; Fotyma et al., 1997). However, there are two major criticisms of such an approach: first, the basic fertilizer response curves were mainly established in the 1940s to 1960s, and databases need to be updated and validated for SSNM, respectively (Hergert et al., 1997; Voss, 1997). Second, established critical thresholds refer to small homogenous plots (Mallarino and Blackmer, 1992) and are therefore counterproductive with view to the target, which is to identify variability, match it with a variable input, and thus improve recommendations (Hergert et al., 1997). In addition, soils and plants were sampled in the past without reference to spatial auto-correlation, so that the whole system of interpretation still used today for recommendation is based on averages (Nowak, 1998), and will therefore best yield average results (Oestergaard, 1997; Simmelsgaard and Djurhuus, 1997). Sims (1998) named it the "one fits all" problem.

The consistent averaging of results by arranging replicates across the variability within a field is demonstrated in Figure 4.6 for the design of the

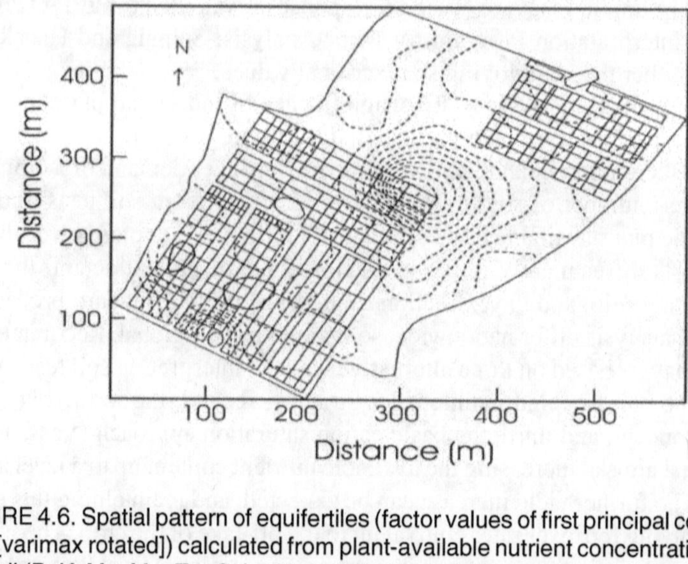


FIGURE 4.6. Spatial pattern of equifertiles (factor values of first principal component [varimax rotated]) calculated from plant-available nutrient concentrations in the soil (P, K, Mg, Mn, Zn, Cu), pH, and plant-available water, and arrangement of experimental plots on an experimental station in Birkenmoor (E 9.575; N 54.245). Adapted from Schnug et al., 1989.

test plots within a research station. This design was superimposed on equifertiles, which are areas with similar soil fertility parameters (calculated from available plant nutrients and soil water across this site; Schnug et al., 1989). Close contour lines reveal rapid changes of soil features or a high degree of heterogeneity. A random distribution of replicates averages these differences. This example stresses that SSNM will be more successful if more variability and extremes are addressed in experimentation. The mismatch of this goal with the existing recommendation systems is obvious. Another, even more serious disadvantage of using common recommendation schemes for decision making in SSNM is the fact that they are not dynamic so that lateral interactions between different management zones will not be addressed in SSNM.

Balanced fertilization using yield maps offers another chance for decision making in SSNM (Blackmore and Larscheid, 1997; Lark, 1997; Larscheid et al., 1997). Although some authors proposed PA measures for yield quality, their results reveal that the variation of the N concentration in barley was extremely small (Stafford, 1999; Thylen et. al., 1999). The variation of N and other minerals, such as P, K, and Mg, was distinctly higher in the generative material of oats than in winter wheat (Table 4.2). This may be explained by the fact that oats are a husk cereal and the share of hulls varies between grains of one panicle (Geisler, 1980). The variation of macronutrients in wheat straw was two- to fourfold higher than in grains, but similar in both plant parts in the case of oats (Table 4.2). Therefore, the off-take of plant minerals may be calculated as a multiple of the yield in the case of huskless cereals if plant residues (straw) remain on the field. In the case of cereals with hulls, such as barley, oats, rice, and millet, this might provide unsatisfactory results with view to a balanced fertilization.

### *Decisions Based on Local Knowledge*

Local knowledge has already been acknowledged as a valuable source of information for evaluating the response of a site to fertilizer inputs, but LK is more than just the knowledge of facts—it also comprises decision-making processes.

Perhaps the simplest way of decision making for SSNM would be based on easy-to-appraise topographic characteristics of the landscape. An example for this procedure was given by Haneklaus, Schroeder, and Schnug (1997).

A more complex strategy involves knowledge about soil parameters. Next a simple example is given for the set up of a fertilizing strategy based

TABLE 4.2. Range of variation of nutrient off-take by oats ($n = 174$) and winter wheat ($n = 132$) determined by 30-meter grid sampling on two fields in Mariensee, Lower Saxony (E 9.49158; N 52.55012).

| | Grain—Winter Wheat | | | | Straw—Winter Wheat | | | |
|---|---|---|---|---|---|---|---|---|
| **Nutrient (percent)** | **min.** | **max.** | **mean** | **CV (percent)** | **min.** | **max.** | **mean** | **CV (percent)** |
| N | 1.88 | 2.17 | 2.06 | 3.1 | 0.41 | 0.79 | 0.57 | 10.4 |
| P | 0.31 | 0.41 | 0.35 | 4.4 | 0.05 | 0.12 | 0.08 | 19.6 |
| K | 0.45 | 0.58 | 0.50 | 5.0 | 1.25 | 2.17 | 1.61 | 11.9 |
| Mg | 0.08 | 0.10 | 0.09 | 3.5 | 0.03 | 0.01 | 0.06 | 14.8 |
| Yield ($Mg \cdot ha^{-1}$) | 5.9 | 9.6 | 8.1 | 9.5 | | | | |
| | **Grain—Oats** | | | | **Straw—Oats** | | | |
| N | 1.50 | 2.03 | 1.82 | 6.6 | 0.16 | 0.61 | 0.30 | 26.9 |
| P | 0.21 | 0.39 | 0.29 | 13.7 | 0.11 | 0.35 | 0.25 | 12.4 |
| K | 0.35 | 0.58 | 0.48 | 10.5 | 2.58 | 4.55 | 3.68 | 10.6 |
| Mg | 0.05 | 0.15 | 0.12 | 18.0 | 0.02 | 0.09 | 0.05 | 19.0 |
| Yield ($Mg \cdot ha^{-1}$) | 4.0 | 8.6 | 7.4 | 8.6 | | | | |

*Note:* CV = coefficient of variation.

on LK: first, the farmer generates by means of self-surveying georeferenced maps of a field for clay and organic matter. In this example six categories were addressed to each variable (Table 4.3).

Next, the farmer calculates the mean N rate to be applied to "average" zones of the field employing standard test methods such as "$N_{min}$" (VDLUFA, 1991). The $N_{min}$ content reflects the amount of plant-available mineral N (ammonia and nitrate nitrogen) in the 0-90 cm soil layer at the start of the vegetation period, and is a common procedure to determine the N supply. The variation of rates from this mean value then depends on the LK of how different zones of the land react to N under consideration of factors such as weather, crop type, and variety. Assuming linear relationships between clay and organic matter content and N rates, the surplus and reduction of the rates may only be developed for the extremes.

TABLE 4.3. Variable-rate scheme for SSNM for the first spring application of nitrogen to winter wheat based on self-surveying of clay and organic matter, and decision making based on local knowledge in Mariensee, Lower Saxony (E 9.49527; N 52.5484).

| Clay content (percent) | Organic matter content (percent) 2.2-2.4 | 2.4-2.6 | 2.6-2.8 | 2.8-3.0 | 3.0-3.2 | 3.2-3.4 |
|---|---|---|---|---|---|---|
| 14-18 | 120 | 110 | 100 | 90 | 80 | 70 |
| 18-22 | 110 | 100 | 90 | 80 | 70 | 60 |
| 22-26 | 100 | 90 | 80 | 70 | 65 | 60 |
| 26-30 | 90 | 80 | 70 | 65 | 60 | 55 |
| 30-34 | 80 | 70 | 65 | 60 | 55 | 50 |
| 34-38 | 70 | 65 | 60 | 55 | 50 | 40 |

*Assumption:* 20 kg·ha$^{-1}$ $N_{min}$ at the beginning of the vegetation period and a target value for the first nitrogen application to winter wheat of 120 kg·ha$^{-1}$ N.

### *On-Site Experimentation*

Most of the rules for applying fertilizers derive from field experimentation carried out on selected sites. The locations were often chosen because of organizational and administrative reasons rather than growth conditions. This may be not a problem if soil-independent factors such as varieties and pesticides are tested, but has great impact on the results obtained for plant-available nutrients. The larger the geographical distance is between the experimental site where recommendations have been elaborated and the farm on which they will be applied, the greater is the possibility of a mismatch of these rules. Before PA technologies were available, usually numerous experimental stations carried out traditional plot experiments in order to take the ecological spectrum of site characteristics into account. Now, using PA technologies for positioning, variable-rate application, and online yield measurements, it is possible to investigate multivariate experimental designs on each farm including the whole bandwidth of ecological variability (Reetz, 1997; Schroeder and Schnug, 1995).

SSNM develops a farm toward a data warehouse, and from this data, by means of strategies known today as data mining, knowledge discovery, and knowledge management (NRC, 1997; Bauer and Guenzel, 2001), rules for spatially variable fertilization can be developed that are genuinely site-specific. A three-step example for evaluating a strategy for spatially variable lime application is given in Figures 4.7 through 4.9.

In the first step the pH values, based on soil sampling in a 30-meter grid, are plotted against the relative grain yield data, and by means of the upper-boundary-line approach (Webb, 1972) the response curve for pH to crop yield is calculated (Figure 4.7). The principle of the boundary-line analysis and the application of the boundary-line development system (BOLIDES) in SSNM are comprehensively discussed by Schnug et al. (1996). A major advantage of using BOLIDES in SSNM is the distinctly higher number of replicates than in common field experimentation, which increases the accuracy of the results obtained (McBratney and Pringle, 1997).

Erroneous extremes were excluded before further processing of the data. The boundary line, usually a fourth-order polynomial function, lies on the upper edge of the body of the data (Figure 4.7). This line describes the response to variation in the test parameter in which all other factors are, within the constraints of the field, as close as possible to nonlimiting in terms of crop yield. Data points below this line relate to samples in which some other factor limited the crop response to the nutrient. The first derivative of the fitted polynomial gives predicted yield response to fertilization in relation to the nutrient content. The optimum nutrient value corresponds with the zero of the first derivative of the upper boundary line and the sign

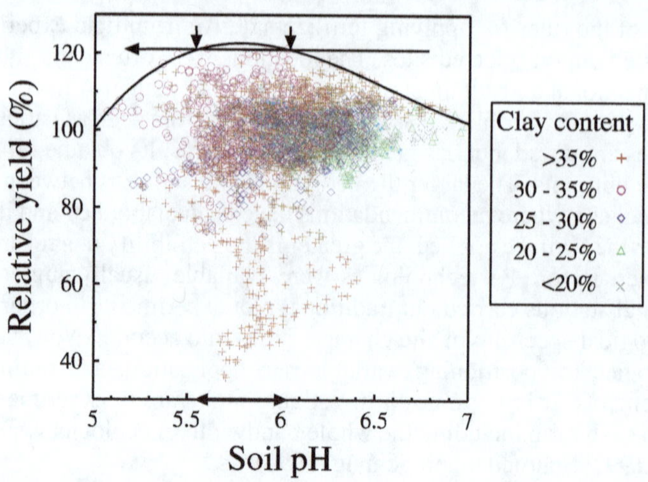


FIGURE 4.7. Evaluation of the optimum range of pH values (5.6-6.1) in the top soil of a clayey loam soil in Northumberland, United Kingdom (W 1.05833; N 55.28333). (See corresponding PLATE 4.7 in the central gallery.)

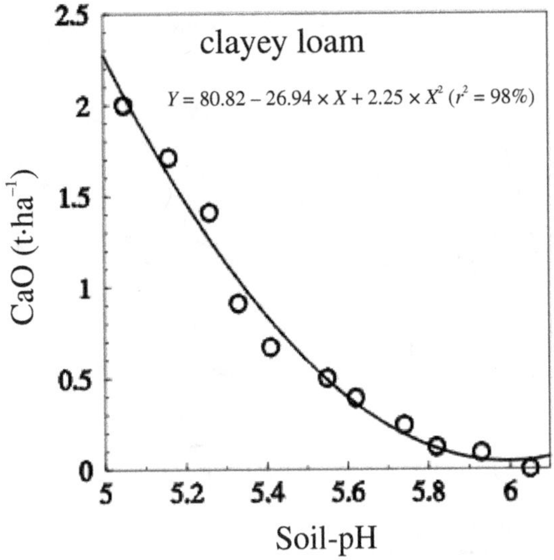


FIGURE 4.8. Determination of the lime requirement by the modified (Haneklaus and Schnug, 2000) Jensen method (Thun et al., 1954) for a clayey loam soil in Northumberland, United Kingdom (W 1.05833; N 55.28333).

of the second derivative at this point. For the determination of optimum ranges, i.e., the range of nutrient concentration that is sufficient to realize 95 percent of the maximum yield, standard, numerical root-finding procedures were used for real polynomials of the fourth degree with constant coefficients. On the field in the example given in Figure 4.7, highest yields can be expected between pH 6.1 and 6.4.

In the second step the response of the soil pH to lime application is analyzed for the individual texture classes of the field by means of the modified Jensen method (Haneklaus and Schnug, 2000; Thun et al., 1955). In Figure 4.8 a typical buffer curve is presented for one texture class (clayey loam).

Local calibration was applied when selecting samples from the complete set of grid samples. The buffer functions were then used to calculate variable lime rates. The last step is the application of the algorithms to the field data and the calculation of a lime application map for navigating and regulating a variable-rate spreader system (Figure 4.9).

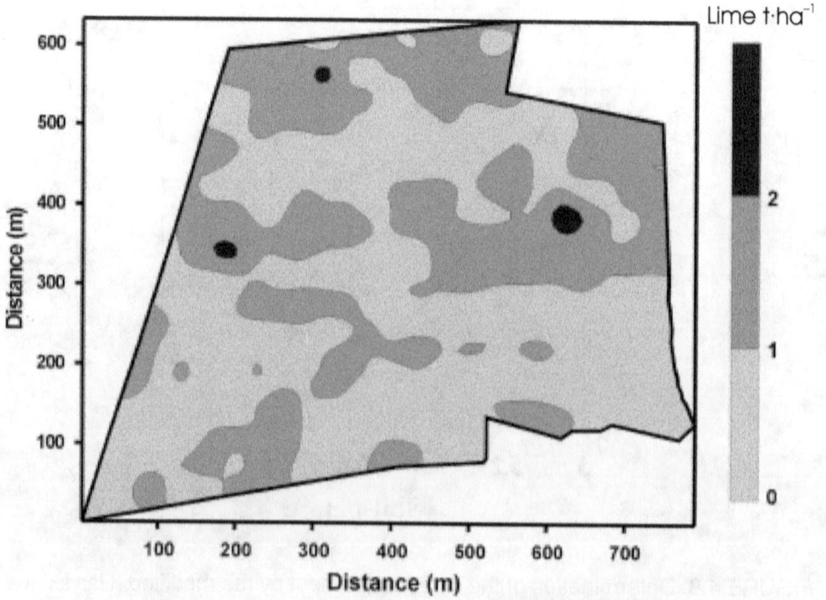


FIGURE 4.9. Lime application map based on the determination of optimum pH ranges by BOLIDES and lime requirement assessment according to Jensen of a clayey loam soil in Northumberland, United Kingdom (W 1.05833; N 55.28333).

### *Monitoring of Temporal Variation*

The validity of plant-available soil nutrient data ranges from days for highly mobile nutrients such as N and S (Bloem, 1998; Haneklaus et al., 1999) to 3-5 years for P, K, and Mg, and 5-10 years for pH (Hergert, 1997).

Not only do physico-chemical soil properties cause changes in the nutritional status through leaching, capillary rise, mineralization, immobilization, and mobilization processes, but climatic conditions and topographic features also influence nutrient states through weathering and soil erosion processes. In addition to these natural conditions, SSNM apparently favors an adjustment of within-field variability of the soil nutrient status as fertilization rates are usually related inversely to the nutrient status. This was shown by Goedeken et al. (1998) and Gassner et al. (2002) for variable-rate fertilization of P. Therefore it is important for SSNM to update temporal changes with minimum sampling efforts. For this purpose already existing spatial information of a site is used to guide sampling to representative plots, referred to in the literature as directed, smart, or targeted sampling in order to reduce sampling efforts drastically (McBratney et al., 1999; Mulla, 1997; Pocknee et al., 1996; Schnug et al., 1998). Locations for directed sampling can be chosen from remote-sensing images (Haneklaus et al., 2000; Hergert, 1997; Mulla, 1997) and from equifertile maps that reflect zones of similar crop productivity, for instance derived from yield maps (Schnug et al. 1992; Schnug et al., 1994) or from different management zones (McBratney et al., 1999). The criteria of defining management zones for SSNM need to have a direct impact on the variation of the nutrient demand as management zones are not static, but flexible in space, because of, for example, pedogenetic factors and time due to management practices. For a further reduction of sampling points, only those locations should be selected for permanent investigations that have proved to be consistent in their yield potential over a validated time and comprise the whole range of variation of soil properties and their temporal fluctuations (Larscheid et al., 1997). This minimum number of control points comprises so-called monitor pedo cells (Haneklaus et al., 2000; Panten et al., 2005).

### *Fuzzy Algorithms*

Several authors addressed the identification of management zones, which are areas in the field with similar yield response to different factors as a basic step for decision making in SSNM. Normalized fuzzy partition entropy proved to be an appropriate tool to distinguish a number of different classes of natural variation, which can be addressed to operational units for site-specific management (Lark, 2001). Sources of variation can, however, not are explained by this procedure (Roubens, 1982). Fuzzy partition was used to define management zones from yield, topographic and soil data (Lark, 2001; Shatar and McBratney, 1999). Available information on soil characteristics may be used directly for the set-up of the classification or class memberships of soil features may be predicted indirectly employing related parameters.

## *SSNM FOR DIFFERENT NUTRIENT SOURCES*

Strategies for SSNM need to take into account different forms of fertilizers, such as single-nutrient mineral fertilizers, compound fertilizers, organic fertilizers, and secondary raw material fertilizers, because either monofactorial or multifactorial rules need to be elaborated for a balanced nutrient input.

### *Soil Nutrients and Liming*

Developing strategies for SSNM of soil nutrients may appear odd, but it is an intriguing approach under exceptional conditions, for example when fertilizers are not available, their use is restricted or banned. In developing countries for example, nutrient inputs are regularly limited because of economic constraints. In organic farming systems, in which legal guidelines exclude or restrict their use (EU guidelines [2091/91/EWG] in Schmidt and Haccius, 1998), the relative significance of soilborne nutrients for the nutrition of plants increases. This stresses the particular value of SSNM for small-scale farming and low-input systems, whereas SSNM usually focuses on large-scale farming. The strategy for SSNM would then be to apply spatially variable measures, which increase the utilization of soil nutrients. Increasing the mobility of soil nutrients or uptake efficiency of the crop could do this. The first option would be feasible for nutrients with an expressed dependency on soil pH, such as micronutrients and P (Schnug, 1992; Schnug and Finck, 1982), the second by promoting root growth in order to improve the source:sink ratio of nutrients. Suitable measures would imply variable tillage and sowing operations, variable-rate applications of lime, or acidifying fertilizers such as elemental S.

SSNM for liming is generally acknowledged as important because pH is highly variable in soils (Bongiovanni and Loewenberg-Deboer, 2000; Heiniger and Meijer, 2000). The soil pH is one of the most important factors influencing the mobility of soil nutrients (Schnug, 1982) and variable-rate liming is not only a very effective way to optimize the nutrient supply, but also to limit the availability of toxic heavy metals such as cadmium (Bruemmer et al., 1986).

The decision-making strategy for the site-specific amount of lime to be applied can be made by means of local calibration in order to adjust target soil pH values. Compared to Viscarra Rossel and McBratney (2001), who used sophisticated modeling, this approach is empiric.

### *SSNM with Mineral Fertilizers*

Fertilization with natural products such as plant residues, lime, gypsum, manure, compost, or sludge for maintaining soil fertility is as old as farming itself. Mineral fertilizers are either available in solid, liquid, or gaseous form, which determines the output technique. The specific problem of SSNM for organic fertilizers and secondary raw material products is the fact that they contain different nutrients in varying concentrations.

### *Single-Nutrient Mineral Fertilizers*

Single-nutrient mineral fertilizers are easiest to use for SSNM, particularly in the case of monocomponent products, such as urea, ammonia nitrate urea, or ammonia, as a linear transformation of nutrient demand into variable rate can be applied for different fertilizer products according to the following general formula:

$$Y = 100/a \times X$$

with $Y$ = variable nutrient rate (kg·ha$^{-1}$), $X$ = variable nutrient demand (kg·ha$^{-1}$), and $a$ = nutrient content in fertilizer product (percent).

In comparison with single nutrient fertilizers, compound products offer the following advantages: reduced expenditure of time for application and flexible, tailor-made nutrient ratios for small quantities. In a case study strategies for SSNM of compound fertilizers were elaborated to prove the general possibility of merging the fine-scale spatial variability of soil features with the variable-rate input of compound fertilizers. The major basic requirements for fertilizer management employing compound fertilizers are that the nutrient ratios provided by the manufacturer are flexible, and access is provided to a geographical information system in order to manage georeferenced soil and crop data. This implies the recording of input and off-take and the corresponding annual adjustment of fertilizer strategies under practical conditions.

### *Compound Fertilizers Without N*

The exclusive use of compound fertilizers is not possible as this would promote an imbalanced nutrient supply, and therefore single-nutrient fertilizers would need to be applied complementarily. The steps for calculating a balanced nutrient input within one crop rotation that matches the spatially variable nutrient demand by use of a PK compound fertilizer are as follows and summarized in Table 4.4:

1. Estimation of the individual P and K demand of each crop in the whole crop rotation
2. Calculation of optimum variable fertilizer rates for one nutrient, e.g., K (usually the most deficient nutrient or that nutrient with a high demand by the crop) for the first year of cropping
3. Determination of the widest P:K ratio (P:$K_{max}$) aiming at a P input that corresponds with the lowest P demand of the crop in the first year, de-

termined in the field in order to avoid surpluses that could not be balanced in the short term

4. Calculation of optimum P fertilizer rates for the second and third year of cropping by subtracting the P input of the first year from the P demand of the whole crop rotation (Thus the P input is already balanced after two years.)
5. Determination of the narrowest P:K ratio (P:$K_{min}$) aiming at a K input that corresponds with the lowest K demand of the winter wheat and barley crop determined in the field
6. Calculation of the optimum K fertilizer rates in the third year of cropping (This is achieved by subtracting the K rates already applied in the first and second year from the sum of K rates determined for the crop rotation.)

Following the approach presented in Table 4.4, compound fertilizers with more than two nutrients can also be applied in variable rates for balanced fertilization (Table 4.5).

*Compound Fertilizers with N*

The variable rates of N-containing compound fertilizers need to follow the prognosed N demand of the crop in order to maintain a high level of productivity. A mere linear optimization of the N:P:K ratio will not lead to a balanced nutrient input as the discrepancy between actual P and K fertilizer demand and applied P and K rates favors an imbalanced input and thus an

TABLE 4.4. Algorithms for a balanced variable-rate application of a PK fertilizer in a three-year crop rotation.

| | | Algorithms for | |
|---|---|---|---|
| **Year** | **Type of fertilizer** | **K** | **P** |
| 1 | PK | $K_{OPT}{}^{1}$ | $P_{P:Kmax}{}^{2}$ |
| 2 | PK | $K_{P:Kmin}$ | $P_{OPT} = P_{tot}{}^{3} - P_{yr1}$ |
| 3 | K | $K_{OPT} = K_{tot} - (K_{yr1} + K_{yr2})$ | |

*Note:* [1]OPT = optimum variable rate application; [2]P:K= variable-rate fertilization according to widest/narrowest (max/min) P:K (1:X) ratio resulting in minimum fertilizer expenditure; [3]tot = variable nutrient rates for crop rotation.

TABLE 4.5. Algorithms for a balanced variable-rate fertilization with triple compound fertilizers in a four-year crop rotation.

| | Algorithms for variable rates | | |
|---|---|---|---|
| Year | P | K | Mg |
| 1 | $P_{OPT}$[1] | $K_{P:K\ (min)}$[2] | $Mg_{P:Mg\ (min)}$ |
| 2 | $P_{P:Mg\ (max)}$ | $K_{K:Mg\ (max)}$ | $Mg_{OPT} = Mg_{tot}$[3] $- Mg_{yr1}$ |
| 3 | $P_{P:K\ (max)}$ | $K_{OPT} = K_{tot} - (K_{yr1} + K_{yr2})$ | — |
| 4 | $P_{OPT} = P_{tot} - (\Sigma\ P_{yr1\text{-}3})$ | — | — |

*Note:* [1]OPT = optimum variable rate application; [2]P:K = variable-rate fertilization according to widest/narrowest (max/min) P:K (1:X) ratio resulting in minimum fertilizer expenditure; [3]tot = total nutrient rates for crop rotation.

increased surplus or deficit in the field (Haneklaus and Schnug, 1999a). This means that fertilizer strategies employing N-containing compound products cannot focus on the optimum nutrient ratio, but rather have to use nutrient ratios that avoid surpluses in combination with the use of single-nutrient products. In the following examples it is assumed that the variation of the N rates remains constant in each year of cropping.

A balanced input of N, P, and K using compound fertilizers can for instance be realized with an alternative use of compound and single-nutrient fertilizers (Haneklaus and Schnug, 1999a). The narrowest N:P:K ratio is chosen in the first year thus avoiding surpluses in the nutrient balance, which could not be balanced within one crop rotation. In the subsequent year the optimum P and K rates are applied after consideration of the amounts added the year before. Instead of an alternative use of NPK and single-nutrient fertilizers, a strategy employing triple-, double-, and single-nutrient fertilizers within a three-year crop rotation may be applied (Table 4.6).

Again, in the first year of cropping the narrowest $N:P_{min}$ and $N:K_{min}$ ratio is chosen for fertilization. In the second year a NK fertilizer product is used, again with the narrowest $N:K_{min}$ ratio observed on the field. P is applied in the form of a single-nutrient fertilizer product whereby the optimum rates result from the difference between the total sum to be applied minus the sum of P applied by the narrowest $N:P_{min}$ ratio in the first and third year of cropping. In the third year of cropping the optimum K rates to be applied in single-nutrient form result from the difference of the total sum of K to be applied minus K applied in the first and second year. The main difference of this strategy compared to the alternative use of compound and single nutri-

TABLE 4.6. Algorithms for a balanced variable-rate application of N, P, and K over a three-year crop rotation for SSNM with N-obtaining compound fertilizers.

| Year | Algorithms for NPK fertilizer | | |
|---|---|---|---|
| 1 | | NPK fertilizer | |
| | $N_{OPT}$[1] | $P_{N:P\ (min)}$ | $K_{N:K\ (min)}$ |
| 2 | | NK fertilizer | P fertilizer |
| | $N_{OPT}$[1] | $K_{N:K\ (min)}$ | $P_{OPT} = P_{tot}$[2] $- (P_{yr1} + P_{yr3})$ |
| 3 | | NP fertilizer | K fertilizer |
| | $N_{OPT}$[1] | $P_{N:P\ (min)}$ | $K_{OPT} = K_{tot} - (K_{yr1} + K_{yr2})$ |

*Note:* [1]optimum variable-rate application; [2]tot = total optimum, spatially variable amount of P and K for the crop rotation.

ent fertilizers is that larger amounts of P and K are applied in the form of compound fertilizers. This gives the farmer more flexibility to react to changing market prices for different products.

### *SSNM for Organic Fertilizers and Secondary Raw Materials*

Strategies for the variable-rate application of farmyard manures and secondary raw material fertilizers are required most urgently because of environmental concerns. The specific problem of using these nutrient sources for SSNM is the fact that farmyard manure shows a particular inhomogeneity resulting in varying nutrient concentrations of the material, and that waste materials for recycling may contain environmentally relevant elements. An important premise in this context is that products used for fertilization need to be free from organic xenobiotics since no bias-free risk assessment is possible for these substances.

Organic fertilizers and secondary raw materials may be of higher significance for a successful development of PA than manufactured fertilizers because of the strong need to control their fate and environmental impacts, and because the costs of applying PA can be covered by the main use of the primary product.

#### *Farmyard Manure and Slurry*

The stocking density on livestock farms is limited to two large livestock units per hectare, and for environmental protection a maximum output of 170 kg·ha$^{-1}$ N by farmyard manure or slurry is permitted in the EU

(Bundesgesetzblatt, 1996). If the mean fertilizer N demand is below this level, SSNM of organic fertilizers on conventional farms could therefore run counter to the interest of the farmer, which is to empty storage capacities. On the contrary, manure is a limited resource in organic farming and SSNM is supposed to improve the N utilization efficiency and thus crop productivity.

The problem of inhomogeneity of organic fertilizers could be overcome either by a standardization process for the materials before use (this could for instance be done by incineration of slurry, whereby, however, N is lost), thorough homogenization of slurry together with an extremely fast analyzing service that provides the input data for calculations of variable rates or an on-the-go sensing system that analyzes relevant nutrient concentrations while the material passes the output equipment (Scotford et al., 1998). Another problem of SSNM with organic fertilizers is that the water content and nutrient ratios (N:P:K) are naturally fixed by animal species so that the problem of an imbalanced P and K supply is even bigger than applying N-containing compound fertilizers (see previous discussion). It is technically possible to vary the output of organic fertilizers as variable-rate spreaders for slurry exist and are under construction for manure (DLG, 2002).

## *INTERACTION OF SSNM WITH OTHER PA MEASURES IN THE FIELD*

Financing SSNM just from reduced fertilizer costs or increased yield benefits is a marginal business. Important for the economic success of SSNM, in fact for the future of PA, will be the identification of beneficial side effects, which can only be realized with SSNM and not with uniform nutrient-application techniques. From the recent state of the art such benefits may derive from weed-control measures and improved logistics and traffic on the field. The spatial variability of the fertilizer demand within a field is also a challenge for optimizing logistics and traffic. This is already common practice in the transport business (Tullberg, 2001). Optimizing places and times for reloading of fertilizer spreaders saves time and money and optimizing the travel routes of the spreader on the field can reduce traffic on the field, which is in line with the aims of soil conservation. This is especially important for bulky fertilizers, such as lime, manures, slurry, and secondary raw materials, which are usually distributed by heavy machinery.

Including site-specific knowledge of pH and weed infestations in the decision-making process is another example for the possibility to reduce or renounce the herbicide input, which can be of great importance in organic farming systems. Another useful side effect of liming could be the control of the availability and carryover of toxic heavy metals, which are known to be strongly dependent on soil pH (Bruemmer et al., 1986).

Heisel et al. (1999) found the population of *Lamium* ssp. to be related to the silt content and showed that co-kriging with soil properties significantly improved the mapping of different species. Figures 4.10 and 4.11 give examples from Vladutu et al. (1988) for the dependency of weed infestations on soil pH in cereals. For SSNM these results stress the significance of site-specific lime applications as the soil pH not only directly influences crop productivity, but also has significant side effects, such as on weed infestations. The optimum pH of a soil is always a compromise, however, between higher pH values to maintain soil structure and lower pH values to enhance the plant availability of micronutrients. Such a compromise can also be seen in the yield response curve in Figure 4.7, where yield reductions below the pH optimum are caused by an unfavorable soil structure, while those above the pH optimum range can be attributed to Mn deficiency.

SSNM offers one more challenge for weed control with the concept of nutrient-induced competition (NIC): the basic idea is to apply strongly concentrated nutrient solutions specifically to weeds. This causes physiologi-

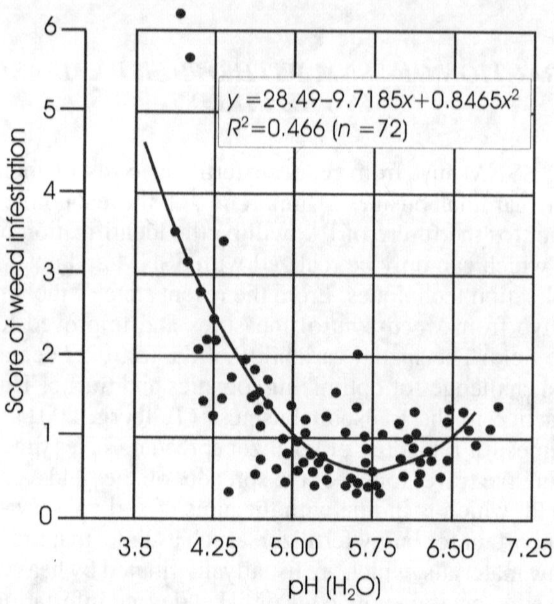


FIGURE 4.10. Relationship between weed infestation in winter wheat and soil pH on a brown luvic soil in Livada, Romania. Adapted from Vladutu et al., 1988.

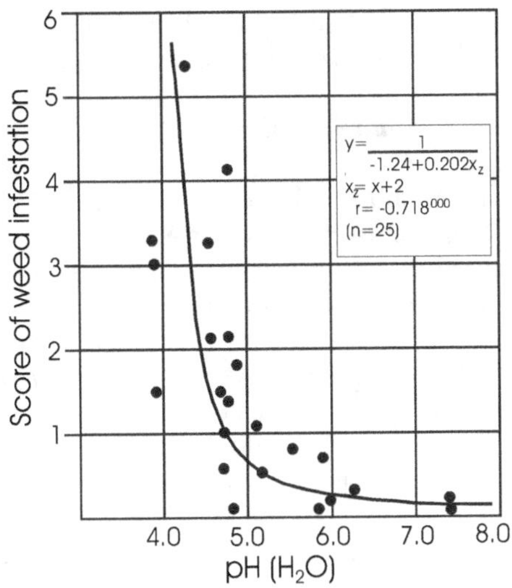


FIGURE 4.11. Relationship between infestation with *Agropyron repens* in winter barley and soil pH on an albic luvisol in Livada, Romania. Adapted from Vladutu et al., 1988.

cal disorders either by extreme nutrient surplus or salt stress, thus killing or at least reducing the vigor of the weed plant. This again enhances the relative competitiveness of the crop plant. Technically this would employ sprayers with optoelectronic control systems, which are able to differentiate between weeds and crop plants (Biller et al., 1997). Suitable nutrient solutions may contain KCl, NaCl, or sulfates of the micronutrients Fe, Mn, Zn, and Cu. The amount of nutrients applied by NIC need to be considered in the site-specific nutrient balance.

## *QUALITY ASPECTS*

In general, the risk that a mismatch of information and field operations causes adverse effects increases the smaller the distance is between two significantly different spatial events. In fact, the discussion of the "null hypothesis" for PA and thus also for SSNM (McBratney and Whelan, 1999) is only sensible after a thorough evaluation of errors involved.

A great problem to define quality criteria for SSNM is the fact that it has variability, which is searched for so that erroneous variability can easily be hidden. Adopting SSNM in practical farming therefore needs quality standards for information input and technical implements. In addition, with SSNM the demands toward the quality of education and advice increase.

### *Accuracy of Positioning and Mapping*

Table 4.7 gives some information of characteristic values for the accuracy of different components employed in SSNM. The data revealed that the accuracy of georeferenced information should be no longer a problem for SSNM. The guaranteed quality standards for purchasable information are also satisfactory (Resnik and Bill, 2000). The greatest source for errors will still be failures by the operator. As pointed out earlier, the quality of analytical data is not the subject of this chapter.

The question whether SSNM should preferably operate in a continuous context or in discrete management zones is still open, though Thrikawala et al. (1998) expect an increase of the fertilizer utilization efficiency with decreasing management unit sizes and an increase in spatial variability. In addition, the scale of temporal variation, its impact on crop yield in the ranking with other growth factors, the bandwidth of the optimum nutrient range, the response in terms of yield, and the scale of variation of the input

TABLE 4.7. Positional accuracy of different data sources and technical equipment in SSNM.

| Source | Positional accuracy (m) |
|---|---|
| *Satellite positioning* | |
| Standard | 2-3 |
| Differential | < 1 |
| With S/A | 5 |
| *Ordnance survey maps* | |
| Maps (scale 1:5000) | 3 |
| Digital elevation data | 0.5 |
| *Soil survey* | |
| Digitizing of map features (sampling points) | 3 |
| Boundaries of choropleth maps | 15-20 |
| *Remote-sensing images* | 10-20 |
| Yield maps | 10-15 |

are other important criteria. In other words, if the spatial and temporal variation of the nutrient are high, and the optimum range is small, yield fluctuations need to be faced when the nutrient supply is slightly below or above the optimum, and if crop performance is primarily dependent on a sufficient supply of this nutrient, a continuous mode is supposedly preferable for matching the site-specific nutrient demand.

Some general remarks on the quality of yield maps are given in this context because techniques for generating yield maps are becoming increasingly available on farms, and manufacturers of harvesters and yield-mapping systems advertise their systems specifically for SSNM. Colvin et al. (1997), Griepentrog (1999), Murphy et al. (1995), and Nissen and Soederstroem (1999) discuss the system-inherent errors of different yield-recording sensors.

The two major problems are the true positioning of the harvested yield and the accuracy of the amount of yield harvested (Haneklaus et al., 2001; Panten et al., 2002). Although satellite positioning allows a precise tracking of the moving harvester there is still a considerable source of error in matching the yield to its true position, which is due to variability of the time the crop needs to travel from the point where it is cut to the point where the yield is recorded by a sensor (Whelan and McBratney, 2002). Whelan and McBratney (2002) developed a transfer function to correct this error and yielded a significant increase of the small-scale variability of crop productivity.

The second source of error is the lack of secure means for the determination of the actual cutting width (Haneklaus et al., 2001; Selcuk and Colvin, 2002; Stafford et al., 1997). Panten et al. (2002) estimate for a combine harvester with a 6 m cutting bar and a GPS-assisted, flow-control, yield-mapping system, the first error to be in the range of 10 m, the second to equal about 8 to 10 percent of the recorded yield.

### *Accuracy of Navigation and Application*

The transfer of application maps for SSNM into field operations requires not only information about fertilizer rate, but requires also data for positioning and navigation. Precise navigation is also crucial for the allocation of monitor pedo cells for resampling.

Even the simplest form of SSNM employed on the smallest scale needs markers for orientation in space such as landmarks. Another option is tramlines, which have been common in northern European farming for 30 years (Parry, 1976) (in other parts of the world called controlled traffic [Tullberg, 2001]). However, the application of this method is limited to decision making and fertilization on small farms (< 20 ha). Under research conditions the accuracy of positioning and navigation with dead reckoning was found to be in a range of 15 to 20 m under optimum conditions. Satellite-aided navigation of fertilizer spreaders is meanwhile a low-cost technology, which meets the same accuracy as positioning.

The precise match of position and variable rate with the data of the fertilizer map is more problematic. The error during variable-rate application of fertilizers has several main components: the error of the navigation system, the hysteretic between the time the system is advised to apply a certain amount of fertilizer (Griepentrog and Persson, 2001), unevenness in the lateral distribution pattern (Persson and Bangsgaard, 1999), and inhomogeneities in the fertilizer material itself (Thirion et al., 1999). In general, the relative size of this error increases in the following order: liquids < solids; manufactured fertilizers < farm sources; slurries < manures; pneumatic spreaders < spin disk spreaders; irrigation < spreading (Cahn, 1995; Chaplin et al., 1995; King et al., 1997; Olieslagers et al., 1995; Persson and Bangsgaard, 1999; and Thirion et al., 1999).

This brief overview reveals that the lack of positive results from applying SSNM can simply be because the techniques, technologies, and sources employed are too inaccurate and not because the concept itself is wrong. The great number of error sources and nonspatial variability supports the hypothesis that successful SSNM requires the development and implementation of quality standards and control measures.

### *Quality of Education and Advice in Agriculture*

The understanding of the farmer as the biological interface adds one more aspect of quality to SSNM, namely that of education. The better the education, the better the biological interface will perform in adjusting fertilizer rates to the spatially variable needs of soils and crops. Excellent education in SSNM is also a strategy to ensure that knowledge is transferred most efficiently from the book to the field (Holt and Sonka, 1995; Krill, 1997; Mangold, 1995). Recent practice is that the decision maker himself or herself either applies the strategies for SSNM, or those responsibilities are handed over to a usually lower-educated machine operator. In the first case, the limitation is storage and recognition rate of the human brain, in the second case these restrictions are extended by the problem of a lower level of education.

Highly educated staff that apply PA technologies usually transfer their skills and knowledge to field workers. The complexity of SSNM compared to common fertilizer practice will open new markets for external services.

## *ECONOMIC, ECOLOGICAL, AND SOCIAL IMPACTS OF SSNM*

SSNM is the first and still dominating application of PA, but after almost 15 years of research and development its implementation and impact on agricultural business is still poor. Continuously decreasing fertilizer costs and product prices explain the birth problem of PA, i.e., the possible profit of SSNM was simply too small to pay the extra costs for the advanced concept. At the same time federal and industrial resources for education and advice in agriculture were reduced or withdrawn so that the backbone for an introduction of the new technology was lost.

### *Economic Impacts*

The crucial question of "Can SSNM be implemented successfully and permanently in future agriculture?" depends only on economic constraints. Although promised since the beginning of PA and still being advocated that SSNM would find widespread adoption, profits have been rarely in the ranges where farmers would immediately adopt the technique (Beguyot, 2002; Bullock, 1999; Daberkov et al., 2002; Kilian et al., 2001; Schmerler and Basten, 1999; Sparovek and Schnug, 2001; Swinton et al., 2001). Fertilizer savings ranged from 10 to 20 percent, but yield increases varied only between 3 to 5 percent (Van Alphen and Stoorvogel, 2000; Bongiovanni and Loewenberg-Deboer, 2000; Fixen and Reetz, 1995; Forcella, 1993; Lenz, 1997; Long et al., 1997; Schnitkey et al., 1997; Snyder et al., 1997; Wollenhaupt and Bucholz, 1993). This indicates that in the majority of these studies the crop has been already overfertilized in the past, which worsens the expected economic benefit of variable-rate application. Only a few papers report that the achieved returns from SSNM were considerably lower than expected (Hergert, 1997; Oestergaard, 1997).

The profitability of SSNM increases clearly when fertilizer prices are constantly high, as it was for P in the former German Democratic Republic (Witter, 1987) or as it is generally for fertilizers in Brazil (See Chapter 22). However the situation is problematic in the case of ammonia-based N fertilizers, whose price depends mostly (90 percent) on the costs for natural gas (TFI, 2002; IFA, 2002). Here, the rapid fluctuations make investments in new technologies on the farm difficult, particularly if the returns are marginal. An economic advantage of SSNM can be expected if contractors provide technologies and services (Luetticken et al., 1997). Although this may also create new jobs, the advocates of "agrarianism" fear that "farmers lose all connection to nature and sink into deeper dependence on the agribusiness juggernaut" (Kimbrell, 2002, p. 2).

It is often thought that the profitability of SSNM could be improved by charging farmers for environmental clean-up (McBratney, 2001). This approach is particularly risky, because environmentally harmful nutrient surpluses are characteristic of livestock farms (Vermeulen et al., 1998), not of arable farms, and the latter clients will most likely implement PA technologies first. A better way to solidify the economic basis of SSNM is to add value to the technologies employed. Intensive livestock farms could for instance implement SSNM with economic success, if by legal order the whereabouts of animal manures and slurries must be proved.

Circumstantial factors of agricultural production might strongly influence beneficial effects of SSNM. A general ban of growth regulators as a precautionary measure for consumer protection would have immense consequences for N-fertilization practices. The impact of growth regulators and N fertilization on crop performance is demonstrated in Table 4.8. The high variability of plant-available N in dependence on geomorphological features was already shown in Figure 4.2a, and values for $N_{min}$, N-fertilizer rate, yield and N content of grain were generalized from sampling points within the same field. Considering the negative effects that overfertilization of N has on crop yields (e.g., by early lodging) and the great variability of the N supply in the soil, the upper limit of N fertilization applied for optimum yield refers to the areas with the highest soil supply. In a system with uniform application rates this would leave the yield potential in areas with low soil N supply unexploited. In comparison, when growth regulators may be used the N input can match the lowest, plant-available N pool in the soil without the risk of lodging or yield losses in zones with a high natural N supply (Table 4.8).

The latter strategy accepts environmentally hazardous N surpluses. The example given in Table 4.8 reveals that a high input of N will cause even stronger losses in yield than a low N input if growth regulators were banned, whereas the site-specific application of N would be a tool to maintain crop productivity. The N utilization efficiency would be nearly twice as high in the site-specific N management than in a low N input system in which growth regulators are applied (Table 4.8). The studies of Whitley et al. (2001) confirm such interactions; they found higher N fluxes in the surface soil and increased N-uptake efficiency after variable-rate N fertilization to potatoes in an elevated landscape. Thus in countries where growth regulators are banned, for example in Sweden for corn (Jordbruksverket, 2001), the economic situation for SSNM substantially improves.

Adopting the concept of SSNM in farming is not too costly. It is rather the way of beginning to think on the spatial aspects of managing natural resources, which have to change first. Economic problems arise when additional PA technologies have to be financed. This includes hardware as well

TABLE 4.8. Influence of uniform and site-specific N fertilization in dependence on topography and use of growth regulators on yield and N-fertilizer efficiency of a winter wheat crop with an N demand of 120 kg·ha$^{-1}$ N grown in northern Germany.

*N management*

| Topography | Share of area (percent) | Plant-avail. N in soil (kg·ha$^{-1}$) | Uniform N rate (kg·ha$^{-1}$) | Total N supply (kg·ha$^{-1}$) | Variable N rate (kg·ha$^{-1}$) | Total N supply (kg·ha$^{-1}$) |
|---|---|---|---|---|---|---|
| Summit | 10 | 10 | 120 | 130[a] | 110 | 120 |
| Upper slope | 20 | 20 | 120 | 140 | 100 | 120 |
| Middle slope | 40 | 40 | 120 | 160 | 90 | 130 |
| Lower slope | 20 | 80 | 120 | 200 | 40 | 120 |
| Depression | 10 | 120 | 120 | 240[b] | 0 | 120 |
| Mean | | 49 | 120 | 169 | 75 | 124 |

*Effect of N management and use of growth regulators on crop yield (relative)*

| Topography | No GR[c] No N | No GR Low N input | With GR Low N input | No GR High N input | With GR High N input | No GR Variable N rate |
|---|---|---|---|---|---|---|
| Summit | 20 | 40 | 40 | 100 | 100 | 110 |
| Upper slope | 30 | 50 | 50 | 90 | 100 | 110 |
| Middle slope | 40 | 60 | 60 | 60 | 100 | 105 |
| Lower slope | 80 | 90 | 90 | 30 | 100 | 100 |
| Depression | 100 | 90 | 100 | 20 | 100 | 100 |
| Mean | 50 | 65 | 66 | 60 | 100 | 105 |
| *Effect of N management and use of growth regulators on fertilizer N efficiency* | | | | | | |
| N off-take[d] | 60 | 104 | 106 | 120 | 200 | 210 |
| N content[e] | 1.3 | 1.6 | 1.6 | 2.0 | 2.0 | 2.0 |
| N utilization efficiency[f] | | 59 | 61 | 60 | 140 | 211 |

*Note:* [a]low N input = 130 kg·ha$^{-1}$ N; [b]high N input = 240 kg·ha$^{-1}$ N; [c]GR: growth regulator; [d]N off-take (kg·ha$^{-1}$); [e]N content of grain (percent); [f]N utilization efficiency (percent) determined by the difference method.

as labor costs for running the systems. Although many glossy promises have been made by companies and researchers over the last decade (e.g., Taylor et al., 2002) the basic funding for PA technologies on a farm will have to come from other sources than saved inputs or prevented environmental costs. The most promising strategy for financing the fundamental costs of PA will most likely be to make the "last man" on the tractor redundant, that is, through the use of automated machine guidance and telematics (Geo-Tec, 2002; Goetting, 2002; Hagras et al., 1999; Schnug et al., 2003; Umeda et al., 1999). This is supposed to save labor costs and pay for the system, and SSNM will become a welcome and added value to the system.

### *Ecological Impacts*

The intrinsic hypothesis for the environmental-friendliness of SSNM is evident—less fertilizer is applied to areas with increased vulnerability to nutrient losses. Until now only a few experimental proofs with direct measurements were available in the literature (e.g., Whitley et al., 2001). Nevertheless, sufficient evidence for the positive effects of SSNM on nutrient discharges to the environment is given when spatial nutrient balances for systems with and without applying SSNM are made up (Haneklaus and Schnug, 1998) and if residual nutrients in soils are considered (Kitchen et al., 1995).

The economy of SSNM would be best supported if, in addition to improving the nutrient efficiency, more value could be added to this basic application. A very important side effect of SSNM could be the control of the whereabouts of recycled wastes, both from animal production and industry outside agriculture. Intensive livestock farming is still the major source for nonpoint nutrient pollution of water bodies and the atmosphere (Isherwood, 1998; Sapek, 1998; Sugiharto et al., 1994). The obligatory use of SSNM and PA technologies would therefore ensure an efficient documentation of the whereabouts of these residues.

Another important area for SSNM may become logistics and field traffic. A major future application for PA, and therefore also beneficial for the economy of SSNM, is plant protection measures in order to minimize the leaching of pesticides (Khakural et al., 1993; Paulsen, 1989). The input of herbicides may be reduced when the impact of soil characteristics on weed distribution is considered and included into decision making for SSNM with lime (Heisel et al., 1999; Vladutu et al., 1988). Taking advantage of the innovative concept of nutrient-induced competition, SSNM may contribute to a reduced input of pesticides or other chemicals used for plant protection.

Another example for environmental benefits from SSNM was given in Table 4.8, where the variation of N fertilizer rates was calculated for management systems operating without growth regulators. It is the use of growth regulators that enables an excessive use of N as undesired side effects of over-fertilization are compensated (Table 4.8). Growth regulators such as chlorocholinchloride (CCC) accumulate in crops in dependence on N fertilization and thus enter the food chain (Bier and Dedek, 1972; Landazuri et al., 1993). Due to the toxic effects of CCC in animals, adverse effects in humans cannot be excluded (Landazuri et al., 1993). Employing SSNM bears a fair chance that growth regulators could be replaced by better concepts, knowledge, and new technologies. Without any doubt, SSNM will be part of future codes of good agricultural practice and PA technologies will be implemented in best management practices (Casell and Clausen, 1993; Park et al., 1994; Yadav and Wall, 1998).

### *Social Impacts*

Every new technology, especially when entering old and very traditional professions such as agriculture, has social impacts that need to be considered in order to avoid social conflicts. Nowack mentioned this aspect in 1993.

When demands toward the quality of work change, the people involved will have to change as well. Higher educational levels are needed to conduct and follow up the more complicated decision making, and extended knowledge and skills are needed to operate sophisticated technology. Agriculture was always a business in which low and even uneducated people could find work, and this aspect needs to be taken into account. Although modern technologies might enable low and uneducated tractor drivers to carry out SSNM operations in the field, the next step is that they are replaced by highly accurate navigating robots (Geo-Tec, 2002; Goetting, 2002; Hagras et al., 1999; Umeda et al., 1999).

There is a trend toward a need for higher qualified staff in agriculture, but this means higher salaries, too. At the given economic situation of the whole sector this implies less jobs and increasing farm sizes. There is also a need for more sophisticated information and advice from outside the farm, which may create new jobs. But as this is also a prominent subject for internet services (Luetticken, 2001), the number of newly created jobs will be much lower than the ones lost. A socially just implementation of SSNM and PA into future agriculture requires that ample attention is paid to the LK of farmers in order to avoid the old problem of alienation (Berry, 2002; Marx, 1844) and that PA becomes "a technology that diminishes the traditional old role of farmers by substituting capital and technology for age-old local knowledge" as Kimbrell (2002) feared in his book *Fatal Harvest.*

## FUTURE RESEARCH NEEDS

SSNM is a concept for improving the sustainability of nutrients in agriculture, and it needs further development before it will be widely implemented. Based on the current status given earlier, the main direction for research and development of SSNM will have seven core targets:

1. Improved access to input data, which provides the required information for effective decision making in SSNM. A major contribution can be expected from real-time crop surveillance, because then dynamic maps of soil and crop features can be established. The further development of on-the-go sensing systems is vital as well. More efforts need to be made for faster, cheaper, and easier access to spatial data, and the Internet seems to be the perfect medium.
2. Development of decision-making strategies and rules including fuzzy logic algorithms.
3. Precise variable-rate application technologies for manufactured fertilizers, but particularly for manures and slurries.
4. For PA and variable-rate technologies quality control measures and standards need to be defined in order to minimize technical errors inherent to the equipment.
5. Interdisciplinary applications for PA technologies need to be developed to support the profitability of SSNM.
6. Education and advice in SSNM need to be strengthened for improving the efficient use of sophisticated PA technologies and for implementing the beneficial concept on farms with low nutrient inputs.
7. Last, but not least, agricultural and environmental politics need to become aware of the potential of SSNM for sustainable agricultural production and need to be encouraged to provide the political backup for its global realization.

## REFERENCES

Adsett, J.F. and G.C. Zoerb. 1991. Automated field monitoring of soil nitrate levels. In *Automated Agriculture for the 21st Century,* Am. Soc. of Agr. Engineers, St. Joseph, Michigan, pp. 326-335.

Barnes, E.M., M.S. Moran, P.J. Pinter, and T.R. Clarke. 1997. Multispectral remote sensing and site-specific agriculture: Examples of current technology and future possibilities. In *Proceedings 3rd Int. Conf. on Precision Agriculture,* eds. P.C. Robert, R.H. Rust, and W.E. Larson, ASA-CSSA-SSSA, Madison, WI, pp. 845-854.

Bauer, A. and H. Guenzel. 2001. *Data Warehouse Systeme*, Dpunkt Verlag, Heidelberg.

Bauer, M. 1994. *Vermessung und Ortung mit Satelliten,* Verlag Wichmann, Heidelberg.

Bauernblatt/Landpost. 1989. Kleine Chips fuer die Optimierung. *Bauernblatt/ Landpost* 8 Juli 27 (43/139):48-49.

Beguyot, P. 2002. Motivations for undertaking or expectations of French farmers regarding precision agriculture. In *Abstracts 6th Int. Conf. on Precision Agriculture,* eds. P.C. Robert, R.H. Rust, and W. E. Larson, ASA/CSSA, SSSA/ ASAE, Madison, WI, p. 95.

Berry, W. 2002. The whole horse—the preservation of the agrarian mind. In *Fatal harvest: The Tragedy of Industrial Agriculture,* ed. A. Kimbrell, Foundation for Deep Ecology, Sausalito, CA, pp. 7-11.

Bier, H. and W. Dedek. 1972. Transport, Verteilung und Rueckstaende von $^{14}$C- und $^{15}$N-CCC (2-Chloroaethyltrimethylammoniumchlorid) in Weizen. *Biochemie und Physiologie der Pflanzen* 163:169-174.

Biller, R.H., A. Hollstein, and C. Sommer. 1997. Precision application of herbicides by use of optoelectronic sensors. In *Precision Agriculture 1997: Proceedings 1st European Conf. on Precision Agriculture,* Vol. 2, ed. J.V. Stafford, BIOS Scientific Publishers, Oxford, UK, pp. 451-458.

Birrell, S.J. and J.W. Hummel. 1993. Multi-ISFET sensors for soil nitrate analysis. In *Proceedings Workshop Soil-Specific Crop Management*, ASA-CSSA-SSSA, Madison, WI, pp. 349-350.

Blackmer, A.M., J.S. Schepers, and G.E. Meyer. 1995. Remote sensing to detect nitrogen deficiency in corn. In *Proceedings Site-Specific Management for Agricultural Systems,* eds. P.C. Robert, R.H. Rust, and W.E. Larson, ASA-CSSA-SSSA, Madison, WI, pp. 505-512.

Blackmer, A.M. and S.E. White. 1997. Remote sensing to identify spatial patterns in optimal rates of nitrogen fertilization. In *Proceedings 3rd Int. Conf. on Precision Agriculture,* eds. P.C. Robert, R.H. Rust, and W.E. Larson, ASA-CSSA-SSSA Madison, pp. 33-42.

Blackmore, B.S. and G. Larscheid. 1997. Strategies for managing variability. In *Precision Agriculture 1997: Proceedings 1st European Conf. on Precision Agriculture,* Vol. 2, ed. J.V. Stafford, BIOS Scientific Publishers, Oxford, UK, pp. 851-859.

Bloem, E. 1998. Schwefel-Bilanz von Agraroekosystemen unter besonderer Beruecksichtigung hydrologischer und bodenphysikalischer Standorteigenschaften. *Landbauforschung Voelkenrode* 192:1-156.

Bongiovanni, R. and J. Loewenberg-Deboer. 2000. Economics of variable rate lime in Indiana. *Precision Agriculture* 2:55-70.

Bouma, J. 1997. From soil survey to a soil database for precision agriculture. In *Precision Agriculture 1997: Proceedings 1st European Conf. on Precision*

*Agriculture,* Vol. 1, ed. J.V. Stafford, BIOS Scientific Publishers, Oxford, UK, pp. 61-69.

Bouma J. 1999. Land evaluation for landscape units. In *Handbook of Soil Science,* ed. M.E. Summer, CRC Press, Boca Raton, FL, pp. E393-E412.

Boydell, B. and A.B. McBratney. 1999. Identifying potential within-field management zones from cotton yield estimates. In *Proceedings 2nd European Conf. on Precision Agriculture,* ed. J.V. Stafford, Sheffield Academic Press, Odense, Denmark, pp. 331-342.

Bruemmer, G., J. Gerth, and U. Herms. 1986. Heavy metal species, mobility and availability in soils. *Zeitschrift Pflanzenernaehrung und Bodenkunde* 149: 382-398.

Bullock, D.S. 1999. The economics of precision farming: A primer for agronomists designing experiments. In *Proceedings 2nd European Conf. on Precision Agriculture,* ed. J.V. Stafford, Sheffield Academic Press, Odense, Denmark, pp. 937-946.

Bundesgesetzblatt. 1996. *Verordnung ueber die Grundsaetze der guten fachlichen Praxis beim Duengen (Duengeverordnung),* Bundesgesetzblatt, Jg. 1996 Teil 1, Nr. 6, Bonn 6.2.1996.

Burrough, P.A. 1993. Soil variability: A late 20th century view. *Soils and Fertilizers* 56:529-562.

Cahn, M.D. 1995. A variable rate applicator for side-dressing liquid N fertilizer. In *Proceedings 2nd Int. Conf. on Site-specific Management for Agricultural Systems,* eds. P.C. Robert, R.H. Rust, and W.E. Larson, ASA-CSSA-SSSA, Madison, WI, pp. 683-690.

Cahn, M.D., J.W. Hummel, and B.H. Brouer. 1994. Spatial analysis of soil fertility for site-specific crop management. *Soil Science Society American Journal* 58: 1240-1248.

Cambardella, C.A., T.B. Moorman, J.M. Novak, T.B. Parkin, D.L. Karlen, R.F. Turco, and A.E. Konopka. 1994. Field-scale variability of soil properties in central Iowa soils. *Soil Science Society American Journal* 5:1501-1511.

Capelle, A. 1999. Die Eignung von Bodenkarten unteschiedlicher Massstaebe fuer die Erfassung der kleinraeumigen Heterogenitaet des Bodens, *KTBL-Arbeitspapier* 264, Erfassung der kleinraeumigen Heterogenitaet, pp. 42-46.

Cassell, E.A. and J.C. Clausen. 1993. Dynamic simulation modeling for evaluating water quality response to agricultural BMP implementation. *Water Science Technology* 28:635-648.

Catt, J.A., A.E. Johnston, and J.N. Quinton. 1997. Phosphate losses in the Woburn Erosion Reference Experiment. In *Phosphorus Loss from Soil to Water,* eds. H. Tunney, O.T. Carton, P.C Brookes, and A.E. Johnson, CAB International, Wallingford, UK, pp. 374-377.

Chaplin, J., E. Roytburg, and J. Kaplan. 1995. Measuring the spatial performance of chemical applicators. In *Proceedings 2nd Int. Conf. on Site-specific Management for Agricultural Systems,* eds. P.C. Robert, R.H. Rust, and W.E. Larson, ASA-CSSA-SSSA, Madison, WI, pp. 651-670.

Claydon, J. 1989. Patent no. 0042 245, Wickhambrook, UK.

Clevers, J.G.P.W. 1994. Imaging spectrometry in agriculture: Plant vitality and yield indicators. In *Imaging Spectrometry: A Tool for Environmental Observations,* Vol. 4, eds. J. Hill and J. Megier, Kluwer Academic Publishers, Dordrecht, The Netherlands, pp. 193-221.

Collins, C. and A.L. Budden. 1998. Soil analysis techniques—the need to combine precision with accuracy. In *Proceedings Int. Fertilizer Society Conf.,* No 418, York, UK.

Colvin, T.S., D.L. Karlen, D.B. Jaynes, S. Arslan, M.A. Al-Mahasneh, and W. Li. 1997. Comparisons of grain yield monitors to scales for mapping. In *Precision Agriculture 1997: Proceedings 1st European Conf. on Precision Agriculture,* Vol. 1, ed. J.V. Stafford. BIOS Scientific Publishers, Oxford, UK, pp. 469-475.

Daberkov, S.G., J. Fernandez-Cornejo, and M. Padgitt. 2002. Precision agriculture technology diffusion. In *Abstracts 6th Int. Conf. on Precision Agriculture,* eds. P.C. Robert, R.H. Rust, and W. E. Larson, ASA/CSSA, SSSA/ASAE, Madison, WI, p. 93.

Dampney, P.M.R., M.A. Froment, and C. Dawson. 1997. The variability of pH and available phosphorus, potassium, and magnesium in soils within arable fields in England. In *Precision Agriculture 1997: Proceedings 1st European Conf. on Precision Agriculture,* ed. J.V. Stafford, BIOS Scientific Publishers, Oxford, UK, pp. 79-86.

Davenport, J.R. and J.D. Jabro. 2001. Assessment of hand held ion selective electrode technology for direct measurement of soil chemical properties. *Communications of Soil Science and Plant Analysis* 32:3077-3085.

Dawson, C.J. 1996. Implications of precision farming for fertilizer application policies. In Proceedings No. 391, The International Fertilizer Society, York, UK.

Dawson, C.J. and J. Johnston. 1997. Aspects of soil fertility in the interpretation of yield maps as an aid to precision farming. In *Proceedings 2nd European Conf. on Precision Agriculture,* ed. J.V. Stafford, Sheffield Academic Press, Sheffield, UK, pp. 87-94.

DLG. 2002. http://presse.dlg.org/workspace/de/page5850.html.

Dobermann, A. and P.F. White. 1999. Strategies for nutrient management in irrigated and rainfed lowland rice systems. *Nutrient Cycling in Agroecosystems* 53:1-18.

Doerge, T. 1998. Management zone concepts. In *Site-Specific Management Guidelines,* eds. D.E. Clay, N.R. Kitchen, C.G. Carlson, J.L. Kleinjan, and W.A. Tjentland, Potash and Phosphate Institute, Site-Specific Management Guidelines No. 2, South Dakota State University, Brookings, pp. 1-4.

Engel, R.E., D.S. Long, G.R. Carlson, and C. Meier. 2000. Method for precision nitrogen management in spring wheat: I fundamental relationships. *Soil Science Society American Journal* 64(1):328-338.

Everett, M.W. and F.J. Pierce. 1997. Variability of corn yield and soil profile nitrates in relation to site-specific N management. In *Proceedings 3rd Int. Conf. on Precision Agriculture,* eds. P.C. Robert, R.H. Rust, and W.E. Larson, ASA-CSSA-SSSA, Madison, WI, pp. 43-53.

FAL. 2002. http://www.pb.fal.de/en/index.htm.

Fixen, P.E. and H.F. Reetz. 1995. Site-specific soil test interpretation incorporating soil and farmer characteristics. In *Proceedings Site-Specific Management for Agricultural Systems,* eds. P.C. Robert, R.H. Rust, and W.E. Larson, ASA-CSSA-SSSA, Madison, WI, pp. 731-742.

Fleming, K.L., D.G. Westfall, D.W. Wiens, and M.C. Brodahl. 2000. Evaluating farmer defined management zone maps for variable rate fertilizer application. *Precision Agriculture* 2:201-215.

Forcella, F. 1993. Value of managing within-field variability. In *Proceedings 1st Workshop Soil-Specific Crop Management,* eds. P.C. Robert, R.H. Rust, and W.E. Larson, ASA-CSSA-SSSA, Madison, WI, pp. 125-132.

Fotyma, M., A. Faber, M. Czajkowski, and K. Kubsik. 1997. Preliminary experience concerning precision agriculture in Poland. In *Precision Agriculture 1997: Proceedings 1st European Conf. on Precision Agriculture,* Vol. 1, ed. J.V. Stafford, BIOS Scientific Publishers, Oxford, UK, pp. 111-118.

Franko, U., G.J. Crocker, P.R. Grace, J. Klir, M. Koerschens, P.R. Poulton, and D.D. Richter. 1997. Simulating trends in soil organic carbon in long-term experiments using the CANDY model. *Geoderma* 81:109-120.

Franzen, D.W., A.D. Halvorson, J. Krupinsky, V.L. Hofman, and L.J. Cihacek. 1998. Directed sampling using topography as a logical basis. In *Proceedings 4th Int. Conf. on Precision Agriculture,* eds. P.C. Robert, R.H. Rust, and W.E. Larson, ASA-CSSA-SSSA, Madison, WI, pp. 1559-1568.

Fromm, H., K. Winter, J. Filser, R. Hantschel, and F. Beese. 1993. The influence of soil type and cultivation system on the spatial distributions of the soil fauna and microorganisms and their interactions. *Geoderma* 60:109-118.

Gassner, A. 2003. Factors controlling the spatial speciation of phosphorous in agricultural soils. *Special issue FAL–Agricultural Research* 244.

Gassner, A., J. Fleckenstein, S. Haneklaus, and E. Schnug. 2002. Spatial speciation—a new approach to assess soil analysis methods. *Communications of Soil Science and Plant Analysis* 33(15-18): 3347-3357.

Geisler, G. 1980. *Pflanzenbau,* Verlag Paul Parey, Berlin.

Geo-Tec. 2002. http://www.geo-tec.de/geo-tec/geo-tec-e/index.htm.

Goedeken, M.W., G.V. Johnson, W.R. Raun, and S.B. Phillips. 1998. Soil test phosphorus crop response projections to variable rate phosphorus application. *Communications of Soil Science and Plant Analysis.* 29(11-14):1731-1738.

Goetting. 2002. http://www.goetting.de/e_home.shtml.

Graichen, G. 1990. *Das Kultplatzbuch,* Bechtermuenz Verlag, Augsburg.

Grenzdoerffer, G. J. 2001. Requirements and possibilities of remote sensing for precision agriculture—current status and future developments. In *Proceedings 3rd European Conf. on Precision Agriculture,* eds. G. Grenier and S. Blackmore, Agro-Montpellier, Montpellier, France, pp. 211-216.

Griepentrog, H.-W. 1999. Ertragsermittlung im Maehdrescher. In *Erfassung der kleinraeumigen Heterogenitaet,* KTBL-Arbeitspapier 262:68-77.

Griepentrog, H.-W. and K. Persson. 2001. A model to determine the positional lag for fertilizer spreaders. In *Proceedings 3rd European Conf. on Precision Agriculture,* eds. G. Grenier and S. Blackmore, Agro-Montpellier, Montpellier, France, pp. 671-682.

Guyot, G. and F. Baret. 1988. Utilisation de la haute resolution spectrale pour suivre l'etat des couverts vegetaux. In *Proceedings 4th. Coll. on Spectral Signature of Objects in Remote Sensing,* Aussois, France, ESA SP–287, European Space Agency, Paris, pp. 279-286.

Hagras, V., M.C. Callaghan, and M. Carr-West. 1999. A behavior based hierarchical fuzzy control architecture for agricultural autonomous mobile robots. In *The Int. Conf. on Computational Intelligence for Modeling, Control, and Automation,* ed. M. Mohammadian, IOS Press, Amsterdam, The Netherlands, pp. 172-177.

Hake, G. 1982. *Kartographie I: Allgemeines, Erfassung der Informationen, Netzentwuerfe, Gestaltungsmerkmale, topographische Karten.* 6. neubearbeitete Auflage, Walter de Gruyter, Berlin.

Han, S., R.G. Evans, S.M. Schneider, and S.L. Rawlins. 1997. Spatial variability of soil properties on two center-pivot irrigated fields. In *Proceedings 3rd Int. Conf. on Precision Agriculture,* eds. P.C. Robert, R.H. Rust, and W.E. Larson, ASA-CSSA-SSSA, Madison, WI, pp. 97-106.

Haneklaus, S., H. Lilienthal, K. Panten, E. Haveresch, and E. Schnug. 2001. Routines for efficient yield mapping. In *Proceedings 5th Int. Conf. on Precision Agriculture,* eds. P.C. Robert, R.H. Rust, and W.E. Larson, ASA-CSSA-SSSA, Madison, WI, CD-ROM.

Haneklaus, S., H.M. Paulsen, D. Schroeder, U. Leopold, and E. Schnug. 1998. Self-surveying—a strategy for efficient mapping of the spatial variability of time constant soil parameters. *Communications of Soil Science and Plant Analysis* 29(11-14):1593-1601.

Haneklaus, S., I. Ruehling, D. Schroeder, and E. Schnug. 1997. Studies on the variability of soil and crop fertility parameters and yields in different landscapes of Northern Germany. In *Precision Agriculture 1997: Proceedings 1st European Conf. on Precision Agriculture,* Vol. 2, ed. J.V. Stafford, BIOS Scientific Publishers, Oxford, UK, pp. 785-792.

Haneklaus, S. and E. Schnug. 1998. Impacts of precision agriculture technologies on fertilization. In *Proceedings 11th Int. Symposium of CIEC Codes of Good Fertilizer Practice and Balanced Fertilization,* eds. E. Schnug and M. Fotyma, CIEC, Pulawy, Poland, pp. 95-107.

Haneklaus, S. and E. Schnug. 1999a. Decision-making strategies for the variable rate application of compound fertilizers. *Communications of Soil Science and Plant Analysis* 31(11-14):1863-1873.

Haneklaus, S. and E. Schnug. 1999b. Self-surveying and variable N rates. *Agroprecise* 1(2):4-6.

Haneklaus, S. and E. Schnug. 2000. Spatial variability of soil acidity and its impact on production technology. In *Proceedings Int. Scientific Conf. on Soil Acidification and Liming: Fundamental and Technological Researches,* eds. Z. Dabkevicius, E. Lapinskas, V. Petraitas, A. Siuliauskas, V. Slapakauskas, A. Rasinskiene, V. Tarvydas and V. Vasiliauskiene, Lithuanian Institute of Agriculture, Klaipeda, Lithuania, pp. 67-73.

Haneklaus, S. and E. Schnug. 2002. An agronomic, ecological, and economic assessment of site-specific fertilization. *FAL–Agricultural Research* 52:123-133.

Haneklaus, S., E. Schnug, and K. Panten. 2000. Evaluation of structural coincidences of patterns in remote sensing images and yield maps for the identification of sampling locations. *Aspects of Applied Biology* 60:37-44.

Haneklaus, S., D. Schroeder, and E. Schnug. 1997. Strategies for fertilizer recommendations based on digital agro resource maps. In *Proceedings 3rd Int. Conf. on Precision Agriculture,* eds. P.C. Robert, R.H. Rust, and W.E. Larson, ASA-CSSA-SSSA, Madison, WI, pp. 361-368.

Haneklaus, S., D. Schroeder, and E. Schnug. 1999. Decision-making strategies for fertilizer use in precision agriculture. In *Proceedings 4th Int. Conf. on Precision Agriculture,* Part A, eds. P.C. Robert, R.H. Rust, and W.E. Larson, ASA-CSSA-SSSA, Madison, WI, pp. 1785-1797.

Hansen, J. 1990. *EDV-Einsatz in der Landwirtschaft unter besonderer Beruecksichtigung des Einsatzes im Betrieb Osterhof.* Jahresarbeit im Fach Pflanzenproduktion an der Hoeheren Landbauschule Rendsburg, BS thesis.

Heiniger, R.W. and A.J. Meijer. 2000. Why variable rate application of lime has increased grower profits and acceptance of precision agriculture in the southeast. In *Proceedings 6th Int. Conf. on Precision Agriculture,* eds. P.C. Robert, R.H. Rust, and W. E. Larson, ASA/CSSA, SSSA/ASAE, Madison, WI, CD-ROM.

Heisel, T., A.K. Ersboell, and C. Andreasen. 1999. Weed mapping with co-kriging using soil properties. *Precision Agriculture* 1:39-52.

Hergert, G.W. 1997. A futuristic view of soil and plant analysis and nutrient recommendations. *Communications of Soil Science and Plant Analysis* 29(11-14): 1441-1444.

Hergert, G.W., W.L. Pan, D.R. Huggins, J.H. Grove, and T.R. Peck. 1997. The adequacy of current fertilizer recommendations for site-specific management in the USA. In *Precision Agriculture 1997: Proceedings 1st European Conf. on Precision Agriculture,* Vol. 1, ed. J.V. Stafford, BIOS Scientific Publishers, Oxford, UK, pp. 371-378.

Heute, A.R. 1998. A soil adjusted vegetation indices (SAVI). *Remote Sensing of Environment,* 25:295-309.

Holt, D.A. and S.T. Sonka. 1995. Virtual agriculture: Developing and transferring agricultural technology in the 21st century. In *Proceedings 2nd Int. Conf. on Site-Specific Management for Agricultural Systems,* eds. P.C. Robert, R.H. Rust, and W.E. Larson, ASA-CSSA-SSSA, Madison, WI, pp. 883-898.

IFA. 2002. International Fertilizer Industry Association; http://www.fertilizer.org

Isherwood, K.F. 1998. Good fertilizer practice and balanced fertilization. In *Proceedings 11th Symposium of CIEC Codes of Good Fertilizer Practice and Balanced Fertilization,* eds. E. Schnug and M. Fotyma, CIEC, Pulaway, Poland, pp. 157-170.

Jackson, W., B. Colman, and W. Berry. 1984. *Meeting the Expectations of the Land: Essays in Sustainable Agriculture and Stewardship,* University of Nebraska Press, Lincoln, NE.

Jordbruksverket. 2001. http://www.sjv.se.

Khakural, B.R., P.C. Robert, and D.J. Fuchs. 1993. Leaching and runoff or pesticides under conventional and soil specific management. In *Proceedings 1st*

*Workshop Soil-Specific Crop Management,* eds. P.C. Robert, R.H. Rust, and W.E. Larson, ASA-CSSA-SSSA, Madison, WI, pp. 363-364.

Kilian, B., T.M. Hurley, and G. Malzer. 2001. Economic aspects of precision agriculture: An economic assessment of different site-specific N-fertilization approaches. In *Proceedings 3rd European Conf. on Precision Agriculture,* eds. G. Grenier and S. Blackmore, Agro-Montpellier, Montpellier, France, pp. 521-532.

Kimbrell, A. 2002. The virtual farm—precision farming and technology take command. In *Fatal harvest—the tragedy of industrial agriculture,* ed. A. Kimbrell, Foundation for Deep Ecology, Sausalito, California, pp. 261-265.

King, B.A., I.R. Mc Cann, J.C. Stark, and D.T. Westermann. 1997. Spatially varied nitrogen application through a center pivot irrigation system. In *Proceedings 3rd Int. Conf. on Precision Agriculture,* eds. P.C. Robert, R.H. Rust, and W.E. Larson, ASA-CSSA-SSSA, Madison, WI, pp. 85-96.

Kitchen, N.R., D.F. Hughes, K.A. Sudduth, and S.J. Birrell. 1995. Comparison of variable rate to single rate nitrogen fertilizer application: Corn production and residual soil NO3-N. In *Proceedings 2nd Int. Conf. on Site-Specific Management for Agricultural Systems,* eds. P.C. Robert, R.H. Rust, and W.E. Larson, ASA-CSSA-SSSA, Madison, WI, pp. 427-442.

Kloepfer, F., S. Haneklaus, and E. Schnug. 1994. Ergebnisse aus dem Projekt "Computer Aided Farming" (CAF). *KTBL-Arbeitspapier* 175:132-175.

Kokaly, R.F. and R.N. Clark. 1999. Spectroscopic determination of leaf biochemistry using band-depth analysis of absorption features and stepwise multiple linear regression. *Remote Sensing of Environment* 67:267-287.

Krige, D.G. 1951. A statistical approach to some basic mine valuation problems on the Witwatersrand. *Journal of the Chemical, Metallurgical, and Mining Society of South Africa* 52:119-139.

Krill, T.L. 1997. Site-specific management: Educating through cooperating farmers. In *Proceedings 3rd Int. Conf. on Precision Agriculture,* eds. P.C. Robert, R.H. Rust, and W.E. Larson, ASA-CSSA-SSSA, Madison, WI, pp. 1165-1172.

Kristensen, K., S.E. Simmelsgaard, J. Djurhuus, and S.E. Olesen. 1995. Spatial variability of soil physical and chemical parameters. In *Proceedings Seminar on Site-Specific Farming,* SP report no. 26, March 20-21, ed. S.E. Olesen, Koldkærgård, Aarhus, Denmark, pp. 39-55.

Lamp, J. and E. Schnug. 1987. Rechnergestuetzte Duengung mit Hilfe digitaler Hofbodenkarten. *Schriftenreihe Agrarwissenschaftliche Fakultaet, Kiel* 69:61-70.

Lamp, J. und E. Schnug. 1990. CAF, Rechnergestuetzte Pflanzenproduktion-Moeglichkeiten, Methoden, Management. *DLG-Tagungsbericht Fulda,* DLG, Frankfurt, pp. 92-105.

Landazuri, J.C., U. ter Meulen, E.A. El Harith, and K.D. Guenther. 1993. Distribution and excretion of N15-chlorcholinchloride by laying hens. *Journal of Animal Physiology and Animal Nutrition* 69:211-216.

Lark, R.M. 1997. Variation in soil conditions and crop performance. In *Precision Agriculture 1997: Proceedings 1st European Conf. on Precision Agriculture,* Vol. 1, ed. J.V. Stafford, BIOS Scientific Publishers, Oxford, UK, pp. 127-135.

Lark, R.M. 2001. Some tools for parsimonius modeling of within-field variation of soil and crop systems. *Soil & Tillage Research* 58(3-4):99-111.

Larscheid, G., B.S. Blackmore, and M. Moore. 1997. Management decisions based on yield maps. In *Precision Agriculture 1997: Proceedings 1st European Conf. on Precision Agriculture,* Vol. 2, ed. J.V. Stafford, BIOS Scientific Publishers, Oxford, UK, pp. 895-903.

Lenz, D. 1997. Calculating profitability of grid soil sampling and variable rate fertilizing for sugar beets. In *Proceedings 3rd Int. Conf. on Precision Agriculture,* eds. P.C. Robert, R.H. Rust, and W.E. Larson, ASA-CSSA-SSSA, Madison, WI, pp. 945-948.

Liebig, J. 1840. *Die Chemie in ihrer Anwendung und auf Agrikultur und Physiologie,* Vieweg, Braunschweig.

Lilienthal, H. 2003. Entwicklung eines bodengestuetzten Fernerkundungssystems für die Landwirtschaft. *Special issue FAL–Agricultural Research* 254.

Lilienthal, H., S. Haneklaus, E. Schnug, and E. Haveresch. 2000. Utilization of hyperspectral data for the evaluation of the spatial variability of the nitrogen status of wheat. *Aspects of Applied Biology* 60:189-194.

Lilienthal, H. and E. Schnug. 2004. Equifertiles—the key to extend the supervised areas of ground based remote sensing systems. In *Proceedings 7th Int. Conf. on Precision Agriculture and Other Precision Resources Management,* ASA-CSSA-SSSA, Madison, WI, CD-ROM.

Lilienthal, H. and E. Schnug. 2005. Eignung spektraler Signaturen zur Lokalisierung von Schwefelmangel in Raps mit Hilfe der Fernerkundung. *Special issue FAL–Agricultural Research* 286:47-54.

Long, D.S., G.R. Carlson, and G.A. Nielsen. 1997. Cost analysis of variable rate application of nitrogen and phosphorus for wheat production in northern Montana. In *Proceedings 3rd Int. Conf. on Precision Agriculture,* eds. P.C. Robert, R.H. Rust, and W.E. Larson, ASA-CSSA-SSSA, Madison, WI, pp. 1019-1032.

Luetticken, R.E. 2001. Development of an Internet-based communication and information network to progress the implementation of precision agriculture. In *Proceedings 5th Int. Conf. on Precision Agriculture,* eds. P.C. Robert, R.H. Rust, and W.E. Larson, ASA-CSSA-SSSA, Madison, WI, CD-ROM.

Luetticken, R.E., W. Koch, and R. Bill. 1997. The potential of yield maps and soil survey data in low cost site-specific farming strategies. In *Precision Agriculture 1997: Proceedings 1st European Conf. on Precision Agriculture,* Vol. 2, ed. J.V. Stafford, BIOS Scientific Publishers, Oxford, UK, pp. 803-810.

Lui, W., S.K. Upadhyaya, T. Kataoka, and S. Shibusawa. 1997. Development of a texture/soil compaction sensor. In *Proceedings 3rd Int. Conf. on Precision Agriculture,* eds. P.C. Robert, R.H. Rust, and W.E. Larson, ASA-CSSA-SSSA, Madison, WI, pp. 617-630.

MacMillan, R.A., W.W. Pettapiece, S.C. Nolan, and T.W. Goddard. 2000. A generic procedure for automatically segmenting landforms into landform elements using DEMs, heuristic rules, and fuzzy logic. *Fuzzy Sets and Systems* 113(1):81-109.

Mallarino, A. P. 1996. Spatial variability patterns of phosphorous and potassium in no-tilled soils for two sampling scales. *Soil Science Society American Journal* 60:1473-1481.

Mallarino, A.P. and A.M. Blackmer. 1992. Comparison of methods for determining critical concentrations of soil test phosphorus for corn. *Agronomy Journal* 84:850-856.

Mangold, G. 1995. How can you transfer site-specific crop management technology to farmers? In *Proceedings 2nd Int. Conf. on Site-Specific Management for Agricultural Systems,* eds. P.C. Robert, R.H. Rust, and W.E. Larson, ASA-CSSA-SSSA, Madison, WI, pp. 921-926.

Marx, K. 1844. Estranged labor. In *Economic and Philosophical Manuscripts of 1844,* Progress Publishers, Moscow, 1959. Available at http://www.marxists.org/archive/marx/works/1844/manuscripts/labour.htm.

Matheron, C. 1971. *The Theory of Regionalized Variables and Its Applications,* Ecole Nationale Supérieure de Paris, Fontainbleau, France.

McBratney, A.B. and M.J. Pringle. 1997. Spatial variability in soil—implications for precision agriculture. In *Precision Agriculture 1997: Proceedings 1st European Conf. on Precision Agriculture,* Vol. 1, ed. J.V. Stafford, BIOS Scientific Publishers, Oxford, UK, pp. 3-31.

McBratney, A.B. and B.M. Whelan. 1999. The null hypothesis of precision agriculture. In *Proceedings 2nd European Conf. on Precision Agriculture,* ed. J.V. Stafford, Sheffield Academic Press, Sheffield, UK, pp. 947-958.

McBratney, A.B., B.M. Whelan, D.J.J. Walvoort, and B. Minasny. 1999. A purposive sampling scheme for precision agriculture. In *Proceedings 2nd European Conf. on Precision Agriculture,* ed. J.V. Stafford, Sheffield Academic Press, Sheffield, UK, pp. 101-110.

McBratney, A.B. 2001. Environmental economics & precision agriculture: A simple nitrogen fertilization example. In *Proceedings 3rd European Conf. on Precision Agriculture,* eds. G. Grenier and B. S. Blackmore, Agro-Montpellier, Montpellier, France, pp. 539-543.

Miller, M.P., M.J. Singer, and D.R. Nielsen. 1988. Spatial variability of wheat yield and soil properties on complex hills. *Soil Science Society American Journal* 52:1133-1141.

Modern Farming. 1991. Satellite link combine leads quest for lower crop costs. *Modern Farming* 7(3):8-10.

Moran, S. 2000. New imaging sensor technologies suitable for agricultural management. *Aspects of Applied Biology* 60:1-10.

Mueckenhausen, E., H. Mertens, and H.J. Dubber. 1988. *Die Bodenkarte 1:5000 auf der Grundlage der Bodenschaetzung,* No. 151, Ed. Landesausschuss fuer landwirtschaftliche Forschung, Erziehung und Wirtschaftsberatung beim Ministerium fuer Umwelt, Raumordnung und Landwirtschaft des Landes Nordrhein-Wetfalen, Fourth Edition, Duesseldorf, pp. 9-14.

Mulla, D.J. 1988. Estimating spatial patterns in water content, matric suction, and hydraulic conductivity. *Soil Science Society American Journal* 52:1547-1553.

Mulla, D.J. 1997. Geostatistics, remote sensing, and precision farming. In *Precision Agriculture: Spatial and Temporal Variability of Environmental Quality,* eds. J.V. Lake, G.R. Bock, and J.A. Goode, John Wiley & Sons, Chichester, UK, pp. 100-119.

Mulla, D.J. and A.U. Bhatti. 1997. An evaluation of indicator properties affecting spatial patterns in N and P requirements for winter wheat yield. In *Precision Agriculture 1997: Proceedings 1st European Conf. on Precision Agriculture,* Vol. 1, ed. J.V. Stafford, BIOS Scientific Publishers, Oxford, UK, pp. 145-153.

Murphy, D.P., E. Schnug, and S. Haneklaus. 1995. Yield mapping—a guide to improve techniques and strategies. In *Proceedings Workshop Soil-Specific Crop Management,* eds. P.C. Robert, R.H. Rust, and W.E. Larson, ASA-CSSA-SSSA, Madison, WI, pp. 33-47.

Nissen, K. and M. Soederstroem. 1999. Mapping in precision farming—from the farmer's perspective. *Proceedings 2nd European Conf. on Precision Agriculture,* ed. J.V. Stafford, Sheffield Academic Press, Sheffield, UK, pp. 655-664.

Nowak, P. 1993. Social issues related to soil-specific crop management. In *Proceedings 1st Workshop Soil-Specific Crop Management,* eds. P.C. Robert, R.H. Rust, and W.E. Larson, ASA-CSSA-SSSA, Madison, WI, pp. 269-286.

Nowak, P. 1998. Agriculture and change: The promises and pitfalls of precision. *Communications of Soil Science and Plant Analysis* 29(11-14):1537-1541.

NRC. 1997. *Precision Agriculture in the 21st Century: Geospatial and Information Technologies in Crop Management,* Committee on Assessing Crop Yield: Site-Specific Farming, Information Systems, and Research Opportunities, Board on Agriculture, National Academy Press, Washington, DC.

Oestergaard, H.S. 1997. Agronomic consequences of variable N fertilization. In *Precision Agriculture 1997: Proceedings 1st European Conf. on Precision Agriculture,* Vol. 1, ed. J.V. Stafford, BIOS Scientific Publishers, Oxford, UK, pp. 315-320.

Olieslagers, R., J. De Baerdemaeker, L. Bashford, and H. Delcourt. 1995. The accuracy of site-specific fertilizer application by means of a spinning-disc fertilizer spreader. In *Proceedings 2nd Int. Conf. on Site-Specific Management for Agricultural Systems,* eds. P.C. Robert, R.H. Rust, and W.E. Larson, ASA-CSSA-SSSA, WI, Madison, pp. 709-720.

Panten, K. 2002. *Ein Beitrag zur Fernerkundung der raeumlichen Variabilitaet von Boden- und Bestandesmerkmalen.* Landbauforschung Voelkenrode (FAL Agricultural Research), SH232.

Panten, K., S. Haneklaus, J. Rogasik, and E. Schnug. 2005. Predicting sugar beet yield variability using yield maps of combinable crops and the "monitor pedo cell" approach. *Special issue FAL–Agricultural Research* 286:65-70.

Panten, K., S. Haneklaus, and E. Schnug. 2002. Spatial accuracy of online yield mapping. *FAL-Agricultural Research* 52(4): 205-209.

Panten, K., S. Haneklaus, D. Schroeder, and E. Schnug. 1998. Remote sensing as an aid for the spatial management of nutrients. In *Proceedings 4th Int. Conf. on Precision Agriculture,* Part A, eds. P.C. Robert, R.H. Rust, and W.E. Larson, ASA-CSSA-SSSA, Madison, WI, pp.1501-1511.

Park, S.W., S. Mostaghimi, R.A. Cooke, and P.W. McClellan. 1994. BMP impacts on watershed runoff, sediment, and nutrient yields. *Water Resource Bulletin* 30:1011-1023.

Parker, C.G., K. Phelps, R.J. Reader, and C.J. Hinde. 1996. Why computer-based models need more than good science. *Aspects of Applied Biology* 46:297-304.

Parry, J. 1976. Cereals special. *British Farmer and Stockbreeder* 5(118):39-46.

Paulsen, S. 1989. Rechnergestuetzte Pflanzenproduktion: technisch aufwendig *Agrarpraxis* 8(89):42-43.

Peck, T.R. and S.W. Melsted. 1967. Field sampling for soil testing. In *Soil Testing and Plant Analysis,* Third Edition, ed. M. Stelly, SSSA, Madison, WI, pp. 25-35.

Persson, K. and J. Bangsgaard. 1999. Accuracy of fertilizer distribution for site-specific fertilization. In *Proceedings 2nd European Conf. on Precision Agriculture,* ed. J.V. Stafford, Sheffield Academic Press, Sheffield, UK, pp. 815-824.

Pierce F.J., D.D. Warncke, and M.W. Everett. 1995. Yield and mutirent variability in glacial soils of Michigan. *Site-Specific Management for Agricultural Systems: Proceedings Second Int. Conf.,* Minneapolis, pp. 133-151.

Pocknee, S., B.C. Boydell, H.M. Green, D.J. Waters, and C.K. Kvien. 1996. Directed soil sampling. In *Proceedings 3rd Int. Conf. on Precision Agriculture,* eds. P.C. Robert, R.H. Rust, and W.E. Larson, ASA-CSSA-SSSA, Madison, WI, pp. 159-168.

Preusse, U. 1989. Signale aus dem Weltraum. *DLG-Mitteilungen,* pp. 21-22.

Reetz, H.F. 1997. On-farm research opportunities through site-specific management. In *Proceedings 3rd Int. Conf. on Precision Agriculture,* eds. P.C. Robert, R.H. Rust, and W.E. Larson, ASA-CSSA-SSSA, Madison, WI, pp. 1173-1176.

Resnik, B. and R. Bill. 2000. *Vermessungskunde fuer den Planungs-, Bau- und Umweltbereich,* Herbert Wichmann Verlag, Heidelberg.

Reusch, S. 1997. *Entwicklung eines reflexionsoptischen Sensors zur Erfassung der Stickstoffversorgung landwirtschaftlicher Kulturpflanzen.* PhD, thesis Kiel University.

Reuter, D.J., J.B. Robinson, K.J. Peveril, G.H. Price, and M.J. Lambert. 1997. Guidelines for collecting, handling, and analyzing plant material. In *Plant Analysis: An Interpretation Manual,* eds. D. J. Reuter and J. B. Robinson, CSIRO Publishing, Collingwood, Australia, pp. 53-70.

Reyns, P., B. Missotten, H. Ramon, and J. De Baerdemaeker. 2002. A review of combine sensors for precision farming. *Precision Agriculture* 3:169-182.

Robert, P.C., R.H. Rust, and W.E. Larson, eds. 1993. *Proceedings 1st. Workshop Soil-Specific Crop Management,* ASA-CSSA-SSSA, Madison, WI.

Robert, P.C., R.H. Rust, and W.E. Larson, eds. 1995. *Proceedings 2nd Int. Conf. on Site-Specific Management for Agricultural Systems,* ASA-CSSA-SSSA, Madison, WI.

Robert, P.C., R.H. Rust, and W.E. Larson, eds. 1997. *Proceedings 3rd Int. Conf. on Precision Agriculture,* ASA-CSSA-SSSA, Madison, WI.

Robert, P.C., R.H. Rust, and W.E. Larson, eds. 1999. *Proceedings 4th Int. Conf. on Precision Agriculture,* ASA-CSSA-SSSA, Madison, WI.

Robert, P.C., R.H. Rust, and W.E. Larson, eds. 2000. *Proceedings 5th Int. Conf. on Precision Agriculture,* ASA-CSSA-SSSA, Madison WI, CD-ROM.

Robert, P.C., R.H. Rust, and W.E. Larson, eds. 2002. *Proceedings 5th Int. Conf. on Precision Agriculture,* ASA-CSSA-SSSA,Madison, WI, CD-ROM.

Robert, P.C., R.H. Rust, and W.E. Larson. eds. 2004. *Proceedings of the 7th Int. Conf. on Precision Agriculture,* ASA-CSSA-SSSA, Madison, WI, CD-ROM.

Roubens, M. 1982. Fuzzy clustering algorithms and their cluster validity. *European Journal of Operational Research* 10:173-197.

Russell, G. 1996. Spatial aspects of modeling in applied biology. *Aspects of Applied Biology* 46:1-4.

Sabbe, W.E. and W. Marx. 1987. Soil sampling: Spatial and temporal variability. In *Soil Testing: Sampling, Correlation, Calibration, and Interpretation,* ed. J.R. Brown, SSSA Spec. Publ. 21, SSSA, Madison, WI, pp. 1-14.

Sapek, B. 1998. Farm as a source of soil, water, and air pollution with nitrogen, phosphorus, and potassium. In *Proceedings 11th Symposium of CIEC Codes of Good Fertilizer Practice and Balanced Fertilization,* eds. E. Schnug and M. Fotyma, CIEC, Pulawy, Poland, pp. 124-144.

Schepers, J.S., M.R. Schlemmer, and R.B. Ferguson. 2000. Site-specific considerations for managing phosphorus. *Journal Environmental Quality* 29:125-130.

Schmerler, J. and M. Basten. 1999. Cost/benefit analysis of introducing site-specific management on a commercial farm. In *Proceedings 2nd European Conf. on Precision Agriculture,* ed. J.V. Stafford, Sheffield Academic Press, Sheffield, UK, pp. 959-968.

Schmidt, H. and M. Haccius. 1998. *EU Regulation "Organic Farming."* A legal and agro-ecological commentary on the EU's Council Regulation (EEC) No. 2091/91, Margraf Verlag, Weikersheim, Germany.

Schnug, E. 1982. *Untersuchungen zum Einfluß bodenversauernder Duengung auf die Spurennaehrstoff-Versorgung von Kulturpflanzen.* PhD Agrarwissenschaftliche Fakultaet, Kiel.

Schnug, E. 1992. Mobilità e biodisponibilità dei microelementi. In *L'azoto la pianta l'ambiente—I microelementi nel terreno e nella pianta,* Agenzia per i Servizi di Sviluppo, Quaderni, 9:131-143.

Schnug, E., F. Achwan, and J. Heym. 1996. Establishing critical values for soil and plant analysis by means of the Boundary Line Development System (BOLIDES) *Communications of Soil Science and Plant Analysis* 27:2739-2748.

Schnug, E. and A. Finck. 1982. Trace element mobilization by acidifying fertilizers. In *Proceedings 9th Int. Plant Nutrition Coll.,* Vol. 2, ed. A. Scaife, Commonwealth Agricultural Bureaux, Warwick, UK, pp. 582-587.

Schnug, E., J. Fleckenstein, and S. Haneklaus. 1997. Coca Cola is it! The ubiquitous extractant for micronutrients in soils. In *Soil and Plant Analysis in Sustainable Agriculture,* eds. T.M. Hood and J. Benton Jones Jr., Marcel Dekker Inc., New York, pp. 349-358.

Schnug, E. and S. Haneklaus. 1992. PIPPA: Un programme d'interprétation des analyses de plantes pour le colza et les céréales. *Supplément de Perspectives Agricoles* 171:30-33.

Schnug, E. and S. Haneklaus. 1997a. A rapid method for the indirect determination of the clay content by X-ray fluorescence spectroscopy of rubidium in soils. In *Soil and Plant Analysis in Sustainable Agriculture,* eds. T.M. Hood and J. Benton Jones Jr., Marcel Dekker Inc., New York, pp. 577-590.

Schnug, E. and S. Haneklaus. 1997b. A rapid method for the indirect determination of the organic matter content of soils. In *Soil and Plant Analysis in Sustainable*

*Agriculture,* eds. T.M. Hood and J. Benton Jones Jr., Marcel Dekker Inc., New York, pp. 563-576.

Schnug, E. and S. Haneklaus. 1998. Diagnosis of sulphur nutrition. In *Sulphur in Agro-Ecosystems,* eds. E. Schnug and H. Beringer, Kluwer Academic Publishers, Dordrecht, The Netherlands, pp. 1-38.

Schnug, E., S. Haneklaus, H. Lilienthal, and K. Panten. 2000. LASSIE—an innovative approach for the continuous remote sensing of crops. *Aspects of Applied Biology* 60:147-154.

Schnug, E., S. Haneklaus, and D. Murphy. 1994. Equifertiles—an innovative concept for efficient sampling in the local resource management of agricultural soils. *Aspects of Applied Biology* 37:63-72.

Schnug, E., J. Lamp, and F. Knoop. 1985. Regionale Variabilitaet von Merkmalen der Bodenfruchtbarkeit. II. Beispiele, Ursachen und praktische Bedeutung. *Mitteilungen Deutsche Bodenkundliche Gesellschaft* 43(II):679-684.

Schnug, E., H. Lilienthal, K. Panten, and S. Haneklaus. 2003. Smart navigation—Salvation for precision agriculture? In *Proceedings 11th IAIN (International Association of Institutes of Navigation) World Congress,* Berlin, CD-ROM.

Schnug, E., D. Murphy, E. Evans, and S. Haneklaus. 1992. Advanced approaches to local resource management in crop production. In *Proceedings 2nd ESA Conf.,* ed. J.V. Stafford, ESA, Warwick, UK, pp. 372-373.

Schnug, E., D. Murphy, E. Evans, and S. Haneklaus. 1993. Local Resource-Management in computer aided farming: A new approach for sustainable agriculture. In *Optimization of Plant Nutrition,* eds. M.A.C. Fragoso and M.L. van Beusichem, Kluwer Academic Publishers, Dordrecht, The Netherlands, pp. 657-663.

Schnug, E., D. Murphy, E. Evans, S. Haneklaus, and J. Lamp. 1993. Yield mapping and application of yield maps to computer aided local resource management. In *Proceedings Workshop Soil-Specific Crop Management,* eds. P.C. Robert, R.H. Rust, and W.E. Larson, ASA-CSSA-SSSA, Madison, WI, pp. 87-93.

Schnug, E., K. Panten, and S. Haneklaus. 1998. Sampling and nutrient recommendations—the future. *Communications of Soil Science and Plant Analysis* 29(11-14):1455-1462.

Schnug, E., H.M. Paulsen, and J. Lamp. 1989. Beruecksichtigung der kleinraeumlichen Variabilitaet von Merkmalen der Bodenfruchtbarkeit bei der Anlage von Feldversuchen. *Mitteilungen Deutsche Bodenkundliche Gesellschaft.* 59(II): 795-796.

Schroeder, D., S. Haneklaus, and E. Schnug. 1997. Information management in precision agriculture with LORIS. In *Precision Agriculture 1997: Proceedings 1st European Conf. on Precision Agriculture,* Vol. 2, ed. J.V. Stafford, BIOS Scientific Publishers, Oxford, UK, pp. 821-826.

Schroeder, D. and E. Schnug. 1995. Application of yield mapping to large scale field experimentation. *Aspects of Applied Biology* 43:117-124.

Schumann, M., M. Kuecke, and E. Schnug. 1997. Fallstudien und Konzeption zur Einfuehrung bilanzorientierter Duengung in der deutschen Landwirtschaft. *Landbauforschung Voelkenrode,* SH 180.

Scotford, I.M., T.R. Cumby, P.A. Richards, and L. Han. 1998. Development of a prototype nutrient sensing system for livestock slurries. *Journal of Agricultural Engineering Research* 69:217-228.

Selcuk, A. and T.S. Colvin. 2002. Grain yield mapping: Yield sensing, yield reconstruction, and errors. *Precision Agriculture* 3:135-154.

Sérélé, C.Z., Q. Hugh, J. Gwyn, J.B. Boisvert, E. Pattey, N. McLaughlin, and G. Daoust. 2001. Detection of corn nitrogen status from airborne VIS-NIR imagery using artificial neural networks. In *Proceedings 5th Int. Conf. on Precision Agriculture,* eds. P.C. Robert, R.H. Rust, and W.E. Larson, ASA-CSSA-SSSA, Madison, WI, CD-ROM.

Shatar, T.M. and A.B. McBratney. 1999. Empirical modeling of local yield response and identification of areas within uniform yield response. In *Proceedings 2nd European Conf. on Precision Agriculture,* ed. J.V. Stafford, Sheffield, UK, Sheffield Academic Press, pp. 581-592.

Shatar, T.M. and A.B. McBratney. 2001. Subdividing a field into contiguous management zones using k-zone algorithm. In *Proceedings 3rd European Conf. on Precision Agriculture,* eds. G. Grenier and B.S. Blackmore, Agro-Montpellier, Montpellier, France, pp. 115-120.

Shi, Z., K. Wang, J.S. Bailey, C. Jordan, and A.J. Higgins. 2000. Sampling strategies for mapping soil phosphorous and soil potassium distributions in cool temperate grassland. *Precision Agriculture* 2:347-357.

Simmelsgaard, S.E. and J. Djurhuus. 1997. The possibilities of precision fertilization with N, P, and K based on plant and soil parameters. In *Precision Agriculture 1997: Proceedings 1st European Conf. on Precision Agriculture,* Vol. 1, ed. J.V. Stafford, BIOS Scientific Publishers, Oxford, UK, pp. 179-187.

Sims, A. 1998. Phosphorus soil testing: Innovations for water quality protection. *Communications of Soil Science and Plant Analysis* 29(11-14):1471-1490.

Snyder, C., J. Havlin, G. Kuitenberg, and T. Schroeder. 1997. An economic analysis of variable rate nitrogen management. In *Proceedings 3rd Int. Conf. on Precision Agriculture,* eds. P.C. Robert, R.H. Rust, and W.E. Larson, ASA-CSSA-SSSA, Madison, WI, pp. 989-998.

Sogbedji, J.M., H.M. van Es, S.D. Klausner, D.R. Bouldin, and W.J. Cox. 2001. Spatial and temporal processes affecting nitrogen availability at the landscape. *Soil & Tillage Research* 58(3-4):233-244.

Sparovek, G., M. Cooper, D. Dourado-Neto, R.F. Maule, P.L.F. de Vidal-Torrado-Pimenta, S. Martins, P.R. Teramoto, A.C. Silva, J. Van de Steeg, and E. Schnug. 2002. The Land Resource Information and Suitability System for Family Agriculture (LARISSA), developed for the Brazilian agrarian reform. *Der Tropenlandwirt* 103(1):47-59.

Sparovek, G. and E. Schnug. 2001. Soil tillage and precision agriculture—a theoretical case study for soil erosion control in Brazilian sugar cane production. *Soil & Tillage Research* 61:47-54.

Stafford, J.V. 1999. An investigation into the within-field spatial variability of grain quality. In *Proceedings 2nd European Conf. on Precision Agriculture,* ed. J.V. Stafford, Sheffield Academic Press, Sheffield, UK, pp. 353-361.

Stafford, J.V., B. Ambler, and H.C. Bolam. 1997. Cut width sensors to improve the accuracy of yield mapping systems. In *Precision Agriculture 1997: Proceedings 1st European Conf. on Precision Agriculture,* Vol. 2, ed. J.V. Stafford, BIOS Scientific Publishers, Oxford, UK, pp. 519-527.

Stewart, C.M. and A.B. McBratney. 2001. Using bare soil imagery to determine management zones for the variable-rate application of inputs for cotton. In *Proceedings 3rd European Conf. on Precision Agriculture,* eds. G. Grenier and B.S. Blackmore, Agro-Montpellier, Montpellier, France, pp. 319-324.

Stewart, C.M., A.B. McBratney, and J.H. Skerritt. 2002. Site-specific durum wheat quality and its relationship to soil properties in a single field in northern New South Wales. *Precision Agriculture* 3:155-168.

Sudduth, K.A. and S.C. Borgelt. 1993. Sensing for variability management. In *Proceedings 1st Workshop Soil-Specific Crop Management,* eds. P.C. Robert, R.H. Rust, and W.E. Larson, ASA-CSSA-SSSA, Madison, WI, pp. 375-376.

Sugiharto, T., T.H. McIntosh, R.C. Uhrig, and J.J. Lardinois. 1994. Modeling alternatives to reduce dairy farm and watershed nonpoint source pollution. *Journal of Environmental Quality* 23(1):18-24.

Swinton, S.M., K.Q. Jones, N.R. Miller, O. Schabenberger, R.C. Brook, and D.D. Warncke. 2001. Comparison of site-specific and whole-field variability management in Michigan soybeans and corn. In *Proceedings 5th Int. Conf. on Precision Agriculture,* eds. P.C. Robert, R.H. Rust, and W.E. Larson, ASA-CSSA-SSSA, Madison, WI, CD-ROM.

Syers, J.K. and T.W. Walker. 1969. Fractionation of phosphorus in two cultivated soils and particle size separates. *Soil Science,* 108:283-289.

Taylor, J. C., Godwin, R. J. and Wood, G. A. 2002. Management guidelines for precision farming: An operational framework for cereal production in the UK. In *Abstracts 6th Int. Conf. on Precision Agriculture,* eds. P.C. Robert, R.H. Rust, and W. E. Larson, ASA/CSSA, SSSA/ASAE, Madison, WI, p. 94.

TFI. 2002. The Fertilizer Institute; http://www.tfi.org.

Thirion, F., P. Zwaenepol, and F. Chabot. 1999. Integrated weighing platform for manure spreader regulation. In *Proceedings 4th International Conference on Precision Agriculture,* eds. P.C. Robert, R.H. Rust, and W.E. Larson, ASA-CSSA-SSSA, Madison, WI, pp.1219-1226.

Thun, R., R. Herrmann, and E. Knickmann. 1955. Methode der elektrometrischen Neutralisation nach Jensen. In *Die Untersuchung von Boeden,* eds. R. Thun and E. Knickmann, Neumann Verlag, Berlin, pp. 149-151.

Thrikawala, S., A. Weersink, and G. Kachaoski. 1998. Management unit size and efficiency from nitrogen fertilizer applications. *Agricultural Systems* 56(4): 513-531.

Thylen, L., P.A. Algerbo, and C.G. Pettersson. 1999. Grain quality variations within fields of malting barley. In *Proceedings 2nd European Conf. on Precision Agriculture,* ed. J.V. Stafford, Sheffield Academic Press, Sheffield, UK, pp. 287-296.

Tullberg, J.N. 2001. Controlled traffic for sustainable cropping. *Proceedings 10th Austr. Agronomy Conf.,* Hobart, Australia. Available at http://www.regional .org.au/asa/2001/plenery/ 2/tullberg.htm.

Umeda, M., S. Kubota, and M. Iida. 1999. Development of "Stork," a watermelon-harvesting robot. *Artif. Life Robotics* 3:143-147.

Van Alphen, B.J. and J.J. Stoorvogel. 2000. A methodology for precision nitrogen fertilization in high-input farming systems. *Precision Agriculture* 2:319-332.

Vauclin, M., S.R. Vieira, G. Vachaud, and D.R. Nielsen. 1983. The use of cokriging with limited field soil observations. *Soil Science Society American Journal* 47(2):175-184.

VDLUFA. 1991. *Bestimmung von mineralischem (Nitrat-) Stickstoff in Bodenprofilen ($N_{min}$-Labormethode),* Methodenbuch I., ed. G. Hoffman, VDLUFA Verlag, Darmstadt, Germany, Kap. A.9.1.4.1.

Vermeulen, S., I. Steen, and E. Schnug. 1998. Nutrient balances at the farm level. In *Proceedings 11th Symposium of CIEC Codes of Good Fertilizer Practice and Balanced Fertilization,* eds. E. Schnug and M. Fotyma, CIEC, Pulawy, Poland, pp. 108-123.

Viscarra Rossel, R.A. 2001. *Development of a Proximal Soil Sensing System for the Continuous Management of Acid Soil.* PhD thesis, Australian Centre for Precision Agriculture, Dept. of Agric. Chemistry and Soil Science, The University of Sydney, Australia.

Viscarra Rossel, R.A. and A.B. McBratney. 2001. A two-factor empirical deterministic response surface calibration model for site-specific predictions of lime requirement. *Precision Agriculture* 2:163-178.

Vladutu, I., T. Fritea, and P. Kurtinecz. 1988. *Contributii privind integrata a buruienilor pe agriloiluvisolurils din nord-vestul Transilvaniei. II. Efectul altor metode agrotechnice asupra combaterii buruienlior.* Probleme de agrofitotehnie teoretica si aplicata, Institutul de cercetari pentru cereale si plante tehnice Funduela, Romania, X, 2:153-177.

Voss, R. 1997. Fertility recommendations: Past and present. *Communications of Soil Science and Plant Analysis* 29(11-15):1429-1440.

Walters, C.S., B.E. Frazier, B.C. Miller, and W.L. Pan. 1997. Remotely sensed organic matter mapping for site-specific N management in dryland winter wheat. In *Proceedings 3rd Int. Conf. on Precision Agriculture,* eds. P.C. Robert, R.H. Rust, and W.E. Larson, ASA-CSSA-SSSA, Madison, WI, p. 579.

Webb, R.A. 1972. Use of the boundary line in the analysis of biological data. *Journal of Horticultural Science* 47:309-319.

Webster, R. and C. de la Cuanalo. 1975. Soil transect variograms of North Oxfordshire and their interpretation. *Journal of Soil Science* 26:176-194.

Webster, R. and A.B. McBratney. 1987. Mapping soil fertility at Broom's Barn by simple kriging. *Journal of the Science Food and Agriculture* 38:97-115.

Whelan, B.M. and A.B. McBratney. 2000. The "null hypothesis" of precision agriculture management. *Precision Agriculture* 2:265-278.

Whelan, B.M. and A.B. McBratney. 2002. A parametric transfer function for grain-flow within a conventional combine harvester. *Precision Agriculture* 3:123-134.

Whitley, K.M., J.R. Davenport, and S.R. Manley. 2001. Difference in nitrate leaching under variable and conventional nitrogen fertilizer management in irrigated potato systems. In *Proceedings 5th Int. Conf. on Precision Agriculture,* eds.

P.C. Robert, R.H. Rust, and W.E. Larson, ASA-CSSA-SSSA, Madison, WI, CD-ROM.

Witter, B. 1987. Hoehere Effektivitaet beim Einsatz der Mineralduengerfonds durch differenzierte Schlag- und Teilschlagduengung. *Feldwirtschaft* 28:120-124.

Wollenhaupt, N.C. and D.D. Buchholz. 1993. Profitability of farming by soils. In *Proceedings 1st Workshop Soil-Specific Crop Management,* eds. P.C. Robert, R.H. Rust, and W.E. Larson, ASA-CSSA-SSSA, Madison, WI, pp.199-212.

Wollring, J. and S. Reusch. 1999. Variable N-Duengung mit Hilfe eines On-line Sensors. In *Sensorsysteme im Precision Farming,* eds. R. Bill, G. Grenzdoerffer, and F. Schmidt, Interner Bericht Nr. 12, Institut fuer Geodaesie und Geoinformatik, Universitaet Rostock, Deutschland, pp. 93-104.

Yadav, S.N. and D.B. Wall. 1998. Benefit-cost analysis of best management practices implemented to control nitrate contamination of groundwater. *Water Resource Research* 34(3):497-504.

# Chapter 5

# Precision Water Management: Current Realities, Possibilities, and Trends

Carl R. Camp
E. John Sadler
Robert G. Evans

## *INTRODUCTION*

In order to establish a common understanding, we first need to clarify or explain our characterization of the term *precision water management.* The traditional meaning of precision water management has been that of either applying precise amounts of water to crops or applying water at precise locations (e.g., within the soil profile) or at precise times. Perhaps a good example of this traditional definition is drip irrigation, which is generally accepted as a very precise irrigation method because water can be precisely controlled with regard to application rate, timing, and location with respect to the plant. This definition continues to be used today in some areas. However, in this chapter we define precision water management as site-specific water management, specifically the application of water to a given site in a volume and at a time needed for optimum crop production, profitability, or other management objectives at that specific site. This is in contrast with a simultaneous application of a single amount to the entire area of the irrigation system. Moving irrigation systems have been used to accomplish this in all known applications thus far, and will be the primary focus in the bulk of this chapter.

---

Contribution of the U. S. Department of Agriculture, Agricultural Research Service. Mention of a trade name, proprietary product, or specific equipment does not constitute a guarantee or warranty by the USDA and does not imply approval of a product to the exclusion of others that may be suitable.

*Handbook of Precision Agriculture*
Published by The Haworth Press, Inc, 2006. 
doi:10.1300/5627_05

Limited literature exists on this topic, with the first known dating to 1992. Most is concerned with hardware development, and only a few discuss economics of site-specific irrigation (Evans, Han, et al., 1996; Watkins et al. 1999; Lu et al., 2004a,b; Lu et al., 2005). This method of water management continues to be very much a research issue. Development of the hardware to provide variable-rate applications and the hardware and software to control these applications is in its infancy. However, existing commercial self-propelled systems such as center-pivot and lateral-move machines are particularly amenable to site-specific approaches because of their current level of automation and large area of coverage with a single pipe lateral. This is reflected in some commercial irrigation systems that have recently been modified for precision irrigation. In addition to irrigation, these machines offer an outstanding platform for mounting sensors that can provide real-time monitoring of plant and soil conditions and serve as a transport device for separate agro-chemical application systems.

## CURRENT STATUS

### Research Sites

Four research efforts in precision irrigation were initiated in the early 1990s. The Idaho, Colorado, and South Carolina efforts started almost concurrently, and the Washington effort started shortly thereafter. Although the approaches used in these efforts were different in both technique and scale, the objectives were similar. To explore these similarities and differences in more detail, each effort will be described briefly and later by component in more detail. While reading these descriptions, it may be helpful to refer to the summary of system characteristics by location in Table 5.1.

#### *Idaho*

In 1993, McCann and Stark received a U.S. patent for a precision irrigation control system (McCann and Stark, 1993). They initially worked on a 100 m lateral-move system with a 9.1 m control zone and two commercial sprinklers of different size at each location. The two sprinkler sizes (one three times the flow rate of the other one) allowed them to be used separately or together to achieve a range of water application rates (0, 1/3, 2/3, and full). The next year, they scaled up to a medium-sized center pivot (210 m long, seven towers), and used three commercial sprinklers, but water application rates were controlled separately only on one complete span. Two years later, they reported work on another center pivot (354 m long, nine

TABLE 5.1. Summary of characteristics of precision irrigation systems developed by different groups across the United States.

| Location | Type (#)[a] | Length (# spans) x trvl (linear) | Control element size along boom, then along travel | Nozzle type (# for multiple) | Wetted radius | Water supply method[b] | Control type[c] | Nutrient injection | Position Method[d] |
|---|---|---|---|---|---|---|---|---|---|
| Ft Collins, CO<br>Fraisse et al., 1992, Duke et al., 1992 | LM (1) | 176 m (4) x 640 m | 22 m x 53 m | Pulsed | 3.8 m | STP, PR | PLC | | GPS, DR |
| Aberdeen, ID<br>King et al., 1995 | LM (1) | 100 m x 180 m | 9.1 x 9.1 m, then 18.2 x 18.2 m | Multiple (2) stepwise | 9 m | STP | Bus | Yes | |
| Aberdeen, ID<br>King et al., 1995 | CPF (1) | 210 m (7) | 28.7 m x 5.6 degree | On/off | 9.1 m | STP | Bus | | R |
| Aberdeen, ID<br>King et al., 1996 | CPF (1) | 354 m (9) | 38.1 m x 6 degree (more for inner rings) | Multiple (2) stepwise | 9 m | STP | Bus | Yes | R |
| Aberdeen, ID<br>Wall et al., 1996 | CPF(1) | 392 m (10) | 38 m x 1 degree ctrl | Multiple (2) stepwise | 10 m | VRP | Bus | Yes | R |
| Florence, SC<br>Camp and Sadler, 1994 | CPF (2) | 140 m (3) | 9.1 m x 7.5 degree (#1) 4-16 deg (#2) | Multiple (3) stepwise | 2.5 m | MP | PLC | Yes | R |
| Prosser, WA<br>Evans et al., 1996 | CPF (1) | 390 (8) | 6-12 m x 0.5 degree ctrl | Pulsed | 9 m | PSM | RS-485 custom | Yes | EC |
| Paterson, WA<br>Harting, 1999 | CPF (3)<br>CPP (3) | 360 (7) | 12m x 0.13 degree ctrl (GPS) | Pulsed | 9 m | PSM | PLC | Yes | GPS |
| Tifton, GA<br>Perry, Pocknee, Hart, et al., 2002; Perry, Pocknee, Hansen, et al., 2002; Perry et al., 2003 | CPF (1) | 173 m (3) | 13.7 m x 2 degree ctrl | Pulsed | 9.1 m | STP, PR | RS-485 Canlink 3000 | No | GPS |

TABLE 5.1 *(continued)*

| Location | Type (#)[a] | Length (# spans) x trvl (linear) | Control element size along boom, then along travel | Nozzle type (# for multiple) | Wetted radius | Water supply method[b] | Control type[c] | Nutrient injection | Position Method[d] |
|---|---|---|---|---|---|---|---|---|---|
| Tifton, GA Perry, Pocknee, Hart, et al., 2002; Perry Pocknee, Hansen, et al., 2002; Perry et al., 2003 | CPF (1) | 186 m (3) | 10.9 m x 2 degree ctrl | Pulsed | 15.2 m | STP, PR | RS-485 Canlink 3000 | No | GPS |
| Tifton, GA Perry, Pocknee, Hart, et al., 2002; Perry Pocknee, Hansen, et al., 2002; Perry et al., 2003 | CPF (1) | 303 m (5) | 12.8 m x 2 degree ctrl | Pulsed | 6.1 m | STP, PR | RS-485 Canlink 3000 | No | GPS |
| Tifton, GA Perry, Pocknee, Hart, et al., 2002; Perry Pocknee, Hansen, et al., 2002; Perry et al., 2003 | CPP (1) | 429 m (last 1, end gun) | 11.0 m x 2 degree ctrl | Pulsed | 12.2 m | CP | RS-485 Canlink 3000 | No | GPS |
| Tifton, GA Perry, Pocknee, Hart, et al., 2002; Perry Pocknee, Hansen, et al., 2002; Perry et al., 2003 | CPF (1) | 186 m (4) | 12.2 m x 2 degree ctrl | Pulsed | 8.5 m | CP, PR | RS-485 Canlink 3000 | No | GPS |

| | | | | | | | | | |
|---|---|---|---|---|---|---|---|---|---|
| Tifton, GA<br>Perry, Pocknee, Hart, et al., 2002; Perry Pocknee, Hansen, et al., 2002; Perry et al., 2003 | CPF (1) | 187 m (4) | 10.7 m x 2 degree ctrl | Pulsed | 9.1 m | PSM | RS-485 Canlink 3000 | No | GPS |
| bbock, TX<br>Bordovsky and Lascano, 2003 | CPP (1) | 410 m (last 4) | 16.1 m x 3 degree ctrl | Multiple (3) stepwise | n/a | STM, PR | RS-485 PLC | Yes | |
| Lubbock-Halfway, TX<br>Bordovsky and Lascano, 2003 | LM (1) | 189 (5) | 18.3 m x 18.3 m | Multiple (3) stepwise | n/a | STM, PR | PLC | No | |
| Sidney, MT<br>Evans and Iversen, 2005 | LM | (5) x 250m | 15m | Pulsed | 3 | STM | PLC | Yes | GPS |
| Sidney, MT<br>Evans and Iversen, 2005 | LM | 350 m (7) x 350 m | 15 m x 25 m | Pulsed | 3 | VRP | PLC | Yes | GPS |
| Prosser, WA<br>Pierce et al., 2005 | LM | 74 m (2) x 120 m | 3m | Pulsed | 5 | STM | Digital | No | GPS |

*Source:* Adapted from Sadler, Evans, et al., 2000.
[a]CPF = center pivot, full circle; CPP = center pivot, some spans; LM = lateral move.
[b]STM = single turbine pump; PR = pressure regulated; MP = multiple pumps; VRP = variable rate pump; PSM = pressurized system mainline; CP = centrifugal pump.
[c]PLC = programmable logic controller.
[d]GPS = Global Positioning System; DR = dead reckoning; R = resolver at pivot; EC = electronic compass.

towers) with a control zone size of 38.1 m. In the same year, they reported their last work on another center pivot (392 m long) with emphasis on the control system. This center pivot was divided into 32 segments, each 13 m long with control initiated on 1° increments in the travel direction. Application rates were stepwise variable using multiple nozzles and included both water and chemical applications.

*Colorado*

In 1992, the USDA-ARS (United States Department of Agriculture-Agricultural Research Service) research unit in Fort Collins reported their work with a lateral-move system used for research (Fraisse et al., 1992). The 174 m, four-tower system was divided into eight control zones, each 22 m in length, using a separate manifold for each zone. Variable application rates were achieved by cycling (1-minute base cycle) the manifold pilot-operated solenoid valves, which provided pulses of varying frequency to commercial sprinklers. They also reported additional research on determining location and showed that a combination of a Global Positioning System (GPS) and dead reckoning could achieve an accuracy of 7 m. In 1999, they reported initial results from a large interdisciplinary research project on two large center pivots (72 and 52 ha). Although variable-rate applications were not made during the first two years, extensive soil, crop, and weed data were collected to determine baseline spatial variability of soil water, nutrients, crop yield, soil electrical conductivity, and soil organic matter. They used these data layers to find autoregressive models that explained yield variation. They also modeled soil water spatially and combined these results with other information.

*South Carolina*

In 1992, the USDA-ARS group in Florence, completed the conceptual design for a site-specific center-pivot system and installed modifications to achieve site-specific applications on two commercial center pivots (148 m long, 3 spans) in 1995 and 1999 (Camp and Sadler, 1994). During this time, they also developed a staged-pump, variable-flow water supply system for the two modified center pivots. These center pivots were divided into 13 zones, using manifolds 9.1 m long. Each zone included three manifolds sized to provide seven stepwise application rates. Infrared thermometers were mounted on both center pivots to map soil and canopy temperatures.

### *Washington*

In 1996, the Washington State University/USDA-ARS group reported their developments, which followed extensive feasibility studies on commercial fields (Evans, Hans et al., 1996). They retrofitted both a commercial center-pivot and a lateral-move system with addressable solenoid valves on each sprinkler, but retained the original sprinkler packages. Variable application rates were achieved by pulsing sprinklers (1.0- to 2.5-minute base cycle) in groups or zones of two to four sprinklers using a system that distributed control signals in a carrier frequency, superimposed on existing system electrical power wiring. They also reported various efforts to accurately determine position of the irrigation system. In 1999, they reported a pesticide system that included a separate flow path for pesticides. It used the irrigation machine as a ground-driven application system, possibly avoiding pesticide label issues regarding injection of chemicals into the irrigation stream.

### *Later Efforts*

Since the research at the four sites mentioned earlier, research and demonstration efforts have been conducted at multiple locations, particularly the University of Georgia's Tifton research station and in western Texas. In 2003 and 2004, two new PLC-based control systems have been installed and used for research in Sidney, Montana, one of which has both sprays and LEPA (low-energy precision application) irrigation methods on the same machine for plot research. A digital control system has been developed and tested at Washington State University–Prosser that uses an onboard computer to implement radio-transmitted instructions from a base computer. A second generation of the Prosser system will be installed in Sidney, Montana, in 2006 (see Table 5.1). In addition, starting in January 2006, a commercial firm in Georgia (http://betterpivots.com/) has retrofit, using the Tifton design, several on-farm pivots across the southern United States. Scientists at the Federal German Agricultural Research Center, Braunschweig, Germany, have also conducted an assessment of the feasibility and significance of applying spatially variable irrigation under a center-pivot systems, but have not yet implemented the system.

## ***Commercial Sites***

A few commercial firms have been active in site-specific irrigation development. Perhaps the first was Precision Irrigation Control Systems in

Idaho operating under a license from J. R. Simplot, using technology very similar to that developed by the Idaho group. Another commercial firm, Agri-Northwest in Washington, has installed site-specific control systems on several traveling irrigation systems, especially for potatoes in the northwestern United States, using technology similar to that developed by the Washington group (Harting, 1999). Finally, Valmont Industries, Inc., a manufacturer of center-pivot irrigation systems in Nebraska, has marketed a separate flow path, low-volume pesticide application system that can be mounted either on center-pivot or lateral-move machines. Without modification, each span of this pesticide system can be controlled independently. With modification, it could be adapted to smaller management zones.

### International Sites

Although there have been both considerable interest and some research involving precision irrigation outside the United States, this technology appears to be less developed than other precision agriculture activities such as fertilizer application and yield monitoring. Some locations in Europe are investigating yield response to water and fertilizer application rates within relatively large management zones (e.g., sector of center pivot), but attempts to manage water applications in smaller zones, which would require modification of the water delivery system, have not been reported.

## IRRIGATION APPLICATION AND SYSTEM CONTROL

### Irrigation Application Devices

Traditionally, irrigation application rates have been altered by manually changing either the entire sprinkler or the sprinkler orifice, or in the case of moving irrigation systems, by changing the travel speed. In earlier traveling systems, the travel speed was adjusted manually, but newer systems with programmable controls allow dynamic adjustment of the travel speed. Labor costs to make manual changes were unreasonably expensive. For both manual and programmable systems, the management zone was often a large portion of the total system area and the zone had relatively inflexible boundaries (e.g., sector of circle or annular ring). This made it difficult or impossible to match the variable-rate management zone with the area that needed site-specific management.

Several innovative technologies have been developed to variably apply irrigation water to meet anticipated whole-field management needs in precision irrigation, primarily with center-pivot and lateral-move irrigation systems. In general, the operational criteria for these systems include ease of retrofit to existing commercial irrigation systems, good water application uniformity within and between management zones, robust electronics, compatibility with existing irrigation-system equipment, bi-directional communication, and flexible expansion for future development and functional requirements. In addition, management of the precision water application systems must include interactions between individual sprinkler wetted diameters, the start/stop movement of towers, and solenoid valve cycling. These new precision water application technologies generally can be classified as either (1) a multiple of discrete fixed-rate application devices operated in combination to provide a range of application depths, (2) flow interruption to fixed-rate devices to provide a range of application depths that depend upon pulse frequency, or (3) a variable-aperture sprinkler with time-proportional control. Case studies of existing systems are used to illustrate these points.

*Multiple Sprinklers*

In earlier research on lateral move irrigation systems in Texas, Lyle and Bordovsky (1981, 1983) used three individually-controlled manifolds, each delivering discrete but different flow rates, from various combinations of different orifices sized to achieve a series of discrete incremental application rates. In Arizona, Roth and Gardner (1989) used three lines on a lateral-move irrigation system to apply five different water depths to a series of research plots, but the depth distribution among plots was fixed, which prevented independent control of application depth either along the system length or in the travel direction. Although these two systems had the potential to be used for dynamic control, neither was used in this manner.

Those approaches were extended to full site-specific application by the USDA-ARS Coastal Plains Soil, Water, and Plant Research Unit in Florence, South Carolina, to provide independent control of application depths on two small center-pivot systems (Camp and Sadler, 1994; Omary et al., 1997). For each, the mainline was divided into 13 segments of 9.1 m each, with three independently controlled, parallel manifolds of six spray nozzles spaced 1.5 m apart. To minimize water application to adjacent segments, wide-angle industrial spray nozzles with wetted diameters of 3 to 5 m were used. The manifolds and nozzles were sized to provide 1/7, 2/7, and 4/7 of a base application depth. Thus, combinations of the three manifolds delivered from 0 to 100 percent of the base amount in 1/7 increments.

University of Idaho researchers employed a similar incremental approach, using pairs of sprinklers with different nozzle sizes at each sprinkler location on commercial lateral move and center pivot machines. The sprinklers applied either 1/3 or 2/3 the full application depth. The proper application rate was achieved by switching solenoid valves that directed flow to the appropriate sprinkler (McCann et al., 1997).

*Pulsing Sprinklers*

In Colorado, the USDA-ARS research unit at Fort Collins developed a variable-rate water application system on a four-span lateral-move machine using commercially available equipment to achieve variable application depths when operating in either direction. Commercial irrigation sprinklers (Quad IV LEPA nozzles, Senninger Irrigation Inc., Orlando, Florida) were attached to flexible hose drops spaced 1.5 m apart and mounted on manifolds suspended below the mainline pipe with each manifold supplying half of a span (~22 m). The nozzle package was designed for maximum uniformity, so the orifice size and instantaneous application rate was the same for all nozzles. Different water depths along the mainline were obtained by pulsing the water flow to each manifold so that the time-averaged discharge depended on the pulsing frequency of the valves controlling the flow (Fraisse et al., 1992). Uniformity of closing times for each valve model was very important in obtaining uniform application in a field (Duke et al., 1992). One disadvantage was that the variance in closing time for individual valves was high enough that the control program had to account for the individual valve closing times. Check valves were installed above each LEPA nozzle to prevent drainage from the manifold and drop tubes when the solenoid valve was off in the pulse cycle.

Researchers with Washington State University, USDA-ARS (Prosser), and industry partners developed and tested a prototype precision irrigation system on a commercial 45 ha center-pivot system (Evans, Han, et al., 1996; Evans, Schneider, et al., 1996; Evans and Harting, 1999). This center pivot was divided into zones of two to four heads each. Every rotator spray head (Nelson Irrigation, Inc., Walla Walla, WA) included a pressure regulator and a water-operated, normally open solenoid valve (which could irrigate even if the control system failed) to regulate the amount of water to be applied. Each of the 30 zones were switched on or off by individual, custom-built zone controllers.

*Variable-Orifice Sprinkler*

The water flow rate from an orifice can be reduced by reducing the effective area of the orifice. In Idaho, King and Kincaid (1996) developed a variable-orifice sprinkler in which the flow rate was reduced by insertion of a pin into the orifice. Maximum flow occurs when the pin is withdrawn. When the pin is inserted, flow rate is reduced proportionally to the reduction in area. Time-averaged variable flow rate between the lower limit and maximum flow rate is achieved by moving the pin in and out at an appropriate duty cycle. In this case, the flow rate ranged from about 40 to 100 percent of the maximum flow rate with minimal effect on the sprinkler distribution uniformity (King et al., 1997). Prototype sprinklers were tested on a lateral-move irrigation system with sprinklers spaced 3 m apart and mounted 2 m above the ground.

### *System Control*

#### *Idaho Control System*

Various forms of control systems for self-propelled irrigation systems were developed and tested, starting with off-the-shelf components and, more recently, with in-house electronic designs. The first control system utilized a carrier signal superimposed on the irrigation system wiring. McCann and Stark (1993) obtained a U.S. patent for a computer control system to provide site-specific application of water and chemicals for both lateral-move and center-pivot irrigation systems. This led to the development of a Supervisory Control and Data Acquisition System (SCADA) network to provide variable-rate water and chemical application and to serve as a data acquisition system for decision-support software. The SCADA network utilized the 480-volt, three-phase power line of the irrigation system as the bi-directional communication medium for control and data acquisition. The SCADA network consisted of nodes placed along the system lateral where control and/or data acquisition are required and the 480-volt, three-phase power line are easily accessible (King and Wall, 1997). Each network node provided for 14 lines of digital input/output, one 0-20 ma analog output, one 0-4.096 VDC (volts, direct current) input (12 bit), and one RS-232 serial communication port. King et al. (1999) and King and Wall (1998) provided a more detailed description of the Idaho SCADA network.

Field implementation of the SCADA network required a network node at the master control point (pump or pivot-point location) and one at alternate system towers starting with the first tower. If the master control point were not at the pivot point, an additional network node would be needed there to monitor the location of the system lateral using a digital angular position encoder. Variable-rate water application was accomplished using a double sprinkler arrangement at each active outlet along the system lateral. Stepwise variable flow rate was achieved by controlling operation of each sprinkler using solenoid-activated diaphragm valves on each sprinkler. Each system span was divided into three individual control zones by wiring the valves in corresponding groups. Multiple application maps were stored in the master controller and the user could select which one to use. Maps and system operational data were transferred to and from the master controller over an RS-232 port using a laptop computer (McCann et al., 1997; King et al., 1999; King and Wall, 2000).

*Colorado Control System*

The USDA-ARS research unit in Fort Collins, Colorado, developed a control system for a variable-rate water application system on a four-span, lateral-move machine that was used to irrigate many small research plots with various irrigation amounts while minimizing the areas of transitional depths between plots. A C:A:M:S (Valmont Industries, Inc., Valley, Nebraska) control panel was used to vary system speed and, consequently, application depth in the direction of travel. This microprocessor-based controller could be programmed to change speed or control output relays by time, position, or a status change of an input port. Up to nine independent programs could be stored in the panel memory (Evans et al., 2000).

The solenoids that controlled water flow to each manifold were connected to a General Electric Fanuc Model 90-30 PLC (GE Fanuc model 90-30, Charlottesville, Virginia) via electric cable along the entire length of the system. The programmable logic controller (PLC) program used a 60-s cycle interval, and the duty cycle (time on/cycle interval) could be programmed from 1 to 100 percent. Multiple programs stored in the PLC produced various combinations of duty cycles for the eight half spans. The program running in the C:A:M:S panel selected the PLC program for pulsing by setting the appropriate input ports connected to the PLC. However, because of the experiments under this lateral move, only three programs were routinely used and were controlled by switches parallel to the input ports.

*South Carolina System*

The USDA-ARS research unit in Florence, South Carolina, developed a programmable control system for variable-rate water and nutrient applications on two small center-pivot systems. A computer (PC) was mounted on a GE Fanuc Model 90-30 PLC backplane (GE Fanuc model 90-30, Charlottesville, Virginia) and connected via the PLC system bus. The PC-PLC control system was mounted on the mobile portion of the system about five meters from the pivot center, from which it controlled all manifold solenoid valves via electric cables along the entire length of the system. The computer also controlled irrigation-system travel speed and chemical-injection pump-flow rate, both based on a spatially referenced mapping system. Software written in Visual Basic for DOS (Microsoft Corporation, Redmond, Washington) converted a set of control values to on-off settings in the directly addressable solenoid control registers of the PLC (Camp et al., 1998; Sadler et al, 1996). The on-board PC repeatedly interrogated the stationary C:A:M:S (Valmont Industries, Inc., Valley, Nebraska) computer management system to determine the angular position of the first span and other parameters to provide assurance of proper operation. From the angular position and the fixed radius of each segment along the truss, the segment location (expressed in polar coordinates) was established for that time. For each location, the appropriate table lookup was performed, and the solenoid registers were set accordingly. The software also determined the correct system speed from the table and set the appropriate value in the stationary management system. The software also included a routine, based on position measurements, to correct consistent errors in truss angular position reported by the stationary management system. If one of several system variables (pressure, flow rate, position, voltage, etc.) was reported as out of the specified range, the PLC controller closed all solenoid valves and stopped the center-pivot drive system. The problem was recorded to a log file for user information. Additional details regarding the control system and overall center-pivot operation were reported by Camp et al. (1998).

### *Washington Research System*

Researchers with Washington State University, USDA-ARS (Prosser), and industry partners developed and tested a prototype precision irrigation system on a commercial 45 ha center-pivot system (Evans, Han, et al., 1996; Evans, Schneider, et al., 1996; Evans and Harting, 1999). The center pivot was divided into 30 zones of two to four heads each. The zones were turned on or off by individual custom-built zone controllers that communicated with a computer via a 900 MHz spread-spectrum radio modem. The individually addressable controllers communicated via a two-wire RS-485 bus along the pivot lateral and could also monitor RS-232 devices such as digital compasses or pressure transducers. A computer in the farm office communicated with the pivot controllers, an on-site weather station, rain gauges, and soil water sensors; developed real-time "management maps" or prescriptions; and implemented them depending on the position of the center pivot in the field. Application of the prescribed irrigation amount in a management zone was achieved by cycling the sprinklers on and off at selected intervals over a period of time. This control system theoretically allowed application depths ranging continuously from 0 to 100 percent, but changes were made in about 5 percent increments in practice. A PC-DOS-based center-pivot irrigation simulation model (CPIM) was developed, tested, and linked to the GIS (geographic information system) (Evans et al., 1993) used by the control system. The CPIM model was calibrated for the specific sprinkler packages and hydraulic characteristics of the individual center-pivot system. A site-specific irrigation-scheduling program was written that utilized the output from CPIM and included uniformity considerations (Han et al., 1996).

*Washington Commercial System*

On a large commercial farming enterprise, Harting (1999) installed a precision irrigation control system specifically designed for a central cluster of pressure control valves, shut-off valves, fertilizer storage tanks, and fertilizer injectors located within the triangular area between three adjacent circles. The cluster also contained the electrical panels for on-off and speed control for each of three center-pivot machines. The control system was composed of an off-the-shelf PLC, electric solenoid valves, and hydraulic valves. The sprinklers were controlled in banks of four sprinklers and were pulsed on and off to reduce the amount of applied water in a duty cycle of one minute. Three PLC counters were used to control the sprinkler banks. These created 20 s offsets between banks so that all of the sprinklers did not shut off at once, decreasing the possibility of control-valve problems at the cluster point. Normally open pneumatic valves were used rather than water-assisted electronic solenoid valves because the air was much cleaner than the irrigation water, which meant that foreign material did not plug the orifices. Normally closed electric solenoid valves, placed in banks of five valves, were used to control the pneumatic valves.

The control system compared the position of each sprinkler bank to a map in the memory of the main controller. The program was designed to allow up to 80 individual area definitions per circle with a maximum of eight treatment areas. Each sprinkler bank operated on a 1-minute cycle. The precision irrigation control resided in the PLC at the cluster and consisted of three databases and corresponding programs. The control program consisted of several function blocks that were added to the main program as needed. Associated with the treatment area definitions was another database that contained the distance from the center point to the middle of each sprinkler bank and the total flow rate that the sprinkler bank could apply. A third database contained the application depth percent (treatment) of the base application rate, which was entered by the user for each individual treatment area.

## *AUXILLIARY SYSTEM COMPONENTS*

### *Location and Alignment*

The control systems for most site-specific irrigation application systems use some form of spatially indexed data to determine the appropriate application rate for specific sites. The basis for these spatially indexed data is typically a widely accepted georeference system, such as latitude and longitude. Consequently, it is necessary to know the precise location of all elements of the application system at all times during operation if accurate site-specific applications are expected. Various approaches have been used (cf. Table 5.1), but the greatest challenge is cost. Although it is often desirable to have multiple location sensors along the truss length of a moving irrigation system, cost would be prohibitive. A general solution has been to use one or two sensors to locate one or both ends of the moving system and to calculate the location of intermediate points. Because moving irrigation systems consist of multiple segments or spans, with each end of the span moving independently but within confined limits, the truss is not always linear. This is not a significant problem for small systems, but misalignment can be significant for large systems. Fortunately, in many cases, the shape of the truss is consistent, predictable, and describable for specific operational conditions.

Because only one end of a center-pivot system moves, fewer sensors are needed to determine location. One approach (e.g., Florence) has been to use the angular resolver furnished with many current commercial center-pivot systems to determine the radial location of the first span, but the system alignment can still vary by at least the angular displacement at each tower. The location of the remaining spans can either be assumed based upon an assumed truss shape (e.g., linear) or be calculated based on observations or measurements. Another approach (e.g., Washington research) has been to use multiple electronic compasses to continuously measure the misalignment so that the locations can be calculated more precisely. Others (e.g., Washington commercial) have used Global Positioning Systems at one or more sites along the truss to allow calculation of other locations. Until recently, this option has been cost prohibitive because of the precision required but is now becoming a more viable option. End tower measurements are inherently more accurate than resolvers because the radius of the center pivot amplifies the resolver imprecision.

The determination of precise locations for lateral move systems is similar to that of center pivot systems except that it is more difficult because both ends move. Phene et al. (1985) used a laser alignment system. Although the travel path is constrained by the guidance system, some variation usually exists in repeatability. The issues of truss misalignment are similar for lateral-move and center-pivot systems. In general, more sensors are required for determining locations in lateral-move systems than in center-pivot systems. Most lateral-move precision irrigation systems have used one or more GPS sensors to determine location. A variation of this method, which uses a combination of a less precise GPS (lower cost) and dead reckoning, has also been suggested as a lower-cost alternative (Colorado) location.

### *Management Database and Decision Support*

Any information to be used in the precision management applications must be indexed by its geographical location and stored in an electronic format that can be readily accessed by a computer or computer-based controller. As such, it operates as a special-purpose GIS. This management database houses the data with which the control system operates the irrigation machines, records the actual application amounts for later use or documentation, and provides the framework on which a decision support system can operate. Data that may be stored could include user-entered soil characteristics, cultural operations, or application maps. It could also include historical georeferenced data such as yield maps, past application maps, or cultural histories. Spatial arrays of sensors, either mounted on the system or in the field, could potentially feed information directly into the management database.

The ultimate use of the information in the management database is to contribute to the decision support system. This component lags behind the development of the hardware components. Despite knowledge transferred from uniform, whole-field management, precision water management, as well as the larger area of precision agriculture, remains limited by an insufficient knowledge base for making management decisions. In the case of water management, the basic decisions are when to apply irrigation and how much water to apply. Of course, many factors affect these basic decisions, including adequacy of water supply, expected economic benefit of the added water, benefit of alternate use of the water, regulatory and social issues, and impact on other farm enterprises. Little information is available for some of these factors. Consequently, the knowledge base needed to make intelligent, profitable, and environmentally sound decisions with regard to precision water management is currently not adequate. However, some information is available and significant new information is being developed.

As with precision fertilizer management, initial efforts in precision water management were based on soil map unit or a measured soil physical property (i.e., electrical conductivity or soil texture). In many cases, these efforts were not very successful when crop yield was the primary method for measuring effectiveness. Some have suggested that spatially referenced crop yield for several seasons may be a basis for effective precision water management because it integrates many other effects including weather. Others, primarily in research, have investigated the potential use of canopy temperature or some derivative as a basis for dynamic precision water management, building on whole-field irrigation water management using similar tools (Wanjura et al., 1992; Barnes et al., 2000; Sadler et al., 2002.) The decision support system will likely incorporate all of these elements, and very well may integrate them using computer simulation models.

### *Variable Water Supply*

Most conventional moving irrigation systems are designed for and operate with a constant water flow rate and pressure to the system in which all sprinklers operate most of the time. With precision irrigation, in which variable flow rates are required for several management zones within the total system, water must be supplied to the system at constant pressure but at a variable flow rate. The magnitude of the flow rate variance depends upon the system design and operation characteristics, but in extreme cases it can vary from full-design flow rate to almost zero.

Analysis of field data (Evans, Han, et al., 1996; Evans, Schneider, et al., 1996) has shown that application depths under precision irrigation with self-propelled machines in arid areas are normally reduced by either no more than 15 to 20 percent or by 100 percent (no application) as would be the case of exclusion zones, and that lesser values (i.e., 75 to 90 percent reductions) are not usually needed except for chemigation uses and areas where surface and shallow subsurface flows accumulate. In humid areas, however, a wider range of application rates may be required to accommodate different rainfall amounts and widely varying soil water storage capacities across a field.

At least two possible solutions currently exist for a variable water supply. The first method consists of a multiple-pump plant. For example, one such system used four pumps, each drawing from a reservoir and each connected to a discharge manifold, and used in combinations with pressure bypass to the reservoir to provide constant pressure at flow rates ranging from 0 to 3,000 $Lmin^{-1}$ (Camp et al., 1998). The second, which involves fewer components, is a variable-rate pump (e.g., variable-frequency motor controller) such as that used by Harting (1999). Both of these solutions are expensive compared to constant-rate pumps, but the cost of the variable-rate pump may decrease with time.

### *Nutrient Application Methods*

In all known applications, variable-rate application of nutrients has been accomplished by maintaining a constant concentration of the nutrient in the water supply and using the variable-rate application capabilities of the irrigation machine to vary the amount of water applied. This requires knowledge of the water flow rate; either by direct measurement or by calculation using the design (or individually measured) flow rates of the segments that are operating at a given time. With that information, the controller must signal the nutrient pump to inject the nutrient solution at a flow rate that is proportional to the system water flow rate.

The primary difficulty encountered with this technique is the delay in transporting the nutrients the full length of the machine. This requires a startup zone that either can tolerate a shortage of nutrients on the far end, or stationary operation with the end span operating to move the water/nutrient mixture through the system main line. In the case of center-pivot systems, the startup zone could be traversed again after completing the circle, in the same travel direction, to allow the nutrient to be flushed from the system, thus compensating for the shortage in the first pass during startup. This would cause application of twice the amount of water, but could theoretically meet the nutrient application requirement. Although this would be easily accomplished with a full-circle center pivot, it would be more difficult with partial-circle pivots and lateral-move machines, requiring these systems to deadhead back to the beginning of this startup zone.

Because site-specific applications of fertilizers are often applied in less water than normally applied during a single irrigation application, water deficits may occur in some areas within the system, especially for arid climates in which irrigation machines must operate most of the time. Furthermore, multiple site-specific fertilizer applications during the growing season could further exacerbate this problem. Conversely, site-specific fertilizer applications could cause excess water application in some areas, causing surface accumulation or possible leaching of nutrients toward the ground water. Consequently, site-specific fertilizer applications with moving irrigation machines must be carefully coordinated with all irrigation applications, both before and after fertilizer applications.

### *Pesticide Application Methods*

Site-specific pesticide applications with self-propelled systems are especially attractive for precision agriculture approaches because agro-chemicals (herbicides, insecticides, and fungicides) can be applied after full canopy cover to the entire irrigated area or to problem areas only. Site-specific pesticide applications could

1. reduce pesticide costs to growers, making them more competitive in global markets;
2. reduce the potential for resistant pest populations;
3. reduce total use and minimize negative impacts of pesticides on the environment and nontarget organisms; and
4. reduce potential for leaching of pesticides and thus increase chemical efficacy.

Furthermore, utilization of remote detection technologies (e.g., remote sensing and image analysis) and dynamic control of site-specific variable-rate application devices could allow treatment of discrete areas within the total system ("spot" treatment) with the proper amount of pesticide. This could potentially reduce pesticide use, enhance effectiveness of integrated pest management (IPM) programs, and greatly improve the economics of site-specific technologies. It is also highly probable that future environmental regulations may require spatially-referenced records of all pesticide applications, which this technology could easily provide.

The development of hardware and software for site-specific pesticide applications with center-pivot and lateral-move irrigation systems is proceeding along several fronts. Several agro-chemical and irrigation manufacturing companies have recognized the potential demand for this technology and are either developing or implementing separate pesticide application systems. However, most of these designs have limited site-specific capabilities. Two general approaches are being explored. One is parallel to nutrient injection: to use the existing irrigation sprinkler or spray heads and cycle the water supply on and off (pulsing) to vary the water application rate at very low volumes (e.g., 2.8-3.7 instead of 46-75 $m^3ha^{-1}$ normally applied for irrigation, but will vary with length and capacity of the center-pivot or lateral-move system). The second approach is to install separate flow-path systems (water and pesticide supply pipes, supporting structure, nozzles) on the center-pivot or lateral-move system structure and operate them independently of the water application system. These systems apply the water and chemical mixtures to crops, usually also at much lower water application rates (e.g., 1-2 $m^3ha^{-1}$). Either approach can be modified to have site-specific capabilities. Reliability, portability, and economics of these pesticide systems will be major considerations.

Because insects are sometimes easily mapped and amenable to spot treatments, insect control could also utilize the same system control technology and application systems. However, the real challenge will be to develop ways to coordinate pesticide applications with water applications during peak water use periods so that there are minimal negative impacts on crop yield and quality as well as on the environment.

## MANAGEMENT ZONES

### Size

Management zones are defined here as an area within a field or irrigation system where crops respond somewhat uniformly to cultural inputs such as water, fertilizer, and agro-chemicals. The size and shape of management zones can vary considerably among sites and regions depending upon several sources of spatial variability. Sources of variability include soil properties, cultural practices and features, topography, presence of sensitive zones (e.g., recharge zones, ponded water, diverse crops, differing plant density, etc.), and other factors. For some regions, large contiguous portions of a field or irrigation system may be relatively uniform and may be managed in a similar manner. For other regions, soil map unit or other factors contributing to spatial variability may be reasonably uniform only within much smaller areas, which would probably require more resolved site-specific management. For example, Sadler et al. (1995) found 12 soil map units within a small (6 ha) center-pivot system in the southeastern Coastal Plain of the U.S. and reported significant yield differences within soil map units for sites only 25 to 30 m apart (Sadler, Bauer, et al., 2000). For the northwestern United States, Evans and Harting (1999) concluded that growers probably cannot practically manage areas less than 0.4 to 0.5 ha within a field because of other cultural aspects. In other cases, the management zone size may be determined by the areal extent of a cultural feature such as a ground water recharge zone, a pond or zone of saturated soil, rock outcropping, or a building. The size of the desired management zone may be different for irrigation, nutrients, and pesticides and may change during the growing season.

### *Conflicts and Consequences*

This last point illustrates one important conflict, namely that the size of the desired management zone may be quite different for applications of water, nutrients, and agro-chemicals. Because the design of most current site-specific application systems constrains the minimum management zone size, the desire for conflicting management zones must be resolved. Also, there is a close relationship among the wetted diameter of the water application device (sprinkler, spray, or drip), the need for overlap to achieve acceptable application uniformity, the allowable overspray into adjacent management zones, and the desired management zone size. Conflicts among these factors are normally resolved during the system design phase with a heavy emphasis on system cost, potential profitability, and legal constraints. For example, with sprinklers mounted every 2.5 to 3 m along the system truss and wetted diameters ranging from 6 to 10 m or more, controlling banks of 3 to 5 sprinklers might be acceptable for a management zone of 0.5 ha. For a smaller zone, the sprinklers can be controlled individually, but the application uniformity would probably be lower and the increased cost would reduce the potential economical feasibility of the system. On the other hand, more sprinklers can be controlled together or placed on a separate manifold, which would decrease cost, but the larger control area would reduce the probability of matching the desired management zone size. Consequently, the precision irrigation application system must provide acceptable variable-rate applications within the desired management zone size and shape.

Most water application devices do not provide acceptable application uniformities within the wetted diameter when operated alone or in a fixed position. Although acceptable application uniformities can normally be achieved by overlapping the patterns of adjacent devices or by moving the pattern over a fixed soil location, the uniformity will decrease in any border areas. Consequently, the site-specific water application system can be designed either for maximum application uniformity within the management zone, which means accepting overspray into adjacent zones, or for maximum management zone integrity (no overspray allowed), which means accepting poorer application uniformity in the border areas.

In general, the adverse effects of errors in matching management zone size and shape with the desired area of site-specific management can range from minor to major, depending on the sensitivity of the environmental area to the commodity being applied. In the case of water and nutrients, adverse effects could range from reduced crop yield to environmental damage caused by surface runoff or leaching of nutrients toward the ground water. In the case of pesticides, improper application rate, overspray, or drift could cause reduced crop yield, crop death, or serious environmental damage. In cases where regulations or statutes are violated, errors in application could result in fines.

## APPLICATIONS AND JUSTIFICATIONS

Obviously, precision water management can be used more effectively in situations in which the benefit is greater, whether required by regulation or justified by increased profit or conservation of resources. One such case is where profitable crop production is possible with precision management of water and agro-chemicals, but statutes, rules, or regulatory policy prevent profitable crop production using conventional methods. For example, if ground water recharge zones exist within the irrigated area, nutrients and pesticides (or, in some cases, water) cannot be applied in or near these areas. Similar restrictions may apply to other environmentally sensitive zones and cultural features.

In other cases, soils may exhibit enough local spatial variability to make precision water, nutrient, or pesticide management profitable, both from economic and environmental points of view. Although the total amount of water or chemicals applied to the field may not change, the distribution may change drastically and could significantly increase total crop yield. The use of precision water management may prevent waterlogged soils caused by areas with limited infiltration rate and could improve trafficability for farm equipment. If the water supply is limited, either permanently or temporarily due to reduced rainfall, precision water management allows the manager to allocate the limited resource among the various competing crops to meet management objectives or to optimize profit. In general, precision management of water, nutrients, and pesticides allows the overall optimization of inputs to achieve enterprise goals, which can range from maximum profit to resource conservation and environmental protection.

## CURRENT TRENDS

In many parts of the world, water supplies, especially those available for agricultural use, are declining. A major reason for the declining supply available to agriculture is the increased competition for existing water supplies among the various user groups. In the current food pricing structure, water has more value for industrial, municipal, and sometimes environmental uses than for agriculture. Consequently, as supplies decline and competition increases, the cost of water for agricultural use will probably increase. In certain cases, water supplies may be removed from agricultural use altogether. In either case, agricultural producers will have to manage water more efficiently and increase productivity for each unit of water used. In some cases, agricultural production may shift to geographic areas where water supplies are more available, but some of these areas may have soils that exhibit more spatial variability or require more management to prevent environmental damage.

Declining water supplies combine with other factors that negatively impact agricultural production (i.e., increased regulation, decreased profit margins, and environmental concerns) to create significant challenges for the food and fiber industry. However, a portion of the water supply problem may be alleviated through the use of treated wastewater. Although it is currently used for turf and landscape plants, with proper treatment it might be used for some food and fiber crops in the future. Use of this treated water supply would probably require additional regulation and testing, but precision irrigation, with its reliance on computers for storage of spatially referenced data and for spatial control of applications, could easily provide documentation of each operation. Finally, another factor is the possible change in climate for the long term. Significant atmospheric warming could change rainfall patterns, alter water supplies, require shifts in the specific crops produced in an area, and move agricultural production to other geographic areas.

## *COST-BENEFIT ISSUES*

The cost-benefit issues are certainly not the same throughout the world. The issues vary with size of enterprise, type and size of irrigation system, crops produced, prevailing crop prices and subsidies, and production costs. Therefore, these issues will be discussed for two broad categories—automated irrigation on medium- to larger-sized farms and manually controlled irrigation on smaller farms. For both of these categories, the primary current need is production functions to predict the crop yield response to various incremental changes in manageable inputs, such as water, nutrients, and pesticides.

### *Larger Farms*

Because commercially produced precision water management systems and crop production functions needed for optimization are not yet available, it is difficult to develop accurate cost-benefit scenarios. All current precision irrigation systems are commercial systems that have been modified, which adds expense. If only water and/or fertilizers are managed site-specifically, random variability in irrigation applications caused by wind, sprinkler patterns, and start-stop operations may not be sufficient to justify precision applications when water and fertilizer costs are low. However, when more extensive spatial variation or localized production losses due to management occur, the benefits may justify the cost. Of course, the economic benefits are often much greater for higher-value crops such as potatoes, tomatoes, and other vegetable crops, where improved quality is of equal or greater importance than crop yield (e.g., Watkins et al., 1999). Where the justification is obtained for high-value crops, field crops grown in rotations on the same fields may benefit from the availability of precision irrigation.

A critical factor is that water, fertilizer, and pesticide applications should all be included in the precision management system. Depending upon the crop, savings can be substantial for pesticide applications, especially if custom applications such as aerial applications can be avoided. Some savings may accrue from less water, fertilizer, and pesticide use, from selective and improved accuracy of pesticide applications, and from better utilization of manpower. Also, future environmental regulations may greatly change the feasibility of precision irrigation systems because they may be able to accomplish applications with less environmental impact. For example, application restrictions on aerial chemical application and recent changes in endangered species listings could have a large impact on the adoption of precision irrigation systems in some areas.

Finally, integrated dynamic control of water, nutrient, and agro-chemical applications may be possible in the future using remote or local sensors for early detection of stress or pests that may cause crop yield or quality problems. Precise applications of water, fertilizer, or pesticides could be made on a real-time, site-specific basis to remedy the deficiency. Although some sensors could be mounted on the moving irrigation system, others may be constrained to fixed locations within each management zone (e.g., soil water sensors).

### *Smaller Farms*

Although the popular concept is that precision agriculture technology is appropriate only for large, mechanized farms, primarily because of equipment costs and location positioning systems, it is probably not true. Certainly the scale and type of equipment and, possibly, the technology will not be the same for the two cases. However, the basic concept of precision agriculture, that of optimizing the management of each crop unit (primarily inputs) for maximum profit (or other objective), applies equally to both production scales. In fact, the smaller the parcel, uniform and spatially variable management become more similar. In many developing countries, small parcels have been individually managed for many years, but the degree to which inputs are custom tailored for each parcel is unknown. Keeping this in mind, it still appears that development of technology for smaller farms lags that for larger farms by several years. Because it is currently much less developed, considerable speculation is required to describe a precision water, fertilizer, and pesticide application system for smaller farms.

To reduce equipment costs for determining location and automation, precision water applications will most likely be accomplished through design of a fixed system rather than through operation of a moving irrigation system. Consequently, the source and degree of spatial variability and the cropping systems must be determined earlier in the process and are often related to soil properties or topography. The irrigation system of choice may be some form of drip irrigation because of its high precision and its design adaptability to small, irregularly shaped areas. For instance, the low-cost drip irrigation methods described in Postel et al. (2001) for farmed areas ranging from as small as 100 $m^2$ to approximately a hectare could potentially be managed very precisely in space because the management zone is very small. However, these methods were developed with the objective of bringing rural communities above subsistence levels, essentially by doubling the food supply. In achieving this objective, lower irrigation system costs were a high priority, and uniformity within the management zone may be somewhat less than would be expected with standard drip irrigation practice. Further, the agronomic knowledge available may be a limiting factor for fully adapting to spatially variable soils. If these or standard irrigation systems are used for fertilizer and pesticide management, the spatial extent of chemical applications will probably be much less than for large, moving systems. However, for relatively small areas, these applications could be accomplished by other means. Obviously, this technology must be developed for specific applications and will probably be much more diverse in type than that for larger farms.

## FUTURE DIRECTIONS

For the profitable adoption of site-specific water, nutrient, and agrochemical application systems, crop response information is needed for optimum management of these systems. Initially, most site-specific management decisions, primarily fertilizer applications, were based on soil analyses collected on a grid pattern or on a soil map unit. More recently, field maps of selected measured variables (e.g., electrical conductivity, crop yield) have been used. More specific crop response information is currently being developed and reported (Camp, Sadler, Evans, and Millen, 2000; Camp et al., 2001) for selected crop-soil-site combinations, but more is needed, especially for the different physiographic regions where this technology will be used. Another critical need is an economically priced variable-rate metering device or sprinkler that can deliver the range of water-flow rates required for most moving irrigation systems. The variable-orifice sprinkler (King and Kincaid, 1996), a digitally controlled metering device (Camp, Sadler, Evans, and Usrey, 2000; Sadler et al., 2001), and pulsing commercial sprinklers are examples of attempts to provide this technology. Existing site-specific water application systems appear to satisfactorily accomplish the desired applications but most are not suitable for commercial production and all are more expensive than systems used for uniform application.

Existing control systems are generally adequate for managing site-specific applications of water, fertilizer, and pesticides if the basis for management can be represented by a digital map or file of spatially referenced data. Because of both mass of wire and cost considerations, future systems will probably implement addressable bus technology and may utilize a carrier signal on the system wiring instead of separate control wiring. Management systems that respond dynamically to real-time data collected by remote or local sensors mounted on the irrigation system should be available in the near future. These data could include water and nutritional status of the crop, pest infestations, and water stored in the soil.

Although certain combinations of cost and availability of water may justify precision application of water alone, precision water application systems must usually incorporate fertilizer and pesticide applications to be cost effective. Initially, this technology will probably be used predominantly on higher-value crops, such as vegetables, but will cascade to other crops as the cost decreases. For example, the cost of precise geographical positioning has decreased considerably during the past few years. Although most of this cost reduction resulted from a unique government action that eliminated the need for more expensive technology, other costs should decrease as new technology becomes commercially available.

Future adoption of precision water management concepts in developing countries will depend upon the development of appropriate technologies for applications in which most operations are performed manually and investment capital is low. In these cases, site-specific management concepts will most likely be accomplished through irrigation system design rather than through system operation. While the irrigation water supply may be pressurized via pumping, the irrigation system will probably be stationary (e.g., drip, sprinkler) rather than moving (e.g., center pivot, lateral move), which reduces or eliminates the need for dynamic location determination (e.g., GPS). Fertilizer and pesticide management will most likely be accomplished via conventional application techniques with application rates determined by manual interpretation of maps showing spatial variability and/or application rates.

## *CONCLUSIONS*

Beginning in 1992, research and development for site-specific management of water have been reported for four locations in the United States. Although all activities have similar goals, the techniques to achieve variable-rate applications have been different. All of these systems have used moving irrigation systems, center pivot or lateral move, with the system control based on stored databases of spatially referenced data. However, the method of implementing valve control to achieve the desired application rate ranged from programmable logic controllers to addressable solenoid valves. The method for determining location of discrete parts of the irrigation system at any time has also varied and has included GPS, electronic compass, dead reckoning, and system angular position obtained from the center-pivot control system. Location determination is more complex for lateral-move systems than for center pivots.

The variable-rate water application systems employed include multiple sprinklers or groups of sprinklers at each location that were switched to provide discrete rates, cycling sprinklers or groups of sprinklers for time-proportional pulsing, and a variable-aperture sprinkler with time-proportional control. The size of management zones for these systems varied, due, at least in part, to the interaction between the sprinkler wetted diameter and management zone size, along with the associated application uniformity issues. However, the size and shape of areas that require similar but unique input management are the major factors that determine management zone size, with due consideration of practical cultural constraints at that site.

Integration of site-specific water, fertilizer, and agro-chemical applications is usually important for the profitable and sustainable use of this technology. With the capital investment in a moving irrigation system for irrigation, the cost of adding fertilizer and agro-chemical application technology should be less costly than performing these operations separately by ground or aerial application techniques. Although initial applications of site- specific water, fertilizer, and agro-chemical applications will be more profitable for high-value crops, they should also become profitable for lower-value crops as system costs decrease. Geographically referenced documentation of all applications to crops, potentially to be required to ensure environmental quality, should be easily accomplished using precision irrigation systems. Although the adoption of precision water management technology using automated control of moving irrigation systems will probably continue to be limited to larger farm enterprises, the basic technology could be applied to smaller farms, even in developing countries, where irrigation systems are stationary. In these cases, the spatial variability would be managed primarily through the water-management-system design and manual control.

## REFERENCES


Barnes, E.M., P.J. Pinter Jr., B.A. Kimball, D.J. Hunsaker, G.W. Wall, and R.L. LaMorte. 2000. Precision irrigation management using modeling and remote sensing approaches. In *Proceedings of the 4th Decennial National Irrigation Symposium,* eds. R.G. Evans, B.L. Blenham, and T.P. Trooien, St. Joseph, MI: ASAE/IA, pp. 332-337.

Bordovsky, J.P., and R.J. Lascano. 2003. Variable rate irrigation using low energy precision application (LEPA). 2003 Electronic Proceedings of the Beltwide Cotton Conferences, Nashville, TN.

Camp, C.R. and E.J. Sadler. 1994. *Center pivot irrigation system for site-specific water and nutrient management.* ASAE Paper #94-1586. St. Joseph, MI: ASAE, p. 9.

Camp, C.R., E.J. Sadler, D.E. Evans, and J.A. Millen. 2000. Corn yield response to N fertilizer for a range of water regimes and soil mapping units in the southeastern Coastal Plain. In *Proceedings of the 5th International Conference on Precision Agriculture,* eds. P.C. Robert, R.H. Rust, and W.E. Larson, Madison, WI: ASA-CSSA-SSSA, CD-ROM.

Camp, C.R., E.J. Sadler, D.E. Evans, and J.A. Millen. 2001. Irrigation and nitrogen management with a site-specific center pivot. In *Third European Conference on Precision Agriculture,* Vol. 2, eds. G. Grenier and S. Blackmore. Montpellier, France: Agro Montpellier, pp. 635-640.

Camp, C.R., E.J. Sadler, D.E. Evans, and L.J. Usrey. 2000. Variable-rate, digitally controlled metering device. *Applied Engineering in Agriculture* 16:39-44.

Camp, C.R., E.J. Sadler, D.E. Evans, L.J. Usrey, and M. Omary. 1998. Modified center pivot system for precision management of water and nutrients. *Applied Engineering in Agriculture* 14:23-31.

Duke, H.R., D.F. Heermann, and C.W. Fraisse. 1992. Linear move irrigation system for fertilizer management research. In *Proceedings of International Exposition and Technical Conference,* Fairfax, VA: The Irrigation Association, pp. 72-81.

Evans, R.G., G.W. Buchleiter, E.J. Sadler, B.A. King, and G. Harting. 2000. Control systems for precision irrigation with self-propelled systems. In *Proceedings of the 4th Decennial National Irrigation Symposium,* eds. R.G. Evans, B.L. Blenham, and T.P. Trooien, St. Joseph, MI: ASAE/ IA, pp. 322-331.

Evans, R.G., S. Han, L.G. James, and M.W. Kroeger. 1993. *CPIM—a computer simulation program for center pivot irrigation systems.* ASAE Paper #93-3065. St. Joseph, MI: ASAE.

Evans, R.G., S. Han, S.M. Schneider, and M.W. Kroeger. 1996. Precision center pivot irrigation for efficient use of water and nitrogen. In *Proceedings of the 3rd International Conference on Precision Agriculture,* eds. P.C. Robert, R.H. Rust, and W.E. Larson Madison, WI: ASA-CSSA-SSSA, pp. 75-84.

Evans, R.G. and G.B. Harting. 1999. Precision irrigation with center pivot systems on potatoes. In *Proceedings of ASCE 1999 International Water Resources Engineering Conference,* eds. R. Walton and R.E. Nece, Reston, VA: ASCE, CD-ROM.

Evans, R.G. and W.M. Iversen, 2005. Combined LEPA and MESA irrigation on a site-specific linear move system. In *Proceedings of the 26th Annual International Irrigation Show,* November 6-8, Phoenix, AZ: Irrigation Association, CD-ROM.

Evans, R.G., S.M. Schneider, R.A. Boydston, and S. Han. 1996. *Precision agriculture for center pivot irrigated wheat and potatoes in Washington.* ASAE Paper PNR-967502. Presented at 51st Annual Meeting of Pacific Northwest Section of ASAE, Yakima, WA, September 22-24, p. 15.

Fraisse, C.W., D.F. Heermann, and H.R. Duke. 1992. Modified linear move system for experimental water application. In *Advances in Planning, Design, and Management of Irrigation Systems as Related to Sustainable Land Use,* Vol. 1, Leuven, Belgium, pp. 367-376.

Han, S., R.G. Evans, and S.M. Schneider. 1996. *Development of a site-specific irrigation scheduling program.* ASAE Paper #96-2076. St Joseph, MI: ASAE.

Harting, G.B. 1999. As the pivot turns. *Resource,* 6:13-14.

King, B.A., R.A. Brady, I.R. McCann, and J.C. Stark. 1995. Variable rate water application through sprinkler irrigation. In *Site-Specific Management for Agriculture Systems,* eds. P.C. Robert, R.H. Rust, and W.E. Larson, Madison, WI: ASA-CSSA-SSSA, pp. 485-493.

King, B.A. and D.C. Kincaid. 1996. *Variable flow sprinkler for site-specific water and nutrient management.* ASAE Paper #962074. St. Joseph, MI: ASAE.

King, B.A, I.R. McCann, C.V. Eberlein, and J.C. Stark. 1999. Computer control system for spatially varied water and chemical application studies with continuous-move irrigation systems. *Computers and Electronics in Agriculture* 24: 177-194.

King, B.A., J.C. Stark, I.R. McCann, and D.T. Westermann. 1996. Spatially varied nitrogen application through a center pivot irrigation system. In *Proceedings of the 3rd International Conference on Precision Agriculture,* eds. P.C Robert, R.H. Rust, and W.E. Larson, Madison, WI: ASA-CSSA-SSSA, pp. 85-94.

King, B.A. and R.W. Wall. 1997. *Digital power line carrier control system for optimum operation of variable speed pumping plants with center pivots.* ASAE Paper #972191. St. Joseph, MI: ASAE

King, B.A. and R.W. Wall. 1998. Supervisory control and data acquisition system for site-specific center pivot irrigation. *Applied Engineering in Agriculture* 14(2):135-144.

King, B.A. and R.W. Wall. 2000. Distributed instrumentation for optimum control of variable speed electric pumping plants with center pivots. *Applied Engineering in Agriculture* 16:45-50.

King, B.A., R.W. Wall, D.C. Kincaid, and D.T. Westermann. 1997. *Field scale performance of a variable rate sprinkler for variable water and nutrient application.* AAE Paper #972216. St. Joseph, MI: ASAE.

Lu, Y.-C., C.R. Camp, and E. J. Sadler. 2004a. Efficient allocations of irrigation water and nitrogen fertilizer in corn production. *Journal of Sustainable Agriculture* 24(4):97-111.

Lu, Y.-C., C.R. Camp, and E.J. Sadler. 2004b. Optimal levels of irrigation in corn production in the southeast coastal plain. *Journal of Sustainable Agriculture* 24(1):95-106.

Lu, Y.-C., E.J. Sadler, and C.R. Camp, 2005. Economic feasibility study of variable irrigation of corn production in southeast Coastal Plain. *Journal of Sustainable Agriculture* 26(3):69-81.

Lyle, W.M. and J.P. Bordovsky. 1981. Low energy precision application (LEPA) irrigation system. *Transactions of the ASAE* 24:1241-1245.

Lyle, W.M. and J.P. Bordovsky. 1983. LEPA irrigation system evaluation. *Transactions of the ASAE* 26:776-781.

McCann, I.R., B.A. King, and J.C. Stark. 1997. Variable rate water and chemical application for continuous-move sprinkler irrigation systems. *Applied Engineering in Agriculture* 13:609-615.

McCann, I.R. and J.C. Stark. 1993. Method and apparatus for variable application of irrigation water and chemicals. U.S. Patent No. 5,246,164. September 21, 1993.

Omary, M., C.R. Camp, and E.J. Sadler. 1997. Center pivot irrigation system modification to provide variable water application depths. *Applied Engineering in Agriculture* 13:235-239.

Perry, C., S. Pocknee, and O. Hansen. 2003. A variable rate pivot irrigation control system. In *Proceedings of the Fourth European Conference on Precision Agriculture,* eds. J. Stafford and A. Werner, Wageningen, The Netherlands: Wageninger, pp. 539-544.

Perry, C.D., S. Pocknee, O. Hansen, C. Kvien, G. Vellidis, and E. Hart. 2002. Development and testing of a variable-rate pivot irrigation control system. ASAE Paper No. 02-2290. St. Joseph, MI: ASAE.

Perry, C.D., S. Pocknee, E. Hart, G. Vellidis, D. Thomas, N. Wells, and C. Kvien. 2002. Precision pivot irrigation. In *Proceedings of the 6th International Conference on Precision Agriculture,* Minneapolis, MN: ASA-CSSA-SSSA, pp. 969-983.

Phene, C.J., T.A. Howell, and M.J. Sikorski. 1985. Traveling trickle irrigation, a new approach to irrigation. In *Advances in Irrigation,* Vol. 13, ed. D. Hillel, New York: Academic Press, Inc., pp. 1-49.

Pierce, F., J. Chavez, T. Elliot, G. Matthews, R. Evans, and J. Kim (2005). An integrated wireless approach for remote, real-time irrigation monitoring and control. Paper No 323-3. Salt Lake City, UT: ASA-CSSA-SSSA.

Postel, S., P. Polak, F. Gonzales, and J. Keller. 2001. Drip irrigation for small farmers: A new initiative to alleviate hunger and poverty. *Water International* 26(1):3-13.

Roth, R.L. and B.R. Gardner. 1989. Modified self-moving irrigation system for water-nitrogen crop production. *Applied Engineering in Agriculture* 5:175-179.

Sadler, E.J., P.J. Bauer, and W.J. Busscher. 2000. Site-specific analysis of a droughted corn crop. I. Growth and grain yield. *Agronomy Journal* 92:395-402.

Sadler, E.J., W.J. Busscher, and D.L. Karlen. 1995. Chapter 11. Site-specific yield histories on a SE Coastal Plain field. In *Site-Specific Management of Agricultural Systems,* eds. P.C. Robert, R.H. Rust, and W.E. Larson, Madison, WI: ASA-CSSA-SSSA, pp. 153-166.

Sadler, E.J., C.R. Camp, D.E. Evans, and J.A. Millen. 2002. Corn canopy temperatures measured with a moving infrared thermometer array. *Transactions of the ASAE* 45(3):581-591.

Sadler, E.J., C.R. Camp, D.E. Evans, and L.J. Usrey. 1996. A site-specific center pivot irrigation system for highly-variable Coastal Plain soils. In *Proceedings of the 3rd International Conference on Precision Agriculture,* eds. P.C. Robert, R.H. Rust, and W.E. Larson, Madison, WI: ASA-CSSA-SSSA, pp. 827-834.

Sadler, E.J., C.R. Camp, D.E. Evans, and L.J. Usrey. 2001. Variable-rate, digitally-controlled fluid metering device. U.S. Patent No. 6,293,429. September 25, 2001.

Sadler, E.J., R.G. Evans, G.W. Buchleiter, B.A. King, and C.R. Camp. 2000. Design considerations for site specific irrigation. In *Proceedings of the 4th Decennial National Irrigation Symposium,* eds. R. G. Evans, B. L. Benham, and T. P. Trooien, St. Joseph, MI: ASAE, pp. 304-315.

Wall, R.W., B.A. King, and I.R. McCann. 1996. Center-pivot irrigation system control and data communications network for real-time variable water application. In *Proceedings of the 3rd International Conference on Precision Agriculture,* eds. P.C. Robert, R.H. Rust and W.E. Larson, Madison, WI: ASA-CSSA-SSSA, pp. 757-766.

Wanjura, D.F., D.R. Upchurch, and J.R. Mahan. 1992. Automated irrigation based on threshold canopy temperature. *Transactions of the ASAE* 35(5):1411-1417.

Watkins, K.B., Y.C. Lu, and W.Y. Huang. 1999. Economic returns and environmental impacts of variable rate nitrogen fertilizer and water applications. In *Proceedings of the 4th International Conference on Precision Agriculture,* eds. P.C. Robert, R.H. Rust, and W.E. Larson, Madison, WI: ASA-CSSA-SSSA, pp. 1667-1679.

Chapter 6

# Site-Specific Weed Management

Roland Gerhards
Svend Christensen

## *INTRODUCTION*

Farmers have increased post-emergent herbicide use since the 1960s, thereby replacing mechanical weeding methods and prophylactic practices according to the principles of integrated pest management. The use of herbicides, however, has not eliminated weed problems in agriculture (Haas and Streibig, 1982; Firbank and Watkinson, 1986) but has changed the weed composition and reduced the total population density (Chancellor, 1977). The frequency of herbicide-resistant or tolerant biotypes of weeds and the demands for lower herbicide residues in food, water, and soil have all increased since the 1970s, further reinforcing the need to develop weed management systems that minimize the use of herbicides. These concerns have given impetus in Europe to research and development into weed control measures that can reduce pesticide use. Site-specific weed management is an option with positive effects on the environment and economy. With a large within-field variation in weed occurrence, patch spraying based on the need for weed control may reduce treatment cost as well as herbicidal loading to the environment.

In the past two centuries, field sizes have changed dramatically in many countries. Figure 6.1 shows a 62 ha field from a large Danish farm, which 200 years ago consisted of 22 subfields partitioned presumably on the basis of spatial variation in soil fertility and drainage. The poor fields were used as grassland whereas cereals were grown on the fertile parts. Further amalgamation occurred 100 years ago, reducing the number of subfields to five. The entire area is now managed as one field, although large within-field variation still exists. The development of Differential Global Positioning Systems (DGPS) and yield-monitoring technologies in the late 1980s inspired several scientists, advisors, and farmers to study the spatial variation

*Handbook of Precision Agriculture*

doi:10.1300/5627_06

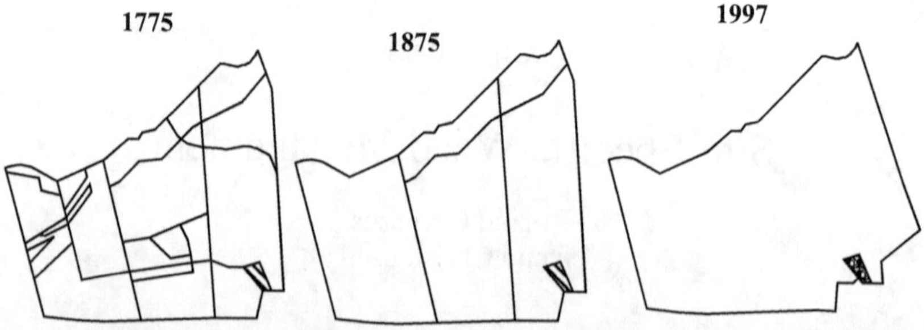


FIGURE 6.1. The history of increasing field size during the past two centuries of a 62 ha field at Bramstrup, Odense, Denmark.

of crop and soil parameters in conjunction with weed occurrence and density within and between fields.

This chapter gives an overview of spatial and temporal weed studies and research and development on spatial weed management and automatic systems for weed detection.

## *WEED DISTRIBUTION IN THE FIELD*

Weed seedling distributions have been found to be spatially and temporally heterogeneous within agricultural fields. Weeds often occur in aggregated patches of varying size or in stripes along the direction of cultivation (Marshall, 1988; Thornton et al., 1990; Wiles et al., 1992; Cardina et al., 1995; Johnson et al., 1996; Gerhards, Sökefeld, et al., 1997; Christensen and Heisel, 1998) (see Figure 6.2). In certain areas of the United States, weeds were found along the field borders and progressed inward. The variation in weed seedling populations has so far been ignored for weed management decisions since techniques to assess the weed seedling distribution in acceptable time have not been available. In most studies, weed seedling distribution in fields was assessed manually prior to postemergent herbicide application at the intersection points of a regular sampling grid that was established in the field (Clay et al., 1999). Irregular sampling grids were used when the person counting the weeds documented the GPS coordinates of the sampling points. The square grid size has varied from a few meters up to approximately 50 m, and depended on the width of the spray boom used for site-specific herbicide application. The density of emerged weed seedlings has been counted prior to and after postemergence herbicide application.

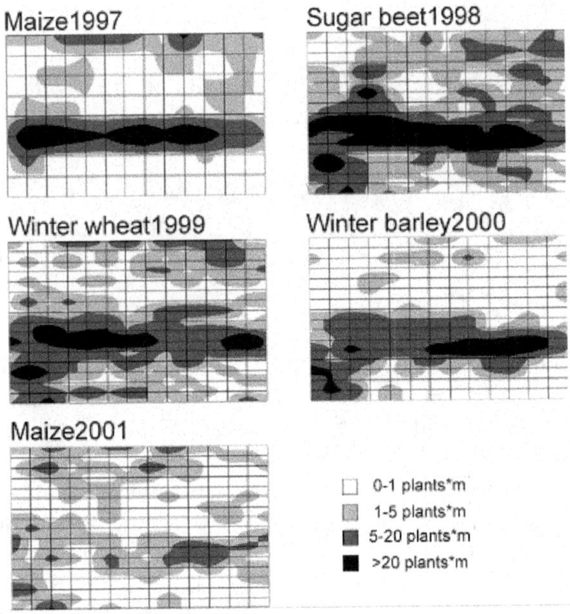


FIGURE 6.2. Distribution of blackgrass *(Alopecurus myosuroides)* prior to post-emergence weed control in a 2.3 ha arable field at Dikopshof Research Station near Bonn, Germany. *Source:* Adapted from Krohmann et al., 2002.

This enabled one to calculate the efficacy of weed control. Weed seedlings were counted in a sampling frame of usually less than one 1 $m^2$ placed at all grid intersection points.

Different mapping methods have been applied to characterize spatial distribution of weeds and soil parameters within fields. Generated maps differ depending on the interpolation method that is applied and the sample spacing (Dille et al., 2003). Johnson et al. (1995) used geostatistical methods to quantify spatial dependence of weed seedling populations. Kriging methods were then applied to estimate and map weed density at unsampled positions in the field. Gerhards, Sökefeld, et al. (1997) found that a triangulation interpolation method was more accurate than ordinary kriging to characterize weed seedling populations with a directional pattern. This method overcomes the problem of discontinuities between adjacent sampling points that result from grid sampling. A plane is fitted through three sampling points that surround the point being estimated. The equation of the plane can be expressed as

$$Z = ax + by + c \tag{6.1}$$

where $Z$ is the observed seedling density, $x$ and $y$ are the coordinates in east-west and north-south directions, and $a$, $b$, and $c$ are parameters that can be calculated by solving the following system of equations:

$$Z_1 = ax_1 + by_1 + c$$
$$Z_2 = ax_2 + by_2 + c$$
$$Z_3 = ax_3 + by_3 + c$$

Starting with the density count at the first x, y-intersection, weed density is calculated for each pixel between the sampling points. This interpolation method differs from ordinary kriging, in that equal weight is given to all sampling points (Isaaks and Srivastava, 1989).

Interpolated weed maps were reclassified based on weed infestation levels (Gerhards, Sökefeld, et al., 1997) using a geographic information system. Density classes were equal for all species in this study to facilitate the analysis of overlay maps. A weed treatment map was created to provide a decision rule for the patch sprayer. The decision rule was based on the growth stage of the occurring weed species and their competitiveness. One approach to assess the competitiveness of weed species was by yield-loss functions (Cousens, 1985). A different approach used competitive indices for the weed species observed (Pallut, 1992).

## *STABILITY OF WEED POPULATIONS*

Few studies have attempted to quantify spatial stability of weed patches in agricultural fields. If weed patches are consistent in density and location over years, maps from one year could be used to direct sampling plans and to regulate weed control methods in subsequent years. Wilson and Brain (1991) found that the pattern of blackgrass (*Alopecurus myosuroides* Huds.) patches persisted during a ten-year study. Persistence of patches was attributed to the poor ability of blackgrass to colonize new locations when effective herbicides were applied. The pattern of blackgrass patches was most stable in fields planted to cereals. Pester et al. (1995) observed significant stability for velvetleaf (*Abutilon theophrasti* Medik.) populations using Pearson, Spearman rank, and chi-square correlation analysis to quantify year-by-year relationships between weed density at individual x, y-coordinates of a sampling grid in four fields. Walter (1996) also used the chi-square correlation method and found that field violet (*Viola arvensis* Murr.), common lambsquarters (*Chenopodium album* L.), and prostrate knotweed (*Polygonum aviculare* L.) distributions were stable in cereal grain fields over three years.

Gerhards, Wyse-Pester, et al. (1997) studied the spatial stability of velvetleaf, hemp dogbane (*Apocynum cannabinum* L.), common sunflower (*Helianthus annuus* L.), yellow foxtail (*Setaria glauca* L.), and green foxtail (*Setaria viridis* L.) over four years (1992-1995) in two fields in eastern Nebraska. The first field was planted to soybean in 1992 and corn in 1993, 1994, and 1995. The second field was planted to corn in 1992 and 1994, and soybean in 1993 and 1995. Weed density was counted prior to postemergence herbicide application at approximately 800 locations per year in each field on a regular 7 m grid. The same locations were sampled every year. Weed density at locations between the sample sites was determined by linear triangulation interpolation. Weed seedling distribution was significantly aggregated with large areas being weed free in both fields. Common sunflower, velvetleaf, and hemp dogbane patches were very persistent in the east-west and north-south directions and in location and area over the four years in the first field. Foxtail distribution and density continuously increased in each of the four years in the first field and decreased in the second field. Foxtail species are mostly wind dispersed, velvetleaf and common sunflower have very large seeds that have mostly dropped on the soil before maize is harvested, and hemp dogbane is mostly dispersed by vegetal organs. A GIS was used to overlay maps from each year for a species. This showed that 36 percent of the sampled area was free of common sunflower, 62.5 percent was free of hemp dogbane, and 11.5 percent was free of velvetleaf in the first field, but only 1 percent was free of velvetleaf in the second field. The stability of broadleaf weed patches observed in this study suggests that weed seedling distributions mapped in one year are good predictors of future seedling distributions.

The dynamics of weed populations are influenced by farming practices, soil parameters, and climatic conditions (Firbank and Watkinson, 1986; Nordmeyer and Niemann, 1997; Mortensen et al., 1998; Dieleman et al., 2000; Timmermann et al., 2002; Jurado-Esposito et al., 2004). Still, only a few results are available to quantify those effects. An understanding of fundamental weed population biology would improve our ability to develop site-specific management decisions.

Krohmann et al. (2002) studied the dynamics of weed seedling distribution over five years in a rotation of maize, sugar beet, winter wheat, and winter barley, and in continuous maize. In one field, blackgrass was the dominating weed species. Blackgrass was controlled site-specifically in all five years based on weed distribution maps and an economic weed control threshold model. High-density blackgrass populations persisted at the same locations in the field although effective herbicides were sprayed every year. However, population density did not increase at locations where no or few grass weeds were found and thus no herbicides were applied (Figure 6.2). In a different field *Viola arvensis* and *Chenopodium album* were the dominating species. *Viola arvensis* was observed in one half of the field in 1998 maize, 1999 sugar beet, 2000 winter wheat, and 2001 winter barley. *Chenopodium album* was found in the other half of the field in 1998 maize and 1999 sugar beet, but this species did not cause any competition in winter wheat and winter barley. Therefore, weed distribution maps obtained in maize and sugar beet were suitable for site-specific weed control in winter wheat and winter barley (Figure 6.3).

Nordmeyer and Niemann (1992) found that blackgrass populations mostly occurred at locations in the field where the clay content was high. Timmermann et al. (2002) reported that the crop rotation had a long-term effect on weed density and weed species composition. In fields that had

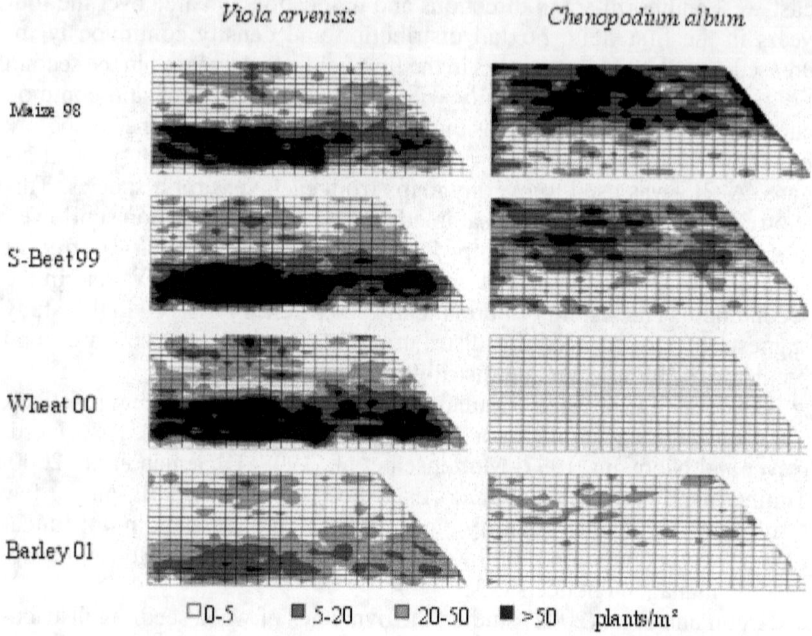


FIGURE 6.3. Distribution of *Viola arvensis* and *Chenopodium album* in maize 1998, sugar beet 1999, winter wheat 2000, and winter barley 2001 in a 5 ha arable field at Dikopshof Research Station near Bonn, Germany. *Source:* Adapted from Krohmann et al., 2002.

been planted with 50 percent maize in the rotation more than 20 years ago, the density of *Chenopodium album* was still much higher than in fields with a high percentage of winter annual grains in the rotation. Tillage practices combined with certain crop rotation had a very strong effect on the organic matter content. Fields that had been planted with potatoes were lower in organic matter content than fields where mostly grains were planted. The reason is that more intensive soil tillage is practiced in potatoes than in grains and therefore, mineralization of organic matter is increased. The difference in organic matter content again had a strong influence on weed species composition. *Galium aparine* mostly occurred in fields with high organic matter contents (Timmermann et al., 2002).

Weed population models have been applied to quantify the effects of site-specific weed management practices (Paice et al., 1997). However, the mechanism of weed patch stability is largely unknown. Some reports suggest that efficacy of weed control methods was lower in weed patches with a high density than at low-density locations (Mortensen et al., 1998). One reason for the density-dependent herbicide efficacy would be that the leaves of the weeds protect each other. Wartenberg and Dammer (2000) also found that the efficacy of postemergent herbicides was lower at high weed density locations. He assumed that this dependency is caused by the irregular distribution of spray droplets. The higher the weed density, the greater the probability that individual weeds do not receive a lethal herbicide dose. Krohmann et al. (2002) found that the persistence of weed populations was also attributed to weed seedlings that emerged after weed control methods had been applied and that survived herbicide application. Those late-emerging and surviving weeds were able to produce viable seeds in maize and sugar beet but not in winter wheat and winter barley. The authors assume that competition of the crop was higher in winter annual grains and therefore late emerging weed seedlings were suppressed.

## *WEED MONITORING*

Several monitoring concepts have been suggested for patch spraying. Bajwa and Tian (2000) showed promising results of mapping weed density (ratio of the weed-covered area to the total area) from relatively high–resolution aerial images of a crop field with an actively growing crop. Heisel et al. (1996) were mapping weeds from weed counting in grids. It is obvious, that manual weed counting is not feasible because it is time consuming. In contrast, real-time monitoring and spraying is a concept that involves simultaneous weed detection, decision-making, and spraying (Gerhards and Christensen,

2003). One advantage of the weed-mapping concept is the possibility of choosing a herbicide appropriate for treating the whole field weed infestation.

One major step toward a practical solution for site-specific weed control is the development of precise and powerful sampling techniques to automatically determine in-field variation of crop cover and weed seedling populations. Real-time optoelectronic sensors and digital image analysis systems for plant species discrimination have been investigated. Vrindts and de Baerdemaeker (1997), Biller (1998), and Moshou et al. (1998) used optoelectronic sensors to measure the reflectance in the green, red, and near-infrared light wave bands. Green leaves were characterized by a high reflectance in the green and near-infrared spectrums and a low reflectance in the red spectrum compared with the reflectance curve of a bare soil. Felton and McCloy (1992) developed a spot-spraying system for nonselective herbicides based on the information of real-time reflectance sensors. The nozzles of the sprayer were turned on automatically when the normalized difference vegetation index (NDVI) exceeded a set threshold indicating a higher proportion of green vegetation in the field of view of the sensor.

Gerhards et al. (1998), Chapron et al. (1999), and Sökefeld et al. (2000) used digital image analysis systems to identify plant species based on characteristic shape-, color-, and texture-features for each individual object in the image. Those systems were suitable for real-time weed and crop identification and allowed for site-specific application of selective herbicides in various crops.

For automatic weed detection, Gerhards and Sökefeld (2001) mounted three digital cameras in the front of the sprayer. Each camera was equipped with a near-infrared band pass filter (780 nm-1150 nm). Under cloudy conditions, the reflectance of green plants in the near-infrared spectrum was higher than the reflectance of the soil. The cameras were triggered with an exposure time of 1/4000 s to achieve well-focused images at a speed of 7 km/h. Approximately every 2 m a set of three images was taken and stored on an onboard computer of the vehicle together with the DGPS coordinates of the picture. A histogram of the gray-level intensity was calculated for all images. A gray-level threshold was set automatically between the peak of dark pixels representing the soil and the peak of light pixels representing the plants. In a binary picture, the soil was displayed in black and the plants in white. In the next step, the contours of all white objects in the picture were extracted (Figure 6.4). Objects that were smaller or bigger than plants were automatically removed from the image-setting thresholds for the contour length. The contour of plants was then transformed into a chain code of standard vectors from pixel to pixel. Each vector was represented by a number of 0 to 7 indicating the orientation of the vector in relation to the previous pixel. The chain code was already characteristic of the plant species in

FIGURE 6.4. Original image, binary picture, and contours of maize and *Chenopodium album* obtained from a monochrome CCD-shutter camera with a near-infrared band pass filter under cloudy conditions and dry soil. *Source:* Adapted from Gerhards and Sökefeld, 2001.

the image but it was still dependent on size, rotation, and position of the plant within the picture. Therefore, the chain code was transformed into a function with a standardized contour length of $2\pi$ on the abscissa and the variation of the angle on the ordinate. The result of this transformation was that the contours of the plants were independent of the height of the camera above the ground (zooming invariance). A Fourier transformation was then applied to obtain characteristic parameters of the plant contours that were also independent of rotation and position of the plant within the image. It was found that the amplitudes of the Fourier function were most suitable to describe the shape of weed seedlings. In addition to the Fourier descriptors, two geometric parameters (compactness and the quotient of minimum Ferrets diameter and maximum Ferrets diameter) and the area of the plants were calculated.

Both sensor systems, the optoelectronic systems and the image analysis system have been successfully tested under field conditions. Optoelectronic systems have been applied to control weeds precisely and in real-time on fallow land, in urban areas, along rail tracks, and between crop rows. Image analysis systems are currently being tested to identify weeds in cereals, sugar beet, and maize. However, both systems need further adaptations. One problem is the variation of soil reflection and sunlight intensity, which can result in misclassification of weed and crop. Consequently, the sensors have to be protected against direct sunlight or multispectral cameras have to be used to increase the contrast between soil and green plants. A second problem arises from overlapping leaves of different plant species, which makes the classification more difficult. Image analysis algorithms are currently being used to identify plant species with overlapping leaves (Chapron et al., 1999).

Sensor data can be used for real-time control of the sprayer or can be taken to create weed distribution and application maps (Mortensen et al.,

1998; Sökefeld et al., 2000). Combinations of real-time sensors and historic weed maps have also been investigated (Christensen and Heisel, 1998).

## DECISION MAKING

Computerized decision support systems offer an ideal means of achieving economical-and environmentally safe weed management. Christiansen et al. (2003) used a decision algorithm for patch spraying broad-leaved weeds in cereals (DAPS). Weed species competitiveness, weed density, crop and herbicide price, and dose-response parameters are included in the decision algorithm to find the most economic herbicide dose. Kriging and a GIS are used to produce a treatment map. A preliminary trial with site-specific weed control using DAPS reduced herbicide input by 41 percent compared with whole-field spraying with PC-Plant Protection recommendation (Heisel et al., 1997).

The long-term effect of using DAPS and three other decision algorithms was evaluated with a population model that simulated blackgrass propagation and dispersal in space and time. A population of blackgrass in a subfield measuring 140 m × 140 m was used as the starting point of the model. This area was divided into cells of 1 $m^2$ and the population model was employed in each cell using the three decision algorithms. The model simulated the density of blackgrass in each of the different stages of its annual lifecycle. The dispersal of seeds to and from the neighboring cells was also simulated, i.e., the model kept track of the "import" and "export" of seeds between the cells. Model simulations over eight years (Figures 6.5 through 6.7) showed that site-specific control of blackgrass using an economic threshold model leads to the greatest variation in plant density as well as propagation into areas where only a few plants existed in the first year. There was a tendency over the years for overall herbicide use to increase when using the economic threshold model. A uniform dose recommended by the Danish advisory system PC-Plant Protection (Rydahl and Thonke, 1993) reduced the stand of blackgrass, though not enough to prevent yield loss following control. Over a period of eight years the decision algorithm DAPS resulted in the lowest treatment index as well as the lowest yield loss following control and the lowest seed bank.

## SITE-SPECIFIC HERBICIDE APPLICATION

A few sprayers have been developed to apply herbicides intermittently (Stafford and Miller, 1993; Paice et al., 1995; Gerhards et al., 1997; Chris-

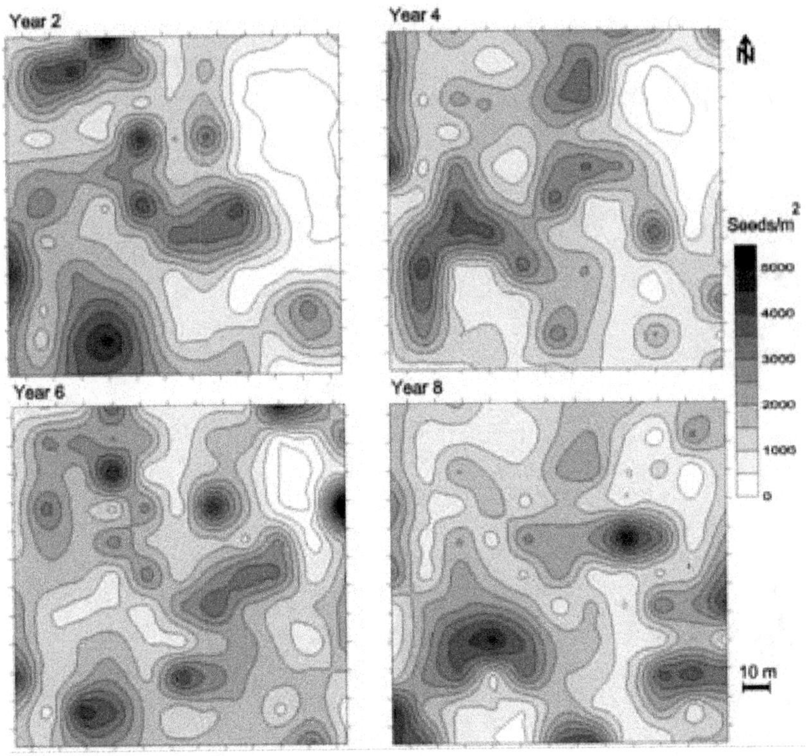


FIGURE 6.5. The seed bank of blackgrass in years 2, 4, 6, and 8 in a field treated with a normal dose of Isoproturon herbicide as determined by the threshold model.

tensen and Heisel, 1998). Selective control of small sections of the spray boom is needed to treat highly aggregated weed patches. In addition, the herbicide dosage needs to be varied according to the size of weeds and the species occurring in the field. Herbicide dosage is mostly regulated by the pressure in the hydraulic system. With standard nozzles, the pressure can only be varied in a limited range without affecting the droplet size and distribution. Air-assisted nozzles allow a variation of the herbicide rate by approximately 3+/- 50 percent of the standard rate, which for many cases is sufficient in plant protection (Audenaert, 2001). The variation does not significantly affect the droplet size and distribution.

A basic requirement for patch sprayers is the variation of herbicide mixture or type of herbicide according to the current weed population and density. So far, a practical system for real-time herbicide variation has not been

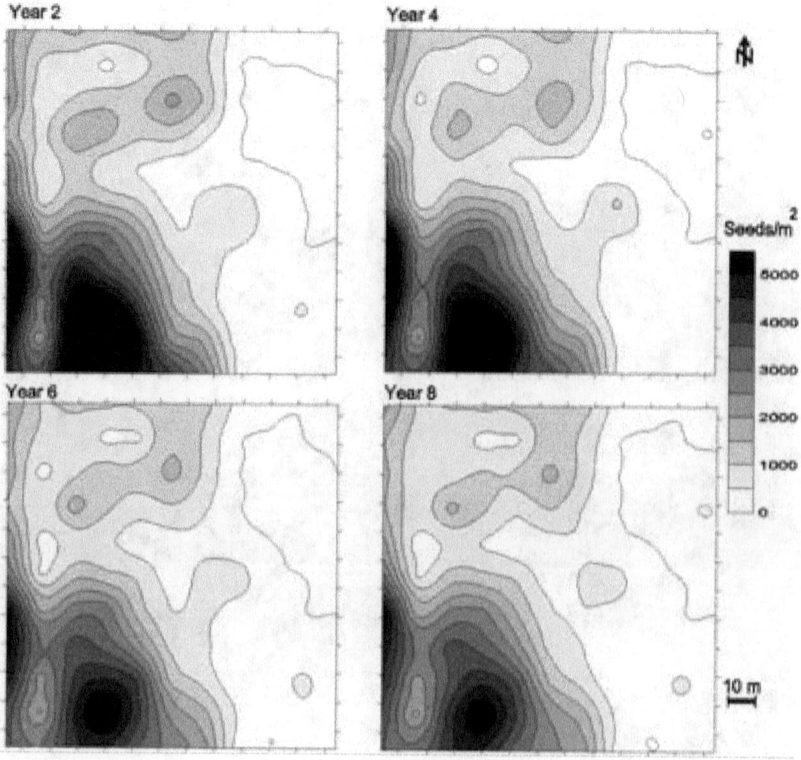


FIGURE 6.6. The seed bank of blackgrass in years 2, 4, 6, and 8 in a field treated with Isoproturon herbicide as recommended by PC-Plant Protection.

found. Several companies have developed direct injection systems that inject concentrated pesticide solutions in a water stream (MSR Agroinject, Dosatron, Öko-Lok, Biotronic, MidTech, and Dos-Intro). Chemical injection pumps have mostly been placed in front of the carrier pump. Therefore, a reaction time of several seconds was required until the pesticide mixture reached the nozzles. Currently, injection systems for sprayers are being investigated that apply concentrated pesticide solutions directly into the nozzles of the sprayer so as to allow online variation of herbicide type and dosage.

Gerhards et al. (1998) developed with Rau-Company in Germany an experimental patch sprayer with a 15 m boom divided into five sections of 3 m each (Figure 6.8). Each section was separately turned on and off from a control unit via solenoid valves. The herbicide dose for the full spray boom was

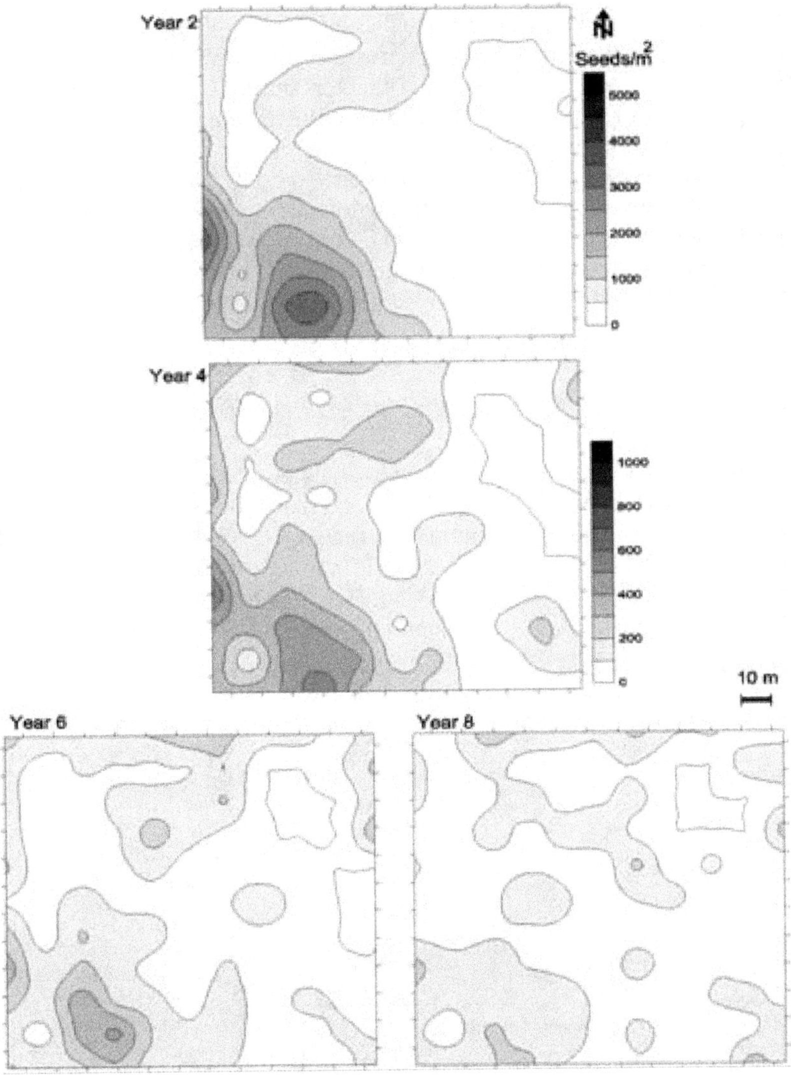


FIGURE 6.7. The seed bank of blackgrass in a field treated with site-specific herbicide doses calculated using DAPS in years 2, 4, 6, and 8.

regulated by the same control unit via a spray computer. The herbicide rate was varied by changing the pressure in the system. The full rate was applied with 300 L water per ha. The nozzles allowed a reduction down to 200 $Lha^{-1}$ with almost constant distribution of droplet size. The main hydraulic circuit on the sprayer was similar to that on a conventional sprayer with an output from the main pump fed to a pressure control valve which regulated the rate that was set with the spray computer. During herbicide application, the spray control system was linked to an onboard computer loaded with the weed treatment map. A GPS in the differential mode was used for real-time location of the patch sprayer. The onboard computer compared the actual position of the sprayer with the information in the weed treatment map and signals were transmitted to the control unit via a data bus to open each individual solenoid valve when herbicide application was warranted. In the same way, the herbicide dose was adjusted to the recommended rate from the treatment map (Figure 6.8).

## SITE-SPECIFIC WEED CONTROL

Some studies have been conducted to apply postemergence herbicides in winter wheat, winter barley, sugar beet, and maize based on georeferenced maps of weed seedling distribution (Gerhards, Sökefeld, et al., 1997; Williams et al., 2000; Tian et al., 1999). In winter wheat and winter barley, weed distribution maps for grass weeds, *Galium aparine,* and other broadleaf species were plotted and an economic weed threshold model (Gerhards

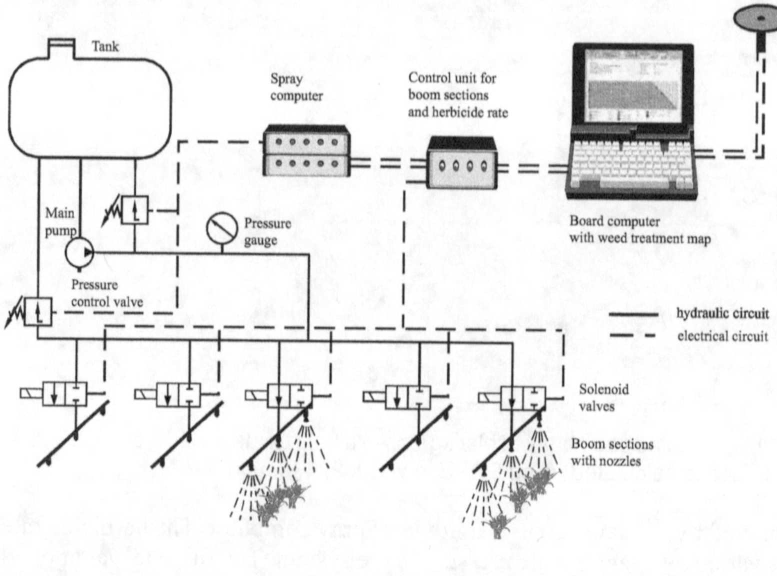


FIGURE 6.8. Sprayer control system for site-specific herbicide application. *Source:* Adapted from Gerhards et al., 1998.

and Kühbauch, 1993) was applied to characterize areas in all three maps where herbicide application was warranted. A GIS was used to overlay all three binary maps in one weed treatment map. This map was divided into 3 m strips, since each 3 m section of the spray boom could be controlled separately. The total width of the spray boom was 15 m.

In sugar beet and maize, different decision algorithms were used for site-specific herbicide application (LIZ, 1994; Williams et al., 2000). Because of the wider spacing between crop rows, weed competition is higher in these crops than in winter grains. Therefore, the sprayer was turned off only at locations where no or very few weeds were observed. The herbicide rate in sugar beet and maize was reduced compared with conventional broadcast applications by adjusting the herbicide dose to the spatial variability of weed species composition and weed density (Williams et al., 2000). In winter wheat, an economic threshold model (Gerhards and Kühbauch, 1993) linked control decision algorithms with weed maps. Herbicide use with this map-based approach was reduced some 40 to 50 percent. In maize, site-specific herbicide applications reduced herbicide load by 11.5 percent for broadleaf-weeds and 98 percent for grass weeds (Williams et al., 2000; Goudy, 2001) (Figure 6.9).

## *FUTURE DIRECTIONS*

Site-specific weed management in agricultural crops offers a great potential to reduce herbicide use. This will increase the acceptance of precision farming practices in the society. In addition, site-specific weed management will also improve our knowledge on the spatial and temporal dynamics of weed populations and on the efficacy of weed control methods. In some studies, it was shown that even the application of effective herbicides in every year did not reduce population density in weed patches, suggesting that a combination of chemical, mechanical, and cultural weed control method would be beneficial at these locations. On the other hand, several studies on site-specific weed control have proved that herbicide input at locations with no or few weeds can be reduced without causing any weed problem in the following years.

Weed monitoring systems are a critical component in the utilization of the ideas and knowledge developed in research projects on site-specific weed control. Novel systems have been developed but there is still a need for improvements. Cost-effective methodologies that combine sensor-based and manual weed surveying may also be a direction for mapping permanent and semipermanent weed management zones, e.g., mapping areas with consistently high weed pressure, areas with consistently low weed pressure

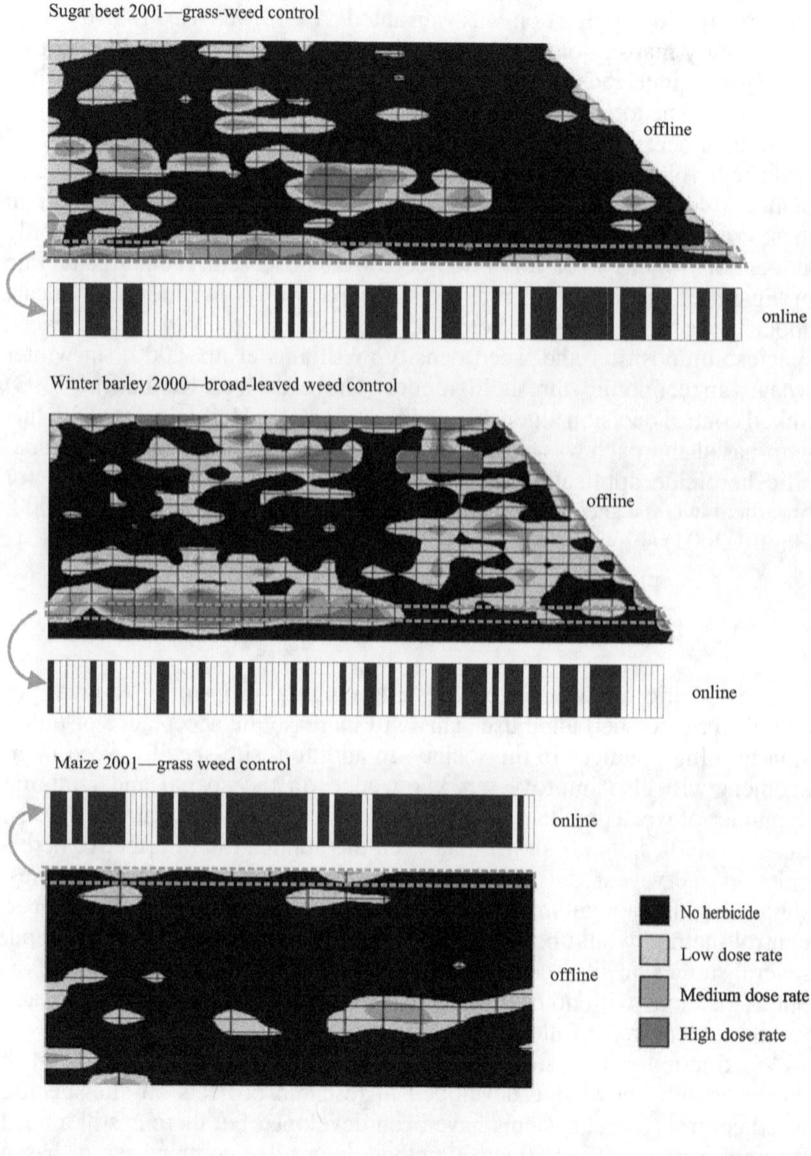


FIGURE 6.9. Application maps for site-specific control of grass weeds and broadleaf weeds in sugar beet, winter barley, and maize. *Source:* Adapted from Gerhards and Sökefeld, 2001.

or areas with unstable weed pressure. Knowledge about the temporal and spatial stability of weed patches may be used to help achieve cost-effective weed mapping.

New technologies and demands for reduced agro-chemical input to benefit the environment and the farm economy have given impetus to research and development of advanced weed control measures, including weed sensing, decision making, and weed control. One of the promising technologies for site-specific weed management is robotic weeding. The introduction of small autonomous vehicles (Marchant et al., 1997; Bak and Jakobsen, 2004) makes it possible to integrate highly accurate identification and positioning of crop and weed seedlings with computer vision (Soegaard, 2004) and highly accurate devices for mechanical and chemical targeting of weed seedlings (Griepentrog et al., 2004). Lee et al. (1999) designed a micro-spray system for applying herbicides on weeds on a centimeter scale in an effort to avoid spraging herbicide on the crop plants and the soil. The precision in targeting the single-weed seedlings was investigated by Søgaard and Lund (2005), and it was documented that the spray liquid was applied at subcentimeter accuracy. This investigation also showed potentials for even higher accuracy. Graglia (2004) investigated the biological efficacy in laboratory by treatment of *Solanum nigrum* L. It was concluded that it was possible to obtain an efficacy of 95 percent by applying 1.0 μg glyphosate per plant. Development of more precise systems working at higher treatment accuracy levels and the possibility of using sensors and data-sampling systems may lead to the implementation of traceability systems in agriculture.

## REFERENCES

Audenaert, J. 2001. Applikationstechnik. In *Precision Farming in Pflanzenschutz*, ed. M. Kunisch, Darmstadt, Germany: KTBL 402, pp. 22-24.

Bajwa S. G. and L.F. Tian. 2001. Aerial CIR remote sensing for weed density mapping in a soybean field. *Transactions of the ASAE* 44(6):1965-1974.

Bak, T. & H. Jakobsen, 2004. Agricultural robotic platform with four-wheel steering for weed detection. *Biosystems Engineering* 87(2):125-136.

Biller, R.H. 1998. Pflanzenunterscheidung und gezielter Einsatz von Herbiziden, *Forschungs - Report 1*. 34p.

Cardina, J., D.H. Sparrow, and E.L. McCoy. 1995. Analysis of spatial distribution of common lambsquarters *(Chenopodium album)* in no-till soybean *(Glycine max). Weed Science* 43:258-268.

Chancellor, R.J. 1977. A preliminary survey of arable weeds in Britain. *Weed Research* 17:283-287.

Chapron, M., M. Requena-Esteso, P. Boissard, and L. Assemat. 1999. A method for recognizing vegetal species from multispectral images. In *Precision Agricul-*

*ture1999: 2nd European Conference on Precision Agriculture,* Part 1. ed. J.V. Stafford, Sheffield, UK: Sheffield Academic Press, pp. 239-248.

Christensen, S. and T. Heisel. 1998. Patch spraying using historical, manual, and real-time monitoring of weeds in cereals. *Zeitschrift für Pflanzenkrankheiten und Pflanzenschutz, Sonderheft XVI* pp. 257-263.

Christensen, S., T. Heisel, and M. Paice. 1999. Simulation of long-term *Alopecurus myosuroides* population using three patch spraying strategies. In *Proceedings of the 2nd European Conference on Precision Agriculture,* Part 2, ed. J.V Stafford, Sheffield, UK: Sheffield Academic Press, pp. 977-987.

Christensen, S., T. Heisel, A.M. Walter, and E. Graglia, 2003. A decision algorithm for patch spraying. *Weed Research* 43(4):276-284.

Clay, S.A., G.J. Lems, D.E. Clay, F. Forcella, M. M. Ellsbury, and C.G. Carlson. 1999. Sampling weed spatial variability on a field-wide scale. *Weed Science* 47:674-681.

Cousens, R. 1985. A simple model relating yield loss to weed density. *Annals of Applied Botany* 107:239-252.

Dieleman, J.A., D.A. Mortensen, D.D. Buhler, C.A. Cambardella, and T.B. Moorman. 2000. Identifying associations among site properties and weed species abundance. I. Multivariate analysis. *Weed Science* 48:567-575.

Dille, J.A., M. Milner, J.J. Groeteke, D.A. Mortensen and M.M. Williams. 2003. How good is your weed map? A comparison of spatial interpolators. *Weed Science* 51(1):44-55.

Felton, W.L. and K.R. McCloy. 1992. Spot spraying. *Agricultural Engineering* 11:9-12.

Firbank, L.G. and A.R. Watkinson. 1986. Modeling the population dynamics of an arable weed and its effects upon crop yield. *Journal of Applied Ecology* 23:147-159.

Gerhards, R. and S. Christensen. 2003. Real-time weed detection, decision making, and patch spraying in maize, sugarbeet, winter wheat, and winter barley. *Weed Research* 43:385-392.

Gerhards, R. and W. Kühbauch. 1993. Dynamisches Entscheidungsmodell zur Lenkung von Unkrautkontrollmaßnahmen in Wintergetreide mit Hilfe digitaler Bildverarbeitung. *Journal of Agronomy and Crop Science* 171:329-335.

Gerhards, R. and M. Sökefeld. 2001. Sensor systems for automatic weed detection, In *Proceedings of the British Crop Protection Conference,* Brighton, UK: British Crop Protection Council, pp. 827-834.

Gerhards, R., M. Sökefeld, and W. Kühbauch. 1998. Einsatz der digitalen Bildverarbeitung bei der teilschlagspezifischen Unkrautkontrolle. *Zeitschrift für Pflanzenkrankheiten und Pflanzenschutz,* Sonderheft XVI, pp. 273-280.

Gerhards, R., M. Sökefeld, K. Schulze-Lohne, D.A. Mortensen, and W. Kühbauch. 1997. Site specific weed control in winter wheat. *Journal of Agronomy and Crop Science* 178:219-225.

Gerhards, R., D.Y. Wyse-Pester, and D.A. Mortensen. 1997. Characterizing spatial stability of weed populations using interpolated maps. *Weed Science* 45:108-119.

Goudy, H.J., K.A. Bennett, R.B. Brown, and F.J. Tardif. 2001. Evaluation of site-specific weed management using a direct-injection sprayer. *Weed-Science* 49(3): 359-366.

Graglia E. 2004. Importance of herbicide concentration, number of droplets, and droplet size on growth of *Solanum Nigrum* L, using droplet application of glyphosate. *XIeme Colloque International Sur La Biologie Des Mauvaises Herbes*, Dijon, France.

Griepentrog, H.W., S. Christensen, H.T. Søgaard, M. Nørremark, I. Lund, and E. Graglia. 2004. Robotic weeding. AgEng 2004, Leuven, Belgium. Book of Abstracts, ISBN 90-76019-258, 904-905. Paper on CD.

Haas, H. and Streibig, J.C. 1982. Changing patterns of weed distribution as a result of herbicide use and other agronomic factors. In *Herbicide Resistance in Plants*, eds. L. Lebaron and H. Gressel, New York: John Wiley & Sons, p. 57.

Heisel, T., C. Andreasen and A. K. Ersboll. 1996. Annual weed distribution can be mapped with kriging. *Weed Research* 36(4):325-337.

Heisel, T., S. Christensen and A.M. Walter. 1997. Validation of weed patch spraying in spring barley—preliminary trial. In *1st European Conference on Precision Agriculture,* Vol. 2, ed. J. Stafford, Warwick, UK: BIOS Scientific Publishers, pp. 879-886.

Isaaks, E.H. and R.M. Srivastava. 1989. Point estimation and ordinary kriging. In *Applied Geostatistics,* eds. E.H. Isaaks and R.M. Srivastava, New York: Oxford University Press, pp. 249-322.

Johnson, G.A., D.A. Mortensen, and C.A. Gotway. 1996. Spatial and temporal analysis of weed seedling populations using geostatistics. *Weed Science* 44:704-710.

Johnson, G.A., D.A. Mortensen, and A.R. Martin. 1995. A simulation of herbicide use based on weed spatial distribution. *Weed Research* 35:197-205.

Jurado-Exposito, M., F. Lopez-Granados, J.L. Gonzalez-Andujar, and L. Garcia-Torres. 2004. Spatial and temporal analysis of *Convolvulus arvensis* L. populations over four growing seasons. *European Journal of Agronomy* 21(3):287-296.

Krohmann, P., C. Timmermann, R. Gerhards and W. Kühbauch. 2002. Ursachen für die Persistenz von Unkrautpopulationen. *Journal of Plant Diseases and Protection* Sonderheft XVIII: 261-268.

Lee, W.S., D.C. Slaughter and D.K. Giles. 1999. Robotic weed control system for tomatoes. *Precision Agriculture* 1:95-113.

LIZ (Landwirtschaftlicher Informationsdienst Zuckerrübe) 1994. *Herbizideinsatz in Zuckerrüben,* Jülich, Germany: Pfeifer & Langen, pp. 1-6.

Marchant, J.A, T. Hague, and N. Tillett. 1997. Row-following accuracy of an autonomous vision-guided agricultural vehicle. *Computers and Electronics in Agriculture* 16:165-175.

Marshall, E.J.P. 1988. Field-scale estimates of grass populations in arable land. *Weed Research* 28:191-198.

Mortensen, D.A., J.A. Dieleman, and G.A. Johnson. 1998. Weed spatial variation and weed management. In *Integrated Weed and Soil Management,* eds. J.L. Hatfield, D.D. Buhler, and B.A. Stewart, Chelsea, MI: Ann Arbor Press, pp. 293-299.

Moshou, D., B. de Ketelaere, E. Vrindts, P. Kennes, J. de Baerdemaeker, and H. Ramon. 1998. Local linear mapping neural networks for pattern recognition of plants. In *1st IFAC Workshop on Control Application and Ergonomics in Agriculture,* Athens, Greece: pp. 61-66.

Nordmeyer, H. and P. Niemann. 1992. Möglichkeiten der gezielten Teilflächenbehandlung mit Herbiziden auf Grundlage von Unkrautverteilung und Bodenvariabilität. *Zeitschrift für Pflanzenkrankheiten und Pflanzenschutz* Sonderheft XIII pp. 539-547.

Nordmeyer, H. and P. Niemann. 1997. Patchy weed control in agricultural practice. In *Proceedings of British Crop Protection Council, Weeds,* pp. 649-650. Brighton, UK: British Crop Protection Council.

Paice, M.E.R., W. Day, L.J. Rew and A. Howard. 1997. A simulation model for evaluating the concept of patch spraying. *Weed Research* 38:373-388.

Paice, M.E.R., P.C.H. Miller, and D.J. Bodle. 1995. An experimental sprayer for the spatially selective application of herbicides. *Journal of Agricultural Engineering Research* 60:100-109.

Pallut, B. 1992. Zur Verbesserung der Vorhersagegenauigkeit von unkrautbedingten Korneretragsverlusten bei Wintergetreide. *Zeitschrift für Pflanzenkrankheiten und Pflanzenschutz* Sonderheft XIII:129-137.

Pester, D. Y., D.A. Mortensen and C.A. Gotway. 1995. Statistical methods to quantify spatial stability of weed populations. *North Central Weed Science Society* 50:152.

Rydahl, P. and K.E. Thonke. 1993. PC-plant protection: Optimizing chemical weed control. *Bulletin OEPP/EPPO Bulletin* 23:589-594.

Søgaard, H.T. 2004. Weed classification by active shape models. *Biosystems Engineering* 91(3):271-181.

Søgaard, H.T. and I. Lund. 2005. Investigation of the accuracy of a machine vision based robotic micro-spray system. Proc. 5th ECPA and 2nd ECPLF, Uppsala, Sweden. Wageningen Academic Publishers, ISBN 9076998698, pp. 613-619.

Sökefeld, M., R. Gerhards, and W. Kühbauch. 2000. Teilschlagspezifische Unkrautkontrolle–von der Unkrauterfassung bis zur Herbizidapplikation. *Zeitschrift für Pflanzenkrankheiten und Pflanzenschutz* Sonderheft XVII:227-232.

Stafford, J.V. and P.C.H. Miller. 1993. Spatially selective application of herbicide to cereal crops. *Computers and Electronics in Agriculture* 9:217-229.

Thornton, P. K., R. H. Fawcett, J. B. Dent, and T. J. Perkins. 1990. Spatial weed distribution and economic thresholds for weed control. *Crop Protection* 9:337-342.

Tian, L., J.F. Reid, and J.W. Hummel. 1999. Development of a precision sprayer for site-specific weed management. *Transactions of the American Society of Agricultural Engineering* 42:893-900.

Timmermann, C., R. Gerhards, and W. Kühbauch. 2002. Ursachen für Ertragsunterschiede in Ackerschlägen. *Journal of Agronomy and Crop Science* 187:1-9.

Vrindts, E. and J. de Baerdemaeker. 1997. Optical discrimination of crop, weed, and soil for on-line weed detection. In *Precision Agriculture 1997: 1st European Conference on Precision Agriculture,* Vol. 2, ed. J. Stafford, Warwick, UK: BIOS Scientific Publishers, pp. 537-544.

Walter, M. 1996. Temporal and spatial stability of weeds. In *Proceedings of 2nd International Weed Congress,* ed. H. Brown, Sheffield, UK: Sheffield Academic Press, pp. 125-130.

Wartenberg, G. and K.-H. Dammer. 2000. Unkrauterkennung im Echtzeitbetrieb—Möglichkeiten und Probleme. *Zeitschrift für Pflanzenkrankheiten und Pflanzenschutz* Sonderheft XVII:187-194.

Wiles, L.J., G.W. Oliver, A.C. York, H.J. Gold, and G.G. Wilkerson. 1992. Spatial distribution of broadleaf weeds in North Carolina soybean *(Glycine max)* fields. *Weed Science* 40:554-557.

Williams, M.M. II, R. Gerhards, and D.A. Mortensen. 2000. Two-year weed seedling population responses to a post-emergent method of site-specific weed management. *Precision Agriculture* 2:247-263.

Wilson, B.J. and P. Brain. 1991. Long-term stability of distribution of *Alopecurus myosuroides* Huds. within cereal fields. *Weed Research* 31:367-373.

Chapter 7

# Site-Specific Management of Crop Diseases

Karsten D. Bjerre
Lise N. Jørgensen
Jørgen E. Olesen

## *INTRODUCTION*

Plant pathogens are estimated to cause yield reductions of almost 20 percent in the principal food and cash crops worldwide (Oerke et al., 1994). Yield losses depend very much on climatic conditions and the susceptibility of the varieties grown, but crop rotation and other sanitation practices can also play a major role in reducing inoculum and delaying epidemic development. Fungicides are considered to be essential for the effective control of plant diseases in many crops, and today it is common practice to control diseases based on assessments of whether an attack of economic importance will develop.

Yield reductions due to disease attack can vary depending on the crop grown. Early and persistent attack on the vegetative parts of crops can be critical for growing a healthy crop, but attack at later development stages can also cause significant yield reductions and a crop might loose its market value. Fungicides control diseases and reduce the impact from severe disease attack; they stabilize yields and ensure better quality from the treated crops. In order to feed the earth's increasing population, it is important to use fungicides to avoid large yield fluctuations and to minimize the need for increasing the cultivated area, which might lead to further erosion, salinity, and desertification. Fungicides are also known in several cases to improve food quality and safety by controlling fungi that produce mycotoxins (Schneider and Dickert, 1994; Pohland, 1993).

General public concern regarding the environmental impact from the use of pesticides has put focus on developing crop protection agents with a

*Handbook of Precision Agriculture*

doi:10.1300/5627_07

favorable environmental profile and low toxicity to humans and wildlife. In addition, reduction schemes for pesticide use have been introduced in several countries, including Denmark, Sweden, the Netherlands, and Germany (Jørgensen, 1997; Emmerman, 1997; van Alphen, 1997; Anonymous, 2004). This adds continuous pressure on researchers to develop methods that can help to minimize the use of pesticides. Precision farming is believed to have the potential for reducing pesticide input by optimizing the targeting of the pesticides applied, and in addition may provide documentation of application for improved traceability.

As a form of precision agriculture, site-specific crop management may be defined as "Matching resource application and agronomic practices with soil and crop requirements as they vary in space and time within a field" (Whelan and McBratney, 2000, p. 265). These authors noted that much attention has been paid to technological tools for precision agriculture at present, but that the concept of site-specific crop management will only be accepted by the farming industry based on experimental evidence indicating economic benefits from applying such a system. Research on site-specific disease management of arable crops is in its infancy, and until now only a small number of investigations have been published.

This chapter mainly focuses on site-specific management of leaf pathogens in cereals from a North European perspective using fungicides. Specifically, experiences from winter wheat will be presented, as this crop in this region is a high-yielding crop that typically yields between seven to nine tonnes per ha and receives one to three fungicide applications. Other control strategies using site-specific cultivars, variation in seeding rates, and fertilizer rates can reduce crop susceptibility to diseases, and there is a potential for exploiting this in site-specific management, but few studies have so far been presented.

### *The World Market for Fungicides*

Fungicides have become increasingly important as the intensity of crop production has grown. This has been assisted by development of new, effective, and often systemic fungicides. In 1995, fruits and vegetables accounted for about 40 percent, cereals for 25 percent, rice for 15 percent, and potatoes for approximately 5 percent of the global market value (Oerke, 1999). East Asia accounts for more than 35 percent of the total input in fruits and vegetables and West Europe for 30 percent, followed by North and South America totaling 25 percent. The cereal fungicide market is dominated by Western Europe, where about 80 percent of fungicides used in cereals are used on an area constituting only 10 percent of the total cereal area. Similarly, a very large portion of the rice fungicides are used in East Asia, especially Japan. Summarizing all crops, approximately 80 percent of all fungicides are used in West Europe or East Asia, followed by North America, Latin America, and East Europe with shares of 10, 8, and 3 percent, respectively (Oerke, 1999).

Based on the market potential of fungicides in all crops, leaf spot diseases account for about 30 percent of all diseases of economic impact. The downy mildews (25 percent), cereal diseases (20 percent), rice diseases (12 percent), and fruit crop diseases (10 percent) are the major reasons for which fungicides are applied (Schwinn, 1992).

The importance of wheat diseases in different regions of the world varies considerably (Table 7.1). In Europe leaf diseases such as rust, Septoria leaf and glume blotch (*Septoria tritici,* teleomorph: *Mycosphaerella graminicola,* and *Stagonospora nodorum,* syn: *Septoria nodorum,* teleomorph: *Phaeosphaeria nodorum*), and powdery mildew *(Blumeria graminis)* dominate and trigger off most of the treatments. These diseases are caused by polycyclic pathogens with wind or splash dispersal. Depending on the weather and cultivars grown, one to three fungicide treatments may be needed in many wheat crops in Western Europe. In other regions of the world seedborne diseases such as bunt have a major importance due to lack of certification schemes for seeds and use of seed treatments.

TABLE 7.1. Ranking of diseases in wheat in order of importance, by region.

| Ranking | Europe | North America | East Asia | Rest of the world |
|---|---|---|---|---|
| 1 | Rust (*Puccinia* spp.) | Rust (*Puccinia* spp.) | Bunts (*Tilletia* spp.) | Bunts (*Tilletia* spp.) |
| 2 | Leaf and glume blotch (*Septoria tritici, Stagonospora nodorum*) | Common root rot (*Cochliobolus sativus, Fusarium* spp.) | Rust (*Puccinia* spp.) | Rust (*Puccinia* spp.) |
| 3 | Powdery mildew *(Blumeria graminis)* | Bunts (*Tilletia* spp.) | Smuts (*Ustilago nuda*) | Smuts (*Ustilago nuda*) |
| 4 | Foot rot, eyespot (*Tapesia* spp.) | Take-all *(Gaeumannomyces graminis)* | Leaf and glume blotch (*Septoria tritici, Stagonospora nodorum*) | Foot rot, head blight (*Fusarium* spp., etc) |
| 5 | Smuts (*Ustilago nuda*) | Leaf and glume blotch (*Septoria tritici, Stagonospora nodorum*) | Powdery mildew (*Blumeria graminis*) | Leaf and glume blotch (*Septoria tritici, Stagonospora nodorum*) |

*Source:* Modified after Oerke, 1999.

More than 100 active ingredients are registered as fungicides worldwide. Despite this high number, new fungicides are still being developed, of which many are able to improve level of efficacy and yield and quality benefits (Knight et al., 1997). An important reason to continually stimulate development of new fungicides with new modes of action is to combat pathogens with resistance to existing antifungal compounds. This has lately been seen to be very relevant, as several important pathogens have quickly developed resistance to the major fungicide group, strobilurins (Gisi et al., 2000; Fraaije et al., 2003).

## *THE DISEASE MANAGEMENT ARENA*

Disease management is a highly dynamic field due to several factors: a strong climatic influence on seasonal disease prevalence, evolution in cropping methods, changes in crop choice, introduction of new cultivars, and, finally, shifts in pathogen populations, which may affect host-pathogen interactions and pathogen resistance to fungicides. For a given field of a specific crop, thus, management actions in terms of chemical disease control will vary much between years depending on the prevailing diseases and climatic conductivity to epidemic development (Jørgensen et al., 1996).

For centuries farmers and growers have tried to minimize the crop losses caused by plant pathogenic fungi. Traditionally cultural methods and cropping strategies were employed to minimize disease development. These methods and strategies included cultivar choice, crop rotation, seeding density and timing, fertilization regimes, and so forth. During the past 50 to 100 years, fungicides have become more and more widely used on a great number of crops. Disease control using chemicals began during the 1850s when the Bordeaux mixture was introduced to control downy mildew *(Plasmopara viticola)* in French vineyards (Brent, 1985). In the first half of the 1900s protection of many crops became possible with the introduction of organic fungicides. Since the 1960s, the development of new systemic fungicides able to penetrate the leaf and give curative effects permitted a more flexible timing of application and thus further increased the advantages of chemical disease control (Brent, 1985). In some areas and some crops, farmers have come to entirely rely on the use of fungicides, which has led to environmental problems, development of resistant strains of pathogens, and general worry among the public. This has led to the development of integrated control strategies and decision support systems aiming at reductions in fungicide use during recent decades. New strategies such as site-specific disease control might be the next step toward improving disease management while reducing possible side effects.

## *IPM STRATEGIES FOR DISEASE CONTROL*

Much focus over the years has been put on encouraging different integrated pest management (IPM) schemes. There are different concepts and definitions of IPM. They generally call for a complex decision-making system, and set an ideal of replacing the use of chemical products by other means. In the European Commission Directive 91/414 dealing with plant protection products, integrated control is defined as

> the rational application of a combination of biological, biotechnological, chemical, cultural, or plant breeding measures whereby the use of chemical plant protection products is limited to the strict minimum necessary to maintain the pest population at levels below those causing economically unacceptable damage or loss. (Office for Official Publications of the European Communities, 1991)

Depending on the crops grown, within certain regions of the world specific premiums can be obtained for the marketed product if specific IPM guidelines are followed. Retailers' or growers' organizations may demand that specific guidelines are used, and inspection systems have been organized to check if standard operating procedures have been followed. An example is the system formulated by Safeway in the United Kingdom (Garbutt, 2000).

The use of low levels of nitrogen, split nitrogen strategies, late sowing, or low seed rates in winter wheat can minimize disease development. However it is not always economically feasible to use these strategies, as they might also cause a significant reduction in yield (Jørgensen et al., 1997; Olesen et al., 2003a).

At the operational level, disease control in IPM systems often relies on fungicide application. There is a long research history on the optimization of chemical disease control by testing efficacy on different pathogens, testing dose response and duration in relation to disease development, e.g., preventive and curative control, and thereby making it possible to minimize input. The overall aim is to obtain effective control and the best economical response from the treatments (Jørgensen et al., 2000).

### *Decision Support Systems*

Decision support systems have been developed to optimize fungicide treatment with respect to profitability according to factors such as cultivar, growth stage, observed diseases, the weather conditions, and the anticipated epidemic progress. Decision support systems such as PC Plant Protection and Crop Protection Online in Denmark (Secher, Jørgensen, et al., 1995; Henriksen et al., 2000), Pro-plant in Germany (Frahm et al., 1996), and Dessac in the United Kingdom (Paveley and Clark, 2000) are aimed at optimizing disease control from an economic point of view using margin over fungicide costs as the overall target. The systems are to some extent used by the farmers but are more often regarded as a tool for advisors to consult before giving recommendations in newsletters or by phone.

A barrier for wide use of the Danish decision support system Crop Protection Online by practical farmers has been to fulfill the requirements for providing the necessary input data to the system about disease levels in the actual field. Two major reasons have been given for this hurdle. One is difficulties in assessing the actual disease and the frequency of plants attacked with either mildew or rust. The second reason is the steady increase in farm size, which puts limits on how quickly the fields can be treated following the assessment of a given need. In Denmark, windy conditions are very common; this reduces the number of days when application can take place, thereby minimizing the flexibility in using specifically adjusted and appropriate dose rates.

The limitations in existing decision support systems add to the general goal of developing less labor-intensive and more mechanized methods for assessing the disease level and the need for control. Ideally, this should also include the possibility of adjusting the fungicide input in the individual field, taking notice of the intrafield variation of disease and crop status.

## *SITE-SPECIFIC DISEASE CONTROL: THE NEXT STEP IN THE EVOLUTION OF DISEASE MANAGEMENT*

On top of the existing operational disease management systems, a site-specific level may be added. At this level, the intrafield variation of disease and crop condition is the main concern. Although the optimal timing of spray operations certainly may vary on a field basis, the practical and economic limitations imply that the focus will be on the site-specific optimization of whole-field operations. The three main assumptions are that site-specific disease management may be beneficial

1. if intrafield disease epidemic development varies,
2. if the crop sensitivity to disease varies, and/or
3. if the control effect of the applied fungicide(s) varies spatially.

Relevant management actions may potentially include site-specific adjustment of chemical disease control or cultural practices to the observed or perceived disease risk, site-specific adjustment of chemical control to crop disease sensitivity, or sorting of harvest according to the risk of seed infection. The biological and technological background for these management options, their implementation, and potential benefits will be the main topic of the rest of this chapter.

### *Spatial Disease Variability*

Spatial variation in disease severity is one of the major prerequisites for site-specific disease management. Since disease incidence varies within a field, this may seem a redundant requirement. However, disease variation must have an appropriate magnitude and occur on a spatial and temporal scale making site-specific management relevant. Further, if disease variation is to be manageable, disease must either be directly observable by automated monitoring devices or be predictable from ancillary information in terms of observations of crop status, environmental gradients, or disease occurrence in previous crops. If patterns of intrafield disease variation can be related to such ancillary information, this may be used in modeling of disease risk (Bjerre, 1999).

### *Disease Epidemics at Field Scale*

The essential factors in disease development are the presence of a susceptible host, a virulent pathogen, and environmental conditions favorable for disease. In annual crops, the establishment of disease depends on the introduction of primary inoculum from external sources. The pattern of primary infection may be random for wind-dispersed or seedborne diseases (Gaunt and Cole, 1992). For stem and leaf diseases spread from crop debris or soil, the pattern of primary infection may be influenced by inoculum resulting from previous epidemics in the field, exposure to crop debris, or management actions such as temporal field amalgamation (Boland and Hall, 1988; Schuh, 1990a). For root diseases with soilborne inoculum, primary disease patterns may reflect previous disease occurrence and cultivation patterns in the field (Gavassoni et al., 2001).

The mechanisms of disease spread are of significance for the risk and speed of colonization of unaffected plant parts or crops. For polycyclic diseases, repeated cycles of infection and spore dispersal may lead to the formation of disease foci originating from the points of initial infection (Zadoks and van den Bosch, 1994). The inoculum dispersal gradient may be used to predict the risk of infection under conducive conditions (Jeger, 1999). In a comprehensive study of spore gradient data, it was found that deposition gradients for splashborne spores could best be described by exponential models, while power law models best described deposition gradients for airborne spores less than 10 μm (Fitt et al., 1987). Theoretically, the former leads to focus expansion of constant radial speed while the latter leads to a dispersive wave epidemic with an increasing speed of invasion and the formation of distinct daughter foci far from the primary disease front (Ferrandino, 1993; Aylor, 1999). Investigations of disease spatial patterns in wheat revealed that severities of yellow rust *(Puccinia striiformis)* were correlated up to distances of 10 m, presumably due to the pronounced focus formation by this disease (Parker et al., 1997). For powdery mildew of cereals *(Blumeria graminis),* more dense primary infection by wind-spread inoculum is common and foci tend to overlap after some generations (O'Hara and Brown, 1997). However, Parker et al. (1997) also found spatial correlation up to distances of 10 m, but only at early growth stages. Patterns of Septoria leaf blotch *(Septoria tritici)* were found to be near random, presumably due to widespread primary infection by airborne ascospores (Parker et al., 1997). Schuh et al. (1990b) also found random patterns in the early epidemic, but later clumping of disease occurred in some cases, presumably due to microclimatic factors and effects of variable plant density.

The infection process, the multiplication of inoculum, and inoculum dispersal may all be influenced by crop status and environmental conditions. Once established, the role of external inoculum in epidemics of polycyclic diseases may quickly diminish, as shown for powdery mildew of cereals (O'Hara and Brown, 1996). Crop status may vary in terms of plant density (plants per unit area), crop density (leaf area index [LAI]), plant nutritional status, and physiological predisposition to disease. Environmental gradients in terms of shelter from woody vegetation, topography, and soil type variation may affect disease development directly due to resulting gradients in microclimate or indirectly by affecting crop status. Intrafield disease variation (Figure 7.1) has been found in fungicide-treated fields with gradients in environment and/or in crop status (Koch, 1980; Secher, Murali, and Gadegård, 1995b; Bjerre, 1999).

*Shelter*

Woody vegetation surrounding fields may affect microclimate by reducing wind speed and potential evaporation and by increasing daytime temperatures (Aslyng, 1958; Marshall, 1967). The resulting gradients in crop microclimate may affect disease epidemic development. Thus, increased severities of cereal powdery mildew and in some cases Septoria disease in areas close to shelter have been seen in several investigations (Koch, 1980;

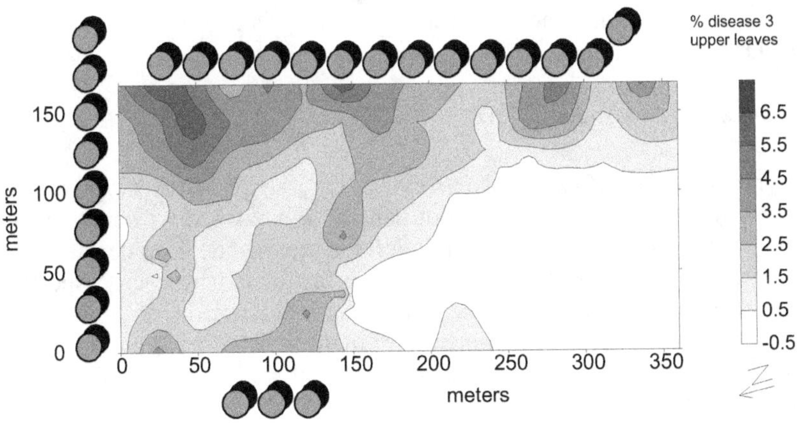


FIGURE 7.1. Severity of powdery mildew *(Blumeria graminis f.sp. tritici)* on the upper three leaves in a winter wheat field (GS 77) at Torbenfeld in Denmark, July 11, 1997. Filled circles outside the map denote shelterbelts *Source:* Bjerre et al., 1998.

Secher, Murali, and Gadegård, 1995; Bjerre, 1999; unpublished results). Woody vegetation may also affect disease development by shading. Progress of black rot in cabbage caused by the bacterium *Xanthomonsa campestris* pv. *campestris* was reduced by shade from trees, presumably due to lower temperatures reducing bacterial growth (Kocks et al., 1998).

### *Drought Stress*

In fields of variable soil type, the occurrence of drought stress may vary on an intrafield basis. This may affect disease development and is also a cause of intrafield variation in canopy development. Prolonged periods of low soil water potential have been shown to increase resistance toward biotrophs, such as powdery mildew in barley and bean rust *(Uromyces phaseoli),* while resistance toward a perthotrophic disease such as spot blotch *(Cochliobolus sativus)* in barley decreased (Ayres and Woolacott, 1980; Oerke and Schönbeck, 1986). Release of drought stress in barley may cause temporary breakdown of *mlo* resistance toward powdery mildew *Blumeria graminis* f.sp. *hordei* (Newton and Young, 1996). Severity of sharp eyespot *(Rhizoctonia cerealis)* in wheat seedlings increased as soil matrix potential decreased from –5 hPa to –200 hPa. Optimal seedling emergence in noninfested soil occurred at –20 hPa to –200 hPa (Kalberer and Gisi, 1997).

*Plant and Crop Density*

The distinction between plant/shoot density and crop density is one of observation method. Plant or shoot density is measured by counting the number of plants or shoots within a defined area, while crop density may be determined by measurement of LAI or estimated from crop spectral reflectance (Wiegand et al., 1992). Intrafield variation in soil conditions may affect seedling establishment and plant density, and/or affect canopy development due to effects of variable soil water supply and nutrient availability.

In a review, Burdon and Chilvers (1982) concluded that increasing plant density in most cases did enhance development of fungal diseases. This could be explained by an increased probability of inoculum interception by plants and a reduced chance of inoculum being lost on the ground or above the crop. However, investigations of plant density effects on cereal powdery mildews and rusts have in several cases shown reduced severities with increasing plant densities (Finckh et al., 1999; Dill-Macky and Roelfs, 2000). These results may be attributed to indirect effects of host density on plant nutritional status, as plants in less dense stands will have higher nutritional status and be more prone to biotrophic diseases. In addition, high-density stands may have negative effects on spore dispersal distances, thus limiting epidemic development (Aylor, 1999). Increases of nitrogen supply enhance canopy development and lead to increase in mildew disease (Tompkins et al., 1992; Olesen et al., 2003a) due to effects on epidemiological parameters and canopy microclimate (Jensen and Munk, 1997; Friedrich, 1994; Tompkins et al., 1993).

For the Septoria diseases of wheat, contrasting results on the effects of plant density (Jørgensen et al., 1997; versus Brocious et al., 1985; Tompkins et al., 1993) and nitrogen level (Lovell et al., 1997; Brocious et al., 1985; versus Johnston et al., 1979) have been found. The contrasting results may be due to opposing effects of canopy density on infection conditions, spore dispersal, and possibly plant physiological predisposition. There may also be differences between the responses of *Septoria nodorum* and *Septoria tritici* to these factors. Microclimate in dense canopies, especially increased duration of leaf wetness, is likely to favor infection and symptom development compared to less dense canopies (Chungu et al., 2001). However, denser canopies are likely to limit spore dispersal by rain splash from lower to upper leaves of the canopy, which is often limiting for disease development here (Lovell et al., 1997). In addition, increased nitrogen levels tend to reduce symptom development of *Septoria nodorum* in growth chamber experiments (Tiedemann, 1996).

Investigation of intrafield variation of Septoria and mildew disease as affected by crop density measured by canopy spectral reflectance have also challenged the notion of higher disease levels in denser crops (Bjerre, 1999; unpublished results). In commercial wheat fields, mildew severities were highest at intermediate crop densities. In thin crops, low mildew levels are expected due to an increased control effect of fungicide spraying and a less disease-conducive microclimate. In dense crops, the reduced mildew severity may be caused by a decrease of canopy N concentration and/or reduced spore dispersal distances.

For Septoria leaf and glume blotch (mainly *Septoria tritici*), a negative relation between disease severity and crop density was often found (Bjerre, 1999; unpublished results), presumably due to a negative effect of dense canopies on splash dispersal of spores from the lower to the upper leaves (Lovell et al., 1997) or due to reduced infection rate at low leaf nitrogen concentrations (Olesen et al., 2003b).

### *Landscape*

In hilly terrain, gradients of soil fertility and moisture may affect crop development and disease levels. In a preliminary investigation, it was found that joint foliar disease ratings for Septoria leaf and glume blotch *(Septoria tritici, Septoria nodorum)* and tan spot *(Pyrenophora tritici-repentis)* in wheat weres higher at upper than at lower slope positions (Kutcher et al., 1999). However, at upper slope positions yield and presumably crop density were reduced, and increased disease levels here may be attributed to a reduced crop nutritional status, drought stress, and/or an effect of crop density on spore dispersal. As yield levels in hilly terrain may vary with slope position according to seasonal weather conditions (Kravchenko et al., 2000), it may consequently be more relevant to use crop density variation than topography in prediction of disease gradients.

## *EFFECTS OF DISEASES AND SPATIAL VARIABILITY ON CROP GROWTH*

Disease is only one of the factors that affect spatial variability of crop growth and yield. The crop is also affected directly by variation in soil and climate conditions, mainly through availability of water and nutrients and the ability of the crop roots to explore the soil. Indirect effects may also affect spatial variability of crop growth. These effects are mainly caused by competition with weeds, predation by insects, or factors affecting crop establishment resulting in reduced density of plants of low vitality.

*Crop Growth and Yield Formation*

Crop dry matter production *(W)* can be taken as the integral of intercepted radiation multiplied by the radiation use efficiency (RUE) (Monteith, 1977):

$$W = \Sigma f_{PAR} \times PAR \times RUE \quad (7.1)$$

where *PAR* is the incident photosynthetic active radiation. The fraction of intercepted photosynthetic active radiation ($f_{PAR}$) can be determined from the green canopy area index *(L)* by use of Beer's law:

$$f_{PAR} = 1 - \exp(-kL) \quad (7.2)$$

The extinction coefficient *(k)* varies between crop species and depends on crop condition; typical values are between 0.4 and 1 (Thorne et al., 1988; Olesen and Grevsen, 1997). This implies a decreasing return from increasing canopy size. In many cereal crops, only a small benefit is obtained from increasing the LAI beyond a value of 5 to 6 (Sylvester-Bradley et al., 1997). The fraction of intercepted photosynthetic active radiation ($f_{PAR}$) may also be determined from spectral reflectance measurements (Christensen and Goudriaan, 1993).

The assimilated dry matter is distributed between vegetative and storage organs with varying proportion depending on the crop development. In cereals, almost all assimilates accumulated after the beginning of grain filling are transferred to the growing grains (Gent, 1994). In addition, stem reserves that are mainly accumulated after flag leaf appearance are transferred to the grains (Olesen, Jensen, and Petersen, 2000). Diseases that mainly occur during grain filling or storage formation in crops will thus greatly affect yields, whereas diseases that occur only during early development often have little effect provided the crop is able to compensate with expansion of the LAI.

As crop dry matter production is related to accumulated canopy light interception, it is likely that many of the factors causing intrafield yield variation are expressed through variation in green canopy size and duration (Whelan et al., 1997). In many crops, green LAI has a linear relationship to nitrogen uptake during vegetative growth (Grindlay, 1997; Lemaire and Gastal, 1997; Olesen et al., 2002). Drought also reduces leaf area considerably (Jamieson et al., 1998). However, many of these soil and climate factors also affect radiation use efficiency (Sinclair and Muchow, 2000).

Relationships defining development of LAI and effects on radiation use efficiency have been used to develop dynamic crop simulation models, which are able to simulate effects of drought and nitrogen availability on crop growth and yield (e.g., Porter, 1993; Jamieson and Semenov, 2000). Provided detailed information on variability in soil conditions and plant density is available, these models may be used to simulate spatial variation in yield potential (Matthews and Cosser, 1997; Basso et al., 2001).

The concept of light interception and radiation use efficiency is also useful for interpreting the effects of foliar diseases. A good correlation between yield and intercepted radiation after flag leaf appearance has thus been found for wheat affected by yellow rust (Bryson et al., 1997). However, diseases may also affect RUE. In winter wheat, mildew, and Septoria disease did reduce both light interception and RUE (Olesen, Jørgensen, and Mortensen, 2000b).

### *Intrafield Crop Variability*

Crop yields often vary considerably over short distances within fields (McBratney and Pringle, 1997; Sadler et al., 2000). Yield maps from multiple years can be used to characterize the stability of the yield patterns (Joernsgaard and Halmoe, 2003). Blackmore (2000) found that the stability of yield patterns was higher for individual crop species than for multiple crops across a field in the United Kingdom. The stability of the yield patterns has in many cases been found to correlate well with rainfall patterns (Bakhsh et al., 2000).

Much of the intrafield crop variability (Figure 7.2) is caused by variation in soil and climate conditions, because management in terms of crop establishment is usually performed in a uniform way across the field. The variability is often closely related to topography which affects both soil and climate conditions (Jones et al., 1989; Kravchenko et al., 2000). The effect of soil conditions on crop growth is mainly mediated through crop water supply (affected by rooting depth and water holding capacity) and nutrient supply (affected by nutrient availability).

Local climatic conditions are affected mainly by shelterbelts and by slope and aspect of the terrain (Rosenberg et al., 1983). These factors can

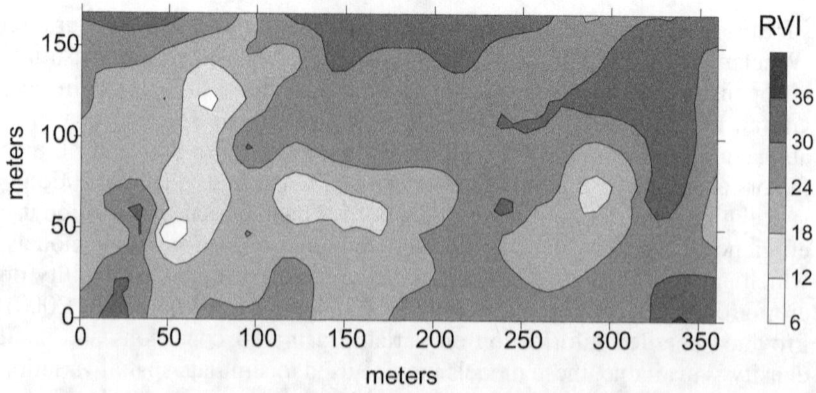


FIGURE 7.2. Crop density in wheat (GS 41) expressed as relative vegetation index for the same field as in Figure 7.1 *Source:* Bjerre et al., 1998.

affect crop growth directly through effects on temperature, radiation, and water availability, and indirectly through effects on soil nutrient turnover and conditions for disease propagation.

*Response to Fungicide Application*

The primary aims for applying fungicides are to delay disease onset and to reduce epidemic development in order to protect the crop from reductions in yield quantity and quality. Increasing levels of disease control and yield returns are normally obtained from applying increasing dosages of fungicides (Figure 7.3). The optimal dose ($F_{opt}$) can be calculated as the point on the yield response curve where the financial gain from the increased yield is equal to the cost of the fungicide needed to obtain the increase in yield. If fungicide dose rates are increased beyond this point, the increase in fungicide costs will be larger than the value of the resulting yield increase. The optimal dose is generally higher in cases where the maximum yield increase for fungicide application is higher, as illustrated in Figure 7.3 for two different application dates. The data originate from two trials, which were treated at two different growth stages (GS 33 and 59) for control of *Septoria tritici.* The late treatment gave the best level of disease control and the largest yield response for disease control (Jørgensen et al., 1999).

In terms of site-specific disease control, the yield response curve emphasizes that optimal inputs may vary according to expected disease severity, crop sensitivity to disease, or the control effect of the applied fungicide(s).

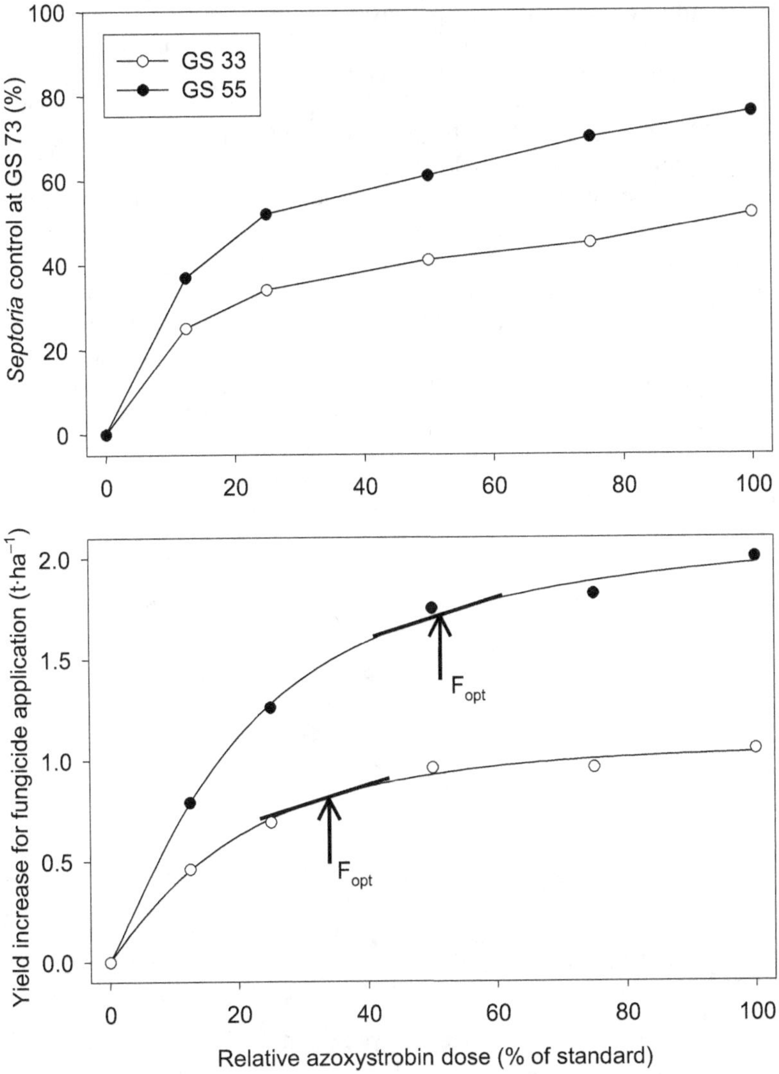


FIGURE 7.3. Percent control of *Septoria tritici* and grain yield of winter wheat for increased dosages of azoxystrobin (Amistar) applied at growth stage 33 or 55. Data from two experiments in Denmark. The optimal dose is shown under the assumption the fungicide cost for a standard rate of azoxystrobin corresponds to 1 ton of grain.

Site-specific fungicide applications will be beneficial only if there is a substantial intrafield variation in $F_{opt}$ (Paveley et al., 1996).

Variable crop sensitivity to disease is somewhat supported by empirical evidence. Thus, yield increase due to conventional disease control was positively correlated with yield level in 42 experiments on winter wheat cv. Riband (Paveley et al., 1996). In Denmark and Germany, positive relations between yield increase and yield level have also been found in winter wheat (Figure 7.4; Oerke and Dehne, 1997). A considerable input of fungicides may be justified on a poorly established crop to make the most of a small canopy. However, optimal fungicide rates will normally be smaller for crops grown under stress conditions, e.g., drought, because these conditions will reduce yields through premature senescence and through reduced radiation use efficiency. The yield response curve and the optimal fungicide dose will therefore depend on the reason for variation in maximum yield. However, variation in yield potential is often linked to variation in soil water holding capacity (Thomsen et al., 1997; Sadler et al., 2000), which again is linked to canopy size. This indicates that spatial variability in canopy size could be an important indicator for optimal fungicide rate.

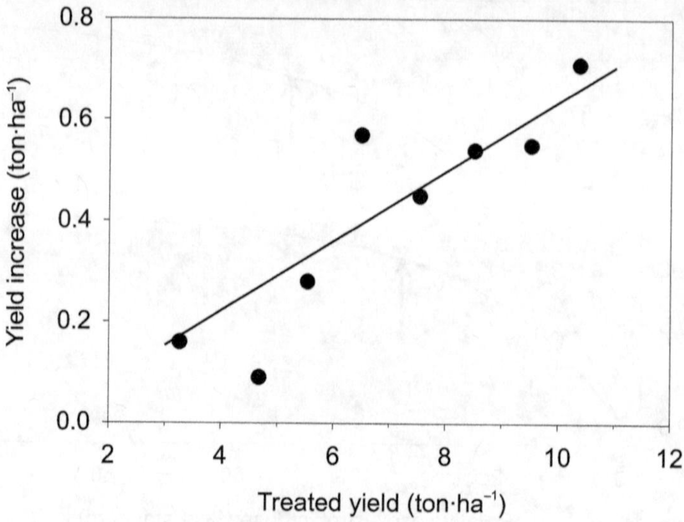


FIGURE 7.4. Relationship between treated yield and yield increase for fungicide application in winter wheat. Data from 481 field experiments in Denmark from 1992 to 1997 were stratified according to yield in treated treatments. The fungicide treatments included at least two applications of 0.5 liter Tilt Top (fenpropimorph 375 g a.i.·$L^{-1}$, propiconazol 125 g a.i.·$L^{-1}$). *Source:* Data from trials carried out by the Danish Agricultural Advisory Centre.

## TECHNOLOGY FOR SITE-SPECIFIC DISEASE MANAGEMENT

Site-specific disease management, as other types of precision agriculture, is information intense. It relies on positioning systems, automated sensing systems, integration of data from several sources, routines for determination of application requirements, and precise application of inputs (Stafford, 2000). Further, monitoring of outcome is essential for evaluation of performance. The technological equipment for precision agriculture has much attention at the present. In the area of disease management, focus is on sensor systems for sensing crop density and disease and on application technology. However, for site-specific management to succeed data from many sources, including the farm manager's knowledge base, must be transformed into information by models and/or expert systems for interpretation and integration. This section will present a concept for fungicide application in crops of variable LAI and discuss the possibilities of implementing site-specific fungicide application with current available technology. Further, advances in application technology and the use of sensor systems for measurement of crop density and disease will be addressed.

### *Canopy Spray Deposition*

The performance of a fungicide application is, among other factors, related to the spray deposition rate on the plant parts to be protected. As the spray droplets are intercepted during their travel through the canopy, this results in a higher deposition in the upper parts of the canopy compared to the lower parts (Permin et al., 1992). The capture efficacy is influenced by the nature of the plant surface, leaf angle distribution, shapes of leaves, stems, and ears, and by the turbulence of the air. The use of air-assisted spraying increases turbulence within the upper leaf layers and may lead to an increased capture efficacy as more pesticide is deposited on the stems and underside of leaves (Hislop et al., 1995). The result is a higher deposition of pesticide in the upper part of the canopy at the expense of deposition in the lower part of the canopy (Gyldenkærne et al., 1999).

The interception of pesticide by the canopy can be described by capture efficiency (Akkerhuis et al., 1998), which is analogous to the extinction coefficient for light used in Equaton 7.2. The effects of crop area index on canopy spray deposition have been addressed in recent research on canopy spray extinction (Akkerhuis et al., 1998) and ground deposition (Gyldenkærne et al., 1999). Akkerhuis et al. (1998) found that the area of stems, leaves, and ears all contributed to canopy spray extinction in spring barley (GS 48 and GS 56) with an extinction coefficient (capture efficiency) of approximately 0.20. In an investigation of ground deposition under wheat and barley canopies, extinction coefficients of 0.49 and 0.42, respectively, were found (Gyldenkærne et al., 1999). The use of a surfactant increased the extinction coefficients in barley and in one of the two wheat cultivars used. Extinction coefficients for light interception are known to vary between crop species, cultivars, and with crop development (Olesen and Grevsen, 1997). It is likely that the capture efficiency for pesticides will also show variation. In addition, the method of leaf area determination and definition may influence the estimates of capture efficiencies.

Other investigations have related spray deposition to canopy size. An inverse relation between spray deposition per leaf area unit and canopy size was found by Miller et al. (2000), as the deposits in winter wheat at GS 39 were 43 percent lower than at GS 32. Canopy density of plots was classified as high or low based on canopy spectral reflectance, but no significant differences in deposits could be found between the groups. However, the reflectance measurements may not have been representative of the crop at the time of spraying, as these were done at an early growth stage (Miller et al., 2000).

The conventional spray application method applies a constant dose rate per ground area irrespective of the variation in canopy size. If the concept of a constant capture efficiency of canopies is assumed in crops of variable canopy area index (CAI) the vertical deposition profiles will be steeper in more dense crops due to a higher extinction rate (Figure 7.5a). By this, the spray deposition, especially on the second to third leaves, is much reduced compared to a more open crop. In order to obtain a more even deposition profile, a site-specific adjustment of dose rate according to crop CAI or some measure related to this might be made (Figure 7.5b). This will result in increased deposition rates in dense crops when compared with the conventional application method, while the opposite is true for less dense crops. This ideally results in a spatially more uniform deposition for the upper leaves of the canopy.

### *Fungicide Application According to Canopy Spectral Reflectance*

Site-specific application according to canopy spectral reflectance in cereals (Figure 7.5b) may be done with technology presently available (Bjerre and Secher, 1998). The decision on fungicide, median dosage, and timing of spraying will be done by the farm manager based on field inspection. Crop density maps can be obtained by tractor-mounted and DGPS-coupled sensors for measurement of canopy spectral reflectance, or

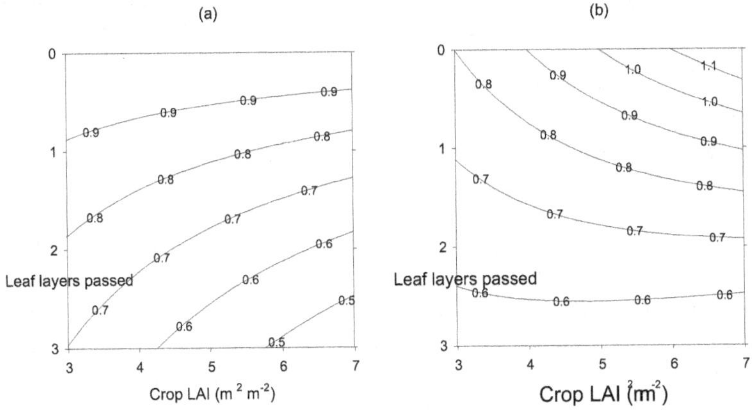


FIGURE 7.5. Relative deposition according to crop leaf area index (LAI) and leaf level at *(a)* conventional and *(b)* site-specific fungicide application. By site-specific application, the conventional dose is varied ±20 percent according to crop LAI. The deposition is calculated relative to the deposition at the top of the conventionally sprayed crop. Deposition is assumed to follow an exponential model with an extinction coefficient of 0.20 (Akkerhuis et al., 1998). *Note:* Each leaf layer has a LAI of 0.2 times the crop LAI, and leaf layers are assumed not to overlap.

alternatively from digitized aerial photographs or high-resolution satellite imagery. Dosage maps may be constructed from crop density maps in a geographical information system for decision support. The spray operation can be done with a conventional field sprayer equipped with a spray controller connected to a DGPS-coupled tractor computer (Bjerre and Secher, 1998). At a dose rate variation within ± 35 percent, a satisfactory application can be obtained by adjustment of sprayer speed and spray pressure. Generally, it is assumed that spray volumes may be adjusted ± 30 percent on conventional flat fan nozzles by changes in spray pressure without negative effects on canopy spray deposition (personal communication, Ivar Lund).

In terms of costs and utilization of machine capacity, a spray operation based on real-time measurement of canopy spectral reflectance would be preferable. However, a robust calibration of sensor readings for the range of crops and the crop physiological and measurement conditions encountered is not likely to be performed by the farm manager. For this reason, it is preferable to obtain a measure of the crop density variation in advance and pragmatically adjust dose rates to the observed variation in sensor reading. When the spray operation is done under good measurement conditions, it may be an option to estimate the range of crop density variation from spot measurements located according to the farmer's knowledge on intrafield soil variation, yield maps, and other clues to crop density variation. In this situation, a relation between sensor readings and dose rate adjustment for the present crop may be determined, and the spray operation may rely on real-time measurements of canopy spectral reflectance. With the development of satellite systems for agricultural purposes, it may be an option to obtain crop density maps from this source (Blakeman et al., 2000; Moran, 2000). Application maps based on satellite imagery may be used directly, or adjusted by use of real-time measurements of crop density. Although crop density information for fungicide application is robust in the time dimension (presumably 10 to 15 days) the use of satellite imagery may, however, be facultative, as an unacceptable low sampling rate may result from cloud cover and logistic constraints inherent in satellite systems. In the present systems, processing time from image acquisition to delivery of georeferenced products is too high for agricultural management purposes (Moran, 2000).

### *Application Technology*

The current application technology has given satisfactory results in winter wheat (Secher, 1997), but little is known about the potential benefits from technological advances in terms of site-specific disease control. Obviously, large adjustments of dose rate with high spatial resolution will be relevant in situations with high intrafield variation in crop cover and may lead to savings of fungicide and environmental benefits in terms of reduced off-target deposition. In theory, such application technology could also be relevant for control of diseases with pronounced disease gradients or discrete disease foci. However, detection and delimitation of such foci would require disease sensor systems not presently available. Further, the high potential for epidemic development and the efficient dispersal mechanisms of airborne pathogens such as cereal powdery mildew *(Blumeria graminis)* and yellow rust *(Puccinia striiformis)* implies that the most efficient control strategy will be whole-field spraying, but with variable dosages.

Spatial resolution may be increased by minimizing the ratio between response time and sprayer velocity, and by subdivision of the sprayer boom, ultimately at nozzle level, into individually controlled sections. Further, the use of air-liquid nozzles or pulsating nozzles (Miller and Combellack, 1997; Lund, 2000; Giles et al., 1999) may enable increased dose rate variation in comparison with conventional flat fan nozzles. This is likely to be advantageous in terms of machine capacity, as large dose-rate variations will be possible without changes in sprayer speed. However, it is not known to what extent dose-rate variation beyond the ± 30 percent range possible with conventional flat fan nozzles is relevant for cereal disease control.

A qualitative advance in application technology may come with the successful development of flow-injection systems, which enables a simultaneous variable-rate application of several substances (Nuspl et al., 1996). Although this possibility is relevant for weed management (Walter et al., 2001), no investigations have yet been made on the potential of this technology for disease management. Many of the presently used fungicides in cereals are to some degree pathogen specific, and at least in winter wheat there is some evidence of differing intrafield disease gradients for mildew and Septoria disease (Bjerre, 1999). Thus, for this crop a site-specific and variable use of fungicides of differing control profiles may be of interest. Differing dispersal and infection biology of co-occurring diseases in other crops may lead to similar differences in intrafield disease risks and may justify separately targeted dose adjustments. However, no investigations have yet been published on this subject.

### *Sensors for Site-Specific Disease Management*

If site-specific disease management is to be practicable, the data needed for treatment maps or other management decisions must be generated by automatic sensor systems sensing crop density, disease, or some suitable surrogate of this. Preferentially, this should be done by real-time, robust, and low-cost mapping systems (Stafford, 2000). In combination with computer-based decision algorithms, this would potentially allow for real-time site-specific application of inputs, e.g., fungicides. However, in the management of soilborne and seedborne diseases historical sensing data may also be used in zoning the field for later diagnosis or site-specific management. Observation of canopy status and disease may be done by remote sensing of canopy light reflection or thermal radiation. In addition, immunological and molecular methods based on physical sampling of air, crop, or soil may be used for identification of the pathogen. The application of remote sensing in crop protection and disease management has been reviewed by Hatfield and Pinter (1993), Nilsson (1995), and West et al. (2003).

Canopy spectral reflectance may be related to LAI and biomass as near-infrared reflectance increases with increasing crop density while visible reflectance decreases. In an investigation of wheat spectral reflectance at multiple sites it was possible to develop general relations between canopy spectral reflectance and LAI values up to 6 (Wiegand et al., 1992). There is in general a linear relationship between crop green area index and relative vegetation index (RVI), which is the ratio of near-infra red to red reflectance (Figure 7.6). However, this relationship is most pronounced in cereals prior to anthesis, because leaf senescence will influence the relationship during grain filling. Canopy spectral reflectance is also influenced by growth habit, water status, nutritional status, and solar angle. Moreover, maximum values of reflectance tend to be reached at LAI values of 3 to 4, which may limit the use of spectral reflectance in determining LAI. However, the relation between canopy interception of photosynthetically active radiation and spectral reflectance seems to be less influenced by these factors (Hatfield and Pinter, 1993).

Remote sensing of leaf and canopy spectral reflectance has been investigated as a tool for diagnosis and quantification of plant disease (Nilsson, 1995; West et al., 2003). For leaves of bean *(Vicia faba)* infected by the necrotrophic fungus *Botrytis fabae* it was found that reflectance in the visible spectrum initially increased and then decreased. However, disease prog-

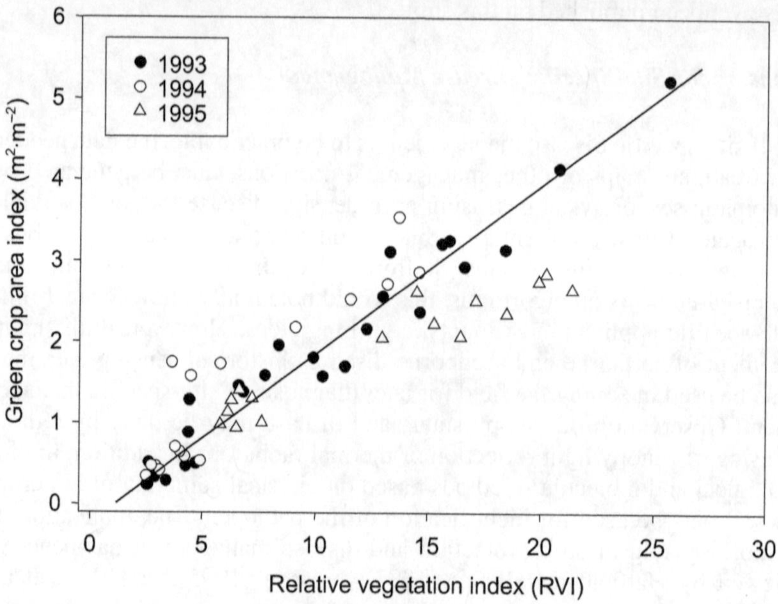


FIGURE 7.6. Linear relationship between relative vegetation index measured using spectral reflectance measurements and green crop area index from destructive sampling. Data from three years in winter wheat prior to anthesis from a field in Denmark with high spatial variability *Source:* Based on data presented by Thomsen et al., 1997.

ress could be related to a flattening of reflectance in the visible spectrum and a consistent decrease of near-infrared reflectance (Malthus and Madeira, 1993). For barley leaves infected by powdery mildew *(Blumeria graminis),* increases in reflectance occurred earlier and were more pronounced in the visible than in the near-infrared part of the spectrum due to sensitivity of the photosynthetic system to stress (Lorenzen and Jensen, 1989). Leaf reflectance may also be influenced by stress factors other than disease. Carter (1993) found that the reflectance of leaves in the visible spectrum increased due to several types of stress agents, presumably due to stress-induced decreases in chlorophyll A content. As foliar diseases typically cause discrete lesions on leaves, these may be detected by image analysis and distinguished from stress symptoms that affect whole plants or plant organs and have relatively uniform patterns, e.g., nitrogen deficiency (West et al., 2003).

In studies of diseased canopies, good correlations between disease occurrence and canopy reflectance in specific spectral bands or with reflectance ratios have been found in many cases (Nilsson, 1995). Canopy spectral reflectance ratios were well correlated with incidence of panicle infection by rice blast caused by the fungus *Magnaporthe grisea* (Kobayashi et al., 2001). In a young wheat crop, areas diseased by yellow rust, healthy areas, and soil background could by distinguished based on four reflectance wavebands measured by spectral line imaging (Bravo et al., 2003). When severities of peanut rust *(Puccinia arachidis)* or defoliation induced by late leaf spot *(Cercosporidium personatum)* have ranged from zero to high levels (>40 percent) in peanut *(Arachis hypogaea)* significant and high correlations with canopy reflectance have been found. However, the relation between percent defoliation and percent reflectance at 800 nm was influenced by the absolute amount of green leaf area present (Nutter, 1989). When severity of Rhizoctonia blight *(Rhizoctonia solani)* in bentgrass was monitored by spectral reflectance during the entire epidemics, the coefficients of determination (0.25-0.55) were lower than those typically found in studies performed on individual assessment dates (Raikes and Burpee, 1998). This may be an effect of increased plant or environment variability over time affecting canopy reflectance and implies that multispectral radiometry may be best suited for monitoring disease severity at discrete points in time. Root and vascular system diseases leading to crop drought stress and increased canopy temperatures may be monitored by IR thermography (Nilsson, 1995). However, drought stress may also be caused by aphid infestation and by poor soil conditions or low N levels leading to reduced root growth and water uptake.

Nilsson (1995) concluded that remote sensing cannot be used in diagnosis of diseases, as other stress agents may lead to similar effects on canopy spectral reflectance. Remote sensing may be used as a supplement to conventional methods, i.e., visual assessment of disease. Although image analysis gives improved capabilities of distinguishing foliar disease from other stress agents, it seems unlikely that a reliable diagnosis of pathogens will result; most diseases symptoms are highly variable according to pathogen strain, host cultivar, and environment. If the cause of stress is identified, differences in spectral reflectance among different parts of a field may be used as an indicator of advanced disease epidemics, which potentially allows for rapid and unbiased assessment of large areas.

Disease sensing based on physical sampling of air, crop, or soil may be used for detection and/or diagnosis of disease. A diagnostic test for *Septoria tritici* based on enzyme-linked immunosorbent assays (ELISA) was able to detect the pathogen at full emergence of upper leaves and gave a good prediction of future disease levels when weather conditions were favorable for disease expression (Lockley et al., 1996). However, under less conducive weather conditions only poor correlations of test results and future disease levels were found. Technological advances in molecular methods, especially real-time polymerase chain reaction (PCR), allow highly specific detection and accurate quantification of plant pathogens (McCartney et al., 2003). Presymptomatic identification and simultaneous identification of several pathogen species may be valuable in disease management. The development of portable real-time PCR instruments allows on-site pathogen detection within one hour or less (Schaad et al., 2003). However, the required processing time for immunological and molecular methods precludes their use in real-time sensing systems (Hollomon, 1998; McCartney et al., 2003). Electronic noses may be used for real-time detection of volatile materials released during pathogen infection, for example in oilseed rape *(Brassica rapa)* (Blight et al., 1995), but the diffusion and dispersal of these materials will be a problem in locating and quantifying disease. In addition, further development of this technology is required before application under field conditions (Schaad et al., 2003). Trapping of airborne spores for identification may be beneficial in determining the timing of disease onset (McCartney et al., 2003), but is not likely to be used for mapping of intrafield disease variation due to the time required for processing the samples.

A rational detection of pathogens in plant or soil material by molecular methods relies on adequate sampling in space and time. Under conditions where substantial intrafield variation is expected, presumptive localization of sampling zones may be of large value. For soilborne diseases and nematodes, stratified sampling zones may be based on yield variation or on remote sensing of crop senescence.

### *Spatial Disease Management*

Intrafield variation in disease occurrence and/or in crop sensitivity to disease suggests that disease management may be optimized by site-specific adjustment of management actions to this variation. If patterns of disease or disease risk are predictable and stable between years, then management may be adjusted to reflect this spatial pattern. This may involve both strategic and operational decisions, where the strategic decisions typically are those aimed at preventing disease (e.g., choice of crop species, variety, seed density, and fertilization), and operational decisions are those aimed at controlling the disease (e.g., adjusting fungicide type and dosage). If patterns of disease variation are not stable between years but are observable or predictable from ancillary information during the growing season, management may be adjusted at the operational level. This may be done by site-specific fungicide applications or by sorting of the harvest (Table 7.2). Presently, fungicide application according to crop density and observed disease incidence is the only method of site-specific disease control that has been tested experimentally. Thus, the discussion of other methods of spatial disease management will be mainly theoretical.

### *Site-Specific Fungicide Application According to Canopy Density*

Chemical control of cereal stem, leaf, and ear pathogens may be adjusted according to variation in crop or in disease (Table 7.2). Variation in cereal crop density at the time of spraying has a direct effect on canopy fungicide deposition (Akkerhuis et al., 1998) but may also be related to crop sensitivity to disease (Paveley et al., 1996; Figure 7.4). The crop density in cereals may be efficiently monitored by remote sensing of canopy spectral reflectance as described in the previous section on technology (Wiegand et al., 1992).

The rationale for fungicide application according to crop density is that a spatially more uniform spray deposition on the leaves to protect from disease can be expected to give overall better disease control (Figure 7.5). If it is assumed that the deposition of fungicide on a leaf area basis must exceed a certain threshold to give adequate protection, then this level will be reached at a certain depth in the canopy depending on the applied dose rate and the extinction coefficient for fungicide deposition in the canopy. This depth should ideally correspond to a specific leaf level irrespective of intrafield variation in crop density. This may be a desirable control strategy as the progress of an epidemic disease is related to crop development and

TABLE 7.2. Implementations and potential implementations of site-specific disease management in annual crops according to management target.

| Management target | Pathogen | Information source | Management action | Status |
|---|---|---|---|---|
| Crop sensitivity to disease and/or canopy spray deposition | Polycyclic leaf and ear pathogens | Measurement of canopy spectral reflectance | Fungicide application | Tested (Secher, 1997; Ewaldz, 2000) |
| | | Site-specific modeling of crop yield potential | Fungicide application | Potential |
| Observed canopy disease status | Polycyclic leaf and ear pathogens | Visual observation of disease + disease forecasting | Fungicide application | Tested (Secher, 1997; Bjerre, 1999) |
| | | Disease sensor + disease forecasting | Fungicide application | Not available |
| Disease risk prediction | Polycyclic leaf and ear pathogens | Modeling disease risk gradients based on ancillary data + disease observation or forecasting | Fungicide application | Potential (Koch, 1980; Secher, Murali, and Gadegård, 1995; Bjerre, 1999) |
| Risk of seed infection | Seedborne pathogens with polycyclic leaf and ear infection | Remote sensing of leaf and ear infection | Sorting of harvest | Potential |
| | Seedborne pathogens | Disease sensor | Sorting of harvest | Not available |
| Disease risk prediction based on previous crops | Inoculum in soil or crop residue | Maps of canopy spectral reflectance and/or yield + disease observation or soil tests | Fungicide application | Potential |
| | | | Solarization, fumigation | Potential |
| | | | Adjustment of general cultural practices | Potential |

leaf emergence (Shaw and Royle, 1993; Paveley et al., 2000) besides climate. Consequently, the optimal timing of fungicide applications is close to the time of full leaf emergence as shown for yellow rust and *Septoria tritici* (Paveley et al., 2000).

Fungicide application according to canopy spectral reflectance has been investigated by the authors. Site-specific dosages were typically adjusted from –20 to +20 percent in comparison with the conventional dosages, and total doses were equal for the two types of treatment. In six trials, a significant yield increase of 0.12 ton·ha$^{-1}$ was found. In addition, disease control tended to be improved as Septoria and mildew disease control were improved in one trial each. Measurements of canopy spectral reflectance were done from two to nine days before spraying, and only minor changes in crop status or density would be expected during this period. For a subset of four trials, crop density variation was found to be smaller than in an observational study of larger parts of commercial fields; thus, it is likely that the results will be transferable to commercial practice. Ewaldz (2000) investigated the effects of constant rate applications in winter wheat classified as having low, medium, or high crop density based on a measurement of spectral reflectance at GS 47. In plots of low crop density, a satisfactory level of Septoria disease control was obtained by a dose of 125 g·ha$^{-1}$ of azoxystrobin at GS 59, while 250 g·ha$^{-1}$ was needed in plots of medium and high crop density. These varying fungicide rates also gave the overall highest profitability.

### *Site-Specific Management According to Disease Incidence*

Variation in disease may influence future epidemic progress and the dose rates needed to achieve a satisfactory control. For research purposes, disease levels may be observed visually, but in commercial practice this route is impracticable. Developments of disease sensors may in the future lead to economically efficient disease monitoring systems, but at present such systems are not available.

In cereals, adjustment of fungicide dose according to observed disease symptoms might be relevant for diseases such as rusts and powdery mildew. For yellow rust *(Puccinia striiformis),* disease progress at the leaf level reflects the disease severity of lower leaves due to spore dispersal within the canopy (Young et al., 2003). As the dose rates required to achieve a satisfactory control and duration of control are related to the rate of epidemic progress (Jørgensen, 1994), it may be rational to adjust these according to observed disease levels. For Septoria disease, the conditions for splash dispersal from lower to upper leaves in the canopy may be more important than the amount of initial inoculum in determining future disease levels. Thus, it was found that only very low disease levels would limit future epidemic development of *Septoria tritici* (Paveley et al., 1997). Consequently, treatment decisions based solely on disease observation may be of limited value. A rational control strategy for this type of disease will also be based on assessment of the risk of vertical transport of inoculum due to rain and canopy density and structure (Lovell et al., 1997).

Significant intrafield variation of mildew disease incidence was found in 12 fields of winter wheat investigated by Secher, Murali, and Gadegård (1995). Site-specific treatment needs, as recommended by the decision support system PC-Plant Protection (Secher, Jørgensen, et al., 1995), were increased for 26 percent of the area and decreased for 28 percent when compared to blanket applications based on average disease incidence. Further, a detailed disease map investigation of winter wheat subfields revealed a considerable variation in mildew and Septoria disease occurrence (Bjerre et al., 1998). Adjustment of fungicide dose according to observed disease levels and canopy size in winter wheat was done by Secher (1997), who found a significant yield response of 0.30 ton·ha$^{-1}$ for site-specific applications measured in relation to a conventional blanket application. Dose rate adjustments at the first application (GS 31) were done according the observed incidence of mildew, as recommended by the decision support system PC-Plant Protection. At the second application (GS 51), dose rates were adjusted according to a visual assessment of crop height and crop density, as mildew incidence was high in all plots at this time. Spectral reflectance measurements performed after the second application revealed a considerable variation in crop density and showed that visual estimates of this had been reliable (Christensen et al., 1997). In another experiment, mildew disease control tended to be improved by site-specific application according to mildew incidence and canopy spectral reflectance, while no effects on yield and Septoria disease control were found (Bjerre, 1999).

The yield increases found for site-specific applications may obviously be caused by better targeting of control to a dominating disease (Secher, 1997) and furthermore to the effect of dose adjustment according to crop density. However, no distinct improvement of disease control was found in any of the investigations, even when doses were adjusted according to disease severity (Secher, 1997; Bjerre, 1999). An alternative explanation is that crop sensitivity to disease is variable and that the dose rate adjustments just traced this variation. This is supported by empirical evidence, as yield response to fungicide application is positively related to yield level (Paveley et al., 1996; Figure 7.4). It is thus likely that the yield increases may be due to a positive relation between the yield effect of fungicide treatment and the canopy size monitored by measurements of spectral reflectance, as the canopy size can be assumed to correlate with yield level. However, more experiments have to be carried out to test this assumption.

Information on crop yield potential can be obtained in other ways, either by directly collecting spatial yield data or by modeling crop yield potential (Table 7.2). Several years of yield maps collected automatically from harvesting machines may be used to generate maps of crop yield potential, and it is possible to classify areas as stable high yielding, stable low yielding, and unstable (Blackmore, 2000; Joernsgaard and Halmoe, 2003). Alternatively, information on spatial variation in soils and local climatic conditions as influenced by slope, aspect, and shelter may be used to simulate crop yield potential (Kitchen et al., 2003). Remote sensing of crop spatial variation can be useful to target sampling and provide spatial inputs for simulation models (Basso et al., 2001; Scharf et al., 2002).

### *Disease Risk Prediction*

Knowledge of intrafield disease variation may be used for site-specific adjustment of fungicide doses (Table 7.2). Based on experience such knowledge is present in the farming community. In Denmark, it is an accepted notion among farmers that the risk of cereal powdery mildew *(Blumeria graminis)* is increased in areas of the field close to shelterbelts or forest edges. This idea is also supported by experience of agricultural advisors and corroborated by scientific evidence from field observations (Koch, 1980; Bjerre, 1999). Based on this, some farmers chose to increase fungicide doses in areas of the field affected by shelter.

Investigations of intrafield disease variation indicate that gradients of mildew and Septoria disease in wheat may be explained by information on location of shelter and intrafield variation in crop density expressed as canopy spectral reflectance (Bjerre, 1999; unpublished results). Modeling of disease risk gradients based on this may be used in decision support systems for site-specific fungicide application. Additional information from climate-based disease risk models and from field observations of disease in high and low risk areas may give information on expected epidemic development and calibration for disease risk maps. Site-specific fungicide application may then be based on expected disease risk and on crop sensitivity to disease. However, no investigations on the benefits from following such a practice have been published. Noticeably, the effects of crop density on powdery mildew and Septoria disease tend to be opposite in some fields (Bjerre, 1999), and the interactions among disease, crop density, and yield potential on crop yield have to be addressed.

### *Sorting of Seed at Harvest*

For seedborne diseases, sorting of harvest according to seed infection level may be an option in disease management (Table 7.2). Sorting may be done by selective harvest (Stafford, 1999) according to a previous zoning of the field with a suitable sensor system able to monitor the risk of seed infection. Alternatively, if rapid and robust sensing systems for seed diagnosis were available, the sorting could be done on the combine harvester and grain could be handled in quality bands according to the level of seed infection. This option could possibly be used for detection of common bunt in wheat caused by the fungi *Tilletia tritici* and *T. laevis* due to the characteristic odor of bunt spores (Borjesson and Johnsson, 1998). However, the spatial variation in seed infection level is unknown. As these diseases have only one yearly generation and are introduced randomly by seeding, a spatially random pattern seems likely. However, in regions where soilborne inoculum is important in the disease cycle (Borgen, 2000) spatial variation in disease incidence may occur. For strictly seedborne and monocyclic pathogens, substantial spatial variation in seed infection levels is not expected and site-specific management of these diseases is not likely to be beneficial.

For polycyclic seedborne diseases, which also infect stems, leaves, and ears, the risk of seed infection may possibly be assessed by remote sensing of canopy spectral reflectance (Table 7.2). For Septoria glume blotch *(Septoria nodorum),* a positive correlation between the incidence of seed infection and disease severity on ear and flag leaf have been reported by Luke et al. (1986). Further, the incidence of seed infection was negatively correlated with plant density, presumably due to a negative effect of dense canopies on splash dispersal of spores (Shah et al., 1995). These results indicate that remote sensing at mid and/or late growing season may be useful in predicting the risk of seed infection. The extent of intrafield variation in seed infection is not known for this disease; as intrafield variation in canopy density is common, variable incidence of seed infection seems likely.

The potential for management of seedborne diseases by sorting of harvest has not been investigated yet. It is of primary interest to evaluate the intrafield variation in the incidence of seed infection and the relationship between this and visual symptoms of disease or other available ancillary information. If the level of seed infection is predictable from maps of canopy spectral reflectance, selective harvest will be possible with current available technologies. However, prediction of the level of seed infection per se may be difficult due to weather effects, among others. Consequently, it may be a more realistic target to predict gradients of seed infection risk. After harvest, the level of seed infection may then be determined for each quality band. Under the current scheme for certification of seed, a tolerance level of 20 to 30 percent infected seed is used, and above this level seed treatments are recommended. If site-specific management of seedborne diseases is to be relevant, intrafield variation of seed infection must have a magnitude that allows the production of seed lots with infection levels lower than this threshold. Naturally, the benefits will be largest in organic seed production where fungicides are not allowed and seedborne diseases are a considerable problem.

### *Disease Risk Prediction Based on Previous Crops*

For pathogens where the inoculum persists in soil or crop residue during the crop-free period, the pattern of disease risk in future crops may potentially be predicted from disease assessments in previous crops (Table 7.2). In fields where chemical disease control is not used, disease levels may be so high that maps of canopy spectral reflectance or yield maps in combination with field verification of disease occurrence may be a practicable route for disease monitoring. If soil tests for disease or nematode occurrence are used, variation in canopy spectral reflectance may potentially be used for division of the field into zones for sampling and management (Heath et al., 2000).

Sclerotinia stem rot caused by the fungus *Sclerotinia sclerotiorum* may potentially be managed based on disease occurrence in previous crops monitored by remote sensing. In a wide range of host plants, this disease causes wilting by blocking of the vascular system. In soybeans *(Glycine max),* disease incidences from 0 to 90 percent were found to be correlated with canopy spectral reflectance at 706 and 760 nm (Dudka et al., 1999). As Sclerotinia stem rot persists as soilborne sclerotia giving rise to ascospore-producing apothecia in coming seasons and disease incidence is correlated to the density of apothecia (Boland and Hall, 1988), zones of high disease risk in future crops may be identified. Mechanical movement of sclerotia in the soil and wind dispersal of ascospores results in aggregated disease patterns of Sclerotinia stem rot (Wegulo et al., 2000; Hartman et al., 1998) and indicate that remedial actions should be based on risk gradients. Management actions may include choice of another crop, fungicide application, or making canopy microclimate less disease conducive by reduced planting rate or increased row spacing (Dudka et al., 1999). Due to the wide host range of sclerotinia stem rot, disease risk may be affected by the history of the field. Temporal amalgamation of fields previously grown with different crops or localized weed problems may result in zones of variable risk of Sclerotinia stem rot (Dudka et al., 1999). Thus, the use of geographical information systems in field management is relevant in avoiding high-risk fields.

Attack of take-all (*Gaeumannomyces graminis* var. *tritici*) in the soil is often present at different levels and intensity in fields. Severe attack will often relate to more sandy areas of the field. If possible, mapping of the previous year's attack should be done based on yield variation or on remote sensing of premature senescence induced by take-all. Actions to be taken to minimize the risk of take-all could be either choosing of a nonsusceptible crop in areas which have previously had severe attack, delaying sowing date (Hornby et al., 1990), or use of seed treatments, which have proved to reduce and delay the outbreak of take-all (Beale et al., 1998). In the case of using seed treatments, only areas of the fields that are known to be at high risk should be treated.

## PERSPECTIVES

Recent results indicate that site-specific disease management is relevant for polycyclic cereal diseases such as powdery mildew and Septoria disease. Yield increases have resulted from adjustment of fungicide doses in winter wheat according to canopy spectral reflectance and disease incidence (Secher, 1997). Dose adjustment according to canopy spectral reflectance may be implemented with currently available technology (Bjerre and Secher, 1998), and the concept may prove relevant in other crop-disease systems, especially rice disease control. Experiments on dose adjustment according to disease incidence in cereals have been based on visual observation of disease. For commercial implementation, however, rapid disease sensing systems or models for prediction of disease risk gradients from ancillary data will be needed. In Table 7.2 a number of other potential applications of site-specific disease management have been listed.

Within the area of site-specific disease management, only a limited number of papers have yet been published, possibly due to a notion that only limited economic or environmental benefits are likely. As many diseases have a highly dynamic and often spatially unpredictable development, this notion may seem justified. The problem is even aggravated by the large differences in life cycles of different pathogens. Therefore, it should be emphasized that we are still in the early stages of validating the potential of site-specific disease management.

The cost-effectiveness of site-specific fungicide application according to canopy spectral reflectance is hard to evaluate, as the future prices for ground-based crop density monitoring and/or satellite imagery, for crop yield, and for increased management are difficult to estimate. However, it seems likely that the crop density maps needed for site-specific fungicide applications may also be used for site-specific fertilization and, assuming a positive profit for the latter (Østergaard, 1997), cost may be divided on two operations. At present, maps of canopy spectral reflectance are expected to cost 6.7 €·ha$^{-1}$ in Denmark if they are delivered by a machine pool as an additional service to fertilization maps at a cost of 13.4 €·ha$^{-1}$. At a wheat grain price of 93 to 134 €·ton$^{-1}$, a yield increase of 0.12 ton·ha$^{-1}$ will result in a positive margin over costs. This will be of interest for larger farms. Continued development of satellite systems for agricultural purposes may decrease costs and simplify management if high-resolution images of canopy spectral reflectance are delivered online at competitive prices.

One of the major constraints in implementing rational disease control is the farm manager's diagnosis of disease and assessment of disease incidence. Disease diagnosis based on visual assessment may be supplemented by molecular techniques based on physical sampling of the crop if processing time can be accepted (McCartney et al., 2003). In contrast, the assessment of disease incidence cannot be facilitated with current available technology; in this area, site-specific disease management will rely on field inspection just as conventional disease control does. However, maps of canopy spectral reflectance and other information layers can serve as input for disease risk models (Bjerre, 1999), and stratified disease observation according to predicted disease risk may reduce workload. Much work still needs to be done before clear suggestions can be made for management strategies that are based on real-time detection of early stages of disease epidemics and subsequent site-specific fungicide application. Fungicide application according to disease level will require sensing systems able to detect very low levels of disease as fungicide application after epidemics have evolved is not rational (Paveley et al., 1996). The position of disease within the crop is often more important than the total amount of disease per se, especially if the crop can tolerate a certain level of disease before economic damage occurs. Determination of disease on individual leaf levels makes an additional demand on sensors for many polycyclic diseases. Disease sensors may be relevant for diseases with very low treatment thresholds. For such diseases (e.g., potato late blight caused by *Phytophtora infestans*), however, it might be most appropriate to monitor the first occurrence of airborne inoculum in the field (McCartney et al., 2003). For the detection of small differences in disease occurrence likely to have an impact on future epidemic development (Paveley et al., 1997), remote sensing of canopy spectral reflectance does not seem to be a suitable candidate (Nilsson, 1995; West et al., 2003).

Fungicide application in cereals and rice is mainly done in Europe and East Asia, which consequently are the regions where site-specific fungicide application in agricultural crops will be of largest interest. In other parts of the world fungicide is mainly used in fruit and vegetable crops (Oerke, 1999). In these crops, however, advances in sensing systems and application methods may result in improved targeting of spray to spray targets and reduced fungicide consumption and environmental impact (Walklate et al., 2002; Miller, 2003).

Precision agriculture including site-specific disease management may be implemented in developing countries if adapted to local conditions and financial capacity (Srinivasan, 2001). Equipment for variable-rate treatment will not be within economic reach of most farmers, but remotely sensed data may potentially be used for delineation of management zones in high-resolution maps. If distributed to consultants or farmers, such maps may be used in site-specific disease management, either in guiding decisions on fungicide application rates or in avoiding sensitive crops in areas with high disease risk (Table 7.2). However, a number of other factors are often more important for crop yields in developing countries than diseases, including water and nutrient availability, weeds, and pests.

## *Research and Development*

For continued development of site-specific disease management several priorities for research and development can be made. Investigations of intrafield variation of disease, crop sensitivity to disease, and disease effects on crop yield and quality are needed. Focus will be on the influence of crop growth conditions and field environment on disease development and the interacting effects of crop growth conditions and disease on yield (Gaunt, 1995; Shtienberg, 1990). The risk of disease epidemic development and crop sensitivity to disease may be modeled based on such constant factors as soil type, topography, and location of shelter, within-season estimates of plant density and canopy development based on remote sensing of canopy spectral reflectance, and weather observations. In an integrated approach to nutrient and disease management, there will be a need for integration and interpretation of data from many sources and development of procedures for determining application requirements. Modeling of disease variation will be relevant only if a substantial part of this can be explained. This may not be the case in all crop-disease systems. In these cases, the modeling of crop sensitivity to disease may prove to be the most important aspect of site-specific disease management.

Development of advanced disease-sensing systems and the use of these in disease management strategies has a high priority. The combined use of several sensing methods may be the most direct route in obtaining biologically relevant information on disease levels and disease patterns. Thus, the use of sensing systems for soilborne diseases may be optimized by rational sampling plans based on analysis of yield maps and/or maps of canopy spectral reflectance. Development of molecular techniques for disease diagnosis and quantification will continue (McCartney et al., 2003) but will be confined to individual crop or soil samples, and require processing time. If these are to be used in site-specific disease management, methods of stratified sampling in zones of variable risk should be developed. Autonomous vehicles may rationalize physical sampling of soil or crop, and reduce the requirement for a short processing time because an autonomous vehicle can work around the clock. Potentially, machine vision and image analysis systems may be used for localization of diseased plants for physical sampling and for rapid assessment of disease severity at a leaf level (West et al., 2003). However, when used on autonomous vehicles this will demand advanced robot systems.

Continued validation of the benefits from site-specific disease management is essential for the acceptance of the concept by the farming industry (Wheelan and McBratney, 2000). Beyond controlled field experiments, this will include whole enterprise investigations, as site-specific disease management must be seen as part of a farmwide implementation of precision agriculture including all field operations and directed toward the management of yield, yield quality, and environmental impact (Stafford, 2000).

Practical testing of equipment for site-specific pesticide application done by regional farmer organizations in Denmark demonstrated that various standards for hardware and software in some cases were an obstacle (personal comm., Ole Møller Hansen). Thus, a continued commercial adaptation to uniform technical standards for tractors and implements is still needed (Stafford, 2000).

### *Drivers for Precision Farming and Site-Specific Disease Management*

Public concern over farming practices and environmental legislation may be a driving force in adaptation of precision farming practices (Stafford, 2000). Pesticide leakage from the root zone and other undesirable environmental effects has caused worry. Site-specific fungicide application in cereals may be part of a strategy for minimizing environmental impact, due to a reduced off-target deposition and potential reductions of fungicide consumption due to a more efficient use. Reductions in fungicide use may result from reductions in the value of crop yield or in some countries pesticide taxes. A shift in the average applied fungicide dose to the more sensitive part of the dose-response curve (Figure 7.3) may increase the yield benefit from an improved targeting of inputs by site-specific application.

At the farm level, an increase in field size and total acreage plus a high frequency of temporal field amalgamations will stimulate the use of geographical information systems in field management. Large fields are likely to stimulate the adoption of site-specific crop management, as benefits from this will increase due to increased intrafield soil and crop variability. A combined implementation of site-specific management of disease, fertilization, weeds, and pests may be relevant due to the possible sharing of data and/or equipment in these applications. As fungicide dose adjustment according to canopy spectral reflectance has proven beneficial, the continued implementation and extension of satellite systems for agricultural purposes may enhance their use, although satellite imagery may be a facultative data source due to cloud cover. In the absence of timely satellite data, conventional application may be chosen or application may rely on ground-based measurements.

## REFERENCES

Akkerhuis, G.A.J.M.J.O., J.A. Axelsen, and C. Kjær. 1998. Towards predicting pesticide deposition from plant phenology: A study in spring barley. *Pesticide Science* 53:252-262.

Anonymous. 2004. Reduktionsprogramm chemisher Pflanzenschutz. Bundesministerium Für Verbrauchershutz, Ernährung und Landwirtschaft. www.verbraucher ministerium.de.

Aslyng, H.C. 1958. Shelter and its effect on climate and water balance. *Oikos* 9:283-308.

Aylor, D.E. 1999. Biophysical scaling and the passive dispersal of fungus spores: Relationship to integrated pest management strategies. *Agricultural and Forest Meteorology* 97:275-292.

Ayres, P.G. and B. Woolacott. 1980. Effects of soil water level on the development of adult plant resistance to powdery mildew in barley. *Annals of Applied Biology* 94:255-263.

Bakhsh, A., D.B. Jaynes, T.S. Colvin, and R.S. Kanwar. 2000. Spatio-temporal analysis of yield variability for a corn-soybean field in Iowa. *Transactions of the ASAE* 43:31-38.

Basso, B., J.T. Ritchie, F.J. Pierce, R.P. Braga, and J.W. Jones. 2001. Spatial validation of crop models for precision agriculture. *Agricultural Systems* 68:97-112.

Beale, R.E., D.P. Phillion, J.M. Headrick, P. Oreilly, and J. Cox. 1998. MON 65500: A unique fungicide for the control of take-all in wheat. In *Brighton Crop Protection Conference, Pest and Diseases.* Farnham, UK: British Crop Protection Council, pp. 343-350.

Bjerre, K.D. 1999. Disease maps and site specific fungicide application in winter wheat. In *Precision Agriculture '99,* ed. J.V. Stafford. Sheffield, UK: Sheffield Academic Press, pp. 495-504.

Bjerre, K.D., L.N. Jørgensen, and B.J.M. Secher. 1998. Sygdomskort og positionsbestemt fungicidanvendelse. In *15th Danish Plant Protection Conference—Pests and Diseases.* Report no. 3. Tjele, Denmark: Danmarks Jordbrugs Forskning, pp. 131-144.

Bjerre, K.D. and B.J.M. Secher. 1998. Field experience with site-specific application of fungicides to winter wheat. In *Brighton Crop Protection Conference—Pests and Diseases 1998.* Farnham, UK: British Crop Protection Council, pp. 987-992.

Blackmore, S. 2000. The interpretation of trends from multiple yield maps. *Computers and Electronics in Agriculture* 26:37-51.

Blakeman, R.H., R.J. Bryson, and P. Dampney. 2000. Assessing crop condition in real time using high resolution satellite imagery. *Aspects of Applied Biology* 60:163-171.

Blight, M.M., C.H. Bock, K.J. Doughty, J.K. Fieldsend, and J.A. Pickett. 1995. Release of isothiocyanates from *Brassica rapa* during infection by *Alternaria brassicae.* In Proceedings of the 9th International Rapeseed Congress: Rapeseed Today and Tomorrow, Cambridge, UK, July 4-7, pp. 604-606.

Boland, G.J. and R. Hall. 1988. Relationships between the spatial pattern and number of apothesia of *Sclerotinia sclerotiorum* and stem rot of soybean. *Plant Pathology* 37:329-336.

Borgen, A. 2000. Perennial survival of common bunt *(Tilletia tritici)* in soil under modern farming practice. *Journal of Plant Diseases and Protection* 107:182-188.

Borjesson, T. and L. Johnsson. 1998. Detection of common bunt *(Tilletia caries)* infestation in wheat with an electronic nose and a human panel. *Journal of Plant Diseases and Protection* 105:306-313.

Bravo, C., D. Moshou, J. West, A. McCartney, and H. Ramon. 2003. Early disease detection in wheat fields using spectral reflectance. *Biosystems Engineering* 84:137-145.

Brent, K.J. 1985. One hundred years of fungicide use. In *Fungicides for Crop Protection, 100 Years of Progress,* ed. I.M. Smith. BCPC Monograph No. 31. Croydon, UK: British Crop Protection Council, pp. 11-22.

Broscious, S.C., J.A. Frank, and J.R. Frederick. 1985. The influence of winter wheat management practices on the severity of powdery mildew and Septoria blotch in Pennsylvania. *Phytopathology* 75:538-542.

Bryson, R.J., N.D. Paveley, W.S. Clark, R. Sylvester-Bradley, and R.K. Scott. 1997. Use of in-field measurements of green leaf area and incident radiation to estimate the effects of yellow rust epidemics on the yield of winter wheat. *European Journal of Agronomy* 7:53-62.

Burdon, J.J. and G.A. Chilvers. 1982. Host density as a factor in plant disease ecology. *Annual Review Phytopathology* 20:143-166.

Carter, G.A. 1993. Responses of leaf spectral reflectance to plant stress. *American Journal of Botany* 80:239-243.

Christensen, S. and J. Goudriaan. 1993. Deriving light intercepted and biomass from spectral reflectance ratio. *Remote Sensing of Environment* 43:87-95.

Christensen, S., T. Heisel, B.J.M. Secher, A. Jensen, and V. Haahr. 1997. Spatial variation of pesticide doses adjusted to varying canopy density in cereals. In *Precision Agriculture '97,* ed. J.V. Stafford. Oxford, UK: BIOS Scientific Publishers Ltd., pp. 211-218.

Chungu, C., J. Gilbert, and F. Townley-Smith. 2001. *Septoria tritici* blotch development as affected by temperature, duration of leaf wetness, inoculum concentration, and host. *Plant Disease* 85:430-435.

Dill-Macky, R. and A.P. Roelfs. 2000. The effect of stand density on the development of *Puccinia graminis* f.sp. *tritici* in barley. *Plant Disease* 84:29-34.

Dudka, M., S. Langton, R. Shuler, J. Kurle, and C.R. Grau. 1999. Use of digital imagery to evaluate disease incidence and yield loss caused by sclerotinia stem rot disease of soybeans. In *Proceedings of the Fourth International Conference on Precision Agriculture,* eds. P.C. Robert, R.H. Rust, and W.E. Larson. Madison, WI: ASA-CSSA-SSSA, pp. 1549-1558.

Emmerman, A. 1997. Program to reduce the risks connected with the use of plant protection products in Sweden. *EPPO Bulletin* 27:212-213.

Ewaldz, N.A.T. 2000. Radiometric readings as a tool for predicting optimal fungicide dose in winter wheat. *Zeitschrift für Pflanzenkrankheiten and Pflanzenschutz* 107:594-604.

Ferrandino, F.J. 1993. Dispersive epidemic waves: 1. Focus expansion within a linear planting. *Phytopathology* 74:969-802.

Finckh, M.R., E.S. Gacek, H.J. Czembor, and M.S. Wolfe. 1999. Host frequency and density effects on powdery mildew and yield in mixtures of barley cultivars. *Plant Pathology* 48:807-816.

Fitt, B.D.L., P.H. Gregory, A.D. Todd, H.A. McCartney, and O.C. MacDonald. 1987. Spore dispersal and plant disease gradients: A comparison between two empirical models. *Journal of Phytopathology* 118:227-242.

Fraaije, B.A., J.A. Lucas, B. Clark, and F.J. Burnett. 2003. QoI resistance development in populations of cereal pathogens in the UK. In *Proceedings of the BCPC International Congress Crop Science and Technology 2003.* Glasgow, Scotland, UK: SECC, pp. 689-694.

Frahm, J., T. Volk, and A. Johnen. 1996. Development of the Pro-Plant decision support system for plant protection in cereals, sugarbeet and rape. *Bulletin EPPO* 26:609-622.

Friedrich, S. 1994. Prognose der Infectionswahrscheinlichkeit durch Ecten Mehltau an Winterweizen (*Erysiphe graminis* DC. f. sp. *tritici*) anhand meteorologischer Eingangsparameter. PhD thesis, Technischen Universität zu Braunschweig, Germany.

Garbutt, N. 2000. Meeting consumers demands for food safety: European retailer protocols for crop protection programs. In *Brighton Crop Protection Conference, Pest and Diseases.* Farnham, UK: British Crop Protection Council, pp. 129-132.

Gaunt, R.E. 1995. The relationship between plant disease severity and yield. *Annual Review Phytopathology* 33:119-144.

Gaunt, R.E. and M.J. Cole. 1992. Spatial analysis of wheat stripe rust epidemics. *Crop Protection* 11:131-137.

Gavassoni W.L., G.L. Tylka, and G.P. Munkvold. 2001. Relationships between tillage and spatial patterns of *Heterodera glycines*. *Phytopathology* 91:534-545.

Gent, M.P.N. 1994. Photosynthate reserves during grain filling in winter wheat. *Agronomy Journal* 86:159-167.

Giles, D.K., M.K. Stone, and K. Dieball. 1999. Distributed network system for control of spray droplet size and application rate for precision chemical application. In *Precision Agriculture '99,* ed. J.V. Stafford. Sheffield, UK: Sheffield Academic Press, pp. 857-866.

Gisi, U., K.M. Chin, G. Knapova, R.K. Farber, U. Mohr, S. Parisi, H. Sierotzki, and U. Steinfeld. 2000. Recent developments in elucidating modes of resistance to phenylamide, DMI and strobilurin fungicides. *Crop Protection* 19:863-872.

Grindlay, D.J.C. 1997. Towards an explanation of crop nitrogen demand based on the optimization of leaf nitrogen per unit leaf area. *Journal of Agricultural Science, Cambridge* 128:377-396.

Gyldenkærne, S., B.J.M. Secher, and E. Nordbo. 1999. Ground deposit of pesticides in relation to the cereal canopy density. *Pesticide Science* 55:1210-1216.

Hartman, G.L., L. Kull, and Y.H. Huang. 1998. Occurrence of *Sclerotinia sclerotiorum* in soybean fields in East-Central Illinois and enumeration of inocula in soybean seed lots. *Plant Disease* 82:560-564.

Hatfield, J.L. and P.J. Pinter. 1993. Remote sensing for crop protection. *Crop Protection* 12:403-413.

Heath, W.L., P.P.J. Haydock, A. Wilcox, and K. Evans. 2000. The potential use of spectral reflectance from potato crop for remote sensing of infection by potato cyst nematodes. *Aspects of Applied Biology* 60:185-188.

Henriksen. K.E., L.N. Jørgensen, and G.C. Nielsen. 2000. PC-plant protection—A tool to reduce fungicide input in winter wheat, winter barley and spring barley in Denmark. In *Brighton Crop Protection Conference, Pest and Diseases.* Farnham, UK: British Crop Protection Council, pp. 835-840.

Hislop, E.C., N.M. Western, and R. Butler. 1995. Experimental air-assisted spraying of a maturing cereal crop under controlled conditions. *Crop Protection* 14:19-26.

Hollomon, D.W. 1998. Diagnostics in modern disease control strategies. In *Brighton Crop Protection Conference, Pest and Diseases.* Farnham, UK: British Crop Protection Council, pp. 771-778.

Hornby, D., G.L. Bateman, and R.J. Gutteridge. 1990. Experiments in England and France on fertilizers, fungicides and agronomic practices to decrease take-all. In *Brighton Crop Protection Conference, Pest and Diseases.* Thornton Heath, UK: British Crop Protection Council. pp. 771-776.

Jamieson, P.D., J.R. Porter, J. Goudriaan, J.T. Ritchie, H. van Keulen, and W. Stol. 1998. A comparison of the models AFRCWHEAT2, CERES-Wheat, Sirius, SUCROS2 and SWHEAT with measurements from wheat grown under drought. *Field Crops Research* 55:23-44.

Jamieson, P.D. and M.A. Semenov. 2000. Modeling nitrogen uptake and redistribution in wheat. *Field Crops Research* 68:21-29.

Jeger, M.J. 1999. Improving understanding of dispersal in crop pest and disease management: Current status and future directions. *Agricultural and Forest Meteorology* 97:331-349.

Jensen, B. and L. Munk. 1997. Nitrogen induced changes in colony density and spore production of *Erysiphe graminis* f.sp. *hordei* on seedlings of six spring barley cultivars. *Plant Pathology* 46:191-202.

Joernsgaard, B. and S. Halmoe. 2003. Intra-field variation over crops and years. *European Journal of Agronomy* 19:23-33.

Johnston, H.W., J.A. MacLeod, and K.S. Clough. 1979. Effects of cycocel (CCC) and fungicide sprays on spring wheat grown at three nitrogen levels. *Canadian Journal of Plant Science* 59:917-929.

Jones, A.L., C.A. Mielke, and C.A. Miller. 1989. Relationship between landscape position and properties to crop production. *Journal of Soil Water Conservation* 44:328-332.

Jørgensen, L.N. 1994. Duration of effects of EBI-fungicides when using reduced rates in cereals. In *Brighton Crop Protection Conference—Pests and Diseases.* Farnham, UK: British Crop Protection Council, pp. 703-710.

Jørgensen, L.N. 1997. The Danish Pesticide Plan—Facts and reality. *EPPO Bulletin* 27:206-209.

Jørgensen, L.N., K.E. Henriksen, and G.C. Nielsen. 1999. Adjusting thresholds for Septoria control in winter wheat using strobilurins. In *Septoria and Stagonospora diseases of cereals: A compilation of global research,* eds. M. Van Ginkel, A. Mc Nab, and J. Krupinsky. El Batan, Texcoco, Mexico: CIMMYT, pp. 173-175. Available online at www.cimmyt.org/english/wps/publs/catalogdb/index.cfm.

Jørgensen, L.N., G.C. Nielsen, and K.E. Henriksen. 2000. Margin over cost in disease management in winter wheat and spring barley in Denmark. In *Brighton Crop Protection Conference, Pest and Diseases.* Farnham, UK: British Crop Protection Council. pp. 655-662.

Jørgensen, L.N., B.J.M. Secher, and G.C. Nielsen. 1996. Monitoring diseases of winter wheat on both a field and at national level. *Crop Protection* 13:383-390.

Jørgensen, L.N., B.J.M. Secher, J.E. Olesen, and J. Mortensen. 1997. Need for fungicide treatments when varying agricultural parameters. *Aspects of Applied Biology* 50:285-292.

Kalberer N. and U. Gisi. 1997. Effect of soil matric potential on sharp eyespot in germinating wheat following seed treatment. *Zeitschrift für Pflanzenernahrung und Bodenkunde* 160:195-199.

Kitchen, N.R., S.T. Drummond, E.D. Lund, K.A. Suddoth, and G.W. Buchleiter. 2003. Soil electrical conductivity and topography related to yield for three contrasting soil-crop systems. *Agronomy Journal* 95:483-495.

Knight, S.C., V.M. Anthony, A.M. Brady, A.J. Greenland, S.P. Heany, D.C. Muray, K.A. Powell, M.A. Schultz, C.A. Spinks, P.A. Worthington, and D. Youle. 1997. Rationale and perspectives on the development of fungicides. *Annual Review of Phytopathology* 35:349-372.

Kobayashi, T., E. Kanda, K. Kitada, K. Ishiguro, and Y. Torigoe. 2001. Detection of rice panicle blast with multispectral radiometer and the potential using airborne multispectral scanners. *Phytopathology* 91:316-323.

Koch, H. 1980. Räumliche Befallsmuster beim echten Mehltau der Gerste. *Zeitschrift für Pflanzenkrankheiten und Pflanzenshutz* 87:731-737.

Kocks, C.G., J. Zadoks, and T.A. Ruissen. 1998. Response of black rot in cabbage to spatial distribution of inoculum. *European Journal of Plant Pathology* 104:713-723.

Kravchenko, A.N., D.G. Bullock, and C.W. Boast. 2000. Joint multifractal analysis of crop yield and terrain slope. *Agronomy Journal* 92:1279-1290.

Kutcher, H.R., S.S. Malhi, A.M. Johnston, and G. Hnatowich. 1999. Impact of topography and management on diseases of canola and wheat. In *Proceedings of the Fourth International Conference on Precision Agriculture,* eds. P.C. Robert, R.H. Rust, and W.E. Larsen. Madison, WI: ASA/CSSA/SSSA, pp. 559-562.

Lemaire, G. and F. Gastal. 1997. N uptake and distribution in plant canopies. In *Diagnosis of nitrogen status in crops,* ed. G. Lemaire. Berlin: Springer-Verlag, pp. 3-43.

Lockley, K.D., N.D. Paveley, A.J. Leadbeater, J.A. Smith, and G.M. Booth. 1996. The value of ELISA diagnostics as tools to optimize fungicide use for the control of *Septoria tritici* in winter wheat. In *Diagnostics in Crop Production, BCPC Symposion Proceedings No 65.* Farnham, UK: British Crop Protection Council, pp. 349-356.

Lorenzen, B. and A. Jensen. 1989. Changes in leaf spectral properties induced in barley by cereal powdery mildew. *Remote Sensing of Environment* 27:201-209.

Lovell, D.J., S.R. Parker, T. Hunter, D.J. Royle, and R.R. Cocker. 1997. Influence of crop growth and structure on the risk of epidemics by *Mycosphaerella graminicola (Septoria tritici)* in winter wheat. *Plant Pathology* 46:126-138.

Luke, H.H., R.D. Barnett, and P.L. Pfahler. 1986. Development of *Septoria nodorum* blotch on wheat from infected and treated seed. *Plant Disease* 70:252-254.

Lund, I. 2000: Sprayer nozzles for precision pesticide application. In *AgEng2000,* Warwick, 00-PA-017, p. 8.

Malthus, T.J. and A.C. Madeira. 1993. High resolution spectroradiometry: Spectral reflectance of field bean leaves infected by *Botrytis fabae. Remote Sensing of Environment* 45:107-116.

Marshall, J.K. 1967. The effect of shelter on the productivity of grassland and field crops. *Field Crop Abstracts* 20:1-14.

Matthews, R.B. and N.D. Cosser. 1997. Using crop simulation models to develop treatment maps in precision agriculture. *Aspects of Applied Biology* 50:181-190.

McBratney, A.B. and M.J. Pringle. 1997. Spatial variability in soils—Implications for precision agriculture. In *Precision Agriculture 1997,* ed. J.V. Stafford. Oxford, UK: BIOS Scientific Publishers Ltd., pp. 3-31.

McCartney, H.A., S.J. Foster, B.A. Fraaije, and E. Ward. 2003. Molecular diagnostics for fungal plant pathogens. *Pest Management Science* 59:129-142.

Miller, P.C.H. 2003. Patch spraying: Future role of electronics in limiting pesticide use. *Pest Management Science* 59:566-574.

Miller, P.C.H. and J.H. Combellack. 1997. The performance of an air/liquid nozzle system suitable for applying herbicides in a spatially selective manner. In *Precision Agriculture '97,* ed. J.V. Stafford. Oxford, UK: BIOS Scientific Publishers Ltd., pp. 651-659.

Miller, P.C.H., A.G. Lane, and H.C. Wheeler. 2000. Matching the application of fungicides to crop canopy characteristics. In *Brighton Crop Protection Conference, Pest and Diseases.* Farnham, UK: British Crop Protection Council, pp. 629-636.

Monteith, J.L. 1977. Climate and the efficiency of crop production in Britain. *Philosophical Transactions of the Royal Society of London* (Series B) 281:277-294.

Moran, M.S. 2000. New imaging sensor technologies suitable for agricultural management. *Aspects of Applied Biology* 60:1-10.

Newton, A.C. and I.M. Young. 1996. Temporary partial breakdown of Mlo-resistance in spring barley by sudden relief of soil water stress. *Plant Pathology* 45:973-977.

Nilsson, H.E. 1995. Remote sensing and image analysis in plant pathology. *Annual Review Phytopathology* 15:489-527.

Nuspl, S.J., W.W. Rudolph, and R. Guthland. 1996. Use of injection for site-specific chemical application. In *Proceedings of the 3rd International Conference on Precision Agriculture,* eds. P.C. Robert, R. H. Rust, and W.E. Larson. Madison, WI: ASA-CSSA-SSSA, pp. 739-744.

Nutter, F.W. 1989. Detection and measurement of plant disease gradients in peanut with a multispectral radiometer. *Phytopathology* 79:958-963.

Oerke, E.C. 1999. The importance of disease control in modern plant production. In *Modern fungicides and antifungal compounds II,* eds. H. Lyr, P.E. Russel, H.W. Dehne, and H.D. Sisler. Andover, UK: Intercept, pp. 11-17.

Oerke, E.C. and H.W. Dehne 1997. Global crop production and the efficacy of crop protection—Current situation and future trends. *European Journal of Plant Pathology* 103:203-215.

Oerke, E.C., H.W. Dehne, F. Schonbeck, and A. Weber. 1994. *Crop protection and crop production.* Amsterdam: Elsevier.

Oerke, E.C. and F. Schönbeck. 1986. Zum Einfluss abiotischer Stressbedingungen auf das Wachstum von Gerste und Bohne und deren Prädisposition gegenüber Schaderregern. *Zeitschrift für Pflanzenkrankheiten und Pflanzenshutz* 93:561-573.

O'Hara, R.B. and J.K.M. Brown. 1996. Immigration of the barley powdery mildew pathogen into field plots of barley. *Plant Pathology* 45:1071-1076.

O'Hara, R.B. and J.K.M. Brown. 1997. Spatial aggregation of pathotypes of barley powdery mildew. *Plant Pathology* 46:969-977.

Office for Official Publications of the European Communities. 1991. Council directive 91/414/EEC: Council directive of 15 July 1991 concerning the placing of plant protection products on the market (91/414/EEC). Available online at europa.eu.int/comm/food/plant/protection/index_en.htm.

Olesen, J.E., J. Berntsen, E.M. Hansen, B.M. Petersen, and J. Petersen. 2002. Crop nitrogen demand and canopy area expansion in winter wheat during vegetative growth. *European Journal of Agronomy* 16:279-296.

Olesen, J.E. and K. Grevsen. 1997. Effects of temperature and irradiance on vegetative growth of cauliflower (*Brassica oleracea* L. *botrytis*) and broccoli (*Brassica oleracea* L. *italica*). *Journal of Experimental Botany* 48:1591-1598.

Olesen, J.E., T. Jensen, and J. Petersen. 2000. Sensitivity of field-scale winter wheat production in Denmark to climate variability and climate change. *Climate Research* 15:221-238.

Olesen, J.E., L.N. Jørgensen, and J.V. Mortensen. 2000. Nitrogen application, irrigation strategy and fungicide control in winter wheat on a sandy soil: II. Radiation interception and conversion. *Journal of Agricultural Science, Cambridge* 134:13-23.

Olesen, J.E., L.N. Jørgensen, J. Petersen, and J.V. Mortensen. 2003a. Effects of rates and timing of nitrogen fertiliser on disease control by fungicides in winter wheat: 1. Crop yield and nitrogen uptake. *Journal of Agricultural Science, Cambridge* 140:1-13.

Olesen, J.E., L.N. Jørgensen, J. Petersen, and J.V. Mortensen. 2003b. Effects of rates and timing of nitrogen fertiliser on disease control by fungicides in winter wheat: 2. Crop growth and disease development. *Journal of Agricultural Science, Cambridge* 140:15-29.

Østergaard, H.S. 1997. Agronomic consequences of variable N fertilization. In *Precision Agriculture 1997,* ed. J.V. Stafford. Oxford, UK: BIOS Scientific Publishers Ltd., pp. 315-327.

Parker, S.R., M.W. Shaw, and D.J. Royle. 1997. Measurements of spatial patterns of disease in winter wheat crops and the implications for sampling. *Plant Pathology* 46:470-480.

Paveley, N.D. and W.S. Clark. 2000. *The wheat disease management guide.* London: Home Grown Cereals Authority.

Paveley, N.D., W.S. Clark, R. Sylvester-Bradley, R.J. Bryson, and P. Dampney. 1996. Responding to inter- and intra-field variation to optimize foliar disease management in wheat. In *Brighton Crop Protection Conference—Pests and Diseases.* Farnham, UK: British Crop Protection Council, pp. 1227-1234.

Paveley, N.D., K.D. Lockley, R. Sylvester-Bradley, and J. Thomas. 1997. Determinants for fungicide spray decisions for wheat. *Pesticide Science* 49:379-388.

Paveley, N.D., K.D. Lockley, T.B. Vaughan, J. Thomas, and K. Schmidt. 2000. Predicting effective fungicide doses through observation of leaf emergence. *Plant Pathology* 49:748-766.

Permin, O., L.N. Jørgensen, and K. Persson. 1992. Deposition characteristics and biological effectiveness of fungicide applied to winter wheat and the hazards of drift when using different types of hydraulic nozzles. *Crop Protection* 11:541-546.

Pohland, A.E. 1993. Mycotoxins in review. *Food Additives and Contaminants* 10:17-28.

Porter, J.R. 1993. AFRCWHEAT2: A model of the growth and development of wheat incorporating responses to water and nitrogen. *European Journal of Agronomy* 2:69-82.

Raikes, C. and L.L. Burpee. 1998. Use of multispectral radiometry for assessment of rhizoctonia blight in creeping bentgrass. *Phytopathology* 88:446-449.

Rosenberg, N.J., B.L. Blad, and S.B. Verma. 1983. *Microclimate: The biological environment.* New York: John Wiley & Sons.

Sadler, E.J., P.J. Bauer, and W.J. Busscher. 2000. Site-specific analysis of a droughted corn crop: I. Growth and grain yield. *Agronomy Journal* 92:395-402.

Schaad, N.W., R.D. Frederick, J. Shaw, W.L. Schneider, R. Hickson, M.D. Petrillo, and D.G. Luster. 2003. Advances in molecular-based diagnostics in meeting crop biosecurity and phytosanitary issues. *Annual Review of Phytopathology* 41:305-324.

Scharf, P.C., J.P. Schmidt, N.R. Kitchen, K.A. Sudduth, S.Y. Hong, J.A. Lory, and J.G. Davis. 2002. Remote sensing for nitrogen management. *Journal of Soil and Water Conservation* 57:518-524.

Schneider, E.F. and K.J. Dickert. 1994. Health costs and benefits of fungicide use in agriculture: A literature review. *Journal of Agromedicine* 1:19-37.

Schuh, W. 1990a. The influence of tillage systems on incidence and spatial pattern of tan spot of wheat. *Phytopathology* 80:804-807.

Schuh, W. 1990b. Influences of tillage systems on disease intensity and spatial pattern of *Septoria* leaf blotch. *Phytopathology* 80:1337-1340.

Schwinn, F.J. 1992. Significance of fungal pathogens in crop production. *Pesticide Outlook* 3:18-25.

Secher, B.J.M. 1997. Site specific control of diseases in winter wheat. *Aspects of Applied Biology* 48:57-64.

Secher, B.J.M., L.N. Jørgensen, N.S. Murali, and P.S. Boll. 1995. Field validation of a decision support system for the control of pests and diseases in cereals in Denmark. *Pesticide Science* 45:195-199.

Secher, B.J.M., N.S. Murali, and K.E. Gadegård. 1995. Site specific control of plant diseases—A great potential. In *Seminar on site specific farming,* ed. S.E. Olesen. Tjele, Denmark: Danish Institute of Plant and Soil Sciences, SP-report 26, pp. 165-169.

Shah, D., G.C. Bergstrom, and P.P. Ueng. 1995. Initiation of *Septoria nodorum* blotch epidemics in winter wheat by seedborne *Stagonospora nodorum. Phytopathology* 85:452-457.

Shaw, M.W. and D.J. Royle. 1993. Factors determining the severity of epidemics of *Mycosphaeerella graminicola (Septoria tritici)* on winter wheat in the UK. *Plant Pathology* 42:882-899.

Shtienberg, D. 1990. Effects of foliar diseases of wheat on the physiological processes affection yield under semiarid conditions. *Plant Pathology* 40:533-541.

Sinclair, T.R. and R.C. Muchow. 2000. Radiation use efficiency. *Advances in Agronomy* 65:215-265.

Srinivasan, A. 2001. Relevance of precision farming technologies to sustainable agriculture in Asia and the Pacific. Proceedings of the Second Asia-Pacific Conference on Sustainable Agriculture, October 18-20, 1999, Phitsanulok, Thailand.

Stafford, J.V. 1999. An investigation into the with-in field spatial variability of grain quality. In *Precision agriculture '99,* ed. J.V. Stafford. Sheffield, UK: Sheffield Academic Press, pp. 353-361.

Stafford, J.V. 2000. Implementing precision agriculture in the 21st century. *Journal of Agricultural Engineering Research* 76:267-275.

Sylvester-Bradley, R., R.K. Scott, D.T. Stokes, and R.W. Clare. 1997. The significance of crop canopies for N nutrition. *Aspects of Applied Biology* 50:103-116.

Thomsen, A., K. Schelde, T. Heidmann, and H. Hougaard. 1997. Mapping of field variability in crop development and water balance within a field with highly variable soil conditions. In *Precision agriculture 1997,* ed. J.V. Stafford. Oxford, UK: BIOS Scientific Publishers Ltd., pp. 189-196.

Thorne, G.N., L. Pearman, W. Day, and A.D. Todd. 1988. Estimation of radiation interception by winter wheat from measurements of leaf area. *Journal of Agricultural Science, Cambridge* 110:101-108.

Tiedemann, A.V. 1996. Single and combined effects of nitrogen fertilization and ozone on fungal leaf diseases on wheat. *Journal of Plant Diseases and Protection* 103:409-419.

Tompkins, D.K., D.B. Fowler, and A.T. Wright. 1993. Influence of agronomic practices on canopy microclimate and Septoria development in no-till wheat produced in the Parkland region of Saskatchewan. *Canadian Journal of Plant Science* 73:331-344.

Tompkins, D.K., A.T. Wright, and D.B. Fowler. 1992. Foliar disease development in no-till winter wheat: Influence of agronomic practices on powdery mildew development. *Canadian Journal of Plant Science* 72:965-972.

Van Alphen, C.A.M. 1997. A strategy for reduced use of plant protection products in the Netherlands. *EPPO Bulletin* 27:210-211.

Walklate, P.J., J.V. Cross, G.M. Richardson, R.A. Murray, and D.E. Baker. 2002. Comparison of different spray volume deposition models using LIDAR measurements of apple orchards. *Biosystems Engineering* 82:253-267.

Walter, A.M., T. Heisel, and S. Christensen 2001. Precision application of herbicides using injection sprayer systems. In *Proceedings of 3rd European Conference on Precision Agriculture,* eds. G. Grenier and S. Blackmore. June 18-20, 2001, Montpellier, pp. 611-616.

Wegulo, S.N., P. Sun, C.A. Martinson, and X.B. Yang. 2000. Spread of Sclerotinia stem rot of soybean from area and point sources of apothecial inoculum. *Canadian Journal of Plant Science* 80:389-402.

West, J.S., C. Bravo, R. Oberti, D. Lemaire, D. Moshou, and H.A. McCartney. 2003. The potential of optical canopy measurement for targeted control of field crop diseases. *Annual Review of Phytopathology* 41:593-614.

Whelan, H.G., R.E. Gaunt, and W.R. Scott. 1997. The effect of leaf rust *(Puccinia hordei)* on yield response in barley (*Hordeum vulgare* L.) crops with different yield potential. *Plant Pathology* 46:397-406.

Whelan, B.M. and A.M. McBratney. 2000. The "null hypothesis" of precision agriculture management. *Precision Agriculture* 2:265-279.

Wiegand, C.L., S.J. Maas, J.K. Aase, J.L. Hatfield, P.J. Pinter (Jr.), R.D. Jackson, E.T. Kanemasu, and R.L. Lapitan. 1992. Multisite analyses of spectral-biophysical data for wheat. *Remote Sensing of Environment* 42:1-21.

Young, C.S., N.D. Paveley, T.B. Vaughan, J.M. Thomas, and K.D. Lockley. 2003. Predicting epidemics of yellow rust *(Puccinia striiformis)* on the upper canopy of wheat from disease observations on lower leaves. *Plant Pathology* 52: 338-349.

Zadoks, J.C. and F. van den Bosch. 1994. On the spread of plant disease: A theory of foci. *Annual Review Phytopathology* 32:503-521.

Chapter 8

# Site-Specific Management of Plant-Parasitic Nematodes

Robert A. Dunn
Jimmy R. Rich
Richard E. Baird

## *INTRODUCTION*

Research into site-specific management of plant-parasitic nematodes through use of GPS technology is in its infancy. For GPS technology to be used successfully in nematode management and other aspects of nematology, a general understanding of nematode biology, spatial distribution, and management procedures is necessary. Thus, the next few pages deal with those topics, followed by a discussion of published research and possible future directions to increase the role of GPS technology in nematology and nematode management.

The overview of biology, diagnosis, and management of plant-parasitic nematodes is a composite of the knowledge and experience of the authors, with literature cited to provide backup for a limited number of statements. For additional detail and different perspectives on general plant nematology, the reader is encouraged to go to any of several general textbooks and references, including Barker et al. (1998), Shurtleff and Averre (2000), Weischer and Brown (2000), and Chen et al. (2004).

## *GENERAL NEMATODE BIOLOGY*

Plant-parasitic nematodes belong in the phylum Nematoda, and one or more species damage all agricultural crops in the world. Worldwide losses from nematodes are difficult to estimate, but in 1987 the cost was placed at more than $77 billion dollars annually (see discussion in Barker et al.,

*Handbook of Precision Agriculture*

doi:10.1300/5627_08

1998). Nematodes also are of great concern to golf course management, ornamental horticulturalists, and home owners, since they can cause severe damage to turfgrass, ornamentals, and home gardens. We often are unaware of losses caused by nematodes, since they are hidden from sight and much of the damage caused by them goes unreported or is attributed to other causes.

Nematodes that attack plants are mostly microscopic in size and cylindrical in shape. Females of a few species lose their worm shape as they mature, becoming greatly enlarged in diameter and assuming varying forms, such as pear, lemon, or kidney shapes. Despite their small size, nematodes are complex in organization. Plant-parasitic nematodes possess all of the major organ systems of higher animals except respiratory and circulatory systems. The body is covered by a multilayered cuticle that is transparent. Thus, sufficient anatomical detail can be seen with the aid of a low-power stereoscopic "dissecting" microscope, generally 20 to 60 X, to identify most nematode specimens to genus. Much higher magnification (900 X or more) often is needed to identify species.

### *Life Cycle and Reproduction*

Plant parasitic nematodes have a simple life cycle, usually six stages: egg, four (in rare cases only three) juvenile stages, and adult. The embryo develops inside the egg to become the first-stage juvenile, which then molts inside the egg to become a second-stage juvenile, which hatches from the egg. The nematode molts three more times to become a fully developed adult. The length of the life cycle varies considerably depending on nematode species, host plant, and the temperature of the habitat. When soil temperatures are near 27°C, many plant nematodes complete their life cycles in about 30 days. The same species may need 90 days to complete a life cycle when soil temperature is near 16°C. Male and female nematodes occur in most species, but reproduction without males is common. The number of eggs deposited by a female varies among species, with most species producing between 50 and 500 eggs.

### *Nematode Feeding and Host-Parasite Relationships*

Plant parasitic nematodes feed on living plant tissues. All have some form of oral stylet or spear somewhat like a hypodermic needle used to puncture the host cell wall. Plant nematodes inject enzymes into the host cell before feeding. These enzymes partially digest the cell contents before they are sucked into the gut. Most of plant injury caused by nematodes is related in some way to the feeding process.

Nematodes may feed on plant tissues from outside the plant (ectoparasitic) or inside the tissues (endoparasitic). If the adult female moves freely through the soil or plant tissues, the species is said to be "migratory." Species in which the adult females become swollen and permanently immobile in one place in or on a root are termed "sedentary."

The feeding/living relationships that nematodes have with their hosts affect sampling methods and the success of management practices. Ectoparasitic nematodes that never enter roots may be recovered only from soil samples. Endoparasitic nematodes often are detected most easily in samples of the tissues in which they feed and live (burrowing and lesion nematodes), but some occur more commonly as migratory stages in the soil (root-knot and reniform nematodes).

Stages of endoparasites which are inside root tissues may be protected from nematicides that do not penetrate into roots, such as some soil fumigants. Root tissues also may shield them from microorganisms that attack nematodes in the soil. Ectoparasites are fully exposed to pesticides and natural control agents in the soil.

### *Ectoparasitic Nematodes*

Ectoparasitic nematodes generally are migratory. Most feed superficially at or very near the root tip or on root hairs, but a few have stylets long enough to enable them to feed deeper in the root. Those that cause the most widespread and severe plant injury are the sting (*Belonolaimus* spp.), stubby-root (*Trichodorus* spp.), awl (*Dolichodorus* spp.), and dagger (*Xiphinema* spp.) nematodes. These feed at or near root tips and usually inhibit root elongation.

Among plant nematodes, only stubby-root nematodes and their close relatives, the dagger and needle (*Longidorus* spp.) nematodes, are known to transmit plant viruses. The corky ringspot disease of potatoes ("spraing" in Europe) is caused by a virus that is carried and transmitted by stubby-root nematodes.

### *Migratory Endoparasites*

Migratory endoparasites can move into, through, and out from host tissues at any stage of development (except the egg). Migratory endoparasites generally live and feed in tender tissues such as the root cortex. They burrow through the tissue, breaking open many cells after feeding on them.

Cells surrounding the feeding area often are killed by toxic materials from the disrupted ones. The relatively large areas of dead cells usually turn brown to become small spots or lesions and often are colonized easily by fungi. Root rot diseases often are associated with infestations of migratory endoparasitic nematodes. Ectoparasitic and migratory endoparasitic nematodes generally deposit their eggs singly as they are produced, wherever the female happens to be in the soil or plant.

Important examples of migratory endoparasites are species of burrowing (*Radopholus* spp.) and lesion (*Pratylenchus* spp.) nematodes. The foliar nematodes (*Aphelenchoides* spp.) are important representatives of a group of migratory nematodes that attack plants at the soil line or above the ground. They feed on or inside the leaves and buds of ferns, strawberries, chrysanthemums, and many kinds of foliage ornamentals.

### *Sedentary Endoparasitic Nematodes*

Sedentary endoparasitic nematodes are typified by the root-knot (*Meloidogyne* spp.), cyst (species of *Heterodera* and related genera), reniform (*Rotylenchulus* spp.), and citrus (*Tylenchulus semipenetrans*) nematodes. These nematodes produce "giant" or "nurse" cells which become specialized food sources for the nematode. At the same time, the nematode becomes immobile, and the body swells to a round, lemon, kidney, or ovoid form.

Mature females of the sedentary endoparasitic nematodes generally produce large numbers of eggs which remain in their bodies or accumulate in masses attached to their bodies. Sedentary endoparasites damage their hosts by redirecting large amounts of energy and nutrients from normal plant activities to development of the nematodes and their special feeding sites.

## *DIAGNOSING NEMATODE PROBLEMS*

It can be very difficult to decide if nematodes are causing or are likely to cause significant crop injury. In a location for which a complete history is lacking, the identities and population densities of nematode pests and the severity of damage that might be expected from them usually can be determined by laboratory assay of soil and/or plant samples.

Another common diagnostic problem is determining the role of nematodes when established plants are making unsatisfactory growth. This task often is difficult because few nematodes cause distinctive diagnostic symptoms. A sound diagnosis should be based on as many as possible of symptoms above and below the ground, field history, diagnostic nematicide tests, and laboratory assay of soil and/or plant samples.

Aboveground symptoms are rarely, if ever, sufficient evidence to diagnose a root nematode problem. Since most plant nematodes affect root functions, most symptoms associated with them are the result of inadequate water supply or mineral nutrition to the tops: chlorosis (yellowing) or other abnormal coloration of foliage, stunted top growth, failure to respond normally to fertilizers, small or sparse foliage, a tendency to wilt more readily than healthy plants, and slower recovery from wilting. Woody plants in advanced stages of decline incited by nematodes will have little or no new foliage when healthy plants have substantial flushes, and eventually exhibit dieback of progressively larger branches.

The distribution of nematodes within any site is very irregular. Therefore, the shape, size, and distribution of areas showing the most severe effects of nematodes will be highly irregular within the field. Nematodes move very few feet per year on their own. In the undisturbed soil of groves, turf, and pastures, visible symptoms of nematode injury normally appear as round, oval, or irregular areas which gradually increase in size year by year. In cultivated land, nematode-injured spots often are elongated in the direction of cultivation, because nematode-laden soil is moved by machinery.

Belowground symptoms are more useful than foliar symptoms for diagnosing many nematode problems. Galls, abbreviated roots, necrotic lesions in the root cortex, and root rotting all may help in diagnosing nematode problems.

### *Field History*

Field history can provide valuable clues to the identity of nematode and other pest problems. If production records show a gradual decline in yields over a period of years, despite no change in cultural practices, nematodes may be indicated. A nematode infestation in a new field usually begins in a small area. It gradually intensifies in the original spot and is spread through the field by cultivation, harvest, erosion, and other factors that spread infested soil or plant parts. Therefore, the total effect of a recently introduced nematode is a gradual production decline for the field, as the percentage of the field that is involved and the severity of damage at any given area in the field increase over the years. Effective use of GPS/GIS technology could make it much easier to track development of such a situation and treat areas within the field according to need as shown by the GIS data.

### *Laboratory Assay*

Sample analysis often is necessary to complete a diagnosis. A variety of methods may be used to extract nematodes from soil and plant tissues representative of the area in question. When they are identified and their relative population densities are known, those data can be compared with experimental data and field observations to estimate whether the crop will be injured by that population under those conditions and, in some cases, to select management tactics specifically suited to the nematodes found.

Accurate characterization of nematode populations becomes critical for purposes of site-specific management, so sample collection, care, and laboratory processing are important. The sample should be prepared from a mixture of at least 10 soil cores taken 20 to 30 cm deep for most row crops. It often is best to discard the top 2 to 3 cm from each core, since nematode numbers may be very low there.

Sample timing is important, since nematodes are most numerous and easiest to detect near the end of a growing season. Predictive samples should be representative (average) of the entire sampling area. Collect cores in a regular pattern over the area. For purposes of mapping nematode populations, a maximum area of one hectare should be sampled. Areas within a field which have different soil types or which were planted to different crops (or varieties) during the past season should be sampled separately. Once collected, mix cores together carefully, then place 0.5 to 1.0 liter of soil in a plastic bag. Care must be taken with soil samples since nematodes will die from overheating, freezing, or drying.

Unfortunately, there are far more crop/nematode combinations for which research data are inadequate to establish thresholds than those for which there are sound thresholds. In these cases, the most important function of the laboratory assay is to determine the kinds of nematodes present. GIS could be very useful to improve such data by providing better site-specific nematode and yield data points to support better correlation of crop loss with nematode populations in combination with other environmental factors (data layers).

### *Alternative Data Collection in the Field*

GPS/GIS technology will be of great value in nematode management by increasing our ability to store nematode data in a very detailed site-specific manner and easily correlate those data with other physical, chemical, and biological factors that can affect crops. To make best use of GIS, we must have good data on nematode presence and intensity. A serious limitation of adapting GIS to nematology is the cost of normal procedures for collecting soil and/or plant tissue samples and processing them in a lab to obtain the data needed for each point on the grid. Identifying nematodes present and quantitatively estimating nematode population density in each sample by normal laboratory procedures is very costly and time-consuming. This could be ameliorated in many cases by recognizing that other kinds of data equally or nearly as good for quantifying nematode populations or their effects on plants often can be collected and recorded directly in the field much more quickly and cheaply. In situations in which visual evaluation of nematodes or their effects on plants can be made in situ, the cost, time delay, and risks of sample damage and loss attendant upon lab sample analysis can be substantially overcome. Far more data points can be collected at less expense, so it is possible to construct a data grid of much finer scale for a given investment with regard to field evidence, such as

- gall ratings of the many species of root-knot nematodes (*Meloidogyne* spp.) on most susceptible hosts;
- for cyst nematodes (species of *Heterodera, Globodera, Punctodera,* etc.), visual recognition and estimation of density of infestation of white females on roots of host plants;
- distinctive neck deterioration of onions, garlic, etc., caused by *Ditylenchus dipsaci;* and
- presence/absence as well as quantitative estimation of percentage of roots involved in distinctive root cortex necrosis of banana roots caused by *Radopholus similis.*

## *PRINCIPLES OF NEMATODE MANAGEMENT—NONCHEMICAL*

Carefully integrating as many as possible of the following components into the management program will help keep most nematode pests below damaging levels and may simplify the decisions that must be made about selecting and applying control agents when they are needed.

### *Crop Rotation*

Rotation is a very old practice for reducing soilborne problems. Many nematodes, soilborne disease organisms, and insects can reproduce and survive on only a few plants. Repeatedly planting a field with the same crop without interruption will enable organisms that reproduce successfully on that crop to continue to increase. Rotating to nonhost crops may stop reproduction of many of those pests and allow natural mortality factors to reduce their numbers. A good rotation program should be a basic component of most land/crop management plans, because of the multiple benefits that can be derived from it (Rhoades, 1976). The key to effective rotation is accurate identification and monitoring of pest populations and accurate mapping of their distributions. GIS clearly improves the ability to do that, with the added benefits of making it easier to correlate nematode data with other factors that might affect choice(s) of rotation crops.

### *Crop Root Destruction*

A cultural practice that receives far less attention than it deserves is destruction of crop roots immediately after conclusion of the cropping season. It is a practice heavily emphasized in some areas for high value crops such as many vegetable crops and tobacco. Fields should be tilled as soon as possible after harvest to kill all plants in the field and destroy their roots. Nematodes, soilborne diseases, and many soilborne insects will continue to feed and multiply on crop root systems as long as they remain alive. When soil temperatures are high, each month that the root system continues to live represents an additional generation and potential increase of about tenfold for root-knot nematodes and many others.

### *Fallowing*

In certain cases, fallowing may be used to help reduce numbers of nematode pests. Fallowing refers to leaving a field with no plants on it for a prolonged period. Most nematodes will decrease after a period of time without an adequate host plant. However, fallowing may be objectionable from a soil management viewpoint, because of the risk of losing soil organic matter, increased danger of erosion, and loss of productive time. If weeds are allowed to grow without control in a "fallow" plot, many kinds of nematodes may be able to survive and reproduce on the weeds, making the practice ineffective.

### *Resistance*

When it is available, host plant resistance to a specific pest usually is the least expensive and most effective means of minimizing losses to that pest (Rich and Olson, 1999). However, using varietal resistance successfully requires knowing the extent of the resistance and which pests are present in a particular situation. There are, for instance, "nematode resistant" varieties of tomatoes, soybeans, southern peas, sweet potatoes, cotton, and tobacco available, but each of these varieties has resistance which is effective against only one, two, or at most three species of nematodes. None are "nematode proof." It still is necessary to use other methods to control any other nematodes that are present, because resistance against one or two species does not inhibit the ability of other nematodes to injure the crop.

## *NEMATICIDES*

Nematicidal chemicals are divided into two main groups, fumigants (volatile) and nonfumigants (nonvolatile). Fumigant nematicides are named for their property of gaseous dispersion after application to soil. Nonfumigants are not volatile and disperse in solution in soil water. No matter how nematicides disperse through the soil profile, an adequate concentration of active ingredient must dissolve in moisture surrounding soil particles to be effective since this is the location of target nematode species (Wright, 1981). Other applications such as foliar sprays (Westerdahl et al., 1991) and seed treatments (Gray and Soh, 1989) may be used in a few cases. Listed in the following section are brief comments on nematicides and their uses, and for a more thorough discussion the reader is encouraged to see Rich et al. (2004).

### *Fumigants*

Chemicals with fumigant properties frequently are divided into the two general groups of multipurpose fumigants and nematicidal fumigants. Multipurpose fumigants have good activity against plant-parasitic nematodes as well as one or more of fungi, weeds, and insects (Lembright, 1990; Dunn and Noling, 1997). These fumigants are general soil biocides and normally are applied to control more than one soilborne pest. Some multipurpose fumigants used commonly now or previously include carbon disulfide, chloropicrin, dazomet, metham, and methyl bromide. Nematicidal fumigants are active primarily against nematodes and soil arthropods at normal use rates but may be active against other soilborne pests at higher rates. Both fumigant types are applied to soil (usually released beneath the surface of the soil through moving or stationary hollow tines, a process called injection).

Multipurpose soil fumigants seldom are used solely for nematode control since they are generally expensive and can be used economically only on crops of high value. Until recently, the most widely used multipurpose fumigant has been methyl bromide due to its broad biocidal spectrum (Methyl Bromide Technical Options Committee, 1995). However, methyl bromide now is scheduled to be totally withdrawn from all soil fumigation uses because of strong evidence that it causes serious deterioration of the ozone layer of the atmosphere which protects the earth from ultraviolet solar radiation. The multipurpose fumigants, chloropicrin and metham sodium, generally are less efficacious nematicides than methyl bromide but will be increasingly used as the methyl bromide phaseout continues.

The only fumigant whose primary function is nematicidal in current use is 1,3-dichloropropene (1,3-D), which is the preferred fumigant when nematode control is the primary consideration. It is comparatively inexpensive, efficacious, and easier to apply than multipurpose fumigants (Taylor, 1959; Dunn and Noling, 1997). The 1,3-D traditionally has been available as a single active ingredient but recently has been formulated more often in combination with chloropicrin or metham to increase spectrum of activity, mainly against fungi. These mixtures are being tested widely for replacement of methyl bromide mixtures in connection with its worldwide phaseout.

Both multipurpose and nematicidal fumigants may be used on a wide range of crops due to both their generally good pest control efficacy and lack of residues in plants.

### *Fumigant Application*

Fumigants generally are injected 15 to 45 cm deep into soil with chisels spaced 10 to 45 cm apart (Lembright, 1990; Dunn and Noling, 1997). Application depth and horizontal spacings are related to specific fumigant vapor pressure, characteristics of the soil, and the width and depth of the soil volume to be treated. Deeper fumigant placement is necessary under conditions such as tree planting sites or nurseries for deep-rooted plant species (Nyczepir, 1991). The point of fumigant application must be sealed by pressing or bedding soil behind the injection knives to prevent escape of the fumigant through the chisel tunnels. Soil injection via metal chisels or disks is most common, but alternative applications have been made with certain fumigants. Several fumigants have been injected as aqueous emulsions or solutions under very high pressure into live bermudagrass turf without metal tools to break the soil/turf surface (Dunn, pers. obs.). Metham may be applied through overhead and drip irrigation systems to manage soilborne fungal and nematode diseases (Noling, 1991; Johnson et al., 1992). Recently, emulsifiable 1,3-D formulations have been applied through drip irrigation systems with good results in finer-texture soil types and mixed results in sandier soils.

Land preparation and correct soil moisture are critical for maximum nematicidal activity of soil fumigants. Fumigants adsorb readily to organic matter; if organic matter content is excessive, application rates must be increased to offset the lost efficacy of material bound by the organic matter (Lembright, 1990). Nematodes also may be harbored in roots, which protect them from sufficient fumigant concentrations for adequate kill. The correct moisture level is important: the gaseous phase must move through pore spaces for distribution within the treated area, but the active ingredient must be dissolved in soil water to reach the target nematodes, which are entirely aquatic organisms. Thus, soil pore spaces should not be occluded by water (Lembright, 1990), but soil that is too dry will allow too much of the fumigant to escape before it can enter the soil moisture in sufficient concentrations to kill nematodes. Waiting periods for fumigant dissipation from soil to avoid phytotoxicity to crop plants may range from three days to more than three weeks. Fine-textured soils, low temperatures, and excessive soil moisture retard escape of the fumes from the soil, requiring longer planting delays (Thomason, 1987).

### *Nonfumigants*

Nonfumigant products may be formulated as either liquid or granular materials and include widely used products such as aldicarb, ethoprop(hos), fenamiphos, and oxamyl. The active ingredients of these nematicides are highly soluble in water, so their redistribution in soil typically is in solution in the soil water phase, normally accomplished by rainfall or irrigation (Smelt and Leistra, 1992). Nonfumigant nematicides are represented by two main classes of chemical compounds, organophosphates and carbamates. The mode of action of both nonfumigant nematicide classes is that of acetylcholinesterase inhibitors (Wright, 1981; Opperman and Chang, 1990). As such, the nonfumigant nematicides are more nemastatic than nematicidal, since nematodes can survive relatively long times without active cholinesterase function. Thus, economically applied dosages of these materials seldom kill nematodes directly, and nematodes may recover and become infective after removal from contact with nonfumigant nematicides. They must remain in contact with nematodes for a suitable period of time (four to eight weeks) to prevent infection and allow for good early plant root growth.

### *Nonfumigant Nematicide Application*

Nonfumigant nematicides usually are applied to soil as in-furrow, band, or broadcast applications (Dunn and Noling, 1997). Band and broadcast applications usually are followed by tilling into the soil for further uniformity of distribution and to prevent runoff. Overhead irrigation may be used to elute them into the soil profile after surface distribution on established turfgrasses and on other surfaces that cannot be disturbed by tillage equipment. Active ingredients move as the soil water moves (Smelt and Leistra, 1992). With excessive rainfall or irrigation, there is a risk of prematurely losing the active ingredient from the root zone and water contamination by leaching or runoff. Organic matter and clay content of soil also affect movement of nonfumigant nematicides. In sandy soils with low organic matter, low cation exchange properties may permit excessive leaching and rapid loss of the material from the root zone (Johnson and Feldmesser, 1987; Rahi et al., 1992).

Unlike fumigant nematicides, some nonfumigants may translocate in living plant tissue either upwardly (apoplastic), downwardly (symplastic), or both (ambimobile) (Wright, 1981). The knowledge of chemical movement and the direction of movement offers flexibility in application strategies for the individual nematicides. For example, oxamyl moves symplastically when applied to plant foliage and has shown activity against soilborne nematodes resulting in crop yield increase (Westerdahl et al., 1991). Alternately, aldicarb moves apoplastically in plants when applied to soil and is used to suppress both nematode populations in soil- and foliar-feeding thrip insect populations (Rhone-Poulenc Ag Company, 1998). The disadvantage of a chemical being mobile in plants may outweigh advantages since chemical residue may persist in plants, limiting crop registrations.

***Biocontrol Agents***

As is true of all living creatures, plant-parasitic nematodes have many natural enemies in the soil where they live. Research has been conducted throughout the world for many years to identify organisms that affect plant nematodes and to make practical use of them for nematode control. An excellent overview of the field is provided by Stirling (1991). Although few products have been offered for commercial biological control of nematodes to date, it clearly is only a matter of time before several kinds of products will become available for use against nematodes in a wide range of growing situations. Those products are likely to include dry and liquid formulations of bacteria, fungi, and other microbes, soil amendments to enhance conditions for effective nematode suppression by both artificially applied and endemic populations of nematode-antagonistic microbes, and perhaps even aqueous suspensions of predatory nematodes or other antagonistic invertebrates. Although these are not traditional chemical nematicides, it is very likely that variable-rate or selective application will be important for focusing the treatments where they are most needed.

## *VARIABLE-RATE NEMATICIDE APPLICATION*

Use of variable-rate application of nematicides for crops is a relatively new concept, and very little research actually is available to determine the logistic and economic potential for this technology. Three studies were conducted recently on use of variable-rate nematicide application for control of cotton (*Gossypium hirsutum*) root-knot nematode, *Meloidogyne incognita. Meloidogyne incognita* was the species of choice because of its characteristic spotty distribution and damage that can be observed in fields.

Standard rates of a granular nonfumigant nematicide, aldicarb, were compared to variable-rate applications at trials conducted in Missouri (Wrather et al., 2001) and Texas (Wheeler et al., 1999). Aldicarb also was used in Georgia field trials (Baird et al., 2001), but the primary interest in that study was comparing standard rates and variable applications of the fumigant 1,3-dichloroproprene (1,3-D). The fumigant was considered the best candidate for study because of the high costs per hectare and its effectiveness in reducing nematode populations in soil. Aldicarb efficacy often is affected by environmental conditions that could impact the variable-rate studies, as discussed in the work by Wheeler et al. (1999).

Based on industry methods for determining nutrient variations within fields, grid sampling also seemed to be the best approach for determining nematode levels in the field studies. The size of the grids, however, may influence the accuracy of the range maps developed for a field. In Texas, plots were subdivided into subplot units and intensively sampled. These samples were then pooled into two composite subsamples per subplot and mean nematode levels determined. The Missouri (Wrather et al., 2001) and Georgia (Baird et al., 2001) studies employed grid sampling methods using 0.1 ha and 0.4 ha grids, respectively, for determining *M. incognita* population levels within the fields of study. In Missouri, 15 soil core samples were randomly collected from an area 4 m × 4 m located at the central point of each grid. In the Georgia study, 10 core samples were randomly collected within a 2 m radius of the center point for each grid square and pooled to obtain the average nematode level.

Results from the Missouri study, using geostatistical analysis of nematode population levels for each sample, showed that taking samples on a 0.1 ha grid was not always acceptable for development of predictive density maps and that the soil sampling grid points might need to be much closer together than economically practical. The study conducted in Georgia, using

0.4 ha grid sizes, was based on 1998 industry standard practices for nutrient sampling in fields to receive variable-rate fertilizer applications (pers. comm. from D. Waters, Southern States Agric. Coop.). The computer program AgLink Professional (AGRIS, Roswell, GA) developed density range maps within each field based on the mean nematode densities determined for each grid sample (Figure 8.1). For grid sampling to be economical, sampling must follow the industry standard methods which minimize labor input and assay costs. Unfortunately, in the Georgia study, no geostatistical comparisons were made to show if spatial dependence between points occurred. The intensive sampling conducted in Texas, although much more accurate, could not be justified economically.

### *Yield Comparisons—Economics*

Overall, variable-rate application of aldicarb in the Texas and Missouri studies produced inconsistent results. In the eight field trials conducted in Texas, variable-rate-treated plots had either greater or similar yields as the standards. In two field trials, no differences occurred, and in three other fields total aldicarb used in the variable-rate treatments was greater than the standard. Overall, the results in the study show much inconsistency with

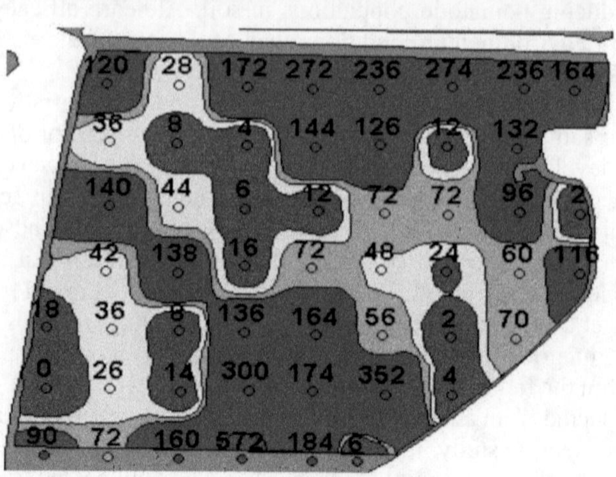


FIGURE 8.1. Numbers shown on the map indicate the *Meloidogyne incognita* juvenile levels per 100 $cm^3$ of soil at each 0.4 ha grid (O = center of grid for sampling); map depicts nematode density ranges, based on the sampling point data, as determined using AgLink Professional software package. *Source:* AGRIS, Roswell, Georgia. (See corresponding PLATE 8.1 in the central gallery.)

variable-rate application of the chemical. Similarly, the variable-rate aldicarb treatments used in the Georgia investigation were inconsistent even when sidedress applications were used to enhance nematode control. In Missouri, Wrather et al. (2001) reported that, with variable-rate application, 46 percent less aldicarb was used in the first year of the study and 61 percent less in the second year. However, no economic comparison was reported for this investigation to show whether the reduced chemical costs resulted in overall economic benefit to growers.

Variable-rate application of 1,3-D produced either similar or greater yields than uniform application of the recommended rates in Georgia. In this study, the cost of soil sampling and assay needed to develop variable-rate application maps was offset by the savings from using less 1,3-D during both years of the study. Additional studies should be conducted to confirm the results for 1,3-D before any recommendations can be made.

### *Integration of GPS/GIS into Nematode Management*

Given a good data matrix of nematode populations in a specific field, the obvious use of GIS technology is to adjust rates of application of a nematicide as one crosses a field, as discussed in the preceding section of this chapter. Other application variations that could be controlled by reference to GIS data might be switching between or using combinations of two nematicides according to species and population levels or delivery of varying quantities and/or mixtures of biological control agents such as different *Pasteuria* populations or mixtures of nematophyllic fungi. Where excellent resistance to critical nematodes exists, as in soybean, the same idea could be applied to multiple-choice seeders, putting different resistant varieties as appropriate for the local nematode populations, or putting resistant varieties only in certain parts of a field.

In the reverse perspective, GPS/GIS technology would be a superb field tool for large-scale correlation of field yields with gridded records of nematode species and populations. The multiple layers of data that are readily manipulated by GIS software would facilitate statistical factoring out of other field conditions, such as soil texture, drainage, cropping history, and other pest/disease occurrence. We still need much better ways to demonstrate the large-scale effects on crop responses of nematodes as they interact with other yield-influencing factors which are as erratically distributed as nematodes. For crops such as grains for which continuous harvest data are easily generated, this would be a good field tool to measure responses to field-scale treatment, perhaps increasing the value of large-scale extension plots for research work.

In some specialized situations, the GPS/GIS technology already is in place for other purposes, so its adaptation to nematological purposes should be a relatively small matter. Certainly, many large-scale farms already are using it for fertility, seeding rate/plant population, etc. Many upper-end golf courses now have GPS systems mounted on golf carts to enable players to ask a computer what their next club should be. It should be a simple matter to add layers of information about the incidence and severity of various pests and diseases including nematodes, as well as agronomic data such as soil characteristics, drainage, slope, etc., once the basic map has been generated.

Another valuable application of GIS to nematology is the ability to maintain long-term records that are readily correlated with other factors of concern regarding nematodes and other organisms of quarantine concern. Once such files are established, the data can be shared easily with appropriate agencies in other nations, states, and local governments that legitimately are concerned about them (Haydock and Evans, 1995; Raisio and Tilikkala, 1993).

GIS technology is ideal for planning long-term management strategies in addition to on-the-spot adjustment of individual tactics such as rates of nematicidal treatments. How better to bring together all of the factors important to producing a crop to help the grower work out long-term rotations in relation to realistic fertilizer rates and plant populations? The ability to overlay records of variations in so many different factors is very powerful, and development of reasonable nematological data/recommendation layers by nematologists may well make it easier to incorporate nematode management into more growers' basic plans and to make growers and managers sensitive to the importance of nematodes in limiting crop performance.

## REFERENCES

Baird, R.E., J.R. Rich, and D. Waters. 2001. Evaluation of variable rate nematicide applications using precision farming methods to manage *Meloidogyne incognita* on cotton. *Nematologia Mediterranea* 29:83-88.

Barker, K.R., G.A. Pederson and G.L. Windham. 1998. *Plant and nematode interactions.* Agronomy Series no. 36. American Society of Agronomy, Madison, WI.

Chen, Z.X., S.Y. Chen, and D.W. Dickson (eds.). 2004. *Nematology: Advances and perspectives,* vols. 1 and 2. CABI Publishing, London, UK.

Dunn, R.A., and J.W. Noling. 1997. *Florida nematode management guide.* Florida Cooperative Extension Service, Gainesville, FL.

Gray, F.A., and D.H. Soh. 1989. A nematicide seed treatment to control *Ditylenchus dipsaci* on seedling alfalfa. *Journal of Nematology* 21:184-188.

Haydock, P.P.J., & K. Evans. 1995. The potential use of global positioning satellite (GPS) technology in the mapping and management of potato cyst nematode populations. *Aspects of Applied Biology* 43:125-128.

Johnson, A.W., A.S. Csinos, A.M. Golden, and N.C. Glaze. 1992. Chemigation for control of black shank-root-knot complex and weeds in tobacco. *Journal of Nematology* 24:648-655.

Johnson, A.W., and J. Feldmesser. 1987. Nematicides—A historical review. Iin *Vistas on nematology,* pp. 448-454. J.A. Veech and D.W. Dickson, eds. Society of Nematologists, Hyattsville, MD.

Lembright, H.W. 1990. Soil fumigation: Principles and application technology. *Supplement to Journal of Nematology* 22:632-644.

Methyl Bromide Technical Options Committee–United Nations Environment Program. 1995. *Report of the Methyl Bromide Technical Options Committee for the 1995 Assessment of the Montreal Protocol on Substances That Deplete the Ozone Layer.* United Nations Environment Program, Australia.

Noling, J.W. 1991. Metham sodium application via drip irrigation systems. Nematology Circular 196. Florida Department of Agriculture and Consumer Services, Division of Plant Industry, Gainesville, FL.

Nyczepir, A.P. 1991. Nematode management strategies in stone fruits in the United States. *Journal of Nematology* 23:334-341.

Opperman, C.H., and S. Chang. 1990. Plant-parasitic nematode acetylcholinesterase inhibition by carbamate and organophosphate nematicides. *Journal of Nematology* 22:481-488.

Rahi, G.S., J.R. Rich, and C.H. Hodge. 1992. Ethoprop depletion from soil as influenced by simulated rainfall. *Journal of Nematology* 24(4S):642-647.

Raisio, R., and K. Tilikkala. 1993. National pest control strategy based on geographical information system (GIS). *SP Rapport* 1993(7):33-37.

Rhoades, H.A. 1976. Effect of *Indigofera hirsuta* on *Belonolaimus longicaudatus, Meloidogyne incognita,* and *M. javanica* and subsequent crop yield. *Plant Disease Reporter* 60(5):384-386.

Rhone-Poulenc Ag Company. 1998. *Temik briefing book.* Research Triangle Park, Raleigh, NC.

Rich, J.R., R.A. Dunn, and J.W. Nowling. 2004. Nematicides: Past and present uses. In *Nematology advances and perspectives: Nematode management and utilization,* pp. 1179-1216, Z.X. Chen, S.Y. Chen, and D.W. Dickson, eds. CABI Publishing, London, UK.

Rich, J.R., and S.M. Olson. 1999. Utility of Mi gene resistance in tomato to manage *Meloidogyne javanica* in north Florida. *Journal of Nematology* 31(4S):715-718.

Shurtleff, M.C. and C.W. Averre III. 2000. *Diagnosing plant diseases caused by nematodes.* American Phytopathological Society, St. Paul, MN.

Smelt, J.H., and M. Leistra. 1992. Availability, movement and accelerated transformation of soil-applied nematicides. In *Nematology from molecule to ecosystem,* pp. 281-293. F.J. Gommers and P. W. Th. Maas, eds. Dekker and Huisman, Wildervank, the Netherlands.

Stirling, G.R. 1991. *Biological control of plant parasitic nematodes.* CAB International, London, UK.

Taylor, A.L. 1959. Progress in chemical control of nematodes. In *Plant pathology—Problems and progress,* pp. 427-434. C.S. Holton, G.W. Fischer, R.W. Fulton, H. Hart, and S.E.A. McCallan, eds. University of Wisconsin Press, Madison, WI.

Thomason, I.J. 1987. Challenges facing nematology: Environmental risks with nematicides and the need for new approaches. In *Vistas on nematology,* pp. 469-476. J.A. Veech and D.W. Dickson, eds. Society of Nematologists, Hyattsville, MD.

Weischer, B., & D. Brown. 2000. *An introduction to nematodes: General nematology.* Pensoft, Sofia-Moscow (in English).

Westerdahl, B.B., D. Giraud, J.D. Radewald, and C.A. Anderson. 1991. Management of *Ditylenchus dipsaci* in daffodils with foliar applications of oxamyl. *Journal of Nematology* 23(4S):706-711.

Wheeler, T.A., H.W. Kaufman, B. Baugh, P. Kidd, G. Schuster, and K. Siders. 1999. Comparison of variable and single-rate applications of aldicarb on cotton yield in fields infested with *Meloidogyne incognita. Journal of Nematology* 31(4S):700-708.

Wrather, J.A., W.E. Stevens, T.L. Kirkpatrick, and N.R. Kitchen. 2001. Effects of site-specific applications of aldicarb on cotton in a *Meloidogye incognita*-infested field. *Journal of Nematology* 32(4S):687-696.

Wright, D.J. 1981. Nematicides: Mode of action and new approaches to chemical control. In *Plant parasitic nematodes,* pp. 421-449. B.M. Zuckerman and R.A. Rohde, eds. Academic Press, New York.

Chapter 9

# Site-Specific Measurement and Management of Grain Quality

Piet Reyns
Josse De Baerdemaeker
Ludo Vanongeval
Maarten Geypens

## *INTRODUCTION*

*Grain quality* is a general term for different variables determined by the final user. Those variables differ greatly, however. For example, the quality requirements of grain for feed differ significantly from those for grain as a food ingredient. In general, grain quality can be divided into two basic categories (Figure 9.1): physical condition and composition (Krischik et al., 1990). Physical condition is further divided into soundness and purity. Soundness includes characteristics that describe the general condition of the grain, such as specific test weight, moisture content, color, and defects such as broken kernels and kernels damaged by mold, insects, moisture, or excessive heat. Purity refers to materials or substances other than the natural kernel that are present in the grain mass. Foreign materials, insects, mycotoxins, and chemical residues are impurities or contaminants.

The composition category contains variables such as protein (and sometimes the fraction of the different amino acids), starch, fiber, oil, and minerals. Despite their importance for the end users, they are not widely used for marketing purposes. One reason is possibly the lack of fast measurement methods and the high cost of instrumentation. Protein content in hard wheat is an exception, in which the protein content directly influences the market value. In general, the grain price paid to the producer is based on the moisture content, specific grain weight, impurities, and sometimes the grain variety (Krischik et al., 1990; Vanorle, 1996). Protein content can affect in some places the price of both wheat and malting barley, while the Hagberg

*Handbook of Precision Agriculture*

doi:10.1300/5627_09

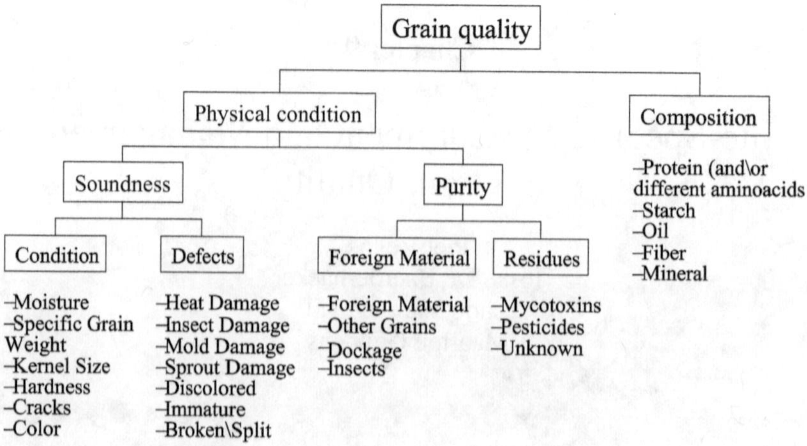


FIGURE 9.1. Different aspects of grain quality.

value affects the price at farm level for milling wheat. In the wholesale trade, the protein content and the Zeleny and Hagberg values sometimes influence the price of bread grain (Bernaerts, 1991).

How can a farmer now use precision agriculture tools in this grain quality business? If it is possible to measure the premium quality parameters during harvest, the on-line sorting could be a tool to optimize the price of the grain. But one can also try to influence the quality through fertilizer uses. Depending on economic and environmental situations, a quantity or quality strategy has to be chosen. If production or soil conditions vary spatially, then quality has to be managed site specifically and an on-line measurement of quality is necessary. In this chapter, the most important quality factors and their classical laboratory test methods are discussed. An overview of the fast measurement techniques and the possibilities for on-line measurements of grain quality during harvest are examined. Based on a literature study of the influence of soil factors and fertilization on various quality factors, different site-specific management strategies are discussed.

## *QUALITY FACTORS AND THEIR MEASUREMENT*

To take into account the possible changes in grain trading standards in the future, the maximum possibilities of on-line measurements will be eval-

uated toward the two most important end uses of the grain—animal feed and bread making.

### *Classical Reference Methods*

#### *Grain for Animal Feed*

The important parameters necessary in animal feed processing are grouped in the so-called Weende analysis. They are the proportion of moisture, ash, protein, fat, crude fiber, starch, and free sugars. At certain reference moisture content, the digestibility of these components of wheat for the different species of animals is known (CVB, 1994). When both nutritional requirement of animals and wheat composition based on the Weende analysis are known, the processor knows how to blend with other products to obtain the appropriate composition of the animal feed (pellets).

The moisture content of grain is mostly determined as the loss in weight of the ground grain when placed in a well-ventilated oven at a temperature of 130°C for 1 to 2 h (ICC standard 109/1, AACC method no. 44-15A). The European reference method uses an oven temperature of 103°C for 4 h. At high moisture content (16 to 17 percent), the grain is dried before grinding at room temperature to avoid moisture loss during grinding. The ash content of the grain is defined as the remaining part of a ground sample when placed for about 12 h in an oven at 600°C (AACC method 08-01). To determine the raw fat content, extraction is done with a paraffin ether solution. The extracted part is the raw fat fraction. The combined cellulose and lignin fraction of the cell walls are defined as the crude fiber content. It is determined by boiling the ground grain for 30 minutes successively in sulfuric acid and potassium hydroxide. The residual part after drying is the cellulose and lignin fraction.

Two different methods are used to determine the raw protein content of grain. In Europe, mostly the Kjeldahl method is used. In this method, an acid breaks down the organic components of the grain with a catalyst in the digestion phase. In a basic environment, all nitrogen in proteins is present as ammonia. This is distilled and the amount of ammonia is then determined by titration with acid. After converting the amount of ammonia to weight of nitrogen, the original amount of protein in the sample is estimated by multiplying by 6.25 (for feed purposes). In the combustion method, nitrogen in the ground grain sample is freed by a pyrolysis in pure oxygen at a temperature of 950°C or higher. The nitrogen gas is isolated from other combustion products and measured by a thermal conductivity detector (Williams et al.,

1998). This method is mostly used in the United States and Canada as the official protein testing method.

Not only the total protein content is important but also the type and amount of different amino acids. Among them methionine, cystine, lysine, threonine, and tryptophan are limiting amino acids for animals, and as such their importance is evident. The others are also important if they are digestible. However, the composition of the amino acids is mostly not determined by the processors because of the time-consuming and difficult wet analysis. Instead, the average composition of the amino acids over different varieties is determined and taken into account to determine the final composition of the feed.

Except for the moisture content and the degree of impurities in the grain, none of the previously mentioned parameters is used as trading standard. Instead, the specific grain weight is a widely accepted standard. It also reflects in part the quality of the grain. It is defined as the weight of a standard volume of grain. The grain is poured in a standardized way in a container with known volume, which is weighed afterward.

The degree of impurities is often part of the trading standards. After sieving the grain, the portion that did not pass the standardized sieves is a measure for the amount of foreign material in the grain.

### *Grain for Bread*

For bread, it is important to obtain flour which provides dough that is easily manipulated, rises well, has a good absorbing capacity, and leads to bread with a large volume after baking. The amount and sort of proteins, starch, and enzymes (amylase activity) play an important role in bread grain quality. Different direct and indirect test methods exist for evaluating the grain composition.

First, the amount of proteins available has to be high enough. The Kjeldahl and combustion methods described earlier are used to measure them. The protein content and quality determine the properties of the gluten. This is a network of proteins that provides the strength, retention of water, and elasticity of the dough.

In comparing grain with the same protein content, the Zeleny index (ICC standard no. 116), similar to the SDS sedimentation index (Axford et al., 1979), gives an idea of the quality of the proteins present. Adding lactic acid to the flour, the volume of flocculating proteins is a measure for their swelling capacity. Higher gluten content and better gluten quality give rise to slower sedimentation and higher sedimentation values.

Sometimes the hardness of the kernels is evaluated as well. The hardness reflects the way starch and proteins are physically built into a matrix. With hard kernels, the protein matrix and starch kernels are more easily damaged, which improves the absorption of moisture and the accessibility of the enzymes. An evaluation method is the particle size index (PSI) (AACC method no. 55-30). The grain is ground using a standard mill and sieved for 10 minutes using a standardized sieve. The throughput proportional to the input is the particle size index.

The elasticity of the dough is determined by the Chopin test (ICC Standard method no. 121). The instrument blows a bubble of dough and measures the pressure during the operation. The alveograph of Chopin represents the pressure in the dough during this test. Two parameters, the P and L values, can be derived: the P value is the maximum pressure attained during inflation of the bubble and is a measure for the resistance to stretch; the L value is a measure of the dough's extensibility. The area under the curve is a measure for the deformation energy and is considered to be a measure of the flour strength (W value of Chopin).

The amount of sprouting of the grain is important. When sprouting is already started in the grain kernel, the protein and starch structure is already partially broken down by enzymatic activities. The loosening of the starch and protein structure leads to a bad dough. The Hagberg falling number expresses the enzymatic activity in the grain (ICC Standard method no.107). To determine it, grain flour is brought in a cup together with a fixed amount of water. The cup is placed in a hot water basin. After a certain time, the viscosity is measured by measuring the sinking speed of a standardized tool in the viscous medium. High starch content increases the viscosity; high enzymatic activity lowers it.

The gluten content is estimated with the glutomatic test (ICC Standard no. 137/1). Dough is prepared from a flour sample by adding a buffered sodium chloride solution. The wet gluten is isolated by washing this dough with sodium chloride solution. The residual water adherent to the gluten is removed by centrifugation and the remainder is weighed.

The specific grain weight and degree of impurities are always measured because of the importance in grain trading.

### *Other Grain Processors and Crops*

Soft grain varieties are also used for starch extraction. The important quality parameters are the same as those used for bread (protein content, Zeleny index, Hagberg falling number, glutomatic test). Notice that the gluten, which is a by-product of the starch extraction, is used in the bread making industries to enrich the flour if the gluten content is too low.

For bakery products other than bread (cookies, cakes, etc.) the same quality parameters as for bread are important. Both the gluten content and the protein content have to be known. The required proportion of these depends largely from one end product to another.

Barley is mainly destined for the feed industry or for the brewing industry. Besides the standard quality demands, the caliber (the percentage of grains with a diameter larger than 2.5 mm), the total protein content, and the starch content also are important for malting barley. Grain selected for malting must have a high capacity for germination, be free of trash and broken kernels, and be relatively free of molds. As the degree of fermentation in the brewing process depends on the available carbohydrates, the concentration and availability (termed the malt extract content) must be as high as possible. A new trend in the quality requirement for barley is the amount of soluble beta-glucans. Their concentration in the final wort at brewing influences the rate of beer filtration and as a result the productivity of the beer.

### *Fast Nondestructive Laboratory Measurement Methods*

Three different measuring techniques are commonly used for grain quality analysis in laboratories: capacitive, microwave, and near infrared reflectance (NIR) or transmittance (NIT) measurements. The standard measurement method of the specific grain weight can also be considered a fast, nondestructive method, since it is based on a simple weighing of a fixed volume of grain. These measuring techniques are discussed in the following paragraphs, together with an overview of the parameters they can measure.

#### *Capacitive Measurements*

With capacitive measurements, the grain forms the dielectric of a capacitance. When an alternating voltage is applied to the circuit with the capacitance, the phase shift of the current and the impedance are strongly correlated to the moisture content (Kandala et al., 1996). The frequency of the circuit, grain density (specific grain weight), temperature, and dimension of the kernels have an influence on phase shift and impedance (Nelson, 1982). When the distance between the electrodes is wide enough, the influence of the kernel dimension is negligible. The temperature dependence is linear and can easily be compensated when measured separately, as is done in most capacitive instruments. The grain density is the most disturbing factor. Sometimes the density is partially kept constant by compressing the grain kernels (many handheld instruments); in other cases the grain density (specific grain weight) is measured with the same instrument. In some scientific research, the density effect is eliminated by measurements at different frequencies (Kandala et al., 1996; Lawrence and Nelson, 1993). In these studies, the standard error of prediction was 0.9 and 0.61 percent moisture content compared with oven determination.

The capacitance measuring technique is widely used by farmers, elevators, and laboratories for the moisture measurement of different kernel products. The accuracy of handheld instruments is typically around 0.7 percent moisture content for wheat compared to the reference oven method. The standard error of a capacitive laboratory instrument was 0.4 percent when used for barley.

### *Microwave Measurements*

In microwave measurements, the attenuation of microwaves emitted to the grain is proportional to the moisture content. However, density and temperature also disturb these measurements. The advantages of this method (Krazewski and Nelson, 1991) compared to capacitive measurements are as follows:

- Physical contact between the sensor and grain is not necessary.
- Penetration of the microwaves in the kernels: not only the moisture content at the kernel surface is measured.
- The measurements are not sensitive to dust and water.

So far not many commercial instruments are known with this technology. Typical accuracy obtained in a research environment is 0.5 percent (standard error of calibration) (Okabe et al., 1973) when compensating only for temperature effects. Calculating a parameter based both on the attenuation and phase shift of the transmitted wave, an accuracy of 0.5 at 95 percent significance level was obtained (Krazewski and Nelson, 1991).

### *Near Infrared Measurements*

Every object has a particular color as a result of its absorption and reflection characteristics for light (electromagnetic waves) at the different wavelengths in the visible region. The color is defined by the detectors in our eyes, which measure the amount of light reflected by the object in the red, green, and blue wavelength region. The light reflected by the object toward our eyes is equal to the incident light minus the absorbed light. In a similar way, every object is characterized by its absorption spectrum in the near infrared wavelength region. The amount of absorption at the different wavelengths depends on the type and amount of chemical bonds between atoms present in the material. This is represented in Lambert-Beer's law stating that at a fixed path length of light through the material, the light absorption is linearly related to the concentration of a particular constituent (Birth and Hecht, 1987).

Most laboratory instruments measure the reflectance of the material, which is defined as the measured reflection divided by the emitted light. Transmission measurements suffer from the low level of the transmitted light, resulting in a low signal-to-noise ratio.

In general, near infrared instruments consist of three major elements: the light source, the wavelength selection mechanism, and the detector element. For grain analysis, mostly two categories of instruments are used in laboratory environment: filter- and monochromator-based instruments (Figure 9.2). They differ in the wavelength selection mechanism. In both instrument types, the different wavelengths are measured sequentially. In the first case, the light coming from the light source passes through a filter wheel with optical filters. Each optical filter represents one wavelength band. The grain sample is illuminated with the filtered light, and the diffuse reflected light is captured by the detector element. This consists often of a gold-coated sphere with one or two PbS (lead sulphide) detectors inside.

Monochromator instruments have a holographic grating to select the different wavelengths. Based on light interference, the wavelength passed to the sample is determined by the position of the grating with respect to the lamp. The advantage of the monochromator instruments is the higher spectral resolution. By turning the grating in small steps, the whole wavelength interval can be measured continuously. For the filter instrument, an optical filter is needed for each desired wavelength.

The most difficult part of NIR technology is the calibration of the instrument and the transfer of these calibrations to other instruments. An accurate prediction can be made only for samples having physical and chemical characteristics close to the samples used for the calibration (Williams, 1996). This means that the calibration set must contain samples of all possible varieties from different cultivating environments, at the whole range of possible values of the parameter to predict. In practice, the calibration has to be evaluated and updated during different years. However, once calibrated the technology has some important advantages compared to the laborious reference methods:

- The analysis time is short (often less than one minute).
- Often the NIR instrument has a higher repeatability compared to the reference method.

The accuracy is always limited by the reference method as the instruments are calibrated with it. The commercial use for measurements on wheat of this type of instruments started around 1971. Initially, only ground wheat could be measured. In Table 9.1, the results of research on different parameters are summarized.

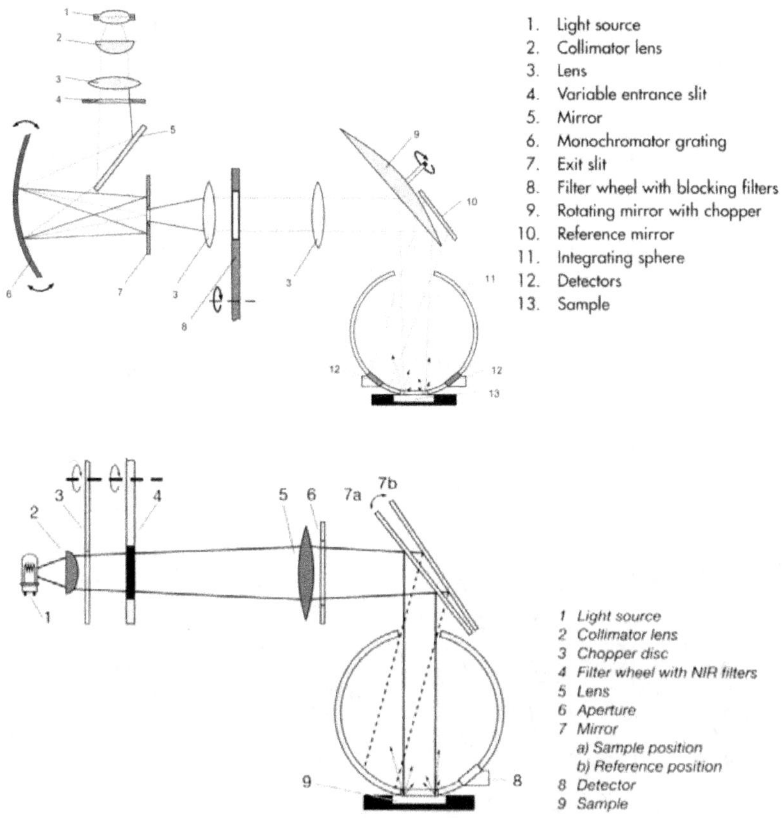


FIGURE 9.2. *Above:* Example of a monochromator-based near infrared reflection instrument (BRAN + LUEBBE InfraAlyzer 500); *below:* Example of a filter-based near infrared reflection instrument (BRAN + LUEBBE InfraAlyzer 500).

TABLE 9.1. Summary of research results for near infrared measurements on ground wheat.

| Parameter | Crop | Accuracy | Measurement range | Reference |
|---|---|---|---|---|
| Moisture | Wheat | 1.5% (CV) | 7-14.5% | Law and Tkachuk (1977) |
| Moisture | Wheat | 0.17% (SEC) | 8-16% | Osborne et al. (1982) |
| Moisture | Wheat | 0.32% (RMSEP) | — | Williams (1979) |
| Protein | Wheat | 0.25% (SEC) | 6-14% | Osborne et al. (1982) |
| Protein | Wheat | 0.98% (CC) | 8-16.6% | Watson et al. (1976) |
| Protein | Wheat | 0.31% (RMSEP) | 10-18% (13.5% MC) | Williams (1979) |
| Protein | Wheat | 0.33% (SEP) | 9.4-18.1% | Williams and Cordeiro (1985) |
| Protein | Wheat | 0.36% (RMSEP) | — | Schäfer and Weinard (1978) |
| Protein | Wheat | 0.28% (RMSEP) | 7-13% | Osborne (1984) |
| Protein | Wheat | 0.28% (RMSEP) | 10-17% (13.5% MC) | Tkachuk (1981) |
| Lysine | Wheat | 1.87 umol·g$^{-1}$ (SEP) 6.2% (CV) | 23.4-39.7 umol·g$^{-1}$ | Williams et al. (1984) |
| Methionine | Wheat | 1.92 umol·g$^{-1}$ (SEP) 13.1% (CV) | 7.6-22 umol·g$^{-1}$ | Williams et al. (1984) |
| Threonine | Wheat | 2.07 umol·g$^{-1}$ (SEP) 5% (CV) | 35.6-55.4 umol·g$^{-1}$ | Williams et al. (1984) |
| Tryptophan | Wheat | 0.97 umol·g$^{-1}$ (SEP) 7.8% (CV) | 10-15.7 umol·g$^{-1}$ | Williams et al. (1984) |
| Water absorption | Wheat | 1.8% (SEP) | 48-76% | Rubenthaler and Pomeranz (1987) |
| Mixing time | Wheat | 0.15 min (SEP) logarithmic values | 0.5-9.5 min | Rubenthaler and Pomeranz (1987) |

| Parameter | Crop | Accuracy | Measurement range | Reference |
|---|---|---|---|---|
| Loaf volume | Wheat | 48 $cm^3$ (SEP) | 568-1328 $cm^3$ | Rubenthaler and Pomeranz (1987) |
| Falling number | Wheat | 73 s (RMSEC) | 62-377 s | Osborne (1984) |
| SDS sed. volume | Wheat | 12.5 mL (RMSEP) | 30-95 mL | Osborne (1984) |

CV = coefficient of variability; SEC = standard error of calibration; RMSEC = root mean square error of calibration; SEP = standard error of prediction; RMSEP = root mean square error of prediction; CC = correlation coefficient.

When estimating the moisture content in ground grain, the reported root mean squared error was always below 0.45 percent moisture content when the grain moisture content was between 7 and 16 percent. The protein content could always be estimated with a mean error below 0.6 percent (measurement range 9.4 to 18.1 percent) and in many cases even below 0.4 percent.

The accuracy obtained in predicting some of the limiting amino acids (lysine, methionine, tryptophan, and threonine) would be high enough for use in wheat processing. However, establishing calibrations for amino acids must be done carefully, as their concentration is often closely correlated to the total protein content. The calibration set of wheat samples should be carefully selected in order to have a wide range of amino acid concentrations at equal protein contents. Osborne (1984) mentioned other studies which indicated the possibility to measure the wheat texture and ash content too.

When trying to establish calibrations for bread grain quality, the same author (Osborne, 1984) concluded that only the protein content could be measured with near infrared technology. Models developed to predict the falling number and SDS sedimentation volume were not significant. In contrast, parameters such as water absorption, mixing time, and loaf volume (similar to Zeleny and SDS sedimentation volume) (Axford et al. 1979) could be predicted (Rubenthaler and Pomeranz, 1987). As loaf volume is highly correlated with the protein content, the calibration set of samples must be carefully chosen to obtain a calibration for loaf volume and not for protein content.

In Table 9.2, the accuracy of a commercial filter instrument for prediction of different parameters of ground wheat is given as an example. Notice

TABLE 9.2. Typical accuracy of a commercial filter-type instrument for the analysis of ground grain (Perten Inframatic 8106).

| Parameter | SEP (%) | Analytical range |
|---|---|---|
| Moisture | 0.35 | 12.8-21.6 |
| Protein | 0.18 | 8-18 |
| Hardness | 3 | 35-65 |
| Zeleny | 6 | 17-66 |
| Water absorption | 0.85 | 49-60 |

*Source:* Cereal Tester, 1997.

SEP = Standard error of prediction.

that a calibration for Zeleny is also provided but with rather low accuracy. The prediction gives only an indication of the magnitude of this parameter.

Research toward measurements on whole grain kernels started around 1980 (Table 9.3). The accuracy for moisture prediction is below 0.4 percent (measurement range 13.6 to 20 percent), which is of the same magnitude as for ground wheat. However, a nonlinear response was observed when measuring on samples with moisture content above 20 percent (Downey, 1985). This moisture content is seldom encountered in grain trading as the first elevator immediately dries this grain.

Protein content seems to be measurable as accurately as with ungrounded grain, with a maximum standard error of 0.6 percent protein. Tkachuk (1981) could even predict the protein content with an RMSEP of 0.3 percent.

Even the relative accurate measurement of the PSI (particle size index) value and degree of insect infestation would be possible according to some authors. The research on insect infestation was focused on both external infestation of *Oryzaephilus surinamensis* (sawtoothed grain beetle) and internal infestation of larvae or pupae of *Sitophilus granarius* (grain weevil). With external infestation, infested grain can be completely distinguished from uninfected starting from an infestation with more than 27 insects per 100 g.

Commercial instruments for measurement on whole grains can typically measure the protein content (SEP 0.2 percent), moisture content (SEP 0.2 percent), and Zeleny (SEP 10 percent) (Cereal Tester, 2000).

Nowadays, NIR measurements are done on a whole range of agricultural products. In Table 9.4, some examples of research on other kernel crops are

TABLE 9.3. Summary of research results for near infrared measurements on wheat (whole grains).

| Parameter | Accuracy | Measurement range | Reference |
|---|---|---|---|
| Moisture | 0.36% (RMSEP) | 13.6-20% | Downey (1985) |
| Moisture | 0.32% (RMSEP) | 9-22% | Spaepen (2000) |
| Protein | 0.3 (RMSEP) | 10-17%<br>(13.5% MC) | Tkachuk (1981) |
| Protein | 0.54 (RMSEP) | 10-16%<br>(0% MC) | Spaepen (2000) |
| PSI | 3.37 (SEP)<br>1.4% (CV)<br>11.7 (RER) | 40.7-50.5 | Williams (1991) |
| Insect infestation | 5.98 insects/100 g (SEC) | 0-50 insects/100 g | Ridgway and Chambers (1996) |

CV = coefficient of variability; SEC = standard error of calibration; SEP = standard error of prediction; RMSEP = root mean square error of prediction; CC = correlation coefficient; RER = (measurement range of reference data)/SEP.

summarized. The accuracy of protein measurement in unground rice is similar to wheat. Prediction of the moisture content is less accurate, but the measurement range is higher and larger compared to wheat. Protein measurement on peas is quite accurate when ground but still only an indication of the magnitude that can be obtained on the whole grains. Results for barley are similar to those of wheat.

Calibrations for commercial instruments (Cereal Tester, 1997) exist for other ground crops, including alfalfa, barley, corn, grass, rapeseed, soya, and sunflower.

### *Summary*

For the most important trading parameters (moisture content, specific test weight, and protein content), only the moisture and protein content are

TABLE 9.4. Results of NIR measurements on kernel crops other than wheat.

| Parameter | Crop | Ground/ whole | Accuracy | Measurement range | Reference |
|---|---|---|---|---|---|
| Protein | Rice | W | 0.24% (SEP) | 7.4-9.3% | Kawamura et al. (1999) |
| Moisture | Rice | W | 0.7% (SEP) | 15.5-33.2% | Kawamura et al. (1999) |
| Protein | Rice | W | 0.28% (SEP) | 6.5-11.5% | Li and Shaw (1997) |
| Amylose | Rice | W | 0.52% (SEP) | 14-25% | Li and Shaw (1997) |
| Protein | Field peas | G | 0.3% (RMSEP) | 14.1-27.6% (0% MC) | Tkachuk et al. (1987) |
| Protein | Field peas | W | 1.3% (RMSEP) | 14.1-27.6% (0% MC) | Tkachuk et al. (1987) |
| Protein | Barley | G | 0.25% (RMSEP) | 9.2-14% | Schmütz and Do (1982) |
| Protein | Barley | G | 0.4% (SEP) 2.6% (CV) | | Williams et al. (1984) |
| Lysine | Barley | G | 2.1% (SEP) 4.6% (CV) | | Williams et al. (1984) |
| Methionine | Barley | G | 1.2% (SEP) 8.4% (CV) | | Williams et al. (1984) |
| Threonine | Barley | G | 2.1% (SEP) 4.7% (CV) | | Williams et al. (1984) |
| Moisture | Barley | W | 0.5% (RMSEP) | 13.6-20% | Downey (1985) |

measurable at the laboratory level in a fast, nondestructive way. Two fast techniques to measure the moisture content exist commercially: capacitive (dielectrically) and near infrared. The protein content can be measured only with the near infrared technology. The typical standard error for measurement of both parameters and techniques is around 0.5 percent.

For the bread-making parameters, an indication of the magnitude of Zeleny (SDS sedimentation value) and hardness can also be obtained.

## *ON-LINE QUALITY MEASUREMENTS*

On-line measurements are done in (at) the process line and are fast enough to enable controlling of the process. Within the on-line measurements, in- and at-line measurements can be distinguished. In-line measurements are done immediately on the flow of the material, while a kind of sampling and sample preparation (e.g., milling) can be involved in at-line measurements. The on-line quality measurements are either dielectric (moisture content) or near infrared.

### *Dielectric On-Line Measurement of Moisture*

In testing three capacitive sensors at three different positions in a combine harvester, the sensitivity was always too low (Mailander et al., 1982). Two flat plate sensors were installed in the grain bin and under the lowest sieve, where the capacitor consisted of a flat plate, the grain (the dielectric material), and the harvester frame. The third sensor was placed in the clean grain auger, which transports the grain from the elevator to the grain bin. Only reasonable results were obtained in corn with moisture content above 23 percent. In this case, correlation coefficient for the auger sensor was 0.87 and 0.75 for the sensor at the sieves.

During laboratory measurements at low flow rates (2 to 14 $kg \cdot s^{-1}$), Berbert and Stenning (1996) could calibrate their sensor with a standard error of 0.5 percent moisture content. The DMC capacitive moisture sensor is used in different commercial "precision farming" packages for combine harvesters. The sensor is mounted on the auger that transports the clean grain to the grain bin. When calibrated with destructive moisture determination of hand-taken samples, the standard error of calibration was 0.88 percent when all different varieties were included (Figure 9.3) (Reyns et al., 2000). The moisture content ranged from 13 to 28 percent. Calibrating for only one field, the standard error was reduced to 0.49 percent (measurement range 15 to 27 percent). RDS,[1] AgLeader,[2] and Claas[3] all include the DMC moisture sensor in their yield mapping system. In the RDS Ceres II system, the sensor is mounted in the center or at the end of the tank-feeding auger. AgLeader and Claas offer the DMC moisture sensor with bypass mounting configuration for the clean grain elevator. The main problem is the need for a uniform density of the grain when presented to the sensor. For this reason the auger flights have to be removed or the bypass-construction is used. However, both of these configurations have problems with wet and unripe grain, causing blocking of the auger and bypass. Claas solved this problem by cleaning the bypass mechanically from time to time.

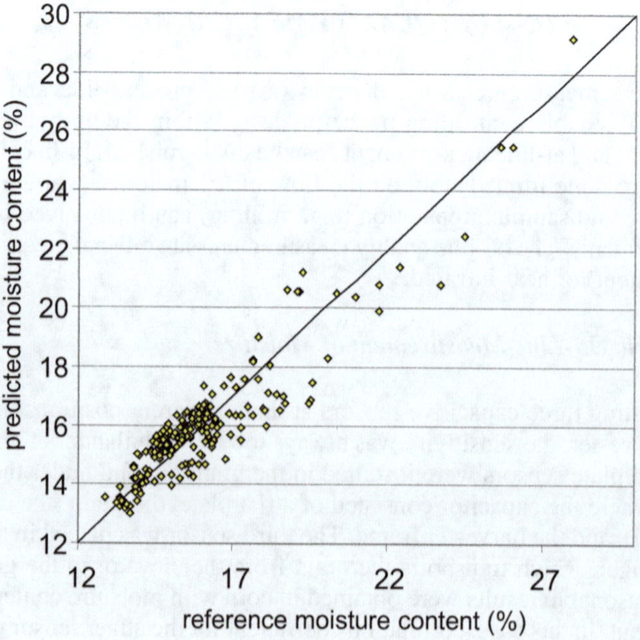


FIGURE 9.3. Evaluation of the DMC moisture sensor.

John Deere[4] developed its own capacitive moisture sensor, similar to the DMC sensor. The sensor is mounted in a bypass at the elevator. In the bypass, an electrically driven wheel is fixed under the sensor to control the amount of grain in the sensor.

### *On-Line Near Infrared Measurements*

#### *General Remarks on Instrumentation and Application*

The types of instruments generally used for on-line near infrared measurements are

- filter-type instruments (as discussed before),
- diode-array instruments, and
- AOTF instruments.

The advantage of diode-array instruments is that the reflection at all wavelengths is measured at once. Instead of splitting the emitted light into its different wavelengths (like with a monochromator instrument), the reflected light is split by a holographic grating and projected on an array of diodes. Each diode represents one wavelength measured. Other advantages are the fast measurement and the absence of moving parts. Those moving parts make filter-type instruments more sensitive for use on a vibrating machine such as a harvester. The measurement of all wavelengths at once makes it easy for use with in-line measurements

AOTF instruments are comparable to monochromator instruments. However instead of using a moving grating for the wavelength selection, an "Acousto-Optical Tunable Filter" is used. In principle, this is a $TeO_2$ crystal, in which the optical properties change with the wavelength of the ultrasonic waves transmitted through it. The AOTF module is also scanning the different wavelengths but at a very high speed. This makes the measurement nearly instantaneous.

For the use in-line, diode-array or AOTF instruments appear most feasible. However, the interface of the measurement with process line must be considered. Typical interfacing problems are shown here:

- Measuring point in the process line to have a uniform density of the material
- Influence of ambient light
- Interface window of instrument with material flow (It should be scratch resistant.)
- Handling of dirt on the interface window

Processing techniques exist to compensate for small changes in the material density, but big changes must be avoided. When no section in the current process line has a uniform density, often a bypass is constructed to control the flow. In some cases the flow in this bypass is also continuous, but often a periodical filling and emptying of the bypass is involved. This requires moving parts that are more sensitive to vibrations and require maintenance. Special care must be taken to the transport mechanism of the grain through the bypass. Often wet material causes blocking of the bypass.

When both the instrument and process line cannot be masked from ambient light, chopping of the sample illumination can be a solution. This allows making a distinction between the reflection of the ambient and chopped light.

Dirt on the measuring window is probably the biggest problem of on-line measuring systems. If the moisture content of the measured material is always low, cleaning with an air punch can be sufficient. The best way is to have a self-cleaning interface. In this way, the friction of the material to the window keeps it clean. Sometimes the friction has to be so high that the window is scratched, even with a hard sapphire window.

When applying NIR sensors on harvesting machines, the calibration of them needs special attention. Skilled people, who are regularly evaluating and updating the instrument calibration, operate laboratory instruments. A calibration works well only if the spectra of the measured material are close to those used for the calibration. As a result, updating is regularly needed with new varieties and grain from different growing conditions. Establishing a stable calibration can take up to five years, and even afterward it must be regularly evaluated. Farmers are not used to doing this, and it should be performed in specialized laboratories. This shifts the problem to the transferability of the calibrations. This can be looked at as transfer from one NIR instrument to another and then transfer from the laboratory to the harvester.

First, the transfer from one near infrared instrument to another is not evident, as every instrument is slightly different. Thermal stabilization and other technical improvements made transfer easier for laboratory instruments. But the NIR technology for on-line measurements is relatively new, and mostly no thermal stabilization is available.

If the calibration equation for a sensor on a particular harvester is exactly known by the responsible laboratory/company, then it must be transferred to the combine. For ease of work, a good communication system must be built into the machines by the manufacturer, to update the calibration on the machine with as little effort as possible for the driver.

To cope with some of these problems, a national or international database of calibration equations could be established, based on an international network of laboratories maintaining this database and driven by the instrument manufacturers.

### *Commercial On-Line Applications for Wheat and Other Grain/Pellet Products*

Different commercial on-line measurement systems exist (Table 9.5) and are mostly used on elevators or conveyer belts in processing plants. The MM55 (Infrared Engineering Ltd.[5]) is a filter-type instrument that can easily be mounted in-line above conveyer belts, etc. A maximum of five wavelengths can be measured (Benson, 1995), and the diameter of the measuring area is 4 cm. A larger area increases the repeatability of the measurement for kernels, as the reflection is averaged over different kernels. The instrument is operating in dual mode, which means the lamp illumination is

TABLE 9.5. Overview of commercial on-line applications.

| Instrument | In/at-line | Parameter | Product | Accuracy | Range |
|---|---|---|---|---|---|
| MM55-SF | in-line | moisture | snack foods | 0.5% (SEP) | 5-25% |
| | | oil content | snack foods | 0.5% (SEP) | 20-45% |
| Infrapowder | at-line | moisture | milk powder | 0.05% (SEP) | 2-4.5% |
| | | fat | milk powder | 0.1% (SEP) | 10-28% |
| | | protein | milk powder | 0.1% (SEP) | 21-28% |
| Prinacon | at-line | protein | barley | 0.3% (SEP) | U |
| | | | wheat | 0.25% (SEP) | U |
| | | | soya | 0.45% (SEP) | U |
| Formal | at-line | U | feedstuff | U | U |
| Brimrose | in-line | protein | | 0.7% (SEP) | U |
| | | oil content | | 0.5% (SEP) | U |
| | | moisture | | 0.42% (SEP) | U |

U = Unknown.

continuously measured and compensated. In predicting moisture content between 5 and 25 percent, a typical error of 0.5 percent is indicated for the MM55-SF. A similar system is available from MSC.[6]

The previously mentioned systems would allow in-line measurements despite the fact that they are not measuring all wavelengths at once, and then the measurements are not on the same grain if this is moving. In contrast, the Prinacon,[7] Formal,[8] and Infrapowder[9] systems all integrate a usual laboratory instrument in an at-line system. The measurement is executed on stationary material in a bypass of the bulk flow. Before measurement, shaking the material controls the density. In the case of the Formal system, the material is even ground. The measuring window is usually cleaned with an air punch.

For the in-line measurements on fish pellets, a Brimrose[10] AOTF module was used (Richmond, 1999). Protein content, oil content, and moisture content could be predicted with a standard error of prediction (SEP) of 0.7, 0.5 and 0.42 percent, respectively. The sensor was measuring through a sapphire window on the free flow of fish pellets. The friction between the pellets and the measuring window kept it clean. To avoid possible influence of ambient light, the light was blocked.

*Research Toward On-Line Measurements on Harvesters*

In 1997, results appeared of tests with an on-the-go protein and moisture sensor produced by Milestone Technology Inc.[11] The manufacturer claims a precision of ± 5 percent, 95 percent of the time, for the measurement of both moisture and protein content. The sensor is mounted on the auger in the grain bin. Today, CNH N.V.[12] is developing similar sensor technology together with Textron Systems.[13] With this sensor, it should be possible to measure moisture, oil, and protein content of corn, wheat, and soybeans on-line as accurately as with laboratory instruments (von Rosenberg et al., 2000). The detecting element incorporated is a CCD array, measuring in the 600 to 1100 short near infrared wavelength range. A reference shutter is included to compensate for drift and other variations of the system components. However, it is not compensating for dirt and scratching of the window. Probably regular update of the calibrations would be needed when commercially applied.

For real-time grain analysis, a sensor has been described (Wright and Hood, 1998) that uses the NIR principle to measure different grain parameters. The sensor is placed on a test machine at the end of an elevator, which conveys samples of clean grain to the sensor. This consists of a test chamber that is periodically refilled. The reflected light is scattered in its different wavelengths by means of a diffraction grating, after which it is detected by a photodiode array. Each photodetector of this array detects one wavelength band of the desired spectrum.

A Perten DA7000 diode array spectrometer was used to measure the wheat protein content on-line (Vigier et al., 2000). The sensor was installed on a chute, capturing part of the grain coming from the grain bin–filling auger. When calibrated in the laboratory with a simulated grain flow, the protein content could be predicted obtaining an RMSEP of 0.4 percent (measurement range 8 to 17 percent protein on dry matter, Kjeldahl conversion coefficient 5.7). However, when tested in the field, the on-line measurement results with this calibration differed largely from the values of samples taken at the same time and analyzed in the lab. This deviation was probably due to the different grain flow pattern in the chute (with variable grain density) compared to the simulation in the laboratory.

Using a portable Zeltex[14] ZX-50 near infrared transmittance instrument, grain moisture and protein content could be measured at-line on a combine harvester (Thylén and Algerbo, 2001). The instrument was installed in a bypass on the elevator, where two electrically driven flaps controlled the material flow. Instead of using a filter wheel, 14 diodes are used to illuminate the sample at different wavelengths according to the optical filter placed before. Those diodes cover the 893 to 1045 nm wavelength range. When calibrated under stationary conditions with samples of known moisture and protein content, the standard error of prediction was 0.185 and 0.358 when compared with the laboratory method. Every five measurements, a reference measurement was executed to compensate for dust and scratching of the sample cup. However, the system got blocked when grain had moisture content above 30 percent.

Measurement of corn moisture content with a Zeiss Corona NIR1.7 sensor (Reyns et al., 2001) resulted in an RMSEP value of 0.32 percent (measurement range 26 to 30.5 percent) (Figure 9.4), compared to the oven reference method. The sensor was installed on the Agleader bypass instead of the DMC moisture sensor. Reference measurements were executed manually from time to time to compensate for instrument drift. However, these measurements were also executed after the grain interfacing window, and thus not compensating for dirt and scratches. The sensor was calibrated and validated with samples taken at the same time as the NIR measurements

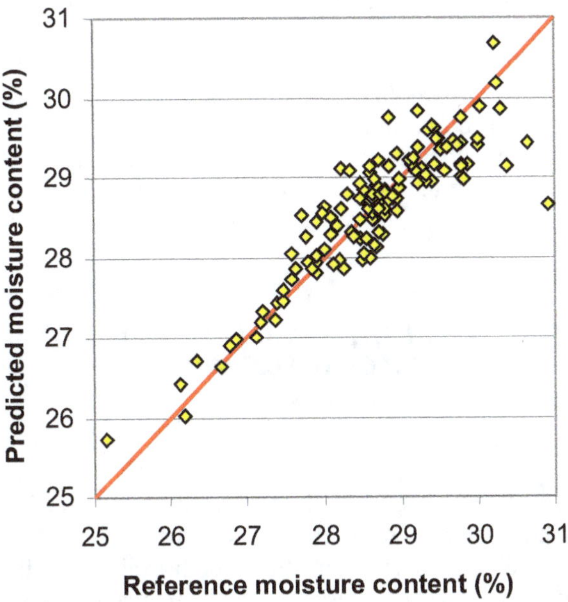


FIGURE 9.4. Results of the on-line measurement of corn moisture content.

during harvest. The same sensor was used to measure on-line the dry matter content of grasses on a forage harvester (Haüsler et al., 2001). In the range from 12 to 42 percent, protein content could be predicted with an SECV (standard error of cross validation) of 1.51 percent. On the used plot harvester, the sensor was installed on a bypass shortly after the cutting drum. Reference measurements were automated by moving the bypass on which a reference was mounted.

In comparing different sensors for on-line moisture measurement of forage maize on a forage harvester, the Perten DA7000 instrument was tested (Kormann, 2001). The instrument was mounted on the upside of the chute of the harvester, looking through a window to the maize flow. Testing on one field, the SECV increased to 1.39 percent. Also the starch, protein, and ash content could be predicted with an SECV of 1.62, 0.28, and 0.25 percent, respectively.

### *Summary*

Different commercial capacitive sensors are available to measure grain moisture content on-line. Accuracy is typically around 0.45 percent (standard error) within one field and 0.8 percent over the whole harvesting season.

Recently, on-line NIR systems on harvesters have been tested (Table 9.6). Measurable parameters would be moisture, protein, and oil content in products such as wheat, corn, and soybeans. Furthermore, measurements on forage harvesters would be possible. To develop these systems for commercial application, problems such as calibration and calibration transferability, dirt on the measuring window, and blocking of the bypasses will have to be solved.

## *INFLUENCE OF PLANT NUTRITION ON THE QUALITY OF CEREAL CROPS*

### *Introduction*

For the optimal management of plant nutrition, a number of factors need to be considered, such as genetic properties of the species, plant nutrients, and weather conditions. The latter affects both available soil N and the growth of the crop. For wheat, barley, and other cereal crops, nitrogen uptake is the main key to yield and quality. The application of nitrogen fertilization at different moments in the growing season can have a positive as

TABLE 9.6. Overview of research on on-line quality measurements on harvesters.

| Crop | Constituent | Accuracy parameter | Range | Calibration | Reference |
|---|---|---|---|---|---|
| Wheat | moisture | 0.19% RMSEP | 12-18% | lab | Thylén et al. (2001) |
| Wheat | protein | 0.36% RMSEP | 8.5-15% | lab | Thylén et al. (2001) |
| Wheat | protein | 0.40% RMSEP | 8-15% | lab | Vigier et al. (2000) |
| Corn | moisture | 0.32% RMSEP | 26-30.5% | on-line | Reyns et al. (2001) |
| Forage | moisture | 1.51% SECV | 12-42% | on-line | Häusler et al. (2001) |
| Forage maize | moisture | 1.39% SECV | 57.4-68.6% | on-line | Kormann (2001) |
| Forage maize | starch | 1.62% SECV | 28-41.2% | on-line | Kormann (2001) |
| Forage maize | protein | 0.28% SECV | 6.5-8.9% | on-line | Kormann (2001) |
| Forage maize | ash | 0.25% SECV | 3.5-4.8% | on-line | Kormann (2001) |

well as a negative effect on the quality of the grain. In this section, it will be discussed how this can be included in precision farming practices.

### *pH*

The acidity of the soil is essential for the assimilation of the mineral elements by the crop. If the pH is suboptimal, symptoms of deficiency and toxicity can occur which obviously influence the yield. Furthermore, the pH has an impact on the mineralization of nitrogen: a low pH of the soil delays this process which affects the supply of mineral nitrogen during the growing season.

The optimal pH zone for barley (Table 9.7) is slightly higher in comparison with wheat because of the higher sensitivity of barley for dissolved Al and Mn ions when the pH is low.

TABLE 9.7. Optimal pH (KCl) zones for winter wheat and winter barley as a function of soil texture.

| Soil texture | Wheat | Barley |
|---|---|---|
| Sandy loam | 5.5-6.0 | 5.9-6.4 |
| Loam | 6.0-6.7 | 6.4-7.1 |
| Clay | 7.1-7.5 | 7.1-7.5 |

In practice, it is impossible to modify the pH for each crop in a crop rotation. Therefore, a compromise is reached: for each soil texture and humus content an optimal pH zone is defined that is favorable for the crops in the crop rotation.

### *Mineral Elements*

Nitrogen is vital to the growth of plants. It is part of all the essential constituents of cells and as such also for chlorophyll which is essential for photosynthesis. However, nitrogen is not the only vital nutrient: phosphorus and potassium also play important roles. Magnesium, calcium, and a number of other elements are also significant.

On one hand the mineral fertilization has to be adapted to the availability of the mineral elements in the soil, and on the other hand to the needs of the crop.

The real demands of the crop are determined on the basis of fertilization experiments in which the effect of the mineral fertilization on the yield is verified. The influence of the main nutrients on the quality of the crop is small, except for nitrogen (Rixhon, 1985). This can be demonstrated based on fertilization experiments performed by the Soil Service of Belgium in Ath (1965-1987). Table 9.8 shows the average protein contents of wheat and barley as a function of the level of phosphorus and potassium fertilization.

### *Nitrogen Fertilization*

For wheat, barley, and other cereal crops, nitrogen is the main key to yield. Crops that have insufficient nitrogen are yellowish. Even a small amount of nitrogen, given at the right time, relieves these symptoms, and larger amounts can increase the yield of grain by a factor of five or six on

TABLE 9.8. Average protein content of wheat and barley as a function of the level of phosphorus and potassium fertilization.

| Level of fertilization | Wheat | Barley |
|---|---|---|
| *Phosphorus* (kg $P_2O_5$/ha) | | |
| 0 | 12.0 | 11.9 |
| 60-80 kg up to 1973, and 80 kg from 1973 | 11.9 | 11.9 |
| 120-140 kg up to 1973, and 160 kg from 1973 | 11.8 | 11.9 |
| *Potassium* (kg $P_2O_5$/ha) | | |
| 0 | 11.9 | 11.9 |
| 80-200 kg up to 1973, and 120 kg from 1973 | 11.9 | 11.9 |
| 160-400 kg up to 1973, and 240 kg from 1973 | 11.8 | 11.9 |

*Source:* Soil Service of Belgium, Ath, 1965-1987.

some occasions (Addiscott et al., 1991). There is a point, however, described as the biological optimum, at which applying extra nitrogen fertilizer ceases to be worthwhile. The extra nitrogen is largely wasted and is likely to contribute to nitrate pollution of the superficial groundwater. Overfertilization can also be disastrous for the crop and may lead to lodging.

Nitrogen may make leaves larger and therefore able to capture more light. Cereal plants have the ability to produce extra shoots, known as tillers. Adding nitrogen increases the number of tillers that survive; extra tillers mean extra leaves, extra stems, extra ears, and more yield potential.

However, fertilizer is not an unmixed blessing to crops. The extra growth of leaves that it brings means extra weight for the stems to bear and extra resistance to wind and therefore stress on the stems. This can increase transpiration and the use of water by the plant. The nitrogen enlarges the cells of stems as well as leaves, making them more succulent but mechanically less strong. One consequence is that well-fertilized cereal crops become more likely to suffer from lodging. Another problem is that more succulent stems are more prone to attack by insects and fungal diseases.

In the absence of stress factors (such as drought, scarcity of nutrients, deficiency, and diseases), the nitrogen response curve (the yields of grain are plotted against the amount of nitrogen supplied) has an optimum, which depends on the total application of nitrogen and the soil capacity to supply nitrogen (Boon, 1988). However, the total production of dry matter (grain and straw) increases when more nitrogen is given. This is due to a stronger vegetative development (larger production of straw). This is illustrated by Figures 9.5 and 9.6.

The response to the nitrogen uptake is positive, even when the amount of supplied nitrogen is higher than the optimum. The extra nitrogen uptake gives rise to the production of storage proteins (gliadin-glutenin) (Biston and Dardenne, 1985). This leads to the conclusion that the application of nitrogen beyond the optimum results in higher protein content because the production of dry matter of the grain is more inhibited in comparison with the nitrogen uptake (Figure 9.5).

In order to increase the protein content together with high yields of grain, a substantial nitrogen uptake is necessary. This requires a high nitrogen uptake capacity of the crop, a healthy crop, a well-developed root system, and a sufficient water supply.

The nitrogen fertilization increases the proportion of water-insoluble proteins, and this influences the Zeleny-sedimentation value. The effect of the nitrogen fertilization on the Zeleny index is to a large extent determined by the variety (Figure 9.7).

The increase in protein content following the application of nitrogen beyond the optimum can account for as much as 80 percent of the increase in gluten content. This explains why nitrogen fertilization indirectly affects the W value of Chopin and the P/L ratio of the dough.

These values are also influenced by the variety. Nevertheless, it can be said that the W value increases as a function of the amount of nitrogen fertil-

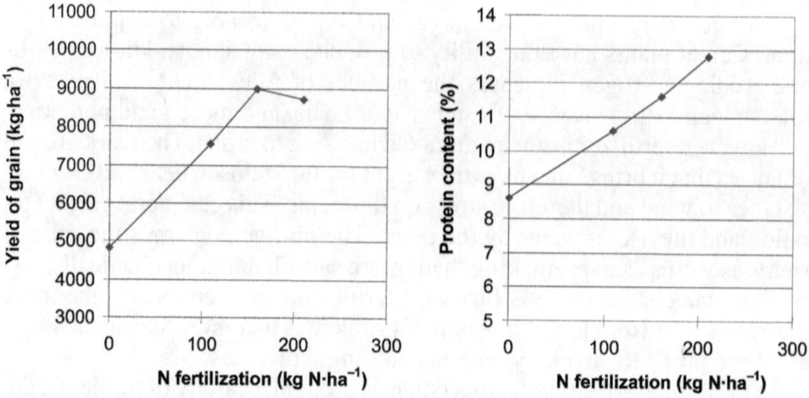


FIGURE 9.5. Effect of the fertilization on the yield of the grains and on the protein content of winter wheat, variety Castell. *Source:* Soil Service of Belgium, Bertem, 1989.

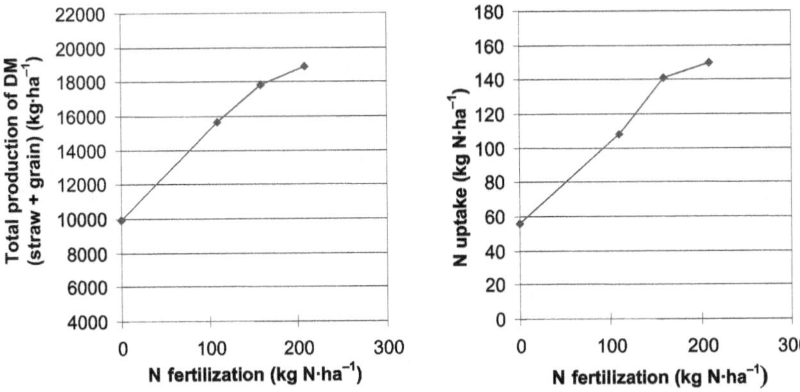


FIGURE 9.6. Influence of the fertilization on the total production of dry matter and on the N uptake by the grain of winter wheat, variety Castell. *Source:* Soil Service of Belgium, Bertem, 1989.

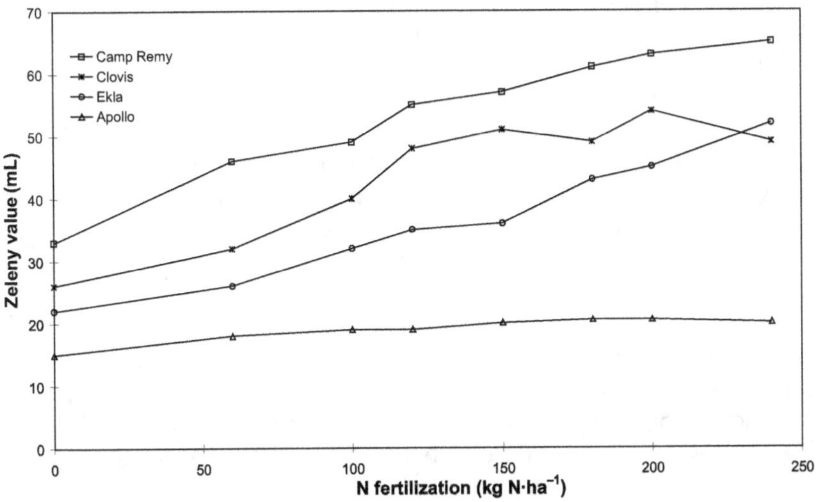


FIGURE 9.7. Effect of the N fertilization on the Zeleny value for different wheat varieties.

izer (Figure 9.8). The P/L ratio decreases and could be under the optimal value, even when high doses of nitrogen are applied (Falisse, 1990).

Under optimal growing conditions, specific grain weight is positively correlated to nitrogen fertilization (Gooding et al., 1999). However, over-

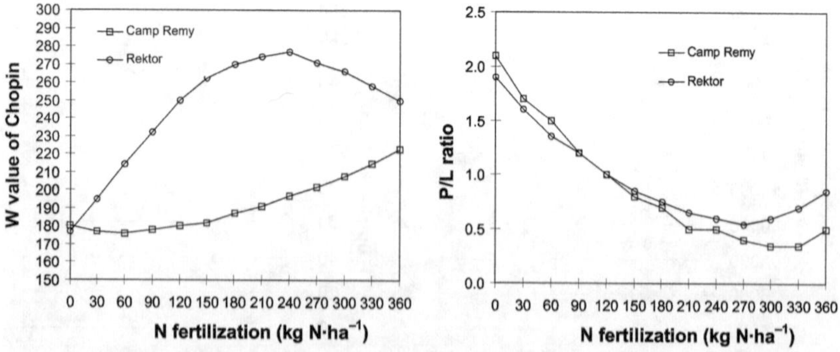


FIGURE 9.8. Influence of the N fertilization on the W value of Chopin and the P/L ratio for the wheat varieties Camp Remy and Rektor. *Source:* Adapted from Falisse, 1990.

fertilization leading to lodging of the crop can diminish it. Also, drought, disease, and other stress factors can adversely affect the specific grain weight. Thylén et al. (1999) reported a negative correlation between grain yield and specific grain weight in malting barley.

To satisfy the quality requirements for malting barley (e.g., a low content of proteins), the choice of variety is again extremely important. The nitrogen fertilization will also play an important role in the final quality of the barley (Table 9.9). Low protein content can be realized only if the nitrogen availability (by the application of fertilization or released by mineralization) is not too high at all times. In this way, the development of a heavy crop is prevented with a smaller risk of lodging and the accumulation of proteins is not stimulated. This can be realized by a good choice of the parcel and by an appropriate fertilization based on the nitrogen supply and the mineralization capacity of the soil.

When nitrogen fertilization is limited, the dose and time of application have limited influence on the quality of the malting barley. In this case, the variety will be the most important factor.

### *Effect of Split Application of Nitrogen Fertilizer on the Quality of Cereals*

The split application of nitrogen fertilizer will help the crop to exploit its full potential, and particularly to ensure that enough nitrogen is available for the crop at all times. We also have to include the environmental objec-

TABLE 9.9. Effect of the nitrogen fertilization on the yield and quality of malting barley, variety Prisma.

| N-fertilization | Yield of grain ($kg \cdot ha^{-1}$) | Protein content (%) | Caliber[a] | Specific grain weight (kg) |
|---|---|---|---|---|
| 40 (40-0)[b] | 7495 | 11.4 | 86.4 | 66.7 |
| 60 (60-0) | 7630 | 11.3 | 87.9 | 67.0 |
| 80 (80-0) | 7881 | 11.4 | 88.3 | 67.0 |
| 100 (100-0) | 7468 | 11.6 | 87.9 | 67.0 |
| 80 (60-20) | 7686 | 11.4 | 88.4 | 66.8 |
| 100 (60-40) | 7837 | 11.7 | 85.2 | 66.6 |

*Source:* Adapted from Gistel, unpublished data.

[a]The percentage of grains with a diameter larger than 2.5 mm (Maddens et al., 1993).

[b]First application at sowing (spring), second application at the end of tillering.

tive of ensuring that there is as little surplus nitrate as possible in the soil at risk of leaching.

The relationships in time between the amount of mineral nitrogen supplied naturally by the soil and the amount of nitrogen that should be available for the winter wheat crop in cool and wet regions show that in autumn the soil supplies enough to meet the needs of the crop. The soil is fairly warm and getting wetter, giving conditions in which soil microbes are able to supply the nitrogen that the crop needs. During the winter, the soil is cool and the microbes produce little mineral nitrogen but the crop does not need much. During the spring, crop growth accelerates and its demand for nitrogen increases. A substantial application of nitrogen fertilizer is usually given at this time. The earlier in the spring the fertilizer is applied, the more likely it is on average to be washed out of the soil. This is not only a waste of money but also an important load on the environment. Some farmers spread the risk by splitting the application into three parts. The development of the crop could be manipulated to the farmer's advantage by using several applications of nitrogen timed to coincide with certain stages in the development of the crop.

An application of nitrogen fertilizer in spring encourages tillering. The application of a very high amount of nitrogen will promote the growth of too many sprouts. These sprouts will compete for light and nutrients in a later stage, which has a harmful impact on the development of the ears.

A period of high nitrogen uptake by the crop starts at the beginning of stem elongation. A sufficient supply of nitrogen in this stage is favorable for the development of the ear, since it will result in a minimal reduction of the number of ears and flowers, which has an influence on the number of grains.

A third dose of nitrogen is applied in the period between the final leaf stage until the initiation of the ear. This application of nitrogen increases the content of nitrogen in the stem and the leaves which extends the life of the green parts of the plant (longer photosynthesis) and which supports the transport of nitrogen to the grain. This third application will have a favorable effect on the nitrogen uptake by the grains.

The third rate also has an important influence on the quality of the wheat destined for the starch industry. Nitrogen fertilization in the final leaf stage has a positive impact on the extraction process: more proteins under the form of gluten are extracted and fewer proteins remain in the starch fraction (Dubois et al., 1995). This increases the extraction yield and the purity of the starch (decrease of the protein content in the starch).

Since the protein content is increased by the application of nitrogen in the final leaf stage, the third application is lowered for wheat destined for the biscuit industry. This reduction of nitrogen fertilization also causes a smaller W value of Chopin, which is favorable for the quality of biscuit wheat (Loison, 1994).

Table 9.10 shows the effect of split application of nitrogen fertilizer on the yield and the quality of winter wheat. The yield of grains, protein content, specific grain weight, and thousand-kernel weight increase after a split application of the fertilizer.

Although properties of the species are determinative, plant nutrition and weather conditions also influence the wheat quality. The complexity of the problem is illustrated on the basis of a fertilization experiment in Middelkerke performed by the Soil Service of Belgium in 1996. In this experiment, a high productive wheat variety (Vivant) was used on a clay texture. Because of the high supply of nitrogen in the soil in the spring of 1996 (after a dry winter), the nitrogen recommendation was 27 kg N·ha$^{-1}$ (27-0-0). Thanks to the favorable growing conditions in 1996, high yields were achieved (Figure 9.9). An increase in nitrogen fertilizer beyond the nitrogen recommendation did not result in an increase in yield. Because of the high reserves of nitrogen in the soil, there was a high nitrogen uptake on the unfertilized object, which resulted in high protein content. However, the protein content increased significantly through nitrogen fertilization (Figure 9.9).

TABLE 9.10. Effect of the application of nitrogen fertilizer on the yield and quality of winter wheat, variety Capitaine.

| N fertilization (kg N·ha$^{-1}$) | Yield of grains (kg·ha$^{-1}$) | Protein content (%) | Specific grain weight (kg) | 1000 kernel weight (g) |
|---|---|---|---|---|
| 209 (not split) | 10885 | 12.2 | 80.4 | 42.4 |
| 159 (not split) | 10708 | 11.2 | 79.1 | 42.0 |
| 109 (not split) | 10531 | 10.3 | 78.6 | 41.9 |
| 209 (57-85-67)[a] | 11639 | 13.1 | 80.6 | 43.0 |
| 159 (41-68-50)[a] | 11476 | 11.7 | 80.4 | 43.5 |
| 109 (25-51-33)[a] | 11389 | 10.9 | 79.1 | 43.2 |
| 0 | 6709 | 8.3 | 74.6 | 39.0 |

*Source:* Soil Service of Belgium, Fontaine Notre Dame, 1989.

[a]Split applications: first application at the beginning of tillering; second application at the beginning of stem elongation; third application in the final leaf stage

However, yield, thousand-kernel weight, and specific grain weight declined when the nitrogen fertilization exceeded the advised dose. Moreover, the extra nitrogen uptake on the fertilized objects was rather small (low nitrogen efficiency), and the higher protein content on the heavy fertilized objects can be explained by a lower yield of grains with the same nitrogen uptake by the grain (Figure 9.10).

The nitrogen residue at harvest is significantly higher on the fertilized objects compared with the unfertilized objects, due to the low nitrogen efficiency and other factors (Figure 9.11).

### *Conclusion*

The question is to what extent nitrogen fertilization management can help to achieve better quality. The specific properties of the variety play a key role, but plant nutrition (mainly by application of nitrogen fertilizer) and weather conditions influence the quality parameters.

The total dose of nitrogen as well as the split application has an influence on the final protein content. Other quality parameters connected to the pro-

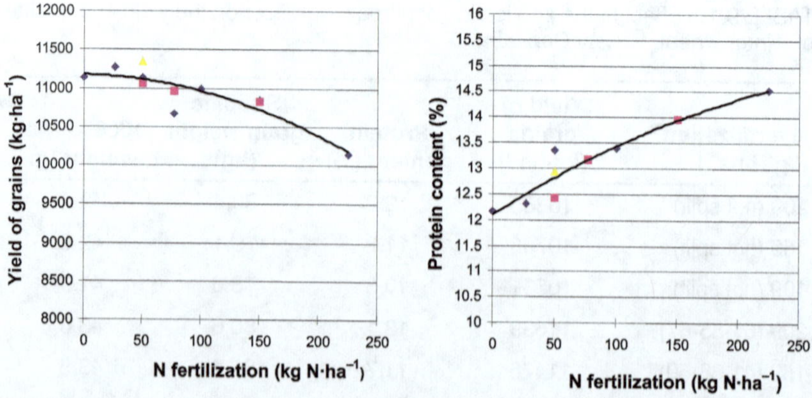


FIGURE 9.9. Influence of the nitrogen fertilization on the yield and protein content of winter wheat, variety Vivant. *Source:* Soil Service of Belgium, Middelkerke, 1996.

tein quality, gluten content, and baking quality are mostly related to the response of the protein content to the nitrogen fertilization.

Nitrogen fertilization beyond the optimum results in higher protein content, especially when this extra application occurs in the final leaf stage.

However, the increase of the global quality is limited because the influence of the extra nitrogen fertilization on the other quality parameters (e.g., falling number of Hagberg, specific grain weight) can be negative. Besides, a high level of nitrogen raises the risk of lodging and diseases and can contribute to nitrate pollution of the superficial groundwater.

As a consequence, an increase in nitrogen fertilization to achieve better quality is sound only when species with a high base quality are used and when the amount of nitrogen fertilizer and the timing of the applications are appropriate.

## GRAIN QUALITY AND CROP MANAGEMENT

### Introduction

Many researchers (Low et al., 1997; Stafford, 1999; Algerbo and Thylén, 1998; Reyns et al., 1999) have shown that not only grain yield but also grain quality often vary site specifically within a field. In the coming years, sen-

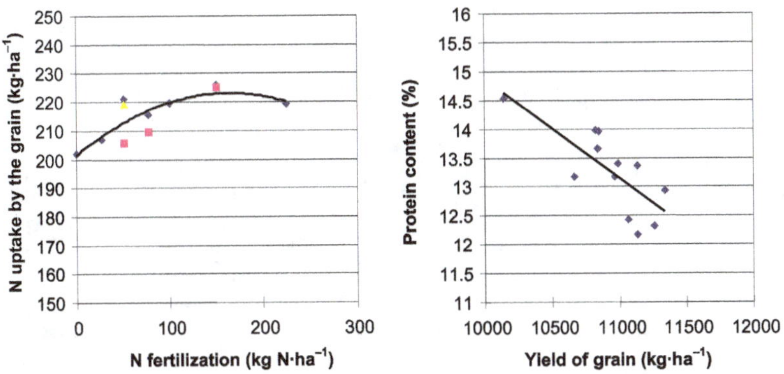


FIGURE 9.10. Influence of the nitrogen fertilization on the nitrogen uptake by the grains *(left)*. Relationship between yield of grains and protein content of winter wheat, variety Vivant *(right)*. *Source:* Soil Service of Belgium, Middelkerke, 1996.

sors will be available to measure grain quality on-line on combine harvesters. This offers new opportunities in site-specific crop management.

The goal of crop management is to maximize the return for the farmer. For wheat this can be defined as follows:

$$\begin{aligned} Return/ha &= yield \times (grain\ price) - fertilization\ cost/ha \\ &\quad - spraying\ cost/ha - fixed\ cost/ha \end{aligned}$$

The fixed cost includes land rent and other expenses not influencing the yield. The cost of pesticides, fungicides, and herbicides will be considered in this discussion as a constant cost per ha, not influencing the yield directly when properly applied. As such, optimization of the return means a maximization of grain yield and price compared to the N fertilization needed to reach them. The price of wheat for feed purposes is negatively influenced by the moisture content and specific grain weight. In Belgium, grains for bread are also tested for the protein content, Zeleny, and the Hagberg falling number, which in turn influence the price. To calculate the optimal fertilization level for a maximum return, the influence of fertilization on these parameters has to be known.

The moisture content is only partially related to the nitrogen supply. Reducing the nitrogen rate can influence it only when lodging or unripe grains caused the high moisture content. Under optimal growing conditions the specific test weight is positively correlated with applied nitrogen, but it is

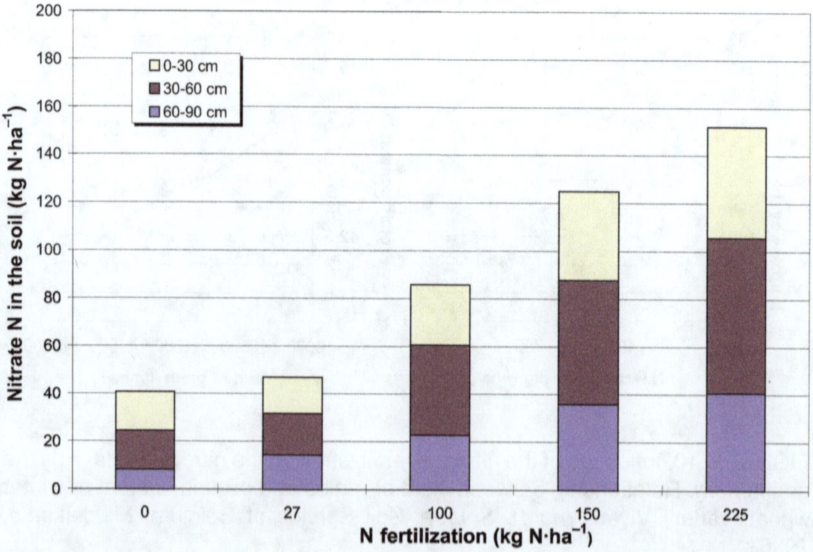


FIGURE 9.11. Influence of the nitrogen fertilization on soil nitrate N at harvest. *Source:* Soil Service of Belgium, Middelkerke, 1996.

influenced by other factors such as weather conditions. Protein content and Zeleny are positively related to the nitrogen content, but no fixed relationship exists for the Hagberg falling number (Gooding et al., 1999). Zeleny is to a large extent determined by the variety, and bringing it to the right level depends on a right choice at the beginning of the season.

To illustrate the influence of grain quality on the price and the optimization process of fertilizer application toward maximum return, simulations are done based on experimental plot data (Falisse, 1990). The variety, Camp Remy, used in the experiments has good quality characteristics for bread making. In the calculation, only grain yield and protein content are included as price-determining factors. To facilitate the simulations, a second-order polynomial was fitted through response data of yield and protein to nitrogen fertilization. The second-order polynomials are chosen as they fit most closely to the data (Figure 9.12).

Optimum N levels are calculated for different situations: optimum yield, maximum financial return when sold as grain for feeding purposes (no protein premium), and maximum return when sold as grain for bread. Prices are taken from the actual Belgian situation for the harvest season 2001. The farmer gets a basic price of 110.3 € per ton and 105.4 € per ton for grain for bread and grain for feeding, respectively, and pays 0.17 € per kg for nitro-

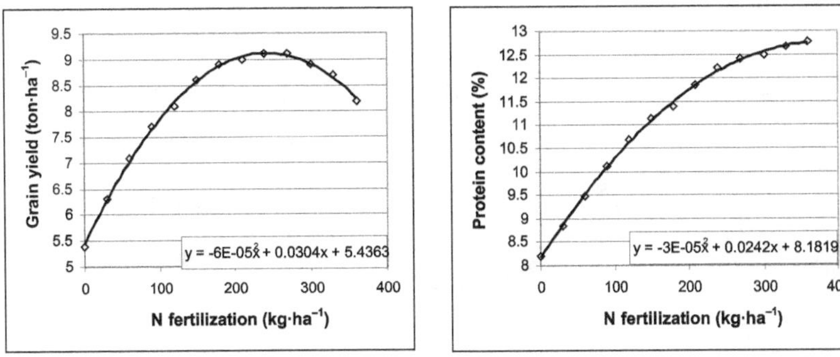


FIGURE 9.12. Grain yield and protein content response curve.

gen. Simulations are calculated for a premium of 5 € per ton and percent protein above 12 percent. The optimum fertilization level is reached when the marginal financial return equals zero (Figure 9.13). The marginal financial return is defined as the additional return (in €) the farmer would have when increasing the nitrogen application with one unit on one hectare.

Because the price of nitrogen is relatively low, the optimal level for animal feeding is close to the level of maximum yield (229 kg·ha$^{-1}$ vs. 242 kg·ha$^{-1}$). A premium of 5€ for protein content (current Belgian situation) shifts the optimum level up to 265 kg N·ha$^{-1}$.

It is clear from these simulations that the economic optimum N level is higher compared to the optimum yield. In practice, the financial optimum can still be somewhat lower, as a high nitrogen rate can require a more intensive use of fungicides, and an increased risk of lodging has to be taken into account. Also, nitrogen losses increase rapidly when applying nitrogen over the optimum yield value.

In the current practice of uniform N application, the application rate is not based on the used curves but calculated mainly on the bases of soil type, available soil organic matter, and nitrogen present as ammonium or nitrate elements.

However, in managing the crop site specifically, the relationships used vary site specifically (Gooding et al., 1999), as they are dependent on the spatially variable soil organic matter, water regime (weather conditions), soil type, and so forth. As a result, the economic optimum fertilization is also variable within the field. In the following paragraphs, some examples of management strategies are discussed, as well as the use of a quality sensor in this management.

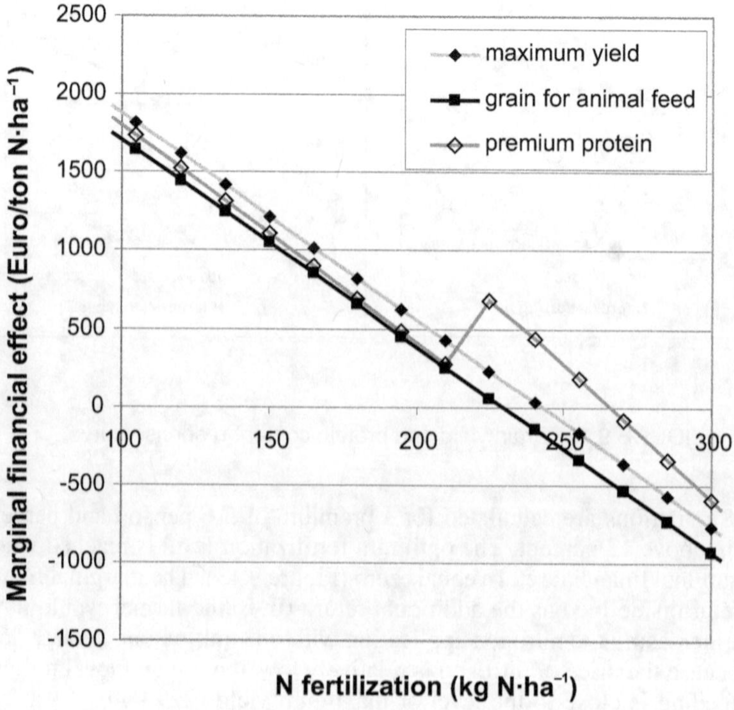


FIGURE 9.13. Optimum N fertilization in different market situations.

## SITE-SPECIFIC CROP QUANTITY AND QUALITY MANAGEMENT

The site-specific grain yield and quality (protein content, specific grain weight, and Hagberg falling number) response curves based on three different nitrogen levels (Gooding et al., 1999) show a significant spatial variation. Test plots with a variation of the latest nitrogen gift (0, 50, and 100 kg·ha$^{-1}$) were created at nine different locations in the field. About one-third of the plots showed an optimum yield at the zero level, one-third at 50 kg N, and in the remaining plots yield still increased up to the 100 kg level. With the exception of one plot, protein increased at all levels, but the responses differed greatly. Particularly in dry parts of the field where the plants were relatively advanced, the response was high. The specific grain weight tended to increase with nitrogen level but with no clear relationship. The reasons for the variation in response were unclear. The same could be concluded for variation of the Hagberg falling number. The economic optimum based on these curves and the European Union grain intervention prices was close to the maximum yield level. The reasons were the low influence of quality on the price as well as a low nitrogen cost. A grain quality premium would have increased the optimum nitrogen level.

Site-specific determination of these response curves is in reality not a feasible strategy. In the case of grain for feeding purposes, the economic optimum could be achieved by fertilization toward maximum yield, since those are close to each other at low nitrogen prices. One possibility is then to apply nitrogen according to the analysis of site-specific soil samples under the supposition that soil organic matter (and present nitrogen under different forms) and soil type are the most important varying factors.

To avoid the high cost of taking many soil samples, some researchers have been trying to optimize fertilization by aid of different sensors. The availability of a sensor to measure the grain protein content could especially offer new possibilities.

Engel et al. (1997) derived from different fertilization trials with different wheat cultivars and water regimes a critical protein level of 13.5 percent. This level equals the point of maximum yield, below which the wheat is underfertilized. The fertilization level of regions in the field with this protein level is calculated as the amount of nitrogen removed through multiplying the N content of the kernels (from protein) with the grain yield. Fertilization of the below- and above-optimal yield regions is based on the grain yield and protein content response curves (Long et al., 1998). When determined under different water regimes, the response would be similar to some extent. Based on the protein content and yield map of the previous year, a fertilization map for the next season is created. In a similar way, Kelly et al. (2001) prescribe fertilization, mentioning a critical protein level of 11.5 percent. Using a linear relationship between the protein content and the nitrogen transfer efficiency, this is taken into account.

Different remarks can be made regarding this approach. First, the critical protein level will be (under optimal growing conditions) dependent on growing conditions such as soil type and variety. This means that this level will have to be determined through fertilization trials by specialists in different climatic regions. Second, because variations can be largely due to variations in soil organic matter content, the approach can be used only if grain is grown in subsequent years. In European conditions, where wheat is mostly cultivated in a crop rotation, the pattern of the soil organic matter content in the field can change over time. Moreover, even if different under- or overfertilized zones can be distinguished in the field, calculation of the nitrogen dose based on response curves will be very inaccurate. It was shown earlier that these curves and their slope cannot be determined accurately enough through fertilizer trials.

Quantifying the optimum fertilization based on harvest observations is not possible because of important year-to-year variations of the nitrogen provided from soil organic matter and the degree in which the plant takes it up. Grain yield and protein sensors are evaluating tools, allowing definition of management zones only to some extent. The protein sensor can help in the analysis of the variation. For example, very low yields together with low protein content indicate a lack of nitrogen. On the other hand, very high protein content with a moderate yield and lodging indicates overfertilization.

The following strategy is then proposed:

1. Mapping of yield and protein content (eventually during consecutive years)
2. Defining of extreme zones (protein content/yield) and zones which are consistent over different years
3. Investigation into the variation source, possibly by soil analysis (soil type, nutritional status), soil maps, and slopes
4. Fertilization prescription based on soil samples in different zones (if they are large enough to compensate for the increased cost of a soil sample)
5. Slight variation of the prescription according to leaf reflectance measurements
6. Agronomical and economical evaluation of the applied variations using yield/protein mapping

Gooding et al. (1999) proved the correlation between the optimal latest dose (third fraction) and values of a Minolta SPAD-meter. Using the reflectance values of a crop-scan MSR5 spectroradiometer just before appearance of the third node, the protein content measured at harvest could be predicted with a RMSEP of 0.87 percent (Figure 9.14). The advantage of this strategy is that variations due to changing weather conditions (changing N supply from soil organic matter, changing amount of leaching) can be taken into account.

When optimizing the price of grain for bread making, the same strategy can be followed. However, a small increase in the fertilization level for each zone has to be considered, as the economic optimum N level is here somewhat higher than for the maximum yield. This must carefully be considered, since overfertilization leads to lodging with risk of a decreased falling number and loss of grain protein premium as a result.

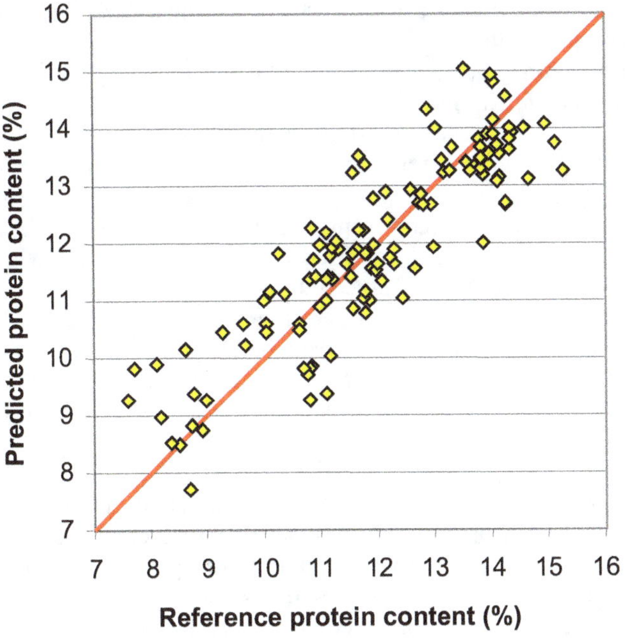


FIGURE 9.14. Prediction of the grain protein content at harvest using reflectance measurements at the growing season. *Source:* Adapted from Dumont, unpublished data.

## *On-Line Grain Sorting*

Economically optimum fertilization varies within the field due to the variation of the response curves of grain quality and yield. When variable-rate fertilization can be applied according to the economic optimum, it will result in a variable quality and yield. To maximize the return, the farmer must get the right price for this quality. But the grain price is currently calculated at the grain stores based on the quality of one grain sample taken from a trailer. To avoid this, four different strategies can be followed (Stafford, 1999):

1. On-combine sorting. The currently developed on-line quality sensors can drive a sorting mechanism, if two or more holding tanks are present on the combine. This system would require different trailers as

well as divisions at the grain store to maintain separation between different quality loads.
2. Harvest zone identification. In case grain quality can be sensed/estimated with a suitable sensing system prior to harvest, the different quality zones can be harvested separately. Alternatively, if quality is measured at harvest, zones that can be separated each year can be harvested separately.
3. In-store sorting. The same sensors as currently developed for combine harvesters can probably be used at the grain store, allowing a relatively crude separation into different quality classes.
4. Uniform grain quality. Site-specific fertilization to get uniform quality over the field can finally give a better return and the extra investment of a grain quality sensor can be avoided. However, this means no sensor is available to help separation of the field into different management zones.

After determining the quality of samples taken at different places in the field, Thylén et al. (1999) could estimate the possible advantage of an on-combine sorting system. Cultivating malting barley during two consecutive years, the value would have increased by $49 per ha the first year, but decreased the second year when sorted. The most important factors for malting barley quality are protein content, germination rate, and grain size. However, only protein content is at this moment measurable on-line.

Sorting of feeding barley according to malting premium limits and on the basis of N content increased the grain value in each of four consecutive years (Stafford, 1999). This increase varied between 34 and 505 £ on a 6.5 ha field if the grain would have been sorted in two quality bands, depending from the premium and the year.

## CONCLUSIONS

Grain quality includes many different parameters, depending on the final use. When used for feeding purposes, this ranges from the protein content to fat, starch, and even the type of amino acids. However, not all these parameters are equally important for every purpose or use, and for some parameters (such as fiber content) the laboratory method to measure them is still not standardized. For bread making, the chemical composition, such as the protein content, is measured, and different physical tests also exist to determine the baking quality. The applied tests differ among companies and among countries. Two common used tests are the Zeleny or SDS sedimentation value, and the Hagberg falling number. These are only indirectly related to the chemical composition, and their determination with fast measuring methods such as near infrared reflection is difficult.

Despite the importance of grain composition for the processing industry, grain trading is mostly based on physical parameters such as specific grain weight, moisture content, and grain impurity. Sometimes the protein content, Zeleny value, and Hagberg falling number also influence price premiums.

Near infrared techniques are widely used in laboratory environments. Sensors are under development that could measure grain moisture and protein content on-line on combine harvesters. There is also a good chance that the on-line instruments will be able to measure the Zeleny value with a limited accuracy. These parameters are also measurable for barley, corn, and oil content in soybeans. Measurements on forage harvesters are under research. To develop these systems for commercial applications, problems such as calibration and calibration transfer, dirt on the measuring window, and the blocking of the bypasses will have to be solved. Different commercial sensors are available to measure grain moisture content on-line based on electrical capacitance. Accuracy is typically around 0.45 percent (root mean squared error) within one field and 0.8 percent over the whole harvesting season. The error (RMSEP) of near infrared laboratory instruments is mostly below 0.5 percent for moisture and protein content prediction.

As of yet, no technique exists to measure on-line the specific grain weight or thousand-kernel weight, although these characteristics are important in grain trading.

The new technology to measure grain quality on-line offers new possibilities for site-specific crop management. Despite much research on the influence of nitrogen supply on grain quality, site-specific fertilization based purely on grain yield and quality sensors will not be possible. Response curves of grain quality differ too much due to conditions that are variable in the field. Such grain quality is positively related to nitrogen uptake; a grain protein sensor can significantly help in defining management zones in the field. The calculation of the optimum fertilization in these zones will still have to be based on soil analysis. Crop reflectance measurements during the growing season will help to adjust this level according to seasonal variations and changing weather conditions.

Another advantage of grain protein sensors is that they open the possibility for on-line grain sorting.

## NOTES

1. RDS Technology Ltd., Cirencester Road, Minchinhampton, Stroud, Gloucestershire, GL6 9HB United Kingdom

2. Ag Leader Technology, 2202 S. Riverside Dr., Ames, Iowa 50010
3. Claas KgaA., Münsterstraße 33, D-33428 Harsewinkel
4. Deere & Company, One John Deere Place, Moline, Illinois 61265-8098; (309) 765-8000
5. Infrared Engineering Limited, Galliford Road, The Causeway, Maldon, Essex CM9 4XD, UK
6. MSC Rütter, P.O. Box 40111 7504 RC Enschede, the Netherlands
7. Prinacon Analyse System A/S., P.O. Box 49 DK-7100 Vejle, Denmark
8. FORMAL SPS, On-line Analysing Systems, Langkaer 72, DK-6100 Haderslev, Denmark
9. Bran+Luebbe Gmbh., P.O. Box 1360, D-22803 Norderstedt, Germany
10. Brimrose Corporation of America, 5024 Campbell Boulevard, Baltimore, Maryland 21236-4968
11. Milestone Technology, Inc., 395 W. Hwy. 39, Blackfoot, Idaho 83221
12. CNH Global N.V., 5729 Washington Avenue, Racine, Wisconsin 53406
13. Textron Systems, 201 Lowell Street, Wilmington, Massachusetts 01887
14. Zeltex Inc., Western Maryland Parkway 130, Hagerstown, Maryland 21740

## REFERENCES

Addiscott, T.M., A.P. Whitmore, and D.S. Powlson. 1991. *Farming, fertilizers and the nitrate problem.* CAB International, Wallingford, UK.

Algerbo, P.A. and L. Thylen. 1998. Variable nitrogen application: Effects on crop yield and quality. In *Proceedings of the 4th International Conference on Precision Agriculture,* eds. P.C. Robert, R.H. Rust, and W.E. Larson (pp. 709-717). ASA, CSSA, SSSA, Madison, Wisconsin.

Axford, D.W., E.E. Mcdermott, and D.G. Redman. 1979. Note on the sodium dodecyl sulfate test of breadmaking quality: Comparison with Pelshenke and Zeleny tests. *Cereal Chemistry* 56(6):582-584.

Benson, I.B. 1995. The characteristics and scope of continuous on-line near infrared measurement. *Spectroscopy Europe* 6:18-24.

Berbert, P.A., and B.C. Stenning. 1996. On-line moisture content measurement of wheat. *Journal of Agricultural Engineering Research* 65:287-296.

Bernaerts, E. 1991. *De afzetkolom van graan in de B.L.E.U.* L.E.I. publikaties nr. 531, studies nr. 40, Brussels, Belgium.

Birth, G.S., and H.G. Hecht. 1987. The physics of near-infrared reflectance. In *Near-Infrared Technology in the Agricultural and Food Industries,* eds. P.C. Williams and K. Norris (pp. 1-15). American Association of Cereal Chemists, Inc., St. Paul, Minnesota.

Biston. R., and P. Dardenne. 1985. Analyse des critères de qualité des blés par spectrométrie de réflextion dans l'infrarouge proche. In *Qualité et teneur en protéines des blés: Note technique du Centre de Recherches Agronomiques de l'Etat* (pp. 17-33). Gembloux, ARC, Belgica.

Boon, R. 1988. *Determinerende factoren bij de stikstofbehoefte van wintertarwe voor maximale opbrengst en kwaliteit.* Bodemkundige Dienst van België, Leuven, Belgium.

Centraal VeevoederBureau (CVB). 1994. *Veevoedertabel, gegevens over chemische samenstelling, verteerbaarheid en voederwaarde van voedermiddelen.* Centraal VeevoederBureau, Lelystad, the Netherlands.

Cereal Tester. 1997. NIR Analyzers INFRAMATIC: Actual application sheets. Cereal Tester, Fleurus, Belgium.

Cereal Tester. 2000. NIR Analyzers INFRAMATIC: Actual application sheets. Cereal Tester, Fleurus, Belgium.

Downey, G. 1985. Estimation of moisture in undried wheat and barley by near infrared reflectance. *Journal of the Science of Food and Agriculture* 36:951-958.

Dubois, A., M. Sindic, and C. Deroanne. 1995. Kwaliteit van het zetmeel en stikstofbemesting. In *Bemesting en plantenbescherming van de graangewassen.* Faculté des Sciences Agronomiques Gembloux en Centre de Recherches Agronomiques de l'Etat, Gembloux, Belgium.

Engel, R.E., D.S. Long, and G.R. Carlson. 1997. On-the-go grain protein sensing is near: Does it have a future in precision nitrogen management for wheat? *Better Crops with Plant Food* 81(4):20-23.

Falisse, A. 1990. Fytotechnie van de granen: 1. Wintertarwe. In *Bemesting en plantenbescherming van de graangewassen.* Faculté des Sciences Agronomiques Gembloux et Centre de Recherches Agronomiques de l'Etat, Gembloux, Belgium.

Gooding, M.J., M. Salahi, and R.H. Ellis. 1999. Spatial variation in the response of wheat grain yield and quality to late nitrogen. In *Proceedings of the 2nd European Conference on Precision Agriculture,* ed. J.V. Stafford. SCI, Odense, Denmark.

Häusler, A., M. Rode, and C. Paul. 2001. New perspectives for sensing crop factors in the field by diode array spectrometers for the NIR range. In *Book of abstracts (Poster) of the 3rd European Conference on Precision Agriculture,* ed. G. Grenier and S. Blackmore (p.14). Agro-Montpellier, Montpellier, France.

Kandala, C.V.K., S.O. Nelson, and K.C. Lawrence. 1996. Determining moisture content in small wheat samples by dual-frequency RF impedance sensing. *American Society of Agricultural Engineers* 39(3):1185-1188.

Kawamura, S., M. Natsuga, and K. Itoh. 1999. Determination of undried rough rice constituent content using near-infrared transmission spectroscopy. *Transactions of the ASAE* 42(3):813-818.

Kelly, R., W. Strong, T. Jensen, and D. Butler. 2001. Strategies to interpret the yield map: Defining yield-limiting factors. *Proceedings of the 10th Australian Agronomy conference, January 28-February 1.* The Australian Society of Agronomy, Hobart, Tasmania, Australia.

Kormann, G. 2001. Untersuchungen zur integration kontinuierlich arbeitender feuchtemesssysteme in ausgewählte Futtererntemaschinen. Dissertation zur Erlangung des akademischen Grades einens Doktor-Ingenieur. Technischen Universität München, fachgebiet für Technik im Pflanzenbau, Weihenstephan.

Kraszewski, A., and S.O. Nelson. 1991. Density-independent moisture determination in wheat by microwave measurement. *Transactions of the American Society of Agricultural Engineers* 34(4):1776-1783.

Krischik, V., D. Shipman, and R. Stuckey. 1990. How grain moves through the marketing system. In *Stored product management* (pp. 13-20). Publication E-912. Oklahoma Cooperative Extension Service, Oklahoma State University, Stillwater, Oklahoma.

Law, D.P., and R. Tkachuk. 1977. Determination of moisture content in wheat by near-infrared diffuse reflectance spectrophotometry. *Cereal Chemistry* 54(4): 874-881.

Lawrence, K.C., and S.O. Nelson. 1993. Radio-frequency density-independent moisture determination in wheat. *Transactions of the American Society of Agricultural Engineers* 36(2):447-483.

Li, W.S., and J.T. Shaw. 1997. Determining the fat acidity of rough rice by near-infrared reflectance spectroscopy. *Cereal Chemistry* 74:556-560.

Loison, M. 1994. Fractionner l'apport d'azote selon le debouché du blé. *Cultivar* 373:44-47.

Long, D.S., G.R. Carlson, and R.E. Engel. 1998. Grain protein mapping for precision management of dryland wheat. In *Proceedings of the 4th International Conference on Precision Agriculture,* eds. P.C. Robert, R.H. Rust, and W.E. Larson (pp. 787-796). ASA, CSSA, SSSA, Madison, Wisconsin.

Low, R., G. Moore, M. Nicolas, and B. Williams. 1997. An investigation of barley grain protein variation. In *Proceedings of the 1st Australian symposium on Precision Agriculture,* ed. B. Whelan. University of Sydney, Sydney, Australia.

Maddens, K., A. Calus, and A. Denoo. 1993. Gerst: Overzicht van het onderzoek in 1991. *Provinciaal Onderzoek- en Voorlichtingscentrum voor Land-en Tuinbouw.* Jaarrapport Provincial, Rumbeke, Belgium.

Mailander, M.P., J.K. Schueller, and G.W. Krutz. 1982. An evaluation of four continuous moisture sensors on a combine. American Society of Agricultural Engineers, Paper 82 1576.

Nelson, S.O. 1982. Factors affecting the dielectric properties of grain. *Transactions of the American Society of Agricultural Engineers* 25(4):1045-1049.

Okabe, T., M.T. Huang, and S. Okamura. 1973. A new method for the measurement of grain moisture content by the use of microwaves. *Journal of Agricultural Engineering Research* 18:59-66.

Osborne, B.G. 1984. Investigations into the use of near infrared reflectance spectroscopy for the quality assessment of wheat with respect to its potential for bread baking. *Journal of the Science of Food and Agriculture* 35:106-110.

Osborne, B.G., S. Douglas, T. Fearn, and K.H. Willis. 1982. The development of universal calibrations for measurement of protein and moisture in UK home-grown wheat by near-infrared reflectance analysis. *Journal of the Science of Food and Agriculture* 33:736-740.

Reyns, P., J. De Baerdemaeker, and H. Ramon. 1999. Site-specific relationship between grain quality and yield. In *Proceedings of the 2nd European Conference on Precision Agriculture,* ed. J.V. Stafford, SCI, Odense, Denmark (pp. 665-676). Sheffield Academic Press, Sheffield, UK.

Reyns, P., K. Maertens, and J. De Baerdemaeker. 2001. On-line measurement of corn moisture content on a combine harvester using NIR technology. In *Proceedings of the 3rd European Conference on Precision Agriculture,* eds. G. Grenier and S. Blackmore (pp. 851-856). Agro-Montpellier, Montpellier, France.

Richmond, J. 1999. Some successful NIR process applications. Applitek On-Line Spectroscopy Seminar, Antwerp (Belgium). Oral presentation.

Ridgway, C., and J. Chambers. 1996. Detection of external and internal insect infestation in wheat by near-infrared reflectance spectroscopy. *Journal of the Science of Food and Agriculture* 71:251-264.

Rixhon, L. 1985. Influence des conditions écologiques et agronomiques. In *Qualité et teneur en protéines des blés: Note technique du Centre de Recherches Agronomiques de l'Etat* (pp. 35-47). Sciences agronomique de Gembloux, Gembloux.

Rubenthaler, G.L., and Y. Pomeranz. 1987. Near-infrared reflectance spectra of hard red winter wheats varying widely in protein content and breadmaking potential. *Cereal Chemistry* 64(6):407-411.

Schäfer, K., and W. Weinard. 1978. Infrarot-Reflektionsspektroskopie: Ein neues verfahren zur gleichzeitigen schnellbestimmung van Eiweiss, Fett und Feuchtigkeit. *Die Mühle + Mischfuttertechnik* 115:240-243.

Schmütz, W., and Q.N. Do. 1982. Weitere verbesserungen bei der Eiweiss-schnellbestimmung in gerste mit der Infrarot-Reflexionsspektroskopie. *Die Mühle + Mischfuttertechnik* 119:258-261.

Spaepen, P. 2000. NIR als meettechniek voor in-line metingen van het vocht- en proteïnegehalte van granen. Eindwerk K.U. Leuven, Faculteit Landbouwkundige en Toegepaste Biologische Wetenschappen, Leuven, Belgium.

Stafford, J.V. 1999. An investigation into the within-field spatial variability of grain quality. In *Proceedings of the 2nd European Conference on Precision Agriculture,* ed. J.V. Stafford, SCI, Odense, Denmark (pp. 353-361). Sheffield Academic Press, Sheffield, UK.

Thylén, L., and P.A. Algerbo. 2001. Development of a protein sensor for combine harvesters. In *Proceedings of the 3rd European Conference on Precision Agriculture,* eds. G. Grenier and S. Blackmore (pp. 869-873). Agro-Montpellier, Montpellier, France.

Thylén, L., P.A. Algerbo, and C.G. Pettersson. 1999. Grain quality variations within fields of malting barley. In *Proceedings of the 2nd European Conference on Precision Agriculture,* ed. J.V. Stafford, SCI, Odense, Denmark (pp. 287-296). Sheffield Academic Press, Sheffield, UK.

Tkachuk, R. 1981. Protein analysis of whole wheat kernels by near infrared reflectance. *Cereal Foods World* 26(10):584-587.

Tkachuk, R., F.D. Kuzina, and R.D. Reichert. 1987. Analysis of protein in ground and whole field peas by near-infrared reflectance. *Cereal Chemistry* 64(6): 418-422.

Vanorle, L. 1996. *La commercialisation des cereales.* Nr. 584, études nr. 74. Publications de l'I.E.A., Brussels, Belgium.

Vigier, F., D. Boffety, A. Marionneau, E. Ollivier, and B. Viallis. 2000. Technique d'échantillonnage et mesure de qualité du grain sur moissonneuse-batteuse. In *Actes du colloque « Agriculture de précision, Avancées de la recherche technologique et industrielle »* ed. P. Zwaenepoel, (pp. 277-287). Cemagref, Enesad, Dijon, France.

von Rosenberg, C.W. Jr., A. Abbate, J. Drake, and D.M. Mayes. 2000. A rugged near-infrared spectrometer for the real-time measurement of grains during harvest. *Spectroscopy* 15(6):34-39.

Watson, C.A., D. Carville, E. Dikeman, G. Daigger, and G.D. Booth. 1976. Evaluation of two infrared instruments for determining protein content of hard winter wheat. *Cereal Chemistry* 53(2):214-222.

Williams, P., D. Sobering, and J. Antoniszyn. 1998. Protein testing methods at the Canadian Grain Commission. Proceedings of the Wheat Protein Symposium, Saskatoon, Saskatchewan, March 9-10. Available online at www.grainscanada.gc.ca/PUBS/confpaper/Williams/ProteinOct98/protein_oct_98.pdf.

Williams, P.C. 1979. Screening wheat for protein and hardness by near infrared reflectance spectroscopy. *Cereal Chemistry* 56(3):169-172.

Williams, P.C. 1991. Prediction of wheat kernel texture in whole grains by near-infrared transmittance. *Cereal Chemistry* 68:112-114.

Williams, P.C. 1996. Recent advances in near-infrared applications for the agriculture and food industries. Presented at the International Japanese Conference on Near-Infrared Reflectance, November 20-21. Available online at www.grainscanada.gc.ca/Pubs/confpaper/nir-e.htm.

Williams, P.C., and H.M. Cordeiro. 1985. Effect of calibration practice on correction of errors induced in near-infrared protein testing of hard red spring wheat by growing location and season. *Journal of Agricultural Science* 104:113-123.

Williams, P.C., K.R. Preston, K.H. Norris, and P.M. Starkey. 1984. Determination of amino acids in wheat and barley by near-infrared reflectance spectroscopy. *Journal of Food Science* 49:17-20.

Wright, S.L., and C.F. Hood. 1998. Near infrared spectrometer used in combination with a combine for real time grain analysis. USA patent no 5.751.421.

# *PART II: APPLICATIONS IN CROPS AND CROPPING SYSTEMS*

## Chapter 10

# Site-Specific Rice Management

Alvaro Roel
G. Stuart Pettygrove
Richard E. Plant

## *INTRODUCTION*

Agricultural producers and researchers have been aware of spatial and temporal yield variability for a long time (Mercer and Hall, 1911; Lark and Stafford, 1997; Lowenberg-DeBoer and Swinton, 1997). Most farmers, when they harvest, recognize the presence of spatial variability in yield, but until recently they have been unable to precisely quantify it. Recently, new technology has become available to measure, reference, and record grain yields to document the location and magnitude of this variability. The introduction of yield monitors, yield mapping software, global positioning systems (GPS), and geographic information systems (GIS) has made it possible to measure crop grain yield within a field at a very high spatial resolution, allowing very fine and precise description of the spatial variability. If the causes of this spatial variability can be identified, then corrective action may possibly increase yield, decrease input costs in low-yielding areas, or both.

Indications suggest that cereal yields will approach the yield potential ceiling in many of the world's most productive cropping systems within the next three decades and that it will be increasingly hard for breeders to increase yield potential any further. If this is indeed the case, it will be essential to ensure that a larger portion of any given field approaches its yield potential to be able to continue increasing production (Matson et al., 1997; Cassman, 1999; Sinclair, 1999). Theoretical estimates of economic and environmental benefits from site-specific deployment are large. In practice, site-specific management (SSM) involves applying the right management, at the right time, at the right place, in the right way (Pierce et al., 1994). It has been suggested that by adoption of SSM techniques significant

*Handbook of Precision Agriculture*

doi:10.1300/5627_10

improvements in nitrogen, phosphorus, and weed and disease management can be achieved as well as important reductions of nonpoint agricultural contamination into the environment (Baker et al., 1996; Mulla and Schepers, 1997).

Rice (*Oryza sativa* L.) is one of the world's most important staple crops. Development of effective site-specific management techniques for rice production could, if they increase production efficiency, contribute dramatically to increased yields. Successful implementation of corrective action by farmers will require accurate data about the spatial variability of biotic and abiotic factors that may influence crop performance. Even more important, an exact understanding of how these factors may spatially and temporally interact to cause the observed variability is essential. Detailed knowledge of the multiple ecophysiological processes governing crop response to interacting environmental factors at different spatial scales is not well understood. This knowledge gap is one of the key limiting factors for the adoption of site-specific management in large-scale agricultural systems (Cassman, 1999).

Research in SSM has primarily been carried out in developed countries and has focused on the management of large fields using mechanized equipment. Even in this context, far more research has been conducted in terrestrial crops than in rice. One of the most active research efforts in the application of SSM concepts to rice production in developing countries has been that of A. Doberman and his co-workers at the International Rice Research Institute. An outline of the general approach of this group is presented by Doberman et al. (1996). They note that rice fields in developing countries are often much smaller than in developed countries, so that a cluster of 10 to 100 fields in a developing country may be considered analogous to a single field in a developed country. In many cases farmers in developing countries do not implement management practices in a manner specific to the conditions of their own field, but rather follow practices (e.g., fertilizer input rates) common to farmers in the region. Thus, SSM in a developing country with small fields may be characterized as management on a field-specific rather than regional basis. Application of technologies such as chlorophyll meters may assist the farmer in assessing the nutrient status of each individual field.

The results of experiments testing the effectiveness of this concept on fertilizer use efficiency are presented by Wang et al. (2001) for trials in southeast China and by Doberman et al. (2002) in a more global context. In both cases the researchers experimentally compared site-specific management of N, P, and K with conventional management practices. They found that SSM provided a significant increase in yield and decrease in fertilizer rate. They note, however, that a barrier to implementation of SSM is the requirement that fertilizer needs be calculated for each individual field, which is not practical. Thus, the analogy between fields in developing countries and portions of a field in developed countries may be extended. In both cases a key component to practical implementation of SSM in rice production is estimating the pattern of field or regional conditions based on a relatively small sample of sites or fields.

In this chapter we describe our experiences with rice production. Our research is conducted entirely within California, whose production system is characterized by a Mediterranean climate, a very high level of mechanization, and a rice monoculture. Many rice production systems used elsewhere share one or more features in common with the California system, but in other ways the California system is unique. We will attempt to interpret our results in a more global context in the Discussion section.

At present, site-specific management is in the initial stages of adoption in California rice production. It is not certain that this technology will achieve widespread use. The most important component of the question of whether precision agriculture technology is appropriate for California rice production, or for any other crop in any other location, is whether the economic benefits of yield mapping, variable-rate input application, and other associated technologies compensate for the costs associated with their use. In order for site-specific management to be appropriate for any crop the following criteria must be met:

1. Yields must vary substantially among the different areas of a field.
2. Yield pattern in a field should be stable or at least predictable over a number of years.
3. The factor or factors responsible for reducing yield must be identified.
4. Growers must be able to either reduce costs or increase yields sufficiently through modification in management practices to offset increased costs associated with these practices.

Many California rice farmers now have the technology to produce yield maps, which arouses interest and leads to anecdotal explanations for the variability each year, but not many farmers really know how to interpret and make use of these maps. Although California rice fields may look uniform at first glance, their properties are actually quite variable. Roel et al. (2001) have documented extensive yield variability in two California rice fields, and unpublished evidence gathered by growers using yield monitors indicates that many rice fields display high levels of variability. Thus, we may consider the first of the four listed criteria to have been met. In the remainder of this chapter we will discuss the other three criteria.

## *QUANTIFYING SPATIAL VARIABILITY AND ITS CAUSES*

### *Patterns of Spatial Variability*

Farmers and researchers have long known, in a relative sense, that crop yields are not always uniform across fields. However, little is known about the spatial structure of these yield patterns or of their consistency from year to year (Jaynes and Colvin, 1997). The consistency of the spatial and temporal structure of crop yield across the field must be investigated before implementing any management strategy (Bakhsh et al., 2000). Most common parametric statistical procedures normally used to characterize variability, such as the coefficient of variation (CV), standard deviation (STD), and confidence interval (CI), are nonspatial methods used to describe variability and, therefore, are potentially misleading when dealing with fields of different sizes. Larger fields will, on average, have a larger CV and STD, and broader CI (McBratney et al., 2000). Therefore, it is necessary to supplement the classical statistical measures with geostatistical methods for the analysis of spatial variability. A convenient means for characterizing spatial structure of any quantity is to assume that it consists of a large-scale deterministic structure or trend and a small-scale stochastic structure (Cressie, 1993).

Environmental effects that are generally relatively permanent within a field, such as soil types, soil textures, and topographic variability, may be assumed to control the large-scale spatial structure. On the other hand, the small-scale structure may be determined primarily by less permanent factors such as weed infestation, disease pressure, and yearly weather variations. Several statistical approaches can be used for the characterization of the deterministic and stochastic spatial structure. Jaynes and Colvin (1997) used a median-polishing technique to separate the large-scale spatial structure (detrending the data) from the total spatial variability. Mulla (1993), Lark et al. (1999), and McBratney et al. (2000) have used variography for describing yield spatial variations. Variograms were calculated using the residuals from the median polishing to characterize the small-scale spatial structure. We can examine some of these methods using a typical California rice field as an example. Figure 10.1 shows the yield pattern within a 40 ha rice field located in Marysville, California, from four growing seasons (1998-2001). The large-scale deterministic structure of yield was investigated by developing the trend surface computed from median-polished data. The trend data were normalized by division with the overall median of the respective year, allowing the visual comparison of trends (large-scale

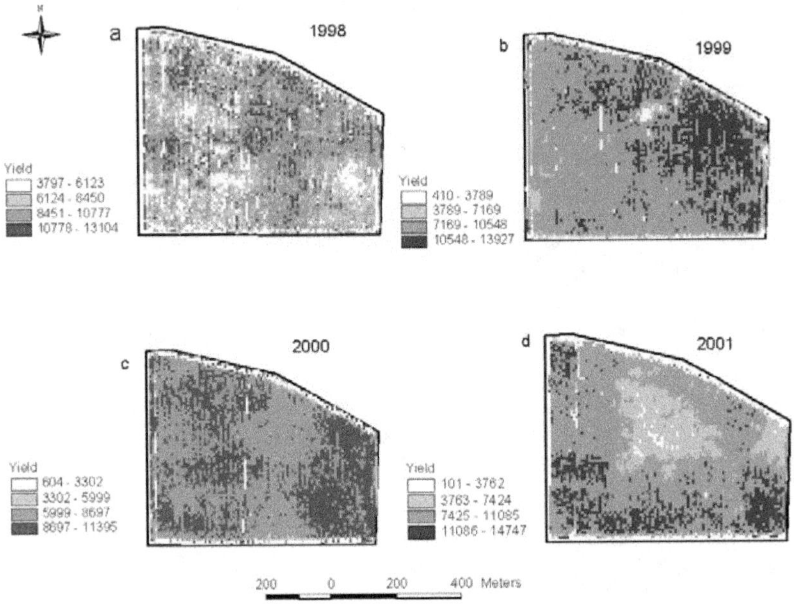


FIGURE 10.1. Rice yield in kg·ha$^{-1}$ in 1998 *(a)*, 1999 *(b)*, 2000 *(c)*, and 2001 *(d)*. (See corresponding PLATE 10.1 in the central gallery.)

deterministic structure) among different years. The trend surfaces for the four years of data are shown in Figure 10.2. Trend surfaces look very dissimilar among years. This is an indication that the spatial behaviors of the main factors controlling this spatial structure are changing from year to year.

Variography was used to characterize the small-scale spatial structure. Figure 10.3 shows the variogram analysis of the yield residuals. A spherical model was fitted to each experimental variogram. Directional variograms computed in different directions verified the assumption of isotropic yield behavior. The spherical models fitted each experimental variogram well, showing determination coefficients ($r^2$) of 0.75, 0.85, 0.93, and 0.93 for 1998, 1999, 2000, and 2001 data, respectively. The ranges of the variograms ($A_0$) of yield residuals varied from 53 m in 1998 to 181 m in 1999, showing clear spatial dependencies and that the dimension of their patches changed from year to year.

As found in other similar studies (Jaynes and Colvin, 1997; Bakhsh et al., 2000; Whelan and McBratney, 2000), the lack of stationarity in

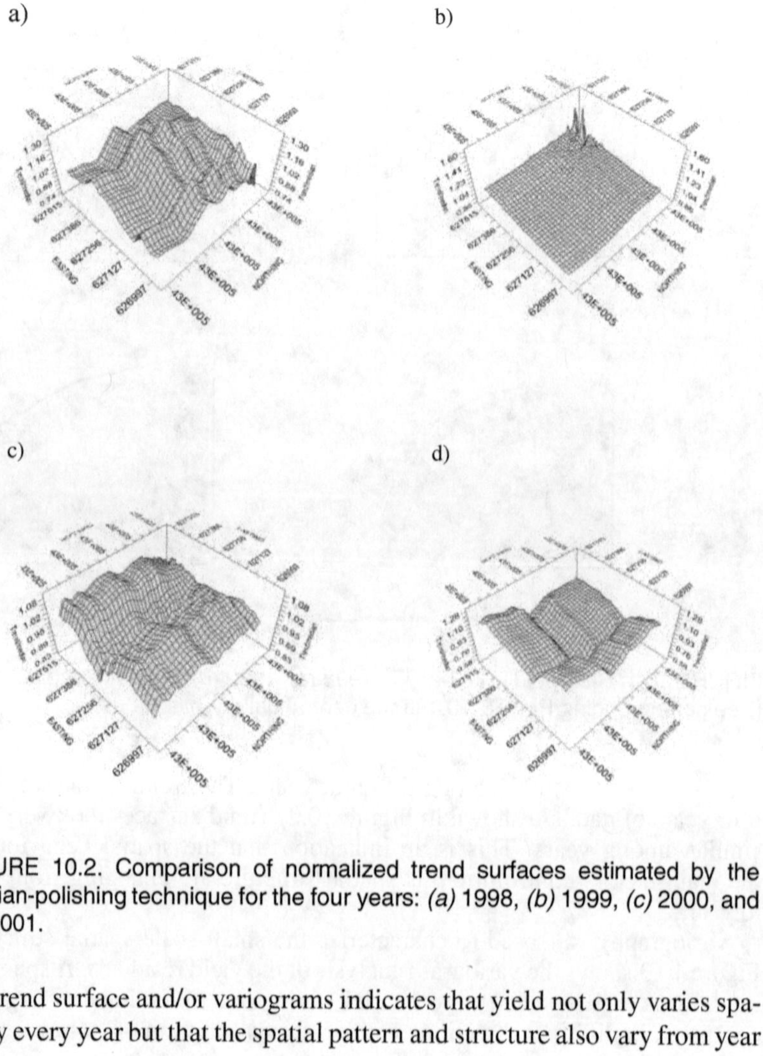

FIGURE 10.2. Comparison of normalized trend surfaces estimated by the median-polishing technique for the four years: *(a)* 1998, *(b)* 1999, *(c)* 2000, and *(d)* 2001.

the trend surface and/or variograms indicates that yield not only varies spatially every year but that the spatial pattern and structure also vary from year to year.

### *Patterns of Temporal Variability*

In order for management decisions based on site-specific information to be effective, the patterns of yield variability must be predictable from one season to the next. Thus, the variability component to be managed, either

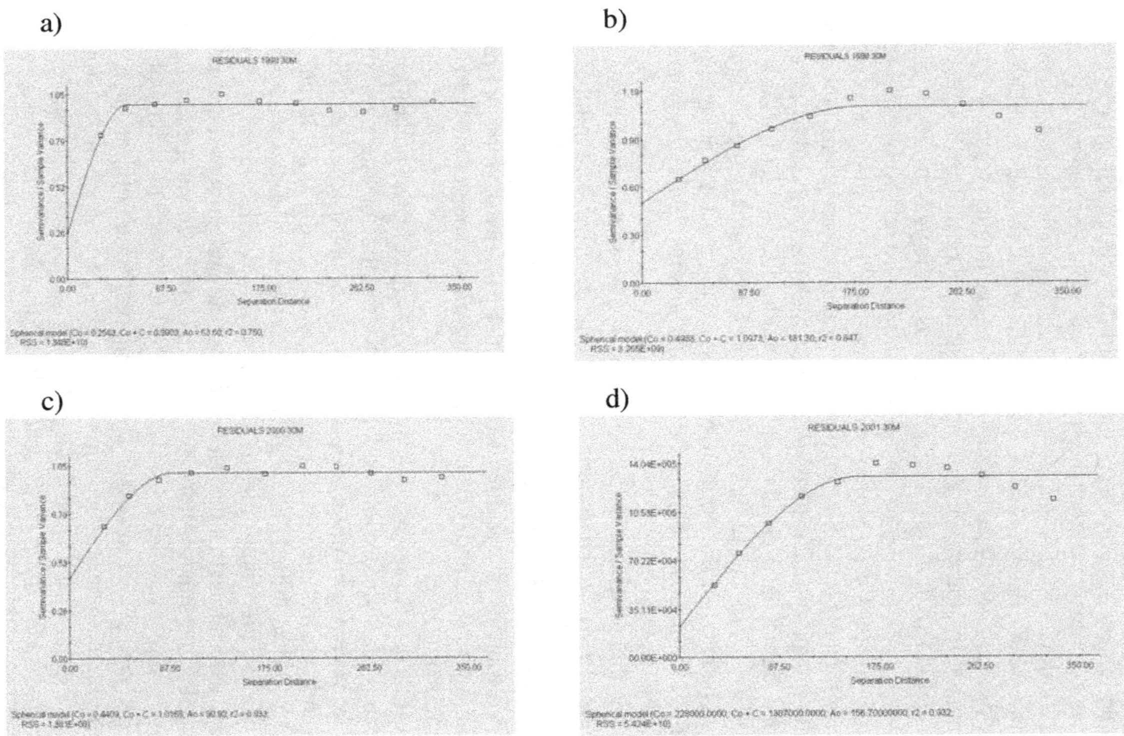


FIGURE 10.3. Experimental and fitted variograms of yield residuals (30 m sampling grid). Values of the fitted variogram sill, $co+c_1$, range, $A_o$ and model fitness, $r^2$, are given in the figures: *(a)* 1998, *(b)* 1999, *(c)* 2000, and *(d)* 2001.

large or small scale, must be either stable in time or be associated with some predictable temporal variability. The stability of the spatial structure over time can also indicate whether the same physical and ecological processes are controlling yields from year to year. To analyze yield temporal stability of the large-scale (deterministic) spatial variability, ranked correlation of grain yields estimated by the median-polishing technique over the years can be calculated. To analyze yield temporal stability of the small-scale (stochastic) spatial variability, ranked correlations of the yield residuals can be used. In many cases, however, these correlations are relatively small (e.g., Jaynes and Colvin, 1997). Moreover, determining the significance of a correlation coefficient is complicated by the spatial autocorrelation of the data.

A different approach can also be used in which the temporal patterns are determined based on a cluster analysis of the annual yields (Lark and Stafford, 1997; Lark et al., 1999). This procedure searches the data (yield values at a pattern of locations across years) for the most distinct groups of observations in terms of yield for the different seasons. Yields are generally standardized to avoid emphasis on data from higher-yielding years. A simple form of this procedure is based on the k-means clustering algorithm (Anderberg, 1973), which optimizes the formation of a fixed number of $k$ groups by minimizing the within-group variance and maximizing the between-group variance. The objective is to attempt to regionalize the field in a limited number of $k$ regions with distinct patterns showing between-season yield variation, and to determine if significant areas of the field have stable production levels.

The results of a cluster analysis may provide an objective quantification of the spatial structure of yield patterns as well as the consistency of these patterns from year to year. Lack of a stationary or consistent spatial structure may indicate that the factors controlling yield are dynamic. Jaynes and Colvin (1997) determined that yield not only varies spatially every year but that the spatial pattern and structure also vary from year to year. Porter et al. (1998) also determined that the interactions of climatic growing conditions, soil properties, and other factors can result in changes in the specific factor controlling yield from year to year and from location to location within a field. The overall evaluation of the spatiotemporal variability can have very important implications to guide future research and production decisions.

We carried out a cluster analysis on the yield trend data shown in Figure 10.2. The standardized yield trend values for each growing season were used as variables. The analysis detected three different behaviors (Figure 10.4): Cluster 1 consisted of points in the field with yield performance similar to or above the annual averages in all years; Cluster 2 consisted of points in the field that presented low yield performance in 1998 and 1999, average performance in 2000, and above average in 2001; and Cluster 3 consisted of

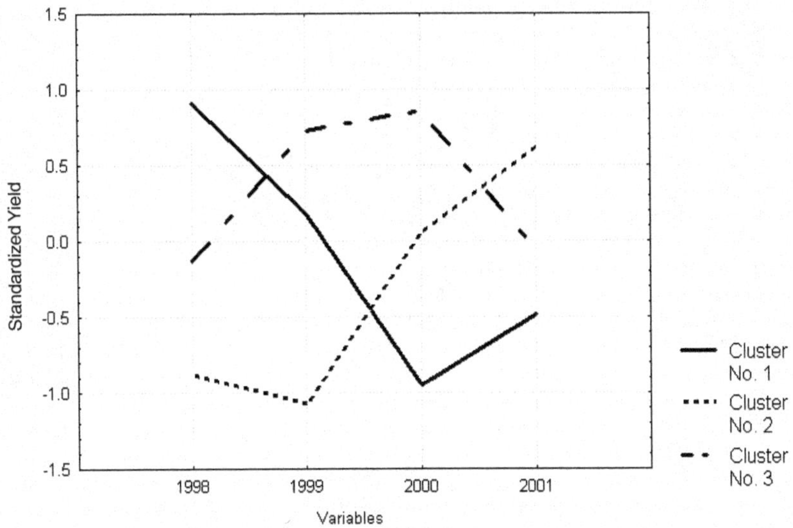


FIGURE 10.4. Cluster behaviors defined by standardized yields in four years.

areas of the field with above average yield performance in 1998, around average in 1999, and below average performance in 2000 and 2001.

Figure 10.5 shows the spatial location of these three different clusters. It is important to highlight that the clustering procedure does not use any information regarding the spatial location of the sampling points. Nevertheless, visual inspection of the maps shows that the three different cluster groups tend to indeed be spatially grouped.

These results indirectly suggest that these are indeed distinct areas of the field that react differently to the interannual or temporal variability. To regionalize these different behaviors the spatial cluster data were interpolated using the Thiessen polygon procedure. This procedure allows the categorization of the entire field to one of the behaviors described. For a better visual representation of the spatial variability of cluster behaviors, the interior of the polygons (squares in this case since they are generated by a uniform grid) were dissolved according to the cluster to which they belong. Figure 10.6 shows this regionalization. Cluster analysis combined with the Thiessen interpolation and the dissolve tool of GIS allowed quantifying the spatial dimension of each temporal behavior and the regionalization of the field according to them. Figure 10.6 shows that the areas of the field

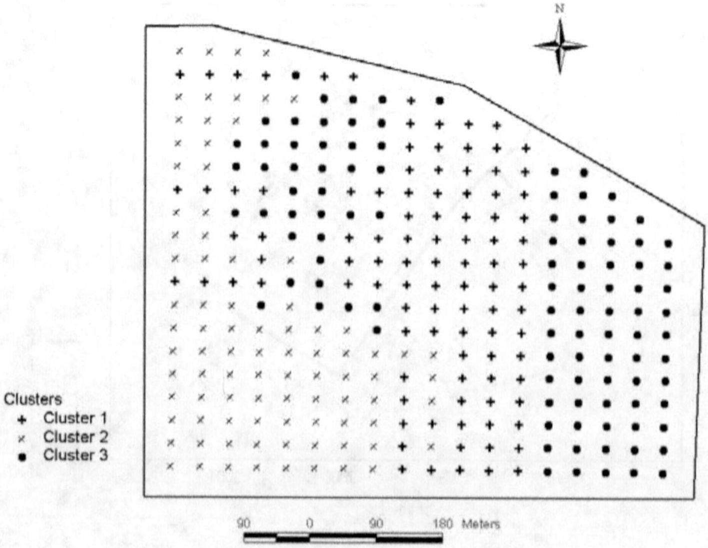


FIGURE 10.5. Spatial distribution of clusters.

in Cluster 3, which corresponded to average or superior yield performances along years, are mostly concentrated in a band along the south-north east portion and near the center of the field. The areas of the field in Cluster 2, which corresponded to points that were increasing their yield performance over time, are mostly concentrated in the south-north west band of the field. Finally, the area of the field that corresponds to Cluster 1 is located mainly in a band along the south-north portion of the center of the field. In summary, the considerable complexity in a sequence of yield maps of one field is reduced to a few general patterns of among-season variations with a particular spatial distribution. The effort may now continue to try to find the factors within the field that are responsible for these behaviors.

### *Determining the Factors Underlying Observed Spatiotemporal Structure*

The previous sections had the purposes of quantifying the spatiotemporal variability, but no causes for this variability were mentioned. The

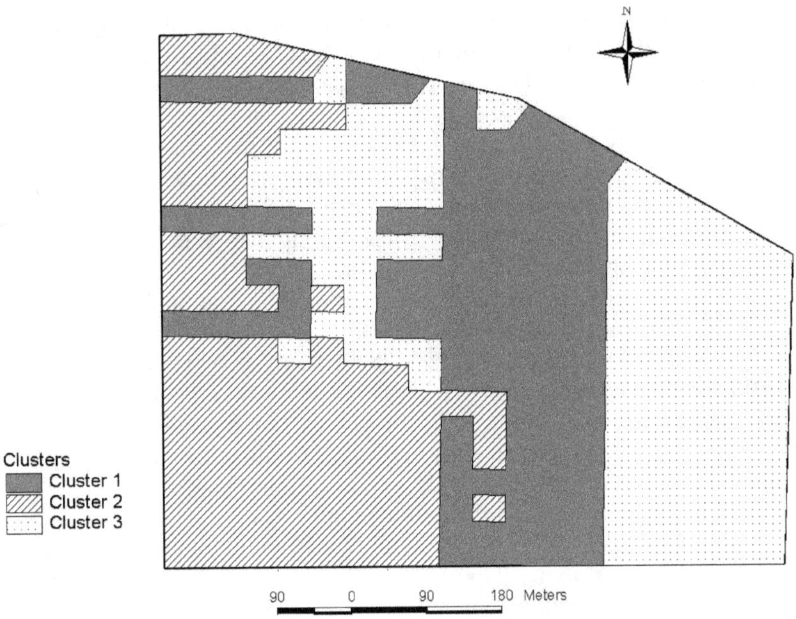


FIGURE 10.6. Regionalized temporal variability.

goal of this section is to indicate how these sources of variability might be identified. In considering how one determines the factors underlying yield variability of rice, it is important to keep in mind those attributes that distinguish rice production from other crops, and in particular from other grain crops. The most obvious distinction is that rice is grown in fields that, for a large part of the season, are flooded. In California, as in other rice-growing areas in the United States, most rice fields are laser leveled. This changes the way in which field topography influences variability. The effect of the "natural" or original topography is manifested through differences in soil quality that result from the cut-and-fill operations associated with laser leveling. The effect of "current" topography is manifested in variations in water depth.

The second principal difference between flood-irrigated rice and upland crops is in the effect of water. In terrestrial crops differences in soil drainage properties can play a major role in determining yield variability, and often the search for factors underlying observed variability is greatly simplified by considering differences in soil water relations (Plant et al., 1999). The differences are much more subtle in rice fields, and this can make the process of determining the factors underlying observed variability much more complex. Making this determination would be greatly simplified in those cases in which large-scale deterministic spatial variation in yield (as determined in the previous section) is due to persistent factors caused by environmental effects that are permanent within that particular field (e.g., soil series, soil textures, topsoil depths, topography) or factors that persist at least for several seasons. Small-scale stochastic spatial variation in yield would then be due to more inconsistent factors in time such as weeds, diseases, crop nutrition, and yearly weather.

Factors controlling yields are often temporally and spatially dynamic. Classical regression analysis is still being used to identify whether two or more agronomic variables are related. Such analysis typically provides low coefficients of determination. The reason for the poor results provided by these analyses is that response functions between crop yield and different variables are neither constant nor consistent across the landscape. Crop response occurs at different levels within a field because the underlying processes and their impact are variable in both space and time (Porter et al., 1998; Nielsen et al., 1999). Sources of intrinsic yield variation are unlikely to be identified simply by looking for regions of consistently high, low, or average yield (Finke and Goense, 1993).

Cahn et al. (1994) were able to determine which soil properties were best or highly correlated with large-scale yield spatial variation in maize (*Zea mays* L.). Factors such as soil water content, $PO_4$-P, and soil organic C presented a spatial structure more similar to the large-scale yield spatial structure than more labile parameters such as $NO_3$-N. These labile parameters were more related to small-scale spatial structure, indicating that some soil properties will be more difficult to use for site-specific soil management than others. Several different statistical approaches may be used for the determination of the underlying factors causing spatial and temporal variability. Scheiner (2001) and Ping and Green (2000) used multivariate analysis of variance (Manova) to test variable differences simultaneously. This approach may be used to search for possible significant differences among persistent factors that may explain the different yield behaviors detected by cluster analysis. This would indicate how these variables differ among these groups and indirectly suggest which factors may be determining these different yield behaviors within the field. At the same time, this analysis would also provide an indirect indication of how much of the variability is not due to the intrinsic factors rather than the interaction of these factors and stochastic factors such as the climatic conditions in each growing season.

Plant et al. (1999) and Buchleiter and Brodahl (2000) have used classification and regression trees (CART) to identify the factors underlying grain yield spatial variability. CART is a nonparametric statistical method (Breiman et al., 1984) that uses a tree-structured approach to classification and regression problems. Using CART for regression has the principal advantage that it can avoid the typical problems of the traditional regression approach already mentioned (response functions between crop yield and different variables are neither constant nor consistent across the landscape). The CART algorithm searches through all values taken by all predictors (persistent and variable factors) to find the split that divides the set into two subsets that are the most homogeneous in terms of the response variable (yield). This approach allows the identification of significant regression coefficients between the response and predictor variables in certain regions that cannot be detected when the entire field is analyzed. Principal component analysis (PCA) can be also used to reveal the relationship among predictor variables, and their effects in the response variable, that otherwise are difficult to reveal. Ping and Green (2000) applied PCA to combine variables to help identify the relationship between yield and soil parameters. This analysis is also relevant for the identification of the potential problem of colinearity among predictor variables that are important to screen before applying traditional multilinear regression (MLR) analysis.

After removing the effect of the deterministic structure (large-scale effect), Jaynes and Colvin (1997) used variogram parameters (range, nugget, and sill) to characterize factors controlling small-scale yield spatial variability. They found that range and sill variogram values were different from year to year. However, for the rainfed terrestrial crops they studied, the range was significantly correlated to the total amount of precipitation during the growing seasons. In agronomic research, intensive soil sampling is commonly used to assist in the determination of the underlying causes of spatiotemporal variability. Although this approach is suitable for research, it is far too expensive to be used in commercial crop production. Therefore, less expensive means of determining soil and plant properties must be found. A promising approach is the use of remote sensing. False color infrared photographs can be taken in all fields at different growth stages of the rice crop (e.g., emergence/seedling stage, panicle differentiation, and heading) to address spatial variability of crop growth. Aerial images cannot be used directly as a part of the data set of potential factors causing spatial variation in yield since remotely sensed crop data can be regarded as covariant with yield. However, visual and analytical interpretation of the images can be used to detect within-field spatial differences that can be useful in identifying the actual sources of yield spatial variability.

### *Economic Analysis*

The economic analysis of the consequences of adoption of site-specific crop management is complex since many of the economic gains, as well as many of the costs, are subtle and not easy to allocate in a partial budget. One way to approach the task is to determine the profit margin in different areas of the field. As a first approximation one can assume that under field-level management, all the field inputs (fertilization, herbicide, seed rate, etc.) are done at a fixed rate, so that the production costs do not have a spatial variability component as does yield. The relationship between the crop value (price × yield) and the total cost of production at each location in the field can be used to generate a map that illustrates the spatial distribution of the net profit or loss across the field. This can be used by the grower to assess the economic consequences of the variability.

Roel and Plant (2001) were able to quantify the economic return of a laser-leveling operation for the field used in our previous examples. The cooperator stated that the total cost (including land cost and drying) in kilograms of rice (break-even point) was 6,500 $kg \cdot ha^{-1}$. For this study we used a price of 0.19 US dollars per kg of rice. Although yield varied across the field, as was shown in Figure 10.1, inputs were applied uniformly, so the costs were the same at each point. The break-even yield value was subtracted from the grain yield, and the resultant number was multiplied by the rice price. This resulted in the profit maps shown in Figure 10.7. The laser leveling was in fact carried out in the spring of 1999. The 1998 profit map was superimposed with a digital elevation map of the field to identify the high and low elevation areas of the field (Figure 10.8). Figure 10.8 shows that there is a relationship between the topographic location of the field and profit. Areas of the field located at low topographic levels tend to present negative profit or loss, while areas located at high levels tend to present higher profits. It is evident, by comparing 1998 and 1999 profit maps, that much of the variability in yield that was present in 1998 was gone in 1999 after the laser-leveling operation was carried out. By computing the difference between mean economic yield in the high and low areas of the field and multiplying this by the total low area we were able to estimate the return of the laser-leveling operation. Not all of this reduction can be attributed to the laser leveling, of course, since other factors such as differences in weather may have also been important. There is still much to be learned about the use of site-specific management tools in rice production; however, this example and others like it show that by keeping the focus on the profit margin these tools can be used to improve economic productivity.

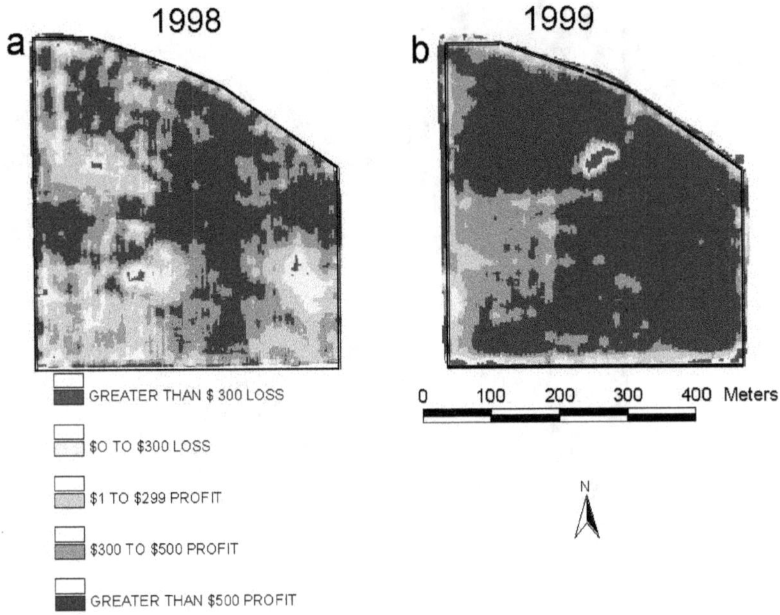


FIGURE 10.7. Dollars per hectare lost or earned in both growing season: *(a)* before leveling, *(b)* after leveling. *Source:* Redrawn from Roel and Plant, 2001. (See corresponding PLATE 10.7 in the central gallery.)

## *DISCUSSION*

Adoption of site-specific management techniques is not a foregone conclusion in California or, for that matter, in the rest of the United States. The decision to adopt will be made by growers who feel the increased costs of equipment, labor, and time are more than offset by the potential to increase profits by reducing production costs or increasing yields. Equipment costs may be expected to decline, possibly in a dramatic way, as the cost of development and production of the equipment is amortized by the manufacturers. Precipitous declines in price are common as information technology matures. Indeed, it is likely that the cost of devices such as yield monitors and global positioning systems may decline sufficiently that it no longer is a factor in an adoption decision in developed countries.

In evaluating the adoption of site-specific management techniques in any crop production system one should consider all the potential trade-offs.

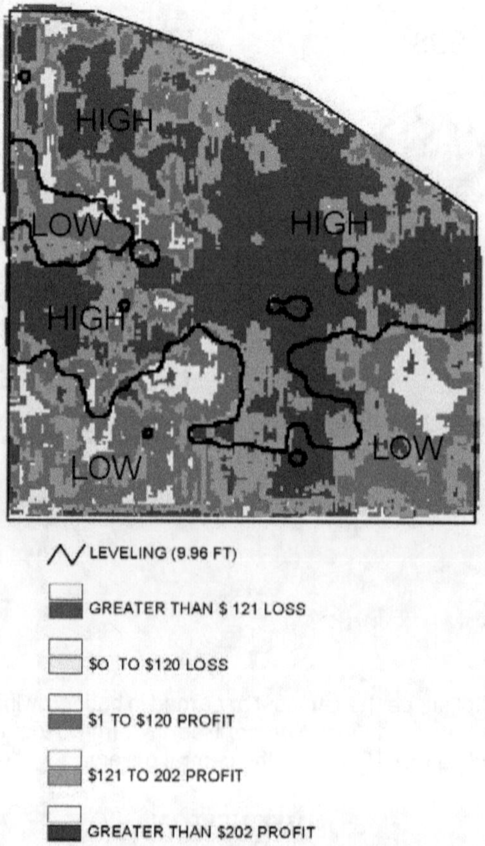


FIGURE 10.8. Dollars per acre lost or earned in 1998 (before leveling the field) and the delineation of the topographic low and high areas of the field. *Source:* Redrawn from Roel and Plant, 2001. (See corresponding PLATE 10.8 in the central gallery.)

Many studies have shown that crop uniformity and yield stability are two concepts that do not necessarily go together. Most of the studies in crop production that relate diversity and stability are concerned with the use of cultivar mixtures or intercropping. These studies have shown that by adoption of variety mixtures farmers have reduced disease pressure and achieved a higher level of plasticity of the crop to negative environmental conditions (Finck et al., 1999; Finck and Wolfe, 1997; Mundt and Browing, 1985). Zhu et al. (2000) and Wolfe (2000) confirmed these results in rice. It may be possible that by the adoption of this new technology farmers will be able to decrease within-field spatial yield variability and in the long term also increase yield stability. Yield stability is particularly important in developing countries where rice is a subsistence crop grown in relatively small fields. In this situation, optimization of production conditions may be less important than ensuring a reasonable production level with a high level of security.

Site-specific management techniques are commonly used with the goal of decreasing yield spatial variability. Through the elimination of heterogeneity at the field, farm, and regional level, the application of simple rules and universal inputs becomes easy. This, however, neglects the possibility of making use of this heterogeneity (Rabbinge, 1997). For example, Zhu et al. (2000) performed a study at the farm level (total of 3,342 ha) to compare the susceptibility of rice to blast disease *(Magnaporthe grisea)* between field seeded with a single cultivar or several different cultivars. This study showed that disease-susceptible rice cultivars planted in mixtures with resistant varieties had 89 percent greater yield, and blast was 94 percent less severe than when a single cultivar was grown. One of the most interesting facts that this study also pointed out was that canopy microenvironmental conditions within a single cultivar crop were more uniform and more conductive for blast.

Wolfe (2000) stated that cultivar mixtures can buffer against unpredictable abiotic variables such as low temperature due to their phenological variability. Indeed, it is likely that the stability of yields from cultivar mixtures over different environments, compared with yields from their components grown as monocultures, results partly from combined restriction of biotic and abiotic stress. Under California conditions, cool air temperatures during the period of panicle development (PD) may lead to pollen sterility and hence to blanking (Hill et al., 1992). The relationship between crop spatial phenological heterogeneity and level of risk due to unfavorable abiotic conditions should also be quantified when precision farming recommendations are implemented. In regions where climatic variables are not as stable as in a Mediterranean climate, like that of California, many of the crop inputs (e.g., fertilizers, herbicide, fungicide, etc.) that must be applied during the growing season may be delayed due to unfavorable climatic conditions. This is especially important in developing countries where the timing and amounts of crop inputs are much less controlled. Many of these inputs must be applied at a specific phenological stage, otherwise their efficacy is reduced. In some regions the possibility of timely harvest is not always guaranteed due to adverse climatic conditions. When unstable climatic conditions exist it is possible that fields may be ready to harvest but frequent rainfalls do not allow entrance into the fields. Once rice grain moisture drops below 16 percent (160 mg·kg$^{-1}$), grain quality may be reduced. All of these factors are indications that a certain degree of crop heterogeneity may act as insurance or risk management strategy against variable abiotic conditions. It is possible that the impacts of losing some degree of this heterogeneity may vary among different agricultural production systems.

Another factor to consider in attempting to analyze the potential effects of site-specific management is its effect on farm size. Plant et al. (2000) argue that one possible outcome of widespread adoption of SSM is an increase in farm size, since the effect of the technology is to reduce the inefficiencies associated with large-scale production. Moreover, a second potential consequence is the replacement of relatively low-skilled jobs with a smaller number of relatively high-skilled jobs. Such replacement is common with increasing mechanization of any industry. In farming systems consisting of a mosaic of relatively small individual holdings, the introduction and adoption of SSM may increase the economic efficiency and competitiveness of large landholdings to the point where small landholders are unable to compete. One way for small landholders to avoid this outcome is to band together as cooperatives, either on their own or through government intervention, to acquire and use the technology of SSM in a way that treats each small field as a portion of a larger system and optimizes production practices in each small field.

In summary, site-specific management at its core is the application of information technology to crop production. As with any new application of information technology, or any technology, the potential for unanticipated negative consequences is a real one. At present the cost of these technologies may make them seem remote for much of the world's rice-producing areas. However, the history of information technologies from television to cellular telephones has shown that these technologies can achieve widespread adoption with amazing speed.

## REFERENCES

Anderberg, M.R. 1973. *Cluster analysis for applications.* New York: Academic Press.

Baker, J., T.S. Colvin, and D.B. Jaynes. 1996. Potential environmental benefits of adopting precision agriculture. In P.C. Robert, R.H. Rust, and W.E. Larson, eds., *Proceedings of the Third International Conference on Precision Agriculture.* Madison, WI: ASA-CSSA-SSSA, pp. 1051-1052.

Bakhsh, A., D.B. Jaynes, T.S. Colvin, and R.S. Kanwar. 2000. Spatio-temporal analysis of yield variability for corn-soybean field in Iowa. *Transactions of the ASAE* 43(1):31-38.

Breiman, L., J.H. Friedman, R.A. Olshen, and C.J. Stone. 1984. *Classification and regression trees.* New York: Chapman and Hall.

Buchleiter, G.W. and M.K. Brodahl. 2000. Analysis of precision farming data with classification trees. In P.C. Robert, R.H. Rust, and W.E. Larson, eds., *Proceedings of the Fifth International Conference on Precision Agriculture.* Madison, WI: ASA-CSSA-SSSA.

Cahn, M.D., J.W. Hummel, and B.H. Brouer. 1994. Spatial analysis of soil fertility for site-specific crop management. *Soil Sci. Soc. Am. J.* 58:1240-1248.

Cassman, K.G. 1999. Ecological intensification of cereal production systems: Yield potential, soil quality, and precision agriculture. *Proc. Natl. Acad. Sci.* 90:5952-5959.

Cressie, N.A.C. 1993. *Statistics for spatial data.* New York: John Wiley and Sons.

Dobermann, A., K.G. Cassman, S. Peng, P.S. Tan, C.V. Phung, P.C. Sta.Cruz, J.B. Bajita, M.A.A. Adviento, and D.C. Olk. 1996. Precision nutrient management in intensive irrigated rice systems. In *Proceedings of the International Symposium on Maximizing Sustainable Rice Yields Through Improved Soil and Environmental Management,* November 11-17, 1996, Khon Kaen, Thailand. Bangkok: Dept. of Agriculture, Soil and Fertilizer Society of Thailand, Dept. of Land Development, ISSS, pp. 133-154.

Dobermann, A., C. Witt, D. Dawe, G.C. Gines, R. Nagarajan, S. Satawathananont, T.T. Son, P.S. Tan, G.H. Wang, N.V. Chien, et al. 2002. Site-specific nutrient management for intensive rice cropping systems in Asia. *Field Crops Res.* 74:37-66.

Finck, M.R., E.S. Gacek, H.J. Czembor, and M.S. Wolfe. 1999. Host frequency and density effects on powdery mildew and yield in mixtures of barley cultivars. *Phytopathology* 48:807-816.

Finck, M.R. and M.S. Wolfe. 1997. The use of biodiversity to restrict plant diseases and some consequences for farmers and society. In L.E Jackson, ed., *Ecology in agriculture.* San Diego; CA: Academic Press, pp. 203-237.

Finke, P.A. and D. Goense. 1993. Differences in barley grain yields as a result of soil variability. *Journal of Soil Science* 40:477-492.

Hill, J., S.R. Roberts, D.M. Brandon, S.C. Scardaci, J.F. Williams, C.M. Wick, W.M. Canevari, and B.L. Weir. 1992. *Rice production in California.* Oakland: University of California Division of Agriculture and Natural Resources.

Jaynes, D.B. and T.S. Colvin. 1997. Spatiotemporal variability of corn and soybean yield. *Agron. J.* 89:30-37.

Lark, R.M., H.C. Bolam, T. Mayr, R.I. Bradley, R.G.O. Burton, and P.M.R. Dampney. 1999. Analysis of yield maps in support of field investigations of soil variation. In J.V. Stafford (ed.), *Precision agriculture '99.* Sheffield, UK: Sheffield Academic Press, pp. 151-161.

Lark, R.M. and J.V. Stafford. 1997. Classification as a first step in the interpretation of temporal and spatial variation of crop yield. *Ann. Appl. Biol.* 130:111-121.

Lowenberg-DeBoer, J. and S.M. Swinton. 1997. Economics of site-specific management in agronomic crops. In F.J. Pierce and E.J. Sadler, eds., *The state of site-specific management for agriculture.* Madison, WI: ASA-CSSA-SSSA, pp. 369-396.

Matson, P.A., W.J. Parton, A.G. Power, and M.J. Swift. 1997. Agricultural intensification and ecosystem properties. *Science* 277:504-509.

McBratney, A.B., B.M. Whelan, J.A. Taylor, and M.J. Pringle. 2000. A management index for precision agriculture. In P.C. Robert, R.H. Rust, and W.E. Larson, eds., *Proceedings of the Fifth International Conference on Precision Agriculture.* Madison, WI: ASA-CSSA-SSSA.

Mercer, W.B. and A.D. Hall. 1911. The experimental error of field trials. *Journal of Agricultural Science, Cambridge* 4:107-132.

Mulla, D.J. 1993. Mapping and managing spatial patterns in soil fertility and crop yield. In P.C. Robert, et al., eds., *Soil specific crop management.* Madison, WI: ASA-CSSA-SSSA, pp. 16-26.

Mulla, D.J. and J.S. Schepers. 1997. Key processes and properties for site-specific soil and crop management. In F.J. Pierce and E.J. Sadler, eds., *The state of site-specific management for agriculture.* Madison, WI: ASA-CSSA SSSA, pp. 1-18.

Mundt, C.C. and J.A. Browning. 1985. Genetic diversity and cereal rust management. In A. P. Roelfs and W.R. Bushnell, eds., *The cereal rusts.* Orlando, FL: Academic Press, Vol. 2, pp. 527-559.

Nielsen, D.R., O. Wendroth, and F.J. Piercew. 1999. Emerging concepts for solving the enigma of precision farming research. In P.C. Robert, R.H. Rust, and W.E. Larson, eds., *Proceedings of the Fourth International Conference on Precision Agriculture.* Madison, WI: ASA-CSSA-SSSA, pp. 308-318.

Pierce, F.J., P.C. Robert, and G. Mangold. 1994. Site-specific management: The pros, cons and the realities In *Proc. Intr. Crop Management Conf.* Ames: Iowa State University, pp. 17-21.

Ping, J. and C.J. Green. 2000. Spatial variability of yield and soil parameters in two irrigated cotton fields in Texas. In P.C. Robert, R.H. Rust, and W.E. Larson, eds., *Proceedings of the Fifth International Conference on Precision Agriculture.* Madison, WI: ASA-CSSA-SSSA.

Plant, R.E., A. Mermer, G.S. Petygrove, M.P. Vayssieres, J.A. Young, R.O. Miller, L.F. Jackson, R.F. Denison, and K. Phelps. 1999. Factors underlying grain yield spatial variability in three irrigated wheat fields. *Transactions of the ASAE* 42(5):1187-1202.

Plant, R.E., G.S. Pettygrove, and W.R. Reinert. 2000. Precision agriculture can increase profits and limit environmental impacts. *California Agriculture* 54(4): 66-71.

Porter, P.M., J.G. Lauer, D.R. Huggins, E.S. Oplinger, and R.K. Crookston. 1998. Assessing spatial and temporal variability of corn and soybean yields. *J. Prod. Agric.* 11:359-363.

Rabbinge, R. 1997. Chairman's introduction. In *Precision agriculture: Spatial and temporal variability of environmental quality.* CIBA Foundation Symposium 210. New York: John Willey & Sons, pp. 1-5.

Roel, A. and R. Plant. 2001. More than a pretty picture. *Rice Farming* 35(4):10-11.

Roel, A., R.E. Plant, J.A. Young, G.S. Pettygrove, J. Deng, and J.F. Williams. 2001. Interpreting yield patterns for California rice precision farm management. In P.C. Robert, R.H. Rust, and W.E. Larson, eds., *Proceedings of the Fifth International Conference on Precision Agriculture* (Unpaginated CD). Madison, WI: American Society of Agronomy.

Scheiner, S.M. 2001. MANOVA: Multiple response variables and multispecies interactions. In S.M. Scheiner and J. Gurevitch, eds., *Design and analysis of ecological experiments.* Oxford, UK: Oxford University Press, pp. 99-115.

Sinclair, T.R. 1999. Crop scientist seek a new revolution. *Science* 283:310-314.

Wang, G.H., A. Dobermann, C. Witt, Q.Z. Sun, and R.X. Fu. 2001. Performance of site-specific nutrient management for irrigated rice in southeast China. *Agronomy Journal* 93:869-878.

Whelan, B.M. and A.B. McBratney 2000. The "null hypothesis" of precision agriculture management. *Precision Agriculture* 2:265-276.

Wolfe, M.S. 2000. Crop strength through diversity. *Nature* 406:681-682.

Zhu, Y., H. Chen, J. Fan, Y. Wang, L. Yan, J. Chen, X. Fan, S. Yang, L. Hu, H. Leung, T. Mew, P. Teng, Z. Wang, and C. Mundt. 2000. Genetic diversity and disease control in rice. *Nature* 406:718-722.

## Chapter 11

# Precision Agriculture Management Progress and Prospects for Corn/Soybean Systems in the Midwestern United States

Thomas S. Colvin

## *INTRODUCTION*

Precision farming with automatic control of inputs and monitoring of corn and soybean yields began in the early 1990s in the Midwest. Automated technologies were developed to permit variable rate fertilizer application in response to variable soil nutrient test levels across fields. Innovators had been using measuring tapes or wheels and flags to break fields into grids before electronic guidance tools, such as the Global Positioning System (GPS), were publicly available. As GPS technology became available, it quickly replaced earth-based electronic location systems. The earth-based systems used triangulation to electronic repeating-beacons placed in fields by operators as they entered. Early adopters, such as Ted Macy in Indiana, used components to put systems together in the early 1990s that allowed them to vary seed and fertilizer inputs and record the results (Macy, 1993). This resulted in savings of almost $35 per ha ($14 per acre), primarily for corn, by optimizing inputs of fertilizer and seed, with small savings on herbicide as well. Macy was interested in site-specific management for almost ten years before the concept could be implemented with automatic controllers in 1991.

## *EXPERIMENTATION IN CENTRAL IOWA*

Research on crop yield variability in farmer-managed corn and soybean fields has been underway at the United States Department of Agriculture

*Handbook of Precision Agriculture*

doi:10.1300/5627_11

(USDA) Agricultural Research Service (ARS) National Soil Tilth Laboratory (NSTL) since 1989. This chapter primarily focuses on the progress in precision agriculture for corn/soybean through this research. Other topics such as the place of technology in precision farming and the critical role of humans are also addressed.

Over time, the definition of precision agriculture that has evolved from this research is "doing the right thing at the right place at the right time," in other words, good management. It is important to notice that this definition says nothing about technology. Therefore, if it is possible to keep all the essential information in one's head, that form of precision agriculture is as satisfactory as keeping the plans and records on paper, or in a computer equipped with a Geographic Information System (GIS). No mention is made of GPS, controllers, regulators, or other specialized tools including computers. Nothing is said about the level of mechanization, or presence or absence of agrochemicals within the production system. Our three Rs apply equally well to an "organic system" producing vegetables with human or horsepower as to a cropping system using the largest tractors and harvesters and the most intense chemical program imaginable. Our definition of precision agriculture is about management (decision making and implementation), not the tools that are used to achieve the desired practices.

To actually do the right thing at the right place at the right time, we must understand the entire system that we are managing and the scale at which actions need to be taken. If an entire state or nation's soils react similarly to an input such as tillage, then it is not necessary to consider variability at a smaller scale. On the other hand, if individual plants or varieties react differently to an input such as an application of fungicide, then we must at least consider managing at that level or at least understand the consequence of not taking that variability into account.

Precision agriculture research at the NSTL began with a study of crop yield variability measured at 224 locations within a 16 ha (40 acre). This was a privately owned field managed by a cooperating farmer in central Iowa. The plots were placed end-to-end (Figure 11.1) and arranged along eight, 28-plot transects. Corn and soybean yields were taken using a small commercial combine equipped with a scale and moisture meter in the grain tank of the combine (Figure 11.2). For these initial studies, the field was treated uniformly each year. The objective was to determine yield variation at various points on the field with respect to each other and time. Figure 11.3 shows the yield records of plots on the fifth transect from the bottom for four years of corn (1991, 1993, 1995, and 1997). Management practices including seeding and fertilizer rates as well as tillage and weed control practices were fairly constant during this period (Colvin et al., 1997). No obvious weed, insect, or disease problems occurred in any year. It is inter-

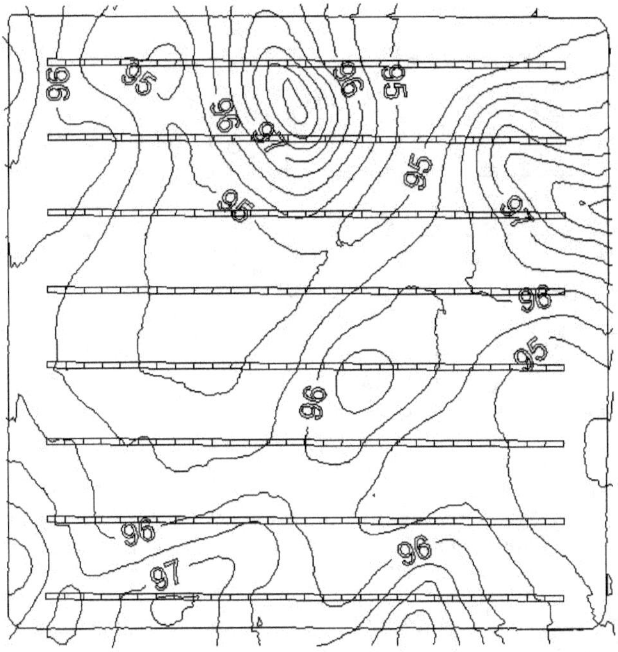


FIGURE 11.1. Cooperators field showing elevation contours and eight transects of 28 yield and soil property measurement plots.

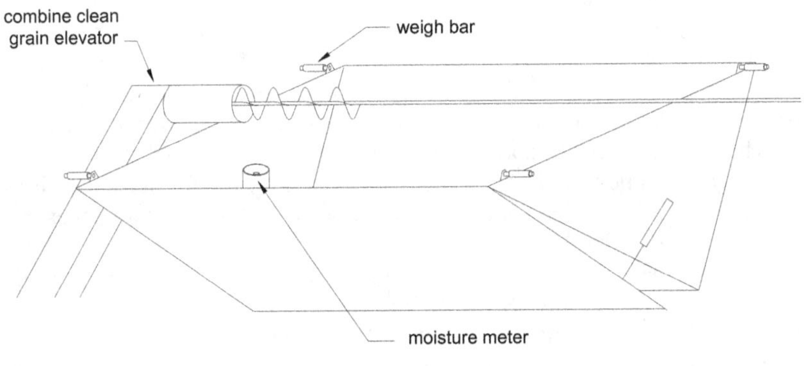


FIGURE 11.2. Diagram of weigh tank and moisture meter constructed in a combine harvester grain tank and used to measure weight and moisture of field transect plots. *Source:* After Colvin, 1990.

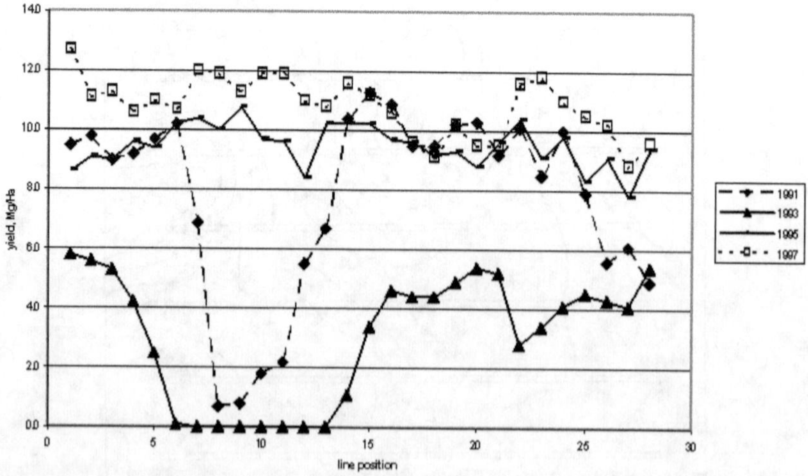


FIGURE 11.3. Graph of corn yields for four years for 28 individual plots on the fifth transect from the bottom for the field shown in Figure 11.1.

esting to note the area from position 6 to 13 on the graph. That area is in a closed depression that is subject to flooding under some conditions that occurred in 1991 and 1993, leading to little or no yield. Flooding occurred in 1991 even though the precipitation for the growing season, as a whole, had been below normal (Kaspar et al., 2003). In the other years, however, the same area had among the highest yields. We have not yet totally accounted for the variability at the other locations.

Jaynes and Colvin (1997) published the first results from that effort based on six years (1989-1994) of data (three years of corn and three years of soybean). They concluded that more time would be needed to firmly establish the pattern of yield variability in both space and time. A few of the locations had predictably high, low, or medium yields relative to the average for the field in any one year, but even then, the specific yield for those locations moved up or down with respect to a specific year and its associated weather patterns. Throughout the midwestern United States, 1993 was an unusually wet year and overall corn yields were cut almost in half compared to the other two corn years in the study. Soybean yields were more stable over years on a field average basis, even though 1990 did have somewhat above average rainfall. The amount of variability in yields was about the same from a statistical standpoint for corn and soybean.

There was not sufficient evidence from the data taken on site-specific yield in the initial six-year period to confidently divide the field into management zones or to begin variable management to either reduce or enhance variability, optimize profit, or minimize environmental impacts. This is not surprising and is not unlike conclusions reached by Wyciskalla et al. (2000) for a two-year study in southern Illinois. In that case, they were not able to find good correlations between a number of soil measurements and yield.

Efforts to quantify and understand crop response variability on the NSTL field continued through the 2001 crop year. Kaspar et al. (2003) combined field surface terrain information (Figure 11.1) with crop yield for six years of corn. This required 11 years of data, tracking variability, for this corn-soybean rotation. Even with that long yield record and the additional information, they still needed to group the years into above-normal and below-normal growing season precipitation categories to develop an accurate statistical model for predicting yield. The wetter-than-normal years were 1993 and 1999, while the four dryer-than-normal years were 1989, 1991, 1995, and 1997. This model was able to account for 78 percent of the spatial variation seen in the four dryer-than-normal years and was then used to determine management zones. When the model developed from the plot data (224 points per year) was compared to combine yield monitor data taken in one of the dry years (1997) as shown in Figure 11.4, it was concluded that the model agreed with the variability seen throughout the field with the yield monitor. It is important to notice that using this model for management, with the level of confidence indicated, would require a good forecast of weather conditions for the growing season.

Bakhsh et al. (2000), studied a corn-soybean system on another farm in central Iowa for three years and reached a similar conclusion that more time would be required to sufficiently quantify variability so that it would be useful for managing production inputs. They did report that the amount of rainfall in any one year was very important, with more rain leading to less diversity in yield response throughout the field. Morgan et al. (2000), reporting on work in Wisconsin, reinforced the importance of annual precipitation when they singled out data from dry years as possibly being helpful in finding relationships between plant available water measurements and yield.

As with other studies throughout the U.S. (Sadler et al., 1995a,b; and Karlen et al., 1990), the initial work on variability at NSTL was done by measuring yields with scales for small plots along transects throughout the field. Subsequent work (Perez-Munoz and Colvin, 1996; Colvin and Arslan, 1999: Al-Mahasneh and Colvin, 2000; and Arslan and Colvin, 2002) has shown that grouping the measurements taken over 6 to 10 seconds with a well-managed yield monitor can produce yield results similar to those taken with a scale over the same area. This provides an opportunity to conduct experiments over entire fields using the yield monitor as the tool to measure crop yields. The study by Bauer et al. (2000) of the potential for

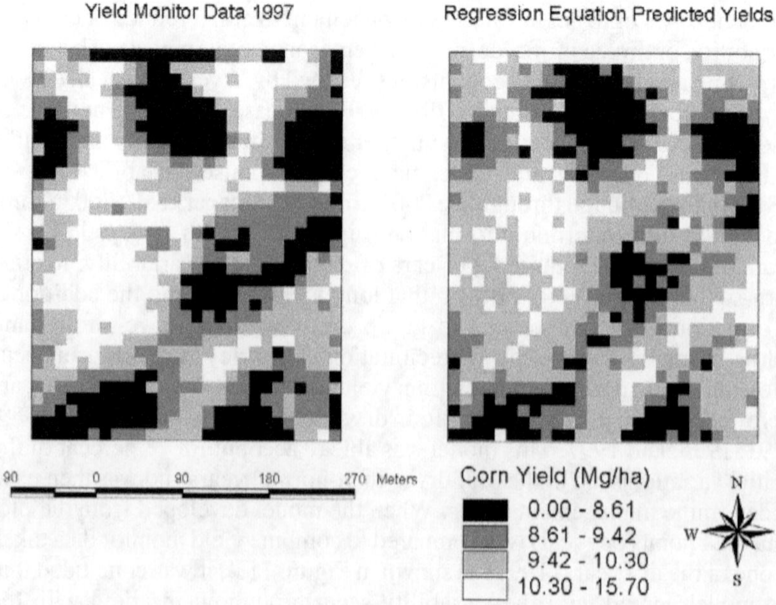


FIGURE 11.4. Corn yield maps of a 16 ha (40 acre) field in central Iowa in 1997 developed from combine yield monitor data and a regression model of the relationship of corn yield and terrain attributes in drier-than-normal years. This is the same field shown in Figure 11.1. *Source:* After Kaspar et al., 2003.

variable rate seeding in Missouri is an example of the use of yield monitors in large field experimentation.

## *AVAILABILITY OF YIELD MONITORS AND SITE-SPECIFIC SOIL TESTING*

In central Iowa, combine yield monitors are available for all major line combines; either as factory installed options on new combines or as aftermarket attachments for combines in the field. In an informal survey of three different dealers, two reported no new combine sales in 2001 without yield monitors, even though they are optional rather than standard equipment. The other dealer reported that few if any yield monitors had been sold, but that customers were equipping grain carts with scales. This informal survey does not suggest that only two-thirds of combines are equipped with yield monitors, but it does indicate that yield mapping may not be a universally accepted practice at this time.

## *OTHER BENEFITS OF YIELD MONITORS*

Yield monitors have uses other than as tools to develop yield maps in the traditional sense for precision agriculture. Some farmers report using yield monitors as the basis for splitting the harvest with landlords. Others use yield monitors as the basis for government or crop insurance reporting, while still others use the yield monitors as a way to optimize loads on trucks leaving the field.

Tracking food, feed, or fiber products to their place of production may require certificates with GPS coordinates for regulatory or other reasons. Yield monitors with GPS might provide part of the story but location information may also be needed from planters, sprayers, and other implements. This might be especially true if plants are used to produce items that we normally think of as pharmaceuticals (or nutriceuticals) and the products must meet Food and Drug Administration (FDA) requirements for trace-ability to their source.

## *STATUS OF SOIL SAMPLING*

The concept of precision agriculture for Corn Belt crops in the commercial sector was initially based on the variable application of N-P-K fertilizer in a site-specific manner. Discussions in late 2001 with central Iowa fertilizer dealers indicated that 40 to 90 percent of the soil sampling being done for farmers is being done on a site-specific basis. This provides information that might be used for variable rate application. One dealer reported on their Web site that they had applied lime in a site-specific manner on approximately as many acres as N-P-K fertilizer.

One problem that has hampered many farmers' ability to make sound decisions relative to the economic use of variable rate application of fertilizer is the absence of site-specific soil-test response curves. It may turn out that generalized response curves developed by soil type or by geographic region are all that is necessary to optimize the use of fertilizer, but this has not been demonstrated at this time. Research is proceeding at a number of locations to determine how different field segments respond to either soil-test levels or applied nutrients. In other words, are other factors such as terrain, management history, or something else needed to fully understand crop response to a specific soil-test value or level of applied plant nutrients?

## PROFITABILITY

Managers may move to more precisionlike management for a number of reasons including "doing a better job for landlords to gain or keep rented land," environmental stewardship, or increased profit. However, except for regulatory requirements, the long-run reason to adopt this type of management would be to improve the profitability (bottom line). Few, if any, reports indicate substantial profit made by adopting precision agricultural management techniques on corn or soybean throughout the U.S. Corn belt (Figure 11.5). For example, publication titled "Precision Farming Profitability" edited by Lowenberg-DeBoer (2000), lists only one experimentally based report of increased profit due to the use of variable-rate fertilizer application and variable-rate seeding. Even in that example, the calculations of profit relied primarily on assumptions and other techniques needed to complete competitive partial budgets for the systems.

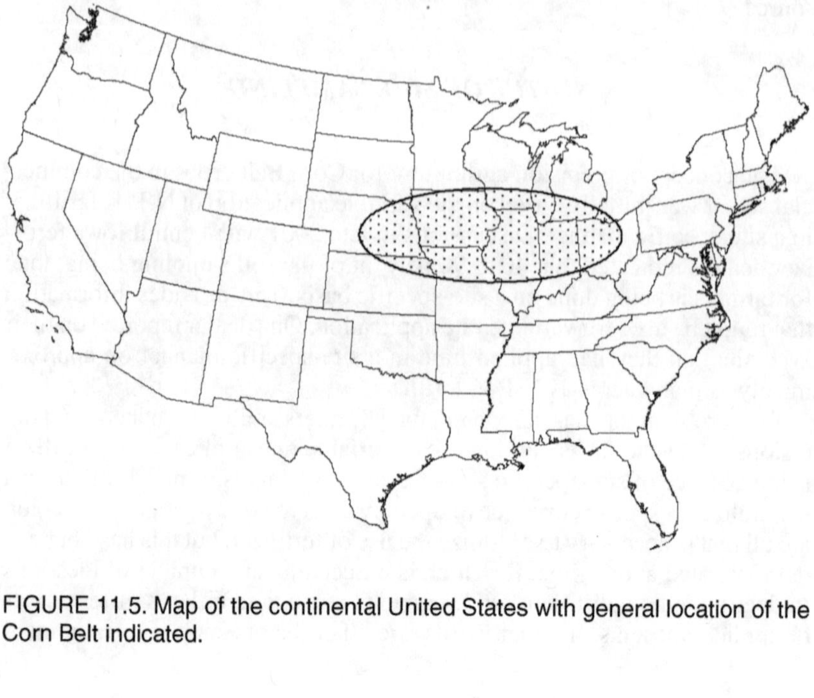

FIGURE 11.5. Map of the continental United States with general location of the Corn Belt indicated.

## *ENVIRONMENTAL ISSUES*

Environmental or other factors may influence the adoption of various concepts or tools currently associated with precision agriculture. Jaynes et al. (2001) showed that different rates of nitrogen fertilizer application can lead to different levels of nitrogen loss in tile drainage water. Similarly, Hanna et al. (2001) and Chaplin et al. (2000) demonstrated the need to ensure that equipment distributes inputs evenly across the entire width of the machine to be sure that application targets are met and that rates can be as low as possible. These studies confirm that the work of balancing nitrogen loss, optimum crop yield, and soil fertility maintenance has only begun.

## *THE HUMAN SIDE OF PRECISION AGRICULTURE*

Previously, it was pointed out that precision agriculture does not depend on technology, even though many readers may find the two inseparable. Precision agriculture does depend on people, along with their commitment to stewardship, their abilities as managers and their commitment to good workmanship. Ideally an activity such as precision agriculture starts with people and adds technology as appropriate. A person can conduct precision agriculture with no special technology, but a computer with an operator who does not care about the results obtained, or has no ability to make quality judgments can soon produce unusable or dangerous results.

I have hired custom applicators to spray herbicides and fertilizers on commercial crops for many years. In one instance, I was watching the operation and noticed that one of the nozzles on a herbicide applicator was not working. I called it to the attention of the operator, but was told that nothing could be wrong because the sprayer monitor had not sounded an alarm. Within a day or two it was noticeable that there was a problem because the burndown herbicide in the mix turned the weeds brown except in the strips where the nozzle had not been working. I called the company and was again told that nothing could be wrong. Finally after an argument, an inspection took place and an attempt was made to correct the problem. This is not the approach that will achieve the results needed for precision agriculture, and in this case, technology was available to assist the operator. Unfortunately, the technology either failed, or the operator failed to notice the output of the system. In the end, a management attitude that is unresponsive or uncaring will lead to problems, not to precision agriculture.

There are also reports of soil samples being sent to multiple laboratories for analysis. Although, the analyses from the samples were reasonably close (Whitney, 1986), fairly wide variations occurred in the fertilizer recommendations based on these test results. The recommendations were implemented on the tested fields with no direct evidence that justification existed for the wide difference in recommendations based on crop yield response. As the report of the work indicated, different time periods were assumed and different philosophies, were presented such as maintenance versus buildup, but it would be helpful for this human reasoning to be made plain. This again is not a recipe for precision agriculture (at least without an explanation of the reasoning), but simply poor economics at best. or potentially bad environmental or economic outcomes at worst.

## THE NEED FOR FUTURE RESEARCH

The underlying principles required to implement precision agriculture methods and management strategies are not likely to be unique to the U.S. Corn Belt. It is likely that local or individual on-farm research will be needed to determine nutrient or tillage response curves that can then be used to optimize crop yield and environmental impact. One possible approach may be for scientists to develop methods of determining yield variability and response curves. These methods then might be used by farmers to develop specific crop response information, based on their own experimentation, for individual farms or fields. If it is discovered that variability at or below some level is not important, then site-specific management will not be needed at that scale.

To develop and promote efficient agricultural systems, we need accepted methods for determining requirements for every production input (seed, fertilizer, pesticide, irrigation, etc.) and the potential risks and benefits associated with each input. Even then, the manager will still make the difference with regard to the various outcomes based on timing and choices that are made. The most successful agriculture managers will strive to achieve the most return with the least input at an optimum and acceptable level of risk.

## REFERENCES

Al-Mahasneh, M.A., and T.S. Colvin. 2000. Verification of yield monitor performance for on-the-go measurement of yield with an in-board electronic scale. *Transactions of the American Society of Agricultural Engineers,* 43(4):801-807.

Arslan, S., and T.S. Colvin. 2002. Grain yield mapping: Yield sensing, yield reconstruction and errors. *Journal of Precision Agriculture,* 3:135-154.

Bakhsh, A., D.B. Jaynes, T.S. Colvin, and R.S. Kanwar. 2000. Spatio-temporal analysis of yield variability for a corn-soybean field in Iowa. *Transactions of the American Society of Agricultural Engineers,* 43(1):31-38.

Bauer, M.G., J.G. Davis, K.A. Sudduth, and S.T. Drummond. 2000. Agronomic and economic evaluation of variable-rate corn seeding on Missouri soils. In *Proceedings of the 5th International Conference on Precision Agriculture,* Pierre Robert (ed.). CD-ROM prepared by ASA/CSSA/SSSA, Madison, WI.

Chaplin, J., J. Kaplan, T. Hustrulid, and B. Hetchler. 2000. A method for testing the effectiveness of fertilizer spreaders for precision agriculture. In *Proceedings of the 5th International Conference on Precision Agriculture,* Pierre Robert (ed.). CD-ROM prepared by ASA/CSSA/SSSA, Madison, WI.

Colvin, T.S. 1990. Automated weighing and moisture sampling for a field plot combine. *Applied Engineering in Agriculture* 6:713-714.

Colvin, T.S., and S. Arslan. 1999 Yield monitor accuracy. Publication SSMG-9 in Site-Specific Management Guidelines published by The Potash and Phosphate Institute. (Available on the Web http://plantsci.sdstate.edu/precisionfarm/paper/papers/SSMG/SSMG-09.pdf).

Colvin, T.S., D.B. Jaynes, D.L. Karlen, D.A. Laird, and J.R. Ambuel. 1997. Yield variability within a Central Iowa field. *Transactions of the American Society of Agricultural Engineers,* 40(4):883-889.

Hanna, M., M. White, P. Boyd, T. Colvin, and J. Baker. 2001. Improving the uniformity of anhydrous ammonia application. PM 1875 Iowa State University Extension, Ames, IA.

Jaynes, D.B., and T.S. Colvin. 1997. Spatiotemporal variability of corn and soybean yield. *Agronomy Journal* 89(1):30-37.

Jaynes, D.B., T.S. Colvin, D.L. Karlen, C.A. Cambardella, and D.W. Meek. 2001. Nitrate loss in subsurface drainage as affected by nitrogen fertilizer rate. *Journal of Environmental Quality* 30:1305-1314.

Karlen, D.L., E.J. Sadler, and W.J. Busscher. 1990. Crop yield variation associated with Coastal Plain soil map units. *Soil Science Society of America Journal* 54:859-865.

Kaspar, T.C., T.S. Colvin, D.B. Jaynes, D. L. Karlen, D.E. James, D.W. Meek, D. Pulido, and H. Butler. 2003. Relationship between six years of corn yields and terrain attributes. *Journal of Precision Agriculture* 4:87-101.

Lowenberg-DeBoer, J. 2000. Precision farming Profitability. Publication #SSM-3. Agricultural Research Programs, Purdue University, West Lafayette, IN.

Macy, T.S. 1993. Macy Farms—Site-specific experiences. In *Soil-specific crop management* (pp. 229-244). ASA-CSSA-SSSA, Madison, WI.

Morgan, C.L.S., J.M. Norman, R.P. Wolkowski, B. Lowery, G.D. Morgan, and R. Schuler. 2000. Two approaches to mapping plant available water: EM-38 measurements and inverse yield modeling. In *Proceedings of the 5th International Conference on Precision Agriculture,* Pierre Robert (ed.). CD-ROM prepared by ASA/CSSA/SSSA, Madison, WI.

Perez-Munoz, F., and T.S. Colvin. 1996. Continuous grain yield monitoring. *Transactions of the American Society of Agricultural Engineers,* 39(3):775-783.

Sadler, E.J., P.J. Bauer, and W.J. Busscher. 1995a. Spatial corn yield during drought in the SE Coastal Plain. In *Site-specific management for agricultural systems* (pp. 365-381). ASA-CSSA-SSSA, Madison, WI.

Sadler, E.J., W.J. Busscher, and D.L. Karlen. 1995b. Site-specific yield histories on a SE Coastal Plain field. In *Site-specific management for agricultural systems* (pp. 153-166). ASA-CSSA-SSSA, Madison, WI.

Whitney, D.A. 1986. *Comparison of fertilizer recommendations made by several laboratories.* Internal report. Kansas State University Agronomy Department.

Wyciskalla, T.D., E.C. Varsa. S.K. Chong, S.A. Ebelhar, S.J. Indorante, M. McCauley, and G.K. Robertson. 2000. An evaluation of soil-site factors responsible for yield variations in two Southern Illinois Farm Fields. In *Proceedings of the 5th International Conference on Precision Agriculture,* Pierre Robert (ed.). CD-ROM prepared by ASA/CSSA/SSSA, Madison, WI.

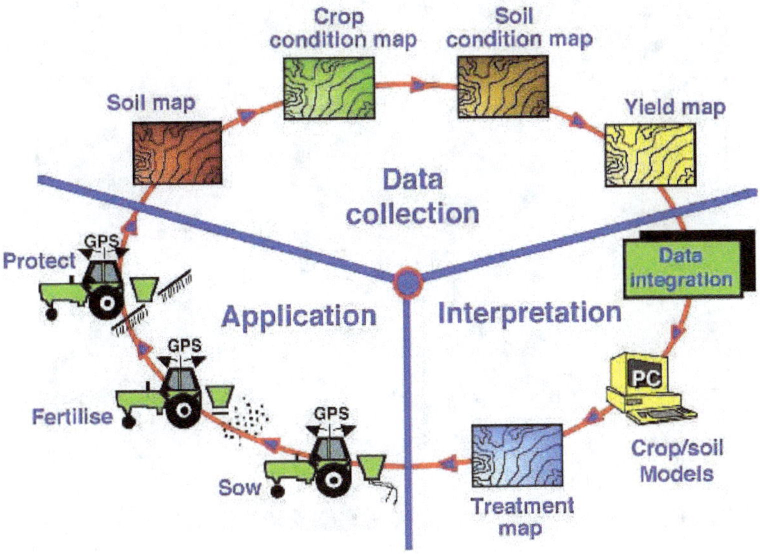


PLATE 2.3. The precision agriculture cycle.

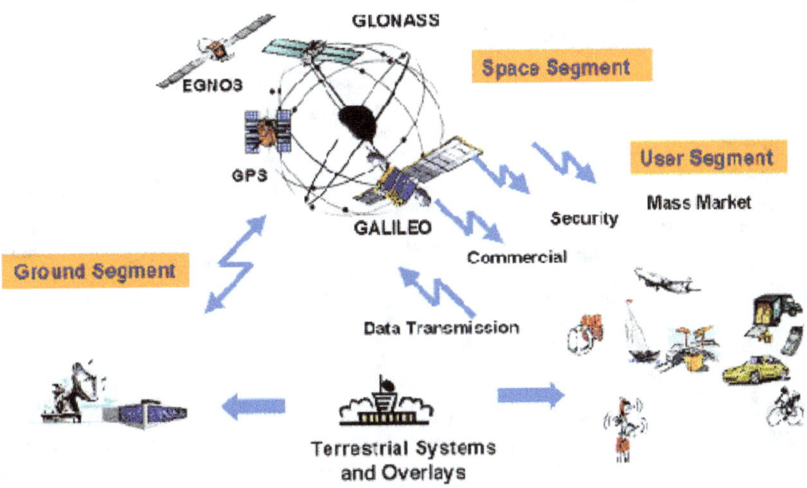


PLATE 2.4. The planned European Galileo GNSS.

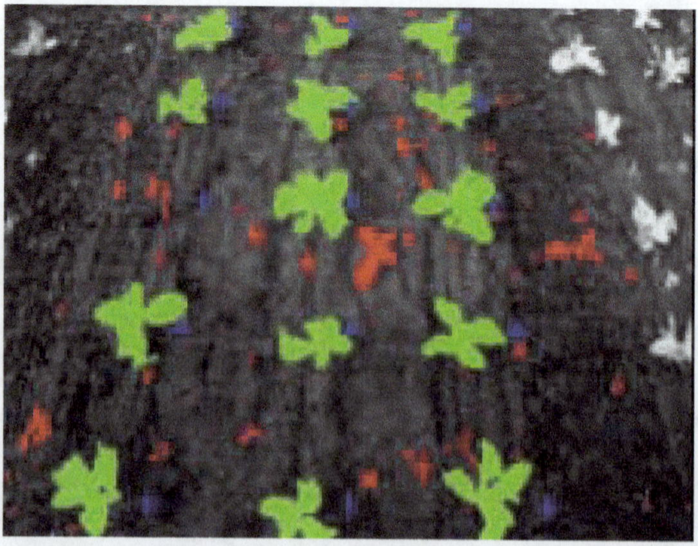

PLATE 2.12. Discriminating individual crop plants. *Source:* After Marchant et al., 1998.

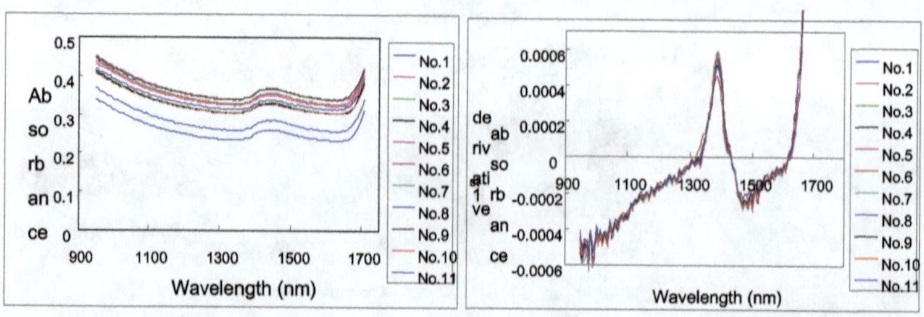


(a) Original soil absorbance

(b) The 1st derivative of the soil absorbance

PLATE 3.8. Effect of soil surface roughness on light energy absorbance in the NIR range. The surface conditions are flat smooth (No. 1), stripes of 3 mm depth at 5 mm spacing (No. 2 and No. 3), scattered hollows with 5 mm diameter and 3 mm depth (No. 4, 10, 11), ditching of 5 mm depth at 5 mm spacing (No. 5, 6, 8), convex of 5 mm elevation (No. 7), concave of 5 mm depth (No. 9). Test soil was Kanto loam (25 percent sand, 34 percent silt, 20 percent clay) in a constant moisture content of 25 percent dry basis. *Source:* Shibusawa et al., 2000.

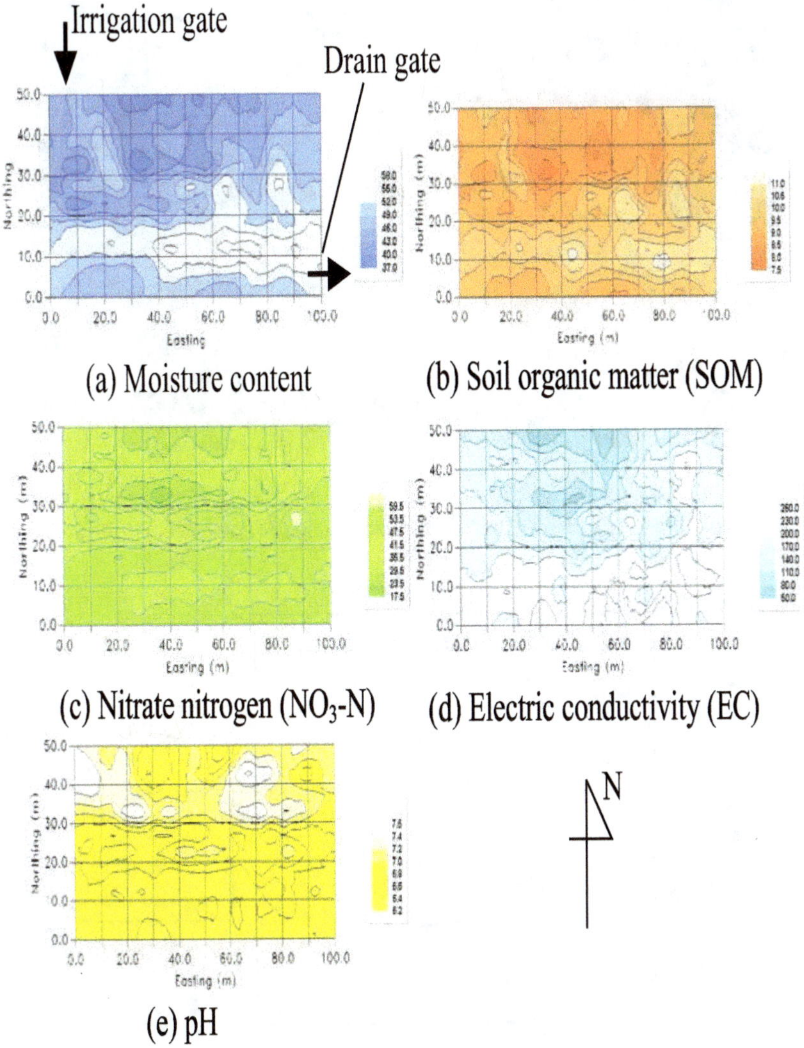


PLATE 3.9. Kriged maps of a 0.5 ha paddy field of 50 m width and 100 m length using 860 data of soil reflectance collected by the real-time soil spectrophotometer. The working depth was about 200 mm, and spacing was 1 m on north-south segments and 5 m on east-west segments.

PLATE 4.1. Spatial variability of macroscopic sulfur deficiency symptoms visible in different colored patches within a flowering oilseed rape field in northern Germany.

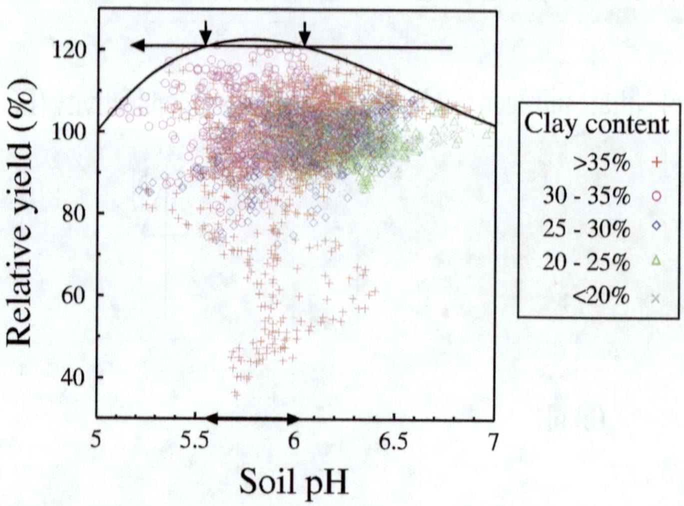


PLATE 4.7. Evaluation of the optimum range of pH values (5.6-6.1) in the topsoil of a clayey loam soil in Northumberland, United Kingdom (W 1.05833; N 55.28333).

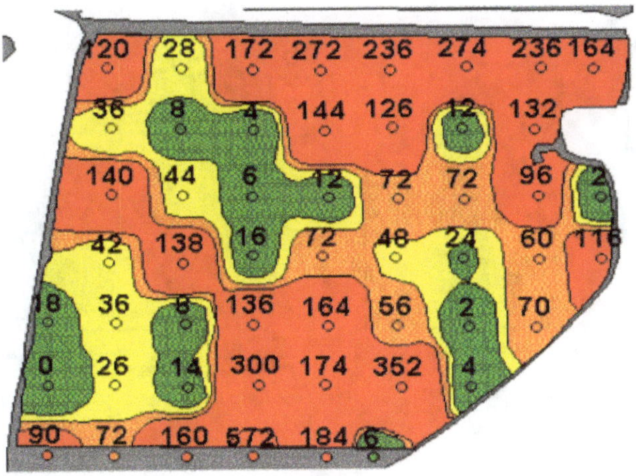


PLATE 8.1. Numbers shown on the map indicate the *Meloidogyne incognita* juvenile levels per 100 $cm^3$ of soil at each 0.4 ha grid (O = center of grid for sampling); map depicts nematode density ranges, based on the sampling point data, as determine using AgLink Professional software package. *Source:* AGRIS, Roswell, Georgia.

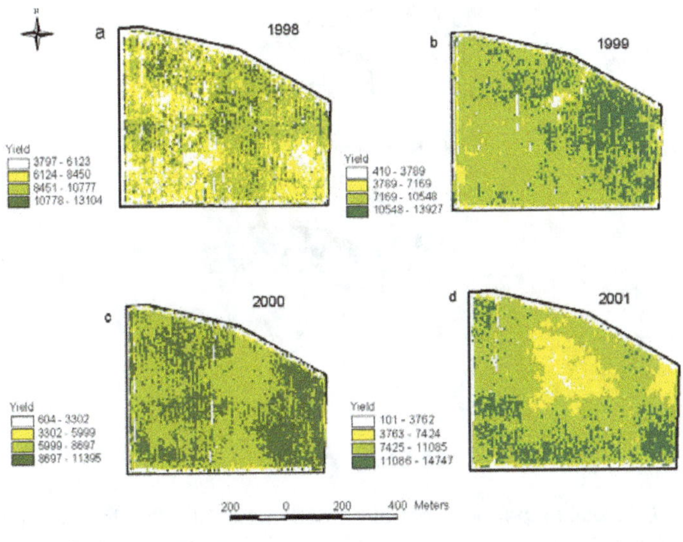


PLATE 10.1. Rice yield in kg·$ha^{-1}$ in 1998 *(a)*, 1999 *(b)*, 2000 *(c)*, and 2001 *(d)*.

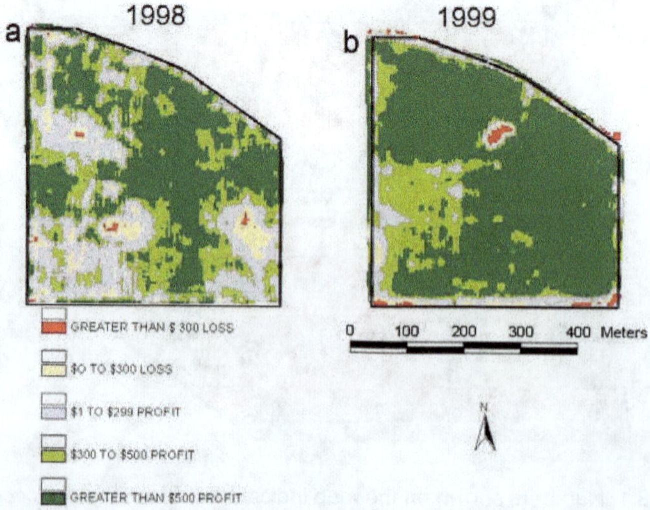


PLATE 10.7. Dollars per hectare lost or earned in both growing season: *(a)* Before leveling, *(b)* after leveling. *Source:* Redrawn from Roel and Plant, 2001.

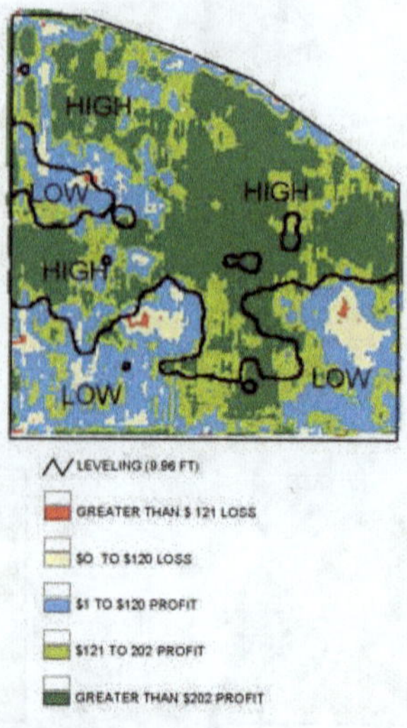


PLATE 10.8. Dollars per acre lost or earned in 1998 (before leveling the field) and the delineation of the topographic low and high areas of the field. *Source:* Redrawn from Roel and Plant, 2001.

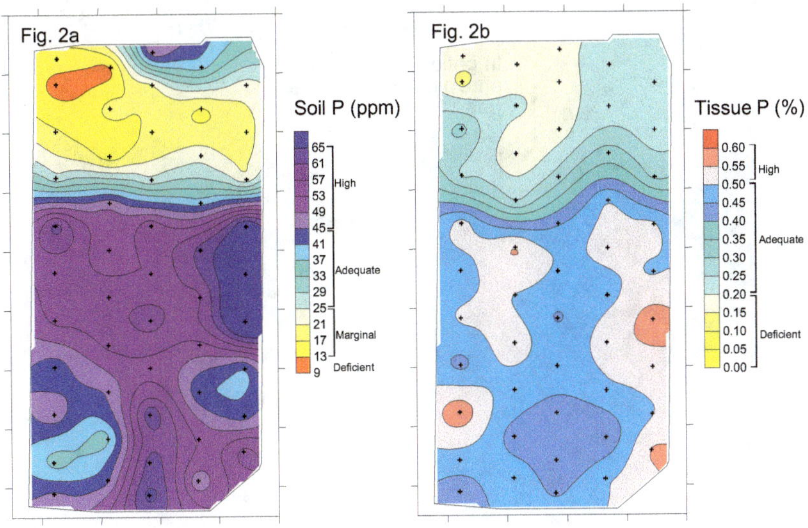


PLATE 13.2. Soil Kelowna phosphorus (0-15 cm) (October 1998) and petiole phosphorus (July 28, 1999) for FL1625 potatoes.

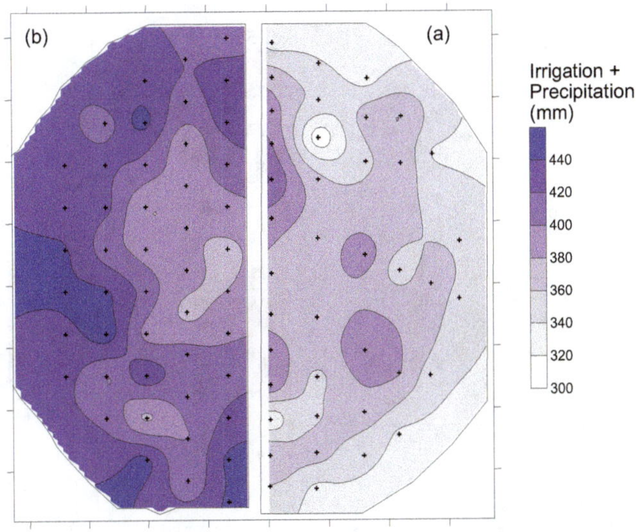


PLATE 13.3. Total irrigation and precipitation (mm) for Snowden potatoes *(b)* June 23-Sept. 4, 1997; *(a)* June 25-Sept. 9, 1996.

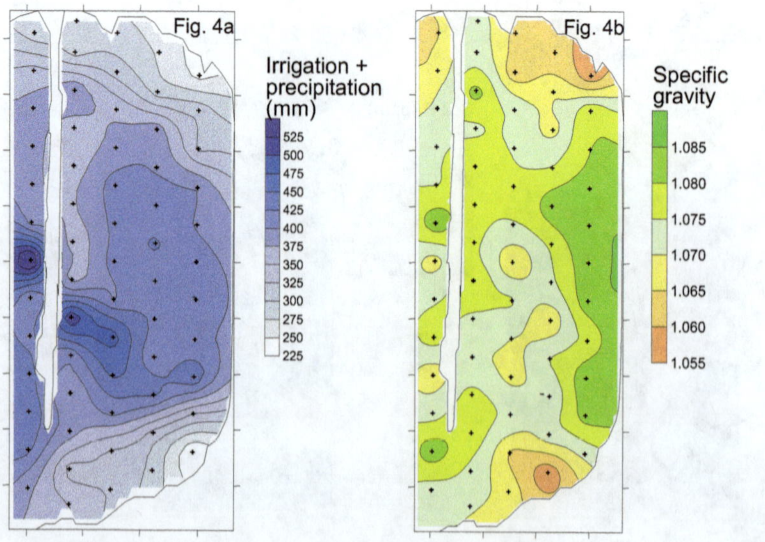


PLATE 13.4. Total irrigation and precipitation (mm) June 17-August 31, 1999, and specific gravity (September 1998) for Russett Burbank potatoes.

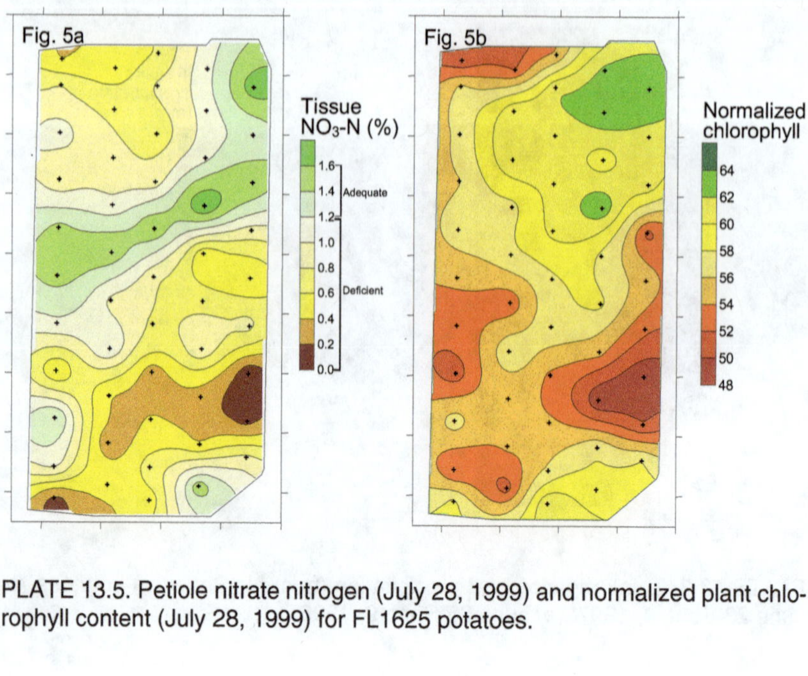


PLATE 13.5. Petiole nitrate nitrogen (July 28, 1999) and normalized plant chlorophyll content (July 28, 1999) for FL1625 potatoes.

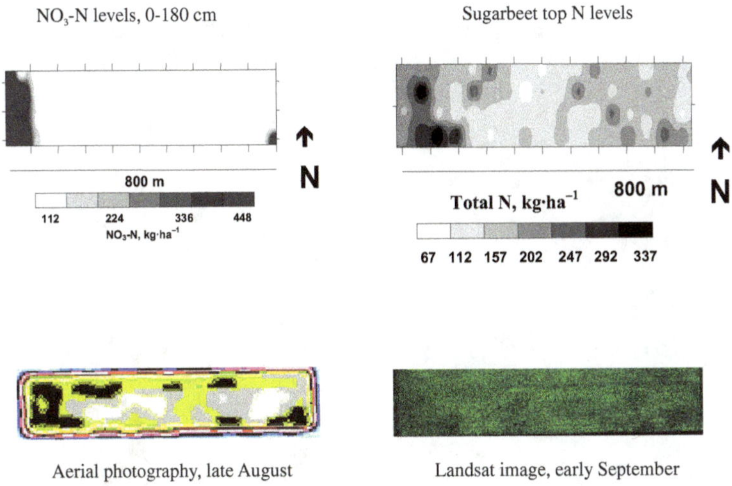


PLATE 14.5. Soil N levels to 180 cm following sugarbeet, sugarbeet top N content at harvest, aerial photography and Landsat 5 satellite imagery showing areas of high N. Darker colors in each image indicate areas of higher N, and lighter colors correspond to lower N.

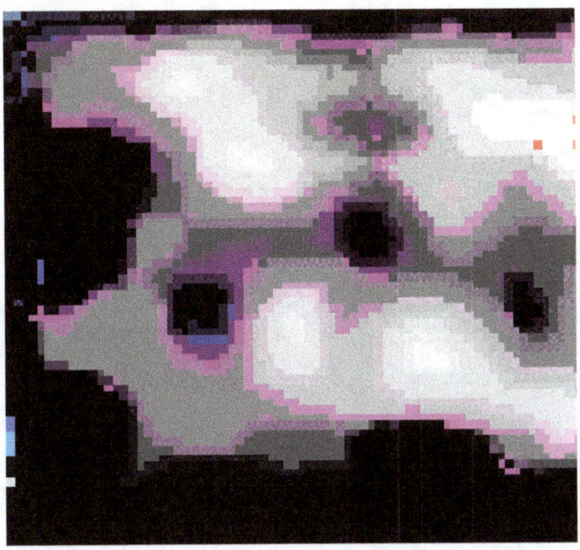

PLATE 14.8. Landsat image of potato field three weeks prior to desiccation. Darker areas correspond to higher amounts of N in foliage.

PLATE 15.2. Leaf area index of vineyards is mapped at 2 m resolution using the relation shown in Figure 15.1 and satellite imagery obtained from IKONOS. Significant variations in LAI are evident even within a single vineyard as a result of spatial differences in resource availability, management practices and presence of pest/diseases. Understanding and properly managing these small scale variations is the key to successful implementation of remote sensing technology in vineyard management.

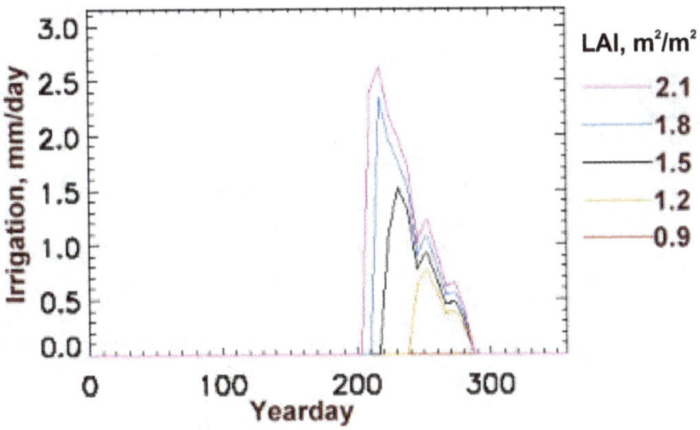


PLATE 15.6. Sensitivity analysis of Ag-BGC showing the amount of irrigation required to keep the vines at a water stress of –1.0 Mpa (predawn) as a function of leaf area index with climate and soils held constant. Dense canopies require greater amounts of water starting earlier in the growing season. Similar analysis can be performed for various combinations of climate, soil, and leaf area. The results of such a comprehensive sensitivity analysis can be used to determine the choice of varietals and plating density.

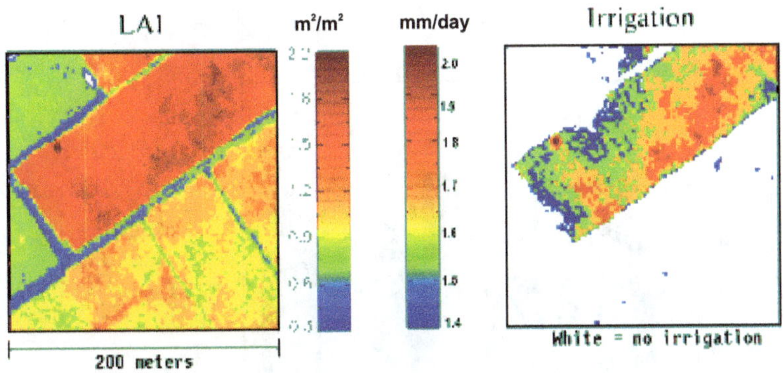


PLATE 15.8. Spatial implementation of Ag-BGC parameterized with LAI from IKONOS data (Figure 15.2) for estimating daily irrigation requirements for a 200 m × 200 m area in the Napa Valley, California. While traditional crop coefficient based irrigation approaches predict uniform irrigation for entire vineyards, spatial BGC simulations show a wide range of water requirements within a vineyard. Variable Rate Technology (VRT) can be brought to bear to address such spatially varying water demands.

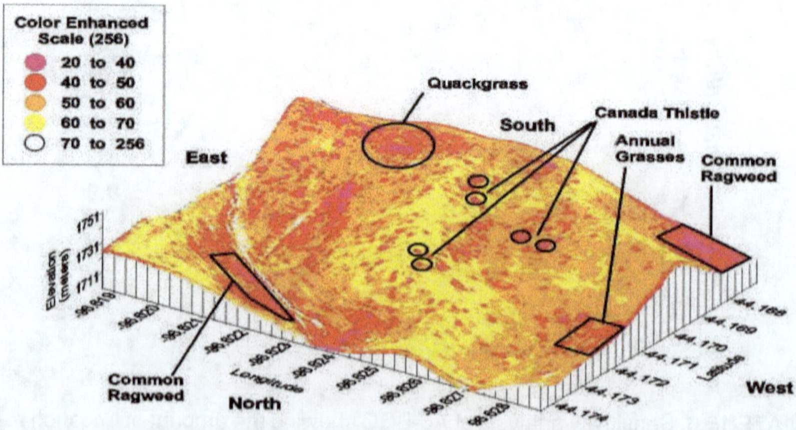


PLATE 16.3. The remote sensed image (NIR) of a 65 ha soybean field collected in October 1996. Patches of Canada thistle are evident in the summit position while common ragweed was found in footslope areas. The color enhanced scale ranges from 0 (minimum reflectance) to 256 (maximum reflectance).

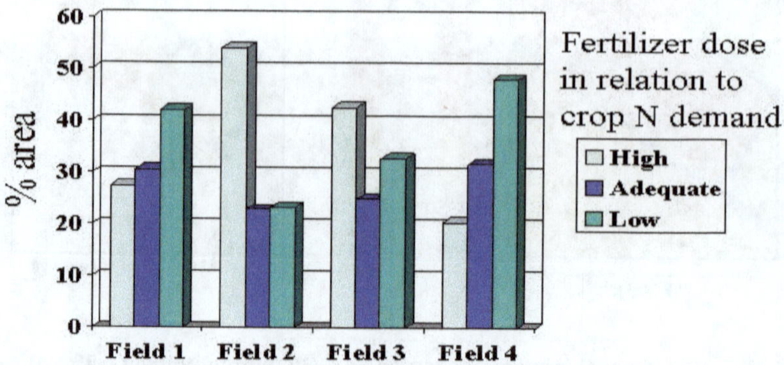


PLATE 18.1. Proportion of area where N fertilizer rates are high, adequate, and low as compared to N demand by rice in four adjacent fields of Hokkaido.

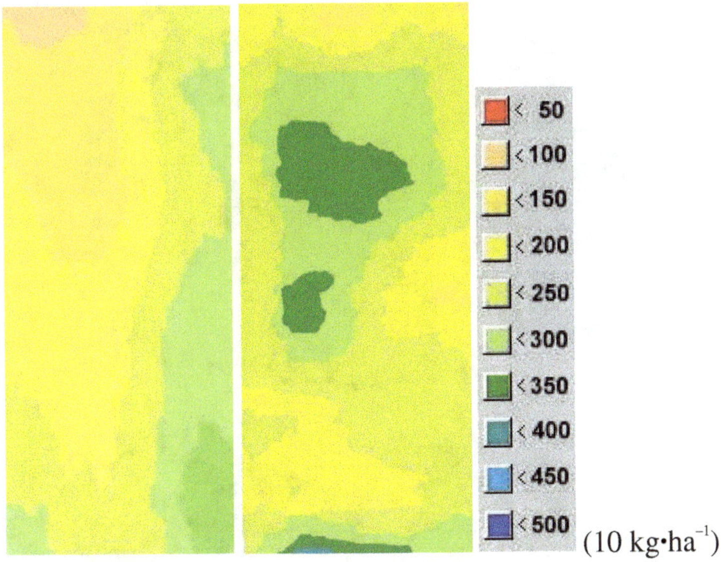


PLATE 20.1. Yield maps of small cotton field in Greece.

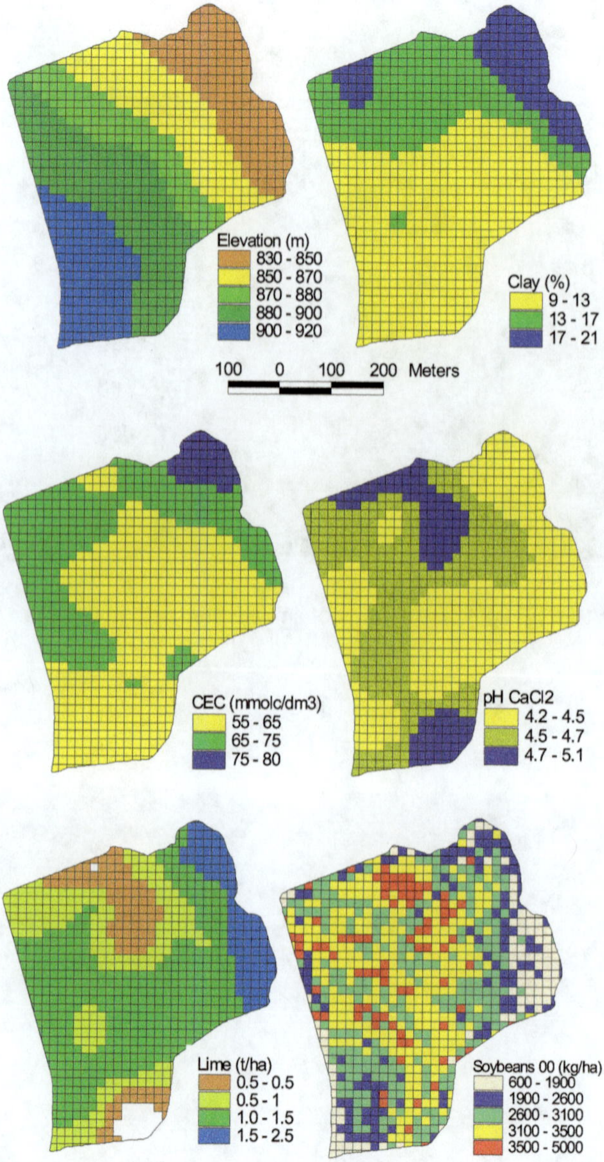


PLATE 22.1. Soil topographic, physical, chemical, and productivity attributes of a commercial field with a rotation of grain crops under no-tillage; coordinates 24°54'S and 50°17'W, State of Parana, Brazil.

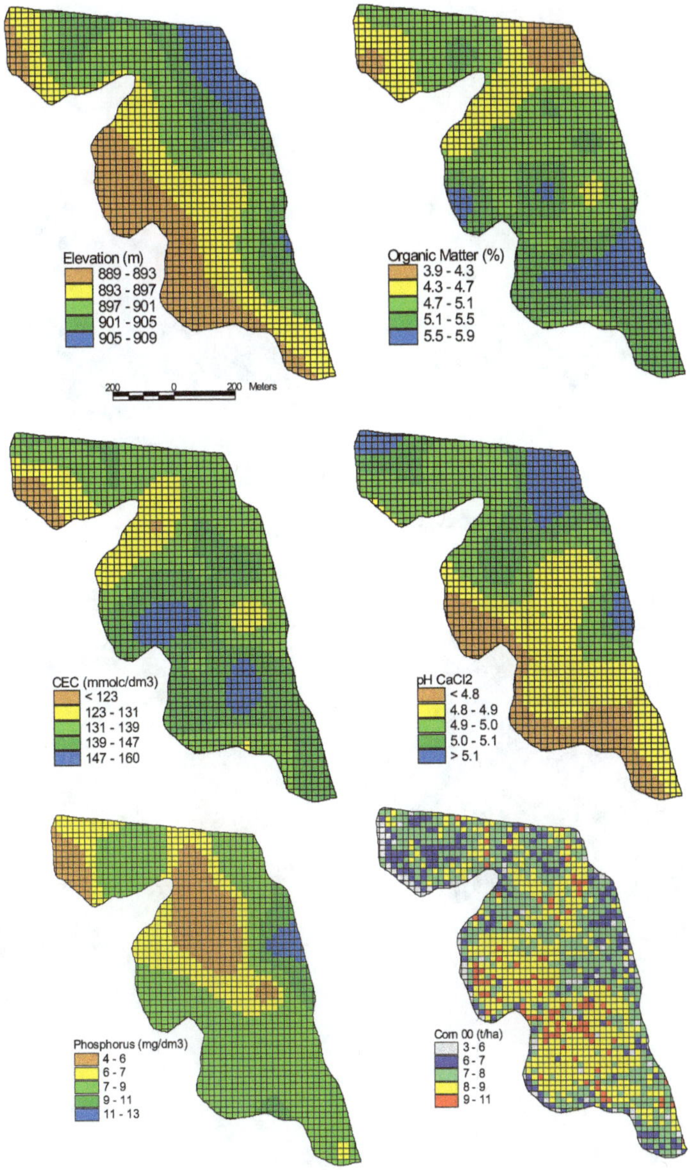


PLATE 22.2. Soil topographic, physical, chemical, and productivity attributes of a commercial field with a rotation of grain crops under no-tillage; coordinates 25°44'S and 51°54'W, State of Parana, Brazil.

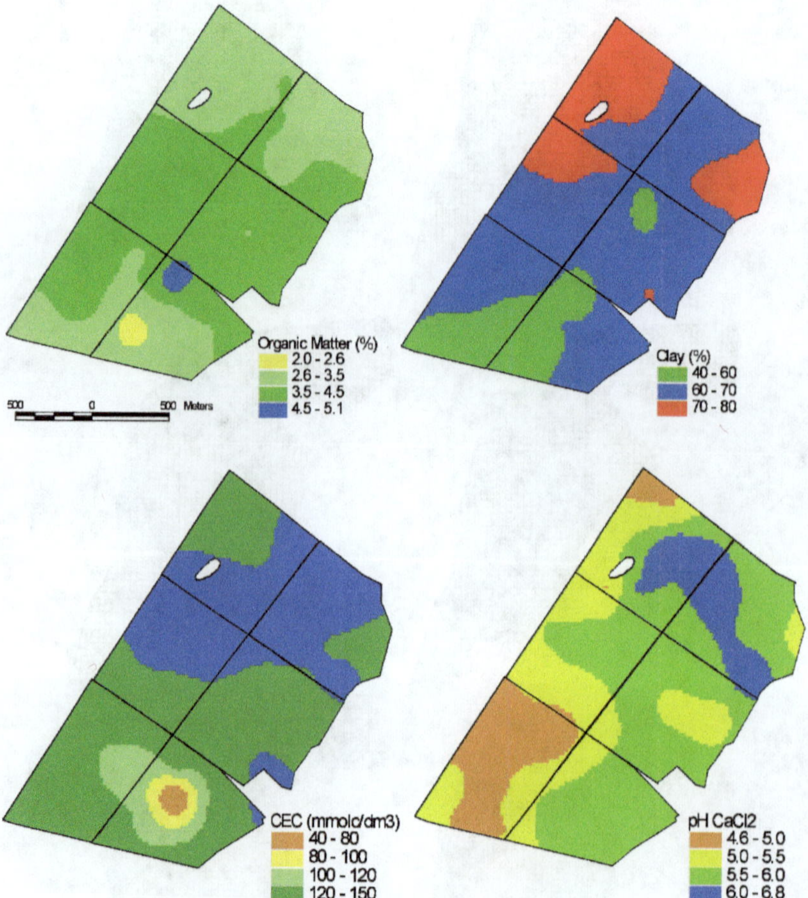


PLATE 22.3. Soil physical and chemical attributes of a commercial field with a rotation of grain crops and pasture under no-tillage; coordinates 21°46'S and 54°24'W, State of Mato Grosso do Sul, Brazil.

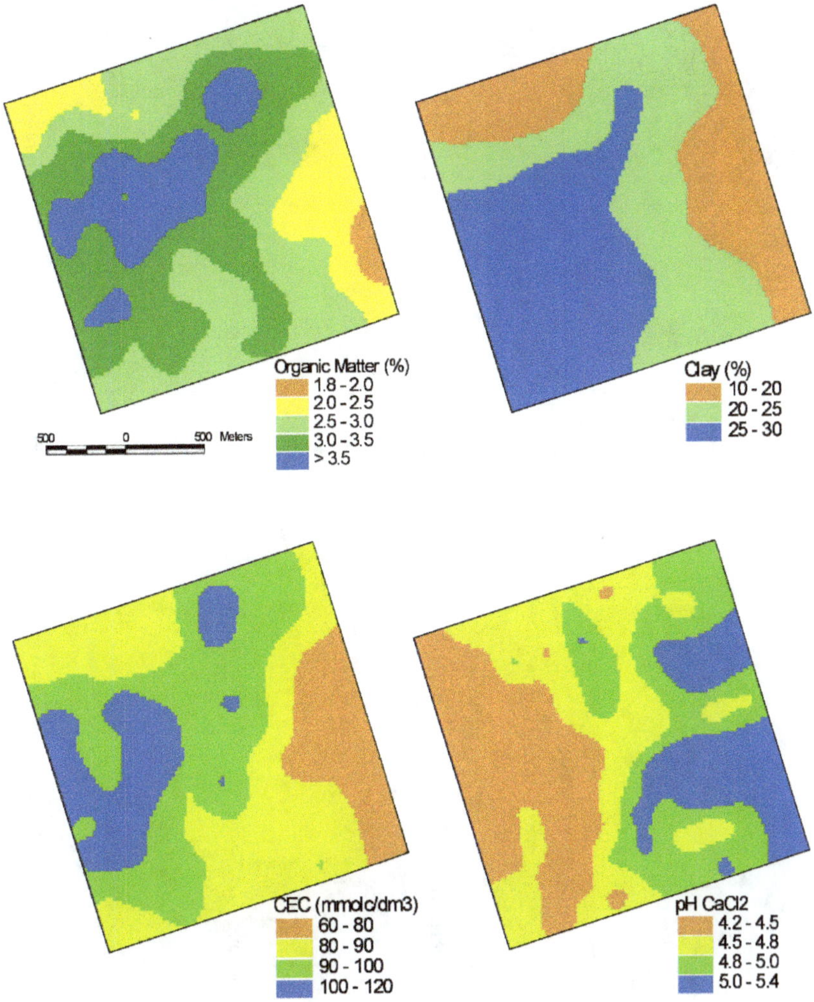


PLATE 22.4. Soil physical and chemical attributes of a commercial field with a rotation of grain crops under no-tillage; coordinates 8°38'S and 46°10'W, State of Piaui, Brazil.

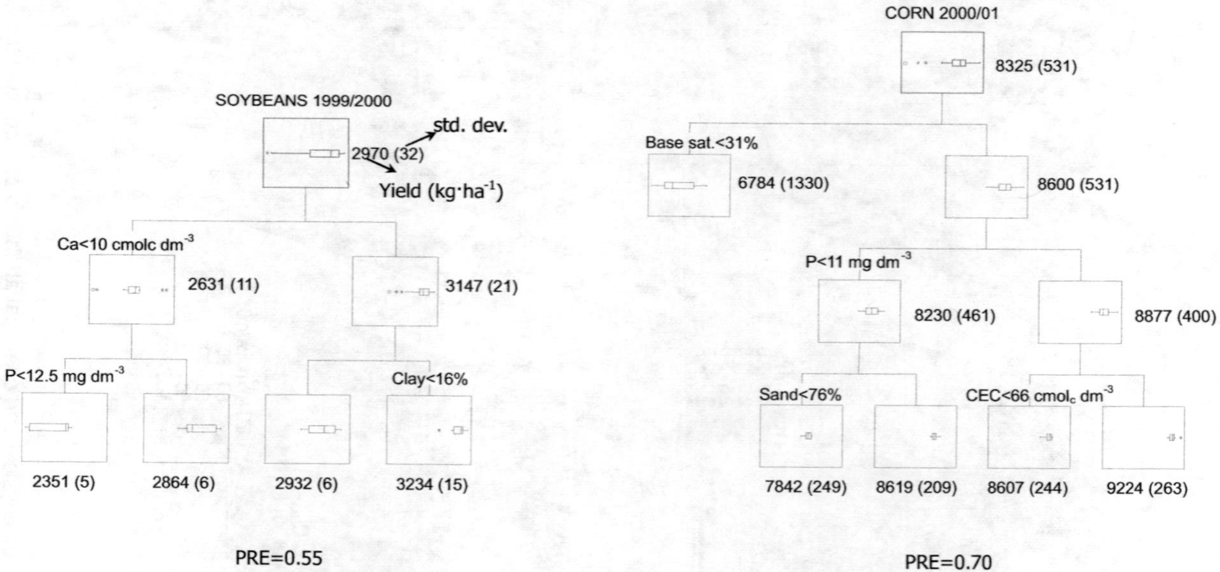


PLATE 22.5. Tree diagrams of the relations between crop yield and significant soil attributes, for field shown in Figure 22.1. PRE is Percent Reduction in Error. Units for P are mgdm$^{-3}$, for Ca and CEC are mmol·dm$^{-3}$.

## Chapter 12

# Site-Specific Management of Cotton Production in the United States

Richard M. Johnson
Judith M. Bradow
Anne F. Wrona

## INTRODUCTION

Site-specific management or precision agriculture, as it is evolving in large-scale crop production, promises new methods for managing cotton production for optimized yields, maximized profitability, and minimized environmental pollution. However, adaptation of site-specific theory and methodology, which have been developed primarily for large-scale grain production, entails overcoming some serious obstacles related to regional production practices and crop valuation methods prevalent in U.S. cotton production. For example, when major cotton-producing counties in the U.S. Cotton Belt are grouped according to

1. genotypes grown,
2. planting dates,
3. growing-season lengths,
4. harvest methods,
5. quality of fiber produced, and
6. fiber end use,

five or six general regions become apparent: Coastal, Humid Southeast Upland, Delta, Texas High Plains, California/Arizona Desert, and Short-Season Border. Because of the genotypes grown, the region-related soil types and landforms, and the weather conditions that vary significantly among and within these regions, fiber properties and end use also differ greatly from region to region and within these generalized areas. Production practices also

*Handbook of Precision Agriculture*

doi:10.1300/5627_12

vary significantly among these regions, and a site-specific strategy that works well in the laser-leveled, irrigated fields of Arizona or California may prove ineffective in the hilly fields where nonirrigated cotton is grown in the Humid Southeast.

Like grain producers, cotton farmers must cope with spatial variability in yield, but the development of cost-effective and accurate cotton yield monitors took longer than comparable developments in grain production. It has only been the past several years that yield monitors have been tested and proven reliable in the field (Sui et al., 2004; Thomasson and Sui 2003; Vellidis et al., 2003) and many producers still question if purchasing a yield monitor is a wise business decision (Larson et al., 2005). In the absence of the geospatial information afforded by fully functional yield monitors, particularly in the early years of site-specific management, cotton yield maps have been constructed from time- and labor-intensive grids and transects that are limited to describing relatively small sectors of any given field. Such high-resolution mapping provides valuable research data, but cost-effective mapping methodology on a broader scale is necessary if even the most efficient, forward-thinking cotton producers are going to invest in site-specific management under the depressed economic conditions now prevailing.

Still further away in the developmental pipeline are the affordable, dependable cotton quality monitors that can quantify the spatial variability in economically important fiber properties such as length, strength, and maturity. Moisture content and seed cotton weight, which could be measured during harvest, do not provide estimates of intrinsic fiber characteristics. Those properties can be determined only after the crop has been harvested and has moved through several steps in post harvest processing. At present, the USDA Agricultural Marketing Service (AMS) regional classing offices determine cotton fiber-quality properties on a bale-by-bale or, in some cases, module-by-module basis after the crop is harvested and the fiber is cleaned and ginned. Currently, no provision is made for documenting or preserving the site-specific relationships that describe the source of the fibers within the bales or modules being classed. Both harvesting and ginning repeatedly mix and blend bolls and fibers. Thus, postharvest procedures totally obliterate the within-crop spatial relationships that are needed for application of some of the most promising site-specific technologies developed for improving fiber quality and, therefore, the economic value of the cotton crop.

Fortunately, researchers and producers have been treating these and other factors perceived as obstacles to the application to site-specific production of cotton as challenges to be met *and* overcome. Regional differences and similarities in cotton production have already been considered in-depth during the development of regional and generalized mechanistic crop models. Now, those extensive databases and the soil/plant/atmosphere system models based on those data are being utilized when reasons are sought for yield variability and when prescriptions for precision crop management are formulated.

Other approaches to site-specific management of cotton production are advancing from the uppermost informational layer downward. The cotton-growing regions of the United States are well covered by GPS satellites and the differential beacon system that facilitate construction of submeter geopositional maps of soils, yields, and fiber properties. In addition, the Wide Area Augmentation System (WAAS) was implemented by the U.S. Federal Aviation Administration (FAA) in 2000 to assist in aircraft navigation and serves as another source of GPS differential correction. Application of remote sensing technology is progressing rapidly in several regions of the U.S. Cotton Belt where forward-thinking growers have established alliances with sectors of the U.S. space program. Remote sensing, the grower's detailed historical knowledge of his own fields, and geopositional mapping of yields and fiber properties are being combined in the formulation of zoned-management prescriptions that optimize cultural inputs and predict and segregate higher-quality fiber during harvest.

Naturally, not every challenge in site-specific management is being met with the same speed or degree of success throughout the U.S. Cotton Belt, but the computer-intensive character of both research in precision agriculture and the application of geospatial technology allow rapid and easy communication and collaboration among consortiums of researchers, contractors, and producers. In this chapter we provide an overview of the current status of site-specific management in the various cotton-producing regions of the United States while identifying further opportunities for collaboration and research on a regional and more generalized basis. Site-specific management of cotton production will be examined by moving through the geospatial (GIS) layers that constitute the soil-to-satellite continuum, and emphasis will be given to current applications and potential uses in the near future.

## SOIL INFORMATIONAL LAYER

An effective site-specific management plan begins with an evaluation of the degree of soil spatial variability present in the field in question. Soil type (i.e., soil series and/or mapping unit) and soil physico-chemical properties may vary considerably over relatively short distances, depending on the soil factor being considered and scale of the field being managed (Addiscott and

Wagenet, 1985; Birrell et al., 1996; Karlen et al., 1996; Flynn, 2001). For example, in some production areas variations in soil pH are minimal and do not contribute to spatial variation in cotton yield production (Flynn, 2001). However, researchers elsewhere (Ebelhar and Ware, 2001) have indicated that topsoil acidity is an important factor in cotton lint yield and that variations in soil pH should be considered in quantitative descriptions of field spatial variability.

Significant spatial variability can occur in soil fertility, moisture content, texture, topography, salinity, crusting, and soil temperature (Han et al. 1996; Ward and Cox, 2000). Singly and interactively, these and other soil factors can affect crop yields and quality sufficiently for spatial variability in soil factors to translate into quantifiable variations in cotton yield and fiber properties across space. Thus, descriptions of the soil informational layer are the frequent bases and origination points for strategies for site-specific management of cotton production.

Evaluation of the degree and distribution of soil spatial variability requires that an effective soil-sampling plan be developed or adapted (Peterson and Wollenhaupt, 2001). The literature reveals that soil-sampling strategies differ among soil-variability studies according to the objectives of those studies and, sometimes, according to the region in which a study is performed. Researchers may choose grid sampling with or without additional transects, topography-based sampling, some form of directed soil sampling, or management-zone based approaches (Pocknee et al., 1996; Sudduth et al., 1997; Clark and Lee, 1998; Franzen, et al., 1998; Sadler et al, 1998; Johnson et al., 1999b; Doerge, 2001; Flynn, 2001). Location of the grids and transects can depend on the presence of characteristic landforms and/or the producer's knowledge of crop histories and cultural problems specific to the fields being mapped (Pocknee et al., 1996; Sadler et al., 1998; Johnson, et al., 1999b; Bronson et al., 2000; Fleming et al., 2000; H. Li et al, 2000; Doerge, 2001; Flynn, 2001).

In previously unmapped fields and/or where historical knowledge is limited, both grid sampling and sampling along transects may be done (Johnson, et al., 1999b; Bronson et al., 2000). For research purposes, the factors governing sampling strategy are density of the data and scale of the data. The interval, or lag distance, between grid points should be relatively close together if spatial variables are to be defined and described adequately (Clay et al., 1999; Doerge, 2001). The sampling frequency required to obtain a "representative" sample depends on the degree of variability and upon the stability of the factor being sampled (Doerge, 2001).

Investigators relating cotton yield to variations in soil properties have used the standard deviation of field parameters such as soil pH, electrical conductivity, organic matter, soil nitrate, bicarbonate-extractable phosphorus, and ammonium-acetate-extractable potassium to determine how many grid samples are needed (Flynn, 2001). The sampling-frequency requirement varied according to the soil parameter. For example, adequate estimation of the field means for soil pH and electrical conductivity required fewer than five subsamples per 16 hectares, but effective nitrate mean estimates were based on 74 subsamples or more, depending on the confidence interval specified.

Where geostatistical approaches are used, it is often critical to obtain at least some samples at relatively small intervals (fine scale and high density) to properly characterize the nugget effect and the shape of the variogram near the origin (Gotway et al., 1996). In recent studies that utilized grid-sampling methods to study the influence of soil variability on cotton yield and quality (Johnson et al., 1999b; Johnson and Bradow, 2000; Ping and Green, 2000; Ward and Cox, 2000; Ping et al., 2004a; Ping et al., 2004b; Ping et al., 2005), the grid-sampling intervals ranged from 7.5 to 101 meters with costs and practical production considerations weighting the choice toward larger intervals and fewer samples.

Site-specific management of any crop depends on the division of large-scale field variability into smaller zones with relatively uniform soil and topological characteristics (Sadler et al., 1998; Johnson et al., 1999b; Ward and Cox, 2000). Soil-variability mapping methods used by most researchers have been borrowed from the mining engineering field and rely on geostatistics to analyze and map the spatial dependence between soil properties (Warrick et al., 1986; Crawford and Hergert, 1997; Goovaerts, 1998). However, other researchers have investigated alternative mapping techniques, such as Inverse-distance and triangulation methods, to describe soil variability (Gotway et al., 1996; Whelan et al., 1996). Examples of some of these approaches are provided in studies investigating variable-rate fertilization (Franzen and Peck, 1995; Franzen et al., 1998; Sadler et al., 1998). Effective soil variability maps are readily included in geographic information systems and can be used to develop variable rate application plans for crop fertilizers, herbicides, pesticides, or plant growth regulators (Thurman and Heiniger, 1998).

The use of intensive grid sampling of soil variability is normally limited to researchers, who require fine-scale quantitative data maps. The costs associated with repeated sampling and replicated analyses are unacceptable to producers, who must reduce, not increase, production costs. Therefore, other less expensive soil mapping approaches have been sought. Soil electrical conductivity mapping was recently utilized to identify soil properties that influence cotton yield (Corwin et al., 2003) and aerial photos were also utilized to provide estimates of cotton yield (Vellidis et al., 2004). Soil survey maps of U.S. cotton-producing counties are available through the

National Soil Survey Center of the USDA Natural Resources Conservation Service, USDA-NRCS (2000b). However, some maps may be outdated, and the maps vary in accuracy and usefulness depending upon which organizational entity collected and compiled the information. More detailed, digitized soil survey maps of some areas are available in the Soil Survey Geographic Database (SSURGO) from the USDA-NRCS Soil Survey Division (2000a). These maps are intended for use by landowners and planners and managers of natural resources, and SSURGO data are designed for use in a Geographic Information System (GIS) like those basic to precision agricultural applications. Some attempts at predicting yields from maps created from SSURGO data alone have been disappointing (Stermitz et al., 1999), but digitized maps and aerial images of bare soil color and foliar reflectance are being used by producers to develop management-zone maps that can be used to formulate precision prescriptions for optimizing and, therefore, reducing production inputs (Fleming et al., 1999; Dupont et al., 2000; Mallarino and Wittry, 2000; Doerge, 2001).

Most available reports relate soil variability to yield variations. Significant positive correlations have been reported between cotton yield and soil texture (Ward and Cox, 2000; Ping and Green, 2000; Ping et al., 2004a), soil phosphorus (Johnson et al., 1999b; Ward and Cox 2000; Ping et al., 2004a; Ping et al., 2005), soil organic matter (Johnson et al., 1999b; Johnson and Bradow, 2000), soil pH (Johnson et al, 1999b; Ping et al., 2004a), and soil calcium and potassium (Ping and Green, 2000; Ping et al., 2004a). However, very few researchers have investigated the relations between soil physico-chemical properties and both cotton yield and quality. Cotton yield and fiber quality were spatially correlated with ranges varying from 23 to 40 meters for yield and from 15 to 106 meters for fiber quality (Johnson et al., 1999b; Johnson and Bradow, 2000; Johnson et al., 2002).

The degree of spatial variation exhibited in both cotton yield and fiber properties suggests that precision management of cotton yield and quality would be both possible and beneficial. Ultimately, however, economic realities in the form of available labor and financial resources become the main factors determining grid size and sampling frequency and, thus, the scale and density of the soil-layer database. Every soil sample taken has a cost in manpower and soil analysis, and objectives must be carefully formulated to provide the maximum amount of reproducible, quantified information with the least expense. Agricultural economics affect the experimental designs underlying research into site-specific crop management, and costs of implementation are the primary limiting factor for crop production applications. Consequently, the grid-sampling frequency commonly recommended for production agriculture, rather than research, is one sample per one to two hectares (Pocknee et al., 1996). Variable-rate fertilizer recommendations, however, set management zones and the related sampling scale at no larger than nine square meters (Doerge, 2001). Thus, the basic characteristics of the data scale and density in the soil informational layer require additional study and development.

## *CROP INFORMATIONAL LAYER*

Application of data from the soil layer requires construction of the additionally complex crop data layer. However detailed and accurate the soil data layer is made, application of site-specific management methods still requires identification and quantization of cause-and-effect relationships between the soil data layer and the corresponding "plant" layer, i.e., crop yield and crop quality. Cotton producers know which parts of their fields have historically produced poor stands and low yields, and some producers have begun identifying management zones on the basis of such historical knowledge combined with soil-type zones. Historical and current characterizations of such management zones are then used in correlating zone-specific soil information with yield data on an annual basis (Elms and Green, 1998; Reed, 2001; Ping et al., 2004b).

At the current stage in the development of management-zone methodology, research approaches are most likely to link soil maps with yield maps (Swindell, 1997; Elms and Green, 1998; Ping and Green, 1999, 2000, 2001; Ping et al., 2004a; Johnson et al., 2002). Multivariate analysis techniques (cluster analysis and discriminant analysis) are being used to identify potential management zones based on yield and soil parameters such as calcium, soil pH, and clay and sand contents (Ping and Green, 2001). These mathematically derived management zones were similar to management zones defined on the basis of soil conductivity, elevation, and slope (Fridgen et al., 2000; Ping and Green, 2001). Alternatively, soil spatial statistics are used to expand and refine mechanistic plant growth simulation models and fiber-quality plant maps (Plant, 1997; Johnson et al, 1998; Akins et al., 1999; Bradow et al., 1999a,b; Johnson et al., 1999a; Landivar, 1999; McCauley, 1999; Johnson et al., 2002).

However, a growing consensus believes that yield mapping in relation to soil and topographic site-specific information is inadequate for predicting yield potential (Pachepsky and Acock, 1997; Werner et al, 2000). Yield mapping of cotton production, in particular, is further complicated by the lack of yield monitors comparable to those being used in site-specific management of grain crops. Development of effective yield monitors for cotton is additionally complicated by the regional use of two different picking systems, stripper and spindle pickers, and by the different row spacing used in the various regions and within specific regions.

The first commercially available cotton yield monitors varied significantly in responsiveness to yield changes, yield map distortions, and estimation of yield (Perry et al., 2001). The monitors tested provide yield maps adequate for production use, but maps produced by the commercial software were subject to different degrees of "smoothing" or "averaging" so that researchers employing any of the monitors have to extract raw yield data and plot new yield maps using software packages compatible with those used in the soil-layer databases. Yield monitoring systems for stripper harvesters were slow to provide the accuracy necessary for use in site-specific management, but progress was made (Roades et al., 2000; Gvili and Brightbill, 2001; Searcy et al., 2001) and eventually adequate systems were developed (Thomasson and Sui, 2003; Sui et al., 2004). Regardless of harvesting method, widespread use of cotton yield monitors and, therefore, site-specific yield management in cotton continues to be hindered by significant problems in maintaining sensor cleanliness and calibration in a "real world" harvest environment (Sassenrath-Cole et al., 1999; Wolak et al., 1999; Perry et al., 2001; Sui and Thomasson, 2001).

The most meaningful factors for both management-zone and growth-simulation models are those most closely related to crop yield. Unfortunately, crop yield patterns extracted from yield maps, even when effective yield monitors are available, may vary so much across crop years that management zones and models based on yield alone are inaccurate and misleading (Kitchen et al., 1995; Ferguson et al., 1997; Porter et al., 1998). Surrogate data such as soil electrical conductivity and color may also be inadequate for defining management zones if the surrogates are not closely correlated to yield. Any factor used to delineate zones (or expand models) must, of course, be reproducible and quantifiable (Doerge, 2001). If a quantified factor is correlated with yield and is also stable over time, it is particularly valuable because it needs to be measured only once.

Alone, the most accurate yield maps would still be insufficient for site-specific management of cotton because the economic return from a cotton crop is determined not only by fiber weight (yield) but also by the quality of the fiber. Both boll flow-rate measurements at the picker level and remote sensing at any stage of crop development cannot be used in predicting or mapping fiber quality because all economically important fiber properties can be measured only after harvesting and ginning (Bradow and Davidonis, 2000). However, on the research level, fiber quality can be mapped on an individual plant basis (Bradow, Bauer, et al., 1997; Bradow, Wartelle et al., 1997) or according to management zone or field spatial coordinates (Bradow et al., 1999; 2000; Sassenrath et al., 2005). Fiber-quality mapping on an individual plant basis is more easily correlated with mechanistic growth models, but maps of spatial variability in fiber quality can be constructed and correlated with soil spatial variability (Bradow et al., 2000; Johnson and Bradow; Johnson et al., 2002; Ping et al., 2004a; Sassenrath et al., 2005; 2000). Both the individual-plant and grid-mapping techniques produce multiyear site-specific databases that can be integrated with corresponding environmental data to provide site-specific recommendations for producers whose fields have been characterized according to management zones based on soil layer or remotely sensed geospatial data.

Depending on geographical region, U.S. cotton is grown under dryland (rainfed) conditions or under irrigation by furrow or center-pivot or drip irrigation, the last water-delivery method being the least frequently employed. Spatial variability in both yield and fiber properties has been found when cotton was grown in irrigated fields (Elms and Green, 1997; DeTar et al., 1999). In one study, drip and furrow irrigation were compared by integration of yield monitoring, remote sensing, and electronic soil survey of electrical conductivity (DeTar et al., 1999). The yield monitor and the mobile measurement of electrical conductivity were effective in detecting abrupt changes in soil properties, but problems with blockage of the yield monitor sensors decreased accuracy and precision. Recent cotton irrigation studies based on historical data have not yet been integrated with GIS technology (Wanjura et al., 2000) since, traditionally, these studies have been designed for use with mechanistic growth models. Integration of long-term environmental data, e.g., temporal and spatial variability in rainfall, with site-specific databases and spatial models is a necessary, but rarely available, tool for understanding seasonal weather and crop production models over time and space (Bosch et al., 1999; Chaubey et al., 1999). Progress has been made, however, in applying site-specific management techniques to the management of cultural inputs on an extremely fine scale in surface-irrigated cotton (Adamsen et al., 2000), and GIS interfaces are being developed for incorporation of midseason adjustments in cotton growth models (Ruble and Searcy, 2000).

Because cotton production demands a great deal of both water and nitrogen, many spatial variability studies focus on these crop resources (Li et al., 2000; Wanjura and Upchurch, 2001; Corwin et al., 2003). Development of variable-rate fertilization in cotton often depends on adaptation of theory and methodology designed for other crops, but some variable-rate studies have been conducted in cotton fields both inside (Clark and McGuckin, 1996; Stevens et al., 1997; Stevens and Moylan, 1998) and outside the United States (Stewart and McBratney, 2000; Bronson et al., 2003).

Cotton insect and weed damage management can depend on mechanistic plant growth models (Jallas et al., 2000). However, progress is being made toward developing a position-sensitive, variable-rate spray tillage tool that can be used in farming systems irrigated with Low Energy Precision Application (LEPA) systems (Bordovsky and Keeling, 2000). Although no short-term economic advantage over traditional cultural methods was found during the two-year duration of that study, the potential of precision cultivation for control of dense weed populations was considered quite high. Site-specific weed management practices developed for other crops in other countries incorporate decision support with patch or site-specific spraying (Heisel et al., 1999), but remote sensing seems to be the most rapidly emerging methodology in U.S. site-specific crop protection.

## REMOTE SENSING INFORMATIONAL LAYER

Increasingly, remote sensing is the preferred methodology used in site-specific management of U.S. crop production (Peterson and Christensen, 2001). Strategic uses of remote sensing are apparent in cotton producers' identification of management zones by soil-color classes (Campanella and Hood, 2000; Mulla and Sekely, 2000; Bloodworth et al., 2001; Horney et al., 2001; Reed, 2001). Also tactical applications, usually sequential remote sensing, are found in the management of pests (Fitzgerald et al., 1999a; 1999b; Willers et al., 1999; Dupont et al, 2000; Fitzgerald et al., 2000; LaMastus et al., 2000; Reynolds et al., 2001; Seal et al., 2001; Sudbrink et al., 2001; Yang et al., 2001; Fitzgerald et al., 2004; Willers et al., 2005), and in irrigation scheduling (DeTar et al., 1999; Plant and Munk, 1999; Yu et al., 2001) and weeds (Alchanatis, et al., 2005; Cohen et al., 2005). Remote sensing technology is reported to facilitate and improve crop scouting (Blumhoff and Johannsen, 2000; Fitzgerald et al, 2000; LaMastus et al., 2000; Sanders et al., 2001 and also to aid in herbicide-based cotton regrowth control (Yang et al., 2005). However, some of the methodology described by these authors would require extensive adaptation to be useful in cotton production and against cotton pests. Remote sensing is also being used to improve cotton yield predictors (Campanella and Hood, 2000; Thomasson et al., 2000; Thomasson et al., 2001, Vellidis et al., 2004, Yang, Everitt, and Bradford, 2004, Yang et al., 2004). Researchers in this area are in the process of determining which reflectance data are most closely correlated with yield without the relationships being confounded by underlying soil type and which data are better descriptors of the soil types (Clevers et al., 1994). The level of correlation differs with the images and indices chosen and with the growth stage of the crop, e.g., canopy closure, and time to harvest when the data were collected (Thomasson et al., 2000).

Remote sensing depends heavily upon electromagnetic reflectance data (Plant et al., 2000). Such data are generally expressed in the form of vegetative indices, i.e., algebraic combinations of the measured canopy reflectance of different wavelength bands. One of the most commonly used vegetative indices is the normalized difference vegetative index or NDVI (Tucker, 1979; Plant et al., 2000, 2001). The NDVI has also been combined with leaf reflectance at 550 nm and red edge shift in an attempt to predict nitrogen status of cotton (Lough and Varco, 2000). Vegetative indices have also been developed to predict crop growth stages (Reddy et al., 2003; Zhou et al., 2005a) leaf nitrogen (Zhou et al., 2005b) and leaf chlorophyll (Boggs et al., 2003; Zhou et al., 2005b). Other researchers who are developing cotton nitrogen-sensing units (Wilkerson et al., 1998) have used a spectral index that is a ratio of the blue, green, amber, and near-infrared reflectance values from the cotton canopy (as early as pinhead square) in predicting plant nitrogen content and expected yield. Multispectral images obtained from a height of 3.7 km were effective in detecting variations in cotton growth related to nitrogen availability, but full canopy development was needed to minimize the effect of soil radiance on the accuracy of crop nitrogen status estimates (Peterson and Varco, 2001).

In an integrated, multidisciplinary study (Barnes et al., 2000), a prototype remote sensor mounted on a linear-move cotton irrigation system was used to collect multispectral (visible, near infrared, and thermal) data. These "remotely sensed" data were combined with soil layer neutron probe data quantifying soil/plant water state and crop layer petiole samples analyzed for crop nitrogen status. Analyses of these data indicated that remotely sensed data could be used to develop maps of water stress, nitrogen status, and canopy density when variations in all these factors occurred simultaneously.

The single-year study cited (Barnes et al., 2000) provided a "snapshot" of crop status at single temporal point during the crop year, but repetition over time is needed to determine how accurately the various interactive factors might be quantified across the growing season. Repetition over time and space is also essential for the integration of multifactor data into functional strategic and tactical site-specific management practices. Without knowledge of the temporal variability among and within crop years, such single-shot descriptions of spatial variations may have some utility in strategic planning and decision making, but the descriptions are only marginally useful in tactical crop management. This limitation can be addressed by obtaining temporal patterns of cotton reflectance, as was done in a project designed to relate NDVI-days, temporal reflectance patterns, and yield (Li et al., 2001).

A detailed description of multiple spatial factors without repetitions over time has limited value in tactical situations (Plant et al., 2001). However, empirical methods are being developed for other crops so that yield or other data collected over several years or by several remote-sensing passes can be compared statistically and used to define site-specific zones within fields (Layrol et al., 2000; Rydberg and Söderström, 2000). Again, significant adaptation would be necessary before such methodology could be applied in cotton production management.

Perhaps the greatest current challenge in developing tactical applications for remote sensing imagery is the need for extensive ground-truthing at and even below the surfaces of the imaged fields. Bare-soil images can be excellent descriptors of drainage patterns and major variations in soil type (or color), but cotton is normally grown with row spacings that promote canopy closure. Reflectance data from a closed canopy do not yet adequately differentiate among nutrient deficiencies, pest infestations, and crop growth stage. Remotely sensed images of ultra-narrow-row fields look quite different from images collected the same day over conventionally spaced plants. Aerial images collected early in the morning when cotton flowers are open differ significantly from images collected later in the day after the flowers have senesced and disappeared from the aerial view. Zonal color shifts have also been detected when bolls began to open in sections of fields where incomplete canopy closure may have accelerated maturation. Of course, the aerial images change yet again after defoliation, between defoliation and harvest, and during harvest if several passes of the picker are performed.

Strategic and tactical applications of information from both the soil and crop layers are similarly dependent on the remote-sensing data layer. Satellite and aerial images are particularly useful when spatial variability in the soil or the crop is related to drainage patterns and related changes in soil species or to identifiable landforms like ridges and poorly drained zones. Zones of soil variation persisting from forgotten roads and field boundaries may be clearly detectable in aerial images and in unexplained anomalies in yield and quality data collected across such zones, which are invisible to observers on the ground. "Sky-truthing" appears to be as integral a part of site-specific crop management as "ground-truthing" has proven to be.

## *INTEGRATION OF INFORMATIONAL LAYERS*

One of the greatest challenges in site-specific management of any crop is assembling and integrating a critical mass of data from all geospatial data layers over time. The relatively small size of researchers listed as authors of the early, i.e., twentieth century papers, in the reference section of this chapter demonstrate how small research teams gathering the "low hanging fruit" in the early days of precision agriculture were able to make rapid progress in developing site-specific methodology and technology, in assembling databases describing geospatial variations, and in delineating management zones for one or two crop inputs. However, those "early" papers frequently describe how resource limitations restricted those databases to incomplete descriptions of one or two data layers, depending on the expertise and experience of the researchers responsible for the underlying experimental designs. In the case of cotton crop management, the differences between growing regions further compartmentalize research efforts and limit the generalized utility of the databases collected in any region. Those regional differences will probably persist, but larger clusters of researchers with diverse expertise and pooled resources are emerging. Larger, more complete "second-generation" methodology is to be expected as resources are pooled within geographical regions to expand the scale of site-specific management research. Some groups (e.g., Barnes and colleagues, 2000; Horney et al., 2001; Seal et al., 2001; Gogineni et al., 2004), have initiated multispecialty studies, and the successes and limitations apparent in these *preliminary* studies suggest some of the challenges and benefits inherent in integrating the various layers and enlisting the participation of a critical mass of researchers from different disciplines.

## *ACCEPTANCE OF SITE-SPECIFIC MANAGEMENT BY COTTON PRODUCERS*

Another challenge faced by advocates of site-specific management of cotton production is the reluctance of producers to adopt new, expensive technology when the price received for a cotton crop is significantly lower than the cost to produce it with the most efficient, cost-effective production methods available. Another significant restriction to acceptance of site-specific management of cotton is the relatively short length of time that has elapsed since the introduction of site-specific management for any crop. For example, the earliest literature citation in the following reference section is 1994 (Clevers et al., 1994). Most economists prefer to consider ten-year horizons when evaluating new management practices (Yu et al., 2001). Thus, site-specific management for cotton is truly in its infancy and is still largely the domain of researchers and the forward-thinking producers who collaborate with them.

Furthermore, producers cite the following major impediments to adoption of site-specific management: high investment cost for equipment, much of which can be used for cotton only; lack of return on investment; and availability of a labor force trained in the use of computers, software, and complex GPS/GIS equipment (English and Roberts, 2001; Roberts et al., 2001). However, adoption of site-specific management is increasingly likely as more early adopters report their positive experiences with crop input savings and increased yields (English and Roberts, 2001; Reed, 2001). Producers who are using the site-specific management tools available now report that diagnostic data can be used successfully and profitably to make permanent drainage improvements, to negotiate rental agreements, and to make variable rate applications of calcium, phosphorus, potassium, and nitrogen (English and Roberts, 2001). Now that effective cotton yield monitors are available, the implementation levels of site-specific technologies should steadily increase in the future (Barnes, 2004). The pioneers in site-specific management of cotton are also creating field-record databases documenting soil quality, productivity, variety responses, and input effectiveness and cost. However, these essential multi-year databases have not yet reached full functionality with respect to temporal variability and economic predictions. Cotton producers recognize the need for downloadable digitized soil maps, farm workers trained in the use of equipment and software, skilled, reliable personnel representing firms providing contract precision farming services, and for standardized equipment and software. The most recent research reports suggest that many of these producers' needs will be the next crop of "low-hanging fruit" as researchers work with producers toward the integration of the soil, crop, and remote-sensing informational layers into a whole that will allow realization of the potential of site-specific crop management.

## REFERENCES

Adamsen, F.J., D.J. Hunsaker, E.M. Barnes, A.J. Clemens, and E. Bautista. 2000. Surface irrigation and precision crop management. (CD-ROM) *Proceedings, 5th International Conference on Precision Agriculture and Other Resource Management;* July 16-19, 2000; Radisson Hotel South, Bloomington, MN, USA. Agronomy Society of America, Crop Science Society of America, Soil Science Society of America, Madison, WI, USA.

Addiscott, T.A. and R.J. Wagenet. 1985. A simple method for combining soil properties that show variability. *Soil Science Society of America Journal,* 49: 1365-1369.

Akins, D.C., J.L. Willers, J.M. McKinion, S.B. Turner, F.D. Whisler, and J. Varco. 1999. Integrating GIS into precision crop management. *Proceedings, 1999 Beltwide Cotton Conferences.* January 3-8, Orlando, FL, USA. National Cotton Council of America, Memphis, TN, USA, pp. 395-399.

Alchanatis, V., L. Ridel, A. Hetzroni, and L. Yaroslavsky. 2005. Weed detection in multi-spectral images of cotton fields. *Computers and Electronics in Agriculture,* 47 (3), pp. 243-260.

Barnes, E.M. 2004. Current implementation of site-specific technologies in U.S. cotton production *Proceedings of SPIE-The International Society for Optical Engineering,* 5544:12-19.

Barnes, E.M., T.R. Clarke, S.E. Richards, P.D. Coilaizzi, J. Haberland, M. Kostrezewski, P. Waller, C. Choi, E. Riley, T. Thompson, R.J. Lascano, H. Li, and M.S. Moran. 2000. Coincident detection of crop water stress, nitrogen status and canopy density using ground-based multispectral data. (CD-ROM) *Proceedings, 5th International Conference on Precision Agriculture and Other Resource Management;* July 16-19, 2000; Radisson Hotel South, Bloomington, MN, USA. Agronomy Society of America, Crop Science Society of America, Soil Science Society of America, Madison, WI, USA.

Birrell, S.J., K.A. Sudduth, and N.R. Kitchen. 1996. Nutrient mapping implications of short range variability. *Proceedings, 3rd International Conference on Precision Agriculture and Other Resource Management;* July 23-26, 1996; Radisson Hotel South, Bloomington, MN, USA. Agronomy Society of America, Crop Science Society of America, Soil Science Society of America, Madison, WI, USA, pp. 207-216.

Bloodworth, K.M., L.M. Bruce, C.D. Rowland, and D.B. Reynolds. 2001. Detection, classification, and quantification of herbicide drift utilizing spectral data. *Proceedings, 2001 Beltwide Cotton Conferences,* January 9-13, Anaheim, CA, USA. National Cotton Council of America, Memphis, TN, USA, p. 1230.

Blumhoff, G.K. and C.J. Johannsen. 2000. Use of remote sensing technology for enhanced crop scouting. (CD-ROM) *Proceedings, 5th International Conference on Precision Agriculture and Other Resource Management;* July 16-19, 2000; Radisson Hotel South, Bloomington, MN, USA. Agronomy Society of America, Crop Science Society of America, Soil Science Society of America, Madison, WI, USA.

Boggs, J.L., T.D. Tsegaye, T.L. Coleman, K.C. Reddy, and A. Fahsi. 2003. Relationship between hyperspectral reflectance, soil nitrate-nitrogen, cotton leaf chlorophyll, and cotton yield: A step toward precision agriculture Journal of Sustainable Agriculture, 22:5-16.

Bordovsky, J.P. and J.W. Keeling. 2000. Perennial weed management with position sensitive, multiple rate spray applicator. *Proceedings, 2000 Beltwide Cotton Conferences,* January 4-8, San Antonio, TX, USA. National Cotton Council of America, Memphis, TN, USA. pp. 399-404.

Bosch, D.D., J.M. Sheridan, and F.M. Davis. 1999. Rainfall characteristics and spatial correlation for the Georgia Coastal Plain. *Transactions of the ASAE* (American Society of Agricultural Engineers), 42:1637-1644.

Bradow, J.M., P.J. Bauer, O. Hinojosa, and G. Sassenrath-Cole. 1997. Quantitation of cotton-fiber-quality variations arising from boll and plant growth environments. *European Journal of Agronomy,* 6:191-204.

Bradow, J.M. and G.H. Davidonis. 2000 Quantitation of fiber quality and the cotton production-processing interface: A physiologist's perspective. *Journal of Cotton Science,* 4:34-64. Available online at http://www.jcotsci.org.

Bradow, J.M., R.M. Johnson, P.J. Bauer, and E.J. Sadler. 1999a. Modulation of economically important cotton fiber properties by field spatial variability. *Proceedings, 1999. Beltwide Cotton Conferences.* January 3-8, Orlando, FL, USA. National Cotton Council of America, Memphis, TN, USA. pp. 512-514.

Bradow, J.M., R.M. Johnson, P.J. Bauer, and E.J. Sadler. 1999b. Site-specific management of cotton fiber quality. *Precision Agriculture '99,* J.V. Stafford (ed.), Sheffield, UK, SCI, Sheffield Academic Press, pp. 677-686.

Bradow, J.M., R.M. Johnson, P.J. Bauer, and E.J. Sadler. 2000. Zoned management of cotton fiber yield and quality. (CD-ROM) *Proceedings, 5th International Conference on Precision Agriculture and Other Resource Management;* July 16-19, 2000; Radisson Hotel South, Bloomington, MN, USA. Agronomy Society of America, Crop Science Society of America, Soil Science Society of America, Madison, WI, USA.

Bradow, J.M., L.H. Wartelle, P.J. Bauer, and G.F. Sassenrath-Cole. 1997. Small-sample cotton fiber quality quantitation. *Journal of Cotton Science,* 1:48-58. Available online at http://www.jcotsci.org.

Bronson, K.F., J. Wayne Keeling, J.D. Booker, T.T. Chua, T.A. Wheeler, R.K. Boman, and R.J. Lascano. 2003. Influence of landscape position, soil series, and phosphorus fertilizer on cotton lint yield. *Agronomy Journal,* 95:949-957.

Bronson, K.F., R.J. Lascano, J.D. Booker, A.B. Onken, J.W. Keeling, T.L. Archer, S. Machado, E.D. Bynum, Jr., D.M. Nesmith, and H. Li. 2000. Grid soil sampling: Comparison of grid size with landscape- and soil texture-based sampling strategies in the Southern Plains. (CD-ROM) *Proceedings, 5th International Conference on Precision Agriculture and Other Resource Management;* July 16-19, 2000; Radisson Hotel South, Bloomington, MN, USA. Agronomy Society of America, Crop Science Society of America, Soil Science Society of America, Madison, WI, USA.

Campanella, R. and K.B. Hood. 2000. Patterns among seeding rates, remotely sensed data, and yield on a Mississippi Delta cotton farm. *Proceedings, 2000 Beltwide Cotton Conferences,* January 4-8, San Antonio, TX, National Cotton Council of America, Memphis, TN. pp. 421-426.

Chaubey, I., C.T. Haan, S. Grumwald, and J.M. Salisbury. 1999. Uncertainty in the model parameters due to spatial variability in rainfall. *Journal of Hydrology,* 220:48-61.

Clark, R.L. and R. Lee. 1998. Development of topographic maps for precision farming with kinematic GPS. *Transactions-ASAE* (American Society of Agricultural Engineers), 41:909-916.

Clark, R.L. and R.L. McGuckin. 1996. Variable rate application equipment for precision farming. *Proceedings, 1996 Beltwide Cotton Conferences,* January 9-12, Nashville, TN, USA. National Cotton Council of America, Memphis, TN. pp. 186-197.

Clay, D.E., C.G. Carlson, J. Chang, S.A. Clay, and D.D. Malo. 1999. Systematic evaluation of precision farming soil sampling requirements. *Precision Agricul-*

*ture, Proceedings, 4th International Conference on Precision Agriculture,* July 19-22, 1998. St. Paul, MN, USA. Agronomy Society of America, Crop Science Society of America, Soil Science Society of America, Madison, WI, USA. pp. 253-265.

Clevers, J.G.P.W., C. Buker, H.H.C. van Leeuwen, and B.A.M. Bouman. 1994. A framework for monitoring crop growth by combining directional and spectral remote sensing information. *Remote Sensing of the Environment,* 50:161-170.

Cohen, Y., V. Alchanatis, M. Meron, Y. Saranga, and J. Tsipris. 2005. Estimation of leaf water potential by thermal imagery and spatial analysis *Journal of Experimental Botany,* 56:1843-1852.

Corwin, D.L., S.M. Lesch, P.J. Shouse, R. Soppe, and J.E. Ayars. 2003. Identifying soil properties that influence cotton yield using soil sampling directed by apparent soil electrical conductivity, *Agronomy Journal,* 95:352-364.

Crawford, C.A. and G.W. Hergert. 1997. Incorporating spatial trends and anisoptropy in geostatistical mapping of soil properties. *Soil Science Society of America Journal,* 61:298-309.

DeTar, W.R., S.J. Maas, and G.J. Fitzgerald. 1999. Drip vs. furrow irrigation of cotton on sand soil with ¼ mile runs—Includes yield monitoring, remote sensing, and electronic soil survey. *Proceedings, Beltwide Cotton Conferences.* January 3-8, Orlando, FL, USA. National Cotton Council of America, Memphis, TN, USA. pp. 375-380.

Doerge, T. 2001. Defining management zones for precision agriculture. Available online at http://www.pioneer.com/usa/technology/precision_farming_management_zones.htm.

Dupont, J.K., R. Campanella, M.R. Seal, J.L. Willers, and K.B. Hood. 2000. Spatially variable insecticide applications through remote sensing. *Proceedings, 2000 Beltwide Cotton Conferences,* January 4-8, San Antonio, TX, USA. National Cotton Council of America, Memphis, TN, USA. pp. 426-429.

Ebelhar, M.W. and J.O. Ware. 2001. Continued evaluation of the influence of soil parameters on cotton and corn yields as determined with GPS/GIS technology. *Proceedings, 2001 Beltwide Conferences,* January 9-13, Anaheim, CA, USA, National Cotton Council of America, Memphis, TN, USA. pp. 614-617.

Elms, M.K. and C.J. Green. 1997. Spatial variability of yield in irrigated cotton. *Proceedings, 1997 Beltwide Cotton Conferences.* January 6-10, New Orleans, LA, USA. National Cotton Council of America, Memphis, TN, USA. pp. 598-600.

Elms, M.K. and C.J. Green. 1998. Cotton yield variability and correlations between yield, previous yield, and soil properties. *Proceedings, 1998 Beltwide Coton Conferences,* January 5-9, San Diego, CA, USA. National Cotton Council of America, Memphis, TN, USA. pp. 630-636.

English, B.C. and R.K. Roberts. 2001. Findings from a focus group of precision farming technology adopters. *Proceedings 2001 Beltwide Cotton Conferences,* January 9-13, Anaheim CA, USA. National Cotton Council of America, Memphis, TN, USA. pp. 291-292.

Ferguson, R.B., G.W. Hergert, and J.S. Schepers. 1997. Strategies for site-specific nitrogen management. *Precision in Agriculture '97,* The 1st European Conference on Precision Agriculture, pp. 387-395.

Fitzgerald, G.J., S.J. Maas, and W.R. DeTar. 1999a. Detection of spider mites in cotton using multispectral remote sensing. *Proceedings, 17th Biennial Workshop on Color Photography and Videography om Research Assessment,* May 5-7, Reno, NV.

Fitzgerald, G.J., S.J. Maas, and W.R. DeTar. 1999b. Early detection of spider mites in cotton using multispectral remote sensing. *Proceedings, 1999 Beltwide Cotton Conferences,* January 3-7, Orlando, FL, USA, National Cotton Council of America, Memphis, TN, USA. pp. 1022-1024.

Fitzgerald, G.J., S.J. Maas, and W.R. DeTar. 2000. Multispectral multitemporal remote sensing for spider mite detection in cotton. (CD-ROM) *Proceedings, 5th International Conference on Precision Agriculture and Other Resource Management;* July 16-19, 2000; Radisson Hotel South, Bloomington, MN, USA. Agronomy Society of America, Crop Science Society of America, Soil Science Society of America, Madison, WI, USA.

Fitzgerald, G.J., S.J. Maas, and W.R. Detar. 2004. Spider mite detection and canopy component mapping in cotton using hyperspectral imagery and spectral mixture analysis. *Precision Agriculture,* 5:275-289.

Fitzgerald, G.J., P.J. Pinter Jr., D.J. Hunsaker, and T.R. Clarke. 2005. Multiple shadow fractions in spectral mixture analysis of a cotton canopy. *Remote Sensing of Environment,* 97:526-539.

Fleming, K.L., D.G. Westfall, D.W. Wiens, L.E. Rothe, J.E. Cipra, and D.F. Heerman. 1999. Evaluating farmer developed management zone maps for precision farming. *Precision Agriculture '99,* J.V. Stafford (ed.) Sheffield, UK. SCI, Sheffield Academic Press, pp. 917-926.

Fleming, K.L., D.G. Westfall, and W.C. Bausch. 2000. Evaluating management zone technology and grid soil sampling for variable rate nitrogen application. (CD-ROM) *Proceedings, 5th International Conference on Precision Agriculture and Other Resource Management;* July 16-19, 2000; Radisson Hotel South, Bloomington, MN, USA. Agronomy Society of America, Crop Science Society of America, Soil Science Society of America, Madison, WI, USA.

Flynn, R. 2001. Impact of grid sampling on traditional agronomic practices. *Proceedings, 2001 Beltwide Cotton Conferences,* January 9-13, Anaheim CA, National Cotton Council of America, Memphis, TN, USA. pp. 507-508.

Franzen, D.W., L.J. Cihacek, V.L. Hofamn, and L.J. Swenson. 1998. Topography-based sampling compared with grid sampling in the Northern Great Plains. *Journal of Production Agriculture,* 11:364-370.

Franzen, D.W., and T.R. Peck. 1995. Field soil sampling density for variable rate fertilization. *Journal of Production Agriculture,* 8:568-574.

Fridgen, J.J., C.W. Fraisse, N.R. Kitchen, and K.A. Sudduth. 2000. Delineation and analysis of site-specific management zones. Available online at http://www.fse.missouri.edu/mpac/pubs/zones.htm.

Gogineni, S., J.A. Thomasson, J. Iqbal, J.R. Wooten, B.M. Kolla, and R. Sui. 2004. Remote sensing input to GIS-integrated cotton growth model: Preliminary results. *Proceedings of SPIE—The International Society for Optical Engineering,* 5544, pp. 186-197.

Goovaerts, P. 1998. Geostatistical tools for characterizing the spatial variability of microbiological and physico-chemical soil properties. *Biol. Fertil. Soils,* 27: 315-334.

Gotway, C.A., R.B. Ferguson, G.W. Hergert, and T.A. Peterson. 1996. Comparison of kriging and inverse-distance methods for mapping soil parameters. *Soil Science Society of America. Journal,* 60:1237-1247.

Gvili, M. and J. Brightbill. 2001. Stripper cotton yield sensor produces yield map. *Proceedings, 2001 Beltwide Cotton Conferences,* January 9-13, Anaheim, CA, USA. National Cotton Council of America, Memphis, TN, USA. pp. 345-346.

Han, S., R.G. Evans, S.M. Schneider, and S.L. Rawlins. 1996. Spatial variability of soil properties on two center-pivot irrigated fields. *Proceedings, 3rd International Conference on Precision Agriculture and Other Resource Management;* July 23-26, 1996; Radisson Hotel South, Bloomington, MN, USA. Agronomy Society of America, Crop Science Society of America, Soil Science Society of America, Madison, WI, USA. pp. 207-216.

Heisel, T., S. Christiansen, and A.M. Walter. 1999. Whole-field experiments with site-specific weed management. *Precision Agriculture '99,* J.V. Stafford (ed.). Sheffield, UK. SCI, Sheffield Academic Press, pp. 759-768.

Horney, R.D., B. Taylor, D.S. Munk, B.R. Roberts, R.E. Plant, and S.M. Lesch. 2001. Development of site specific management for reclamation of salt affect soil utilizing preliminary #M-38 and core soil sample data. *Proceedings, 2001 Beltwide Cotton Conferences,* 9-13 January, Anaheim, CA, USA. National Cotton Council of America, Memphis, TN, USA. pp. 319-322.

Jallas, E., S. Turner, P. Martin, M. Cretenet, and R. Sequeira. 2000. Cottons-IMS: An insect and plant damage management system for cotton. *Proceedings, 2000 Beltwide Cotton Conferences,* January 4-8, San Antonio, TX, USA. National Cotton Council of America, Memphis, TN, USA. pp. 437-439.

Johnson, R.M. and J.M. Bradow. 2000. Potential for precision management of cotton fiber quality. (CD-ROM) *Proceedings, 5th International Conference on Precision Agriculture and Other Resource Management;* July 16-19, 2000; Radisson Hotel South, Bloomington, MN, USA. Agronomy Society of America, Crop Science Society of America, Soil Science Society of America, Madison, WI, USA.

Johnson, R.M., J.M. Bradow, P.J. Bauer, and E.J. Sadler. 1998. Spatial variability of cotton fiber properties. *Proceedings, 1998 Beltwide Cotton Conferences.* January 5-9, San Diego, CA, USA. National Cotton Council of America, Memphis, TN, USA. pp. 1465-1466.

Johnson, R.M., J.M. Bradow, P.J. Bauer, and E.J. Sadler. 1999a. Influence of soil spatial variability on cotton fiber quality. *Proceedings, 1999 Beltwide Cotton Conferences.* January 3-8, Orlando, FL, USA. National Cotton Council of America, Memphis, TN, USA. pp. 1319-1320.

Johnson, R.M., J.M. Bradow, P.J. Bauer, and E.J. Sadler. 1999b. Spatial variability of cotton fiber yield and quality in relation to soil variability. *Precision Agriculture, Proceedings 4th International Conference on Precision Agriculture,* July

19-22, 1998. St. Paul, MN, USA. Agronomy Society of America, Crop Science Society of America, Soil Science Society of America, Madison, WI, USA. pp. 487-497.

Johnson, R.M., R. Downer, J.M. Bradow, P.J. Bauer, and E.J. Sadler. 2002. Variability in cotton fiber yield, fiber quality and soil properties in a south eastern coastal plain. *Agronomy Journal* 94: 1305-1316.

Karlen, D.L., C.A. Cambardella, and T.S. Colvin. 1996. Soil-test variability in adjacent Iowa fields. *Proceedings, 3rd International Conference on Precision Agriculture and Other Resource Management;* July 23-26, 1996; Radisson Hotel South, Bloomington, MN, USA. Agronomy Society of America, Crop Science Society of America, Soil Science Society of America, Madison, WI, USA. pp. 237-243.

Kitchen, N.R., K.A. Sudduth, D.F. Hughes, and S.J. Birrell. 1995. Comparison of variable rate to single rate nitrogen application. Corn production and residual soil NO3-N. *Site-Specific Management of Agricultural Systems.* Agronomy Society of America, Crop Science Society of America, Soil Science Society of America, Madison, WI, USA. pp. 427-442.

LaMastus, F.E., C.M. Smith, D.R. Shaw, and R.L. King. 2000. Potential for weed species differentiation using remote sensing. (CD-ROM) *Proceedings, 5th International Conference on Precision Agriculture and Other Resource Management;* July 16-19, 2000; Radisson Hotel South, Bloomington, MN, USA. Agronomy Society of America, Crop Science Society of America, Soil Science Society of America, Madison, WI, USA.

Landivar, J.A. 1999. Use of simulation models in precision cotton farming. *Proceedings, 1999 Beltwide Cotton Conferences.* January 3-8, Orlando, FL, USA. National Cotton Council of America, Memphis, TN, USA. pp. 1384-1385.

Larson, J.A., R.K. Roberts, B.C. English, R.L. Cochran, and B.S. Wilson. 2005. A computer decision aid for the cotton yield monitor investment decision. *Computers and Electronics in Agriculture,* 48:216-234.

Layrol, L., E. Hedoin, D. Lepoutre, and O. François. 2000. Matching multitemporal yield and images data. (CD-ROM) *Proceedings, 5th International Conference on Precision Agriculture and Other Resource Management;* July 16-19, 2000; Radisson Hotel South, Bloomington, MN, USA. Agronomy Society of America, Crop Science Society of America, Soil Science Society of America, Madison, WI, USA.

Li, H., R.J. Lascano, J. Booker, K.F. Bronson, E.M. Barnes, L.T. Wilson, and E. Segarra. 2001. Temporal patterns of cotton reflectance and NDVI-days line yield modeling. *Proceedings, 2001 Beltwide Cotton Conferences,* January 9-13, Anaheim, CA, USA. National Cotton Council of America, Memphis, TN, USA. pp. 590-594.

Li, H., R.J. Lascano, J. Booker, K.F. Bronson, L.T. Wilson, and E. Segarra. 2000. Underlying field heterogeneity on water and nitrogen use in cotton: State-space analysis. (CD-ROM) *Proceedings, 5th International Conference on Precision Agriculture and Other Resource Management;* July 16-19, 2000; Radisson Hotel South, Bloomington, MN, USA. Agronomy Society of America, Crop Science Society of America, Soil Science Society of America, Madison, WI, USA.

Lough, J.L. and J.J. Varco. 2000. Effects of varying N and K nutrition on the spectral reflectance properties of cotton. (CD-ROM) *Proceedings, 5th International Conference on Precision Agriculture and Other Resource Management;* July 16-19, 2000; Radisson Hotel South, Bloomington, MN, USA. Agronomy Society of America, Crop Science Society of America, Soil Science Society of America, Madison, WI, USA.

Mallarino, A.P. and D.J. Wittry. 2000. Identifying cost-effective soil sampling schemes for variable-rate fertilization and liming. (CD-ROM) *Proceedings, 5th International Conference on Precision Agriculture and Other Resource Management;* July 16-19, 2000; Radisson Hotel South, Bloomington, MN, USA. Agronomy Society of America, Crop Science Society of America, Soil Science Society of America, Madison, WI, USA.

McCauley, J.D. 1999. Simulation of cotton production for precision farming. *Precision Agriculture,* 1:81-94.

Mulla, D.J. and A.C. Sekely. 2000. Evaluation of remote sensing and targeted sampling for variable rate application of lime. (CD-ROM) *Proceedings, 5th International Conference on Precision Agriculture and Other Resource Management;* July 16-19, 2000; Radisson Hotel South, Bloomington, MN, USA. Agronomy Society of America, Crop Science Society of America, Soil Science Society of America, Madison, WI, USA.

Pachepsky, T. and B. Acock. 1997. Holes in precision farming: Spatial variability of essential soil properties. *Precision Agriculture '97,* Proceedings, 1st European Conference on Precision Agriculture, September 7-19, 1997, Oxford, UK. BIOS Scientific Publishers, pp. 163-170.

Perry, C.D., G. Vellidis, N. Wells, and C. Kvien. 2001. Simultaneous evaluation of multiple commercial yield monitors in Georgia. *Proceedings, 2001 Beltwide Cotton Conferences,* January 9-13, Anaheim, CA, USA. National Cotton Council of America, Memphis, TN, USA. pp. 328-339.

Peterson, T. and N. Christensen. 2001. Remote sensing and its link to precision farming: Today's reality and tomorrow's potential. Available online at http://www.pioneer.com/usa/technology/remotesum.htm.

Peterson, T. and N. Wollenhaupt. 2001. Considerations in mapping soil properties or what about this grid-sampling? Available online at http://www.pioneer.com/usa/technology/i961004.htm.

Peterson, W.M. and J.J. Varco. 2001. Utilization of remote sensing to detect fertilizer nitrogen effects on cotton growth. *Proceedings, 2001 Beltwide Cotton Conferences,* January 9-13, Anaheim CA, National Cotton Council of America, Memphis, TN, USA. pp.568-569.

Ping, J. and C.J. Green. 1999. Spatial analysis of agronomic properties in two production cotton fields in West Texas. *Proceedings, 1999 Beltwide Cotton Conferences.* January 3-8, Orlando, FL, USA. National Cotton Council of America, Memphis, TN, USA. pp. 1286-1290.

Ping, J. and C.J. Green. 2000. Spatial variability of yield and soil parameters in two irrigated cotton fields in Texas. (CD-ROM) *Proceedings, 5th International Conference on Precision Agriculture and Other Resource Management;* July 16-19, 2000; Radisson Hotel South, Bloomington, MN, USA. Agronomy Society of

America, Crop Science Society of America, Soil Science Society of America, Madison, WI, USA.

Ping J. and C.J. Green. 2001. Identification of management zones based on cotton yield and soil parameters. *Proceedings, 2001 Beltwide Cotton Conferences,* January 9-13, Anaheim CA, National Cotton Council of America, Memphis, TN, USA. pp. 571-574.

Ping, J.L., C.J. Green, K.F. Bronson, R.E. Zartman, and A. Dobermann. 2004a. Identification of relationships between cotton yield, quality, and soil properties. *Agronomy Journal,* 96:1588-1597.

Ping, J.L., C.J. Green, K.F. Bronson, R.E. Zartman, and A. Dobermann. 2005. Delineating potential management zones for cotton based on yields and soil properties. *Soil Science,* 170:371-385.

Ping, J.L., C.J. Green, R.E. Zartman, and K.F. Bronson 2004b. Exploring spatial dependence of cotton yield using global and local autocorrelation statistics. *Field Crops Research,* 89:219-236.

Plant, R.E. 1997. A methodology for qualitative modeling of crop production systems. *Agricultural Systems,* 53:325-348.

Plant, R.E. and D.S. Munk. 1999. Application of remote sensing to irrigation management in California cotton. *Precision Agriculture, Proceedings 4th International Conference on Precision Agriculture,* July 19-22, 1998. St. Paul, MN, USA. Agronomy Society of America, Crop Science Society of America, Soil Science Society of America, Madison, WI, USA. pp. 1511-1522.

Plant, R.E., D.S. Munk, B.R. Roberts, R.L. Vargas, D.W. Rains, R.L. Travis, and R.B. Hutmacher. 2000. Relationships between remotely sensed reflectance data, Acala cotton growth, and yield. *Proceedings, 2000 Beltwide Cotton Conferences,* January 4-8, San Antonio, TX, USA. National Cotton Council of American, Memphis, TN, USA. pp. 429-432.

Plant, R.E., D.S. Munk, B.R. Roberts, R.N Vargas, D.W. Rains, R.L. Travis, and R.B. Hutmacher. 2001. Application of remote sensing to strategic cotton management. *Journal of Cotton Science,* 5:30-41.

Pocknee, S., B.C. Boydell, H.M. Green, D.J. Waters and C.K. Kvien. 1996. Directed Soil Sampling. *Proceedings, 3rd International Conference on Precision Agriculture and Other Resource Management;* July 23-26, 1996; Radisson Hotel South, Bloomington, MN, USA. Agronomy Society of America, Crop Science Society of America, Soil Science Society of America, Madison, WI, USA. pp. 159-168.

Porter, P.M., J.G. Lauer, D.R. Huggins, E.S. Oplinger, and R.K. Crookston. 1998. Assessing spatial and temporal variability in corn and soybean yields. *Journal of Production Agriculture,* 11:359-363.

Reddy, K.R., D. Zhao, V.G. Kakani, J.J. Read, and K. Sailaja. 2003. Estimating cotton growth and developmental parameters through remote sensing. *Proceedings of SPIE—The International Society for Optical Engineering,* 5153, pp. 277-288.

Reed, J. 2001. Dollars in and dollars out. *Cotton Farming.* 45(36):38-39. Available online at http://www.cottonfarming.com/home/2001-MayCF-precisionag.html.

Reynolds, D.B., L.M. Bruce, J.C. Sanders, and E.L. Sanders. 2001. Spectral classification and identification of weed and crop species. *Proceedings, 2001 Beltwide*

*Cotton Conferences,* January 9-13, Anaheim, CA, USA. National Cotton Council of America, Memphis, TN, USA. p. 1230.

Roades, J.P., A.D. Beck, and S.W. Searcy. 2000. Cotton yield mapping: Texas experiences in 1999. *Proceedings, 1999 Beltwide Cotton Conferences,* January 4-8, San Antonio, TX, USA. National Cotton Council of America, Memphis, TN, USA. pp. 404-410.

Ruble, M.K. and S.W. Searcy. 2000. A GIS interface for cotton growth models that incorporates midseason adjustments. *Proceedings, 2000 Beltwide Cotton Conferences,* January 4-8, San Antonio, TX, USA. National Cotton Council of America, Memphis, TN, USA. pp. 456-459.

Roberts, R.K., B.C. English, and J.A. Larson. 2001. Cotton production in relation to the probability of precision farming technology adoption in Tennessee counties. *Proceedings, 2001 Beltwide Cotton Conferences,* January 9-13, Anaheim, CA, USA. National Cotton Council of America, Memphis, TN, USA. pp. 292-294.

Rydberg, A., and M. Söderström. 2000. Potential crop growth assessment from remotely sensed images compared to ordinary yield maps. (CD-ROM) *Proceedings, 5th International Conference on Precision Agriculture and Other Resource Management;* July 16-19, 2000; Radisson Hotel South, Bloomington, MN, USA. Agronomy Society of America, Crop Science Society of America, Soil Science Society of America, Madison, WI, USA.

Sadler, E.J., W.J. Busscher, P.J. Bauer, and D.L. Karlen. 1998. Spatial scale requirements for precision farming: A case study in the southeastern USA. *Agronomy Journal,* 90:191-197.

Sanders, J.C., L.M. Bruce, and D. B. Reynolds. 2001. Crop monitoring and defoliation timing via spectral data. *Proceedings, 2001 Beltwide Cotton Conferences,* January 9-13, Anaheim, CA, USA. National Cotton Council of America, Memphis, TN, USA. pp. 1230-1231.

Sassenrath, G.F., E.R. Adams, and J.R. Williford. 2005. Rapid sampling system for determination of cotton fiber quality spatial variability. *Applied Engineering in Agriculture,* 21(1):9-14.

Sassenrath-Cole, G.F., S.J. Thomson, J.R. Williford, K.B. Hood, J.A. Thomasson, J. Williams, and D. Woodward. 1999. Field testing of cotton yield monitor. *Proceedings, 1999 Beltwide Cotton Conferences,* January 3-8, Orlando, FL, USA. National Cotton Council of America, Memphis, TN, USA. pp. 364-372.

Seal, M.R., K. Dupont, M. Bethel, D. Lewis, J. Johnson, J.L. Willers, K.Hood, J. Hardwick, R. Leonard, and R. Bagwell. 2001. Utilization of remote sensing technologies in the development and implementation of large-scale spatially-variable insecticide experiments in cotton. *Proceedings, 2001 Beltwide Cotton Conferences,* January 9-13, Anaheim, CA, USA. National Cotton Council of America, pp. 1010-1018.

Searcy, S.W., A.D. Beck, and J.P. Roades. 2001. Cotton yield mapping: Texas experiences in 2000. *Proceedings, 2001 Beltwide Cotton Conferences,* January 9-13, Anaheim, CA, USA. National Cotton Council of America, Memphis, TN, USA. pp. 342-344.

Stermitz, R.J., G.A. Nielsen, and D.S. Long. 1999. Testing quality of soil survey geographic (SSURGO) database for precision farming. *Precision Agriculture,*

*Proceedings 4th International Conference on Precision Agriculture,* July 19-22, 1998. St. Paul, MN, USA. Agronomy Society of America, Crop Science Society of America, Soil Science Society of America, Madison, WI, USA. pp. 319-326.

Stevens, G, S. Hefner, and C. Moylan. 1997. Variable rate lime application with global positioning system. *Proceedings, 1997 Beltwide Cotton Conferences* January 6-10, New Orleans, LA, USA. National Cotton Council of America, Memphis, TN, USA. pp. 603-605.

Stevens, G., and C. Moylan. 1998. Effect of variable rate lime on cotton. *Proceedings, 1998 Beltwide Cotton Conferences,* January 5-9, San Diego, CA, USA. National Cotton Council of America, Memphis, TN, USA. p. 670.

Stevens, G. and C. Moylan. 1999. Effect of variable rate lime on cotton yields. *Proceedings, 1999 Beltwide Cotton Conferences,* January 3-8, Orlando, FL, USA. National Cotton Council of America, Memphis, TN, USA. pp. 1285-1286.

Stewart, C.M. and A.B. McBratney. 2000. Development of a methodology for the variable-rate application of fertilizer in irrigated cotton fields. (CD-ROM) *Proceedings, 5th International Conference on Precision Agriculture and Other Resource Management;* July 16-19, 2000; Radisson Hotel South, Bloomington, MN, USA. Agronomy Society of America, Crop Science Society of America, Soil Science Society of America, Madison, WI, USA.

Sudbrink, D.L. Jr., F.A. Harris, J.T. Robbins, and P.J. English. 2001. Remote sensing and site specific management of cotton arthropods in the Mississippi Delta. *Proceedings, 2001 Beltwide Cotton Conferences,* January 9-13, Anaheim, CA, USA. National Cotton Council of America, Memphis, TN, USA. pp. 1145-1147.

Sudduth, K.A., S.T. Drummond, S.J. Birrell, and N.R. Kitchen. 1997. Analysis of spatial factors influencing crop yield. *Precision Agriculture '97, Proceedings, 1st European Conference on Precision Agriculture,* September 7-19. pp. 439-447.

Sui, R. and J.A. Thomasson. 2001. Field testing of Mississippi State University cotton yield monitor. *Proceedings, 2001 Beltwide Cotton Conferences,* January 9-13, 2001, Anaheim, CA, USA. National Cotton Council of America, Memphis, TN, USA. pp. 339-342.

Sui, R., J.A. Thomasson, R. Mehrle, M. Dale, C. Perry, and G. Rains. 2004. Mississippi cotton yield monitor: Beta test for commercialization. *Computers and Electronics in Agriculture,* 42(3):149-160.

Sui, R., J.B. Wilkerson, W.E. Hart, L.R. Wilhelm, and D.D. Howard. 2005. Multispectral sensor for detection of nitrogen status in cotton. *Applied Engineering in Agriculture,* 21:167-172.

Swindell, J.E.G. 1997. Mapping the spatial variability in the yield potential of arable land through GIS analysis of sequential yield maps. *Precision Agriculture '97, Proceedings, 1st European Conference on Precision Agriculture,* 7-19 September 1997, Oxford, UK. BIOS Scientific Publishers, pp. 827-834.

Thomasson, J.A., J. Chen, J.R. Wooten, and S.A. Shearer. 2001. Remote sensing for predicting cotton yield. *Proceedings, 2001 Beltwide Cotton Conferences,* January 9-13, Anaheim, CA, USA. National Cotton Council of America, Memphis, TN, USA. pp. 317-318.

Thomasson, J.A., J. Chen, J.R. Wooten, S.A. Shearer, and D.A. Pennington. 2000. Cotton yield prediction improvement with remote sensing. *Proceedings, 2000 Beltwide Cotton Conferences,* January 4-8, San Antonio, TX, USA. National Cotton Council of America, Memphis, TN, USA. pp. 419-421.

Thomasson, J.A., and R. Sui. 2003. Mississippi cotton yield monitor: Three years of field-test results. *Applied Engineering in Agriculture,* 19:631-636.

Thurman, M.E. and R.W. Heiniger. 1998. Using GPS to scout cotton for variable rate pix (mepiquat chloride) application. *Precision Agriculture, Proceedings 4th International Conference on Precision Agriculture,* July 19-22, 1998. St. Paul, MN, USA. Agronomy Society of America, Crop Science Society of America, Soil Science Society of America, Madison, WI, USA. pp. 953-961.

Tucker, C.J. 1979. Red and photographic infrared linear combinations for monitoring vegetation. *Remote Sensing of Environment,* 8:127-150.

USDA Natural Resources Conservation Service. 2000a. National Soil Survey Center. Available at http://www.ftw.nrcs.usda.gov/ssur_data.html.

USDA Natural Resources Conservation Service. 2000b. Soil Survey Geographic Database. Available at http://www.statlab.iastate.edu/soils/nssc/.

Vellidis, G., C.D. Perry, G.C. Rains, D.L. Thomas, N. Wells, and C.K. Kvien. 2003. Simultaneous assessment of cotton yield monitors. *Applied Engineering in Agriculture,* 19:259-272.

Vellidis, G., M.A. Tucker, C.D. Perry, D.L. Thomas, N. Wells, and C.K. Kvien. 2004. Predicting cotton lint yield maps from aerial photographs. *Precision Agriculture,* 5:547-564.

Wanjura, D.F., and D.R. Upchurch. 2001. Remote sensing of cotton grown under multiple water and nitrogen levels. *Proceedings, 2001 Beltwide Cotton Conferences,* January 9-13, Anaheim, CA, USA. National Cotton Council of America, Memphis, TN, USA. pp. 311-317.

Wanjura, D.F., D.R. Upchurch, and J.R. Mahan. 2000. Yield and applied water relationships in multiyear drip irrigated cotton production. *Proceedings, 2000 Beltwide Cotton Conferences,* January 4-8, San Antonio, TX, USA. National Cotton Council of America, Memphis, TN, USA. pp. 442-448.

Ward, B.D., and M.S. Cox. 2000. Influence of soil chemical and physical properties on site-specific cotton production. (CD-ROM) *Proceedings, 5th International Conference on Precision Agriculture and Other Resource Management;* July 16-19, 2000; Radisson Hotel South, Bloomington, MN, USA. Agronomy Society of America, Crop Science Society of America, Soil Science Society of America, Madison, WI, USA.

Warrick, A.W., D.E. Myers, and D.R. Nielsen. 1986. Geostatistical methods applied to soil science. *Methods of soil analysis, part 1, 2nd Ed., Physical and mineral logical methods.* Agronomy monograph, 9:53-82.

Werner, A., S. Döllin, A. Jarfe, J. Kühn, J. Pauly, and R. Roth. 2000. Deriving maps of yield potentials with crop models, site information and remote sensing. (CD-ROM) *Proceedings, 5th International Conference on Precision Agriculture and Other Resource Management;* July 16-19, 2000; Radisson Hotel South, Bloomington, MN, USA. Agronomy Society of America, Crop Science Society of America, Soil Science Society of America, Madison, WI, USA.

Whelan, B.W., A.B. McBratney, and R.A. Viscarra Rossel. 1996. Spatial prediction for precision agriculture. *Proceedings, 3rd International Conference on Precision Agriculture and Other Resource Management;* July 23-26, 1996; Radisson Hotel South, Bloomington, MN, USA. Agronomy Society of America, Crop Science Society of America, Soil Science Society of America, Madison, WI, USA. pp. 331-342.

Wilkerson, J.B., R. Sui, W.E. Hart, and D.E. Howard. 1998. Development of a variable-rate nitrogen applicator based on spectral reflectance characteristics of cotton plants. *Proceedings, 1998 Beltwide Cotton Conferences,* January 5-9, San Diego, CA, USA. pp. 439-443.

Willers, J.L., J.N. Jenkins, W.L. Ladner, P.D. Gerard, D.L. Boykin, K.B. Hood, P.L. Mckibben, and M.M. Bethel. 2005. Site-specific approaches to cotton insect control. Sampling and remote sensing analysis techniques. *Precision Agriculture,* 6:431-452.

Willers, J.L., M.R. Seal, and R.G. Luttrell. 1999. Remote sensing, line intercept sampling for tarnished plant bugs (Heteroptera; Miridae) in Mid-South cotton. *Journal of Cotton Science* 3:160:176. Available online at http://www.jcotsci.org.

Wolak, F.J., A. Khalilian, R.B. Dodd, Y.J. Han, M. Keshkin, R.M. Lippert, and W. Hair. 1999. Cotton yield monitor evaluation, South Carolina—Year 2. *Proceedings, 1999 Beltwide Cotton Conferences.* January 3-8, Orlando, FL, USA. National Cotton Council of America, Memphis, TN, USA. pp. 361-364.

Yang, C., J.H. Everitt, and J.M. Bradford. 2004. Using high resolution QuickBird satellite imagery for cotton yield estimation. *ASAE Annual International Meeting 2004,* pp. 893-904.

Yang, C., J.H. Everitt, J.M. Bradford, and D. Murden. 2004. Airborne hyperspectral imagery and yield monitor data for mapping cotton yield variability. *Precision Agriculture,* 5:445-461.

Yang, C., S.M. Greenberg, and J.H. Everitt. 2005. Evaluation of herbicide-based cotton regrowth control using remote sensing. *Transactions of the American Society of Agricultural Engineers,* 48:1987-1994.

Yang, C., S.M. Greenberg, J.H. Everitt, T.W. Sappington, and D.E. Escobar. 2001. Assessment of simulated boll weevil damage using airborne digital imaging. *Proceedings, 2001 Beltwide Cotton Conferences,* January 9-13, Anaheim, CA, USA. National Cotton Council of America, Memphis, TN, USA. pp. 1196-1198.

Yu, M., E. Segarra, S. Watson, H. Li, and R.J. Lascano. 2001. Precision farming practices in irrigated cotton production in the Texas High Plains. *Proceedings, 2001 Beltwide Cotton Conferences,* January 9-13, Anaheim, CA, USA. National Cotton Council of America, Memphis, TN, USA. pp. 201-208.

Zhao, D.H., J.L. Li, and J.G. Qi. 2005. Identification of red and NIR spectral regions and vegetative indices for discrimination of cotton nitrogen stress and growth stage. *Computers and Electronics in Agriculture,* 48:155-169.

Zhao, D., K.R. Reddy, V.G. Kakani, J.J. Read, and S. Koti. 2005. Selection of optimum reflectance ratios for estimating leaf nitrogen and chlorophyll concentrations of field-grown cotton. *Agronomy Journal,* 97:89-98.

Chapter 13

# Potential of Precision Farming with Potatoes

Colin McKenzie
Shelley A. Woods

## *INTRODUCTION*

Global Positioning (Systems) technology (GPS) makes it possible for farmers to measure yields and responses to management practices within their own fields. With GPS it is possible to take numerous measurements of yield or environmental conditions, combine them with computer-aided data collection, and prepare maps of yield or site characteristics. Precision-farming technology developed in North America and Australia because of a need to optimize returns from inputs such as fertilizers (Davenport and Hattendorf, 2000; Hoskinson et al., 2000) and pesticides. In northwestern Europe, precision farming was further driven by environmental concerns, such as excess nutrients and chemicals resulting from intensive agricultural practices affecting water quality.

Potato production involves costly inputs, such as seed, fertilizer, irrigation, and pesticides, as well as labor and equipment for land preparation, seeding and harvesting, and thus is ideally suited to precision farming. Potato production has often been associated with environmental problems such as soil erosion and water contamination with nutrients (Williams et al., 1999) and pesticides. Potatoes are a high-value crop and this value is based on yield and quality that are both sensitive to crop management and environmental conditions. Precision agriculture provides a means to integrate production system information for potatoes (Hess and Hoskinson, 1996; Schneider et al., 1997) and to apply best management practices (Williams et al., 2000) to areas within one potato field.

---

M. Bunney, S. Day, L. Hingley, R. Howard, M. Konschuh, and C. Murray assisted with preparation and editing of the manuscript.

*Handbook of Precision Agriculture*

doi:10.1300/5627_13

The approach to precision farming can be either high tech or low tech (Bouma et al., 1995). The high tech-approach involves GPS and field sampling to apply different rates of inputs to various parts of the field according to a GPS-controlled prescription map. The low-tech approach requires sampling of various parts of the field and subdividing the field so different rates of inputs are supplied by conventional means.

In Canada precision farming with potatoes has not been widely adopted; however, there are some examples. In the province of New Brunswick, Canada, about 3,000 ha of potatoes were planted in fields that received variable-rate fertilizer applications based on 0.4 ha grid soil samples, and about 100 ha were yield monitored in 2000. Variable-rate fertilizer application has also been used in the province of Manitoba and other parts of Canada.

In the province of Alberta, Canada, the area seeded to potatoes has increased from 8,000 ha in 1984 to 22,000 ha in 2001. The size of farms has also increased with about 20 farms growing over 250 ha of potatoes and some with over 500 ha. These larger farms often hire staff to operate the field equipment. A farm with a 500 ha crop of potatoes requires 1,500 to 2,000 ha of land to provide for a three-to four-year rotation and some of this land may be rented. This means the decision maker may not have a detailed knowledge of every field. Aids such as maps of yield, soil fertility, texture, salinity, pH, and weed and pest incidence can be used to develop a crop management program for these potato farms.

## *NUTRIENT MANAGEMENT*

Farmers frequently add more fertilizer than the potatoes remove from the soil in one season. This results in inefficient use of fertilizer and may contribute to soil and water contamination from excess nutrients leached or eroded from the soil. The goal of nutrient management is to obtain optimum economic production while minimizing environmental damage. This may require synchronizing nutrient supply with crop demand by use of controlled-release fertilizers or by adding fertilizers as needed, during the season through banding or fertigation. Hoskinson et al. (2000) described a project to develop an economically optimum variable-rate fertilizer application system for potatoes.

### *Nitrogen*

Nitrogen (N) in the nitrate form ($NO_3$) does not bind to soil particles and readily leaches from the root zone to groundwater. On coarse-textured soils, N deficiencies in crops can develop quickly. For site-specific management of N, soil texture and soil hydrologic characteristics need to be determined (Verhagen et al., 1995). Verhagen (1997) and Watkins et al. (1998), state that temporal variability of N is much larger than spatial variability, making split applications of N fertilizer a prerequisite for site-specific N management. By using soil texture to establish management zones in a barley- potato rotation, precision farming techniques prevented nutrient contamination of water in Colorado (Delgado and Duke, 2000). Whitley et al. (2000), and Watkins et al. (1998), also found that nitrogen leaching to groundwater in potato fields could be reduced by using site-specific crop management and variable-rate application technology.

Spatial variability of potato petiole N was of sufficient magnitude to warrant development of in-season N management zones and this variability was correlated with sand, clay, and organic matter content (King et al., 1999). In experiments at Brooks in southern Alberta, Canada, McKenzie et al. (2000) found potato tissue $NO_3$-N in the first week of July to be significantly correlated with soil $NO_3$-N (0.0-0.60 m) from the previous October on three of six irrigated fields (Figure 13.1). Those sites where tissue N was significantly correlated with soil N should have benefited from site-specific applications of N, however, they were not easy to identify before adding fertilizer. When available nitrate N from all six fields was compared to tissue N, there was no significant relationship (Figure 13.1b). However, tissue N was highly significantly positively correlated to soil clay content (Figure 13.1c). This occurred on soils which ranged from 6 to 36 percent clay. The clay restricted loss of nitrogen by leaching. On coarse-textured soils, knowledge of clay content will assist in determining precision application rates of N and timing of N application to improve N efficiency and reduce N losses.

### *Phosphorus*

Potatoes are sensitive to phosphorus (P) deficiencies because they are inefficient in absorbing P from the soil. Phosphorus fertilizer application rates for potatoes vary according to soil test P and to free lime or active iron and aluminum. In Idaho, a soil with less than 5 percent $CaCO_3$ with 15 ppm P (Na $HCO_3$ extractant) for 0.0-0.30 m requires no addition of P fertilizer. If the soil has 10 percent $CaCO_3$, it requires 59 $kg{\cdot}ha^{-1}$ P and with 15 percent or more $CaCO_3$, 117 $kg{\cdot}ha^{-1}$ P will be required (Tindall et al., 1991). In a site-specific experiment in eastern Washington State, Davenport et al. (1999), found that soil pH was more closely correlated to P content of potato petioles than soil P or soil $CaCO_3$, and that soil pH in conjunction with soil P may provide the best basis for P fertilizer recommendations for potatoes. Johnson et al. (2000), in Australia, found high rates of P needed to

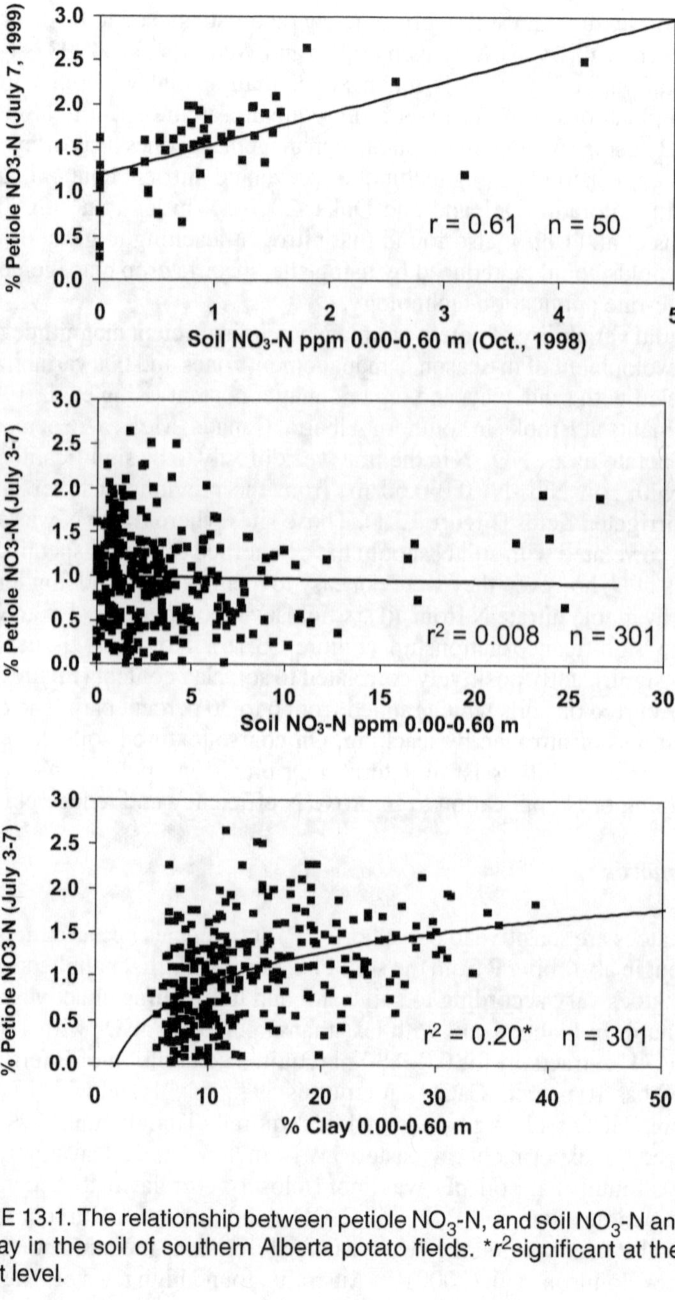


FIGURE 13.1. The relationship between petiole $NO_3$-N, and soil $NO_3$-N and percent clay in the soil of southern Alberta potato fields. *$r^2$significant at the 0.01 percent level.

be applied to potatoes (up to ten times crop removal) on red ferresol soils. On all types of soils, it is necessary to know the P sorption characteristics of the soil, as well as soil P content, when designing precision application rates of phosphorus.

Phosphorus and potassium fertilizers on potatoes were applied in strips at variable rates, conventional rates, and a control without any fertilizer in the province of Quebec, Canada (Cambouris et al., 1999). The researchers obtained a 6 percent decrease in yield in the first year and an increase of 7 percent in the second year for variable-rate application as compared to conventional rate. They attributed the increase in the second year to a more detailed soil sampling grid, which permitted more precise application of fertilizers than in the first year. The differences between soil series in yield potential were significant in both years.

In southern Alberta, Canada, in a precision agriculture experiment in a commercial potato field, correlations between soil P in October (before uniform applications of P fertilizer) and petiole P in the following July were determined and found to be significant in three fields out of six (McKenzie et al., 2000). On an 800 m long, 32 ha field, hog manure had been applied in 1997 to all but the northern 250 m (Figure 13.2). The entire field received

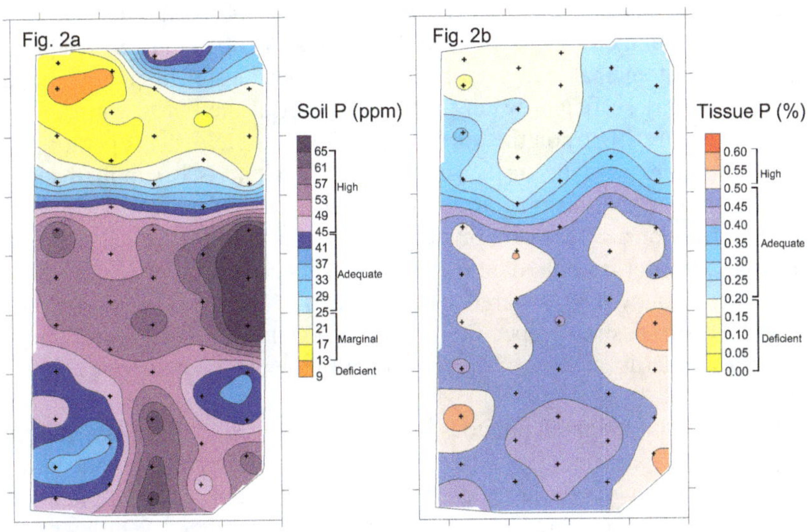


FIGURE 13.2. Soil Kelowna phosphorus (0-15 cm) (October 1998) and petiole phosphorus (July 28, 1999) for FL1625 potatoes. (See corresponding PLATE 13.2 in the central gallery.)

39 kg·ha$^{-1}$ of P fertilizer in the fall of 1998 and 29 kg·ha$^{-1}$ in the spring of 1999. The portion that did not receive hog manure had mostly adequate petiole P levels while the portion which had received hog manure had mostly high petiole P levels. A precision fertilizer application would have been useful as 68 kg·ha$^{-1}$ P was not needed on the southern part of the field and was a sufficient amount on the northern part of the field.

### *Water*

In three years of a site-specific management project with six different fields of potatoes in southern Alberta, Canada, McKenzie et al. (2002) found uniformity of irrigation applications to be a significant factor influencing yield of potatoes. In three fields, the amount of irrigation and precipitation were highly significantly correlated with increasing tuber yield, and in one field it was significantly correlated with decreasing tuber yield. Increasing irrigation and precipitation on one field was highly significantly correlated with increasing mean tuber weight, and on another field highly significantly correlated with decreasing mean tuber weight.

Problems with lack of uniformity of application of water were encountered with irrigation systems (McKenzie et al., 2000) (Figures 13.3 and 13.4). The example in 1996 (Figure 13.3a) shows a 400 m long center-pivot irrigation system that was operating at less than designed pressure and failed to supply sufficient water to the outer part of the field. A new pivot in 1997 (Figure 13.3b) on the other half of the same field, was operating at excess pressure or with an incorrect system design, as it supplied more water to the outside portion of the pivot than the center. A problem with lack of uniformity on a corner-arm pivot system is shown in Figure 13.4a where insufficient water is applied on the corners. This caused reduced quality of the tubers as shown by lowered specific gravity (1.055-1.070) on the corners and more desirable specific gravity (1.065-1.080) to meet minimum contract standards on most of the remainder of the pivot potato area (Figure 13.4b).

In a report on irrigated potatoes in Idaho, King et al. (1999), concluded, "Irrigation management has a big influence on petiole $NO_3$-N concentration. Thus, soil water-holding capacity and cumulative water application may be useful in defining in-season N management zones." Whitley et al. (2000) in eastern Washington, similarly concluded that, "Precision water application (both irrigation and N fertigation) have the potential to further reduce N leaching potential while maintaining crop yield and quality in irrigated potato."

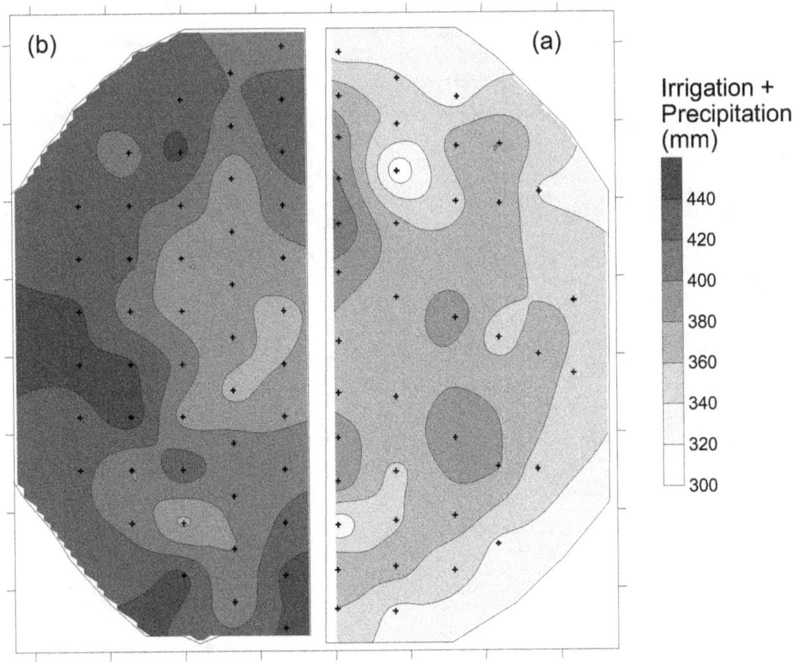


FIGURE 13.3. Total irrigation and precipitation (mm) for Snowden potatoes *(b)* June 23-Sept. 4, 1997; *(a)* June 25-Sept. 9, 1996. (See corresponding PLATE 13.3 in the central gallery.)

An engineering firm (Precision Irrigation) in Idaho, is developing a programmable variable-rate irrigation system that can be attached to a center-pivot system (King et al., 1995). This system would permit an application of variable-rate irrigation according to a prescription map. The prescription map could be based on soil texture, topographic slope, and aspect information or could be developed from thermal infrared imagery of the crop canopy. Such a system, if successful, should increase yield, improve quality, reduce nitrogen losses to groundwater, and improve the efficiency of crop uptake of nitrogen fertilizer applications.

## *REMOTE SENSING*

According to Deguise et al. (1999), with the launch of high-spectral and spatial resolution satellites, it will be possible to use the information from

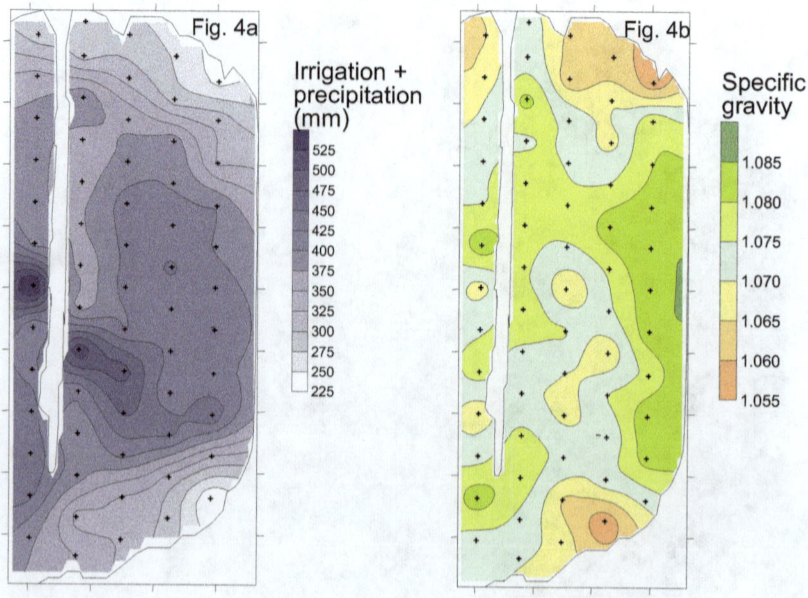


FIGURE 13.4. Total irrigation and precipitation (mm) June 17-August 31, 1999, and specific gravity (September 1998) for Russett Burbank potatoes. (See corresponding PLATE 13.4 in the central gallery.)

these satellites in a timely fashion for the detection of stress, the prediction of yield, and as a diagnostic tool for recurring low productivity areas for numerous crops. Deguise et al. (1999) found that geocorrected airborne imagery of potatoes taken 2,600 m above the ground gave a good correlation with wet biomass of potatoes. This was also closely correlated with dry biomass and leaf area. They state that this method would be more valuable if the location of ground sampling sites in the noncorrectal image could be improved. Field sampling should also match the pixel size.

McKenzie et al. (2002) found a highly significant correlation between potato petiole nitrate N and compact airborne spectral images (CASI) (Figure 13.5) collected in the visible and near-infrared spectrum at 51 locations on a pivot-irrigated field. Petiole $NO_3$-N was more closely correlated ($R^2 =$ 0.55) to leaf area index, a measure of crop vigor, than to chlorophyll concentration ($R^2 = 0.28$).

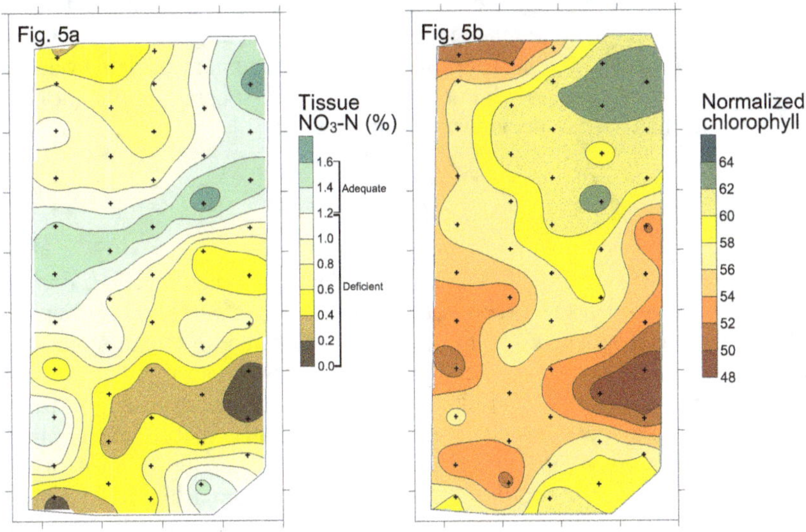


FIGURE 13.5. Petiole nitrate nitrogen (July 28, 1999) and normalized plant chlorophyll content (July 28, 1999) for FL1625 potatoes. (See corresponding PLATE 13.5 in the central gallery.)

## *NEMATODES*

In some areas, plant parasitic nematodes are a major problem facing potato growers. In the United Kingdom, about 40 percent of the potato fields are infested with the potato cyst nematodes *Globodera pallida* and *Globodera rostochiensis,* and about 20 percent of all fields are treated with nematicides each year at a cost of 350 to 500 £$ha^{-1}$ (Stephens et al., 2000). The variability of distribution of nematodes is sufficient to justify spatially varied application of nematicides controlled by GPS systems. This would serve to reduce cost and environmental risk from application of nematicides. The occurrence of the nematodes requires intensive and expensive ground sampling of a grid of 20 m or less (Evans et al., 1998). Nematodes reproduce rapidly when their density is low, which means threshold values for application of nematicides must be set very low (Stafford and Evans, 2000). In Wisconsin, root lesion nematodes *(Pratylenchus penetrans)* were found to require intensive sampling to identify critical densities for site-specific management (Morgan et al., 2000). Remote sensing can be used to identify nematode-damaged crops and select areas for further intensive sampling for site-specific management (Stephens et al., 2000). The potential in the United Kingdom for site-specific management of nematodes in potatoes is being evaluated but large-scale site-specific application of nematicides has not yet been carried out.

## INSECTS

Major insect pests of potato include the Colorado potato beetle *(Leptinotarsa decemlineata)* and green peach aphid *(Myzus persicae).* Foliar applications of chemicals, up to 11 or 12 per season, are the main method of controlling Colorado potato beetle in some areas of the northeastern United States (Weisz et al., 1996). Populations of these species of insect pests are aggregated. Site-specific management might reduce insecticide applications (Weisz et al., 1996). Pesticide resistance is making control of the Colorado potato beetle and the green peach aphid more difficult. Pest management techniques that preserve susceptible genotypes and allow them to interbreed with resistant genotypes have the greatest potential for creating nonresistant insect populations. One way to accomplish this is through the preservation of refugia for susceptible individuals by leaving some habitats untreated (Roush and Daly, 1990). A site-specific pest management program might accomplish this by limiting pesticide applications to specific in-field areas where pest populations are above economic thresholds. This might be relevant to Colorado potato beetle resistance management in the northeastern United States where there is an absence of noncrop hosts. It might also preserve natural pests of the beetle (Weisz et al., 1996). Fleischer et al. (1999) in Pennsylvania, discuss the possibilities of using site-specific applications of insecticide for Colorado potato beetle because the beetles tend to occur first on the edges of potato fields.

## WEED CONTROL

Eberlein et al. (1999) in Idaho, describe the use of an automated irrigation control system in which eight different herbicides were applied to various parts of a field through the irrigation system (herbigation). Herbicide rates applied were all within 5 percent or less from the desired rate. Excellent weed control was achieved when various rates of weeds were seeded into a potato field and herbicide rates were adjusted according to the weed population. This suggests a good potential for site-specific herbigation in fields divided into management zones according to weed population.

## *HARVEST AND SEEDING EQUIPMENT*

Yield mapping equipment for a commercial harvester is described by Rawlins et al. (1995). This equipment efficiently weighs, by means of load cells, the material coming over the harvester belt and positions each weight to the part of the field it came from with GPS (Schneider et al., 1996). A correction is applied to each position to account for spatial displacement by the harvester. This method also weighs the soil and rocks that are carried on the belt along with potato tubers affecting the reliability of the data. Hess and Hoskinson (1996) describe how field weights were credited to a truck, and each truckload was also weighed. They collected about two samples per ha to measure "tare rocks and dirt" and also to grade potatoes for quality, size, frost damage, disease, and specific gravity. Campbell (1999) describes developments such as real time and transfer of data. This permits the operator to have early feedback on yields and the opportunity to recognize equipment malfunctions. He also points out the need for sensors to measure tare such as soil and rocks. Measurements of mass flow on a potato harvester obtained with a bouncing plate were compared with measurements obtained with a digital camera and optical sensor (Algerbo and Ehlert, 2000). Both methods worked but failed to discriminate between potatoes and additional material on the harvester belt. They are continuing with evaluation of the camera in Sweden and Germany.

Godwin et al. (1999) describe the use of a trailer equipped with load cells to collect material from a potato or sugar beet harvester. The equipment recorded the rate of mass flow into the trailer as well as the total weight of each load. Accurate yield maps were prepared. A yield monitor for sweet potatoes that has an image acquisition system consisting of a digital camera is described by Wooten et al. (2000). The system has the ability to distinguish sweet potato images from background material with an accuracy as high as $R^2 = 0.97$. Culls were differentiated from marketable sweet potatoes with an accuracy as high as $R^2 = 0.84$. This machine has not yet been incorporated into a sweet potato yield-monitoring system. Such a machine has the potential to improve yield monitoring and to include quality monitoring. Other types of machines that identify yield by means of bulk density of material on the harvester belt have been tested but are not yet marketed.

Hess et al. (1999) describe a seed-placement system that is monitored by GPS which was able to detect 0.05 m deviations from a planned seed placement of 0.20 m. This system offers the possibility for adapting seed place adjustment to suit each soil type, seed lot, and planter. Uniform seed placement should improve yield and quality.

## SOIL SALINITY

In areas where soil salinity is a hazard, marginal levels of salinity may reduce the growth of potatoes or make harvesting difficult due to poor soil structure. Salinity can be mapped in detail using a sensor which does not require direct contact with the soil such as an electromagnetic induction meter (EM38) and GPS methods (Cannon et al., 1994). This information can be used to adapt cropping choices to salinity characteristics of the field (McKenzie et al., 1997). Saline portions of a potato field could be planted to barley or other crops that are more tolerant to salinity and less expensive to grow than potatoes.

## FIELD SCALE EXPERIMENTATION

Soil fertility has been widely recognized and tested in field scale experiments of precision farming. Soil conservation experiments in Prince Edward Island, Canada, were evaluated by yield monitoring the potato crop (DeHaan et al., 1999). Significant relationships between tuber yield and tillage method, cation exchange capacity, depth of solum, and soil-loss index were determined. These types of relationships would be difficult to establish through conduct of conventional replicated plot experiments.

## PROBLEMS HINDERING THE ADOPTION OF PRECISION FARMING BY THE POTATO INDUSTRY

- It is often difficult to collect yield information that is not appreciably affected by contamination with soils and rocks. This makes yield maps unreliable or only suitable for certain soils. New techniques offer promise of improving the reliability of data collection.
- When rotations of three or four years are used for potatoes, this means information collected on a potato field is not useful until the next cycle of the rotation reoccurs on that field. Potato information may need to be combined with information collected on other crops grown in the rotation. These other crops may not respond to inputs in the same way that potatoes do.
- Fertilizer applications can be done with site-specific techniques which are now available. It is more difficult to collect the soil information needed to prepare the prescription maps.
- Irrigation systems that are able to supply different rates of water to meet variable soil conditions such as drainage, runoff, and water-

holding capacity are not commonly available; although prototypes have been used.

- The need for people with both the computer and diagnostic skills required to make and interpret site-specific information maps limits adoption of precision farming.
- Industry is most interested in measurements of quality such as size, shape, and specific gravity of tubers, but it is difficult to collect this quality data with available site-specific technology.
- The value of precision agriculture techniques in nutrient management is found in increased efficiency, improved production, and improved management of environmental risks from excess fertilizer use. The value of improved efficiency by optimum rather than excess fertilizer use is often small (Killian, 2000). The environmental risks associated with excess fertilizer use are often not recognized in economic terms so response to this receives little attention except in areas where major environmental problems have developed or where laws have been put in place to control fertilizer use, e.g., areas of the Netherlands.
- Remote sensing with accurate georeferencing and quick feedback is not readily available on a commercial scale. If available, this would offer a source of information on which to base site-specific management practices.

## *CONCLUSIONS*

Precision farming with potatoes has developed slowly. The problems that hinder its development are being resolved. The potential for use of precision farming with potatoes is large because of the high inputs in pesticides, fertilizers, water, and tillage. Site-specific application of these inputs may improve yield, quality, or profit. These same inputs also pose risks to environmental contamination and these risks can be reduced by use of precision farming techniques.

## REFERENCES

Algerbo, P.A. and D. Ehlert. 2000. Evaluation of yield sensing systems for potato harvesters. In *Proceedings of the 5th International Conference on Precision Agriculture.* ASA-CSSA-SSSA, Madison, WI, USA. CD ROM.

Bouma, J., J. Brouwer, A. Verhagen, and H.W.G. Bootink. 1995. Site specific management on field level: High and low tech approaches. In *Eco-regional approaches for sustainable land use and food production.* J. Bouma et al. (eds.) Netherlands, Kluwer Academic Publishers. pp. 453-473.

Cambouris, A.N., M.C. Nolin, and R.R. Simard. 1999. Precision management of fertilizer phosphorus and potassium for potato in Quebec, Canada. In *Proceedings of the 4th International Conference on Precision Agriculture.* ASA-CSA-SSSA, Madison, WI, USA. pp. 847-857.

Campbell, R.H. 1999. Recent advances in yield monitoring of conveyor harvested crops. In *Proceedings of the 4th International Conference on Precision Agriculture.* ASA-CSSA-SSSA, Madison, WI, USA. pp. 1101-1106.

Cannon, M.E., R.C. McKenzie, and G. Lachapelle. 1994. Mapping soil salinity with electromagnetic induction and satellite navigation methods. *Canadian Journal of Soil Science* 74: 335-343.

Davenport, J.R. and M.J. Hattendorf. 2000. Using site-specific approaches to advance potato management in irrigated system. *Horticulture Technology* 10 (3): 452-457.

Davenport, J.R., M.J. Hattendorf, R.G. Evans, S. Han, and S.M. Schneider. 1999. Phosphorus fertilizer management strategies for variable rate fertilizer applications in center pivot irrigated potato rotations. In *Proceedings of 4th International Conference on Precision Agriculture.* ASA-CSSA-SSSA. Madison, WI, USA. pp. 745-752.

Deguise, J.C., H. McNairn, K. Staenz, and M. McGovern. 1999. Spatial high resolution crop measurements with airborne hyperspectral remote sensing. In *Proceedings of 4th International Conference on Precision Agriculture.* ASA-CSA-SSSA, Madison, WI, USA. pp. 1603-1608.

DeHaan, K.R., G.T. Vessey, D.A. Holmstrom, J.A. MacLeod, J.B. Sanderson, and M.R. Carter. 1999. Relating potato yield to the level of soil degradation using a bulk yield monitor and differential global positioning systems. *Computers and Electronics in Agriculture* 23: 133-143.

Delgado, J.A. and H.R. Duke. 2000. Potential use of precision farming to improve nutrient management on an irrigated potato-barley rotation. In *Proceedings of 5th International Conference on Precision Agriculture.* ASA-CSSA-SSSA. Madison, WI, USA. CD ROM.

Eberlein, C.V., B.A. King, and M.J. Guttieri. 1999. Weed management with site specific herbigation. In *Proceedings of the 4th International Conference on Precision Agriculture.* ASA-CSA-SSSA, Madison, WI, USA. pp. 869-877.

Evans, K., J. Stafford, R. Webster, P. Halford, M. Russell, A. Barker, and S. Griffin. 1998. Mapping potato cyst nematode populations for modulated applications of nematicide. *Aspects of Applied Biology* 52: 101-108.

Fleischer, S.B., P. Blam, D. Emmen, and H. Hower. 1999. *Dynamics in the spatial continuity of insect density.* ASA-CSA-SSSA, Madison, WI, USA. pp. 133-139.

Godwin, R.J., P.N. Wheeler, M.J. O'Dogherty, C.D. Watt, and T. Richards. 1999. Cumulative mass determination for yield maps of non-grain crops in computers and electronics. *Agriculture* 23: 85-101.

Hess, J.R. and R.L. Hoskinson. 1996. Methods for characterization and analysis of spatial and temporal variability for researching and managing integrated farming systems. In *Precision Agriculture.* ASA-CSA-SSSA, Madison, WI, USA. pp. 641-650.

Hess, R.J., J.M. Svoboda, R.L. Hoskinson, D.W. Hempstead, and W.B. Jones. 1999. Spatial potato seed piece placement monitoring system. In *Proceedings of the 4th International Conference on Precision Agriculture.* ASA-CSA-SSSA, Madison, WI, USA. pp. 653-660.

Hoskinson, R.L., R.K. Fink, R.J. Hess, R.A. Oborn, T.A. Tindall, and L.D. Robinson. 2000. Using the decision support system for agriculture (DSS4Ag) for potato fertilization. In *Proceedings of 5th International Conference on Precision Agriculture.* ASA-CSSA-SSSA, Madison, WI, USA. CD ROM.

Johnson, P., L. Sparrow and R. Doyle. 2000. Improving phosphorus efficiency on red soils. In *Proceedings of the Australian Potato Research Conference,* Potatoes 2000, SARDI, Adelaide, SA. Australia. pp. 121-124.

Killian, B. 2000. Economic aspects of precision farming: A German viewpoint. In *Proceedings of the 5th International Conference on Precision Agriculture.* ASA-CSA-SSSA, Madison, WI, USA. CD ROM.

King, B.A., R.A. Brady, L.R. McCann, and J.C. Stark. 1995. Variable rate water application through sprinkler irrigation. In *Site Specific Management for Agricultural Systems.* ASA-CSSA-SSSA. pp. 485-493.

King, B.A., J.C. Stark, and J.P. Taberna Jr. 1999. In season spatial variability of potato petiole nitrogen. In *Proceedings of 4th International Conference on Precision Agriculture.* ASA-CSSA-SSSA, Madison, WI, USA. pp. 55-65.

McKenzie, R.C., R.J. George, S.A. Woods, M.E. Cannon, and D.L. Bennett. 1997. Use of the electromagnetic induction meter as a tool in managing salinization. *Hydrology Journal* 5(1): 37-50.

McKenzie, R.C., C. Schaupmeyer, T. Goddard, D.C. Penney, T. Harms, and M. Green. 2002. Site specific management of potatoes. Unpublished report. Alberta Agriculture Food and Rural Development, Brooks, AB, Canada.

McKenzie, R.C., S.A. Woods, L.E. Hingley, T.W. Goddard, J. Rodvang, and T. Harms. 2000. Site specific management of potatoes. In *Proceedings of 5th Annual Conference on Precision Agriculture.* ASA-CSSA-SSSA, Madison, WI, USA. CD ROM.

Morgan, G.D., L.K. Binning, and A.E. MacGuidwin. 2000. Spatial variability of root lesion nematodes *Pratylenchus penetrans* in potatoes. In *Proceedings of 5th International Conference on Precision Agriculture.* ASA-CSSA-SSSA, Madison, WI, USA. CD ROM.

Rawlins, S., G.C. Campbell, R.H. Campbell, and J.R. Hess. 1995. Yield mapping of potato in site specific management for agricultural systems. In *Site specific management for agricultural systems.* ASA-CSSA-SSSA, Madison, WI, USA. pp. 59-68.

Roush, R.T. and J.C. Daly. 1990. The role of population genetics in resistance research and management. In *Pesticide resistance in arthropods.* R.T. Roush and B.E. Tabishnik (eds.). New York: Chapman and Hall. pp. 97-152.

Schneider, S.M., R.A. Boydston, S. Han, R.C. Evans, and R.H. Campbell. 1997. Mapping of potato yield and quality. In *Precision agriculture.* BIOS Scientific Publishers Ltd. pp. 253-261.

Schneider, S.M., S.L. Rawlins, S. Han, R.G. Evans, and R.H. Campbell. 1996. Precision agriculture for potatoes in the Pacific Northwest. In *Precision agriculture.* ASA-CSSA-SSSA, Madison, WI, USA. pp. 443-452.

Stafford, J.V. and K. Evans. 2000. Spatial distribution of potato cyst nematode and the potential for varying nematicide application. In *Proceedings of 5th International Conference on Precision Agriculture.* ASA-CSSA-SA, Madison, WI, USA. CD ROM.

Stephens, M., G.A. Wood, and J.C. Taylor. 2000. Opportunities for targeted control of potato cyst nematode using remote sensing. *Aspects of Applied Biology* 60: 213-218.

Tindall, T.A., D.T. Westerman, J.C. Stark, J.C. Ojala, and G.E. Kleinkopf. 1991. *Phosphorus nutrition of potatoes.* Current Information Series #903. Univ. Idaho College of Agric. Co-op Extension System, USA. 4 pp.

Verhagen, A., H.W.G. Booltink, and J. Bouma. 1995. Site specific management: Balancing production and environmental requirements at farm level. *Agricultural System,* 49: 369-384.

Verhagen, J. 1997. Modeling soil and crop responses in a spatially variable field. In *Precision agriculture.* BIOS Scientific Publishers Ltd. pp. 197-203.

Watkins, K.B., Y.C. Lu, and W.Y. Huang. 1998. Economic returns and environmental feasibility of variable rate nitrogen fertilizer and water applications with carry-over effects. *Journal of Agricultural and Resource Economics* 23(2): 401-426.

Weisz, R., S. Fleischer, and Z. Smilowitz. 1996. Site specific integrated pest management for high-value crops: Impact on potato pest management. *Journal of Economic Entomology* 89(2): 501-509.

Whitley, K.M., J.R. Davenport, and S.R. Manley. 2000. Differences in nitrate leaching under variable and conventional nitrogen fertilizer management in irrigated potato systems. In *Proceedings of the 5th International Conference on Precision Agriculture.* ASA-CSSA-SSSA, Madison, WI, USA. CD ROM.

Williams, C.M.J., K. Jackson, and N.A. Maier. 2000. Agronomy overview—Best management practices and future directions. In *Proceedings of the Australian Potato Research Conference.* Potatoes 2000, SARDI, Adelaide, SA Australia. pp. 125-134.

Williams, C.M.J., M.L. Vitosh, N.A. Maier, and D.K.L. MacKerron. 1999. Nutrient management strategies for sustainable potato (*Solanum tuberosum* L.) production systems in the Southern and Northern hemispheres. In *Solanaceae IV: Advances in biology and utilization.* M. Nee, D.E. Symon, R.N. Lester and J.P. Jessop (eds.). Kew, UK. Royal Botanic Gardens, pp. 443-458.

Wooten, J.R., J.G. White, J.A. Tomasson, and P.G. Thompson. 2000. *Yield and quality monitor for sweet potato with machine vision.* Paper # 001123, ASAE Annual Meeting. Milwaukee, WI, USA. July 9-12, 2000.

Chapter 14

# Site-Specific Nitrogen Management in Sugarbeet

David W. Franzen

## *PROPERTIES OF SUGARBEET FAVORABLE TO SITE-SPECIFIC NUTRIENT MANAGEMENT*

In some sugarbeet growing areas, such as the intermountain areas of the United States, successful sugarbeet production means high root yield. In the Red River Valley of Minnesota and North Dakota, growers also have an economic stake in the profitability of sugar processing, since they own the cooperative processing plants that produce sugar in addition to providing raw sugarbeet inputs. Successful sugarbeet production is the result of a combination high sugarbeet yield, high beet sugar content, and low levels of impurities resulting in high recoverable sugar. Undesirable impurities are amino-N content, sodium, and potassium, resulting in increased loss of sugar to molasses. Loss to molasses may be as low as 2 percent or as high as 20 percent of sugar yield. Molasses is about 50 percent sucrose (Hobbis, 1978).

Nitrogen management is key to higher root and recoverable sugar yields and low levels of impurities, (Draycott, 1972; Hobbis, 1973). Several mathematical models have been used to illustrate the response of sugarbeet yield with N application, including a Mitscherlich approach (Crowther and Yates, 1941), a parabolic form (Boyd, 1961), and the form used in the Red River Valley, the linear plateau response (Boyd et al., 1970; Dahnke and Johnson, 1990; Cattanach et al., 1992). From North Dakota and Minnesota databases, sugarbeet yield response is best characterized as linear to a point at which the environment dictates no further response to N is warranted. The current sugarbeet N recommendation equation in the region is as follows:

*Handbook of Precision Agriculture*

doi:10.1300/5627_14

N recommendation (kg·ha$^{-1}$) = 6.1 × Yield Goal (Mg·ha$^{-1}$), less credits due to residual soil N and previous crop

High soil N results in greater sugar beet yield, but it also results in less recoverable sugar and more unrecoverable sugar and amino-N content, which increases processing costs. Table 14.1 illustrates the reduction in recoverable sugar and increase in amino-N content as N rate increases. It is therefore very important that enough N is available for high yields, but the N levels are low enough that late-season plant energies encourage sugar storage and not continued foliar growth, which results in lower sugar content in the roots.

One method to determine whether high N levels are present in the plant is the use of a petiole nitrate test. Ulrich (1950) suggested a critical level, early in the season, of 1,000 ppm in the petiole, which should be allowed to drop below 1,000 ppm about 11 weeks (in California) prior to harvest. In the Northern Plains, the rule is to let petiole nitrate levels drop below 1,000 ppm within about 6 weeks of harvest (Cattanach, 2001, personal communication). Essentially, this threshold means that sugarbeet needs to be slightly N deficient during the final weeks of growth in order to increase sugar storage prior to harvest. To limit the amount of N to the sugarbeet crop, but supply enough to provide for root and sugar yield, soil sampling and nitrate-N analysis is recommended prior to sugarbeet planting. In North Dakota and northwest Minnesota, sampling is usually conducted in the fall because, during the long winter, the soils are deeply frozen, limiting N movement and transformations. Some spring sampling is also conducted, especially on sandy loam or coarser soils and on soils that periodically flood in the early spring. Nitrogen is typically determined from the 0-120 cm depth (Franzen

TABLE 14.1. Sugarbeet yield, recoverable sugar and amino-N content with N rate—Mean of six sugarbeet varieties.

| N rate (kg·ha$^{-1}$) | Yield (Mg·ha$^{-1}$) | Sugar (percent) | Gross sugar (kg·ha$^{-1}$) | Recoverable sugar (kg·ha$^{-1}$) | Amino-N (mg·kg$^{-1}$) |
|---|---|---|---|---|---|
| 168 | 38.3 | 15.4 | 6,233 | 4,949 | 935 |
| 280 | 38.3 | 15.1 | 6,263 | 4,924 | 825 |
| 392 | 39.4 | 14.7 | 6,055 | 4,628 | 1,116 |

*Source:* Adapted from Smith et al., 1980.

and Cihacek, 1998). However, significant N can also be extracted from even deeper depths. Evidence of deep soil N extraction is provided by Rudolph et al. (1980) in which recoverable sugar yield decreased as soil nitrate levels increased at the 0-150 cm depth. Smith (1980) suggests that deep soil sampling may be required to explain the influence of residual N in about 25 percent of Red River Valley fields.

Sugarbeet leaves may contain large amounts of N. Moraghan and Smith (1994) presented a range of N levels present in sugarbeet tops from 72-336 $kg \cdot ha^{-1}$ at harvest. Draycott (1972) points out the N releasing value of sugarbeet tops due to their high N content. Crohain and Rixhon (1967) showed that crops subsequent to sugarbeet might benefit from N released from sugarbeet top residue. The entire crop rotation is important nutritionally, given that certain crops might contribute N to the sugarbeet crop not seen in normal soil sampling, and crops subsequent to sugarbeet may benefit from N mineralized during sugarbeet residue decomposition. It is possible from these studies to envision a soil test report from a field following sugarbeet with low levels of residual soil nitrate-N, but with sugarbeet leaves containing hundreds of kilograms of N.

Abshahi et al. (1984) followed the application of labeled fertilizer N through a sugarbeet crop and into the following wheat crop. Using sugarbeet residues, application of fertilizer N to wheat was optimized with 62 $kg \cdot ha^{-1}$ N compared to 124 $kg \cdot ha^{-1}$ with no sugarbeet residues. They concluded that sugarbeet residues may contain high levels of N and should be considered a possible N source for subsequent crops.

Moraghan and Smith (1996) applied sugarbeet tops as one would apply a manure application. Two types of sugarbeet top residue applications were made prior to a spring wheat crop. One residue type contained lower sugarbeet N content (14.8 $mg \cdot kg^{-1}$), and the other a higher N content (34.8 $mg \cdot kg^{-1}$) applied at rates of 64 $kg \cdot ha^{-1}$ and 271 $kg \cdot ha^{-1}$, respectively. The high sugarbeet N residues contributed N similar to the yield response of 135-180 $kg \cdot ha^{-1}$ of N as urea, illustrating that sugarbeet tops should be considered a potential N source to subsequent crops. The study suggested that yellow or light green tops at harvest be considered as contributing 22 $kg \cdot ha^{-1}$ N to subsequent crops and green canopied sugarbeet at harvest as contributing at least 70 $kg \cdot ha^{-1}$ N. From these studies, it is clear that nitrogen nutrition in sugarbeet must consider the entire rotation given the cyclic nature of nitrogen uptake, utilization, and mineralization of residues from and to the sugarbeet crop.

## ZONE MANAGEMENT OF NUTRIENTS

Patterns of nutrient levels can be revealed through grid sampling or by using zone sampling. Grid sampling reveals patterns through dense, systematic, unbiased sampling. Several studies (Wollenhaupt et al., 1994; Franzen and Peck, 1995; Ferguson et al., 1996) have studied the density of grid sampling needed to reveal areas of more homogeneous nutrient levels that would make variable-rate fertilizer application with commercial applicators having swaths of around 20-30 m practical. Wollenhaupt et al. (1994) and Franzen and Peck (1995) both concluded that a grid size of 60-66 m was needed to reveal these areas. A grid density of 60-66 m is impractical for sugarbeet production. In the U.S. Corn Belt, sampling is conducted once every three to four years using a core depth of 0-15 cm and analysis for P, K, and soil pH. Nitrogen sampling for sugarbeet must be conducted each year after most crops to ensure that high N levels are not accumulated prior to beet production. Sample core depth is 0-120 cm. This makes the time spent in the field much greater than Corn Belt sampling and the type of equipment required substantially more expensive. Most consultants use hydraulic samplers outfitted onto specialized pickup trucks. Sampling costs for a 0-120 cm core and up to eight cores per sample in the Red River Valley have recently been estimated at $45 US per grid point.

While studying sampling density required in North Dakota for variable-rate fertilizer application (Franzen et al., 1998), it was noticed that similar patterns of nitrate-N appeared in each year, often regardless of crop or rate of N applied (Figure 14.1). This finding was contrary to that which was expected. There was a previous misconception that patterns of $NO_3$-N would not be stable over years due to the mobile nature of the nutrient. If high areas were found in one place in the field one year, they might be found somewhere else the next.

The similar patterns suggest an underlying factor controlling the patterns of $NO_3$-N in the field. After determining the topography of the field through elevation measurement, it was found that the $NO_3$-N patterns were similar to topography (Figure 14.2). Higher N values were found in depressions and swales, while lower N values were found on ridges and slopes. Evidently, mobile nutrients move in the soil in response to water flow, both surface and subsurface, but the water always moves in the same manner (Figure 14.3). If the field had been sampled based on topography zones, only four to six samples would have been required to give similar residual N information as a 33 m grid (Franzen et al., 1998).

Zone sampling assumes that nutrient patterns are related to some logical, predictable reason. It also makes site-specific nutrient management, espe-

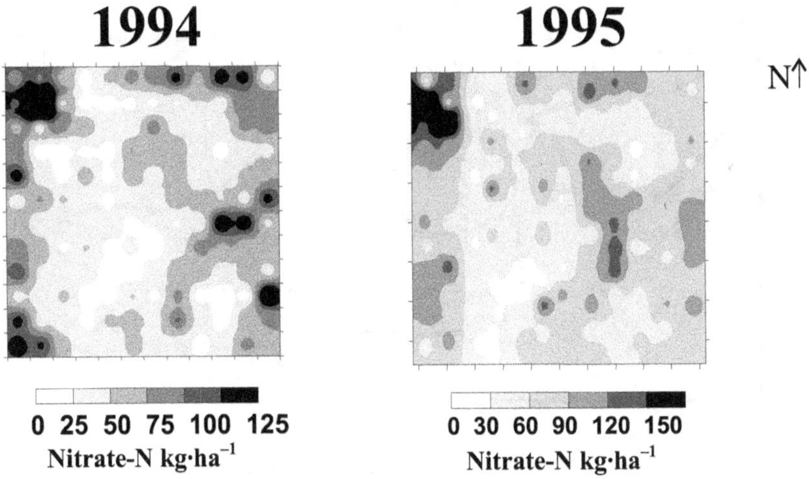


FIGURE 14.1. Soil $NO_3$-N levels 1994 following spring wheat and 1995 following sunflower, Valley City, North Dakota. Patterns expressed through a 33 m sampling grid to 60 cm in depth.

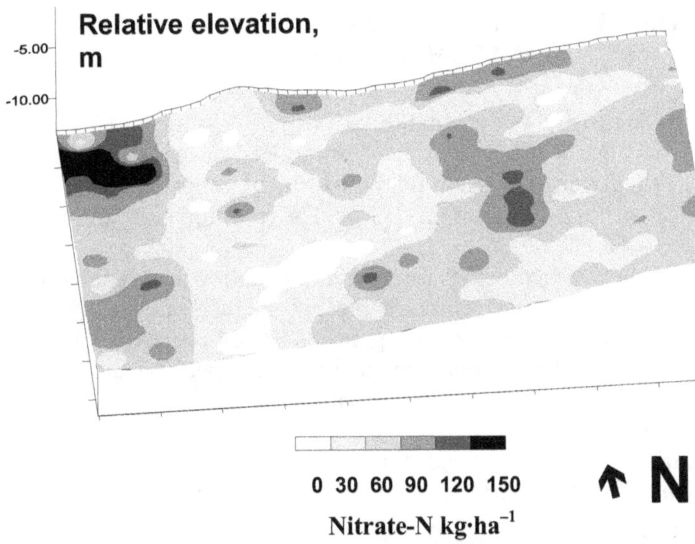


FIGURE 14.2. Soil $NO_3$-N levels overlain onto topography, Valley City, North Dakota.

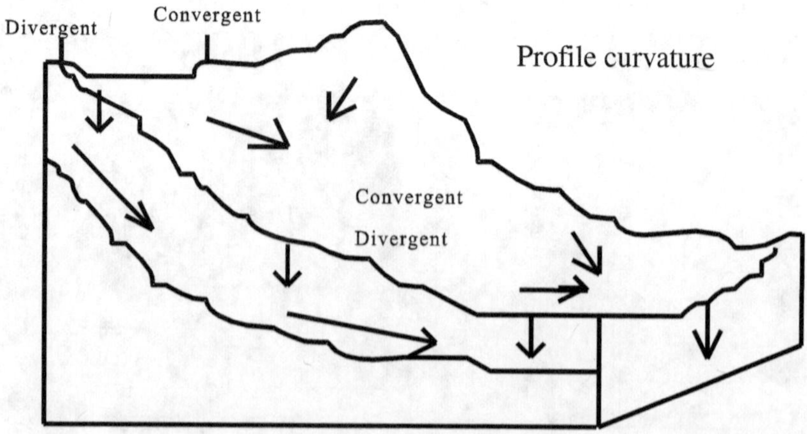


FIGURE 14.3. Basic water flow over and through a landscape. *Source:* Adapted from Ruhe, 1960.

cially nitrogen management, practical in the Northern Plains of the United States.

## *PROFITABILITY OF USING SITE-SPECIFIC NITROGEN MANAGEMENT IN SUGARBEET*

Studies comparing the profitability of using site-specific nitrogen management on sugarbeet began in 1994 in the Red River Valley. Smith and Rains (1995) divided a 60-acre field into 15-acre strips. Each strip was divided into five three-acre sampling grids. The resulting field study is shown in Figure 14.4.

Subtracting estimated costs for grid soil sampling and variable-rate application from the gross income of the two methods of fertilization, the study resulted in a mean net income gain of $177.84 per ha over the conventional, uniform fertilizer application method. Increased revenue for variable rate was the result of increased gross tonnage, sugar content, and decreased loss to molasses.

A similar study was repeated in 1995 (Smith and Rains, 1996). This study also utilized strip trials on a 27 ha field, divided into four alternate strips as in 1994, variable and conventional, but with only one target rate of N, not two as in 1994. Each variable-rate strip was divided into 6, 1.1 ha grid cells and fertilized according to the grid sample analysis from that grid. The conventional was fertilized based on forty soil cores taken randomly from the entire field. The result was an increase of \$111.15 per ha for the variable-rate strips over the net return from the conventional method. The increased revenue resulted from increases in gross tonnage as well as sugar content. A summary of Smith and Rains (1995; 1996) is given in Table 14.2.

Another strip trial in Minnesota was conducted in 1995 by Cattanach et al. (1996). Alternate strips of conventional and variable with two replications were divided into 1.4 ha grids in a 33 ha field. Conventional application was based on the mean $NO_3$-N level from the grid sampling. Net return for the variable-rate method was \$125.97 per ha compared to the conven-

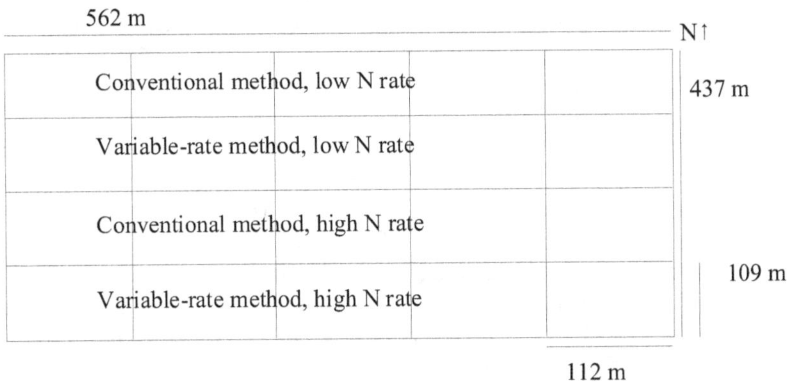


FIGURE 14.4. 60-acre field divided into test strips and 3 acre sampling grids. *Source:* Adapted from Smith and Rains, 1995.

TABLE 14.2. Summary of strip trials, 1994-1996, advantage of variable-rate compared to conventional methods of nitrogen application.

| Year | Yield (Mg·ha$^{-1}$) | Sucrose (percent) | Loss to molasses percent | Recoverable sugar (kg·ha$^{-1}$) | Net return, (\$ per ha) |
|---|---|---|---|---|---|
| 1994 | 2.71 | 0.33 | 0.11 | 642 | 177.84 |
| 1995 | 2.11 | 0.40 | 0.04 | 486 | 118.56 |
| 1996 | 5.87 | –0.08 | 0.03 | 838 | 111.15 |
| Mean | 3.56 | 0.23 | 0.06 | 655 | 135.85 |

*Source:* Smith and Rains, 1995, 1996.

tional method. The increased revenue was the result of increased sugar content (14.9 percent compared to 14.8 percent), yield (18,891 $kg \cdot ha^{-1}$ compared to 17,782 $kg \cdot ha^{-1}$), and decreased loss to molasses (1.43 percent compared to 1.70 percent).

The first studies to show that a simple grid using soil sampling and recommendations in the traditional manner may not be adequate for some fields was conducted in 1995 by Cattanach et al. (1996). Two fields were examined. The first was a strip trial near Felton, Minnesota, on a 9.7 ha field. The field was divided into six strips, 134 m long by 22 m wide, with three treatments: variable rate based on soil testing alone, variable rate based on soil testing with an organic matter adjustment for the grid, and a conventional application based on 30 random soil cores taken from within the study field. The organic matter adjustment was based on a 11 $kg\ N \cdot ha^{-1}$ addition for each percent organic matter less than 3 percent, or a decrease of 11 $kg\ N \cdot ha^{-1}$ for each percent organic matter greater than 3 percent. There was little difference between variable-rate methods in net return. The mean net advantage to variable return was $32.11 per ha. The grower cooperator yield and quality (from data obtained from full truck weights of sugarbeets taken to the factory and the quality samples from those deliveries) also showed a $32.11 per ha advantage of variable-rate over conventional, but also showed a $133.38 per ha net advantage of variable-rate modified by organic matter credits over conventional.

The second field studied by Cattanach et al. (1997) was near Foxholme, Minnesota. This field was grid sampled in a 1.8 ha grid and fertilized using a simple variable rate compared to conventional method. The soil samples were only obtained from the 0-60 cm depth, as opposed to the 120 cm depth usually taken. Although the soil sample nitrate results supported the rates used in the study, the yield and quality at harvest showed a net loss of $602.68 per ha due to variable-rate application. Plant populations were lower, sugar content was similar, and yields were lower in the variable-rate treatments. Apart from the plant population differences, there was concern that the use of relatively large grids may not be appropriate for some fields in directing variable-rate N application. These strip trials showed that improvements in sugarbeet quality were possible through site-specific technology, but only if patterns of N availability were revealed and all N available to the sugarbeet crop was considered.

The important principle binding nitrogen nutrition with sugarbeet production is to provide enough N fertilization to keep root yields high, but avoid excess N that might increase impurity levels and lower sugar content. In the past, N fertilization was based solely on a 0-120 cm soil test. Recently, it has become apparent that controlling and considering the nitrogen supply throughout the sugarbeet rotation both in residue and in all rooting depths of the soil is important if yield and quality are to be maximized.

In one region in the northern end of the Red River Valley, in a sugarbeet, wheat, and potato rotation, sugarbeet yields were high, recoverable sugar levels were low, and impurities were high in sugarbeet in a sugarbeet, wheat, and potato rotation. Growers soil tested, but did not see high levels of N to 120 cm. Variable rate N application based on the 120 cm sampling in a 1.6 ha grid did not improve recoverable sugar content (Reitmeier et al., 1999), showing again that grids of this size may not explain the variability in the field, nor is the use of variable rate N application always beneficial if all sources of N are not considered.

To determine the cause of the problem, the fields were sampled to 180 cm in depth in a 0.2 ha grid and sugarbeet tops were obtained at harvest at the same grid points and analyzed for N content. The result was that following sugarbeet, soil N levels were low. However, sugarbeet tops contained 220-440 kg N·ha$^{-1}$. Aerial photography and satellite imagery (SPOT and Landsat 5) revealed areas that were high in N (see Figure 14.5).

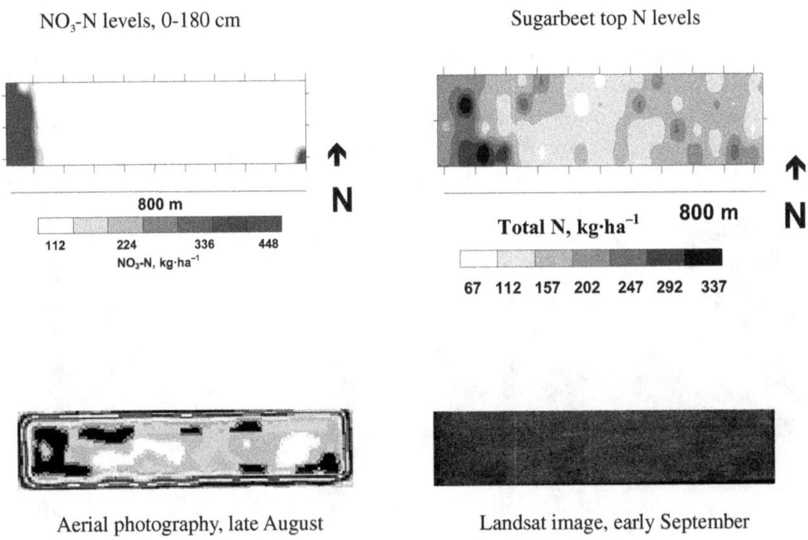


FIGURE 14.5. Soil N levels to 180 cm following sugarbeet, sugarbeet top N content at harvest, aerial photography and Landsat 5 satellite imagery showing areas of high N. Darker colors in each image indicate areas of higher N, and lighter colors correspond to lower N. (See corresponding PLATE 14.5 in the central gallery.)

Sampling in a 0.2 ha grid was also conducted on two 12.5 ha potato fields in the same areas as the sugarbeet fields. These fields were scheduled to be planted to sugarbeet the following year. The sampling indicated that some credit would need to be given for the 0-120 cm $NO_3$-N levels. However, there were also high levels of N at the 120-180 cm depth that normally would not be considered in a nitrogen fertilizer recommendation (Figure 14.6). The amount of N in the 120-180 cm depth was between 10 and 515 $kg \cdot ha^{-1}$.

Additional N was considered from the potato tops (Figure 14.7). Although the amount of total N in the tops only ranged from 11-70 $kg \cdot ha^{-1}$, compared to much higher levels after sugarbeet, the mineralization of a portion of the tops would be expected considering the N content of the tops and estimated C:N ratio of about 10:1. After consideration of the N in potato tops and the N present below 120 cm, it was decided to apply no supplemental N for sugarbeet for the following year.

The following year, sugarbeets were grown in the field with no supplemental N. Yields were 20 $Mg \cdot ha^{-1}$ with 17.2 percent sugar content, which was over 1 percent higher in sugar than the field had achieved in recent memory of the cooperator, and, according to American Crystal agriculturists, 1.5 percent higher sugar than any fields in a similar rotation in the area. To gather the correct residual N information following potatoes, a 120-180 cm depth increment should be added to the sampling recommendation in areas with historic poor sugarbeet quality. However, sampling need not be intensive or dense to obtain the necessary deep N information. Figure 14.8 shows areas in the west with very high vigor. If a small number of cores

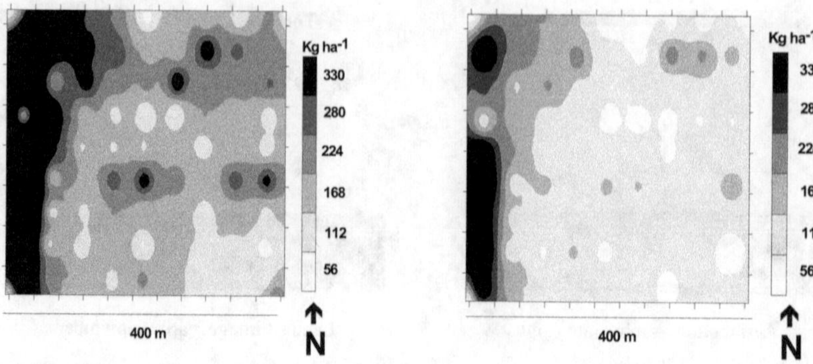


$NO_3$-N, 0-120 cm $NO_3$-N, 0-180 cm

FIGURE 14.6. Soil NO3-N levels following potatoes preceding sugarbeet in a sugarbeet-spring-wheat-potato rotation. *Source:* Adapted from Reitmeier et al., 1999.

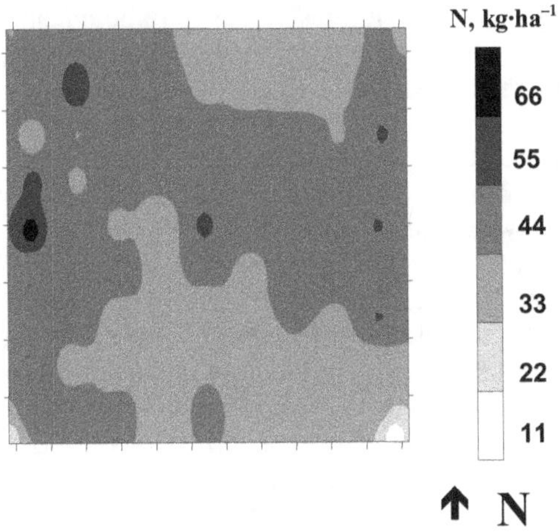


FIGURE 14.7. Potato top N content, prior to desiccation.

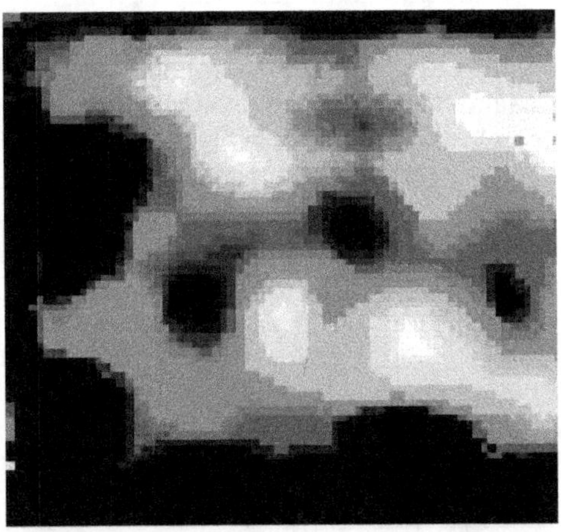

FIGURE 14.8. Landsat image of potato field three weeks prior to desiccation. Darker areas correspond to higher amounts of N in foliage. (See corresponding PLATE 14.8 in the central gallery.)

were composited from the high vigor area and a small number from a low vigor area, appropriate credits would be obtained for N crediting the sugarbeet crop. Although potatoes do not root to 180 cm, the presence of high vigor, greener tops over these areas containing high levels of deep N, suggests that organic matter in these areas may be enhanced with mineralizable N through a history of N recycling by sugarbeet within the rotation.

Following sugarbeet, 80 kg·ha$^{-1}$ credits were given for the most vigorous areas of sugarbeet, 56 kg·ha$^{-1}$ credits were given for intermediate vigor areas, and no N credits were given areas of least vigor and yellow-green color on the aerial images. Wheat was grown the following year, in which a ten foot square of wheat was killed with glyphosate at the 2-leaf stage at each point in the 0.2 ha grid, and the area was sampled to 60 cm every two weeks until harvest to track soil nitrate levels and N mineralization from the sugarbeet residues. The determined amounts were a combination of N mineralization and nitrate as a result of fall ammonia applications applied variably as demanded by the map of credits based on image color. The results showed no differences in wheat yield or soil $NO_3$-N levels at the early June sampling date (see Table 14.3). This showed that issuing credits for sugarbeet color and vigor was appropriate and that savings in N fertilizer and lower residual soil N levels could result.

From this work (Reitmeier et al., 1999; Franzen et al., 2000), it is evident that the problem with poor beet quality stemmed from the failure to account for sugarbeet top N mineralization and deep N following potatoes at the 120-180 cm depth. Following sugarbeet, only the soil test was considered, so high amounts of N, 150-200 kg·ha$^{-1}$ were applied for wheat. During the wheat year in this sandy loam-textured soil, rainfall tended to move large amounts of nitrate below the 60 cm depth, so that when soil sampling was conducted prior to potatoes, much of this N was deemed unavailable for the potato crop. During the potato year, the nitrate leached to deeper depths, with a considerable portion found below 120 cm. Sampling prior to sugarbeet only accounted for the 0-120 cm depth, so the additional N available to the sugarbeet crop at the 120-180 cm depth and the N contribution of the

TABLE 14.3. $NO_3$-N levels by zone, June 1 sampling date and subsequent spring wheat yield.

| Zone | NO3-N, kg·ha$^{-1}$ | Yield, kg·ha$^{-1}$ |
|---|---|---|
| 168 kg N·ha$^{-1}$ | 149 | 3790 |
| 112 kg N·ha$^{-1}$ | 129 | 3820 |
| 90 kg N·ha$^{-1}$ | 127 | 3700 |
| Significance | NS | NS |

*Source:* Franzen et al., 2000.

potato crop were not considered. Sugarbeet would recover and recycle the high N from deep depths and the cycle resulting in poor quality would repeat itself during the next three years.

By adjusting management, sugarbeet quality improved and deep soil N levels were depleted. Management adjustment begins following sugarbeet, when N credits against the amount normally recommended for wheat are based on the satellite imagery of the previous sugarbeet crop. At this time, there is only a rudimentary link between sugarbeet canopy intensity reading by the satellite and mineralizable N. However, with minimum ground truthing compared to the satellite imagery, credits from 0-80 kg N·ha$^{-1}$ are regularly given.

Following wheat, potatoes are fertilized following a 0-60 cm depth soil sampling as has historically been the practice, but following potatoes, the 0-120 cm depth sampling for sugarbeet is conducted, together with a small number of 120-180 cm depth sample cores as indicated by satellite imagery obtained prior to vine desiccation. Credits are given for deep N and an estimate of potato top canopy.

Other researchers have also found that N mineralization of organic matter and residues is important to modify sugarbeet N recommendations. Shock et al. (1998) found that consideration of organic matter can be used to reduce N recommendations to sugarbeet. Shock et al. (2000) also found that there may be little need to fertilize sugarbeet with N following heavily fertilized onions.

## *USE OF IMAGERY FROM SUGARBEET TO MODIFY NITROGEN RECOMMENDATIONS TO SUBSEQUENT CROPS*

The use of imagery from sugarbeet to modify N recommendations was built on earlier work by Crohain and Rixhall, (1967) and by a series of studies by Moraghan and Smith (1994; 1995a, 1995b; 1996), and Moraghan et al. (1997). The work concluded that yellow sugarbeet tops, if the yellowing was the result of N nutrition and not disease, would be expected to have less than 11 kg·ha$^{-1}$ soil N in the top 60 cm of soil for subsequent crops, and the tops would contribute no additional N to the next crop. Areas containing yellow-green color leaves at harvest would also be depleted in soil N to a 60 cm depth, but the decomposition would likely contribute 22-35 kg N·ha$^{-1}$ to subsequent crops. Areas containing green sugarbeet leaves at harvest may or may not contain considerable soil N in the 0-60 cm depth, and the leaves would contribute at least 80 kg N·ha$^{-1}$ to subsequent crops.

Further studies by Reitmeier et al. (1999) and Franzen et al. (2000) showed that additional mineralization beyond the numbers proposed by

Moraghan et al. (1997) were possible and suggested that some small number of tops be collected and analyzed for N in very robust, green areas to help further modify N recommendations to subsequent crops. Mineralization studies by Hapka et al. (2001) showed that release of N from residues is very fast from sugarbeet residues, with up to 90 percent of the N disappearance occurring prior to June in the year following sugarbeet harvest.

## CONCLUSIONS

Site-specific nitrogen management is an important tool to improve quality and maintain yield in sugarbeet production. Although the concept is less than ten years old in the Red River Valley, it is being profitably used to increase recoverable sugar content in sugarbeet and to modify N rates in other crops within the sugarbeet rotation (Franzen, 2004). Nitrogen management for sugarbeet is being conducted using zone soil sampling, sometimes at deeper depths than previously recommended. Nitrogen management for subsequent crops is being modified through the use of satellite imagery zones showing degrees of crop canopy linked to N nutrition of sugarbeet. The entire rotation is considered when managing N for sugarbeet, which is a concept that arguably would not be possible without site-specific management. Further work is needed to better link imagery with potential N mineralization. Additional work is needed to provide a commercially available, practical and affordable yield monitor with sugar sensor, so that sugarbeet producers may map yield and quality to aid in future improvements in production.

The principles of site-specific N management on sugarbeet might be transferred to many crops in other areas. The research on sugarbeet has been one of cropping systems, and it is clear that N in both sugarbeet and other rotational crops influence the availability of N to every member of the rotation. Remote imagery has been useful in limiting the amount of plant and soil testing required to arrive at a recommendation for N rate modification. Other crops also influence one another in modifying N availability and similar tools as those used in sugarbeet research would be useful in identifying important N management zones. The Red River Valley is generally cooler than other sugarbeet growing areas, so the principles of residue decomposition would be expected to be even more appropriate when decomposition rates increase with increasing temperature.

## REFERENCES

Abshahi, A., F.J. Hills, and F.E. Broadbent. 1984. Nitrogen utilization from residual sugarbeet fertilizer and soil incorporated sugarbeet tops. *Agronomy Journal,* 76:954-958.

Boyd, D.A. 1961. Current fertilizer practice in relation to manorial requirements. *Proceedings of the Fertilizer Society of London.* No. 65.

Boyd, D.A., P.B.H. Tinker, A.P. Draycott, and P.J. Last. 1970. Nitrogen requirement of sugar beet grown on mineral soils. *Journal of Agricultural Science, Cambridge* 74:37-46.

Cattanach, A.C., W.C. Dahnke, and C. Fanning. 1992. Fertilizing Sugarbeet. NDSU Extension Circular SF-714. North Dakota State University Extension Service, Fargo, ND.

Cattanach, A., D. Franzen, D. Bernhardson, and D. Braaten. 1997. Red River Valley site-specific nitrogen management experiences in 1996. In *1996 Sugarbeet Research and Extension Reports.* Sugarbeet Research and Education Board of Minnesota and North Dakota. pp. 92-100.

Cattanach, A.C., D.W. Franzen, H. Stanislowski, and L. Sax. 1996. Effect of grid soil testing and variable rate fertilizer application on sugarbeet yield and quality. In *1995 Sugarbeet Research and Extension Reports.* Sugarbeet Research and Education Board of Minnesota and North Dakota. pp. 105-112.

Crohain, A., and L. Rixhon. 1967. Practical fertilizer value of sugar-beet leaves and crowns. *Bull. Rech Agron. Gembloux.* 2:397-428.

Crowther, E.M. and F. Yates. 1941. Fertilizer policy in war-time: The fertilizer requirements of arable crops. *Emperial Journal of Experimental Agriculture.* 9:77-97.

Dahnke, W.C. and G.V. Johnson. 1990. Testing soils for available nitrogen. In *Soil testing and plant analysis,* Third edition. R.L. Westerman, (ed.) Soil Science Society of America, Madison, WI. pp. 127-140.

Draycott, A.P. 1972. *Sugar-beet nutrition.* Applied Science Publishers, Ltd., London.

Ferguson, R.B., C.A. Gotway, G.W. Hergert, and T.A. Peterson. 1996 Soil sampling for site-specific nitrogen management. In *Precision Agriculture, Proceedings of the 3rd International Conference.* June 23-26, 1996. Minneapolis, MN. ASA-CSSA-SSSA, Madison, WI. pp. 13-22.

Franzen, D.W. 2004. delineating nitrogen management zones in a sugarbeet rotation using remote sensing—A review. *Journal of Sugar Beet Research,* 41: 47-60.

Franzen, D.W. and L.J. Cihacek. 1998. Soil sampling as a basis for fertilizer application. NDSU Extension Circular SF-990. North Dakota State University, Fargo, ND. Available online www.ext.nodak.edu/extpubs/soilfert.htm/.

Franzen, D.W., L.J. Cihacek, V.L. Hofman, and L.J. Swenson. 1998. Topography-based sampling compared with grid sampling in the Northern Great Plains. *Journal of Production Agriculture.* 11:364-370.

Franzen, D.W., A.J. Landgraff, J.F. Giles, N.R. Cattanach, and L.J. Reitmeier. 2000. Nitrogen availability and movement within wheat fields following sugarbeet. In *1999 Sugarbeet Research and Extension Reports.* Sugarbeet Research and Education Board of Minnesota and North Dakota. pp. 117-126.

Franzen, D.W. and T.R. Peck. 1995. Field soil sampling density for variable rate fertilization. *Journal of Production Agriculture.* 8:568-574.

Hapka, A.J., D.W. Franzen, J.F. Giles, and N.R. Cattanach. 2001. Timing and release of nitrogen from residues. In *2000 Sugarbeet Research and Extension Reports.* Sugarbeet Research and Education Board of Minnesota and North Dakota. pp. 114-121.

Hobbis, J.K. 1973. What do the nitrate N and conductance reading from the tare laboratory sugar samples mean? In *1973 Sugarbeet Research and Extension Reports.* North Dakota State University, Fargo, ND, and University of Minnesota, St. Paul, MN. pp. 113-199.

Hobbis, J.K. 1978. Methods for estimating recoverable sugar in sugarbeets. In *1977 Sugarbeet Research and Extension Reports.* Sugarbeet Research and Education Board of Minnesota and North Dakota. pp. 247-251.

Moraghan, J., K. Horsager, L. Smith, and A. Simms. 1997. Canopy reflectance and GPS technology in relation to sugar production. In *1996 Sugarbeet Research and Extension Reports.* Sugarbeet Research and Education Board of Minnesota and North Dakota. pp. 121-133.

Moraghan, J.T. and L.J. Smith. 1994. Influence of sugarbeet tops on growth of a subsequent wheat crop. In *1993 Sugarbeet Research and Extension Reports.* Sugarbeet Research and Education Board of Minnesota and North Dakota. pp. 262-272.

Moraghan, J.T. and L.J. Smith. 1995a. Early season release of inorganic N from sugarbeet tops. In *1994 Sugarbeet Research and Extension Reports.* Sugarbeet Research and Education Board of Minnesota and North Dakota. pp. 209-214.

Moraghan, J.T. and L.J. Smith. 1995b. Influence of sugarbeet tops on growth of wheat in 1994. In *1994 Sugarbeet Research and Extension Reports.* Sugarbeet Research and Education Board of Minnesota and North Dakota. pp. 199-208

Moraghan, J.T. and L.J. Smith. 1996. Nitrogen in sugarbeet tops and the growth of a subsequent wheat crop. *Agronomy Journal.* 88:521-526.

Reitmeier, L.J., D.W. Franzen, J.F. Giles, A.C. Cattanach, and N.R. Cattanach. 1999. Nitrogen management in a wheat/potato/sugarbeet crop rotation. In *1998 Sugarbeet Research and Extension Reports.* Sugarbeet Research and Education Board of Minnesota and North Dakota. pp. 125-134.

Rudolph, G., R. Dahl, and J. Coffman. 1980. Soil sampling and the effect of nitrogen on quality. In *1979 Sugarbeet Research and Extension Reports.* Sugarbeet Research and Education Board of Minnesota and North Dakota. 10:134-137.

Ruhe, R.V. 1960. Elements of soil landscape. In *Transactions of the 7th International Congress of Soil Science.* Vol. 4. Int. Soc. Of Soil Science. Madison, WI. pp. 165-170.

Shock, C.C., B.G. Feibert, and D. Westermann. 1998. "On farm" implementation of lower nitrogen fertilizer inputs through nitrogen accounting and validation of organic matter mineralization. Oregon State University Agricultural Experiment Station, Special Report 988:126-134. Online at http://www.cropinfo.net/Annual Reports/1997nitrogen.wq.html.

Shock, C.C., M. Seddigh, L.D. Saunders, T.D. Steiber, and J.G. Miller. 2000. Sugarbeet nitrogen uptake and performance following heavily fertilized onion. *Agronomy Journal.* 92:10-15.

Smith, L.J. 1980. Deep soil testing—Does it pay? In *1979 Sugarbeet Research and Extension Reports.* Sugarbeet Research and Education Board of Minnesota and North Dakota. 10:138-141.

Smith, L.J. and D. Rains. 1995. Grid soil testing and variable rate spreading for profitable sugarbeet production—Fact or fiction? In *1994 Sugarbeet Research and Extension Reports.* Sugarbeet Research and Education Board of Minnesota and North Dakota. pp. 181-185.

Smith, L.J. and D. Rains. 1996. Grid soil testing and variable rate fertilization for profitable sugarbeet production—Part 2. In *1995 Sugarbeet Research and Extension Reports.* Sugarbeet Research and Education Board of Minnesota and North Dakota. pp. 113-116.

Smith, L.J., G. Varvel, J. Lensing, and G. Malzer. 1980. The effects of 3 nitrogen levels on sugarbeet variety performance. In *1979 Sugarbeet Research and Extension Reports.* Sugarbeet Research and Education Board of Minnesota and North Dakota. 10:142-149.

Ulrich, A. 1950. Critical nitrate levels of sugar beets estimated from analysis of petioles and blades, with special reference to yields and sucrose concentrations. *Soil Science.* 69:291-309.

Wollenhaupt, N.C., R.P. Wolkowski, and M.K. Clayton. 1994. Mapping soil test phosphorus and potassium for variable-rate fertilizer application. *Journal of Production Agriculture.* 7:441-448.

Chapter 15

# Application of Remote Sensing and Ecosystem Modeling in Vineyard Management

Ramakrishna R. Nemani
Lee F. Johnson
Michael A. White

## *INTRODUCTION*

Proper site selection, historically, has been one of the most important aspects of fine wine production. Decades, and in some cases, centuries of experience with climate-topography-soil combinations allowed vintners around the world to identify areas that are most suitable for a given variety of winegrapes. Recent research and technology have considerably simplified the site-selection process (Gladstones, 1992). At the same time, wine production has evolved from stemming "out of passion" to a production management system. Large capital investments in viticulture now dictate that maximum potential of the entire vineyard be exploited, paying proper attention to marginal areas (Bramley and Proffitt, 1999). Winegrape yield as well as various measures of wine quality have been shown to vary substantially within a single block (Lamb, 1999). Inability to characterize such variability often leads to poor yield prediction and the acceptance of lower-quality wines from whole vineyards. Current management practices tend to follow the "average" approach, i.e., managing entire blocks with one prescription. In this context, precision viticulture (PV) may be defined as monitoring and managing spatial variation in productivity-related variables such as yield and quality within a single vineyard (Cook et al., 2000). To be an economically viable tool, PV should provide not only information to the vineyard managers about the performance of the cropping system over a variable land surface but also control technology to manage variations more precisely.

*Handbook of Precision Agriculture*

doi:10.1300/5627_15

A number of high-tech tools are available to the vineyard manager (i.e., remote sensing, geographic information systems, Global Positioning Systems, etc.), that complement his or her intuition and knowledge in identifying and paying special attention to the marginal areas. In this chapter, we discuss our experiences of applying some of these tools to vineyard management (VM) in the Napa Valley, California. First, we discuss the role of remote sensing in identifying problem areas in a vineyard, and then introduce simulation modeling that may help in finding proper solutions.

## *THE VINEYARD AS AN ECOSYSTEM*

Although a number of high-tech tools are available for mapping variability across a landscape, understanding this requires an ecosystem approach with a view to identifying variables that are most suited for management. In a natural ecosystem, plant productivity is a result of complex interactions among climate, soil, and leaf area displayed. An equilibrium generally exists between climate-soil and leaf-area index (LAI, area of leaves per unit area of ground, $m^2/m^2$), i.e., for a given climate, LAI increases or decreases depending on soil water-holding capacity and nutritional status (Nemani and Running, 1989). Agronomic practices aim to not only maximize productivity, but also optimize the quality and quantity of harvestable products. Genetic engineering is now routinely used in various annual crops to modify the patterns of productivity as well as resistance to pests and diseases. However, because winegrapes are a perennial crop, fewer such opportunities are readily available. Since climate is to a large extent beyond control, modifying soil and leaf area offer the best options to obtain a desired management goal. Even within a single vineyard, small differences in topography or soil properties can add up to significant differences in growth potential. Independent of management, such differences in growth potential are manifested as changes in leaf area. Similarly, impairment of a plant's ability to take up water or nutrients from soils either due to pest or disease infestation also manifests as reductions in leaf area (Peterson and Johnson, 2000). Since LAI is a key variable controlling the water loss, photosynthesis and radiation penetration through the canopy, and the ability to estimate and map LAI across vineyards would be valuable in various management decisions relating to pruning, irrigation, and regulating crop loads (Winkler, 1958; Smart, 1985). A number of viticultural operations (planting density, trellising, and pruning, etc.) are aimed at regulating canopy leaf area for entire blocks.

## *TOOLS IN VINEYARD MANAGEMENT*

If we accept LAI as a key variable in the vineyard production management system, which can be used to regulate yield and quality, then accurate characterization of its spatiotemporal variability is a key first step. Once an accurate map of LAI is derived either from ground observations (for smaller areas) or from remotely sensed data (for larger areas), one must find ancillary data (soil's physical and chemical properties, topography, and climate) that could potentially explain the LAI variations. Collection of such data may itself involve a number of technologies, including Global Positioning Systems (GPS) and personal data assistants (PDAs). A GIS is then required to bring the disparate data sources into a uniform projection and spatial resolution. Often, models are required to study the subtle relations that exist between different variables. For example, at a given soil water-holding capacity one may find a large variation in LAI values because of seasonal changes in LAI and microclimatic differences. Simulation models are also an important tool for forward-looking decision support as in the case of estimating irrigation requirements and harvest scheduling. A brief description of each of the technologies is given, with regard to their use in vineyard management.

### *Remote Sensing*

An aerial view of a vineyard often gives a better perspective of vineyard canopy conditions than ground observations. This perspective, when translated to quantitative information, allows for better management prescriptions. Remote sensing, observing an object from a distance, has graduated from aerial photography to highly sophisticated imaging sensors. These sensors can collect information at multiple wavelengths of the solar spectrum, and at spatial resolutions as fine as half a meter. Much of the remotely sensed data used in precision agriculture are collected in the red (RED, chlorophyll absorption) and near-infrared (NIR, internal leaf scattering) wavelength bands (Moran et al. 1997). Various combinations of the two bands, called spectral vegetation indices, have been shown to be highly sensitive to vegetation cover and density. For example, a widely used index called the Normalized Difference Vegetation Index (NDVI), computed as (NIR-RED)/(NIR+RED), has been shown to be a useful measure of vine health (Johnson et al., 1996; Lamb, 2000).

Along with the advances in our understanding of canopy reflectance measures and their relationship to underlying scene components, the technology of collecting the information has dramatically improved. Various public and private firms offer technologies ranging from multispectral (a few wavelength bands) to hyperspectral (as many as 256 wavelength bands) sensors (Lebof, 2000). For most practical purposes, a simple system collecting information in the red and NIR wavelengths seems to be sufficient (Escobar et al., 1998). However, for remote sensing to be useful, the information has to be timely and cost effective (Fritz, 1996; Lamb, 1999).

During the 1990s, vegetation indices were used in vineyards to delineate spatial differences in canopy conditions as a result of pest infestation and/or resource availability (Baldy et al., 1996; Lobitz et al., 1997). Similarly, vegetation indices have also been used to segment the vineyards into uniform zones of canopy vigor. Such segmentation was found to be helpful in maximizing the production of quality wines (Johnson et al., 1996). Although the qualitative use of remote sensing is indicative of its underlying potential, we must strive to extract quantitative information to exploit its full potential. Quantitative measures such as LAI allow comparisons from one year to the next or from one vineyard to the next. A number of methods are available to measure LAI in the field. They include destructive methods (leaf removal and pruning weights) and nondestructive methods based on light interception and shoot lengths. If a relationship can be established between remote sensing data and field-measured LAI, spatial maps of LAI can be created without much effort.

### *Remote Sensing of Leaf Area in Vineyards*

With the goal of quantifying the relationship between LAI and NDVI, we chose 20 plots in the Robert Mondavi vineyards of the Napa Valley in California where LAI was measured/estimated using a number of techniques (Grantz and Williams, 1993). Remotely sensed data, acquired from IKONOS data, were used to compute NDVI for each 2 $m^2$ area. A strong linear relationship was observed between NDVI and field-estimated LAI (Figure 15.1). Though theory suggests that the NDVI-LAI relation becomes asymptotic at high canopy LAIs (Myneni et al., 1997), LAI of vineyards tends to be in the linear range of the NDVI-LAI relationship (< 3 LAI). Interestingly, the NDVI-LAI relationship does not seem to be sensitive to the trellising types as the field-measured LAIs represent a wide range of spacings and trellising types. A number of other factors such as sensor calibration, soil color, soil moisture, sun angle, and viewing geometry do influence the NDVI-LAI relationship (Myneni et al., 1997). Therefore, one must pay attention to these factors when imagery from multiple dates are used. Cover crops in various stages of development or senescence can also create problems for extracting quantitative information about the vine canopies. Use of multiple images during a growing season may help alleviate this problem.

### *Mapping Leaf Area Index in Vineyards*

We applied the NDVI-LAI relationship shown in Figure 15.1 to the entire scene shown in Figure 15.2 and estimated LAI for each 2 $m^2$ area. LAI of vineyards ranged from 0.2 (in newly planted blocks) to over 2.0 (in mature vineyards). The cost of producing such a map is less than one dollar per acre in California. Therefore, LAI maps based on remote sensing are a cost-effective way of making decisions regarding pruning, irrigation, and managing crop loads. On closer examination of the LAI map, one can see signif-

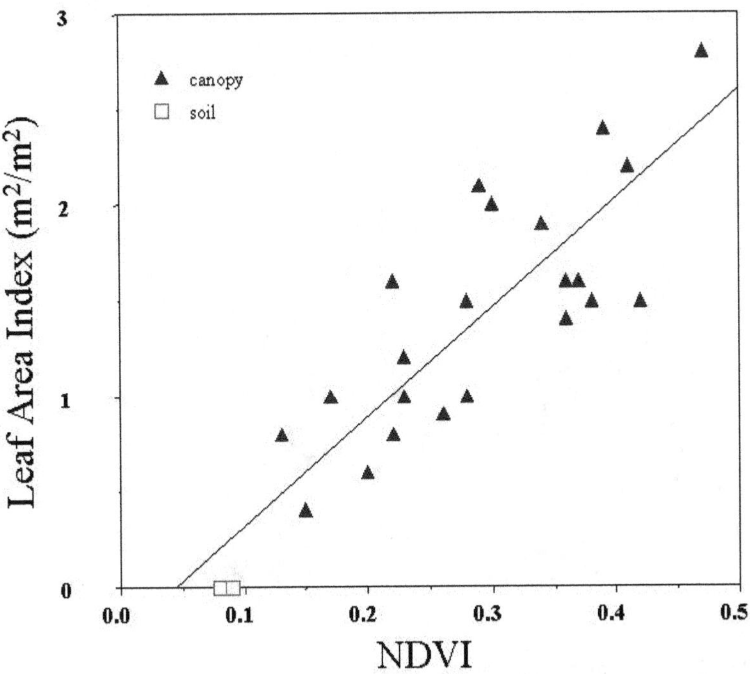


FIGURE 15.1. Observed relation between vegetation index from satellite data (IKONOS) and ground-based leaf area index measurements of winegrape canopies in Napa Valley, California ($y = -0.25198 + 5.7085x$; $R^2 = 0.737$; Johnson et al., 2001). Results are consistent with canopy radiative transfer theory (Myneni et al., 1997) explaining the linear relationship between NDVI and LAIs under 3.0.

FIGURE 15.2. Leaf area index of vineyards is mapped at 2 m resolution using the relation shown in Figure 15.1 and satellite imagery obtained from IKONOS (Nemani et al., 2001). Significant variations in LAI are evident even within a single vineyard as a result of spatial differences in resource availability, management practices and presence of pest/diseases. Understanding and properly managing these small scale variations is the key to successful implementation of remote sensing technology in vineyard management. (See corresponding PLATE 15.2 in the central gallery.)

icant variations in LAI within a single block. Information on such spatial heterogeneity, caused by differences in resource availability or pest/disease infestations, can be used in two different ways. In the short term, the blocks may be divided into uniform zones that can be managed differently for fertilizer application, irrigation, and time of harvesting, etc. In the long run, such information is useful in planning the vineyard: variety, spacing, and trellising type. Once the vineyard is planted with the knowledge of underlying spatial heterogeneity, year-to-year management becomes more efficient and cost effective.

One remaining impediment to the widespread use of remote sensing in VM is turnaround time of data collected from air/spaceborne sensors. Though acquisition costs have come down considerably, reliable and timely delivery has so far been elusive. Part of the problem is size of the market, as remote sensing gains wider acceptance one would hope that commercial-sector development would rise to the occasion.

### *Climate, Topography, and Soils*

Although remote sensing is useful for detecting the spatial and temporal variations in canopy conditions, it cannot explain the underlying causes for a given canopy condition. As discussed earlier, among the variables that can influence leaf-area distribution are the changes in microclimate, topography, and soil properties. Differences in microclimate as a result of topography (slope, aspect, and elevation) and prevailing wind direction can significantly alter solar radiation, temperature, and water demand. For example, in the mid-latitudes a vineyard facing south can receive as much as 30 percent more radiation when compared to a northfacing vineyard. A paucity of weather stations creates another problem for accurately mapping climatic parameters such as rainfall, temperature, humidity, and wind conditions (O'Neal et al., 2000). Although a weather station in the vineyard provides the best source of climate information, a number of numerical models are available that extrapolate/interpolate weather variables from the point of observation to various points around the landscape using topography (Thornton et al., 1997). An illustration of such mapping is shown for wine producing areas of California using daily weather observations from over 200 stations and 1 km topographic data (Figure 15.3). Similarly, such models can also be used to map microclimate variations at much finer resolutions.

Topographic conditions also influence water flow around the landscape, creating dry conditions at the top of the hill and wetter conditions at the bottom. Over time, such changes in water flow create differential soil depths

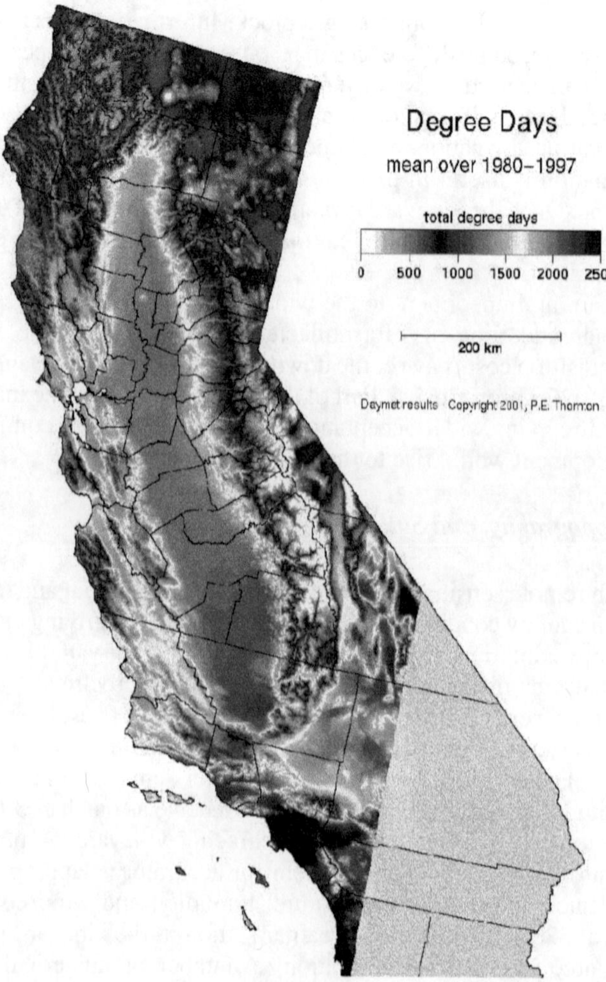


FIGURE 15.3. Mapping the spatial variability in microclimate with DAYMET model (Thornton et al., 1997) using weather station observations and topography. Accurate spatial representation of weather conditions depends on the station density and topographic data used in the interpolation process. Once the basic weather elements (temperature, rainfall, humidity, and wind) are interpolated to a finer grid, then a number of viticulturally sensitive parameters (number of frosts, degree days, total rainfall, etc.) can be estimated at each point. Here, spatial distribution of growing degree days computed with a base temperature of 10°C is shown for wine-producing regions of California.

around the landscape as a result of erosion and sedimentation. Even small variations in topography can create substantial variations in water availability around the vineyard, which may result in excess or poor canopy growth. It is now possible to map topographic variations at submeter resolutions using laser altimetry and stereophotography (Figure 15.4).

Spatial mapping of soil parameters is by far the most important and difficult problem in vineyard management. A number of advanced techniques are being developed for mapping soil physical and chemical properties. Among them, microwave remote sensing and measurements of electrical conductivity (Kitchen et al., 1999) offer the potential to map soil texture and depth with a reasonable accuracy. The best option for mapping soils at

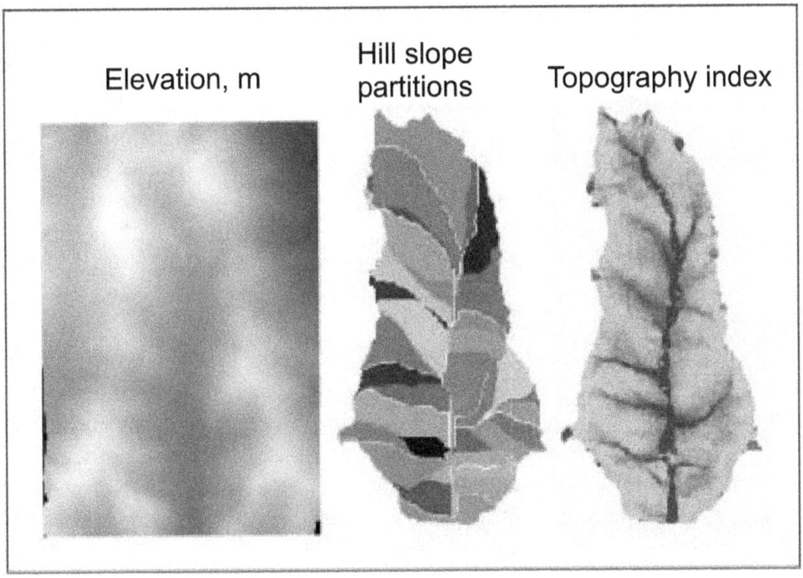


FIGURE 15.4. Topographic variations in Carneros area of Napa Valley, California, mapped using stereophotography at 2 m resolution. Once a digital elevation model (DEM) is created (brighter shades represent higher elevations), the data can be used with terrain analysis software to produce viticulturally significant variables such as slope, aspect, and a topography index (TI) indicating water flow routes. The slope and aspect data can then be combined to produce hillslope partitions that have similar terrain characteristics such as incident radiation. The topography index data can be a key indicator of water stress and canopy development as low TI areas (lighter shades) tend to produce high water stress and low LAI, while high TI (dark shades) areas tend to accumulate more water resulting in low water stress and high canopy densities.

the present time would be to use traditional soil sampling methods, along with a sampling scheme that captures the inherent variability developed using aerial imagery (Varvel et al., 1999).

### Ecosystem Modeling

Simple overlays of various information fields may not be sufficient to unravel the complex soil-plant-atmosphere interactions that take place in the vineyard. For example, what looked like a uniform canopy in the spring may turn out to be quite variable in the late summer as a result of differential water availability. Here we need tools that explore the complex interactions among climate, soil, vegetation, and management practices. Ecosystem models are designed to combine daily weather, soil physiochemical properties, and vegetation physiological responses to explain the differences in vegetation stress, canopy growth, and productivity. Such models are available for almost all agricultural crops. Our model is an adaptation of a generic ecosystem model, BIOME-BGC (Figure 15.5; White et al., 2000), customized for grape canopies. Given daily weather conditions (temperature, wind, humidity, rainfall, and radiation) and LAI, the model was able to reproduce measured soil moisture and vegetation stress (Nemani and Running, 1989). Plant moisture status is actively managed in vineyards through pruning and irrigation.

The model is also capable of estimating photosynthesis and respiration, which are useful for estimating productivity of vineyards. Estimates of the photosynthesis-respiration balance can be used to calculate optimum crop loads, by choosing the proper leaf area-to-fruit ratio. Once the models are calibrated with field observations, they can be used in a number of ways to explore various management options both in the site-selection process as well as growing-season vineyard operations.

#### *Models in Vineyard Establishment*

An example of how models can be useful in vineyard management is shown in Figure 15.6, where the amount of irrigation required to maintain a given level of vegetation stress is simulated at different LAIs. To run the model, we used daily climate data from the Napa Valley for 1985-1999, and assumed the soils to be of sandy-loam with a depth of 120 cm. As LAI increases, under same soil and climatic conditions, the vines require more water on a daily basis with irrigation starting earlier in the growing season. A sensitivity analysis of the model using various combinations of soil-

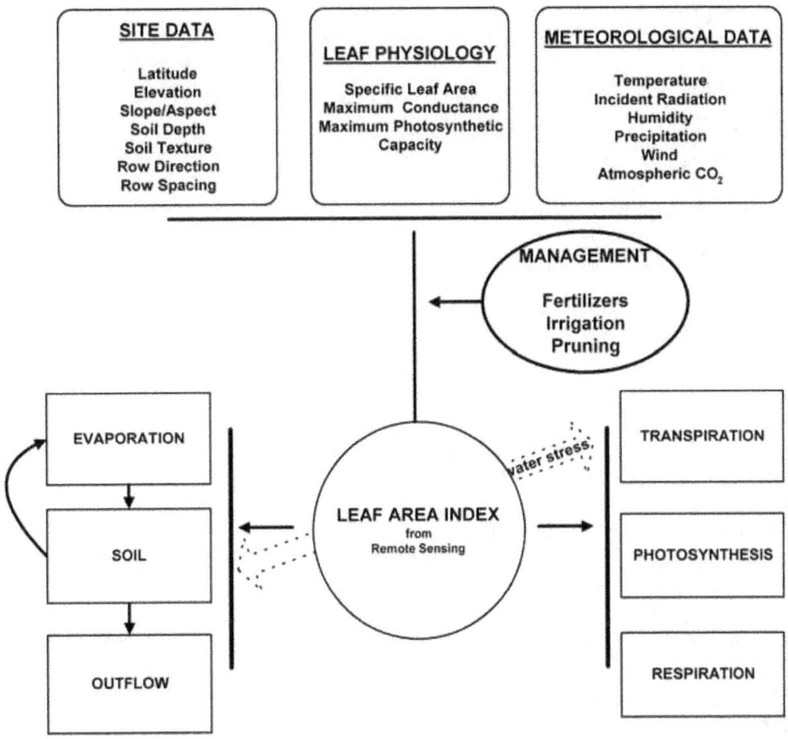


FIGURE 15.5: Conceptual flowchart of the Ag-BGC model (adaptation from the BIOME-BGC ecosystem model, White et al., 2000) useful for estimating water and carbon fluxes between vineyards and the atmosphere. Although it is relatively straightforward to translate variables such evaporation, transpiration, and water stress into useful information like irrigation requirements and scheduling, more research is needed to understand how photosynthesis-respiration estimates relate to productivity-based variables such as grape yield and wine quality.

climate-LAI conditions can provide valuable information for vineyard establishment.

### *Models in Operational Management*

Models can also be used operationally to estimate irrigation requirements on a daily basis (Grimes and Williams, 1990). Since weather conditions play a dominant role in irrigation requirements, the model can be set

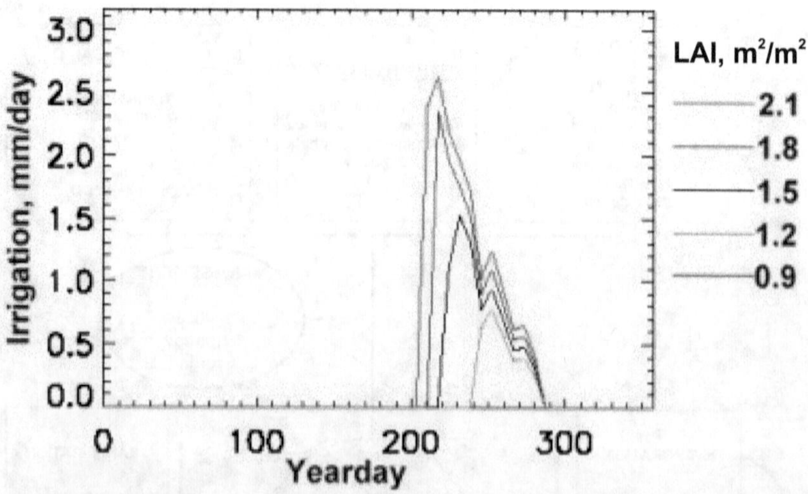


FIGURE 15.6. Sensitivity analysis of Ag-BGC showing the amount of irrigation required to keep the vines at a water stress of −1.0 Mpa (predawn) as a function of leaf area index with climate and soils held constant. Dense canopies require greater amounts of water starting earlier in the growing season. Similar analysis can be performed for various combinations of climate, soil, and leaf area. The results of such a comprehensive sensitivity analysis can be used to determine the choice of varietals and plating density. (See corresponding PLATE 15.6 in the central gallery.)

up to run on a daily basis with weather data collected in/around the vineyard. Efforts are underway to implement the model spatially, combining the remotely sensed LAI maps, soil texture and depth, and interpolated weather conditions (Figure 15.7; Ford et al., 1994). Spatial estimates of soil moisture and evapotranspiration can then be used to assess where and how much water needs to be applied (Figure 15.8). Such detailed maps of irrigation requirements are a considerable improvement over current methods. At present, irrigation requirements are calculated for the entire vineyard using a crop coefficient and potential evapotranspiration estimates from a nearby weather station. However, the crop coefficients do not change to reflect the variable density of canopies in the vineyard. Using remotely sensed LAIs would significantly improve the estimation of irrigation requirements (Booltink and Epinat, 2000).

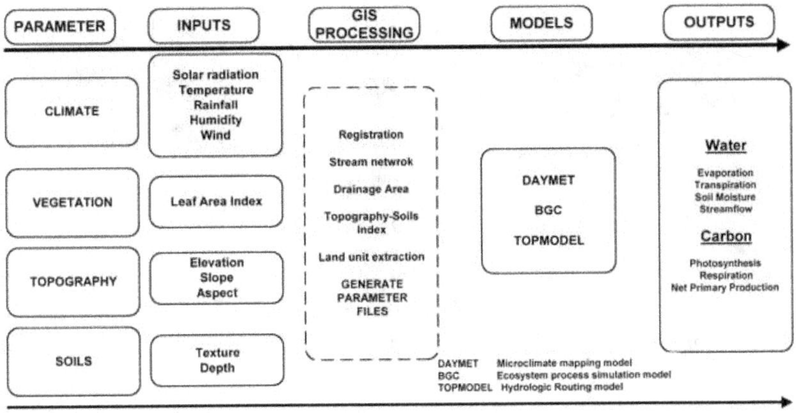


FIGURE 15.7. Schematic diagram showing model and data integration for estimating spatial variability in ecosystem processes in a vineyard. The TOPMODEL (Ford et al., 1994) is used to capture lateral flow of water around the landscape.

### *Information Technology*

Of the many high-tech tools available for VM, the most important is a GIS, which allows the grower to incorporate the disparate data sources into a uniform database for analysis and display. Commercial GIS packages come in various configurations and complexity, and most packages seem to offer the capabilities required for VM. The problem is one of data collection rather than software functionality. Many current GIS packages also allow interfacing with various database packages that allow sophisticated statistical and search functions. A recent development in GIS technology is the ability to ingest information collected in the field with PDAs using a wireless interface. Along with GPS technology for recording location information, the PDAs will radically alter the way information is gathered and stored for later processing. The wireless interfacing also allows the vineyard manager to access information ondemand from the host computer, thus allowing near real-time decisions.

As we have seen in the case of information technology, there is bound to be information overload for growers using advanced technologies. Much of the current focus is on gathering information about the variability present in the vineyard. Although mapping the variability presents an opportunity for improvement, this does not translate to profitability unless the links between control strategies and the resulting outcome are worked out. A log-

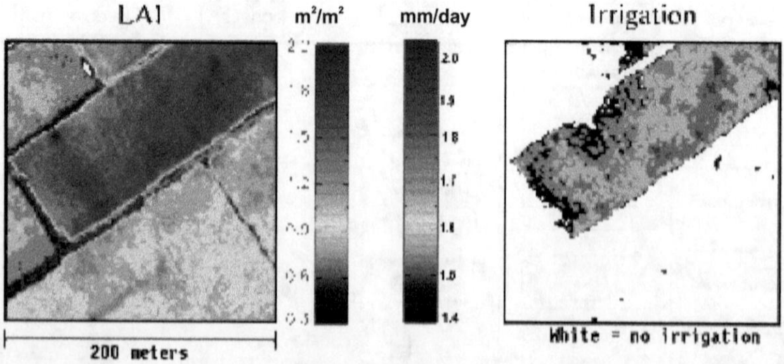


FIGURE 15.8. Spatial implementation of Ag-BGC parameterized with LAI from IKONOS data (Figure 15.2) for estimating daily irrigation requirements for a 200 m × 200 m area in the Napa Valley, California. Although traditional crop coefficient-based irrigation approaches predict uniform irrigation for entire vineyards, spatial BGC simulations show a wide range of water requirements within a vineyard. Variable Rate Technology (VRT) can be brought to bear to address such spatially varying water demands. (See corresponding PLATE 15.8 in the central gallery.)

ical next step would be developing tools to organize and process the information so as to identify opportunities for change. Such opportunities need to be diagnosed in a way that action is possible and profitable. The wine industry already has yield-monitoring technology, and on-the-go quality monitoring is expected soon (Bramley and Proffitt, 1999). Predictive models need to be improved in the context of decision-support systems that can provide a range of solutions with associated returns on investments. Developing such a robust decision-support system involves growers and a number of specialists in viticulture, agricultural economics, soil science, entomology, and pathology who are familiar with the production system. Formation of such teams would bring the credibility needed to enhance the utilization of advanced technologies among growers (Schepters and Francis, 1998).

## CONCLUSIONS

Improving profitability while minimizing environmental impacts is the main goal of precision agriculture (PA). How this is achieved depends on the problems at hand, available tools, and the capacity to implement technology. Impressive progress has been made in developing or implementing tools for collecting information about the spatial variability in the vineyards. However, the process of diagnosing problems and prescribing proper solutions that are economically viable is still in its rudimentary stages. Without the certainty of the effects of management changes on production and profit, it is difficult to convince growers of the need to implement PA practices. Few studies have addressed the cost effectiveness of PA technology. We need to design experiments with the participation of various specialists and growers to address the problem using an "end-to-end" approach dealing with agro-ecology as well as economics (Bullock and Bullock, 2000). Such experiments would result in enhanced grower skills, development of standard software tools to analyze data and prescribe solutions, and rapid development of the commercial sector. Although French vintners developed the concept of "terroir," it is the vintners in California and Australia that are leading the conceptual development and implementation of PA tools in viticulture. The health and vitality of the wine industry in the two countries are likely to reflect the success or failure of precision viticulture.

## REFERENCES


Baldy, R., L. DeBenedictis, L. Johnson, E. Weber, M. Baldy, B. Osborn, and J. Burleith. 1996. Leaf color and vine size are related to yield in a *Phylloxera*-infested vineyard. *Vitis,* 35(4):201-205.

Booltink, H.W.G. and V. Epinat. 2000. Using remote sensing data to improve simulation modeling for precision agriculture. In *Proceedings of the 5th International Conference on Precision Agriculture,* American Society of Agronomy, Madison, WI. CD-ROM.

Bramley, R.G.V. and A.P. Proffitt. 1999. Managing variability in viticultural production. *Grapegrower and Winemaker,* July:11-16.

Bullock, D.S. and D.G. Bullock. 2000. From agronomic research to farm management guidelines: A primer on the economics of information and precision technology. *Precision Agriculture,* 2:71-101.

Cook, S.E., M.L. Adams., and R. Bramley. 2000. What is obstructing the wider adoption of precision agriculture technology? In *Proceedings of the 5th International Conference on Precision Agriculture,* American Society of Agronomy, Madison, WI. CD-ROM.

Escobar, D.E., J.H. Everitt, J. Noreiga, I. Cavazos, and M. Davis. 1998. A twelve band airborne digital video imaging system (ADVIS). *Remote Sensing of Environment,* 66:122-128.

Ford, R., S. Running, and R. Nemani. 1994. A scalable system for modular ecological modeling. *IEEE Computational Science and Engineering,* 1:32-44.

Fritz, L.W. 1996. The era of commercial earth observation satellites. *Photogrammetric Engineering and Remote Sensing,* 62:39-45.

Gladstones, J. 1992. *Viticulture and environment.* Wine titles, Adelaide, Australia.

Grantz, D. A. and L. Williams. 1993. An empirical protocol for indirect measurement of leaf area index in grape (*Vitis vinifera* L.). *HortScience,* 28:777-779.

Grimes, D. and L. Williams. 1990. Irrigation effects on plant water relations and productivity of Thompson seedless grapevines. *Crop Science,* 30:255-260.

Johnson, L., B. Lobitz, R. Armstrong, R. Baldy, E. Weber, J. DeBenedctis, and D. Bosch. 1996. Airborne imaging aids vineyard canopy evaluation. *California Agriculture,* 15(4):14-18.

Johnson, L., D. Roczen, S. Youkhana, R. Nemani, and D. Bosch. 2001. Mapping vineyard leaf area with multispectral satellite imagery. *Computers and Electronics in Agriculture,* 381(1):37-48.

Kitchen, N.R., K.A. Sudduth, and S.T. Drummond. 1999. Soil electrical conductivity as a crop productivity measure for claypan soils. *Journal of Production Agriculture,* 12:607-617.

Lamb, D.W. 1999. Monitoring vineyard variability from the air. *Australian Viticulture,* 3(6):22-23.

Lamb, D.W. 2000. The use of qualitative airborne multispectral imaging for managing agricultural crops—A case study in southeastern Australia. *Australian Journal of Experimental Agriculture,* 40:725-738.

Lebof, J. 2000. Practical applications of remote sensing technology—An industry perspective. *Hortechnology,* 10(3):475-480.

Lobitz, B., L. Johnson, C. Hlavka, R. Armstrong, and C. Bell. 1997. Grapevine Remote Sensing Analysis of Phylloxera Early Stress (GRAPES): Remote sensing analysis summary, *NASA Technical Memorandum* #112218.

Moran, M.S., Y. Inoue, and E.M. Barnes. 1997. Opportunities and limitations for image-based remote sensing in precision crop management. *Remote Sensing and Environment,* 6:319-346.

Myneni, R., R. Nemani, and S. Running. 1997. Algorithm for the estimation of land cover, LAI, and FPAR based on radiative transfer models. *IEEE Transactions on Geoscience and Remote Sensing,* 35:1380-1392.

Nemani, R., L. Johnson, and M. White. 2001. Adding science to intuition: Application of remote sensing and ecosystem modeling to vineyard management. *Australian Grapegrower and Winemaker,* Technical Issue #29:45-47.

Nemani, R., and S. Running. 1989. Testing a theoretical climate-soil-leaf area hydrologic equilibrium of forest using satellite and ecosystem simulation. *Agricultural and Forest Meteorology,* 44:245-260.

O'Neal, M.R., J.R. Frankenberger, and D.R. Ess. 2000. Spatial precipitation variability in the choice of nitrogen fertilization rates. In *Proceedings of the 5th International Conference on Precision Agriculture,* American Society of Agronomy, Madison, WI. CD-ROM.

Peterson, D.L. and L. Johnson. 2000. The application of earth science findings to the practical problems of growing winegrapes. *Geographic Information Sciences,* 6(2):181-187.

Schepters, J.S. and D.D. Francis. 1998. Precision agriculture—What's in our future? *Communications in Soil Science and Plant Analysis,* 29:1463-1469.

Smart, R. 1985. Principles of grapevine canopy microclimate manipulation with implications for yield and quality. *American Journal of Enology and Viticulture,* 36:230-239.

Thornton P.E., S.W. Running, and M.A. White. 1997. Generating surfaces of daily meteorological variables over large regions of complex terrain. *Journal of Hydrology,* 190:214-251.

Varvel, G.E., M.R. Schlemmer, and J.S. Schepters. 1999. Relationship between spectral data from an aerial image and soil organic matter and phosphorus levels. *Precision Agriculture,* 1:291-300.

White, M., P.E. Thornton, S.W. Running, and R.R. Nemani. 2000. Parameterization and sensitivity analysis of the BIOME-BGC terrestrial ecosystem model: Net primary production controls. *Earth Interactions,* 4, Paper No. 3.

Winkler, A. 1958. The relation of leaf area and climate to vine performance and grape quality. *American Journal of Enology and Viticulture,* 9:10-23.

# Chapter 16

# Site-Specific Management from a Cropping System Perspective

David E. Clay
Sharon A. Clay
Gregg Carlson

## *INTRODUCTION*

Cropping systems are composed of many individual components that consider rotations, tillage, fertilizer efficiency, and pest management. All of these factors interact to influence yield, which can be defined by the following equation:

$$\text{Yield} = f(WS, W, I, D, N, C, T, R, O) \quad (16.1)$$

where *WS* is water stress, *W* is weeds, *I* is insects, *D* is diseases, *N* is nutrients, *C* is climate, *T* is tillage, *R* is rotations, and *O* is other factors. Some of these factors can be managed while other ones cannot. Given the large number of options available within each component in a cropping system, a producer must be very careful to select management strategies that increase profitability. If Equation 16.1 could be uniquely determined, then the cost effectiveness of different management scenarios could be determined using enterprise analysis (Swinton and Lowenberg-DeBoer, 1999). However, solving Equation 16.1 is difficult because (1) most field experiments only investigate one or two of the various parameters at any given time, and (2) spatially dependent soil properties, such as the soil water-holding capacity, ability to supply nutrients, drainage, texture, water infiltration rates, structure, cation exchange capacity, electrical conductivity, pH, organic matter, and temperature, interact to influence yield (Timlin et al., 2001). An

*Handbook of Precision Agriculture*

doi:10.1300/5627_16

understanding of the interactions among spatially dependent soil properties and the various management options are needed to improve site-specific management. These, in turn, are needed to predict the impact of different management options and climatic scenarios on crop productivity within agricultural systems. If interactions are not taken into account management recommendations may be less than optimal.

## UNDERSTANDING YIELD VARIABILITY

### Water Stress

Water availability is the most important factor limiting yield in many areas of the world. For example, Clay, Engel, et al. (2001) reported that in the Great Plains of the United States (Montana) irrigation increased wheat (*Triticum aestivum* L.) yields from 1,540 to 5,260 $kg \cdot ha^{-1}$ in 1996. Understanding the methods used to measure water stress and the problems associated with these measurements can lead to an appreciation of the difficulty to quantify water stress in production fields. A traditional measurement of water stress is water use efficiency (WUE), i.e., the ratio between biomass production and evapotranspiration (Hatfield et al., 2001). Evapotranspiration (ET) can be estimated by the following equation:

$$\begin{aligned} ET = {} & \text{Rainfall} + \text{Initial water} - \text{End water} \\ & - \text{Water loss} + \text{Water gain} \end{aligned} \tag{16.2}$$

where initial water and end water are the amounts of water contained in the plant root zone prior to and after a specified time interval. For crops grown in semiarid environments in nearly level fields, many studies assume that the amount of water lost through deep drainage or runoff or gained through run-in is insignificant (Angus and van Herwaarden, 2001). If this assumption is accepted, then biomass yields, rainfall, and the change in the amount of water contained in the plant root zone can be used to calculate WUE for many points in the field. Clearly, this assumption is not valid for fields with topographic relief. In sloping land, runoff or runon can be measured, modeled, or estimated from topographic information. For example, Halvorson and Doll (1991) addressed the runoff and runon problem by using topography to calculate a topographic factor. If the factor was positive, then a net gain in run-in was expected and if negative, then a net loss from runoff was expected. Halvorson and Doll (1991) used the topographic factor to predict areas were runoff was expected and move appropriate amounts to areas where runon was expected. This approach worked better in wet than dry years.

Water stress also can be evaluated by measuring $^{13}C$ discrimination ($\Delta$) in $C_3$ and $C_4$ plants. A detailed discussion why $^{13}C$ discrimination can be used as an estimator of water stress is beyond the scope of this chapter and is available in Boutton, (1991), Farquhar and Lloyd (1993), and Clay, Clay et al. (2001). Prior to using $\Delta$ a few definitions are needed. The ratio between $C^{13}$ and $C^{12}$ is the *R* value. The *R* value is used to calculate $\delta^{13}C$ using the following equation:

$$\delta^{13}C = [R(sample)/R(standard)-1] \times 1000‰ \tag{16.3}$$

where, R(sample) is the $^{13}C/^{12}C$ ratio of the sample and R(standard) is the $^{13}C/^{12}C$ ratio of PDB, a limestone from the Pee Dee formation in South Carolina (Farquhar and Lloyd, 1993). Typically, $\delta^{13}C$ values for air, $C_3$, and $C_4$ plants are –8, –27, and –13 per mille, respectively. A negative sign indicates that the sample has a lower $^{13}C/^{12}C$ ratio than PDB. In many cases it is convenient to report $^{13}C$ discrimination ($\Delta$), which is calculated using this equation:

$$\Delta = (\delta^{13}C_a - \delta^{13}C_p)/(1 + \delta^{13}C_p/1000) \tag{16.4}$$

where, $\delta^{13}C_a$ is the $\delta^{13}C$ value of air (–8 per mille) and $\delta^{13}C_p$ is measured value of the plant.

Research has shown that $\Delta$ can be used to assess water stress in plants (Clay et al., 2005). For example, Clay, Clay, et al. (2001) reported that for corn, $\Delta$ contained spatial structure and was higher in summit areas and lower in footslope areas, and $\Delta$ was able to explain 90 percent of the corn yield variability across landscapes. These results suggest that corn grown in summit areas was more water stressed than corn plants in footslope areas. In a different study, Clay, Engel, et al. (2001) reported that $\Delta$ explained 84 percent of wheat yield variability in plots subject to only water stress, and that $\Delta$ explained 85 percent of the yield variability in plots subject to both water and N stress. These studies suggest that $\Delta$ can be used as a diagnostic tool for characterizing water stress in both $C_3$ and $C_4$ plants. However, care must be used in using $\Delta$ as a diagnostic tool for water stress because any factor that has the potential to influence water use or photosynthesis efficiency can impact the ratio between $CO_2$ concentrations inside and outside the leaf and thereby impact $\Delta$ (Condon et al., 1992).

The impact of water stress on yield spatial variability has been reported in many field studies (Clay, Clay, et al., 2001; Clay, Englel, et al., 2001; Timlin et al., 2001). Relatively low yields in shoulder and summit areas (Figure 16.1b) have been attributed to

1. water redistribution following precipitation;
2. eroded shoulder and summit soils holding less plant available water than footslope soils; or
3. shoulder soils having a lower fertility level than footslope soils (Malo and Worcester, 1975; Hanna et al., 1982; Halvorson and Doll, 1991; Rockström et al., 1999; Clay, Clay et al., 2001).

Water redistribution does not just occur in rolling landscapes containing relative large slopes, it also occurs in coarse-textured soils with relatively small slopes (<2 percent) (Delin et al., 2000).

The impact of water availability at different landscape positions on yield was measured in two multiple-year field-scale experiments conducted in eastern South Dakota. In the first experiment, corn and soybean *(Glycine max)* yields were measured using a calibrated yield monitor between 1995 and 2000. Yield at each sampled point was converted to relative yield by dividing the observed yield by the maximum yield observed between 1995

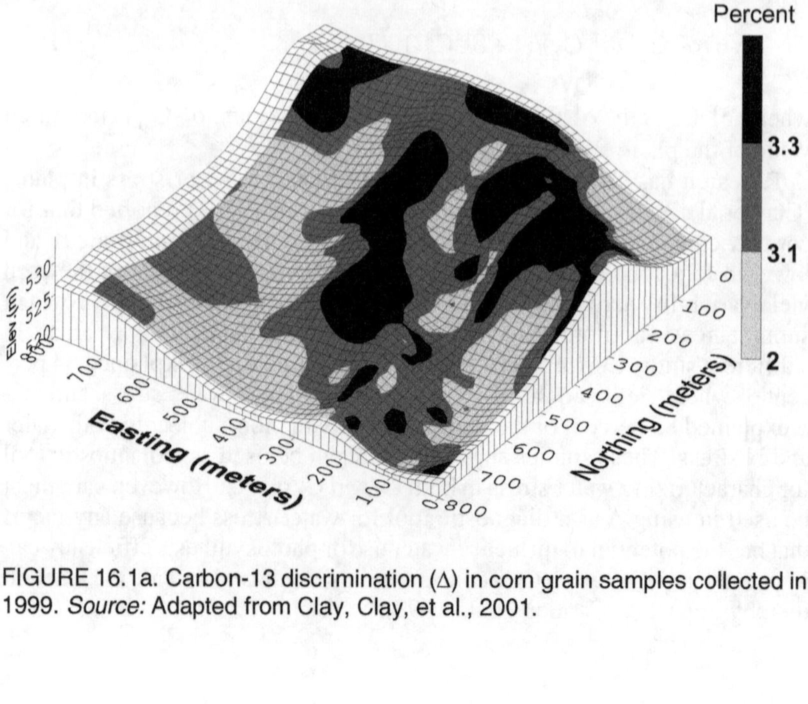


FIGURE 16.1a. Carbon-13 discrimination (Δ) in corn grain samples collected in 1999. *Source:* Adapted from Clay, Clay, et al., 2001.

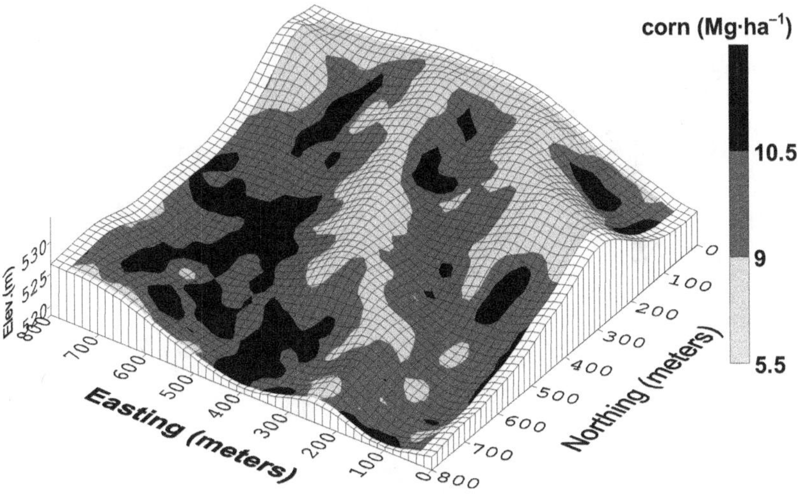


FIGURE 16.1b. Corn yield in 1999 superimposed on a topography map. *Source:* Unpublished data.

and 2000. Summit area yields were relatively stable and produced a single population histogram with a mean of 60 percent of the maximum yield (Figure 16.2). However, in the footslope areas, two peaks were observed. One peak was centered on 80 percent of maximum yield, while the other peak was centered on 40 percent of maximum yield. High yields were associated with years when water did not pond, while low yields were associated with years when water ponding did occur (Figure 16.2).

The second experiment evaluated if water stress was responsible for corn yield spatial variability during a year when water ponding did not occur. In this experiment, 12 randomly selected corn plants located in the summit and footslope areas were watered with 2 L (1.35 cm) of water six times in July and August 1999. In the summit areas, unwatered plants had lower yields (87 g/plant, 95 percent confidence interval [CI] = 37) than watered plants (145 g/plant, 95 percent CI = 31). However, in footslope areas, watered (154 g/plant, 95 percent CI = 46) and unwatered plants (163 g/plant, 95 percent CI = 30) had similar yields. In the summit area, unwatered plants had higher $\Delta$ (3.35‰, 95 percent CI = 0.071) than watered (3.11‰, 95 percent CI = 0.069) plants and in the footslope area watered and unwatered plants had similar $\Delta$ (2.9‰). This experiment was repeated with soybeans in 2000. Plants in summit areas had lower yields and $\Delta$ in unwatered (5.4 g per plant; 18.2‰) than watered (9.2 g per plant; 18.6‰) areas. However, in the footslope area, watering had a limited impact on

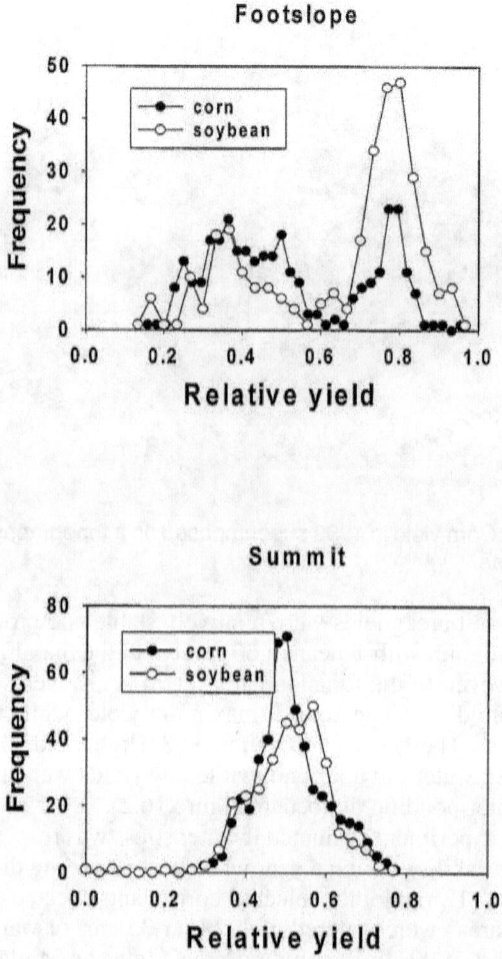


FIGURE 16.2. The influence of landscape position on corn and soybean yield histograms (1995 through 2000) harvested from a field located in eastern South Dakota. *Source:* Unpublished data, 2001.

yields and Δ and were 10 g per plant and 20.0‰, respectively. These data indicate that water stress reduced corn and soybean yields in summit areas and had a limited impact on yields in footslope areas. Halvorson and Doll (1991) had similar results in North Dakota and reported that wheat yields from summit areas were the lowest; yields from small depressions were the largest; and the ability to explain yield variability was improved by accounting for water runoff and runon. Bakhsh et al. (2000) had similar results in Iowa and reported that higher elevation areas had less corn yield, whereas footslope and toeslope areas had more variable yields.

Water stress does not just limit yields in semiarid areas such as the Great Plains. Wyciskalla et al. (2000) reported that in two southern Illinois corn fields, yields were reduced by unseasonable warm temperatures and drought in mid-to-late summer. Camp et al. (2000) reported that in the southeastern Plains of the United States, corn yield responses to irrigation varied among the various soils' map units, with some soils responding more to irrigation than others. Stone et al. (1985) reported that in North Carolina, landscape differences in plant available water was a better predictor of corn yield than erosion phase. Timlin et al. (1998) reported that in New York increasing the soil's water-holding capacity tended to increase corn yield. In Kentucky, corn yields were reduced 12 to 21 percent on eroded soils when compared to uneroded soils (Frye et al., 1982). In Iowa, Spomer and Piest (1982) compared yields for 16 years and found that footslope positions produced higher crop yields than summit or backslope positions. In Illinois, Olson and Nizeyimana (1988) reported that soil formed in loess without root-restricting subsoils showed only a slight (5 percent) corn yield reduction with increasing erosion. However, in crops growing in soil with root-restricting subsoils (clay pans, fragipans, or glacial tills), moderate (24 percent) yield reductions were observed.

Erosion of summit and shoulders areas, which reduces soil fertility and water-holding capacity, may partially be responsible for low yields in shoulder areas. Larney et al. (2000) reported yield losses due to erosion can be restored and that the best amendment for restoring productivity was manure (158 percent grain yield increase) and the least effective amendment was N and P fertilizers (40 percent grain yield increase). They also reported that the largest yield increase from manure was observed in soil with the greatest amount of topsoil loss and the removal of crop water stress by irrigation did not completely offset the loss of topsoil, even when adequate N and P fertilizers were applied. In Kentucky, Frye et al. (1982) had similar results and reported that yield potential decreases due to soil erosion were linked to reduced plant available water resulting from increased clay content of the Ap horizon. However, Bauer and Black (1992) reported that in the northern Great Plains, productivity loss was due to a loss in soil fertility rather than a loss in water-holding capacity.

### *Weeds*

Many producers believe that high weed populations always reduce yields. However, weeds may or may not cause yield reductions. For example, if an area was waterlogged, drought stressed, or ravaged by disease, crop growth may have been poor thus allowing weeds to flourish. When these areas are harvested, the producer may incorrectly assume that the yield loss resulted from weed pressure, rather than relating the yield loss to some other factor that led to less than desired yields.

Weeds occur in patches, and although there are exceptions, rarely does one weed dominate an entire field (Figure 16.3) (Cardina and Doohan, 1999; Clay S.A. et al., 1999). Weed patches have been reported to be fairly stable in location even in crop rotations, although densities often fluctuate from season to season depending on the crop and environmental conditions (Cousens and Mortimer, 1995; Clay S.A., Kruetner, et al., 2006).

Weed patch locations are likely the result of weeds being better adapted to one location over others. Interactions between landscapes, temperature, and germination can influence weed populations. For example, Forcella (1999) reported slower emergence of common ragweed *(Ambrosia artemisiifolia)* and green foxtail *(Setaria viridis)* in sandy versus clay-loam soils. These differences were attributed to the clay soil holding more water and remaining cooler during the spring than the sandy soil. Lems (1998) re-

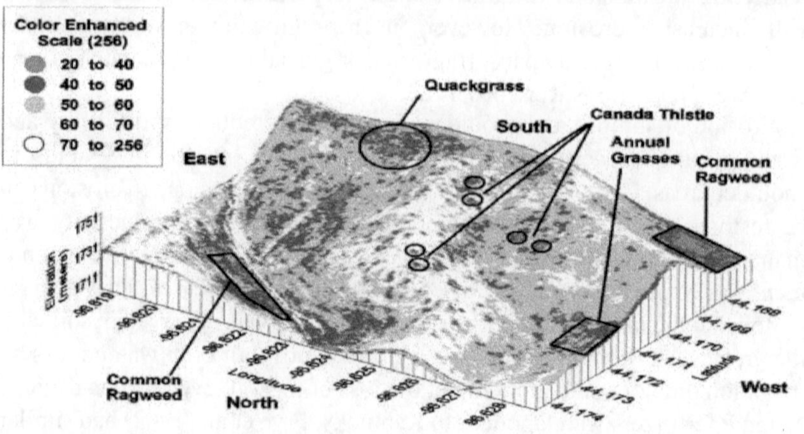


FIGURE 16.3. The remote sensed image (NIR) of a 65 ha soybean field collected in October 1996. Patches of Canada thistle are evident in the summit position while common ragweed was found in footslope areas. The color-enhanced scale ranges from 0 (minimum reflectance) to 256 (maximum reflectance). (See corresponding PLATE 16.3 in the central gallery.)

ported that Canada thistle *(Cirsium arvense)* populations were higher in well-drained (summit) than poorly drained (toeslope) soils (Table 16.1). These differences were attributed to a high water table in toeslope areas that reduced Canada thistle root growth (Holm et al., 1991). In contrast, common ragweed and kochia *(Kochia scoparia)* had populations highest in wet areas and lowest in dry summit areas.

Common ragweed segregation can also result from common ragweed entering secondary dormancy when soil temperature exceeds about 25°C (Forcella, 1999). The summit areas may warm more quickly than footslope areas. Therefore, in summit areas, if temperatures exceed 25°C, then common ragweed may have only a single germination event followed by dormancy. However, in footslope areas, if soil temperatures do not exceed 25°C until midsummer, then a number of germination events can occur. Understanding these biological features will aid in interpreting remote sensed imagery, defining management zone boundaries, and making appropriate weed-management decisions.

Weeds reduce yields by competing with the crop for water, nutrients, and light resources. The availability of these resources may be influenced by landscape position.

As discussed earlier, landscape positions influence water availability, and therefore crops growing in summit areas may be more water stressed than plants in footslope and toeslope areas. Wilson (1992) reported that in South Dakota, weeds reduced wheat yields more under high than under low water stress conditions. Similar results were reported for corn competing

TABLE 16.1. The influence of landscape position on Canada thistle, common ragweed, and kochia population distribution on May 8, 1998.

| Landscape position | Weeds (average population per m$^2$) | | |
|---|---|---|---|
| | **Canada thistle** | **Common ragweed** | **Kochia** |
| Toeslope | 0.9 (0.5) | 43 (14) | 111 (28) |
| Footslope | 1.2 (0.7) | 17 (6) | 118 (23) |
| Backslope | 1.2 (0.5) | 5 (2) | 33 (12) |
| Shoulder | 3.1 (1.1) | 4 (2) | 7 (4) |
| Summit | 4.1 (2.5) | 3 (1) | 1 (1) |

*Source:* Adapted from Lems, 1998.

*Note:* The 95 percent confidence interval of the means are in parentheses.

with velvetleaf *(Abutilon theophrasti)* (Clay and Brix-Davis, 1996), and between soybean and perennial sowthistle *(Sonchus arvensis)* (Zollinger and Kells, 1993).

In an unpublished study conducted by S. A. Clay, the impact of landscape position, weeds, and water stress on soybean yield was investigated (Table 16.2). In this study, soil water contents under natural rainfall conditions, as expected, were lower in summit than footslope positions. Soil water content differences were reflected in soybean yields. Under natural rainfall conditions, controlling weeds did not increase yields in footslope areas but did increase yields in summit areas. Under irrigated conditions, controlling weeds increased yields at both landscape positions. This research suggests in areas where crop yields are reduced by water stress, the crop may be more susceptible to water use by weeds than areas where the crop was not water stressed.

Water used by weeds reduces the water available to the crop. The WUE of different plants can be used to estimate how water extraction by weeds influences crop growth. Geddes et al. (1979) reported that water use efficiencies for common cocklebur *(Xanthium strumarium)* were 2.04 mg biomass·g$^{-1}$ water. Norris (1996) compiled water use efficiencies for various crops and weeds reported by Shantz and Piemeisel (1927) and Dillman (1931) (Table 16.3). Generally, WUEs were higher in $C_4$ than $C_3$ plants. These results are important and if we assume a linear relationship between water availability and yield, then 120 mm of water is required to produce 2,000 kg·ha$^{-1}$ of a $C_3$ weed with a WUE of 1.66 mg biomass·g$^{-1}$ water (average of $C_3$ plants) and 60 mm of water is required to produce 2,000

TABLE 16.2. The influence of weeds, landscape positions, and water stress on soybean yield in an eastern South Dakota field.

| | | Landscape position (g per plant) | |
|---|---|---|---|
| **Treatment Water** | **Weed** | **Summit** | **Footslope** |
| Natural rainfall | None | 5.45 (1.42) | 7.01 (1.51) |
| Natural rainfall | Weeds | 3.25 (1.51) | 6.68 (1.77) |
| Natural and watered | None | 9.36 (1.92) | 9.60 (1.69) |
| Natural and watered | Weeds | 4.42 (1.01) | 6.24 (1.85) |

*Source:* Unpublished data, S. A. Clay.

*Note:* The 95 percent confidence interval for each mean is in parenthesis.

TABLE 16.3. Water use efficiencies for several plants in mg plant/g water.

| Species | Photosynthesis system | Shantz and Piemeisel (1927) WUE | Dillman (1931) WUE |
|---|---|---|---|
| Oats | $C_3$ | 1.65 | 1.87 |
| Soybean | $C_3$ | 1.55 | |
| Alfalfa | $C_3$ | 1.15 | 1.25 |
| Cotton | $C_3$ | 1.76 | |
| Sugarbeet | $C_3$ | 2.65 | 3.29 |
| Oilseed rape | $C_3$ | 1.40 | |
| Cocklebur | $C_3$ | 2.41 | |
| Smooth bromegrass | $C_3$ | 1.02 | 1.28 |
| Lambsquarter | $C_3$ | 1.52 | 2.30 |
| Prostrate knotweed | $C_3$ | 1.47 | |
| Corn | $C_4$ | 2.77 | |
| Sorghum | $C_4$ | 3.51 | 3.71 |
| Foxtail millet | $C_4$ | 3.65 | 3.98 |
| Witchgrass | $C_4$ | | 3.94 |
| Common purslane | $C_4$ | 3.65 | 3.47 |
| Redroot pigweed | $C_4$ | 3.28 | 3.83 |
| Russian thistle | $C_4$ | 3.18 | 4.46 |

*Source:* Adapted from Norris, 1996.

$kg \cdot ha^{-1}$ of a $C_4$ weed with a WUE of 3.33 mg biomass$\cdot g^{-1}$ water (average of $C_4$ plants). Based on only the WUE, a loss of 120 mm of water will reduce soybean (WUE = 1.55 mg biomass$\cdot g^{-1}$ water) and corn (WUE = 2.77 mg biomass$\cdot g^{-1}$ water) biomass 1,860 $kg \cdot ha^{-1}$ and 3,324 $kg \cdot ha^{-1}$, respectively.

Weeds compete with crops for nutrients. Vengris et al. (1955) reported that corn grown with weeds had 10 percent lower N content and 23 percent lower K content than corn grown under weed-free conditions. Studies conducted with old (hybrid released in 1959) versus newer hybrids (hybrid released in about 1990) under low N and weedy conditions indicated that yield of the older hybrid was reduced much more than yield of newer hybrids (Tollenaar et al., 1994). Selecting hybrids that can withstand nutrient and/or weed stress may become an important part of designing weed-management systems.

Adding more nutrients without weed control can stimulate more weed than crop growth, resulting in greater crop loss (Zimdahl, 1980). Under hydroponic conditions, it was found that Palmer amaranth *(Amaranthus palmeri)* had 3.7 times more roots that were 5 times greater in length and much thinner than soybean roots (Wright et al., 1999). Differences in root morphology can contribute to weeds exploring a much greater volume of the surface soil than the crop, which can provide the weed with a competitive advantage for utilizing water and nutrient resources.

The type of fertilizer applied can influence the type of weed pressure. Experiments were conducted in wheat with N, P, and K applied alone or in combination (Das and Yaduraju, 1999). In this study, grass weeds dominated in plots treated with N, whereas in plots where N was excluded and only P and K was applied broadleaf weeds dominated. The amount of competition, interference, and crop yield loss will depend on the limiting nutrient, the crop, and the weed species and infestation density. For example, weed competition for N was much greater in sorghum than soybean (Everaarts, 1992; Everaarts, 1993). Different densities of common lambsquarters *(Chenopodium album)* and littleseed canarygrass *(Phalaris minor)* reduced N uptake to varying degrees in wheat but this nutrient reduction was not the major cause for the observed yield reduction (Iqbal and Wright, 1999).

Many producers take pride in having clean fields. One reason frequently given for zero weed tolerance is to control the weed seed bank. It is important to point out that

1. the seed bank contains tens of millions of seeds per ha and therefore subtracting or adding a few seeds of some weed species will have little influence on weed seed numbers;
2. controlling weeds when their population is below the economic threshold does not increase profitability;
3. some of the new technologies provide excellent weed control, and therefore it is easier to clean up escapes; and
4. as stated earlier, yield threshold values may vary with landscape position.

Of course, there are some weed species such as new invasive weeds, or those that are extremely troublesome for which growers should have zero tolerance. In the U.S. Midwest these weeds include woolly cupgrass *(Erio-*

*chloa villosa),* black nightshade (*Solanum* spp.), common and tall waterhemp (*Amaranthus rudis* and *A. tuberculatus,* respectively), giant foxtail *(Setaria faberii),* wild proso millet *(Panicum miliaceum),* and state-designated noxious weeds.

If weed decisions are based on the infestation level, yield reduction models for each crop/weed should be developed. Numerous studies have been conducted related to yield reduction potential of specific weeds in specific crops, how time of emergence influences crop yield and growth (Clay S.A., et al., 2005), and how weeds compete between themselves. For example, one giant ragweed *(Ambrosia trifida)* competing in corn causes a similar yield loss as 20 green foxtail plants (Zimdahl, 1980). The cost of control, yield potential of the crop at a specific field site, and anticipated selling price for the crop must be accounted for in some manner to determine the economic threshold of specific weed-management strategies which would be cost effective (Bussan and Boerboom, 2001; Clay S.A., Banken, et al., 2006; Clay S.A., Kreutner, et al., 2006).

### *Soil Nutrients*

Numerous studies have investigated nutrient spatial variability and its relationship to crop yields (Khakural et al., 1996; Sudduth et al., 1996; Bakhsh et al., 2000; Kravchenko and Bullock, 2000). A common feature of most studies is low linear correlation coefficients between yield and soil nutrient concentrations. Poor correlation can result from several factors. First, yields are not correlated to extractable soil nutrients because soil nutrient concentrations are higher than the critical level, the amount of a nutrient extracted by the soil test is not correlated to nutrient availability, and yields are limited by a factor other than soil nutrients (Black, 1993). Second, the yield response relationship [y = f(nutrients)] is nonlinear. This hypothesis has been demonstrated in numerous soil fertility studies (Black, 1993). Third, as discussed, yields are the result of interactions between the crop and many other factors, and therefore an analysis that considers only a single factor should explain only a small portion of the yield variability (Timothy et al., 1994; Paz et al., 1999; Clay, Engel, et al., 2001). For example, if water stress reduces the corn yield potential in summit areas from 5 to 3.5 $Mg \cdot ha^{-1}$, then the plants in the summit area may require less nutrients than plants in the footslope areas. Under these conditions, the relationship between yield and soil-nutrient may be landscape dependent. Interactions among such factors as yield potential and soil nutrient supplying power can show some areas of a field being responsive to nutrients and other areas not being responsive.

Interactions among soil nutrient availability and the soil's physical, biological, and chemical properties might be responsible for crops that respond to nutrients differently at different landscape positions. For example, Malzer et al. (2000) reported that in Minnesota, yield and crop response to N and P were impacted by drainage and compaction. The highest potential returns from P were observed in poorly drained compacted soils. Timothy et al. (1994) reported that winter wheat yields in the Palouse region of eastern Washington were greater on the southern-facing backslope than shoulder positions and that different N response functions were observed for different landscape positions. Nolan et al. (1998) reported that simple landscape classification delimited areas of the field that responded differently to N fertilizer.

Yields and N recommendations in footslope areas can also be impacted by interactions between N and water cycles. Stout and Schnabel (1997) showed that for perennial ryegrass (*Lolium perenne* L.) water use efficiency decreased with poor drainage because of the loss of N through denitrification. Clay et al. (1997) reported that in poorly drained areas of a field there was a net loss of 95 kg $N \cdot ha^{-1}$, while in adjacent tile-drained areas there was a net gain of 98 kg $N \cdot ha^{-1}$. The yield loss was attributed to denitrification and the N gain was attributed to N mineralization.

In dryland conditions, understanding interactions between N, water, and plant characteristics is critical to maximize yield. For example, Angus and van Herwaarden (2001) reported that in dryland wheat grown in southeastern Australia, yield reductions from N occurred if N stimulated early vegetative growth and water use, which in turn, reduced available water during grain fill. Clearly, to minimize the over- and underapplication of fertilizers, landscape position and its yield potential must be considered.

### *Quantifying Yield Loss*

One approach to quantify the effect of different factors on yield at different landscape positions is to use simulation models such as CROPGRO-Soybean (Hoogenboom et al., 1994); and CERES-Maize (Jones and Kiniry, 1986) or CERES-Wheat to solve equation 16.1. These models integrate the effects of soil properties, drainage, pest levels, crop genetics, management, and weather information on a daily basis to predict the impact of different management options and climatic conditions on yield at specific points in the field (Batchelor et al., 1993; Shen et al., 1998; Batchelor and Paz, 1999; Calmon et al., 1999). Over the past 25 years, many of the individual components of these models have undergone continuous testing and improvement.

For example, the water flow components of these models were tested (Shen et al., 1998) and modified to account for landscape processes (Basso, 2000).

Currently, these models are being used to conduct systems analysis of plots, specific areas within fields, and specific areas within regional areas (Paz et al., 2003). On the plot scale (0.07-0.2 ha each), the CROPGRO-Soybean model was evaluated in semiarid tropical conditions (Alagarswamy et al., 2000). The model was used to develop a yield-evaporation relationship and to assess the influence of soil water storage on soybean yield. Yield was linearly related to ET and reduced nonlinearly with decreases in soil depth. The model predicted that severe yield reduction occurs if the soil depth was less than 45 cm.

On the regional scale, the CERES-Wheat (wet) model was used to evaluate factors limiting yields in 10 by 10 km grids in Finland (Saarikko, 2000). The model was able to correctly predict that at site and regional scales, climatic conditions influenced wheat yield variation.

On the field scale, Paz et al. (1999) evaluated the causes of corn yield variation in an Iowa field. In this study, the field was separated into grids and the CERES-Maize model was used to solve equation 16.1 for each grid. This study determined that corn plant population was a major factor contributing to yield variability. In a different study conducted in Iowa using the same approach, Batchelor and Paz (1999) used CROPGRO-Soybean to separate the effects of soybean cyst nematode *(Heterodera glycine)* (SCN), weeds, and water stress on soybean yield. This study showed that by accounting for water stress, SCN, and weed interactions, 77 percent of the soybean yield variability could be explained and that the influence of individual yield-limiting factors could be evaluated.

In summary, use of experimental approaches makes it difficult to quantify the magnitude of the impact of different yield-limiting factors on yield. Quantifying yield reductions is further complicated by numerous interactions between the factors. Simulation modeling within a geographic information system (GIS) provides a mechanism to predict the relative importance of different yield-limiting factors on causing yield variability.

## *MANAGING YIELD VARIABILITY*

Fertilizer applicators, planters, sprayers, and tillage implements controlled by GIS and GPS provide the hardware required to change varieties on-the-go or apply variable-rate applications of agrichemicals used to correct problems caused by many of the factors in Equation 16.1. For example, a producer may choose to do any of the following:

- Plant drought-resistant cultivars in summit areas and disease-resistant cultivars in footslope areas
- Plow footslope areas and use no-tillage in summit areas
- Preferentially apply manure to shoulder and summit areas
- Use precision farming techniques to develop weed and nutrient response functions
- Develop a database that monitors the extent of pest and disease problems
- Apply different rates or blends of fertilizers to correct nutrient problems
- Use remote sensing to identify weed patch locations and nutrient stresses (Figure 16.3)

Given the large number of options available, a producer must be very careful to select management strategies that increase profitability. Using variable-rate technology does not necessarily reduce chemical costs and/or increase profits. Instead, precision farming uses information to improve decisions and does not require variable-rate applications. D.E. Clay et al. (1999) showed that precision phosphorous (P) management has the potential to increase profits more in some fields than others. Simulation modeling showed that P management was more likely to increase profitability in: (1) fields where the crop responded to P; (2) fields containing a substantial amount of variability; and (3) fields where the traditional P recommendations were relatively low. This study pointed out that in conducting an enterprise analysis of individual decisions, accurate fertilizer response functions are needed. For example, in fields with a relatively high level of P variability, precision farming techniques will be more profitable in fields with high expected yield responses than low expected responses from P. Response functions can be obtained by: (1) using precision farming techniques to determine response functions for the different soils (Malzer et al., 2000); (2) conducting a literature review of published papers and state progress reports; or (3) visiting with experts. Response functions also can be estimated using experimental approaches described by Black (1993). However, in most cases, soil-specific response functions for most agronomic inputs are not available.

### *The Use of Tillage to Manage Yield Variability*

Tillage impacts water budgets, weeds, diseases, soil nutrient availability, soil temperature, and soil productivity (Frye et al., 1982; Olson and Nizeyimana, 1988; Gollany et al., 1992; Khakural et al., 1992). We hypothesize that there are portions of the same field that would benefit from increasing and other portions that would benefit from decreasing tillage intensity. Soil properties that must be evaluated when considering site-specific tillage options are soil wetness, water infiltration, soil temperature, soil structure, and the soil's susceptibility to compaction, crusting, drainage, and erosion.

Tillage impacts water budgets by at least three different mechanisms. First, water infiltration and water flow are impacted by tillage. Initially, tillage may increase water infiltration, however, as raindrops destroy aggregates, water infiltration rates can decrease (Pikul and Aase, 1995). Higher infiltration capacity in no-tillage than tillage systems has been attributed to more macropores and biochannels resulting from higher organic matter contents in the soil surface and the protection of the soil surface from raindrop impact by surface residue.

Second, tillage changes water flow in the soil profile. Zachmann et al. (1986) investigated the impact of earthworms, tillage, and residue on macroporous water flow and redistribution. In this study, water redistribution was evaluated by quantifying the fate of $Br^-$ applied to soil. The no-tillage surface-residue treatments more than doubled the number of burrows open to the soil surface than the incorporated tillage treatment, and the percentage of applied $Br^-$ found below 19 cm was higher in the no-tillage (33 percent) than the incorporated residue treatment (23 percent). Tillage differences were attributed to an unbroken network of continuous macropores, cracks, and fissures in the no-tillage system.

Third, tillage reduces the amount of residues on the soil surface, which in turn impacts soil temperature, infiltration, evaporative losses, and water content (Greb, 1966; Johnson et al., 1984; Azooz and Arhad, 1995; Sauer et al., 1996). Reduced evaporative losses in no-tillage were attributed to surface residue having a higher reflectance than bare soil. Under these conditions, reduced evaporative losses in no-tillage can result in higher surface water contents. For example in South Dakota, Khakural et al. (1992) reported that surface volumetric water contents and temperatures were lower in ridge and no-tillage than moldboard plow systems.

Changes in soil water budgets and soil temperatures will impact plant growth and microbial activity. In semiarid regions, no-tillage-induced increases in soil water can increase the cropping frequency and reduce wind and water erosion (Anderson et al., 1999; Norwood, 1999; Schillinger et al., 1999). For example, Gibson et al. (1992) reported that when compared with conventional tillage, leaving sorghum stubble on the soil surface increased sorghum yields 390 $kg \cdot ha^{-1}$. Halvorson et al. (2000) reported that in North Dakota under non-N limiting conditions, no-till grain yields were greater than conventional tillage in years when plant available water was greater than 260 but less than 400 mm. When total plant available water exceeded 400 mm, conventional tillage yields exceeded no-tillage yields.

Scientists have also hypothesized that tillage may impact root distribution. However, the evidence indicating that no-tillage actually changes root distribution is weak. Barber (1971) reported that in the 0 to 30 cm soil depth, no-tillage and chisel plow had similar root lengths. Hilfiker and Lowery (1988) reported that root distribution in no-tillage and tilled systems were similar. Rasse and Smucker (1998) reported that for one out of two years, the number of roots present in the C horizon were greater for a tilled than a no-tilled system.

### *The Use of Crop Rotations to Manage Yield Variability*

Crop rotations influence many components of the production system including weed growth, weed reproduction, diseases, nutrient availability, and water budgets (Ball and Miller, 1990; Blackshaw, 1994; Buhler et al., 1994; Singer and Cox, 1998; Shirima et al., 2000; Manley et al., 2001). Single weed or weed complexes are observed most often in a monoculture cropping system in which the same or similar herbicide strategies are used (Ball and Miller, 1990; Buhler, 1992; Kapusta and Krautz, 1993). This type of selection pressure eliminates susceptible species and promotes tolerant or resistant weeds to dominate the field. For example, the first widely used herbicide was 2,4-D (2,4-dichlorophenoxyacetic acid), which controls broadleaf plants. Growers using 2,4-D dramatically reduced broadleaf weeds in corn and wheat fields but saw a dramatic shift to grass weed problems such as yellow, green, and giant foxtail and wild oat *(Avena fatua)*. Another popular herbicide, still in high use in U.S. Midwest corn production, is atrazine (6-chlor-N-ethyl-N'-(1-methylethyl)-1,3,5-triazine-2,4,diamine). This herbicide can be applied pre- or postemergence and controls both broadleaf and grass weeds. At times, atrazine was applied at rates exceeding 3.3 $kg{\cdot}ha^{-1}$. The consequences of using atrazine yearly at this high rate were (1) the development of atrazine-resistant weeds; (2) carryover problems; and (3) in many areas the atrazine contamination of surface and groundwater supplies through leaching or wind erosion (DeSutter et al., 1999; Clay et al., 2000). If the crop was corn, no carryover problems resulted. However, crops such as soybean, wheat, or alfalfa could not be planted after corn due to injury or total crop failure. The application of atrazine resulted in a multitude of hectares to be planted in corn/corn rotations, which in turn influenced fertilizer, pest, and disease control strategies. Today, atrazine is typically applied at rates near 1.1 $kg{\cdot}ha^{-1}$ and carryover problems have been lessened considerably.

In contrast, in cropping systems that utilized a variety of crops, a diversity of herbicides, and reduced tillage, grass and broadleaf seed production can be less than in a monoculture system with high tillage inputs (Clay and Aguilar, 1998; Kegode et al., 1999). For example, velvetleaf plant characteristics differ when grown in corn, soybean, and alfalfa *(Medicago sativa).* In a corn field not treated with herbicide, velvetleaf had 33 percent less biomass and produced 60 percent less seed than velvetleaf grown in soybean (Bussan and Boerboom, 2001). Yield losses of soybean have been estimated at about 10 percent per velvetleaf plant present per meter of row (Stoller et al., 1987), compared to losses in corn of about 4.5 percent (Scholes et al., 1995). When alfalfa was cut for hay three times during a growing season, velvetleaf plants remained small and no seed was produced (S. A. Clay, unpublished data, 2001). These data indicate that velvetleaf control is most critical in the soybean years of a corn/soybean/alfalfa rotation.

Crops planted in rotations typically have higher yields than crops not planted in rotations. Yield increases due to rotations can be partially attributed to disease control. For example, in southern Australia wheat following canola *(Brassica napus)* on average yielded 19 percent better than wheat following wheat (Angus et al., 1999). Duczek et al. (1999) reported that a minimum of two years is required to control wheat septoria leaf spot, caused by *Septoria tritici* and *S. avenue,* and one year between wheat crops is required to control tan spot *(Pyrenophora tritici-repentis).* Fernandez et al. (1998) reported that the best rotation to reduce wheat leaf spot was two consecutive years of wheat followed by two years of a noncereal crop. Shirima et al. (2000) reported that in Tabora, Tanzania, using a fallow-corn-tobacco rotation could control nematodes in tobacco.

Plant diseases also may be influenced by the crop nutrient and soil water status. Pacumbaba et al. (1997) studied the influence of complete fertilization (N, P, and K), N only (ammonium nitrate), P only (0-46-0), and K only (0-0-60) on soybean stem canker (*Diaporthe phaseolorum* Sacc.), phytophthora root rot (*Phytophthora megasperma* sp. glycinea), bacterial blight (*Pseudomonas syringae* pv. glycinea), soybean mosaic virus (SMV), and soybean cyst nematode *(Heterodera glycines).* They reported that (1) soybean plants fertilized with only K had a higher incidence of phytophthora root rot and SMV; (2) soybean plants fertilized with up to 100 $kg \cdot ha^{-1}$ of P had a lower incidence of SMV; and (3) damage from soybean cyst nematodes or bacterial blight was not influenced by fertilizers. Giesler and Powers (1999) reported that rotations could reduce but not eliminate soybean cyst nematodes. Fernandez et al. (1998) reported that in southwestern Saskatchewan, Canada, N deficiencies in wheat increased leaf spot in dry summers and P deficiencies decreased leaf spot in cool wet springs. Kutcher et al. (1999) reported that at Prince Albert, Canada, septoria and tan spot in wheat were observed to be more severe on upper as opposed to lower slope positions and there were no relationships observed between N and P fertility and foliar diseases. At this site, yields at lower landscape positions were greater than yields on upper slopes. At Watrous, Canada, Kutcher et al. (1999) reported that common root rot *(Cochliobolus sativus)* was less severe on upper slope positions than mid and lower slopes.

Crop rotations also influence soil nutrient status. For example, Carpenter-Boggs et al. (2000) reported that in South Dakota, rotations and N fertilizer influence N mineralization and the inclusion of a legume in the rotation increases net N mineralization. In South Dakota the N fertilizer credit for corn when soybeans are the previous crop is about 1 kg N$\cdot$ha$^{-1}$ for every 60 kg soybeans$\cdot$ha$^{-1}$ produced. The sequence of the rotation may also impact the plant N requirement. Scharf et al. (2000) reported that in Missouri, where corn was planted into sod, corn in the zero N treatments showed N deficiency symptoms and the economic N rate was much higher than in the sod-soybean-corn cropping sequence.

### *Plant Genetics*

The plant genetics selected by a producer is one of the most important decisions made during any given year. Many producers will or do have the ability to plant crops with different genetic characteristics in different areas of the field. In order to target plant with specific genetic characteristics to specific problems, producers need to know the plants' genetic characteristics.

Genetics influence the plant's ability to withstand diseases, water stress, pests, and use nutrients (Hall and Cholick, 1989). In the past, most producers were limited to planting a single cultivar in a field. Using site-specific farming, producers have the capability to plant different areas of the field with different cultivars. To match cultivars with soil characteristics producers need

- simple techniques for identifying management zones, and
- information on how different cultivars respond to different environmental conditions.

Many producers have been using strip trials to obtain this information (Doerge and Gardner, 1999).

Producers and scientists have been experimenting with different approaches for characterizing management zones. Soil apparent electrical conductivity ($EC_a$), as measured by an electromagnetic sensor is one such approach (Kitchen et al., 1999; Clay, Chang et al., 2001). Apparent electrical conductivity represents an integrated measure of many soil properties (clay content, water content, varying depths of conductive soil materials, salinity, metals, bulk density, and temperature). Many of the soil properties integrated by $EC_a$ measurement are related to crop productivity. In Missouri, $EC_a$ can be used to estimate topsoil depth (Kitchen et al., 1999) and in South Dakota, $EC_a$ can be used to characterize soil drainage (Clay, Chang et al., 2001).

Reese et al. (2000) in South Dakota investigated the use of $EC_a$ as a tool for matching soybean cultivar to soil characteristics. In South Dakota, $EC_a$ is typically low in summit areas and high in footslope and toeslope areas (Clay, Chang et al., 2001). Differences in $EC_a$ result from footslope soils containing more salts (gypsum and calcium carbonate), organic matter, and water than summit soils (Malo and Worcester, 1975). In one South Dakota field, Reese et al. (2000) planted phytophthora-resistant and susceptible soybeans, the resistant soybeans yielded better in high $EC_a$ areas (poorly drained) and the susceptible soybean cultivar yielded better in low $EC_a$ areas (well-drained). However, in a different field, the susceptible soybean cultivar did better at all landscape positions. Differences between the two field studies were attributed to the suitability of cultivars that were selected. At one site, the phytophthora-resistant cultivar was both iron-chlorosis and phytophora-resistant, while at the other site the phytophthora-resistant cultivar was not resistant to iron chlorosis.

Different cultivars also have different abilities to withstand water stress. The ability of plants to withstand water stress is influenced by the volume of soil explored by the roots, matching the plants' phenology to water supply, osmotic adjustments, individual leaf size, and the rate of leaf appearance. Field- and model-simulated research has shown that the larger the soil volume explored by the roots the better adapted they are to water stress conditions (Jordan et al., 1983; Jones and Zur, 1984; Sinclair, 1994; Oikeh et al., 1999). Ludlow and Muchow (1990) suggested that other traits may influence water use efficiency (WUE): $CO_2$ assimilation activity, early stomata closure, or delayed stomata closure. Sinclair and Muchow (2001) conducted a simulation analysis of these traits in corn using the Muchow and Sinclair (1991) model and reported that

- corn yield and WUE for the control plant (80 cm rooting depth) was 409 $g \cdot m^{-2}$ and 0.99 $mg \cdot g^{-1}$, respectively;
- increasing the rooting depth from 80 to 120 cm increased corn yield from 409 to 563 $g \cdot m^{-2}$ and the WUE from 0.99 to 1.23 mg biomass$\cdot g^{-1}$ water;

- slowing the rate of leaf appearance reduced the rate of vegetative development and resulted in increased yield (540 $g \cdot m^{-2}$ ) and WUE (1.23 mg biomass$\cdot g^{-1}$ water);
- reducing leaf size reduced both yield (370 $g \cdot m^{-2}$) and water use efficiency (0.9 mg biomass$\cdot g^{-1}$ water);
- early stomatal closure slightly increased yield (445 $g \cdot m^{-2}$) and WUE (1.08 mg biomass$\cdot g^{-1}$ water); and
- delaying stomatal closure reduced yield (381 $g \cdot m^{-2}$) and water use efficiency (0.93 mg biomass$\cdot g^{-1}$ water).

These findings are important because they suggest which genetic traits should be selected for in plant breeding programs. In selecting for drought-tolerant plants, it may be possible to use $^{13}C$ discrimination ($\Delta$) as a diagnostic tool. For example, if early stomata closure is the desired attribute, then $C_3$ plants grown under moderate water stress should have relatively low $\Delta$ and if delayed stomatal closure is the desired attribute, then the plant should have relatively high $\Delta$ (Condon et al., 1992).

Genetically modified organisms (GMO) have changed pest control. Managing pests through genetics rather than chemicals has the potential to reduce the total amount of pesticides applied to soil. However, in many areas of the world consumers are concerned with how GMOs will influence human health and nontarget organisms. Irrespective of the safety question, GMO seeds are expensive, may require refuge areas, and can result in pest population shifts (Anderson and Hellmich, 2001). One example of a GMO crop is Bt corn (Anderson and Hellmich, 2001). Bt corn, available since 1996, has a resistance to European corn borer *(Ostrinia nubilalis).* The gene with corn borer-resistance was obtained from the soil bacterium *Bacillus thuringiensis.* In 2000, approximately 25 percent of the total corn acreage in the United States contained the Bt gene. Many scientists are concerned that planting too much Bt corn can lead to Bt resistance in European corn borer populations. To prevent the loss of this management tool, the current recommendation is that growers should plant 20 to 25 percent of their crop with non-Bt corn. Refuge areas can be the breeding ground for non-Bt-resistant European corn borer. The refuge areas must be planted close enough to Bt corn to encourage random mating of susceptible and resistant moths. A rule of thumb is that the refuge should be within 800 m of the Bt corn. Refuges can be planted as blocks, strips, and perimeter plantings. Clearly, precision farming techniques can help design effective refuge programs and monitor shifts in pest pressure.

A second example of a GMO crop is Roundup Ready soybean. Transgenic soybean is a relatively new use of genetic modification technology. In

1996, Roundup-(glyphosate) resistant transgenic soybean varieties were not available in South Dakota. In 1997, about 40 percent of the soybean acres were planted with Roundup Ready varieties and in 2000 this amount rose to 67.5 percent of South Dakota's 4.3 million acres (South Dakota Ag Statistics Service, 2000). By 2004, over 89 percent of the soybean acreage and about 50 percent of the corn acreage were treated with glyphosate (South Dakota Ag Statistics Service, 2005). The shift to this technology is due to the complete control of all emerged plants that do not carry the resistant gene. However, even with this technology, shifts in weed species across fields have been observed (Hart and Wax, 1999). Weeds that have some tolerance to the herbicide, such as velvetleaf and common cocklebur, have become problematic in some fields (Shaner, 2000). Other species, such as common and tall waterhemp have become more prevalent, not because they are tolerant of the chemical, but the emergence of these species is later in the season after row closure when herbicide and tillage treatments have been completed. If use patterns continue with this herbicide in the future, weeds truly resistant to glyphosate such as rigid ryegrass *(Lolium rigidum)* may develop (Powles et al., 1998).

In summary, many producers will or do have the ability to plant crops with different genetic characteristics in different areas of the field. To minimize errors in matching genetics to environments, producers need to know the genetic characteristics of the crop.

## CONCLUSIONS

Site-specific farming is not about using gadgets and gizmos but using information and knowledge to solve problems. Many producers and scientists are frustrated by the difficulty in converting information into knowledge. Both biological knowledge and technical feasibility are needed to integrate precision farming technologies into current production systems (Kennedy, 2000). Economic benefits and wide-scale adoption of site-specific management practices by small- and medium-sized producers will be realized only if (1) costs for obtaining spatial information are minimal; (2) we can accurately predict the impact of different management options on yield; (3) we can develop relatively simple decision rules that reduce the economic risk associated with adopting precision technologies; and (4) we can transfer decision tools to end users. Developing products that meet these goals is complicated by spatial and temporal yield variability. Simulation models such as CROPGRO-Soybean and CERES-Maize when linked to soil property information and remote sensing through GIS have been used to predict the causes of yield variability, assess the economic consequences of using different information layers in making decisions, and determine the impact of different management decisions and climatic scenarios on profitability. Failure to account for interactions between the various components of a system can cause less than optimum recommendations.

## REFERENCES


Alagarswamy, G., P. Sigh, G. Hoogenboom, S.P. Wani, P. Pathak, and S.M. Virmani. 2000. Evaluation and application of the CROPGRO-Soybean simulation model in a Vertic Inceptisol. *Agricultural Systems,* 63:19-32.

Anderson, R.L., R.A. Bowman, D.C. Nielson, M.F. Vigil, R.M. Aiken, and J.G. Benjamin. 1999. Alternative crop rotations for the central Great Plains. *Journal of Production Agriculture,* 12:95-99.

Anderson, P.L. and R.L. Hellmich. 2001. Bt corn and insect resistance management: What are they? In *Site Specific Management Guidelines.* eds. D.E. Clay, D. Armstrong, B. Dartst, K. Griffin, C. Reese and K. Reitsma. Available online at www-ppi-far.org/ssmg. Potash and Phosphate Institute, Norcross, GA. SSMG-33.

Angus, J.F., J.M. Desmarchelier, P.A. Gardner, A. Green, P.J. Hocking, G.N. Howe, J.A. Kirkegaard, S. Marcroft, J.A. Meed, G.D. Pitson, et al. 1999. Canola and Indian mustard as break crops for wheat. *Proceedings of 10th International Rapeseed Congress, Canberra,* ACY Australia. Sept. 26-29, 1999. Groupe Consultatif Int.de Recherche sur le Colza, Paris. CD-ROM computer file.

Angus J.F. and A.F. van Herwaarden. 2001. Increasing water use efficiency in dryland wheat. *Agronomy Journal,* 93:290-298.

Azooz, R.H. and M.A. Arhad. 1995. Tillage effects on thermal conductivity of two soils in northern British Columbia. *Soil Science Society of America Journal,* 59:1413-1423.

Bakhsh, A., T.S. Colvin, D.B. Jaynes, R.S. Kanwar, and U.S. Tim. 2000. Using soil attributes and GIS for interpretation of spatial variability in yield. *Transactions of American Society of Agricultural Engineers,* 43:819-828.

Ball, D.A. and S.D. Miller. 1990. Weed seed population response to tillage and herbicide use in three irrigated cropping sequences. *Weed Science,* 38:511-517.

Barber, S.A. 1971. Effect of tillage practices on corn *(Zea mays)* root distribution and morphology. *Agronomy Journal,* 63:724-726.

Basso, B. 2000. Digital Terrain Analysis and Simulation Modeling to Assess Spatial Variability of Soil Water Balance and Crop Production. Doctoral dissertation, Michigan State University, East Lansing, Michigan.

Batchelor, W.D., J.W. Jones, K.J. Boote, and H.O. Pinnschmidt. 1993. Extending the use of crop models to study pest damage. *Transactions of American Society of Agricultural Engineers,* 36:551-558.

Batchelor, W.D. and J.O. Paz. 1999. Crop growth models and yield variability. *Integrated Crop Management Newsletter.* May, 1999 edition. IC-482, Available online at www.ipm.iastate.edu/ipm/icm/1999/5-5-1999/cgmod.html (accessed 4/8/02).

Bauer, A. and A.L. Black. 1992. Organic carbon effect on available water capacity of three textural groups. *Soil Science Society of America Journal,* 56:248-254.

Black, C.A. 1993. *Soil fertility evaluation and control.* Boca Raton, FL: Lewis Publishers.

Blackshaw, R.E. 1994. Rotation effects downy brome *(Bromus tectorum)* in winter wheat *(Triticum aestivum). Weed Technology,* 8:728-732.

Boutton, T.W. 1991. Stable carbon isotope ratios of natural materials: 1: Sample preparation and mass spectrometer analysis. In *Carbon isotope techniques,* eds. D.C. Coleman and B. Fry. New York: Academic Press. pp. 155-171.

Buhler, D.D. 1992. Population dynamics and control of annual weeds in corn *(Zea mays)* as influenced by tillage systems. *Weed Science,* 40:241-248.

Buhler, D.D., D.E. Stoltenberg, R.L. Becker, and J.L. Gunsolus. 1994. Perennial weed populations after 14 years of variable tillage and cropping practices. *Weed Science,* 42:205-209.

Bussan, A.J. and C.M. Boerboom. 2001. Modeling the integrated management of velvetleaf in a corn-soybean rotation. *Weed Science,* 49:31-41.

Calmon, M.A., W.D. Batchelor, J.W. Jones, J.T. Ritchie, K.J. Boote, and L.C. Hammond. 1999. Simulating soybean root growth and soil water extraction using a functional crop growth model. *Transactions of the American Society of Agricultural Engineers,* 42:1867-1877.

Camp, C.R., E.J. Sadler, D.E. Evans, and J.A. Millen. 2000. Corn yield response to fertilizer for a range of water regimes and soil mapping units in the southeastern coastal plain. *Proceedings of the 5th International Conference on Precision Agriculture.* eds. P.C. Robert, R.H. Rust, and W.E. Larson, July 16-19, 2000, Bloomington Minnesota, American Society of Agronomy, Madison WI. CD-ROM.

Cardina, J. and D.J. Doohan. 1999. Weed biology and precision farming. In *Site specific management guidelines.* eds. D.E. Clay, D. Armstrong, B. Dartst, K. Griffin, C. Reese and K. Reitsma. Available online at www-ppi-far.org/ssmg. Potash and Phosphate Institute, Norcross, GA. SSMG-25.

Carpenter-Boggs, L.C., J.L. Pikul Jr., M.F. Vigil, and W.E. Riedell. 2000. Soil nitrogen mineralization influenced by crop rotation and nitrogen fertilization. *Soil Science Society of America Journal,* 64:2038-2045.

Clay, D.E., J. Chang, S.A. Clay, M.M. Ellsbury, C.G. Carlson, D.D. Malo, D. Woodson, and T. DeSutter. 1997. Field scale variability of nitrogen and delta 15-N. *Communications in Soil Science and Plant Analysis,* 28:1513-1527.

Clay, D.E., J. Chang, D.D. Malo, C.G. Carlson, C. Reese, S.A. Clay, M. Ellsbury, and R. Berg. 2001. Factors influencing spatial variability of soil apparent electrical conductivity. *Communications in Soil Science and Plant Analysis,* 32: 2993-3008.

Clay, D.E., S.A. Clay, Z. Lui, and C. Reese. 2001. Spatial variability of C-13 Isotopic discrimination in corn *(Zea mays). Communications in Soil Science and Plant Analysis,* 32:1813-1827.

Clay, D.E., S.A. Clay, D.J. Lyon and J.M. Blumenthal. 2005. Can $^{13}C$ discrimination in corn *(Zea mays)* grain be used to characterize intra-plant competition for water and nitrogen? *Weed Science,* 53:23-29.

Clay, D.E., R.E. Engel, D.S. Long, and Z. Liu. 2001. Nitrogen and water stress interact to influence C-13 discrimination in wheat. *Soil Science Society of America Journal,* 65:1823-1828.

Clay, D.E., D.D. Malo, C.L. Reese, and C.G. Carlson. 1999. Identifying good candidates for precision phosphorus management. In *Site specific management guidelines* eds. D.E. Clay, D. Armstrong, B. Dartst. K. Griffin, C. Reese and K. Reitsma. Available online at www-ppi-far.org/ssmg. Potash and Phosphate Institute, Norcross, GA. SSMG 13.

Clay, S.A., and I. Aguilar. 1998. Weed seed banks and corn growth following continuous corn and alfalfa. *Agronomy Journal,* 90:813-818.

Clay, S.A., K.R. Banken, F. Forcella, M.M. Ellsbury, D.E. Clay, and A.E. Olness. 2006. Influence of yellow foxtail *(Setaria pumila)* on corn *(Zea mays)* growth and yield. *Communications in Soil Science and Plant Analysis,* (in press).

Clay, S.A. and K. Brix-Davis. 1996. The interaction between the spatial variability of velvetleaf population and corn grain yield. *Proceeding of the 3rd International Conference of Precision Agriculture.* eds. P.C. Robert, R.H. Rust, and W.E. Larson. Minneapolis, MN American Society of Agronomy, Madison, WI. p. 565.

Clay, S.A., R.H. Dowdy, J.A. Lamb, J.L. Anderson, B. Lowery, R.E. Knighton, and D.E. Clay. 2000. Herbicide movement and dissipation of atrazine in four midwestern sites. *Journal of Environmental Science Health Part B,* 35:259-278.

Clay, S.A., J. Kleinjan, D.E. Clay, F. Forcella and W. Batchelor. 2005. Growth and fecundity of several weed species in corn and soybean. *Agronomy Journal,* 97:294-302.

Clay, S.A., B. Kreutner, D.E. Clay, C. Reese, J. Kleinjan and F. Forcella. 2006. Spatial distribution, temporal stability, and yield loss estimates for annual grasses and common ragweed (*Ambrosia artimissifolia*) in a corn/soybean production field over nine years. *Weed Science,* (in press).

Clay, S.A., G.J. Lems, D.E. Clay, F. Forcella, M.M. Ellsbury, and C.G. Carlson. 1999. Sampling weed spatial variability on a field-wide scale. *Weed Science,* 47:674-681.

Condon, A.G., R.A. Richards, and G.D. Farquhar. 1992. The effect of variation in soil water availability vapour pressure, and nitrogen nutrition on carbon isotope discrimination in wheat. *Australian Journal of Agricultural Research,* 43: 935-947.

Cousens, R. and M. Mortimer. 1995. *Dynamics of Weed Populations.* University Press, Cambridge, Great Britain. p. 332.

Das, T.K. and N.T. Yaduraju. 1999. Effect of weed competition on growth, nutrient uptake, and yield of wheat as affected by irrigation and fertilizers. *Journal of Agricultural Science,* 133:45-51.

Delin, G.N., R.W. Healy, M.K. Landon, and J.K Bohlke. 2000. Effect of topography and soil properties on recharge at two sites in an agricultural field. *Journal of American Water Research Association,* 36:1401-1416.

DeSutter, T., S.A. Clay, and D.E. Clay. 1999. Agrochemical concentration on wind blown sediments. *Journal of Environmental Science Health Part B.* 33:683-691.

Dillman, A.C. 1931. The water requirement of certain crop plants and weeds in the northern Great Plains. *Journal of Agricultural Research,* 42:187-235.

Doerge, T.A. and D.L. Gardner. 1999. The pioneer split-planter comparison method. SSSMG-11. In *Site specific management guidelines.* eds. D.E. Clay, D. Armstrong, B. Dartst, K. Griffin, C. Reese and K. Reitsma. Available online at www-ppi-far.org/ssmg. Potash and Phosphate Institute, Norcross, GA.

Duczek, L.J., K.A. Sutherland, S.L. Reed, K.L. Bailey, and G.P. Lafond. 1999. Survival of leaf spot pathogens on crop residues of wheat and barley in Saskatchewan. *Canadian Journal of Plant Pathology,* 21:165-173.

Everaarts, A.P. 1992. Effects of competition with weeds on growth, development and yield of soybeans. *The Netherlands Journal of Agricultural Science,* 40:91-107.

Everaarts, A.P. 1993. Effects of competition with weeds on growth, development and yield of sorghum. *Journal of Agricultural Science,* 120:187-196.

Farquhar, G.D. and L. Lloyd. 1993. Carbon and oxygen isotope effects in the exchange of carbon dioxide between terrestrial plants and the atmosphere. In *Stable isotopes and plant carbon-water relations.* eds. J.R. Ehleringer, A.E. Hall, and G.D. Farquhar, Academic Press., New York. pp. 47-70.

Fernandez, M.R., R.P. Zentner, B.G. McConkey, and C.A. Campbell. 1998. Effect of crop rotations and fertilizer management on leaf spotting diseases of spring wheat in southwestern Saskatchewan. *Canadian Journal of Plant Science,* 78: 489-496.

Forcella, F. 1999. Estimating timing of weed emergence. In *Site specific management guidelines.* eds. D.E. Clay, D. Armstrong, B. Dartst. K. Griffin, C. Reese, and K. Reitsma. Available online at www-ppi-far.org/ssmg. Potash and Phosphate Institute, Norcross, GA. SSMG-20.

Frye, W.W., S.A. Ebelhar, L.W. Murdock, and R.L. Blevens. 1982. Soil erosion effects on properties and productivity of two Kentucky soils. *Soil Science Society of America Journal,* 46:1051-1055.

Geddes, R. D., H.D. Scott, and L.R. Oliver. 1979. Growth and water use by common cocklebur *(Xanthium pensylvanicum)* and soybean *(Glycine max)* under field conditions. *Weed Science,* 27:206-212.

Gibson, G., B.J. Radford, and R.G. H. Nielsen. 1992. Fallow management, soil water, plant–available soil nitrogen and grain sorghum production in south west Queensland. *Australian Journal Experimental Agriculture,* 32:473-482.

Giesler, L.J. and T.O. Powers. 1999. Soybean cyst nematode in Nebraska. In *Nebraska Guide, Nebraska Cooperative Extension Service,* University of Nebraska, Lincoln, Nebraska. G99 1383-A, Available on-line at www.ianr.unl.edu/pubs (accessed 4/8/02).

Gollany, H.T., T.E. Schumacher, M.J. Lindstrom, P.D. Evenson, and G.D. Lemme. 1992. Topsoil depth and desurfacing effects on properties and productivity of a typic argiustoll. *Soil Science Society of America Journal,* 56:220-225.

Greb, B.W. 1966. Effect of surface applied wheat straw on soil water losses by solar distillation. *Proceeding of the Soil Science Society America,* 30:786-788.

Hall, E.F. and F.A. Cholick. 1989. Cultivar x tillage interactions of hard red spring wheat cultivars. *Agronomy Journal,* 81:789-792.

Halvorson, A.D., A.B. Black, J.M. Krupinski, S.D. Merrill, B.J. Wienhold, and D.L. Tanaka. 2000. Spring wheat response to tillage and nitrogen fertilization in rotation with sunflowers and winter wheat. *Agronomy Journal,* 92:136-144.

Halvorson, G.A. and E.C. Doll. 1991. Topographic effects on spring wheat yields and water use. *Soil Science Society of America Journal,* 55:1680-1685.

Hanna, A.Y., P.W. Harlan, and D.T. Lewis. 1982. Soil available water as influenced by landscape position and aspect. *Agronomy Journal,* 74:999-1004.

Hart, S.E. and L.M. Wax. 1999. Review and future prospectus on the impacts of herbicide resistant maize on weed management. *Maydica,* 44:25-36.

Hatfield, J.L., T.J. Sauer, and J.H. Prueger. 2001. Managing soils to achieve greater water use efficiency: A review. *Agronomy Journal,* 93:271-280.

Hilfiker, R.E. and B. Lowery. 1988. Effect of conservation tillage systems on corn root growth. *Soil and Tillage Research,* 12:269-283.

Holm, L.G., D.L. Plucknett, J.V. Pancho, and J.P. Herberger. 1991. *The world's worst weeds: distribution and biology.* Malablar, FL: Krieger Publishing Co.

Hoogenboom, G.J., J.W. Jones, P.W. Wilkens, W.D. Batchelor, W.T. Bowen, L.A. Hunt, N.B. Pickering, U.Singh, D.C. Godwin, B. Baer, K.J. Boote, J.T. Ritchie, and J.W. White. 1994. Crop models. In *DSSAT* v3., vol 2-2. eds. G.Y.Tsuji, G. Vehara, and S. Balas. University of Hawaii, Honolulu, Hawaii. pp. 95-244.

Iqbal, J. and D. Wright. 1999. Effects of weed competition on flag leaf photosynthesis and grain yield of spring wheat. *Journal of Agricultural Science,* 132:23-30.

Johnson, W.C., B. Lowery, and T.C. Daniel. 1984. Soil moisture regimes of three conservation tillage systems. *Transactions of the American Society Agricultural Engineers,* 27:1385-1390.

Jones, C.A. and J.R. Kiniry. 1986. *CERES-Maize: A simulation model of maize growth and development.* College Station, TX: Texas A&M University Press.

Jones, J.W. and B. Zur. 1984. Simulation of possible adaptive mechanisms in crops subjected to water stress. *Irrigation Science,* 5:251-264.

Jordan, W.R., W.A. Dugas, Jr. and P.J. Shouse. 1983. Strategies for crop improvement for drought-prone regions. *Agricultural Water Management,* 7:281-299.

Kapusta, G. and R.F. Krautz. 1993. Weed control and yield are equal in conventional, reduced, and no-tillage soybean *(Glycine max)* after 11 years. *Weed Technology,* 7:443-451.

Kegode, G., F. Forcella, and S.A. Clay. 1999. Influence of crop rotation, tillage, and management inputs on weed seed production. *Weed Science,* 47:175-183.

Kennedy, G.G. 2000. Perspectives on progress in IPM. In *Emerging technologies for integrated pest management. Concepts, research, and implementation.* eds. G.G. Kennedy and T.B. Sutton APS Press. St. Paul, MN, pp. 2-11.

Khakural, B.R., G.D. Lemme, T.E. Schumacher, and M.J. Lindstrom. 1992. Effect of tillage systems and landscape on soil. *Soil and Tillage Research,* 25:43-52.

Khakural, B.R., P.C. Robert, and D.J. Mulla. 1996. Relating corn/soybean yield to variability in soil and landscape characteristics. *Precision Agriculture: Proceedings of the 3rd International Conference.* eds. P.C. Robert, R.H. Rust, and W.E. Larson, June 23-26. 1996. Minneapolis, MN, American Society of Agronomy, Madison WI. pp. 117-128.

Kitchen, N.R., K.A. Sudduth, and S.T. Drummond. 1999. Soil electrical conductivity as a crop productivity measure for claypan soils. *Journal of Production Agriculture,* 12:607-617.

Kravchenko, A.N. and D.G. Bullock. 2000. Correlations of corn and soybean grain yield with topography and soil properties. *Agronomy Journal,* 92:75-83.

Kutcher, H.R., S.S. Malhi, A.M. Johnston, and G. Hnatowich. 1999. Impact of topography and management on diseases of canola and wheat. *Proceedings of the 4th International Conference on Precision Agriculture.* eds. P.C. Robert, R.H. Rust, and W.E. Larson, St. Paul MN, American Society of Agronomy, Madison WI. pp. 559-562.

Larney, F.L., B.M. Olson, H. Janzen, and C.W. Lindwall. 2000. Early impact of topsoil removal and soil amendment on crop productivity. *Agronomy Journal,* 92:948-956.

Lems, J. 1998. *Weed spatial variability and management on a field-wide scale.* South Dakota State University, master's thesis. Brookings, SD. p. 105.

Ludlow, M.M. and R.C. Muchow. 1990. A critical evaluation of traits for improving crop yield in water-limited environments. *Advances in Agronomy,* 43:107-153.

Malo, D.D. and B.K. Worcester. 1975. Soil fertility and crop responses to selected landscape positions. *Agronomy Journal,* 67:397-401.

Malzer, G.L., D.L. Mulla, and T.S. Murrel. 2000. Getting specific with site-specific management. In *Site specific management guidelines.* eds. D.E. Clay, D. Armstrong, B. Dartst. K. Griffin, C. Reese, and K. Reitsma. Available online at www-ppi-far.org/ssmg. Potash and Phosphate Institute, Norcross, GA. SSMG-23.

Manley, B.S., H.P. Wilson, and T.E. Hines. 2001. Weed management and crop rotations influence populations of several broadleaf weeds. *Weed Science,* 49: 106-122.

Muchow, R.C. and T.R. Sinclair. 1991. Water deficit effects on maize yields modeled under current and greenhouse climates. *Agronomy Journal,* 83:1052-1059.

Nolan, S.C., T.W. Goddard, D.C. Penney, and F.M. Green. 1998. Yield response to nitrogen within landscape classes. In *Precision Agriculture: Proceedings of the 4th International Conference* July 19-22, 1998, St Paul Minnesota, eds P.C. Robert, R.H. Rust, and W.E. Larson, American Society of Agronomy Madison WI. pp. 479-485.

Norris, R.F. 1996. Water use efficiency as a method for predicting water use by weeds. *Weed Technology,* 10:153-155.

Norwood, C.A. 1999. Water use and yield of dryland row crops as affected by tillage. *Agronomy Journal,* 91:108-115.

Oikeh, S.O., J.G. Kling, W.J. Horst, V.O. Chude, and R.J. Carsky. 1999. Growth and distribution of maize roots under nitrogen fertilization in plinthite soil. *Field Crops Research,* 62:1-13.

Olson, K.R. and E. Nizeyimana. 1988. Effect of soil erosion on corn yield of seven Illinois soils. *Journal of Production Agriculture,* 1:13-19.

Pacumbaba R.P., G.F. Brown, and R.O. Pacumbaba. 1997. Effect of fertilizers and rates of application on soybean diseases in northern Alabama. *Plant Disease,* 81:1459-1460.

Paz, J. O., W.D. Batchelor, D.E. Clay, S.A. Clay, and C. Reese. 2003. Characterization of soybean yield variability using crop growth models and $^{13}C$ discrimination. ASAU meeting presentation #033044.

Paz, J.O., W.D. Batchelor, T.S. Colvin, S.D. Logsdon, T.C. Kaspar, D.L. Karlen, and B.A. Babcock. 1999. Model-based techniques to determine variable rate nitrogen for corn. *Agricultural Systems,* 60:69-75.

Pikul J.L. and J.K. Aase. 1995. Infiltration and soil properties as affected by annual cropping in the northern Great Plains. *Agronomy Journal,* 87:656-662.

Powles, S.B., D.F. Lorraine-Colwill, J.J. Dellow, and C. Preston. 1998. Evolved resistance to glyphosate in rigid ryegrass *(Lolium ridigum)* in Australia. *Weed Science,* 46:604-607.

Rasse, D.P. and A.J.M. Smucker. 1998. Root recolonization of previous root channels in corn and alfalfa rotations. *Plant and Soil,* 204:203-212.

Reese, C.L., D.E. Clay, C.G. Carlson, R. Berg, and D. Diedrich. 2000. Identifying soybean Phytopthora tolerant and non-tolerant varieties management zones using soil electrical conductivity. In *Proceedings of the 5th International Conference on Precision Agriculture.* eds P.C. Robert, R.H. Rust, and W.E. Larson, July 16-19, 2000, Bloomington Minnesota, American Society of Agronomy, Madison, WI.

Rockström, J., J. Barrow, J. Brouwer, S. Galle, and A. de Rouw. 1999. On-farm spatial and temporal variability of soil water in pearl millet cultivation. *Soil Science Society of America Journal,* 63:1308-1319.

Saarikko, R.A. 2000. Applying a site based crop model to estimate regional yields under current and changed climates. *Ecological Modeling,* 131:191-206.

Sauer, T.J., J.L. Hatfield, and J.H. Prueger. 1996. Corn residue age and placement effects on evaporation and soil thermal regime. *Soil Science Society of America Journal,* 60:1558-1564.

Scharf, P.C., D.B. Quarles, and J.A. Lory. 2000. Nitrogen response of no-till corn in first and second years following conservation reserve program. *Communications Soil Science and Plant Analysis,* 31:2501-2508.

Schillinger, W.F., R.J. Cook, and R.L. Papendick. 1999. Increased dryland cropping intensity with no-till barley. *Agronomy Journal,* 91:744-752.

Scholes, C., S.A. Clay, and K. Brix-Davis. 1995. Velvetleaf effect on corn growth and yield in South Dakota. *Weed Technology,* 9:665-668.

Shaner, D.L. 2000. The impact of glyphosate-tolerant crops on the use of other herbicides and on resistance management. *Pest Management Science,* 56:320-326.

Shantz, H.L. and L.N. Piemeisel. 1927. The water requirement of plants at Akron, *Colorado Journal Agricultural Research,* 34:1095-1190

Shen, J., W.D. Batchelor, R. Kanwar, J.T. Ritchie, and J.W. Jones. 1998. Validation of the water balance model in CROPGRO-Soybean. *Transactions of the American Society Agricultural Engineers,* 41:1305-1313.

Shirima, D.S., R. Otsyina, W.P. Mwagemi, and J. Bridge. 2000. Effect of natural and *Sephania* fallows and crop rotations on the incidence of root-knot nematodes and tobacco products in Tabora, Tanzania. *International Journal of Nematology,* 10:49-54.

Sinclair, T.R. 1994. Limits to crop yield? In *Physiology and determination of crop yield.* ed. K.J. Boote, American Society of Agronomy, Madison, WI. pp. 509-532.

Sinclair, T.R. and R.C. Muchow. 2001. System analysis of plant traits to increase grain yield on limited water supplies. *Agronomy Journal,* 93:263-270.

Singer, J.W. and W.J. Cox. 1998. Agronomics of corn production under different crop rotations in New York. *Journal of Production Agriculture,* 11: 462-468.

South Dakota Ag Statistics Service. 2000. South Dakota 2000 Soybean varieties. Online at www.nass.usda.gov/sd/releases/soyvar00.pdf. (accessed May 2001).

South Dakota Ag Statistics Service. 2005. South Dakota Agriculture. Bulletin No. 65. June 2005.

Spomer, R.G. and R.F. Piest. 1982. Soil productivity and erosion of Iowa loess soils. *Transactions of the American Society Agricultural Engineers,* 25:1295-1299.

Stoller, E.W., S.K. Harrison, L.M. Wax, E.E. Regnier, and E.D. Nafziger. 1987. Weed interference in soybean *(Glycine max). Reviews in Weed Science,* 3:155-181.

Stone, J.R., J.W. Gilliam, D.K. Cassel, P.B. Daniels, L.A. Nelson, and H.J. Kleiss. 1985. Effect of erosion and landscape position on the productivity of the Piedmont soils. *Soil Science Society America Journal,* 49:987-991.

Stout, W.L. and R.R. Schnabel. 1997. Water-use efficiency of perennial ryegrass as affected by soil drainage and nitrogen fertilization on two floodplain soils. *Journal of Soil and Water Conservation,* 52:207-211.

Sudduth, K.A., S.T. Drummond, S.J. Birrell, and N.R. Kitchen. 1996. Analysis of spatial factors influencing crop yield. In *Precision Agriculture: Proceedings of the 3rd International Conference.* June 23-26, 1996. eds. P.C. Robert, R.H. Rust, and W.E. Larson, Minneapolis, MN, American Society of Agronomy, Madison WI. pp. 129-140.

Swinton, S.M. and J. Lowenberg-DeBoer. 1999. Profitability of site specific farming. In *Site specific management guidelines.* eds. D.E. Clay, D. Armstrong, B. Dartst, K. Griffin, C. Reese and K. Reitsma, Available online at www-ppi-far.org/ssmg. Potash and Phosphate Institute, Norcross, GA. SSMG-3.

Timlin, D.J., Y. Pachepsky, V.A. Snyder, and R.B. Bryant. 1998. Spatial and temporal variability of corn grain yield on a hillslope. *Soil Science Society American Journal,* 62:764-773.

Timlin, D.J., Y. Pachepsky, C. Walthall, and S. Loechel. 2001. The use of a water budget and yield map to characterize water availability in a landscape. *Soil Tillage Research,* 58:219-231.

Timothy E.F., B.C. Miller, and W.L. Pan. 1994. Winter wheat yield and grain protein across varied landscape positions. *Agronomy Journal,* 86:1026-1032.

Tollenaar, M., S.P. Nissanka, A. Aguilera, S.F. Weise, and C.J. Swanton. 1994. Effect of weed interference and soil nitrogen on four maize hybrids. *Agronomy Journal,* 86:596-601.

Vengris, J., W.G. Colby, and M. Drake. 1955. Plant nutrient competition between weeds and corn. *Agronomy Journal,* 47:213-216.

Wilson, J.A. 1992. *The effect of foxtail* (*Setaria* spp.) *interference in spring wheat.* South Dakota State University, master's thesis. Brookings, SD.

Wright, S.R., M.W. Jennette, H.D. Coble, and T.W. Rufty, Jr. 1999. Root morphology of young *Glycine max, Senna obtusifolia,* and *Amaranthus palmeri. Weed Science,* 47:706-711.

Wyciskalla, T.D. E.C. Varsa, S.K. Chong, S.A. Ebelhar, S.J. Indorante, W.M. McCauley, and G.K. Robertson. 2000. An evaluation of soil site factors responsible for field variation in two southern Illinois farm fields. In *Proceedings of the 5th International Conference on Precision Agriculture.* July 16-19, 2000, Bloomington, MN, eds. P.C. Robert, R.H. Rust, and W.E. Larson. American Society of Agronomy, Madison, WI. CD-ROM.

Zachmann, J., D.R. Linden, and C.E. Clapp. 1986. Macroporous infiltration and redistribution as affected by earthworms, tillage, and residue. *Soil Science Society of America Journal,* 51:1580-1586.

Zimdahl, R.L. 1980. *Weed-crop competition: A review.* International Plant Protection Center. Oregon State Univ. Corvallis, OR. p. 195.

Zollinger, R.K. and J.J. Kells. 1993. Perennial sow thistle (*Sonchus arvensis*) influence on soybean *(Glycine max)* and dry edible bean *(Phaseolus vulgaris). Weed Technology,* 7:52-57.

# *PART III: CURRENT STATUS*

Chapter 17

# Africa

W.T. (Wimpie) Nell
Ntsikane Maine
P. M. Basson

## *INTRODUCTION*

New technological developments such as precision agriculture have revolutionized the traditional approach of farming. Precision agriculture provides highly sophisticated information to assist farmers/managers in their normal, everyday planning and strategic decisions, which can expand farmers' management capacity and management time. Used efficiently and correctly, precision agricultural techniques and site-specific management can assist the farm manager in identifying opportunities to optimize input applications, cut input costs, increase production, reduce production risks, and promote greater profitability within the strategic plan of the farming business (Nell and Napier, 2005). Precision agriculture in this chapter refers to the management of variability by means of Geographic Positioning Systems (GPS), and site-specific management refers to the management of variability without using GPS-based systems.

Many types and phases of sustainable technologies are needed to steer agriculture along a sustainable path. The development of new technologies such as precision agriculture and site-specific management can contribute to the successful management of agriculture in a changing environment. Precision agriculture is a complementary process of different kinds of man-

Acknowledgment is given to New Holland South Africa for the financial assistance for the compiling of this chapter.

*Handbook of Precision Agriculture*

doi:10.1300/5627_17

agement practices, including environmental management. Harris (1997) stated that precision agriculture intends to influence the cost of production by optimizing inputs. As an effective technology in the optimal use of resources, it can be the long-term solution and the ultimate choice in improving agricultural productivity in Africa.

Farmers in South Africa are currently moving toward the use of precision agricultural techniques to combat the ongoing price-cost squeeze in South Africa and to lower production risk. They further regard it as an opportunity to increase the profitability of their farming businesses, to increase yields, to improve quality, to lower manpower costs, and to conserve the natural resources for their descendants. The prediction is that more farmers will adopt precision agricultural techniques when more scientific research results on the profitability of precision agriculture become available.

The objective of this chapter is to investigate the status of the adoption of precision agriculture and site-specific management technologies in Africa and South Africa. The climatic conditions of Africa and South Africa as well as a short background of agriculture in Africa and South Africa will first be attended to, followed by a discussion of the current status of site-specific management and precision agriculture in Africa and South Africa. The research on precision agriculture in South Africa will further be addressed as well as constraints in adoption of precision agriculture and site-specific management technologies. The observations for Africa are based on research done by Nell (1998) on the transfer and adoption of new technologies, as well as Internet searches and by means of knowledge gained from agricultural companies involved in Africa. (The observations for South Africa are based on research in precision agriculture done by Maine for a master's degree in agricultural management [2001].) The chapter will be closed by a summary on the future prospects and challenges for precision agriculture in Africa and South Africa.

## *CLIMATIC CONDITIONS*

### *Africa*

Most of Africa has a warm or hot climate, but the humidity and rainfall vary from area to area. Africa has a very uneven rainfall distribution; some areas have either excessive rainfall, whereas other areas receive insufficient rainfall to support agriculture. More than half of Africa receives less than 500 mm rain annually. Africa's climate has made agricultural improvements and the adoption of modern technologies difficult. In areas with limited and unreliable rainfall, the choice of crops and production strategies is difficult. On the other hand, in areas that receive too much rainfall, the leaching away of soil nutrients and soil erosion become a problem. As for the hot, humid climate in some areas of Africa, the spreading of insects and diseases that destroy crops and livestock, as well as diseases affecting people, is prevalent. The highest temperatures in Africa occur in the Sahara and in parts of Somalia. The coolest regions are the northwest, the highland areas of the east, and parts of the south (Walter et al., 2001).

The larger part of the African continent is arid and semiarid (Sanders et al., 1996), farming is relatively difficult and will require good management skills. Sound dryland crop management practices can produce high yields, but by using new technologies and proper irrigation practices, production can be significantly improved and the risk of crop losses during seasons with low rainfall can be reduced to a large extent. The correct use of water retention technologies and production systems will therefore be a necessity for sustainable success.

### *South Africa*

Agricultural production in South Africa is practiced under various climatic and topographical conditions. The diverse topography of South Africa gives rise to vastly different climates. The weather is generally pleasant, and ranges from mild, cold winters to hot summers. Temperatures can drop as low as –17°C (20°F), with a humidity of 5 percent during extreme winter cold spells, and can increase to as high as 40°C (104°F) during hot summers.

The climatic conditions are to a large extent dependent on the altitude of the area in question and its proximity to one of the oceans. Winter rainfall and beautiful summers are experienced on the West Coast and the Cape, while the inland regions enjoy cold, dry winters and summer rainfall in the form of almost daily afternoon thunderstorms. Humid and warm to hot conditions prevail on the East Coast and Durban.

The aridity index of South Africa (Figure 17.1 and Table 17.1) indicates that 90 percent of the country is arid and semiarid, which is perhaps one of the main reasons why agricultural development experienced difficult stages in the past. The western part of the country is arid with higher temperatures throughout the year than the semiarid and subhumid eastern part of the country (Figure 17.1). The country is also divided into two distinct rainfall areas, namely winter rainfall in the south and summer rainfall in the north.

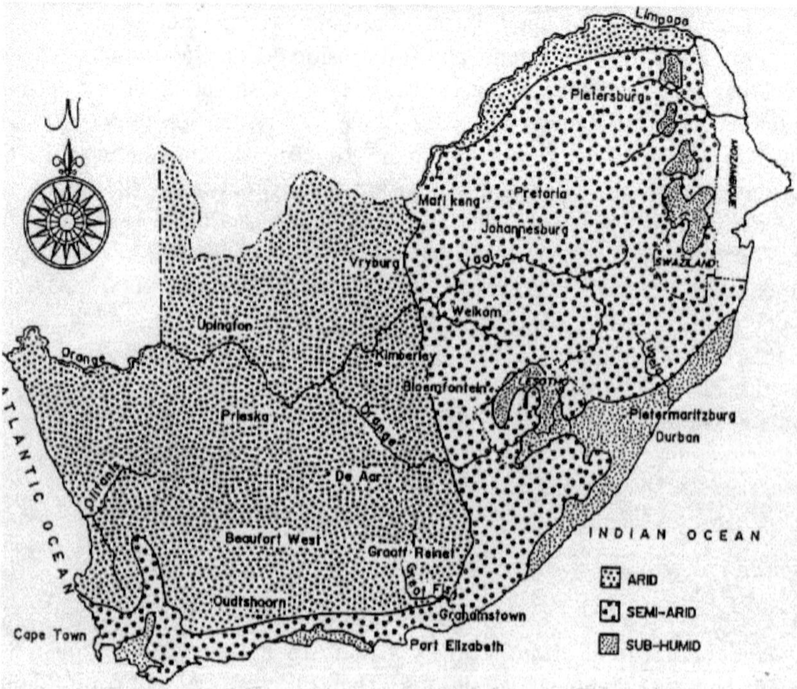


FIGURE 17.1. Bio-climatology zones of South Africa. *Source:* Adapted from UNESCO, 1977.

## BACKGROUND OF AGRICULTURE

### Agriculture in Africa

Africa is a continent of amazing contrast, from deserts to subhumid areas (Figure 17.1), with good potential for the marketing of services and products, and on the other hand, poor economic growth and severe poverty. Africa covers approximately 30.19 million km$^2$; about one-fifth of the earth's total land area. Africa has a population of about 825 million people, with over 600 million people living south of the Sahara. About two-thirds of all Africans live in rural areas, where they make a living by growing crops or raising livestock (Walter et al., 2001).

Africa raises more than two-thirds of the world's camels, nearly one-third of its goats, and about one-seventh of its sheep and cattle. Most of the

TABLE 17.1. Bioclimatology of South Africa.

| Climatic zone | Area (percent) | Annual rainfall (mm) | Annual evaporation (mm) | Aridity Index |
|---|---|---|---|---|
| Arid | 50 | < 500 | > 2500 | < 0.2 |
| Semiarid | 40 | 500-750 | 2,500-1,500 | 0.2-0.5 |
| Subhumid | 10 | > 750 | < 1500 | > 0.5 |

*Source:* Adapted from UNESCO, 1977.

world's cassava, cocoa beans, and yams are produced in Africa, but Africa has the least developed economy of any other continent with the exception of Antarctica (Walter et al., 2001).

Concern over Africa's economic growth is one of the important topics in international development. A large section of the continent continues to show dismal performance on almost all the human development measuring instruments (Center for International Development, 1999). Performance indicators of human development are associated with, amongst others, low agricultural productivity and environmental degradation. According to the Natural Resources Institute (NRI) (1994), the poor performance of agriculture corresponds with the overall economic performance as reflected in an annual decrease of approximately 1 percent per capita income during the 1980s. The NRI (1994) furthermore indicates that an increase in agricultural production in Africa has been achieved mainly by the expansion of crop fields (area). "During the 1980s, there was a 28% increase in area under cereals in Africa, but only a 12% increase in yields" (NRI, 1994). It is thus obvious that the strategy of expansion of the area under production will not be sustainable, and better production and technological strategies will thus have to be implemented to promote sustainable success.

Walter et al. (2001) and Sanders et al. (1996) identified several reasons that account for the low overall agricultural productivity in Sub-Saharan Africa:

- Too small production areas and inefficient production methods
- Thin and poor soils
- Erosion of the precious topsoil in parts of western and central Africa as a result of heavy rains and poor resource management
- Periodic droughts, floods, or insect infestations that destroy crops or sharply reduce crop yields

- Unfavorable climate in many parts of Africa hinder cattle production
- Lack of infrastructure
- Lack of technologically induced agricultural policies
- Political instability with accompanying wars as well as faction and tribal fights
- Low literacy with the accompanying low management level of farmers

The transfer and adoption of site-specific management and production technologies can therefore contribute substantially to the growth in agricultural production.

### *Agriculture in South Africa*

South Africa lies at the southern bottom end of the African continent and stretches between the twenty-second and the thirty-fourth degrees of the southern latitude and hence is part of the subtropical zone. South Africa has an area of 1,223 million km$^2$ with a population of 43.6 million. According to the *Abstract of Agricultural Statistics* (2004), 82.3 percent of the total area is used for agricultural purposes. South Africa is a country of contrasts. It varies from arid in the west to subhumid in the east. Approximately 90 percent of the country is arid or semiarid (UNESCO, 1977), with erratic rainfall that can result in long drought spells as well as devastating floods. Aridity seriously affects grazing capacity. The carrying capacity of natural veld, for example, varies from two hectare per livestock unit (LSU) in the east to 96 hectare per LSU in the west (Richtersveld) (RSA, Department of Agriculture, 1993).

South Africa generally has a scarcity of water resources, and only 13.7 percent of the total area is used for intensive agriculture (arable land), relying mostly on rainfall (*Abstract of Agricultural Statistics,* 2000). This situation makes irrigation-based agriculture very important. Irrigation farming is practiced on 7.2 percent of the arable land (Dennis and Nell, 2001).

Agriculture accounts for about 5 percent of the country's gross domestic product (GDP). The major crops in the country's agriculture include citrus and deciduous fruits, grapes, maize (corn), wheat, oilseeds, dairy products, sugarcane, tobacco, wine, meat, and wool. South Africa's agriculture is well-developed and can successfully sustain itself on the world markets, especially through products such as wine, fruit, vegetables, and sugar. South Africa is a net exporter of food, and agricultural products are listed high among the country's export products (U.S. Department of State, 2005).

Most of the products mentioned are produced under irrigation in South Africa, and there are a number of well-developed irrigation schemes in the country. Meat and animal fiber production is subject to climatic conditions, with certain livestock adapted to certain regions. Extensive cattle, sheep, and goat farming on natural pastures is predominant in the arid area of the Karoo. Farming in the northwest is restricted to goats and sheep, although cattle farming is possible in areas with greater precipitation. Dairy production is more common close to densely populated areas and also on the east coast because of higher rainfall (South Africa Online Travel Guide, 2005).

Approximately 95 percent of all agricultural production in South Africa is produced by technically developed commercial farmers who operate in a free market. Little or no subsidies are received from the government. The majority of agricultural production in South Africa is practiced under rainfed conditions in the semiarid areas (Figure 17.2). This is an indication that water is a scarce resource in South Africa and must be managed accordingly. Currently, overirrigation tends to occur where there is enough water, while underirrigation is the norm where water is limited. Precision agriculture is one of the technologies that can contribute to the more efficient use of the scarce water resources.

In the South African agricultural system, farmers can be grouped into three levels of technology adoption, namely high technology → high management; high technology → low management; low technology → low management (Nell et al., 1999). Bembridge (1991) stressed that future technology transfer strategies in South Africa should be based on a target approach to reach progressive, low-access, and resource-poor farmers. Emerging and small-scale farmers have to be established and provided (equipped) with new technologies as well as management skills to make the best use of these new technologies to compete in a free national and international market (Kirsten, 1994). Nell et al., (1999) emphasized that emerging farmers need institutional support from the government to increase the management level in order to adopt modern technologies to be competiti ve.

### *South Africa As a Leader in Africa's Agriculture*

The Republic of South Africa is the only African nation generally classified as economically developed. All other African countries have a developing economy (Walter et al., 2001). South Africa has furthermore a productive and industrialized economy that surprisingly exhibits many characteristics associated with developing countries, such as division of labor between the formal and informal sector, and an uneven distribution of wealth and income. The emerging of South Africa in international markets after the lifting of sanctions, is an opportunity for economic growth in the region. As a result, the benefits can spill over to other countries in the continent. South Africa, together with Nigeria, shadow the economies of the rest of Sub-Saharan Africa, as reflected by their share to the region's GDP. In 1990, the GDP of Sub-Saharan Africa was approximately US$254,000 million, of which Nigeria accounted for 14 percent and South Africa for 36 percent (Digges et al., 1997).

"South Africa has sophisticated financial structure with a large and active stock exchange, which ranks 18th in terms of total market specialization" (U.S. Department of State, 2005). South Africa's formal sector, which is based on mining, manufacturing, services, and agriculture, is well-developed.

Farmers in large parts of Africa have only recently started using animal traction in the place of human traction, and in areas where animal traction has been used previously, farmers started using small tractors (Sanders et al., 1996; Lowenberg-DeBoer, personal communication, 2001). South African farmers started using fertilizers in the early 1920s (Fertilizer Society of South Africa, FSSA, 1989), mechanization on a commercial scale in the mid-1930s (Van der Merwe, 2001), and hybrid seed in the late 1940s (Nefdt, personal communication, 2001). The fact that international agencies still struggle to convince the majority of farmers in Africa to use the mentioned technologies is an indication that the South African agricultural sector is technologically more advanced than the rest of Africa and is ahead of Africa by 50 to 80 years.

## *SITE-SPECIFIC MANAGEMENT*

As stated earlier, site-specific management is referred to as the management of variability without GPS-based systems. Many examples can be found of this farming technology. As soon as a farmer starts to differentiate the treatment of two different areas of land, individual livestock or groups of livestock with different potential, it can be seen as the management of variability or site-specific management.

### *The Situation in Africa*

Many farmers in Africa choose different types of soil for the production of different crops as a result of past experience, and this can be seen as a form of site-specific management. There are not many references in the literature of formal site-specific management projects in Africa. Blackmore (1999) referred to tea growing in Tanzania as the best example of precision agriculture (referred to in this chapter as site-specific management) in the world. Blackmore further described tea as a highly structured crop grown in blocks in seven-year cycles. Each block is first surveyed and identified as a unique management zone, after which yield maps are produced (without GPS). Yield maps arise from the recording of the weight of the tea plucked, together with the block from which it came from. Inputs are then precisely applied by hand, with the labor force applying the desired amount of inputs in the designated area. It is alleged that the cost of implementation in this system is very low because of highly structured fields and the recording system already in place. This may sound very simple, but in essence, this is, according to Blackmore (1999), precision agriculture (site-specific management).

Site-specific management in dairy farming is found in some African countries, with three electronic management systems operational in Mozambique, four in Namibia, 25 in Zambia, and no feeding systems are operational in Zimbabwe. Approximately 24 dairy farmers use other site-specific management technologies in Namibia and 278 in Zimbabwe (Van Vuuren, personal communication, 2005).

Site-specific management is, according to Visser (personal communication, 2001), relatively common amongst commercial farming operations in East Africa (Kenya, Uganda, Tanzania, Rwanda, Mauritius, and Zanzibar) (Table 17.2). According to Visser ( personal communication, 2001), no GPS farming practices are used in East Africa amongst commercial farmers. Official figures are very unreliable as data are not upgraded on a regular basis by the different agricultural departments in Africa (Visser, personal

TABLE 17.2. Indicators of site-specific management in East Africa, 2005.

| Enterprise/practice | Number of farmers (Actual numbers) | Percent of farmers (Estimated) |
|---|---|---|
| Electronic milk systems | 5 | |
| Dairy production–feeding as per production | 110 | |
| Coffee cultivation | | 75 |
| Tea cultivation | | 75 |
| Poultry production | | 90 |
| Pork production | | 85 |
| Horticulture | | 80 |
| Cut-flower production | | 100 |

communication, 2001). These figures stayed very much the same until 2005 (Visser, personal communication, 2001).

### *The Situation in South Africa*

Ample examples exist of site-specific management in South Africa. Site-specific management is exercised to some extent in nearly all agricultural enterprises, depending on the management level of the farmer/manager. The higher the management level or capacity of the farmer/manager, the more intense is the management of variability in farming.

Given the results discussed, site-specific management is quite common amongst the high-value agricultural enterprises, such as dairy, fruit, nuts, etc. It is further mainly commercial farmers and projects driven by foreign NGOs and agencies, which make use of site-specific management technologies.

#### *Soil Classification*

Soil classification is in most cases the beginning stage and backbone for site-specific management as well as precision agriculture in crop production. It helps the farmer/manager to distinguish between the better or higher potential soils and the weaker or lower potential soils. It also helps the farmer/manager in field division, crop planning, and the planning of crop rotation systems. Up to 1997, 300,000 hectare arable land was classified by means of soil surveys (Loxton and Thompson, 1978). Another 4.3 million hectare was classified from 1977 to 2001, of which a 100,000 hectare of land was classified in the former homelands and a further 1 million hectare of forest plantations (Nell, personal communication, 2001). Approximately 25 percent of the total arable land was classified, but only a relatively small number of these farmers actually applied the soil classifications in crop production planning. The reason for this situation is, according to Van der Berg (personal communication, 2001), the lack of management abilities and skills to apply and use the highly sophisticated information.

#### *Production of Citrus, Deciduous Fruit, Table Grapes, Vegetables, and Cash Crops*

The form of site-specific management referred to by Blackmore (1999) is quite common in South Africa amongst citrus, deciduous fruit, table grape, and vegetable farmers. It is common practice to record yield levels per block or area and apply inputs accordingly (Karsten, personal communication, 2001). It is usually done by allocating a specific code or number to a laborer designated to a specific block that must be harvested, who then attaches codes to the harvesting bins before it is weighed at the packing sheds. It is then automatically fed into a computer program which analyzes the results. Recommendations for rectifying any identified problems in a specific block or area can then be attended to accordingly.

*Dairy Cattle Feeding*

Another agricultural enterprise that has shown immense technological development over the past 60 years is the dairy farming sector. The term *management of variability* or *site-specific management* in dairy farming is not a new concept in South Africa. The feeding of dairy cows according to their potential or production started in South Africa in the early 1930s.

Scales were initially used to weigh the milk and the individual milk production was recorded manually and written in a milk recording register. The herd was then divided into three or more groups and fed according to the average production of specific groups. The weighing system was replaced by the Wykato milk recording system, which is based on a milk flow meter. This system is based on the same principle as the flow meter used in a combine yield monitor in the precision agricultural process. The information available from this system is much more accurate than the first system mentioned, which improved the level of site-specific management in dairy farming.

Technologically, the most developed system used in South Africa is the electronic identification system. This system links all technical information of the milk to the individual cow, which can be treated accordingly. Information such as weight, liters, milk solids, protein, butter fat, temperature, nitrogen levels, somatic and bacteria count, can be used to improve the efficiency of the individual cow as well as the herd. This can increase the sustainable success of the herd. The higher the accuracy and detail of information available, the higher the farmer's management level must be in order to use and manage the detailed information for the success of dairy farming.

South Africa had 4,500 commercial dairy farmers in 2005 (Grobler, 2005). In 2001 there were 5,700 dairy farmers in South Africa, of which approximately 150 used electronic milking systems. Approximately 1,200 farmers used the so-called Largebore swing-over systems in 2001. Approximately 2,000 of the other farmers used milk parlors, which were equipped with various kinds of milking equipment to improve the level of site-specific management (Marais, personal communication, 2001). No updated information was available on the current systems in South Africa because the number of providers of milking machines has increased significantly.

## PRECISION AGRICULTURE

"Precision agriculture is broadly defined as monitoring and control applied to agriculture, including site-specific application of inputs, timing of operations and monitoring of crops and employees" (Lowenberg-DeBoer and Boehlje, 1996). Crop management decisions are traditionally made on a field-by-field basis, the entire field treated as a homogenous unit. However, the properties of soil within a field are diverse and may vary widely within one field, necessitating precision in crop management to obtain sustainable success. In this section, precision agriculture refers to management of variability, using GPS- and/or GIS-based systems.

### *The Situation in Africa*

In 2001, Bryce-Rogers indicated that precision agricultural activities are practiced in Zimbabwe, and according to agricultural equipment and input suppliers, with the exception of Zimbabwe and South Africa, no precision agricultural activities are currently practiced in Africa. The Nemakonde Agricultural Development Company (NADC), a division of Zimbabwean business Origen Corporation, started with precision agriculture in Zimbabwe in 1999. Precision agricultural hardware and software were purchased from Techniland, a South African-based precision agriculture company. Bryce-Rogers (personal communication, 2001) stated that the main reason for venturing into precision agriculture was to provide Origen Corporation and its agricultural clients in the Chinhoyi area, which is 120 kilometers northwest of Harare, with sophisticated agricultural services. These services included precise Differential Geographic Positioning System DGPS-based land mapping, DGPS-based soil sampling, and DGPS-based variable application of nutrients and lime where necessary.

As a result of this initiative, the view of Bryce-Rogers (personal communication, 2001) was that more accurate information was made available for local agronomy services to make fertilizer recommendations and blends. At the initial stage, a high interest in precision agriculture was observed, with approximately 18,000 hectares mapped in the first six months of the venture, of which 5,000 hectares were sampled. The main clients were producers of row crops (soybean, maize, wheat, and cotton), with tobacco farmers and a small number of intensive flower producers making up the balance (Bryce-Rogers, personal communication, 2001).

According to Bryce-Rogers (personal communication, 2001) NADC sent its soil samples to independent laboratories in Harare and used the results to generate nutrient maps for fertilizer recommendations to its clients.

These nutrient maps, accompanied by recommendations and prescription files generated by NADC, were used for variable-rate applications. Variable-rate applications were done using dolomitic lime and superphosphate as the equipment purchased was only capable of variably applying granular straights. By the middle of 2001, 30,000 hectares had been DGPS mapped and 8,000 hectare DGPS sampled for 58 clients. The main users of precision agriculture at the time of the research were tobacco producers and not row croppers as initially experienced. This was primarily because of the high cost of the precision farming service and the more intensive nature of tobacco farming and its increased return per hectare. Of the 58 clients, 30 farmers used variable-rate applications for fertilizer and lime. The use of yield monitors was very rare due to the high cost of combines fitted with the technology in Zimbabwe (Bryce-Rogers, personal communication, 2001).

Bryce-Rogers (personal communication, 2001) asserted that there were a few independent users of precision agricultural techniques in Zimbabwe. Case International, through its local agent, provided Differential Geographic Positioning System (DGPS) units to customers who had yield monitors fitted in their combines. This was limited to harvesting periods only. The upheaval in the agricultural industry in Zimbabwe with the resultant uncertainty about future land tenures and financial constraints, has led to a decline in the use of precision agricultural services in the Chinhoyi area. NADC ceased its precision farming activities in Zimbabwe in 2005.

Bryce-Rogers (personal communication, 2001) noted that NADC was also involved in precision agriculture in Zambia (2003-2004) where 20,000 hectares were DGPS mapped and 1,000 hectares sampled. Zambia is currently a growth market for precision agriculture due to its recent agricultural industry growth. The larger estates are investing in both GIS and precision agriculture technology for crop growth analysis, yield monitoring, and variable nutrient spreading.

Fraser (personal communication, 2005) reported that Flexicoil Ltd. was the only company currently involved in equipment that can be used for precision agriculture in Kenya. The company sold 12 air seeders to farmers in Kenya, but not for GPS-based variable rate applications. None of the farmers in Kenya used (GPS-based) field maps in the production process.

Nielson (personal communication, 2001) reported that no precision agricultural initiatives have been identified in West Africa. The International Institute for Tropical Agriculture (IITA) has a mandate for poverty alleviation, food security insurance, and enhancement of environmental services in West Africa, which is geared toward severely disadvantaged communities with simple technological opportunities for intervention; precision agriculture schemes are not yet in the pipeline.

IITA does however make extensive use of GPS and GIS technologies in its research. In early 2005, most of IITA scientists used GPS as a matter of course in the field, recording locations of crop trials, gene-bank acquisitions, pest and disease outbreaks, and socioeconomic surveys. Most GPS units are simple Garmin Etrex, and data are normally downloaded directly to PCs via cable links. In a few cases, GPS units are used in conjunction with handheld PDAs for more complex data collection and navigation. IITA maintains an expanding archive of spatial data for Sub-Saharan Africa, including climate, vegetation, infrastructure, population, relief, and administrative maps, and makes extensive use of satellite imagery, mainly Landsat TM. GIS is used in targeting IITA agricultural technologies, in analyzing distributions of genotypes and crop pests and diseases, and in processing socioeconomic data. A Web server is currently being established to disseminate these data within the organization (Nielson, personal communication, 2005).

In countries where plantations are established by large international companies for the production of wood, GPS- and GIS-based systems are used to manage variability in soils (Lowenberg-DeBoer, personal communication, 2001).

### *Reasons for Limited Adoption of Precision Agriculture by Large Commercial Farmers in Africa*

Economies of scale, poor infrastructure, a low level of management skills, and low profit levels are the main reasons why farmers in Africa have not adopted precision agricultural technologies to a larger extent. Three of the most important constraints to agricultural development in West Africa as reported by Sanders et al. (1996) are

- inadequate adoption and diffusion of the substantial achievements of public investment in agricultural technology research over the past 20 years;
- the failure of economic policies to encourage output and investment in the agricultural sector; and
- the inability of farmers to acquire capital either from their own savings or from the private or public sectors to finance the increased input purchases necessary for technological change in agriculture.

These constraints are interrelated. Higher inputs are required to increase yields. An economic environment in which profits can be made at an acceptable level of risk, the provision of adapted agricultural technologies, and the evolution of input and product marketing systems are necessary conditions to encourage farmers to purchase more inputs (McMillan et al., 1997).

According to Penning de Vries and De Wit (1987), the fact that potential food production in Sub-Saharan Africa by far exceeds its current production levels is the main reason why serious agricultural research must be done to develop improved varieties of crops, new husbandry and management technologies, such as site-specific management techniques that in turn can offer farmers new technologies, and better means of production and management. Charreau and Rouanet (1986) pointed out that there are many technologies available for soil and water management as well as agronomic technologies which are not used because of the lack in management skills. Pinstrup-Andersen and Pandya-Lorch (1997) also found that because of the lack of management skills, the farmers in most Sub-Saharan African countries have gained only marginal productivity. If applied and managed correctly, new technologies can have a significant impact on agricultural production.

In 1976 a total of 10,146 families were settled in the Mahaweli Ganga Development Project in Sri Lanka. After three years it became apparent that the project's objectives were going awry. After an assessment, it was found that the settlers had an acute absence of productive and managerial human skills. In other words, they were not able to adopt and manage the new technologies transferred to them (Kahn, 1982).

The availability of appropriate technology and extension services is essential in the establishment of profitable agricultural enterprises (Binswanger and Deininger, 1996). Kirsten (1994) referred to Eicher and Baker who came to the conclusion that over the past 20 years most extension services and officers in Africa have been poorly equipped and undertrained, when compared to their counterparts in Asia or Latin America. Most extension services in Africa are orientated toward technical problems and ill-equipped toward farm management or social aspects that are necessary for technology transfer and adoption. Nagy et al. (1988) found that experienced extension officers are one of the most important components of technology transfer and adoption-support programs.

Walter et al. (2001) pointed out that agriculture was the leading economic activity in Africa, but outdated tools and methods were used for farming on poor and thin soils. Something has to be done to enhance the declining agricultural productivity in Africa. Sanders et al. (1996) pointed out that Sub-Saharan African countries have undertaken the improvement of agricultural production, especially food crops, but this has not been an easy venture and has had several constraints.

Practically, precision agriculture is not without constraints. In Zimbabwe, Bryce-Rogers (personal communication, 2001) identified the cost of differential signal as one stumbling block in the expansion of precision agriculture. "Individual users with combines are not prepared to pay the US$2,000 per annum for DGPS signal when the unit will only be used for six weeks a year" (Bryce-Rogers, personal communication, 2001). Combines are also extremely expensive. Furthermore, the economic climate and political upheaval prevailing in Zimbabwe have affected the farming industry, and have made users cautious of spending money on extra or nonessential operations. Bryce-Rogers' (personal communication, 2001) view was therefore that the future of precision agriculture in Zimbabwe is uncertain and improvement will only occur once there is political change or stability.

### *Precision Agriculture in South Africa*

South Africa has advanced more than other African countries in precision agricultural practices. Farmers in South Africa are involved in precision agriculture activities ranging from the diagnostic through positioning stage to the variable-rate application stage. Precision agriculture is prevalent in three main regions in the country as indicated in Figure 17.2. The first region is the Corn (maize) Triangle which includes the northern Free State, and part of the North West Province and Mpumalanga; the second region includes the irrigation schemes of the Northern Cape; and in the third region, to a lesser extent, is the Western Cape Province (the Swartland and Rûens).

#### *Equipment, Inputs, and Services Offered*

A number of companies play an important role in the advancement of precision agriculture in South Africa, ranging from fertilizer and seed companies to general agricultural companies (former agricultural cooperatives), machinery dealers, and consultants.

- Fertilizer companies include Kynoch, Omnia, Nitrophoska, SSK (Technifarm)
- Current and former cooperatives include SSK (Technifarm), Afgri (Techniland), GWK, Senwes, Suidwes, and the NWK
- Machinery dealers include Case, New Holland South Africa (Flexicoil), John Deere, and Massey Ferguson
- Precision farming electronic suppliers include Astrata and Kee Technologies

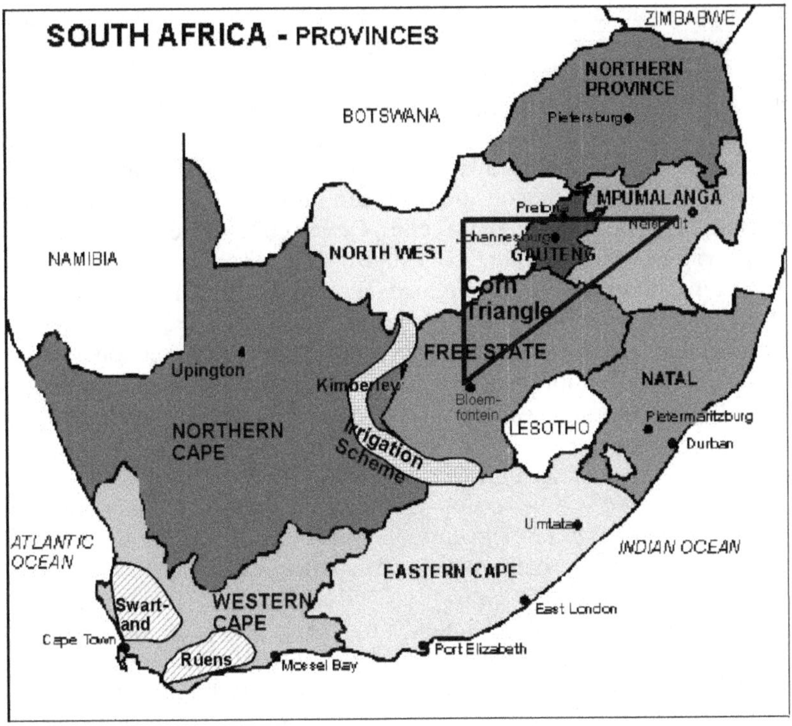


FIGURE 17.2. Regions of South Africa in relation to adoption of precision agriculture.

*Company Profiles*

Kynoch (Yara), a fertilizer company operating in the entire country, has been analyzing and evaluating precision agricultural concepts and products on farm levels, jointly with farmers and producers since 1990 (http://www.yara.com). Kynoch offers precision agricultural services, specializing in remote sensing and focusing on geographic database management. It has a comprehensive soil analysis system. Samples are taken using remote sensed imagery and vegetation indexes, and "Smart Sampling" is used in identifying management zones. For recommendation purposes, the soil analysis results are interpreted and managed with a program called Kynoch Farm Vision that is based on a Geographic Information System (GIS) (www

.yara.com). Yield monitoring is done only in cases where farmers or custom combines are equipped with yield monitors.

Omnia offers precision agricultural services to farmers in collaboration with other companies and cooperations. The services include grid soil-sampling with soil chemical maps, soil classification, creation of VRT application maps, statistical analysis on what causes yield differences with the aid of "Infolab," precision application of fertilizer, and satellite imagery to identify or define management zones. Omnia collaborates with other service providers, whereby other companies gather data (for example, yield mapping), and Omnia does recommendations for fertilizer application and alleviation of soil problems. The company currently serve 250,000 hectare with precision farming practices and does differentiated correction with fertilizers and lime/gypsum on average of 60,000 hectare annually. A fleet of 20 VRT applicators is available to farmers for variable application of fertilizers.

Nitrophoska is another fertilizer company with its base in Western Cape Province. It started with precision agriculture in 1998 and offers precision agricultural services to farmers through grid soil sampling to identify management zones for variable-rate fertilizer application. Teragators with GPS are then used to variably apply fertilizer according to the management zones (De Swart, personal communication, 2001). Nitrophoska is currently in the fertilizer and lime business which mainly applies phosphorus, gypsum, dolomitic and calcitic limes. The area serviced has doubled in the past four years. No yield monitoring was available at the time of this research amongst the company's clients.

Senwes Ltd. is an agricultural company that provides agricultural support and services for precision agricultural activities nationwide. Senwes started offering precision agricultural services to farmers in 1996. However, it has also been involved with soil classification services since 1980. It carries out comprehensive soil classification with emphasis on the physical characteristics to determine different soil potentials within a field. These potentials are generated with a simulation model to determine potential crop yield for different crop types. Chemical analysis for pH and other nutrients is also included in the potential determination, even though the main focus is on physical soil characteristics. GPS is used in taking soil samples to earmark the area concerned and to develop "decision maps." The company established that there is a high correlation between the physical characteristics of the soil and yield, ending up with a potential map. The correlation results indicated that the higher the soil potential, the higher the yield (Muller, 2001).

Senwes mapped many hectares of and through soil classification during the past ten years. Senwes was instrumental in doing the first variable-rate application based on soil potential for row crops (2,000 hectare maize) during the 2000 planting season of the maize triangle in the North West Province. Input costs were cut significantly and the yield harvested in June/July 2001 showed substantial improvement. Senwes is strongly convinced of the necessity of yield monitoring, as it is an effective tool to measure (quantify) the progress made, and cost efficiency of the inputs made during the precision farming process (Helm, personal communication, 2001).

GWK started its first trials on precision farming during November 2004. One farm was identified and the chemical analyses made on a grid sample basis of 100 m × 100 m. At this early stage it looks at the maps provided on a whole-farm basis. By doing this it is able to get an idea of the soil genesis and how it influences the characteristics of the different chemical boundaries.

It is important to investigate the soil physical as well as chemical aspects. By making use of the yield maps, areas will be identified to make decisions on how to approach the fields. In most cases there is a definite correlation between soil chemistry, physics, and the yield. There is a big difference between precision farming management under dryland conditions and irrigation. Because of the small areas of the irrigation fields, chemical and physical problems will be rectified with VRA equipment.

Both grid and smart sample methods will be used in the future to identify management zones. This will be done with the yield data in mind. There are about 24 GPS-equipped yield monitors and 10 non-GPS-equipped yield monitors in the GWK area and only one farmer using VRA equipment.

NWK is an agricultural cooperative that has been offering a wide range of services to farmers in the North West Province since 2000. NWK is at present fully involved in precision agriculture. Grid-sampling services are provided in collaboration with Omnia. A satellite image service is also available to farmers to assist them with affordable crop monitoring at the farm level. Yield monitoring and VRT equipment are available to all farmers in collaboration with the various machinery suppliers. A geographic information system is used to handle different precision farming applications. An Internet-based management system is used as a valuable aid for the modern grain producer to apply precision agricultural principles.

Suidwes Agriculture (Pty.) Ltd. started with precision agriculture at the beginning of 2001 with a group of 20 farmers who were interested in precision agriculture. The company provides the following precision farming services to farmers:

- GPS measurements to accurately determine areas
- Soil classification to determine the physical characteristics of the soil in order to determine long-term expected yield

- Chemical analysis of the soil done on a grid-sampling basis from which a spatial map is generated
- Satellite images and harvester monitor data to help with the interpretation and monitoring of the growth and yield of the crop
- All the mentioned data are conduced into a GIS where the data are interpolated

SSK (Technifarm) (Sentraal-Suid Koöperasie Beperk/Central South Cooperative Ltd.) is an agricultural organization in the Western Cape providing different services to cash-crop farmers in the area. SSK started with precision agricultural services in the farming community mid-2001 with the purchase of five yield monitors rented out to farmers. These yield monitors were used for the first time during the wheat harvesting season in 2001. SSK subsequently purchased a shareholding in Technifarm, a precision farming holding that provides the following services: GPS farm mapping, soil sampling, variable-rate application of lime and fertilizer, supplying and installation of precision farming equipment, i.e., yield monitors, variable-rate controllers, supplying of NIR scanners, etc. Technifarm provides these services to SSK's farmers as well as the rest of the producers in the Western Cape. Technifarm has GPS grid sampled more than 100,000 ha since 2001 and has done VRA of inputs on more than 32,000 ha (Enslin, personal communication, 2005).

Techniland, a division within Afgri, has been involved with precision agriculture since 1998. Techniland offers unique and comprehensive products and services that enable farmers to maximize their effectiveness by making the right choices in terms of inputs and production. It all starts with an accurate quantification of the resources available, presented in an easily understandable format (maps and summary tables) that allows the farmer and agronomist to make quantified decisions in terms of inputs and production objectives. By providing the farmer the facility to accurately determine specific areas within fields that require specific treatments, risk and costs can be drastically reduced. Through this process, clients can make financially sound decisions, thereby lowering the risk to financiers and themselves.

Techniland focuses on incorporating modern technology and quantitative assessments of agricultural resources (climate, soils, and crops) by applying sound scientific principles adjusted to local conditions. Assessments are done by in-field inspections and sampling as well as complementary remote sensing technology. This has resulted in developing a world-class offering at an affordable cost. The services offered by Techniland include the following:

- DGPS mapping of field boundaries and maintenance of GIS layers
- Assessment of soil physical and chemical status through grid sampling
- Data processing (production of user-friendly spatial maps and information)
- Quantifying the in-field agronomic yield potential of farming units
- Providing decision support tools to farmers and agronomists
- Providing agronomic analysis and advice
- Variable rate application services for inputs (fertilizer, herbicide, insecticide)
- Data collection and variable rate equipment sales and support
- Software sales and support

Within the Techniland offering is the Agrimage satellite imagery range of products. These enable near real-time monitoring of crop development through time series of processed imagery and seasonal growth curves. Other applications include early detection of stressed areas, field scouting, smart sampling, early yield estimation, quantification of crop damage for insurance purposes, and comparison of cultivar performances. Techniland is continually striving to develop new applications through research and development programs in conjunction with end users (farmers). Regular contact with international experts also ensures that the latest technology is sourced and adjusted to local conditions to the benefit of the farmer (Badenhorst, personal communication, 2005).

New Holland South Africa is able to supply all the precision agricultural implements and hardware by means of its Flexicoil range, and contracts software to Astrata and Kee Technologies. Flexicoil Ltd. started with precision agriculture as an independent company in 1994, selling precision agricultural equipment in South Africa. The package offered by New Holland currently includes spreaders, air carts for bulk seed and fertilizer handling, seeders, planters, and sprayers. The Task Controller II is also available, which is compatible to other precision agricultural equipment. There are currently 38 Flexicoil systems operational in South Africa. Sales of precision farming equipment have increased substantially over the past four years (Van der Merwe, personal communication, 2005).

John Deere started providing precision agricultural equipment and hardware in the form of yield monitors and yield-mapping systems to agents and dealers in 1997. It does not deal with farmers directly, but through agents such as agricultural cooperatives. They sell Agricultural Management Systems in the Free Sate, North West, as well as Zambia (approximately ten yield monitors and ten guidance systems). Estimates show that approximately 200 John Deere clients are using precision farming equipment, which includes the following products:

- Parallel tracking
- Auto trac
- Field Document Systems (software that documents information on crops/cultivar, etc.)
- Map-based description (seed monitoring variable-rate equipment)
- Harvest Document Systems (yield monitor, mapping, and software)

One silage harvester with a yield monitor is operational in South Africa and a cotton combine with a yield monitor has recently been launched (Hartzenberg, personal communication, 2005).

Massey Ferguson started providing precision agricultural equipment to farmers in 1998 and does not provide any precision agricultural services. Satellite DGPS systems and signals are currently provided by Omnistar and Racal, but Omnistar plays a dominant role in this regard. Astrata South Africa distributes and supports leading precision farming systems from Ag Leader Technology and Trimble Navigation.

Ag Leader Technology has been the pioneer of precision farming technology for more than a decade. The company introduced the industry's first yield monitor in the early 1990s and has continued to develop and market some of the industry's most advanced integrated precision farming systems and software, such as Insight, used throughout the farming process including yield monitoring, variable-rate control, and Trimble's AutoPilot system. Trimble, the world leader in GPS innovation and development, has numerous products that dominate the manual and automated guidance markets with all products supporting patterns for curves, contours, parallel lines, headlands, and pivots. Trimble also has a range of laser- (Spectra Precision) and GPS-based land-leveling products for water management (Lipawsky, personal communication, 2005).

KEE Technologies is an Australian agricultural electronics company (since 1982) focusing on agricultural electronics to better farming performance and profit. KEE Technologies entered South Africa at the end of 2001 as part of its global expansion program that included Brazil, the United States, the United Kingdom, and Europe. KEE markets its products through selected distributors such as Senwes, Suidwes, and Technifarm. KEE's flagship is the ZYNX, a multifunctional controller doing guidance, auto steering (parallel, contour, and race tracking), spraying, spreading, and variable rate control (up to five products individually and simultaneously), as well as auto section control to prevent overapplication of input products (Helm, personal communication, 2005).

### *Current Status*

In this changing agricultural environment, farmers have to be on a constant lookout for an increase in the price of inputs, which rises faster than the value of outputs, the so-called price-cost squeeze that catches up with the profit margins of less proactive farmers. In an endeavor to lower the effect of the price-cost squeeze, farmers are forced to increase productivity, which can be done by spreading fixed costs over larger areas. Olson (1998) stated that to achieve this goal it requires speed in machinery, speed in information processing, and speed in decision making.

Precision agriculture applies inputs according to the soil requirements, thereby lowering input costs to help combat the ever growing price-cost squeeze in South African agriculture. Harris (1997) concluded from his research that cost savings might reasonably be expected to be 15 percent of variable costs, depending on the variation in yield potential. Yield monitoring is currently mainly concentrated in the Corn Triangle, the Western Cape, and the irrigation schemes of the Northern Cape.

As in the rest of the world, precision agriculture is still in its infancy stage in South Africa but showed a significant increase in the past four years. Between 1 and 2 percent (Lipawsky, personal communication, 2005) of its farmers are involved in precision agriculture. Information about the current status of farmers was gathered from the role-players as well as from farmers themselves (Matela, 2001). In 1999, the South African agricultural sector was comprised of 54,000 commercial farmers (*Agribusiness and Farming,* 2000).Twelve thousand (22.2 percent) of these farmers were in the three main regions—the Corn Triangle and the irrigation schemes of the Northern Cape (9,000), and the Western Cape Province (3,000). Almost all farmers involved in precision agriculture practiced in these three regions. Furthermore, precision agriculture in South Africa focused mainly on cash-crop production.

The major role-players in the industry were contacted to obtain the numbers of farmers involved in precision agriculture, as well as their names and contact details with the purpose to draw a random sample from the total population. The number of farmers was easily obtained, but because of the alleged confidentiality between companies and clients it was not possible to obtain the names of the majority of farmers in cases where variable-rate fertilizer and lime were applied by companies or where fertilizer companies only helped farmers in identifying management zones (Matela, 2001).

The information concerning the use of precision agriculture technologies received from the role-players is reflected in Table 17.3. Quite a number of farmers were involved in variable-rate application technologies for lime (244) and fertilizer (251). This emanates from the fact that a greater number of role-players are in the fertilizer business, and with aggressive marketing, these companies were able to win some clients. The acidity problem prevalent in South Africa accounts for a high usage of variable-rate lime application. Farmers using yield monitors numbered approximately 600.

The names of 25 farmers were obtained from the role-players and questionnaires for the research by Matela (2001) were sent to these farmers. The main results from the 25 farmers on the usage of precision agricultural technologies are summarized in Table 17.4. From the results it is evident that there were farmers using only grid soil-sampling without using GPS-based yield monitoring, and also farmers using yield monitors without grid soil-sampling results. Of the 25 respondents, 20 implemented grid soil-sampling on their fields as a result of the grid soil-sampling process being the en-

TABLE 17.3. Estimates of hardware used for precision agriculture in South Africa, 2005.

| Technology | Approximate number of units |
|---|---|
| Yield monitors (GPS linked) | 600 |
| Variable rate applicators | 500 |
| Manual tractor guidance (light bars) | 200 |
| Automated guidance systems in tractors | 50 |

TABLE 17.4. Precision farming technologies adopted by 25 farmers in Northern Free State, the North West Province and Mpumalanga; the irrigation schemes of the Northern Cape, and parts of Western Cape Province (the Swartland and Rûens).

| Technology | Number of farmers |
|---|---|
| Grid soil sampling | 20 |
| Yield monitoring | 12 |
| Differential lime application | 16 |
| Differential fertilizer application | 8 |

trance step and building block into precision application of inputs. Lime application was popular amongst these farmers, with 16 out of the 25 farmers involved in differential lime application, and only 8 out of 25 implementing variable fertilizer application. Yield monitoring was applied to some extent (12 out of 25), though some farmers did not use the yield-monitoring results for identifying management zones and thus variable-rate application of inputs. Yield monitoring is currently mainly concentrated in the Corn Triangle and the irrigation schemes of the Northern Cape. In mid-2001, yield monitoring was nearly nonexistent in the Western Cape.

Some of the 25 farmers produced cash crops on rainfed fields as well as under irrigation. These farmers indicated that they started on irrigation fields with precision agriculture because the potential for profit improvement was higher on irrigation than on rainfed fields. Smith (personal communication, 2001) suggests that farmers are currently only starting to enter into precision agricultural practices. Omnia did potential and chemical classifications for a number of farmers in the Corn Triangle and the irrigation areas of the Northern Cape. Some farmers who had already implemented precision agriculture and were involved in irrigation in Mpumalanga, the Free State, and Northern Cape, indicated that they felt that precision agriculture would be more profitable on irrigation fields as the variation in yields are larger under irrigation than under rainfed conditions. The scope for increment in profitability is also better.

Precision irrigation practices, where the application of water is varied according to the potential of the crops, did not exist in South Africa at the time of the research. In their research, Dennis and Nell (2001) found that there was a great need for precision agriculture technology in South Africa as it is expected that the price of water will increase substantially in future, which will put pressure on the profitability of irrigation farming in South Africa. With regard to plantations, about one-third of the forest plantations in South Africa were planned and managed by means of GPS-based systems and the rest were managed by means of GIS imaging (Schoeman, personal communication, 2001).

## *CONSTRAINTS IN THE ADOPTION OF PRECISION AGRICULTURE AND SITE-SPECIFIC MANAGEMENT TECHNOLOGIES*

Precision agriculture provides highly sophisticated information to the farmer, but the management skills of the majority of farmers are not developed to such an extent to take advantage of this. The farm managers, who have not developed good management skills, will not know how to get the most advantage from the precision agricultural system. It will have a negative effect on the profitability of adopting precision agricultural techniques.

Profitability is envisaged as the main factor that drives precision agriculture in South Africa (Lowenberg-DeBoer, personal communication, 2001). However, profit margins may be lower for crops of past comparative advantage, but higher for high-value crops. Precision agriculture does have some drawbacks. The capital outlay for the whole system, which amounts to approximately R700,000 or US$87,500 (yield monitor, task controller, air cart, seeder or planter, DGPS signal, computer and software) is relatively expensive for South Africa. The decline in the value of the South African rand would make it more expensive to obtain these new precision technologies if the current trend persists.

Furthermore, farmers may not be prepared for the extra costs that come with precision agricultural practices. For example, one contractor (custom operator) in the North West Province has fitted a yield monitor to his combine, but has suspended the satellite signal service as it was found to be unprofitable or too costly. The contractor has mapped fields for 40 farmers, but the farmers were not willing to pay for the extra costs associated with a GPS-based yield monitoring on the combine harvester. Only one farmer used the precision yield data obtained by the contractor. This indicates that farmers in specific areas are not ready to pay for highly sophisticated information to improve the profitability of their farming operations.

In the United States, labor is expensive, but capital is cheaper, and the system is thus designed to minimize labor and increase capital (Lowenberg-DeBoer, personal communication, 2001). This will, however, have a negative impact on South African rural communities that rely on farm jobs. In South Africa the productivity of labor is low and therefore also relatively expensive, and this can motivate farmers to mechanize further by means of precision agriculture. The relatively low profit margins of the main crops (maize and wheat) planted in the Corn Triangle and irrigation areas in the Northern Cape, further contribute to the constraints of adopting precision agricultural technologies.

## *RESEARCH ON PRECISION AGRICULTURE IN SOUTH AFRICA*

The Centre for Agricultural Management at the University of the Free State in South Africa is conducting research on the economic aspects of precision agriculture. The research is funded by New Holland South Africa (NHSA) and assisted technically by Senwes. The Centre also coordinates a one-day short course, "Introduction to Precision Agriculture," which was especially developed for farmers. This course is also financially sponsored by NHSA and technically assisted by Senwes.

Only one postgraduate study in South Africa on precision agriculture could be found, "The Relationship Between the Spatial Variability of Some Physical Aspects of Soil and the Variability of Rain-Fed Maize (Corn) Yields Within 11 Maize Fields in the Theunissen District: A Case Study" (Muller, 2001). The main focus of this study was to determine the correlation between physical aspects of the soil (plant-available water capacity) and the variation in maize yields. The results showed a high correlation, which indicates that a thorough knowledge of the soil can contribute substantially to identifying management zones. The following research projects on precision agriculture are under way:

- The current status of precision agriculture in South Africa
- The profitability of precision agriculture in the Northwest Free State
- The prospects for precision irrigation in South Africa
- Use of technology for more effective production

## *PROSPECTS FOR PRECISION AGRICULTURE*

Technological development in agriculture is part of the development of the modern farming business. If the rest of Africa follows the same development trend as South Africa, site-specific management will increase in Africa to improve yield levels of crops and livestock. This will contribute to the alleviation of poverty in Africa. However, the trend and effect of HIV/AIDS in Africa are unclear. This makes the forecast of the future for site-specific management and precision agriculture in Africa, South Africa included, extremely difficult.

The effect of HIV/AIDS is highly speculative as different sources have different viewpoints on what the future of this disease is going to be. Official sources are very cautious to forecast the trend. A discussion on the general view/outlook regarding different the HIV/AIDS trend will be followed by a discussion of HIV/AIDS in Africa by means of two scenarios and the future challenges for Africa. The section will end with a discussion of the future potential of precision agriculture for South Africa.

### *Synopsis of Future Expectations in Africa*

Agricultural development and human capital capacity development go hand in hand. The human capital capacity of the person adopting the more advanced agricultural technologies will be improved as well as the production and profit of the farming business. This in turn will contribute to the alleviation of poverty in Africa. Training of farmers as well as extension officers in more advanced production systems and management technologies, such as site-specific management, will also contribute to the expansion of human capital capacity and the well-being of the African population. This in turn can contribute to the alleviation of the impact of the HIV/AIDS problem facing Africa, which has grasped the attention of the international community.

*Scenario 1: The AIDS pandemic can be stopped.*

In Kampala, Uganda, the prevalence of HIV dropped from 29 percent in 1992 to 8 percent in 2002, and in Addis Ababa from a peak of 24 percent in 1995 to 11 percent in 2003. This could indicate that the population of Africa will increase, which will put pressure on food resources. This can be counteracted by means of expansion of site-specific management, precision agriculture, and biotechnology.

*Scenario 2: The AIDS pandemic cannot be stopped.*

Agriculture is the largest sector in most African economies, accounting for a large portion of production and a majority of employment. Studies done in Tanzania and other African countries have shown that HIV and AIDS will have adverse effects on agriculture, such as a loss of labor supply and remitted income.

Sub-Saharan Africa has just over 10 percent of the world's population, but accommodates almost two-thirds of all living people with HIV (estimated: 23.1 to 27.9 million). In 2003 alone, between 2.6 and 3.7 million in the region became newly infected with HIV. Approximately 2.5 million related AIDS deaths occurred. In Africa, the pandemic spreads through the general population and not only to populations defined as high risk. This has resulted in the African adult population remaining stable in recent years.

The loss of a few workers at a crucial period during the planting or harvesting season can significantly reduce the size of the harvest. In countries where food security is a continuous issue due to droughts, decline in household production (the first level of farming) will have serious consequences. A loss of agricultural labor is likely to cause farmers to switch to less labor-intensive farming. This may cause switching from export crops to food crops. HIV and AIDS will affect the production of cash crops as well as food crops. This will have a major impact in the second-level higher farming system.

Community authorities often allocate land to families on basis of their size. Sickness and death due to HIV and AIDS can result in the inability of households to cultivate all the land at their disposal. Remote fields tend to be left fallow and the total output of the agricultural unit consequently declines. Declining yields are less immediately visible but important, because poor timing in such essential farming operations as tillage, planting, and weeding can be devastating.

The most vulnerable population to HIV/AIDS are the able-bodied and most productive age groups, and this has a significant impact on labor implications (FAO, 1995). Labor shortages result from lost employees who become ill and die, as well as their family members that have to nurse them. These people not only reduce the work force, but important knowledge and skills are lost. In many areas, the usual way for children to learn the required agricultural skills is by working with their parents. Given the AIDS pandemic, this is often no longer possible, and owing to the gender division of labor knowledge, the surviving parent is not always able to transfer the skills of the deceased one. That causes training costs to increase. Cash crops are therefore often abandoned because of the inability to maintain enough labor for both cash and subsistence crops. Switching from labor-intensive crops, to less labor-intensive ones impacts on the nutritional quality of diets.

The direct impact of HIV/AIDS on crop production in eastern Africa was observed from the resultant reduction in land area cultivated, a decline in crop yields, and a decline in the range of crops grown. Labor-intensive farming systems are particularly vulnerable to the pandemic. Cultivation in most African countries is mainly done by hand, although animal traction, and to a lesser extent tractors, are available to wealthier farmers (FAO, 1995). Livestock serve multiple functions, and frequently represent a form of asset. The medical costs incurred by those affected by HIV and AIDS often require the selling, payment in kind, and sacrifice of livestock as payment for traditional or other forms of medical treatment (Du Guerny, 2005).

The impact of HIV/AIDS on farming communities differs from village to village and country to country. However, the pandemic is undermining the progress made in the past 40 years of agricultural and rural development. This poses enormous challenges to governments, nongovernmental organizations, and the international community. The disease is no longer just a health issue—it has become a major development problem (FAO, 2005).

The prevalence of HIV/AIDS and/or an unstable political environment can have a negative impact on the economies of African countries, as well as on agriculture and the adoption of new agricultural technologies. Sixty-five percent of the world's HIV-infected population lives in Africa (Lewallen, 1997). According to the statistics, 47 million infections had occurred worldwide by the end of 1998, of which 34 million occurred in Sub-Saharan Africa alone. This indicates that 72 percent of world's infections were prevalent in Africa. By the end of 1998, 5.8 million new cases of HIV infections were reported and 4 million of these were in Sub-Saharan Africa. Furthermore, deaths due to HIV/AIDS occurring in Sub-Saharan Africa reached 12 million, compared to 14 million for the whole world, thus Sub-Saharan Africa accounts for 86 percent of total deaths due to HIV/AIDS. The trend of deaths and newly reported infections was even higher for children, with 94 percent and 90 percent, respectively (Stenicki, 1999). It is alleged that the survival of HIV-infected people in Sub-Saharan Africa is shorter than those in developed countries (Maartens and Wood, 1998).

The statistics are appalling and clearly indicate that Africa is hit the hardest, considering that Africa's population represents only a tenth of the world's population. It is therefore apparent that the pandemic will have a severe impact on African economies, including agriculture. According to Stenicki (1999), as skilled adults die, many African economies, which are already weak, will deteriorate further. For example, between 1996 and 1997, Botswana, as one of the highly-infected countries, as a result of the impact of HIV/AIDS, fell 26 places down the Human Development Index, which takes into consideration wealth, literacy, and life expectancy (Stenicki, 1999).

The Food and Agriculture Organization (FAO) projected that the labor force in the ten most-affected African countries will be reduced by as much as 26 percent by 2020, with a devastating effect on rural production and food security (FAO, 2005). According to the report, the virus has a major impact on nutrition, food security, agricultural production, and rural communities in many countries. The view of the FAO (1995) was that the impact of HIV/AIDS, especially where it reached pandemic proportions in some countries, should be taken into account in all steps of a project cycle, such as project identification and design, project appraisal, project implementation, monitoring, and evaluation.

A synopsis of the impact of HIV/AIDS on agriculture, and more specifically precision agriculture, is essential. However, there is no country-wide population-based data that can be used to determine the size and shape of the pandemic in different countries (Ncayiyana, 1998). Stenicki's (1999) view is that the HIV prevalence is expected to rise. The dilemma is whether HIV/AIDS can be curbed or not, and the effects of each scenario.

Agricultural production systems in Africa are generally very labor intensive. If AIDS cannot be stopped, farmers will be forced to adopt modern technologies such as mechanical equipment, site-specific management, and precision agriculture to be competitive in an international market. However, such advanced technologies can also be affected because of the skilled labor needed to operate such sophisticated equipment. If a skilled individual dies because of the disease, someone else has to be trained, which will increase training costs.

### *Future Challenges for Africa*

Kirsten (1994) suggests it is not only necessary that emerging and small-scale farmers be provided with new technologies, but their management skills will also have to be sharpened to ensure that they will make the best use of new technologies to compete in free national and international markets.

To be competitive and to increase their productivity, emerging farmers are in desperate need of government support in order to adopt modern technologies. Nell et al. (1999) emphasized that emerging farmers need institutional support from the government to increase their management levels. According to Sanders et al., (1996) this should also include improvements in resource management (soil fertility, water, veld, and capital management). Binswanger and Deininger (1996) furthermore agreed that the availability of appropriate technologies to improve profitability and extension services is essential in the establishment of profitable agricultural enterprises.

In Sub-Saharan Africa, the most successful technology change took place in Burkina Faso, Mali, and Senegal where the introduction of new cultivars was a great success due to the fact that farmers adopted improved crop management techniques. When farmers start using more complex management systems and the results are positive, they will be more open to further improvements, especially if the overall economic environment for increased crop production is favorable.

There are prospects for increases in precision agriculture technologies in some parts of Africa. The involvement of NADC in precision agriculture in Zimbabwe has stimulated interest in its future expansion. Interest in precision agriculture has been observed in other parts of the country, from tobacco farmers in Kiroi and Birndira, as well as the sugar industry based in the low veld of Zimbabwe. The sugar industry has also shown interest in satellite images, but NADC is not yet involved in this aspect because of prohibitive costs. NADC also observed interest in precision agriculture among Zambian farmers in the Mazubuka and Mkushi areas, but NADC has not yet started any commercial operations in Zambia (Bryce-Rogers, personal communication, 2001).

The countries in Africa with political stability will be an attraction for adventurous farmers from Southern Africa and even farmers from abroad. The influx of commercial farmers with more advanced technologies will inspire local farmers to improve their farming practices, which can accelerate the agricultural production of these countries. Eastern Africa is an example of such an area. Eastern African countries are ahead of other countries in Africa on more advanced management technologies such as site-specific management. The unstable political situation in Zimbabwe led farmers who had already started using advanced management technologies to be reluctant to expand on these technologies Bryce-Rogers (personal communication, 2001).

The expansion of site-specific management techniques will play an important role in the development of agriculture in Africa in the future. These techniques are not as capital intensive as precision agriculture and the economies of scale can be much lower than precision agriculture. This means that site-specific management techniques can be practiced on a much smaller scale than precision agriculture and still be economical. Much still has to be done in Africa in order to help in the advancement of precision agriculture in the continent. Institutional arrangements to facilitate services, such as leasing instead of ownership of equipment, and providing the technical know-how rather than farmers spending time learning, are essential (Lowenberg-De Boer, personal communication, 2001).

### *Future Potential of Precision Agriculture in South Africa*

South Africa is in the beginning phase of precision agriculture. Farmers are divided into two distinct groups, those who are strongly proprecision agriculture and those strongly negative toward precision agricultural technologies. The better managers realize the prospects for increasing profits and lowering risk, while the poorer managers cannot identify these advantages. Therefore, it is highly unlikely that the number of farmers in South Africa involved in precision agriculture will increase to 40 percent of the total number in the next ten years. The reason for this is that precision agriculture is associated with farmers with a high management capacity and skills and it is common knowledge that in all countries all over the world only the top 10 or 20 percent of the farmers fall in this category.

During the collection of data for this chapter it became evident that the number of farmers interested in starting with precision agricultural technologies will increase rapidly in the coming years. The number of farmers using the identification of management zones and/or yield monitoring increases every year. The one-day course on precision agriculture offered by the University of the Free State is expected to contribute substantially to the expansion of precision agriculture in South Africa.

The fact that SSK bought five yield monitors for renting out to member farmers during the November 2001 wheat harvesting season is a clear indication that they identified the need for precision agricultural technologies in the Western Cape. This will increase the profitability of wheat production and the lowering of production risk. Research to prove the profitability of precision agriculture scientifically is very important to increase the adoption of precision agricultural technologies. Such research projects are planned in the future by the Centre for Agricultural Management.

The projections for the availability of water in South Africa, according to Greens (personal communication, 2001), is that if the current population growth rate in South Africa stays the same, the water resources will be used up to full capacity by 2020. Already, 50 percent of all stored water in South Africa is used to irrigate crops. The newly enacted Water Services and National Water Act, which regulate the usage of water, makes judicious usage of water even more important (Mitchell and Thibaud, 2001). Research on precision irrigation will become very important in South Africa in the future. The expected increment in the cost of water will further encourage research.

Different approaches to precision agriculture by different private institutions and consultants in South Africa will have to be standardized. If not, farmers who do not have the necessary experience and knowledge will waste their money and lose interest in precision agriculture. As commercial farming is vital to the economy, the country needs to develop its farmers and to provide information and assistance. Training is therefore needed to get agriculture back on track. This is why independent research centers such as the Centre for Agricultural Management at the University of the Free State will play an important role in expanding precision agriculture in South Africa. also, research projects will have to be done in collaboration with private institutions and consultants.

## REFERENCES

*Abstract of Agricultural Statistics.* 2004. Pretoria. National Department of Agriculture.

*Agribusiness and farming.* 2000. Bryanston: Trade Edge Publications.

Bembridge, T. J. 1991. Technology transfer in small-scale dryland crop production: Future challenges. *Development Southern Africa,* 8(4):479-493.

Binswanger, H. P. and K. Deininger. 1996. South African land policy: The legacy of history and current options. In *Agricultural land reform in South Africa,* edited by J. van Zyl, J. Kirsten and H. P. Binswanger. Cape Town: Oxford University Press, pp. 64-102.

Blackmore, B.S. 1999. Developing the principles of precision farming. In Proceedings of the international conference on agropoles and agro-industrial technological parks-AGROTECH'99, 1999, Barretos: Associação Nacional de Entidades Promotoras de Empreendimentos de Tecnologias Avançadas/Instituto Barretos de Tecnologia, 1999, pp. 75-77.

Center for International Development, Harvard University. 1999. Seminar on Agricultural Research in Africa, May 14, 1999. www.cid.harvard.edu/cidseminars, 3/16/2005.

Charreau, C. and G. Rouanet. 1986. The technology: Is it available? Where is it being developed? In *Development of rainfed agriculture under arid and semi-arid conditions,* edited by T. J. Davis. Proceedings of the Sixth Agriculture Sector Symposium at the World Bank, Washington, DC, May 6-10, 1986. Washington, DC: World Bank.

Dennis, H. J. and W.T. Nell. 2001. *Precision irrigation in South Africa.* Paper presented at the 12th Conference of the International Farm Management Association, Papendal, Netherlands, June 2001, p. 17.

Digges, P., A. Gordon, and A. Marter. 1997. *International markets for African agricultural exports: Agricultural policy reform and agricultural exports.* Marketing series 10. Chatham, UK: Natural Resources Institute.

Du Guerny, J. AIDS and agriculture in Africa: Can agricultural policy make a difference? http://www.fao.org/docrep/X4390e/X4390e02.htm, 3/16/2005.

Fertilizer Society of South Africa (FSSA). 1989. *Fertilizer handbook.* Translated by E. B. Dickinson, G. Greenfield and J. H. Roos. Hennopsmeer, South Africa: FSSA.

Food and Agriculture Organization of the United Nations (FAO). 1995. *The effects of HIV/AIDS on farming systems in eastern Africa.* Farm Management and Production Economics Service. Rome: FAO.

Food and Agricultural Organization of the United Nations (FAO). 2005. *HIV/AIDS: A rural issue.* http://www.fao.org/Focus/E/aids/aids1-e.htm, 3/16/2005.

Harris, D. 1997. *Risk management in precision farming.* United Kingdom: Bios Scientific Publishers.

Kahn, S. S. 1982. Organizing farmer groups in Mahaweli Ganga. *Agricultural Administration,* 11:303-308.

Kirsten, J. F. 1994. *Agricultural support programs in the developing areas of South Africa.* Doctoral thesis, University of Pretoria, Pretoria.

Lewallen, S. 1997. Ocular manifestations of the human immunodeficiency virus and the acquired immunodeficiency syndrome in developing countries. *AIDS Scan,* 10(1):12.

Lowenberg-DeBoer, J. and M. Boehlje. 1996. Revolution, evolution or dead-end: Economic perspectives on precision agriculture. *Proceedings of the Third International Conference on Precision Agriculture.* Madison, WI, pp. 923-942.

Loxton, R. F. and J.G. Thompson. 1978. *Advances in soil survey and land use in Southern Africa, 1953-1978.* Proceedings 8th National Congress, Soil Science Society of Southern Africa. Pietermaritzburg, South Africa.

Maartens, G. and R. Wood. 1998. Survival differences of heterosexual and homosexual HIV infection in South Africa highlights good hospital care (Somerset Hospital). *AIDS Scan,* 10(1):13-14.

Matela, N. 2001. The status of precision agriculture in cash crop production in South Africa. M.Agric Dissertation, University of the Free State, Bloemfontein, South Africa.

McMillan, D. E., J. Sanders, D. Koenig, K. Akwabi-Ameyaw, and T. Painter. 1997. Agricultural performance of new lands settlement in West Africa. Unpublished paper, University of Florida.

Mitchell, F. and G. Thibaud. 2001. No-till multiplies water infiltration by 10. *Farmer's Weekly,* 6 April.

Muller, C. B. 2001. The relationship between the spatial variability of some physical aspects of soil and the variability of rain-fed maize (corn) yields within 11 maize fields in the Theunissen district: A case study. M.Tech. thesis. Technikon, Pretoria.

Nagy, J. G., J.H. Sanders, and H.W. Ohm. 1988. Cereal technology interventions for the West African semi-arid tropics, *Agricultural Economics,* 2(3):197-208.

Natural Resources Institute (NRI). 1994. *Food policy in Sub-Saharan Africa: A new agenda for research and donor assistance.* Chatham: Natural Resources Institute.

Ncayiyana, D. J. 1998. South African HIV/AIDS research—Barking up the wrong tree? *AIDS Scan,* 10(2):13-14.

Nell, W. T.1998. Transfer and adoption of technology: The case of sheep and goat farmers in Qwaqwa. PhD thesis. University of the Free State, Bloemfontein.

Nell, W.T. and S.J. Napier, 2005. *Strategic approach to farming success. Securing competitive advantage in a turbulent global agricultural environment.* Bloemfontein, South Africa: Wim Nell Agricultural Management Consultant.

Nell, W. T., M.F. Viljoen, and M.C. Lyne. 1999. Towards developing financial management strategies for land reform beneficiaries in the Bloemfontein/Thaba Nchu area of the Free State Province of South Africa. *Development Southern Africa,* 16(3), Spring.

Olson, K. 1998. *Precision agriculture: Current economic and environmental issues.* Sixth Joint Conference on Food, Agriculture, and the Environment. University of Minnesota.

Penning de Vries, F. W. T. and C.T. De Wit. 1987. Identifying technological potentials. In *Accelerating food production in Sub-Saharan Africa,* edited by J. W. Mellor, C. L. Delgado, and M. J. Blackie. London: The Johns Hopkins University Press, pp. 109-117.

Pinstrup-Andersen, P. and R. Pandya-Lorch. 1997. Farm management challenges in the developing world. In *Managing into the 21st century,* Proceedings of the 11th International Farm Management Congress at the University of Calgary, Calgary, Alberta, Canada, July 14-19, 1997.

Republic of South Africa (RSA), Department of Agriculture. 1993. *Grazing capacity maps, Act no.* 43(83). Pretoria: Directorate Resource Conservation.

Sanders, J. H., B.I. Shapiro, and S. Ramaswamy. 1996. *The economics of agricultural technology in semiarid Sub-Saharan Africa.* Baltimore: MD: The Johns Hopkins University Press.

South African Online Travel Guide. http://www.southafrica-travel.net/climate/eklima_f.htm, 3/16/2005.

Stenicki, K. A. 1999. Status of HIV/AIDS in Sub-Saharan Africa. *AIDS Scan,* 11(1):7-8.

United Nations Educational, Scientific and Cultural Organization (UNESCO). 1977. *World map of desertification.* United Nations Conference on Desertification Report A/Conf. 74/2. New York: United Nations.

U.S. Department of State. http://www.state.gov/r/pa/ei/bgn/2898.htm, 3/18/2005.

Van der Merwe, D. 2001. *Personal communications and research from the archives of the National Department of Agriculture.* Agricultural Research Council, Institute for Agricultural Engineering—History book on agricultural engineering at Silverton, South Africa.

Walter, H. S., K.A. Jackson, Jr., K.J. Perkins, and S. Decalo. 2001. "Africa," Discovery Channel School, original content provided by World Book Online, http://www.discoveryschool.com/homeworkhelp/worldbookatozgeography/a / 00674 .html 28/05/2001.

# Chapter 18

# Asia

Ancha Srinivasan

## *INTRODUCTION*

Since the early 1960s, food production in Asia increased at rates and to levels that are unprecedented. Total cereal production, for example, increased from 329 to 1059 million tons (MT) between 1961 and 2005 (FAO, 2006). Much of this gain was due to new varieties with high harvest indexes, intensive use of farm chemicals, expanded irrigation, and advances in farm management. Between 1961 and 2002, fertilizer use in Asia increased from 3.8 to 77 MT, irrigated area from 90 to 194 million ha, and the total number of tractors in use (an indicator of farm mechanization) increased from 200,000 to 7.5 million (FAO, 2006). Recent reports, however, indicate many threats to long-term sustainability of cropping systems. For instance, the rice-wheat rotation, which is the basis of food security in much of South Asia, presents such problems as loss of organic matter, nutrient mining, salinity, waterlogging, and buildup of weeds, pests, etc. Ali and Byerlee (2000) reported a decline of total factor productivity of 0.44 percent per annum in the wheat-rice system in Pakistan from 1971 to 1994. High rates of nutrient and pesticide leaching, pesticide resistance in insects and pathogens, and biodiversity loss have become all too familiar in Asian farming. Nitrate pollution of drinking water in tobacco and rice ecosystems of Malaysia (Ahmad et al., 1996) and of groundwater in Japan (Nishio, 1999), India (Sinha, 1997), and China (Zhang, 2001), and increased soil erosion in Nepal (Carson, 1992) are a few examples. Asia is also characterized by dwindling per capita arable land and water reserves, escalating production costs, and growing food demands due to an increasing and urbanizing population with rising incomes and associated changes in lifestyle and food preferences.

Against the backdrop of these problems Asian agriculturists face the daunting challenge of developing and applying methods that can raise pro-

*Handbook of Precision Agriculture*

doi:10.1300/5627_18

ductivity in sustainable ways. Indeed, a strengths, weaknesses, opportunities, and threats (SWOT) analysis of farming systems in Asia (Exhibit 18.1) indicates an urgent need for adoption of innovative approaches. Precision farming (PF)—a holistic management strategy to tailor agronomic management to fit varying soil and crop conditions within a field—offers new opportunities for farmers to optimize and sustain yields and profits and to reduce pressure on natural resource base, thereby leading to total quality management in farming. The objectives of this chapter are to examine spatial and temporal variability found in Asian soils and cropping systems and to identify the potential for PF adoption. The current status and various constraints for PF adoption in Asia are examined and a few ways to overcome such barriers are proposed.

## *SPATIAL VARIABILITY IN ASIAN FARMS*

The natural soil and crop variability in many parts of Asia is substantial, even on the flat, apparently homogeneous delta lands of various rivers. Both soil fertility status and crop yields vary considerably within a region (macrovariation) as well as within a farm or a field (microvariation). Whereas macrovariation is usually related to pedogenic processes, microvariation is due to various labile soil properties. Field surveys confirmed changes in soil-surface topography and quantitatively important yield variations over distances as short as 10 to 20 m in field crops and even from tree to tree in orchard crops (Srinivasan, 2001b).

Enormous variation was found in yields and soil nutrient supply among fields with similar soil types and crop management practices across Asia. For example, yields of rice supplied with similar amount of N (100 to 120 kg $N \cdot ha^{-1}$) ranged from 2.3 to 6.4 $t \cdot ha^{-1}$ in 56 different small rice fields in a coastal village in Andhra Pradesh, India (Table 18.1). Similar variation was found in the Philippines, Indonesia, Thailand, and Vietnam (Cassman, 1998; Olk et al. 1999). Likewise, a large variation in indigenous nutrient supply was reported in irrigated rice farms of Zhejiang province in China (Wang et al., 2001). Wide spatial and temporal variations of $NO_3$ content in rice fields as small as 10 $m^2$ and 0.3 ha in Japan (Li et al., 1998; Shibusawa et al., 1999) and Korea (Chung et al. 2000) were reported. Results from a multisite study in tropical, subtropical, and temperate ecosystems revealed that spatial variation in yields and soil properties was evident irrespective of the farm size and input use (Table 18.2). In a plantation crop such as tea, wide spatial variation in productivity of bushes is commonly observed. Spatial variation in organic carbon content of the topsoil, an important

## EXHIBIT 18.1. SWOT Analysis of Asian Farming Systems

**Strengths**

- Large areas with suitable climate, soil, land, and water combinations
- High yielding crops and input-responsive varieties
- Experienced and hard working farmers
- Availability of inputs (water, fertilizers, and plant-protection chemicals)
- Technical resources and capacity to adapt cropping systems
- Rapid income growth

**Weaknesses**

- Increasing fragmentation and subdivision of land holdings
- Stagnating/declining yields in major crops and increasing land degradation
- Dependence on few crops and varieties
- Inappropriate use of external inputs
- Raising costs of cultivation and declining profit margins
- Fewer opportunities for off-farm employment
- Adverse environmental impacts such as soil erosion, water pollution, greenhouse gas emissions, and reduced biological diversity
- Low farm mechanization and few resources

**Opportunities**

- Increasing demand for food due to raising human populations
- Diversification opportunities
- Scope for productivity and quality improvement
- New markets due to globalization
- IT and biotechnology revolutions leading to dramatic falls in the cost of information processing
- Changing food preferences and diversified diets
- Rapid urbanization and absorption of rural labor in semi-urban areas

**Threats**

- Increasing market competition
- Declining farm and crop subsidies
- Declining water reserves
- Growing shortages in labor supply
- Low market prices due to high competition and globalization
- Tightening legislation on environmental quality
- Growing income disparities among farming and other sectors

TABLE 18.1. Variability in rice yields with largely similar agronomic management in a South Indian village.

| Yield class ($t{\cdot}ha^{-1}$) | No. of farms | Total N Percent | Available soil P ppm | EC[a] (ds/M) |
|---|---|---|---|---|
| <3.0 | 9 | 0.07-0.09 | 11-23 | 0.5-1.7 |
| 3-4 | 11 | 0.06-0.11 | 10-19 | 0.4-2.2 |
| 4-5 | 16 | 0.10-0.16 | 14-22 | 0.8-2.1 |
| 5-6 | 12 | 0.09-0.15 | 20-37 | 0.7-2.4 |
| >6 | 8 | 0.12-0.14 | 24-46 | 0.6-2.2 |

*Notes:* Cropping system: rice-black gram *(Vigna mungo)* rotation; number/size of farms: 56 (0.1-1.2 ha); variety: high yielding rice variety; N fertilizer applied (rice): 150-160 kg $N{\cdot}ha^{-1}$; P fertilizer applied (black gram): 50 $kg{\cdot}ha^{-1}$; yield variation: rice: 2.3-6.4 $t{\cdot}ha^{-1}$; black gram: 0.2-0.8 $t{\cdot}ha^{-1}$

[a]EC = electrical conductivity

determinant of tea yields, was reported in Sri Lanka (Anandacoomaraswamy and Ananthacumaraswamy, 1999).

In addition to indigenous variability, management-induced variability has also become crucial in certain parts of Asia. Given such high field-to-field variation within small areas with similar soil type, the approach of adopting blanket regional recommendations therefore leads to inefficient use of inputs and will limit efforts to enhance yields because of nutrient deficiencies, excesses, and imbalances. It is therefore imperative to manage such variation more effectively through modern approaches such as PF.

## *DRIVERS AND OPPORTUNITIES FOR ADOPTION OF PRECISION FARMING*

The driving forces for adoption of PF vary with each Asian country and cropping system. Environmental threats to ecosystems, labor shortages due to aging populations, and declining farm incomes are the major drivers for PF adoption in countries such as Japan and South Korea. In contrast, sustainable intensification of crop yields to meet the demands of ever-growing populations will be a major driver in China, India, and Southeast Asia. Even within each country some regions are obviously more suitable than others

TABLE 18.2. Variability in rice yields and soil properties in tropical, subtropical, and temperate cropping systems (ten samples per field) with varying field size and external input usage.

| **Climate** | **Tropical** | | | | **Subtropical** | | **Temperate** | |
|---|---|---|---|---|---|---|---|---|
| Cropping system | Rice-black gram rotation | | | | Rice-wheat | | Rice monoculture | |
| Region | Coastal Andhra Pradesh, India | | | | Haryana, India | | Hokkaido, Japan | |
| Input usage | Low | | Medium | | Medium | | High | |
| Field size | 0.7 ha | | 1.3 ha | | 2.1 ha | | 1.8 ha | |
| Variable | Min | Max | Min | Max | Min | Max | Min | Max |
| Yield ($t \cdot ha^{-1}$) | 0.8 | 1.9 | 2.4 | 5.6 | 4.3 | 6.7 | 4.2 | 8.8 |
| pH | 5.1 | 6.2 | 5.5 | 6.4 | 5.3 | 6.6 | 4.9 | 5.7 |
| EC[a] dS/M | 0.4 | 2.2 | 0.7 | 2.3 | 0.5 | 2.2 | 1.1 | 2.1 |
| OM[b] percent | 1.5 | 2.6 | 1.8 | 2.9 | 1.5 | 2.3 | 1.8 | 2.7 |
| Total N percent | 0.12 | 0.19 | 0.14 | 0.20 | 0.11 | 0.22 | 0.17 | 0.24 |
| $P_2O_5$ ppm | 62 | 194 | 59 | 222 | 117 | 245 | 144 | 331 |

[a]EC = electrical conductivity
[b]OM = organic matter

in implementing PF because of the kinds of cropping systems and farm sizes in the region. For example, in Japan, Hokkaido is perhaps more suited to PF adoption than other prefectures because farms are relatively big. Likewise, a good scope exists for introducing PF in crops such as rice and wheat in Indian states of Punjab and Haryana where external input use is higher and farm mechanization is more advanced than in other states.

Modern PF techniques may be relatively new to Asian farmers, but PF concepts are not. For example, it is common to observe farmers in several parts of Asia physically stake or flag areas showing severe stresses in a field and thus apply varying doses of manures. This is possible because most farms are small, and rough distances to identify stressed areas are estimated by measuring, counting rows, pacing, or other relative means. Even subsistence farmers manipulate fields spatially to take advantage of specific areas they know will suit various kinds of crops. However, since introduction of the green revolution technologies, variations in field conditions were often ignored in favor of adopting blanket regional recommendations, which were developed by the provincial governments and extension agencies by considering the average fertility of many fields and the crop nutrient requirements. Huge subsidies on fertilizers and pesticides as well as labor shortages in some areas also forced farmers to use uniform application rates across the entire field(s). Major drawbacks of blanket applications include increased cost of the additional inputs in areas where it is not required, adverse environmental impacts due to leaching and runoff, and decreased profits due to lower yields in areas where application is inadequate. Repeated uniform applications over many years aggravate these problems further. A reversion to variable management interventions aided by modern tools of information is thus desirable.

Because Asia has the largest number of small farms in the world, future prosperity is directly dependent on the success and productivity of small farms. Although it is true that small farms in Asia are severely handicapped in terms of mechanization and application of new technologies due to constraints such as capital, credit, and other facilities, it does not mean that adoption of PF principles requires a large scale farming operation. Recent reports indicate that small farms are in fact more productive and efficient if their multiple functions such as total output, enhancement of diversity, and ecological services are taken into consideration (Rosset, 2001). Some observers even argue that smaller farms could gain from PF because they are often operated under tighter labor constraints in which automation might enhance field-monitoring capabilities (Cowan, 2000).

PF as practiced in the United States, Europe, and Australia, however, may not be pragmatic for small farms in Asia, but the principles are highly applicable. In small farms (0.1 to 5 ha) PF can be adopted uniformly to an entire field by precisely adjusting the use of inputs and cultivation practices regarding that particular field's soil and crop conditions. Such practices are referred to as field specific rather than site specific. Alternatively, PF can be applied to a group of small farms on a village or block basis by adequately characterizing crop and soil needs within a small region. In cases of large farms (> 50 ha), however, PF can be applied by critically diagnosing the variability in soil properties, crop conditions, etc. We must also remember that many large farms in Asia can be greatly helped by PF technologies. This applies especially to large estate farms or corporate plantation farms.

PF can be employed in both irrigated and rainfed agriculture. In irrigated areas, which comprise 35, 15, and 7 percent of agricultural lands in South Asia, Southeast Asia, and East Asia respectively, and where fertilization has had the greatest impact on yields so far, PF can help in optimizing input use efficiency, varietal choice, and crop quality management. In rainfed systems,

PF can improve overall efficiency of available water through improvements in site-specific soil and residue management. PF practices to improve infiltration and reduce runoff in rainfed areas can have great impact on yield and yield stability.

PF principles can be applied in both low-input and high-input cropping systems, biotech and nonbiotech crops, and in both fertile and marginal lands. Fertile areas within a field can be productively used for cultivation of crops with high quality and yields, and marginal areas can be managed to improve native fertility through spot applications of crop residues, farm manure, compost, and other amendments. Variable nutrient management in fertile lands may not be cost-effective, however, if the range in soil test levels is higher than the required amount of nutrients by crops. PF can be extended even to organic cropping systems for both yield monitoring and variable application of manures and soil amendments.

As for specific cropping systems in Asia, Srinivasan (1999, 2001b) identified several potential opportunities for adoption of PF. PF seems to offer the greatest potential benefits in the following cropping systems:

- Cropping systems in which a variety of inputs are used and input costs are high
- Cropping systems with high spatial and temporal variability
- Vegetable-based cropping systems in which nutrient leaching and pollution of water resources is high
- High value, plantation-crop-based cropping systems

A brief description of possibilities for PF adoption in major Asian cropping systems is given in the following section.

### *Rice-Wheat-Legume Rotations*

Rice-wheat systems are among the most widespread cropping systems in Asia, covering nearly 24 million ha. In some areas, wheat is substituted by or intercropped with oilseed crops or grain or fodder legumes. Rotations also include sugarcane, potato, winter maize, and jute. Nearly 40 percent of an area under rice-wheat systems annually receives more than 400 kg $NPK \cdot ha^{-1}$, and substantial nutrient losses due to incorrect management are common in these areas. Despite considerable efforts to improve fertilizer-use efficiency through varying methods and times of application, fertilizer-use efficiency in the rice-wheat systems of Asia remains low, around 30 to 40 percent. In cases of micronutrients such as Zinc, efficiency seldom exceeds 15 percent. Several solutions for site-specific fertilizer management in rice-wheat rotations are available. These include proximal sensing through use of a chlorophyll meter or leaf color chart based on the actual N needs of the plants, remote sensing of leaf color and identifying potential areas of N deficiency, variable management of split applications, and using more sophisticated variable rate application technologies where practical.

Preliminary results in a 5 ha rice field with varying N status revealed that uniform-rate application of N fertilizer resulted in as much as 42 percent of an area receiving excess fertilizer with nearly 33 percent area receiving insufficient amount (Figure 18.1, Field 3). In such situations, precision N management could be particularly useful. Indeed improved N applications depending on crop N status, as revealed through readings of chlorophyll meter or color charts, have been found to enhance N-use efficiency in rice (Singh and Singh, 1998; Ladha et al. 2000).

### *Maize-Based Cropping Systems*

Asian maize yields are about 1.2-1.4 t·ha$^{-1}$ as against 3.9 t·ha$^{-1}$ of global average. Future demand for maize in Asia, however, is expected to double every 15 years. N deficiency, and associated low soil fertility, is a common feature of maize fields in Asia. Adoption of PF principles such as precision N management based on chlorophyll meter readings or other sensing systems can help in the rational application of fertilizers. Indeed, because of its importance in North America, PF principles and technologies are more widely tested in maize than in other crops. Yield mapping, grid soil

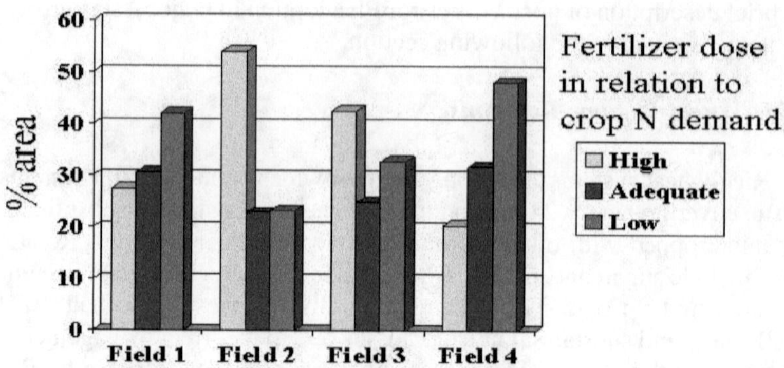


FIGURE 18.1. Proportion of area where N fertilizer rates are high, adequate, and low as compared to N demand by rice in four adjacent fields of Hokkaido. (See corresponding PLATE 18.1 in the central gallery.)

sampling, site-specific pest and weed management, and variable seeding and fertilization may all be relevant in Asia.

### *Sugar Beet–Based Cropping Systems*

Sugar beet is a typical crop grown in temperate parts of Asia. Recent research suggests that many opportunities exist for PF in sugar beet–based systems, especially with regard to N management in rotations. An example from Hokkaido, which grows the entire 100 percent of Japanese sugar beet crop, is examined here. In Hokkaido, N application is highest in sugar beet (130 to 160 kg $N \cdot ha^{-1}$) among the various upland crops. Uniform applications of such high rates without considering soil organic matter and weather conditions often lead to excess N supply, which has deleterious effects on quality of not only sugar beet (low sugar) but also of wheat (delayed maturity) and potato (low starch). Furthermore, sugar beet is unusual in that the N content of residues left in the field is usually greater than that removed in the harvested storage roots. Indeed lack of consideration of the contribution of sugar beet tops to the available soil N pool results in excessive use of N by succeeding crops (Moraghan and Horsager, 1995). Integration of imagery with weather conditions, nutrient uptake, rotation history, and topography in a GIS can help in detecting spatial variation in N status and in determining optimal doses of N. A strategy for precision N management of a three-year rotation of sugar beet–potato–wheat in Hokkaido is given in the following section.

Prior to sowing, field boundaries of sugar beet fields are prepared using GPS and GIS. A farm association acquires aerial photos or satellite imagery at critical stages of sugar beet growth. Imagery taken one month before sugar beet harvest can be used to delineate areas with dark green, light green, or yellow canopy. The dark green areas of the field have large amounts of N stored in the canopy and supply large amounts of N for several succeeding crop years. Therefore, lower rates of N fertilizer must be applied to the succeeding potato in areas of the field that had dark green sugar beet tops, whereas normal rates can be applied to the areas of the field that had a light green or yellow canopy. The imagery with defined crop zones can be delivered to the growers via the Internet after compressing imagery to a reasonable size. The growers can look at such imagery using free GIS viewers (e.g., ArcExplorer) and go to specific areas for N application. GPS is helpful in locating those areas. Simultaneously, petiole sampling is done in various zones for plant N analysis. After harvesting potato in the second year, soil and crop N testing must be done in the areas established in the sugar beet and potato canopies separately so that each zone can be fertilized with the correct amount of N fertilizer for wheat in third year. Potato tops should be considered as a nitrogen credit for wheat. Following harvest of wheat in the third year, soil testing must again be done based on either sugar beet canopy zones or on topography. Satellite images from previous seasons also serve as the best guide to soil sample and fertilize sugar beet in year four. Because excess N even at deeper soil profiles adversely affects sugar beet quality, soils at 60 to 120 cm depth must be sampled to adjust N rates necessary to improve its quality.

### *High Value Crops and Cropping Systems*

Use of external inputs including fertilizers and pesticides is very high in cropping systems involving commercial crops (e.g., cotton, sugarcane, tobacco), vegetables, orchard crops (e.g., grapes, apples, peaches) and plantation crops (e.g., coffee, tea, cocoa, rubber, oil palm). Precision fertilization, and site-specific weed and harvest management practices may be of great help in these crops.

It is reasonable to assume that PF is most likely applied to industrial and plantation crops first prior to widespread introduction in field crops in Asia. However, in Japan, Taiwan, and Korea, where per capita demand for meat could be as much as seven times that of South Asia and where cereal-based feeds are gaining increasing importance, some PF technologies are likely to be applied even in field crops such as rice and maize. It is thus important to make PF approaches accessible and meaningful to smaller diversified operations. Wang (2001) concluded that future efforts in Asia must also focus on integrating available technologies and on developing interface techniques and applied software for farm use.

Based on ongoing research in various countries and local needs, priorities for which PF may be employed in Asian context are given in Table 18.3. The list is not exhaustive but it indicates which PF operations would most likely be relevant to various countries and crops.

## *CURRENT STATUS IN SELECTED COUNTRIES*

Adoption of PF technologies on a commercial scale is virtually nonexistent in Asia. However, as mentioned earlier, research interest on PF is growing rapidly. The Asian Conference on Precision Agriculture was held in August 2005 in Japan and it is planned to be held once every two years in different Asian countries. In addition, in May 2004 the Asian Conference on Precision Farming was held in Kuala Lumpur, Malaysia. An Association of Southeast Asian Nations (ASEAN)-China workshop on precision

TABLE 18.3. Priority areas for potential adoption of precision farming in Asia.

| Crop | Precision farming operation | Countries |
|---|---|---|
| Rice | Yield mapping | Japan, India, China, Korea, Malaysia |
| | Grain-quality mapping | Japan, Korea, China |
| | In-season nitrogen management | Philippines, China |
| | Variable nitrogen fertilization | Japan, Taiwan |
| Wheat | Yield mapping | China |
| | Disease mapping | India |
| | Variable nitrogen fertilization | China |
| Soybean | Soil-compaction mapping | Japan, Thailand |
| Cotton | Pest monitoring | India |
| | Variable nitrogen fertilization | China, India |
| Oil palm | Crop-vigor mapping | Malaysia |
| | Variable nitrogen fertilization | |
| Sugar beet | Nitrogen management in rotation | Japan |
| | Yield and quality mapping | |
| Sugarcane | Variable fertility and harvest management | India |
| Tea | Variable harvest management | Sri Lanka |
| Fruit trees | Tree-specific management | Thailand, Japan |

farming was also held in conjunction with this event. Some ongoing research efforts on regional and country levels are discussed in the following sections.

On a regional basis, researchers at the International Rice Research Institute conducted on-farm experiments at 179 sites in eight key irrigated rice domains of Asia from 1997 to 1999 to evaluate a new approach for site-specific nutrient management (SSNM). Perhaps this is the most extensive PF study conducted in Asia to date. It was found that SSNM increased average grain yield by 0.36 $Mg \cdot ha^{-1}$ (7 percent) compared to the current farmers' fertilizer practice measured in the same cropping seasons (Dobermann et al., 2002).

### Japan

Research emphasis in Japan has been so far on development of machinery suitable for PF including soil nutrient sensors, variable rate fertilizer applicators, compact grain yield monitors, grain moisture sensor, and field robotics (e.g., autonomous tractor using machine vision and GPS, watermelon harvesting robot) (Iida et al., 1998; I Made Anom et al., 2001; Umeda et al., 2001). Most PF research is conducted in Hokkaido University, Kyoto University, Tokyo University of Agriculture and Technology, Bio-oriented Technology Research Advancement Institution (BRAIN) and a few research stations associated with the Ministry of Agriculture, Forestry and Fisheries. The private sector is showing active interest in PF technologies. For example, companies such as Mitsubishi, Japan Space Imaging, Omron, Hitachi, Trimble Japan and many others are looking at opportunities to provide PF services and technologies on a trial basis. Of late, the concept of community-based PF is being promoted in cities such as Toyohashi and Honjo with a view to develop a new regional farming system (Shibusawa, 2004).

### China

The first research center for PF in China was established at China Agricultural University, Beijing, in 1998, and field trials were conducted in a suburb of Beijing, state farms in northeast China, and Inner Mongolia (Wang, 2001). The Chinese Academy of Sciences initiated the first PF program in collaboration with University of Minnesota Precision Agriculture Center, and the Australian Centre for International Agricultural Research with the goal of adapting PF techniques to Chinese conditions. In 2000, a database titled the "US-China Knowledge on Precision Agriculture" was launched at Kansas State University (http://www.lib.ksu.edu/depts/issa/china/precision/). Researchers at East China Normal University in Shanghai initiated PF studies (Kai, 2000). A 15 million yuan ($1.86 million) experiment on PF was started in a modern agricultural zone near Shanghai. Based on analysis of 11 types of nutrients from more than 460 sample fields spreading over 247 hectares of farmland in Songjiang district of Shanghai, researchers found that precision fertilization increased yields of rice (9 to 13 percent), wheat (18 percent), and watermelon (14 to 27 percent) as compared with the common practice of uniform fertilization (China View, 2004). Researchers at the Precision Agriculture Research Center at China Agriculture University, Beijing, and Hebei Agricultural University in Baoding initiated PF studies including variable fertilization in wheat, soil moisture sensing, fertility and yield mapping, etc. (Liu, 2000).

China's continued reforms in agriculture are resulting in a greater number of larger farms, making them conducive for adoption of PF technologies. As part of such reforms, China is planning to introduce PF in western regions and in northern plains where farms are relatively big and farm mechanization reached a threshold level to accept PF (Yu and Zhou, 1999). Currently, utilization efficiency of fertilizers and pesticides in China is only between 20 to 30 percent and researchers expect to improve it with PF up to around 60 percent (Yang, 2001, personal communication). Site-specific nutrient management for irrigated rice in Zhejiang province of Southeast China was shown to improve yields from 5.9 to 6.4 $t \cdot ha^{-1}$ and nutrient efficiency as compared with current farmers fertilizer practice (Wang et al., 2001). Liu et al. (2003) examined applicability of three PF patterns (traditional PF based on traditional technologies, modern PF based on information technology and semiautomation, and high-tech PF based on robotics and information technology) for China and reported that traditional PF is suitable for hilly regions such as Taihang mountain area, and modern PF is suitable for most high-yield producing regions with full or semimechanization such as the Shanghai suburb. They reported that high-tech PF is not suitable for Chinese conditions yet.

### *India*

In India, the National Bank for Agriculture and Rural Development (NABARD) supported a three-year project beginning in 1999 by establishing a resource center for precision farming at JRD Tata Ecotechnology Center of the M. S. Swaminathan Research Foundation (MSSRF). Arava R & D, Israel, provided technical support for this project. The foundation set up five demonstration farms initially in Tamil Nadu and plans to replicate them in other states. In one of the adopted villages a soil spectral variability map showed at least four types of soils in the area, but the entire village was applying a similar fertilizer dose for their chickpea crops. Therefore, a trial on variable rate application technology has been undertaken (Ray et al., 2001). Another project titled "Precision Farming for Sustainable Rice-Wheat Cropping Systems" is in progress at the Indian Agricultural Research Institute. The Space Applications Center of the Indian Space Research Organization, Ahmedabad, initiated a project at Jalandhar to map variability in potato cropping systems. The Project Directorate for Cropping Systems Research in Modipuram initiated a PF project in collaboration with Central Institute of Agricultural Engineering, Bhopal (Shanwad, et al. 2004; Swain et al., 2004). Also, several low-cost, GIS-based decision support systems (e.g., Pragati, Jal chitra) and farm machinery (e.g., GPS-based crop duster capable of spraying an area as small as 16 $m^2$) are attracting wide attention for their use in PF.

### *Republic of Korea*

In the Republic of Korea, research interest in PF has been growing recently for application in crops such as rice and orchard crops (Park, 2003) mainly through the development of technologies (e.g., sensors for soil organic matter content and rice growth, variable rate granular fertilizer controller, rice yield mapping system) suited to Korean cropping systems (Chung et al., 2005). Variable rate fertilization in rice was found to be effective at reducing fertilizer use without adverse impacts on grain yield (Lee et al., 2005).

### *Malaysia*

In Malaysia, research on PF technologies is largely for crops such as oil palm, rubber, and other cash crops, although some research is focused on rice. Many oil palm estates are in the process of preparing digital plantation maps that show road infrastructure, block boundaries, topography, soil and geology, and the location of building. Fertilizers represent a significant portion of total production costs in oil palm plantations, and large-scale estate companies recognize the potential to improve both fertilizer-use efficiency and yield by incorporating appropriate PF techniques in their management strategy. The use of portable GPS and GIS software such as OMP7 (Agrisoft Systems, Germany) is increasingly applied to map oil palm yields and in fertility management. Wahid et al. (2002), for example, developed a variable fertilizer rate map for a field in Ulu Bernam estate in Malaysia. Ibrahim et al. (2000) reported the utility of remote sensing in estimating planting age and leaf area index, which in turn will enable precision management of inputs.

### *Sri Lanka*

On a research scale, efforts are underway to utilize PF techniques for improving yields of tea, which is a highly structured crop grown in blocks in seven-year cycles. Indeed production of yield maps based on records of the weight of tea plucked from each block was considered one of the best examples of precision farming (Blackmore, 1999).

## *CONSTRAINTS AND APPROACHES FOR ADOPTION*

In a presentation at the Fourth International Conference on Precision Agriculture held in 1998, I described in detail the various constraints to adoption of PF in Asia (Srinivasan, 1999). These barriers continue to be relevant now. The principal barriers include the following:

- High cost of obtaining site-specific data on real-time basis
- Lack of willingness to share spatial data among various organizations
- Complexity of tools and techniques requiring new skills
- Cultures, attitudes, and perceptions of farmers including resistance to adoption of new techniques and lack of awareness of agroenvironmental problems
- Small farms, heterogeneity of cropping systems, and land tenure/ownership restrictions
- Infrastructure and institutional constraints including market imperfections
- Lack of success stories of PF adoption and lack of demonstrated impacts on yields
- Lack of local technical expertise
- Variable results and uncertainty regarding benefits and costs of adoption
- Knowledge and technological gaps including
  —Inadequate understanding of agronomic factors and their interaction
  —Lack of understanding of the geostatistics necessary for displaying spatial variability of crops and soils using current mapping software
  —Limited ability to integrate information from diverse sources with varying resolutions and intensities
  —Minimal agronomic support and technical advice due to a chronic shortage of technical manpower. As yet, very few farm advisers or consultants in Asia can offer the specialist services needed for PF

### *Approaches for PF Adoption in Asia*

Many PF technologies are unusual for agriculture in that they were developed outside the traditional agricultural research, development, and dissemination system and were imported from industries not traditionally associated with agriculture (NRC, 1997). From this point, it is reasonable to assume that PF adoption in Asia would occur in phases. Initially, non-GPS dependent variable rate application strategies depending on soil fertility or management zones might be most practical. Such fertility zones can be prepared even with using existing soil test results of individual farms. An information approach using soil tests of each field and making fertilizer rate recommendations can lead to good results in the short to medium term. In the long term, however, efforts to improve the resolution, reliability, and availability of satellite data and to develop cost-effective ways of handling geo-referenced data in current production systems will be increasingly important.

Extensive indigenous knowledge of Asian farmers on local variability of their farmlands confers a comparative advantage in application of PF principles. However, PF technologies require some key changes in mechanisms of their development and delivery to Asian farmers. For example, yield monitoring with combines and VRT are the two most common PF technologies in the United States but they are not yet within economic reach for many Asian farmers. Instead, remote sensing methods to monitor yield variations and combinations of manual and mechanical methods for variable input applications may be more appropriate in the Asian context. It is also important to determine the aspects of PF that are practical at a field scale and to test pragmatic and affordable alternatives that are being developed elsewhere. Indeed, small-scale variability found in Asian fields requires more careful consideration of design parameters and data processing issues (Chung et al., 2005).

In most Asian countries, limited resources of small farms restrict farmers from using PF technologies. However, this limitation can be overcome through devising innovative cost-sharing mechanisms and using freely available mapping software. Many types of PF tools can be considered for adoption in small farms, but no single right way exists for all fields. In some cases, farmers do not need additional information about their fields but only better tools to analyze and interpret existing information. Shibusawa (2002) suggested a regional PF approach applicable to small farms to manage a hierarchy of variability—within-field, between-fields, and between-farmers. He proposed that it is possible to adopt such high-tech approaches as GPS-enabled yield monitoring and mapping for a group of several small farms in a region.

The idea of identifying and distinguishing relatively uniform regions or zones within a field and developing a prescription for management of each zone is one of the most useful PF concepts. Definition of realistic management zones, however, requires considerable skill and knowledge of spatial relationships between yield and other factors. It is possible that some zones show economic response to variably applied herbicides but not to fertilizer. Two common methods for developing management zones are intensive grid soil sampling and mechanical yield monitoring for at least three to four years. However, both these processes are time-consuming and expensive.

Image data, however, can provide a comprehensive view of an area while maintaining spatial connectivity between small fields, thus reducing the need for complex spatial modeling. With the advent of high resolution satellites and decreasing cost of image acquisition, it is expected that aerial remote sensing can help to reduce the costs of mapping yield variations and defining management zones. Multitemporal and multispectral analysis of images acquired during crop growth, and superimposing such images onto a cartographic database can help in identifying specific areas requiring the attention of farm managers and planners. Based on such analysis, the extension staff can advise farmers on how to divide their cropland into uniform zones and perform efficient soil-sampling methods that are less intensive than grid sampling. A software system such as SSToolbox may be used to perform various GIS and statistical analyses necessary to define management zones.

Getting started on a PF strategy in a village or a farm cooperative is not so difficult. The first step is to develop a basic farm plan based on past experience with specific fields, crops, farming practices, and resources available. Using a backpack GPS system, progressive farmers with the guidance of a village extension officer can go around the field boundary and record the coordinates of each field in that village. Recording can be done by obtaining a GPS position every 20 meters or at selected points such as fence posts and known crop boundaries. If fields are reasonably big, GPS information can be recorded while driving in a tractor around the field boundary. A handheld data logger can be used to record soil type and field condition. The extension officer can download the recorded GPS data into a computer and apply differential corrections to the position data before processing attributes or codes. Using a GIS program, a field boundary map can be created and each parcel's total area can be derived. Later, the georeferenced crop data can be overlaid onto existing digital maps of the block. In regions where land fragmentation and subdivision are not common, a single survey is adequate for field boundary definition, which can be used for many years. Once the perimeter map of all fields is complete, the area can be divided into small units and the address of each unit can easily be recorded.

During the crop season, a data logger can be used to record crop type and condition. If a parcel edge is a common boundary between two different crops, more than one attribute can be selected for each GPS position. Remotely sensed imagery obtained during crop growth can be used to detect and monitor any stresses within fields. A GPS receiver can be used later to guide the extension staff back to a specific location for obtaining a soil sample or for collecting insect data from a stressed area. The positional information of sampled sites can easily be entered into a GIS program. Any additional input applications in such stressed areas may be recorded. Farmers can continue to collect data using a spreadsheet template provided by the extension officer to build the farm database. Over time the extension officer can add new layers—crop history, landscape, basic condition of the soil and terrain, fertility, soil composition, weed location, crops and their yield, and other factors—to the farm GIS database. The data can then be used for preparing soil type maps, crop history maps, fertility maps, input recommendation maps, etc.

Another approach, which may prove useful in promoting PF for farmers with limited resources, uses remotely sensed imagery as the starting point. A town office, farmers' association, or PF service bureau can acquire remotely sensed imagery in the form of an aerial photograph or high resolution satellite imagery for the area of its interest. Farmers (in the case of estate farms) or local governments may share costs of acquiring and processing imagery. The first step would be to georeference the image in a desktop GIS such as ArcView or MapInfo by selecting well-known positions as ground control points and then create a base map for all fields. The base maps can be individually delivered to farmers through the Internet and then viewed and printed with free software such as ArcExplorer. As part of this map production, all existing data about the farm can be added. Standard data templates may be distributed to the farmer to enable him to add new data every season. After adding new data, a copy of the spreadsheet is sent by the farmer (or extension staff acting on behalf of farmers) to the farmers' association office for updating the files. Orthorectification and superimposition of cadastral information on images can give a lot of information on fields. It allows accurate calculation of the individual field areas using the tools in GIS. The town office staff can use a GIS as a development tool to create files containing historic information. Problem areas can easily be delineated in a GIS by drawing regions around them. A service provider with GPS facility can sample such problem areas during the crop season so that one can avoid costly sampling of a large area.

In the following year, locations for soil sampling can be chosen by overlaying a grid on both stressed and normal areas of the field. The number of samples may be varied depending on level of stress. After collecting and analyzing soil samples from normal and stressed areas, results can be entered by location in a GIS and a map can be created for each analyzed variable. In order to reduce costs in variable rate application of inputs, it is better to first target problem areas by applying the extra inputs necessary and then apply inputs uniformly across the field. Using a GPS receiver, weak spots in a field can be outlined and spot-treated even before planting. If another image is acquired after crop emergence, it is feasible to identify weak spots that were unnoticed earlier. An image acquired during reproductive growth can show further weak spots that may be corrected before final harvest.

A third approach—perhaps suitable for estate farms or relatively big farms owned or leased by private corporations—uses sophisticated technologies such as a multispectral digital video imaging system and high-end GIS and remote sensing software. The imaging system uses specially filtered video cameras mounted in aircraft to detect selected frequencies of visible light and infrared energy reflected from the ground. Unlike satellite data, video-recorded images based on the green (555 to 565 nm), red (625 to 635, nm), and near-infrared (845 to 857 nm) regions of the electromagnetic spectrum can be viewed and analyzed as soon as the plane lands. Imagery may be captured at altitudes of 1,200 to 1,400 meters under sunny conditions. Analysis of imagery can allow delineation of management zones, based on which crop production decisions may be made.

Since development of PF technologies is largely market driven, the importance of collaboration between the public and private sectors is crucial for its successful adoption in Asia. While the private sector takes up the responsibility of market development, product credibility, and customer satisfaction, the public sector needs to coordinate various development and implementation activities (Mandal and Ghosh, 2000).

## *IMPLICATIONS FOR ADOPTION IN ASIA*

Some researchers consider that PF can help in triggering the next green revolution. The past green revolution, which is now fading, had strong positive and some negative impacts. An assessment of the implications of adoption of PF in the Asian context may help policymakers minimize negative impacts, if any occur. The following issues are worth examining.

### *Adoption Patterns*

As agriculture is an intrinsically conservative business, any major technological development takes much time and effort before it is adopted by a majority of farmers. It took 25 years for hybrid seeds to catch on and more than 30 years to see tractors fully utilized (Robert, 1999). A similar course must be expected for PF. Because diverse cropping systems, land quality, and economic and geographic conditions characterize Asian farming, adoption patterns of PF are likely to be diverse. Even for selected crops, while some farmers may use the Internet to discover marketing opportunities, others may use PF solely for in-field decision making. Owing to relatively high cost and complexity, adoption process will not occur uniformly over time. Individual small farmers in Asia are likely to adopt them slowly, whereas estate owners and farmers' associations may adopt quickly. PF in

Asia is likely to evolve as a combination of services and products mainly taken up by farm associations and the private sector, which can spread the capital costs over large areas and over many years (Srinivasan, 2001a). Diffusion of PF is likely to be more rapid in areas surrounding cities and in areas with larger numbers of farm consultants and dealers.

### *Environmental Implications*

A more precise use of external inputs through PF may alleviate environmental pollution from Asian agriculture in the long term. However, as in all countries, downstream effects of increased soil erosion, water use, and nutrient and pesticide leaching are ordinarily of national concern and of little interest to individual farmers. Only those PF technologies that prove economically profitable are likely to be adopted first. The major challenge, therefore, is to design PF innovations and incentives that are financially rewarding to individual farmers as well as environmentally sound and beneficial to the society as a whole. Synergy between PF and biotechnology may also lead to environmental improvements (NRC, 1997). If yield penalties exist with genetically modified varieties for increased herbicide tolerance, those varieties can be sown in areas with high weed infestation. Likewise, varieties with insecticidal properties may be planted in potential problem areas with pests.

### *Economic Implications*

Although profits for large farms, which are likely to adopt PF technologies first, are expected to be high, small farmers too can get economic benefits over time through adopting proven technologies. Adoption of profitable technologies also should lead to increases in the value of land.

### *Employment Implications*

PF requires significant supporting infrastructure in the form of skilled labor, software development, and hardware availability. Therefore, new industries providing PF services, PF products, and combinations of PF services and products are likely to emerge, which in turn should provide additional employment. For example, agricultural consulting firms specializing in equipment, service, and technology delivery can develop to supply the needs of big farms adopting PF. In cooperation with local governments, they can assist in setting up farm GIS databases as well as process and analyze data to create maps, reports, statistics, and recommendations for farmers in a given area. New opportunities can arise in the form of VRT, equipment sales and recommendations, geoprocessing, etc. All of these services will generate massive amounts of data, which obviously requires skilled personnel with knowledge and experience to interpret the data for guiding the decision process.

## *FUTURE ACTION*

In order to promote PF technologies to foster sustainable agricultural development in Asia, many policy and technological initiatives are necessary at regional, national, provincial, and local levels. Some of those issues are discussed below.

- Appropriate capabilities, policies, and infrastructures must be created in the Asia-Pacific region to exploit PF technologies suitably and effectively. Goal-setting mechanisms must be evolved for identifying the most suitable PF approaches specific to various agroecological and socioeconomic conditions. Partnership arrangements among research and extension agencies must be promoted to develop PF applications that are the most economically beneficial and technically practicable and to devise new methods of cost sharing and fine tuning of technology to make it affordable. Rapid developments in the Internet-based information delivery systems are expected to make some PF technologies affordable for farmers. Regional and intercountry cooperation pacts for conducting research on PF technologies of mutual interest should be reinforced.
- Within each country and various regions within a country, research to identify the most suitable crops and farming systems for applying PF technologies must be intensified. The immediate needs of Asian cropping systems must be considered while formulating PF research projects. Because PF technologies do not come with a guarantee of success in all situations, it is important to determine how to apply them at the level and scale of operation now possible in the Asian context. It is also necessary to localize and modify certain components of technology to meet specific requirements. While it is essential to study the latest innovative PF technologies all over the world, it is more important to determine what is relevant and what is not.
- Inadequate research and institutional support in Asia is a major constraint for adoption of PF technologies. In order to make scientific developments in PF a commercial reality, due attention must be given to the development of capabilities and institutions. Advances in PF ap-

plications will continue to occur in industrial countries in the future. Asian researchers must be aware of those advances and make their own contributions through evaluating available technologies for suitability and removing any existing problems. They also must make cost/benefit assessments and informed judgments on the outcome of introducing PF technologies in social, economic, environmental, and institutional terms and in regard to sustainability.

- Appropriate policies and measures must be established at various levels to enable PF adoption from a "technology push" to an application-driven approach and to create the right environment for active partnerships among government, the private sector, R&D organizations, and NGOs. Because no single establishment can take on the entire PF process—which consists of data logging, point sampling, data analysis, and spatial modeling—it is vital that various agencies join together and contribute their relative advantages in knowledge, skills, and experience. Participatory learning is crucial to effective implementation of PF technologies. Therefore, local knowledge must be actively sought and incorporated during planning. This, however, requires close collaboration among farmers and farmers' associations, community groups, NGOs, spatial data providers, data analysis firms, machinery manufacturers, research and extension agencies, and local governments. Mechanisms to sustain such collaborative activities must be set up.
- Shortage of personnel to conduct research on PF technologies relevant to Asian conditions is another issue to be addressed. As indigenous research capacity is vital for sustained progress in application of PF technologies, governments must allocate adequate budgets and ensure rational utilization of funds per well-chosen priorities. As farmers and associations may not readily agree to share their datasets, government agencies and NGOs should promote models and templates for data sharing and provide examples of benefits of sharing and aggregating data.
- The important role of the private sector and the proprietary nature of many technologies must be recognized in development of PF technologies. Governments, therefore, should help the private sector create suitable applications tailored to the needs of specific agricultural conditions and design ways for effectively integrating technologies in current field operations in Asia. For example, data collection hardware and software, information management software and GPS devices must be user-friendly to be operated by individuals who are novices to these technologies. An urgent need exists to make fully

functional farm GIS easier for nonspecialists to learn and use. Innovative ways on how the private sector can work with farmers, governments, and other institutions in promotion of PF technologies must be explored. Greater investments in PF technology transfer are also necessary, perhaps through promoting support services to NGOs and the private sector.

- Farmers' associations and community-based organizations in each country can play a significant role in promoting sustainable agriculture through shared access to PF technologies. If PF is considered a series of discrete services such as map generation, crop scouting, etc., it is possible to fit these services within the structure of a progressive farm cooperative in each Asian country.
- Remotely sensed high resolution imagery may become more important for assessing crop and soil variability than yield maps in Asia. Therefore, it is important to concentrate on developing robust image analysis methods for farming in the form of, for example, a good quality product of reflectance and an interpreted product based on which farmers can actually make decisions. New cost-effective methods for classifying fields and creating management zones within a village must also be developed.
- As technology for gathering information is far ahead of our understanding of how to use it to help growers make decisions, knowledge-based systems must be developed quickly to realize full potential from PF. In this connection, the development of the ability to use available field information and experience, to form relationships among the various GIS databases and to deduce reliable decision making information is essential. By building farm databases and adding more information to create models, decision support systems that suit the needs and resources of Asian farmers can be developed. Analytical procedures to monitor and manage spatial and temporal yield variation of crops must be simplified. Likewise, research on PF technologies for effective resource management on a watershed basis must be promoted.
- Asian farmers must be assisted in the practical and cost-effective use of PF technologies through
  —conducting pilot demonstration projects and offering training seminars for farmers,
  —providing institutional credit,
  —processing generated data and providing user-friendly summaries, and
  —contributing the technical expertise to integrate various PF components for each farmer's specific application.

## *CONCLUSIONS*

On-site field research across Asia showed the existence of distinct spatial and temporal variation in both crop and soil characteristics. However, the agronomic and economic significance and the causes for such variation have not yet been adequately addressed. Slow but steady progress in PF research has been evident in Asia since 1996, especially in development of machinery suitable to small farms, yield mapping, soil sensing, and variable nitrogen fertilization. Although PF adoption of PF on a commercial scale is virtually nonexistent, research evidence so far suggests that PF could be an attractive option to promote sustainable agriculture with high potential to achieve three interlinked goals: improved crop productivity and yield stability, reduced natural resource degradation, and rehabilitation of environmental goods and services within and beyond farming systems.

Considering the predominance of small farms and fragmented land holdings, heterogeneity of crops and cropping systems, and other cultural variations in terms of technology use, the model of PF denoting the typical Asian agricultural scenario is yet to evolve. Developing such models presents both challenges and opportunities for researchers and policymakers alike. As environmental benefits alone do not create incentives for adoption, PF must be popularized in Asia as an integrated management approach. Another challenge is to reduce the complexity of various PF tools during dissemination to farmers. The nature of PF approach must also be tailored to meet specific conditions in various countries. Ultimately, future success of PF adoption and its potential contributions to sustainable agriculture in Asia would depend on how best we capitalize on existing strengths through collaborative efforts among the various stakeholders.

## REFERENCES

Ahmad, A.R., M. Zulkefli, M. Ahmed, B.Y. Aminuddin, M.L. Sharma, and M. Mohd Zain. 1996. Environmental impact of agricultural inorganic pollution on groundwater resources of the Kelantan plain in Malaysia. In B.Y. Aminuddin, M.L. Sharma, and L.R. Willett (Eds.), *Agricultural impacts of groundwater quality* (pp. 8-21). Australian Centre for International Agricultural Research ACIAR Proceedings No. 61. Canberra: ACIAR.

Ali, M. and D. Byerlee. 2000. Productivity growth and resource degradation in Pakistan's Punjab. Policy Research Working Paper, World Bank. In J.K. Ladha, K.S. Fischer, M. Hossain, P.R. Hobbs, and B. Hardy (Eds.), *Improving the productivity and sustainability of rice-wheat systems of the Indo-Gangetic plains: A synthesis of NARS-IRRI partnership research.* IRRI Discussion Paper, No. 40. Los Baños, the Philippines: International Rice Research Institute.

Anandacoomaraswamy, A. and S. Ananthacumaraswamy. 1999. Precision management of soil organic carbon in tea lands of Sri Lanka. In P.C. Robert, R.H. Rust, and W.E. Larson (Eds.), *Precision agriculture* Part A (pp. 767-772). Proceedings of the 4th International Conference on Precision Agriculture, July19-22, 1998, St. Paul, MN. American Society of Agronomy, Crop Science Society of America, Soil Science Society of America.

Blackmore, B.S. 1999. Developing the principles of precision farming. In *Proceedings of Agrotech 99,* Barretos, Brazil, November 15-19, 1999, Barretos Institute of Technology.

Carson, B. 1992. *The land, farmer and future: A soil fertility management strategy for Nepal.* Kathmandu, Nepal: Department of Soil Conservation.

Cassman, K.G. 1998. Ecological intensification of cereal production systems: Yield potential, soil quality and precision agriculture. Paper presented at the National Academy of Sciences colloquium "Plants and Population: Is there time?" December 5-6, 1998. Beckman Center of the National Academy of Sciences, University of California–Irvine.

China View. 2004. China launches experiments on "precision agriculture." Available online at: http://news.xinhuanet.com/english/2004-06/15/content_1527468 .htm.

Chung, S.O., H.J. Lee, and J.K. Park. 2005. Precision agriculture in the Republic of Korea. *Site Specific Management Center Newsletter* May. Purdue University: 4pp. Available online at: http://www.agriculture.purdue.edu/ssmc/Frames/ SSMC_newsletter_05_2005.pdf

Chung, S.O., J-H. Sung, K.A. Sudduth, S.A. Drummond, and B-K. Hyun. 2000. Spatial variability of yield, chlorophyll content and soil properties in a Korean rice paddy field. Research Publication of Missouri Precision Agriculture Center. Available online at http://mpac.missouri.edu/mpac/pubs/chung_pa2k_fin.PDF.

Cowan, T. 2000. *Precision agriculture and site-specific management: Current status and emerging policy issues.* Congressional Research Service Report RL3063. Washington DC: The National Council for Science and the Environment.

Dobermann, A., C. Witt, D. Dawe, G.C. Gines, R. Nagarajan, S. Satawathananont, T.T. Son, P.S. Tan, G.H. Wang, N.V. Chien, et al. 2002. Site-specific nutrient management for intensive rice cropping systems in Asia. *Field Crops Research* 74: 37-66.

Food and Agriculture Organization of the United Nations (FAO). 2006. FAOSTAT. Statistical databases. Available online at: http://faostat.fao.org.

Ibrahim, S., Z.A. Hasan, and M. Khalid. 2000. Application of optimal remote sensing technology for oil palm management. In C.H. Liu, S. Murai, and A. J. Chen (Eds.), *Proceedings of the 21st Asian Conference on Remote Sensing,* Volume 1 (pp. 302-304). December 4-8, 2000, Taipei, Taiwan. Available online at: http:// www.gisdevelopment.net/ aars/acrs/2000/ps3/ps302pf.htm.

Iida, M., T. Kaho, C. K. Lee, M. Umeda, and M. Suguri. 1998. Measurement of grain yields in Japanese paddy fields. In Robert P.C., R.H. Rust, and W. E. Larson (Eds.), *Proceedings of the 4th International Conference on Precision Agriculture* (pp. 1165-1175). Madison, WI: ASA/CSSA, SSSA/ASAE.

I Made Anom, S.W., S. Shibusawa, A. Sasao, and S. Hirako. 2001. Parameters maps in paddy field using the real-time soil spectrophotometer. *Journal of the Japanese Society of Agricultural Machinery* 63(3): 51-58.

Kai, R. 2000. Research application situation and future developing trend of precision farming. *Acta Agriculturae Shanghai* 16(3): 5-8 (in Chinese with English summary).

Ladha, J.K., K.S. Fischer, M. Hossain, P.R. Hobbs, and B. Hardy. 2000. Improving the productivity and sustainability of rice-wheat systems of the Indo-Gangetic Plains: A Synthesis of NARS-IRRI partnership research. International Rice Research Institute Discussion Paper No. 40. Los Banos, Philippines: IRRI. Available online at: http://www.rwc.cgiar .org/DownloadServer.asp?p=62&F=726&n=IRRIPaper.pdf.

Lee, H.J., C.H. Yi, S.H. Lee, and J.H. Chung. 2004. Reducing fertilizer application and rice growth and yield mapping by variable rate treatment in paddy fields. In T. Fischer, N. Turner, J. Angus, L. McIntyre, M. Robertson, A. Borrell, and D. Lloyd (Eds.) *New directions for a diverse planet: Proceedings of the 4th International Crop Science Congress*, September 26-October 1, 2004, Brisbane, Australia. CD-ROM. The Regional Institute Ltd., Gosford, Australia. Available online at: http://www.cropscience.org.au/icsc2004.

Li, M. Z., A. Sasao, S. Shibusawa, and K. Sakai. 1998. Local variability of soil nutrient parameters in a Japanese small size field. *Journal of Japanese Society of Agricultural Machinery* 61(1): 111-120 (in Japanese).

Liu, G. 2000. Main technique, decision making and application of precision agriculture in China. *Research of Agricultural Modernization* 21(1): 57-59 (in Chinese with English summary). Available online at http://www.bjaeu.edu.cn/keji/pac/incoming/lw12.htm.

Liu, J.O., S. Sun, S. Ninomiya, H. Cai. 2003. Three precision agriculture patterns based on technology and resources in China. In Z. Harnos, M. Herdon, and T.B. Wiwczaroski (Eds.), *Proceedings of the 4th European Federation for Information Technology in Agriculture, Food and the Environment (EFITA)* (pp. 552-558). July 5-9, 2003, Debrecen, Hungary: EFITA.

Mandal, D. and S.K. Ghosh, 2000. Precision farming—The emerging concept of agriculture for today and tomorrow. *Current Science* 79(12): 1644-1647.

Moraghan, J.T. and K. Horsager. 1995. Yield variability in sugarbeet fields. *1995 Sugarbeet Research and Extension Reports* 26: 138-146.

National Research Council (NRC). 1997. *Precision agriculture in the 21st Century: Geospatial and information technologies in crop management.* Committee on Assessing Crop Yield: Site-specific farming, information systems and research opportunities. Washington, DC: National Academies Press.

Nishio, M. 1999. Sustainability and vulnerability: The Japanese agricultural environment in the past and present. In S. Nagavajara (Ed.), *Agriculture and ecological resilience—Striking a balance in the Pacific Rim* (pp. 37-51) The 1st Asia-Pacific High Level Conference on Sustainable Agriculture, October 1998, Beijing, China. Washington DC: American Association for the Advancement of Science.

Olk, D.C., K.G. Cassman, G. Simbahan, P.C. Sta Cruz, S. Abdulrachman, R. Nagarajan, Pham Sy Tan, S. Satawathananont. 1999. Interpreting fertilizer use efficiency in relation to soil nutrient-supplying capacity, factor productivity, and agronomic efficiency. *Nutrient Cycling in Agroecosystems* 53(1): 35-41.

Park, W. P. 2003. Research of precision agriculture in Korea. *Journal of the Korean Society of Precision Agriculture* 2(1): 30-36.

Ray, S.S., S. Panigrahy, and J.S. Parihar. 2001. Precision farming in Indian context. *GIS@Development,* November, 7pp. Available online at http://www .gisdevelopment.net/magazine/gisdev/2001/nov/pfic.shtml.

Robert, P.C. 1999. Precision agriculture: Research needs and status in the USA. In J.V. Stafford (Ed.), *Precision Agriculture' 99* (pp. 19-33). London: Society of Chemical Industry.

Rossett, P. 2001. Small farm agriculture: Meeting future food needs sustainably. In S. Suthipradit, C. Kuntha, S. Lorlowhakarn and J. Rakngan (Eds.), *Sustainable agriculture: Possibility and direction* (pp. 179-193). Proceeding of the 2nd Asia-Pacific Conference on Sustainable Agriculture, October 18-20, 1999, Phitsanulok, Thailand. Bangkok, Thailand: NSTDA.

Shanwad, U.K., V.C. Patil, and H. Honne Gowda. 2004. Precision farming: Dreams and realities for Indian agriculture. In *Proceedings of 7th Annual International Map India Conference,* January 28-30, 2004, New Delhi, India. Available online at: http://www.gisdevelopment .net/application/agriculture/overview/mi04115 .htm.

Shibusawa, S. 2002. Precision farming approaches to small-farm agriculture. *Agro-Chemicals Report* 2(4): 13-20.

Shibusawa, S. 2004. Paradigm of value-driven and community-based precision farming. *International Journal of Agricultural Resources, Governance and Ecology* 3(3/4): 299-309.

Shibusawa, S., A. Sasao, and K. Sakai. 1999. Local variability of nitrate nitrogen in a small field. In J.V. Stafford (Ed.), *Precision Agriculture'99* (pp. 377-386). London: Society of Chemical Industry.

Singh, B. and Y. Singh. 1998. Use of the chlorophyll meter in N management of subtropical wheat following rice. *International Rice Research Notes* 26(3): 37.

Sinha, S. 1997. Global change scenario: Current and future, with reference to land cover changes and sustainable agriculture in the South and South East Asian Context. In V. Ravichandran (Ed.), *Regional land cover changes and sustainable agriculture in South Asia.* Proceedings of the COSTED/EU International Workshop on Regional Land Cover Changes, Sustainable Agriculture and their Interactions with Global Change, December 1996. Chennai: Thailand.

Srinivasan, A. 1999. Precision farming in Asia: Progress and prospects. In P.C. Robert, R.H. Rust, and W.E. Larson (Eds.), *Precision agriculture* Part A (pp. 623-639). Proceedings of the 4th International Conference on Precision Agriculture, July 19-22, 1998, St. Paul, MN. American Society of Agronomy, Crop Science Society of America, Soil Science Society of America.

Srinivasan, A. 2001a. Relevance of precision farming technologies to sustainable agriculture in Asia and the Pacific. In S. Suthipradit, C. Kuntha, S. Lorlowhakarn, and J. Rakngan (Eds.) *Sustainable agriculture: Possibility and direction* (pp. 325-338). Proceedings of the 2nd Asia-Pacific Conference on Sustainable Agriculture, October 18-20, 1999, Phitsanulok, Thailand. Bangkok, Thailand: NSTDA.

Srinivasan, A. 2001b. Site-specific management for selected cropping systems in Asia: Progress and prospects. In Organizing Committee, ICAST (Eds.), *Promoting global innovation of agricultural science and technology and sustainable agriculture development* (pp. 180-190). Proceedings of the International Conference on Agricultural Science and Technology, Session 6: Information technology of agriculture, November 7-9, 2001, Beijing, China.

Swain, K.C., H.P.W. Jayasuriya, and V.M. Salokhe. 2004. Precision agriculture for India: Potential, prospects and strategies. *Presentation at 38th Annual Convention and Symposium of Indian Association of Agricultural Engineers (ISAE),* January 16-18, 2004, Dapoli, Maharashtra, India.

Umeda, M., T. Kaho, M. Iida, and C.K. Lee. 2001. Effect of variable rate fertilizing for paddy field. American Society of Agricultural and Biological Engineers Paper number 011111. ASAE Annual Meeting, 2001, St. Joseph, Michigan: 12pp.

Wahid, O., A. Xaviar, A.M. Tarmizi, and S. Ibrahim. 2002. *Precision agriculture: Fertilizer management map art 1: Spatial and correlation analysis of yield and leaf nutrient.* Malaysian Palm Oil Board Information series TT No. 128. 8pp. Selangor, Malaysia: MPOB.

Wang, G., A. Dobermann, C. Witt, Q. Sun, and R. Fu. 2001. Performance of site-specific nutrient management for irrigated rice in Southeast China. *Agronomy Journal* 93: 869-878.

Wang, M. 2001. Possible adoption of precision agriculture for developing countries at the threshold of the new millennium. *Computers and Electronics in Agriculture* 30: 45-50.

Yu, G. and Y. Zhou. 1999. Precision agriculture and its perspective in *China. Chinese Journal of Natural Resources* 14: 258-264 (In Chinese with English abstract).

Zhang, W-L. 2001. Nitrate pollution of ground water and vegetables in intensive cropping area of North China. Chinese Academy of Agricultural Sciences Forum. ID 306. Available online at: http://www.caas.ac.cn/Forum/Sort_by_Title/sort_by_title_38.html.

# Chapter 19

# Australia

Simon E. Cook
Matthew L. Adams
Robert G. V. Bramley
Brett M. Whelan

## *INTRODUCTION*

Australia presents an interesting case study in the rate and form of adoption of precision agriculture (PA). In terms of the availability of labor and capital to agriculture, it ranks close to the United States as a potential fast adopter of technology. In Australia, labor resources are sparse but sufficient capital is available, and so a relatively rapid adoption of labor-substituting management technology might be expected. However, adoption has been patchy, and has followed a different path to that expected when PA technology first appeared in the early 1990s. As with other complex technologies, adoption has occurred as "steps" in which the processes of acquiring the technology then learning how to use it have overlapped each other. Although we remain at an early stage of development in Australia, this has occurred on a fairly broad front, with several industries taking different parts of the technology and applying them to problems peculiar to each situation.

In Australia, the main driver for adoption is grower demand, encouraged, in this instance, less by the paucity of labor than by the need for a more effective farming process. This chapter describes how PA enables farmers to respond to the need for improved techniques of field management. Since PA in Australia is demand driven, the status of PA is described according to the following:

---

We wish to acknowledge the contribution of Drs. Rob Kelly (Queensland Department of Primary Industries), Robert Corner (Curtin University of Technology), and three anonymous referees for their constructive comments and additions to this chapter.

*Handbook of Precision Agriculture*

doi:10.1300/5627_19

1. The demand for precision agriculture in Australia
2. Precision agriculture methods developed under Australian conditions
3. Applications of precision agriculture in the grains, cotton, wine, and sugar industries
4. Impediments to adoption

However, first, the term *precision agriculture* is defined.

## *WHAT PRECISION AGRICULTURE MEANS IN AUSTRALIA*

Different sectors of Australian agriculture have adopted PA technology in a variety of forms and to varying degrees, but a general understanding of the term *precision agriculture* is that it is a single underlying change in management toward improving the precision of the system used to control crop production. The common objective is to reduce unwanted variation through more precise control of manageable factors. The technology that enables the change is not the focus, nor are the individual practices that farmers may adopt (Cook and Bramley, 1998).

Paralleling observations of improved process control in manufacturing (Kaydos, 1991; Taguchi and Phadke, 1984) and learning cycles of adoption (Kolb, 1984) shows that PA heralds a deep-seated change in the production process itself, not simply in just the tools such a process uses. This improved process is summarized in Figure 19.1 as four phases of observation, which must be interpreted, evaluated for significance, then acted upon for tangible improvement to eventuate, following which the outcome is reobserved and the cycle repeated in a more advanced state. Each phase may

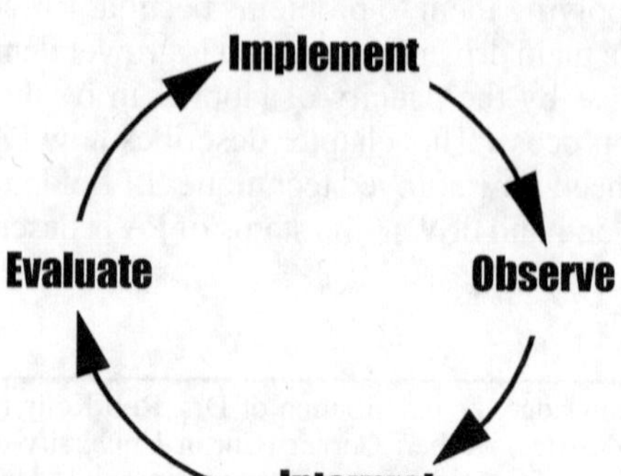


FIGURE 19.1. Four-phase cycle of increasing precision.

introduce new technologies or methods, and, through yield mapping, PA introduces radically improved observation of performance. But the deep-seated change occurs when farmers progress through the *entire* cycle.

Precision agriculture is not about attaching a global positioning system (GPS) to farm equipment and using it drive around fields more precisely, or about getting in an aircraft and taking aerial pictures of fields, or about all of the wallpaper farmers or their advisors have been generating in the guise of yield maps. These are all techniques of PA, but they are not PA in its true sense.

The underlying change enabled by precision agriculture is a shift toward greater certainty in land management. Real change occurs when a farmer knows how to complete the cycle of increasing precision and manage variation more effectively. Until that time, the technology remains little more than a novelty.

## *DEMAND FOR PRECISION AGRICULTURE IN AUSTRALIA: THE BATTLE FOR SUSTAINABILITY NEEDS ACCURATE AND RELEVANT INFORMATION*

This section examines the potential uses of PA in Australia, and the reason why growers might adopt them.

### *Potential Uses of Precision Agriculture in Australia*

Observations indicate that substantial gains could be acquired by managing variation more effectively. This could be achieved by the following:

1. Improved profit through better targeting of inputs
2. Accelerated change through greater certainty of its impact
3. Improving quality assurance through management of inputs or product streaming
4. Reducing the risk to the environment by more precise control of field operations

### *Improving Profit Through Better Targeting of Inputs*

Yield maps often show significant within-field variation in output (Figure 19.2), often substantially greater than that expected by growers. This is an obvious incentive to encourage growers to match inputs to responsiveness, which can occur in at least three ways: continuous variation of inputs

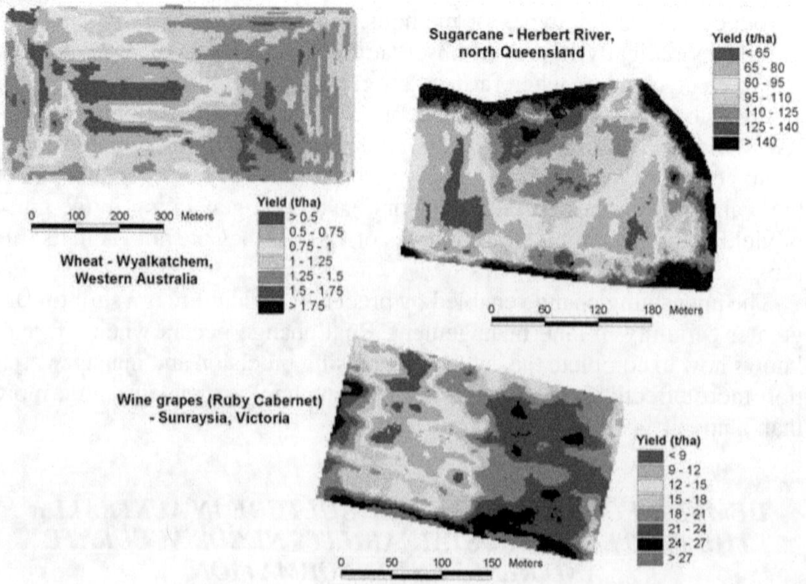


FIGURE 19.2. Paddock-scale yield variation in a range of Australian crops: wheat from a paddock in Western Australia, 1998; first ratoon sugarcane from north Queensland, 1998; and wine grapes from northwest Victoria, 2000.

to match response, the use of management zones, and removal of unprofitable areas from cropping.

Ultimately, growers will want to *maximize* the returns on investment by matching inputs closely with the spatially variable response to the input (for example, by varying the rate of fertilizer according to expected response). However, in Australia, this seems difficult because the uncertainties introduced by other sources of variation often exceed those removed by better targeting of inputs—so that a premature attempt to target fertilizer could actually increase the farmer's exposure to risk (Cook et al., 1996). Two robust alternatives are to define broad management zones or to simply remove from cropping those areas that consistently respond poorly to inputs.

### *Accelerating Change Through Greater Certainty of Its Impact*

Successful export industries need to change more rapidly than their competitors to remain ahead. Cooke (1999) observes that "old" and "new" activities contribute differently to the overall well-being of Australian farm enterprises. Although older, tried-and-tested activities provide relatively "safe" income, their value declines over time. Therefore farmers change by balancing short-term production risks associated with "new" crops against long-term risks of declining value shown by "old" activities. One way to reduce the perceived risk of new activities is to use PA technology to provide detailed objective information to enable growers to learn rapidly about the performance of new crops or varieties, perhaps using techniques of on-farm experimentation within a participatory approach (Bramley et al., 1999, 2005a; Lawrence et al., 2000).

### *Improving Quality Assurance Through Management of Inputs or Product Streaming*

In addition to reducing unit costs, growers can secure revenue by increasing the unit value of their product. This follows the widespread philosophy that uncontrolled variation within *any* part of a processing chain is undesirable and should be reduced at the source (Taguchi and Phadke, 1984). Uncontrolled variation in agricultural produce significantly reduces its downstream value. Low quality wheat makes low quality bread; low quality barley low quality beer; and low quality grapes low quality wine. Establishing an efficient production chain requires certainty of input quality. Since the production chain for Australian agricultural products stretches to importers, the ability to provide quality assurance should be seen as a relatively high priority.

Since all production units (fields or blocks) demonstrate substantial variation in crop quality as well as quantity it is possible to conceive substantial improvements in product streaming by reducing variation in product quality by doing one or more of the following:

1. Segregating the product at the point of harvest (Bramley et al., 2005b) by monitoring quality at or before harvest (on-the-go grain quality-sensors are now commercially available
2. Modifying inputs to reduce the within-field variation of product, for example, by manipulating the application of nitrogen (N) fertilizer to reduce the variation of grain protein content (Stewart et al., 2002)
3. Blending to minimize risk of substandard quality

To segregate grain of different qualities at the point of receival from farmers offers a premium for more desirable grades. The premium offered is moderate, reflecting the small price differential received; hence, few growers segregate *before* delivery even though such a practice would normally be profitable (Stewart et al., 2002). The wine industry, by comparison, receives very large premiums for high quality wine, a good bottle of wine attracting several times the price of a low-grade wine. Consequently, winemakers more carefully select grapes entering their wineries by presampling (Bramley et al. 2005b). Growers who are able to demonstrate adherence to international quality assurance standards may also attract price premiums (Bramley and Proffitt, 1999, 2000).

### *Reducing the Risk to the Environment by More Precise Control of Field Operations*

In Europe, a major purpose of PA is to reduce the risk to the surrounding environment by targeting inputs more effectively, for example, by reducing the use of potentially damaging fertilizers or pesticides. Australian growers have yet to adopt this goal widely and are seen by many as perpetuating an agricultural system that is ill-suited to its environment (Williams et al., 2002). However, all recognize the potential threat to their livelihood that this trend represents, since any industry that fails to demonstrate environmentally sustainable practices risks losing access to lucrative export markets. Although few have yet to adopt the ISO 14000 series of standards, most understand that environmental management systems are an inevitability for which tighter management of field operations will be necessary (Carruthers, 1999).

### *Four Reasons Why Australian Farmers May Adopt Precision Agriculture*

Despite declining importance, Australian agriculture retains a level of significance that is unusual among Organization for Economic Cooperation and Development (OECD) countries. Partially this reflects Australian agriculture's recent history. The Australian economy was founded in the nineteenth century on the basis of a rapid expansion of agricultural and mineral commodities. This expansion was funded by export markets, which enabled wool, meat, and grain industries to be highly profitable. The high prices for agricultural commodities continued, with occasional reversals, through to the mid-twentieth century, at which time the sugar, cotton, and now wine industries rose in importance.[1]

The grains, cotton, wool, sugar, beef, dairy, horticulture, and wine industries occupy diverse environments throughout the continent and appear as a range of forms, from the small winegrower to huge beef cattle stations. However, these sectors share essential characteristics that predispose Australian agriculture to the adoption of PA information technology, including the economically rationalist view of agriculture, the strong export orientation, the need to survive in a risky growing environment, and the need for practical tools to manage large areas. These are described in the following sections.

### *Reason 1: Australian Farmers Are Relatively Responsive to Technological Advances That Can Improve Farm Performance*

Australian agriculture did not arise from indigenous activities and, although many farmers feel strong attachment to the land they now occupy, agriculture could not be described as traditional in the European sense. Agriculture was introduced in the nineteenth century or later as an export-generating industry that was occupied predominantly by immigrants. Many areas of Australia were settled by Europeans relatively recently (for example, large-scale land clearing for agriculture continues in Queensland and ceased in Western Australia only in the 1980s) and the prime motivation for its continuance has been economic rather than cultural. This gives Australian agriculture characteristics that predispose it to information technologies.

First, the relatively short and dynamic history of Australian agriculture fosters an expectation that the industry must change as rapidly as other sectors of the economy in order to survive. Australian agriculture receives substantially less government support than its competitors in other countries (Table 19.1). With the exception of support for R&D and restructuring, which individual farmers cannot support, the government expects farmers to identify alternative means of adapting to global threats and opportunities.

Second, the relative strength of rational economic drivers—compared with social or political forces evident in some of its international competitors—encourages the use of objective information in decision making. Technology has been shown to improve productivity growth rates beyond that expected by classical economists (Martin and Mitra, 1999), and in Australia it is generally expectated that this will happen. Third, as in North America, a relatively strong demand exists for better management of within-field variation, which is introduced by the discordance of field layout with the underlying inherent spatial variation of the land.

TABLE 19.1. Producer support for farmers among selected OECD countries, 1997-1999.

| Location | Producer support: By farmer ($000 per FTE[a]) | By area ($/ha) |
|---|---|---|
| Australia | 3 | 3 |
| Canada | 8 | 47 |
| European Union | 17 | 845 |
| Japan | 23 | 10,671 |
| United States | 18 | 106 |
| All OECD | 10 | 186 |

*Source:* Adapted from OECD, 2000.
[a]FTE = full-time equivalent

### *Reason 2: Australian Agriculture Understands the Need to Improve Profitability and Maintain Export Market Access*

The share of agricultural exports in Australian trade declined from more tha 40 percent in 1980 to 1982 to less than 20 percent in 1994 to 1996. Despite this decline, export agriculture remains an important feature of Australian trade. The Australian wine industry exports 80 percent of its product, and exports from agriculture generally make up a greater proportion of total production than in either the European Union (EU), Canada, or the United States. Although Australia is only the seventh largest producer of grain and the third largest exporter, exports account for substantially more of total production than they do elsewhere (Table 19.2). Consequently, Australian growers recognize the need to maintain profitability in difficult export market conditions.

Rapid change is imposed upon export growers by an apparently inevitable decline in the terms of trade for agricultural produce, which is the value of the commodity traded relative to those of farm inputs (Chisholm, 1992). The implication is that to maintain revenue growers must either increase production per unit cost of input or increase the value of their produce.

TABLE 19.2. Grain exports as a percentage of total production in selected OECD countries.

| Location | 1996-1997 | 1997-1998 | 1998-1999 |
|---|---|---|---|
| United States | 50 | 47 | 50 |
| Canada | 79 | 61 | 75 |
| Australia | 72 | 88 | 89 |

### *Reason 3: Australian Growers Need Precision Agriculture to Cope with a Risky Growing Environment*

Precision agriculture can reduce production risks by providing spatially explicit observations of hazards encountered when agriculture expands into a marginal environment or into one with low resilience. To illustrate this, we use the risks of drought and salinization within the Australian grains industry as examples.

The Australian grains industry occupies a broad arc of approximately 21 million hectares, mainly within the southern half of Australia (AFFA, 2001). Rainfall is the dominant production variable throughout the range of production, such that potential yield is often approximated simply as a function of growing-season rainfall (French and Schultz, 1984). Lack of reliable rainfall determines the geographical limits of production toward the arid center of the continent, and it is likely to become an increasing problem as farmers experience the impact of global climatic change (IPCC, 2001). These limits do not exist as sharp boundaries but fluctuate over space and time. Most if not all farms will experience drought over some parts of their land in some seasons. Some may even have fields in which one part is affected by drought and another by wetness. However, determining the extent and severity of this risk has not been possible without detailed measurement of the effect, and so the tendency has been to manage the risk by hedging. By simply mapping the variation of yield over several seasons and correlating with known soil variations it would be possible to infer intuitively those areas at high risk of drought and so concentrate on the safer areas for investment. More sophisticated modeling techniques may be employed to extrapolate quantitatively into the future, such as the use of coupled simulation models (Hansen and Jones, 2000).

Dryland salinization is a second major environmental hazard faced by Australian grain growers. It threatens a substantial portion of grain production (up to 30 percent of the land in some areas), but is yet to be stemmed (Williams et al., 2002). Dryland salinity is caused by the hydrologic imbalance that results from a switch from perennial vegetation to annual cropping, the latter being less effective at capturing all rainfall. This process has been known for decades and the spread of salinity within catchments has been identified by sophisticated remote sensing techniques. However, the financial risk to cropping—which can eliminate revenue quickly—is more difficult to define. Consequently, one of the main impediments to the collective action necessary to reduce this hazard is the lack of a clear definition of actual impact. Although farmers accept the broad risks, without a delineation of impact many find it difficult to evaluate and remain reluctant to change. Detailed analysis of the impacts and costs—based on actual field measurements using PA—would provide individual managers with powerful analytical tools on which to base more objective decisions.

### *Reason 4: Australian Farmers Need Practical Tools from Precision Agriculture to Ease Field Management*

The fourth contribution of PA to Australian growers is in straightforward practical assistance to farmers to enable them to monitor, analyze and diagnose performance on land about which they would otherwise have very little information. This paucity occurs for a variety of reasons—relating to the development and geography of farms—which results in farms that tend to be very large, remote, and managed by very few people. For example, the "typical" grain farm in Western Australia occupies more than 2000 ha (Connell and Hooper, 2001). It may have a gross annual revenue of two million Australian dollars to three million Australian dollars yet employs just two or three people, normally family members. It is located perhaps 20 to 30 km from the nearest school, possibly 100 km from the nearest hospital, and several hundred kilometers from the nearest university, large bank, retail center, or airport. Land phone lines are likely to rely on relatively old technology and provide relatively slow Internet access. Cellular phone coverage is probably patchy and even nonexistent in many places. A professional farm advisor may live within 100 km or so, but is often located much further away. An agrochemical and fertilizer supplier may be within 50 km, but the associated technical advisor is likely to be much further away. Technical support from government agencies—on which many farmers used to rely—may still exist within the 50 km radius, but government is increasingly harder to access as it withdraws from its role as extension provider.

The information paucity stretches individual farmers to the limit of their experience. This exposes dynamic enterprises to additional risk simply because farmers find it increasingly difficult to keep track of what happens on their farms. Precision agriculture technology can provide assistance in three practical forms:

1. Detailed observations of field operations through automatic recording. This both reduces the time required for supervision and accelerates the rate farmers learn about their business as it changes. The observations also provide objective information for discussion between farmer and farm advisor.
2. Analysis of performance. This might otherwise take years to acquire through trial and error.
3. Rapid diagnosis of malfunction. This could help growers prioritize attention where it is needed.

## *METHODS USED IN AUSTRALIA*

Although some of the PA technology used in Australia has been developed locally, most of the technology is imported from North America and Europe, and therefore discussed in other chapters. This section therefore concentrates on applications of technology peculiar to Australian conditions.

### *Yield Monitoring, Mapping, and Other High-Technology Field Observation Techniques*

Yield monitoring was introduced to Australia in 1993 using a grain yield monitor, GPS, and a base-stationed differential correction transmitter imported from the United States. It revealed a high degree of variation in grain crop yields, which was not so surprising since these yield maps merely confirmed what had been suggested in satellite images of crop biomass variation for years (Wallace et al., 1993). Although sales of imported yield monitors to grains farmers subsequently grew, the market was constrained by the absence of a reliable source of low-cost differential correction to overcome positional errors introduced by selective availability. Satellite-communicated, wide-area differential correction was offered commercially to farmers in 1995, following which yield mapping was extended to irrigated rice, cotton (see Boydell et al., 2000), sugarcane (Harris and Cox, 1997), and grapes (Bramley and Proffitt, 1999). A number of Australian companies are now involved in producing or developing grain, cotton, sugarcane, and grape yield monitors for the world market. Australian research has also led to the detection and site-specific treatment of weeds in fallow fields (Felton and Nash, 1998) and for the accurate guidance of tractors and chemical applicators. Some large international equipment manufacturers also use Australia's hemispheric location to double the number of seasons of field testing each year.

In the mid-1990s intensive grid field soil sampling for grain production was offered in some parts of Australia but faded out quickly because of the high cost relative to perceived benefit. The price of soil testing in Australia is roughly three times the price it is in the United States (Viscarra Rossel and McBratney, 1998) Cook and Bramley (2000) have demonstrated that the benefits to be derived from such soil testing are heavily dependent on the appropriateness to the conditions at specific sites and of models used to assist in soil test interpretation. More often than not these have been developed on the basis of nutrient response experiments conducted over wide biogeographic areas. Nevertheless, interest is increasing in image or map-targeted, georeferenced soil sampling, often following EM38 survey (e.g., Bramley, 2001). A novel research development has been the effort to develop a continual soil pH and lime requirement sensing system (Viscarra Rossel and McBratney, 1999).

Variable rate technology (VRT) has not been widely adopted by Australian growers. This is partly due to lack of reliability in the equipment provided to the Australian market, the cost of the equipment, and the lack of a service provider to develop variable rate nutrient recommendations. In Australia, VRT was used for the first time in 1995 to install on-farm experiments (Adams and Cook, 1997; Cook et al., 1998). The concept was taken further in 1996 and 1997 with the checkerboard and sine wave experimental designs (Figure 19.3; Bramley et al., 1999). The most widely used variable rate equipment in use in Australia is the air seeder and the spreader.

Spray guidance technology has been adopted in Australia more rapidly than either yield mapping or VRT. This reflects the rapid acceptance among farmers for technology that has a rapid and simply acquired benefit. Several domestically developed systems are available. These have driven the development of both hardware and software.

The widespread use of protein monitors is on the short-term horizon (two to three years) and Case New Holland introduced its protein sensor to Australia in late 2001. The ability to understand the complex interactions between yield, moisture, and grain quality offers huge potential for farmers who wish to improve the accuracy of N management (Kelly et al., 2001a).

FIGURE 19.3. Application map and infrared image of the checkerboard experiment, Wyalkatchem, Western Australia, 1997. Each treatment plot measures 30 m × 30 m and is surrounded by an array of other plots containing the other treatments. The speckled nature of the application map arises as a result of this being a map of actual, rather than planned, application of urea.

### *Yield Map Interpretation*

The need for accurate interpretation of yield maps was identified early on as a primary objective of research. Yield maps were perceived to be of great *potential* benefit, which few knew how to capture. Early problems seemed to be the following:

1. Methods were necessary that would add very little to the cost of acquiring a yield map (about two Australian dollars per ha).
2. Methods had to cope with a range of production uncertainties likely to occur in one field, from nutrient deficiencies through to insect pests, drought, and soil degradation. Methods had to be cheap enough to tolerate partial benefit.
3. Methods had to be robust enough to cope with seasonal variation under dryland production systems—which were the earliest applications sought. Reversal of spatial trends of yield variation in dryland crops were not uncommon, resulting either from the interaction of soil and climate or from contrasting phases of the rotation (Cook et al., 1996).
4. Results could not rely on expensive additional field sampling, which appeared impracticable for Australian conditions.

### *Approaches to Modeling Yield Maps*

Precision-agriculture scientists in Australia recognized early on that, given limited resources, research should focus attention where it was needed most urgently. Since PA primarily introduced information, in the form of yield maps, and since the outcome of this information was more precise control, it seemed logical that the most productive use of research was to identify methods to interpret the information and link it to control. For this purpose, new models were deemed to be necessary, if not sufficient, for progress.

Researchers have tried a range of increasingly analytical modeling approaches to interpret yield maps, some of which have entered commercial practice. However, they recognized that a first requirement was to define the basic management problem of uncontrolled variation. This is a more general goal than the attempt to vary specific fertilizers spatially, and enables farmers to consider the *range* of options they have to control their farm as an input:output system. The key features of the system represented in Figure 19.4 are

1. biological processes (a field or farm system) that convert inputs to outputs;
2. variable inputs and outputs, which are partially controllable and predictable and hence manageable;
3. outputs, in the form of yield, revenue, or other outcome, with information from yield maps defining variation in one form of output (crop yield);
4. inputs in three forms: controllable inputs such as fertilizer, seed, or water; uncontrollable but identifiable inputs, such as climate or soil water, which influence the conversion of input to output; and uncontrollable and unidentifiable variation, or noise, which increases the uncertainty in system parameters; and
5. the manager (farmer) who, with guidance, uses knowledge about the input:output relationship to control selected inputs.

Viewed in this way, the problem is generalizable to identifying the relationship between controllable input and output. Yield maps enable output to be measured very precisely, but without information about other factors, the problem of knowing how to modify inputs remains intractable. Various approaches have been applied to this problem (Table 19.3). A fuller treatment is given in Adams et al. (2000b).

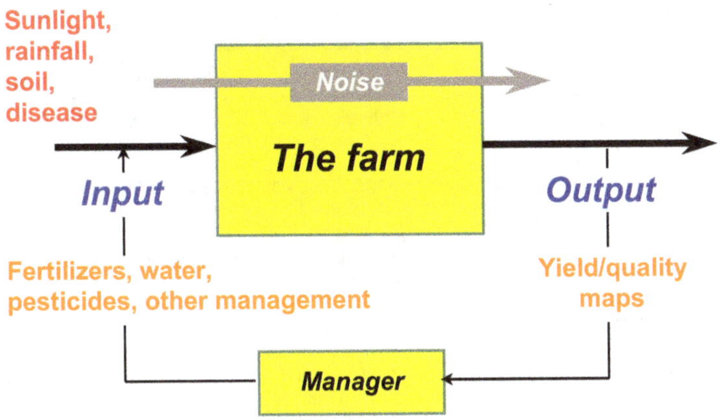


FIGURE 19.4. Simplified concept of a partially controllable system.

### *Simple Intuitive Modeling*

The simplest—assuming that yield maps are interpreted for *some* purpose—and most widely adopted method of interpreting yield maps is simply looking at them and deciding intuitively what the maps mean. This interpretation method is not as illogical as it sounds; farmers have many years' experience on which to base such inference and, in many cases, yield maps provide sufficient objective evidence to confirm a *suspected* problem or identify the extent of a *known* problem and so precipitate appropriate action earlier than might have occurred without the yield map. Naturally, the weakness of such an approach is that the "model" is not specified and so cannot be verified or improved. It is also imprecise, so it cannot be used to instruct variable rate technology.

### *Simple Deterministic Modeling*

Early methods of prescribing variably optimum fertilizer rates relied on simple deterministic models relating soil analytical data to nutrient availability and hence expected response. This is the so-called prescription method. Many standard practices for applying fertilizer were based on these models and it therefore seemed a natural progression to try and adopt them spatially. However, early trials in targeting phosphorus (P) fertilizer (the major variable cost) to dryland wheat by Cook et al. (1996) were not en-

TABLE 19.3. Summary of approaches used to apply PA technology.

| Approach | Description |
|---|---|
| Intuitive | Still widely used by farmers where apparent causes of variation are obvious. |
| Simple deterministic model | Describe explicit mechanisms of crop response to inputs. Widely used to estimate fertilizer requirements. Can be used to show spatial variation if sufficient input data are available. |
| Data-driven and expert systems | Data-driven systems exemplified by cluster analysis, neural networks, and Bayesian knowledge discovery tools. Most appropriate for moderately well defined problems. Requires considerable data input.<br>Expert systems designed to imitate human expert reasoning. Exemplified by expert or rule-based systems. Most appropriate for well-defined or knowledge-rich problems. Capable of accommodating expert knowledge generated using other models (e.g., simulation). |
| Complex deterministic: Simulation models | Simulate fundamental processes of crop growth, water movement, nutrient fluxes, etc. When empirically derived they may also be termed functional models. Possible to apply spatially but normally demanding of input data. |

couraging, because any potential improvements in spatial targeting according to estimated supply were reversed by errors in estimating demand. The spatial errors caused by difficulties in estimating available nutrient, coupled with spatial and temporal errors associated with estimates of demand for nutrient, were deemed at that time to make this approach impracticable for these broadacre conditions.

Subsequent on-farm experimentation enabled further testing of similar methods of estimating the response for N fertilizer (Adams et al., 2000a) and potassium (K) (Cook and Bramley, 2000). The general conclusion is that the demand parameter, that is, the level of yield that would be achieved if the maximum nutrient were applied, was too large to ignore in spatially variable estimates. Corner (2000) attempted to correct this deficiency by developing the Achiever method, which combines three sources of estimates of demand (as yield potential from prior yield maps, satellite imagery, or farmer estimate) with soil-based estimate of nutrient supply. Testing of the Achiever method has been limited due to ongoing issues regarding support of the software, which has prevented general release.

Very simple moving window regression analysis of on-farm trial data was used by Cook et al. (1998) to identify spatially variable responses to fertilizers. Variations of response to P fertilizer could be partially explained by nutrient testing—although tissue analysis proved more capable of explaining response variations than conventional soil analysis. In both cases, the main value appeared to be in retrospective analysis of field crop performance rather than in formal prediction.

Working in the summer-dominant rainfall region of Queensland, Kelly et al. (2001a) used simple deterministic rules, based on the results from prior experimentation, to relate measurements of grain yield and grain protein to the likely requirement for fertilizer N. This approach provides a robust method for farmers to review field data and diagnose likely limitations to acquiring grain of a given protein content for this region.

### *Data-Driven and Expert Systems*

A major source of error in simple deterministic models is their inability to handle data that is merely *associated* with yield response. A more flexible method of using this information is the expert system, which can infer on the basis of known associations which are not necessarily causal—for example, the association between response and prior yield, or "poor" soil. Prior crop yield is not a variable used in the deterministic methods previously described, yet few advisers would dismiss its association with future yield. The Expector method (Corner and Moore, 1999) is a GIS-based application of simple Bayesian networks to the problem of estimating spatially variable yield response. As described by Cook et al. (1996), when applied to the problem of estimating spatially variable P response, it greatly increased the spatial resolution by enabling a wider range of input data (such as a prior yield map). However, for reasons explained previously, with only a few seasons on which to develop rules, this prediction proved to be no more accurate because of unresolved temporal variation.

The dense observations that are provided by yield maps and remotely sensed imagery enable data-driven models, which are models that fit parameters retrospectively to a large data set. Methods that have been tried include Bayesian nets and regression trees (Adams et al., 1998). Bayesian nets explained slightly more of the variation in crop yield than the regression tree. A moving window regression was applied to whole-field fertilizer experiments by Pringle et al. (2003) to identify generalized models.

### *Difficulties of Basing Predictions on Retrospective Analysis*

Information provided by analysis of yield maps indicate precise information about prior crop performance. However, this information acquires practical value *only* when it can help answer the question "What *will* happen?" Retrospective analysis of yield maps is valuable only insofar as it can be used to estimate reliably what *will* happen in the future.

If temporal variability is large, the uncertainty removed by retrospective analysis of spatial crop yield variation is relatively small—the so-called "null hypothesis" commented on by McBratney and Whelan (1999). Spatial precision is improved, but prediction accuracy is less so. This situation can be exacerbated by the variation between cropping phases, especially when successive phases use plants that respond very differently to the same conditions. For example, the Australian grain production system is built on the wheat-legume rotation, which has highly contrasting phenologies and responses to drought or waterlogging.

Temporal variation presents a major problem to accurate prediction of yield response in annual crops, though much less so in perennials such as wine grapes given their temporal stability in patterns of within-vineyard variation (Bramley and Hamilton, 2004). Experience suggests that, in practice, this is less of a problem than one might expect, for although overwhelming evidence shows that farmers are generally risk averse (Antle, 1987), Kingwell (1994) showed that the more successful Australian dryland farmers are actually risk-tolerant. Experience with participatory research using simulation models (Robertson et al., 2000), suggests that many farmers are adept at evaluating risks. Our experience is that in practice a powerful use of PA analysis is to enable a better decision by confirming a preexisting suspicion that is judged in relation to experience of temporal variation.

### *Complex Deterministic Models—Simulation*

Coupled to temporally variable weather data, crop simulation models can be used to try and overcome the difficulties imposed by temporal variation on empirical approaches to prediction. The logical extension of such prediction would be to predict temporal *and* spatial variation by modifying the soil input parameters to account for spatial variability. This has been attempted several times for U.S. conditions (Hansen and Jones, 2000).

Judging by the papers presented on the topic at recent international conferences, crop simulation models have great potential for improvement. However, they have not been used for PA in Australia. This could reflect either a theoretically based disdain for crop simulation models (Passioura,

1996), or the practical difficulties of providing soil input data, when soil maps at scales better than 1:100,000 are rare (Adams et al., 2000a). A more pragmatic concern is that simulation modelers are prone to claim high degrees of accuracy ($r^2 > 0.8$) in the simulations provided by their models for whole paddocks, farms, or regions, yet yield mapping demonstrates coefficients of variation in yield of the order of 35 percent to be typical for a range of crops *within* fields (Pringle et al., 2002). Thus, for a 100 $t \cdot ha^{-1}$ sugarcane crop, the within-field variation in yield can be expected to range between 30 and 170 $t \cdot ha^{-1}$ (Bramley and Quabba, 2001)—a range of variation that casts considerable doubt on the ability of such models to actually simulate the system of manageable interest. Nevertheless, substantial use has been made of crop simulation models in participatory research (Robertson et al., 2000). Here, the role of the model is to provide "cheap experience" and to enable growers to reflect analytically on their observations (Carberry, 2001). Spatial extension of the model to describe temporal *and* spatial variation has been proposed but not tested at field level (Moore and Tyndale-Biscoe, 1999).

### *Remote Sensing*

Given the vast expanse of Australian cropland, techniques of remote sensing have long been seen as a cost-effective source of information about spatial variation of the crop and land resources. A wide range of applications has been developed that can be divided roughly into satellite, airborne, and ground-based techniques. The use of such techniques for land resource assessment in Australia was reviewed in the early 1990s by Johnston (1991), and special attention given to applications in PA by Moran et al. (1997).

#### *Satellite Imagery*

Normalized difference vegetation index (NDVI) has been widely used to estimate the developing crop. For some years, grain farmers in Western Australia have been offered cheap estimates of yield derived from Landsat TM–based imagery under the brand "Agimage" (http://www.dli.wa.gov.au/corporate.nsf/web/Agimage). The correlation between NDVI and grain yield is reported to be up to 70 percent. However, difficulties arise when trying to use satellite imagery to estimate yield variation at the same spatial resolution as yield maps (Corner et al., 1998). Also, problems can arise when biomass fails to convert to grain, as is the case when the plant is affected by late-season drought, certain micronutrient deficiencies, or frost—

which, for example, severely reduced yields in the 1998 season over a substantial area within the Western Australia Wheat Belt. Despite these problems, satellite imagery provides a practical and cost-effective alternative to broadacre farmers who wish to explore spatial variation of crop yield before purchasing a yield monitor. Satellite imagery and derivative products are now being offered as a commercial service (see http://www.agrecon.canberra.edu.au/).

A further difficulty in interpreting satellite imagery under dryland conditions is the interaction between seasonal, soil, and crop type influences, which can confound the reflectance pattern. This problem is reduced in dryland cotton, which appears as a single crop and is normally grown on heavier soils that provide a stabilizing factor of stored water. Consequently, it is possible to use the more easily calibrated imagery to provide an archive of cotton yield (Boydell and McBratney, 1999).

High-resolution imagery, delivered from platforms such as Ikonos, should have applications, particularly under the more intensively managed farmland that predominates in the eastern half of Australia. To date, all such applications are proprietal and have yet to be reported to the wider community.

Bare soil imagery has been used in New South Wales to identify spatial variations in soil condition, especially organic matter content and therefore soil available N (Stewart and McBratney, 2001). Investigations in Western Australia, where such a method would have even greater benefit, proved less encouraging (Corner et al., 1998), principally because soils there were predominantly of low organic matter content and *never* bare. In this situation, all pixels contain mixed signals in which the organic matter influence was dominated by trash.

### *Airborne Sensor Data*

A wide range of airborne remote sensing services is on offer to Australian growers, often as spin-offs from the more lucrative geological prospecting.

Digital multispectral video (DMSV) is probably one of the more widely available services, and it has been used extensively for environmental monitoring. Its routine application to broadacre grain cropping, currently the largest PA market in Australia, is constrained by a relatively high per hectare costs. Such costs are more affordable in higher value crops such as cotton, rice, and grapes. Application of airborne imagery to rice production has been explored, and it is widely used in the wine industry (Hall et al., 2002) as a guide for selective harvesting (Bramley et al., 2005b). It also shows some promise as a means of predicting grape quality (Lamb, 2000; Lamb et al., 2004). The theoretical value of hyperspectral data is known from environmental monitoring and mining exploration applications. Applications in precision agriculture are being developed.

Electromagnetic (EM) surveys of agricultural land are commonplace over much of Australia as part of land resource assessment (Johnston, 1991). Variation of the dielectric constant, on which the technique is based, can also be correlated with soil attributes, in particular with clay content, which has a major influence on soil water availability and hence on crop productivity. EM soil survey is being increasingly used by the wine industry to provide detailed soil information (Bramley et al., 2000, 2001; Lamb and Bramley, 2001; Bramley, 2003), although concerns have been raised about the influence of trellis wires and steel posts (Lamb et al., 2005). Aerial EM imagery is already used in the cotton industry. While some correlations undoubtedly exist, the airborne application of this technique is most suitable for deep profiling. Estimation of surface clay is best undertaken using a ground-based instrument.

A third airborne technique of potential use to PA is sensing of gamma radiometric emission. This technique measures the natural gamma radiation emitted by dry soils, which correlates loosely with soil texture and mineralogy. Emissions from radioactive isotopes of K, thorium, and uranium, as well as total counts are measured. Gamma radiometric emissions have been related to soil physical properties in Australia (Cook et al., 1995) as well as to plant-available K (Wong et al., 2001). Research projects are currently underway toward determining how gamma radiometric data can be used for site-specific management purposes.

## *APPLICATIONS IN THE GRAINS, COTTON, WINE, AND SUGAR INDUSTRIES*

### *Development of Precision Agriculture in the Australian Grains Industry*

Although interest in PA has spread through a range of industries, from broadacre grain, cotton, sugarcane, and potato through to horticulture, wine grapes, and forestry, monitoring and mapping the spatial variation in small-grain crop yields remains the dominant focus of PA in Australia, possibly because research and development is well supported within this industry. The grains industry was the first significant adopter of PA in Australia. Relatively inexpensive imported yield monitors are widely available and are provided as a standard feature on some harvesters. At present, the total number of grain yield monitors operating with GPS is quite small, estimated at about 500. Though some farmers have "disadopted" the technology following initial disappointment at the difficulty of using the results, many are still waiting to see how it can be coupled to management decisions.

### *Management Zones Preferred to Prescription Approach*

The prescription approach to fertilizer application was tried in Australia, but it has been largely eschewed by farmers. One reason for this is that under the climatic uncertainty faced by many dryland farmers—to whom it was initially offered—it is the uncertainty of *demand,* not supply, of nutrient that needs to be managed, which cannot be controlled by prescription. A second is the relatively high cost of soil sampling, which made the costs of within-field analysis impracticable. This was particularly the case given that soil tests are more difficult to interpret in the highly weathered soils that predominate the Australian landscape (Adams et al., 2000a; Cook and Bramley, 2000). A further significant factor is the cost of variable rate technology. Development of PA in the grains industry was initiated more strongly in Western Australia and, consequently, costs of VRT had to be borne by individual farmers rather than harvesting contractors, who are not widely used in that state.

Farmers perceive the following four practical options for change:

- Improved whole-field management, which results from information provided by interpretation of yield maps or on-farm experimentation.
- Removal of sections of fields from cultivation, or conversion into alternative uses other than annual crops—the so-called "culling" or "alternative use" option. This is a practical option for parts of fields that are likely to lose money or present serious risks of leaching.
- Subdivision of fields into more homogenous "management zones" (McBratney et al., 2000; Cupitt and Whelan, 2001). This option is being developed strongly in eastern Australia through the Australian Centre for Precision Agriculture (ACPA). Research in this strategy is developing robust spatial prediction software (e.g. Minasny et al., 1999) and techniques to determine potential management zones within fields, which are confirmed subsequently by soil sampling (Whelan, 2001). An increased interest in varying fertilizer rates by management zones occurred among Western Australian farmers in 2000 and 2001.

This appears to have been driven by successive average to below-average seasons from 1997 to 2000 as well as by a desire to improve fertilizer-use efficiency.
- Continuous variation of fertilizer or sprays using variable rate technology. Australian farmers have generally not adopted this option.

### *Use of Satellite Imagery to Reduce Mapping Costs*

Although few Australian farmers consider the cost of yield monitors to be a major impediment to yield mapping, many balked at the earlier cost of GPS. Until the removal of selective availability, farmers were obliged to use satellite-communicated differential correction signals or local base stations, and for many this proved too expensive to justify. Consequently, the proportion of farmers with yield monitors who map yield is quite small. A commercial alternative to yield mapping has been locally calibrated NDVI (http:// www.dola.wa.gov.au/corporate.nsf). A commercial provider—Wesfarmers CSBP—also sells this product (http:// www.csbp.com.au), which is part of a commercial process for estimating spatially variable yield potential—normally the main source of uncertainty in estimating fertilizer response.

### *On-Farm Experimentation*

An early recognition that generalized recommendations for fertilizer were inappropriate for specific regions led to the preference for locally developed solutions based on empirical on-farm experiments. Methods of using variable-rate technology to automatically install sophisticated designs have been developed and promulgated in a form suitable for leading farmers and their advisors (Bramley et al., 1999). This trend for what is sometimes called *action research* parallels the general move in agricultural research away from the notion of "prescribed" solutions, which are believed to have resulted in dismal record of adoption of decision support by farmers (McCown, 2001).

A significant problem with on-farm experimentation is the sheer volume of information that is generated in the form of yield maps, treatment maps, and other data that are acquired to interpret yield variation (Figure 19.5). Methods of analysis are likely to be daunting to most farmers, and more commercial advisors are needed with sufficient spatial data analytical skills to interpret the data for farmers (Cook et al., 2000).

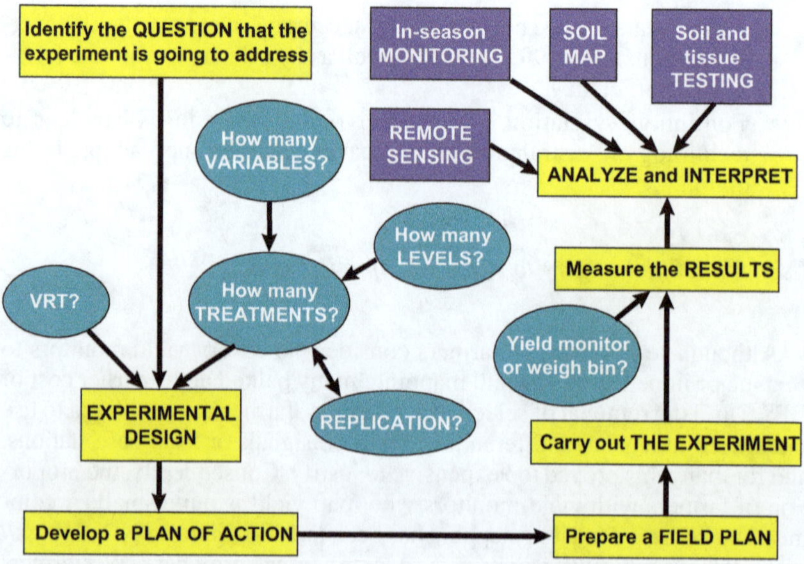


FIGURE 19.5. Scheme of implementing and analyzing on-farm experiment using yield maps. *Source:* Adapted from Bramley et al., 1999.

### *Need for a Wide Range of Managed Variables*

The early uses of PA target better application of N fertilizer, however this capacity alone would not interest many Australian growers because of the high range of nutrient deficiencies likely on highly weathered Australian soils. Large areas are intrinsically deficient in available P, K, and many micronutrients. Most soils are poorly buffered and liable to rapid acidification in the surface and subsurface. A large portion of the grain-growing region of southern Australia has a history of both boron deficiency and toxicity. Many soils are liable to subsoil sodicity.

Given the generally poor buffering capacity of Australia's highly weathered soils, the risks of leaching, acidification, and denitrification from fertilizer not required by the crop are significant. A model has been developed by the ACPA to extend the usual fertilizer-application economics to deal with the environmental costs of overapplication of N fertilizer. This model suggests that the combined improvement of increased profitability plus reduced environmental risk should be taken into account when assessing the benefits of precision application.

### *Machine Guidance Supports the Development of PA*

Although not PA per se, the adoption of machine guidance has supported the expansion of PA. It has enabled spatial technology providers to expand and stay in business while more complex, PA applications develop. The hope is that when the agronomy and information systems are able to provide PA users with functional decision support, experienced technology providers will be available to help deliver the services.

Sprayer guidance technology, some of which is produced in Australia, has a very simple objective: to prevent overlaps and spray misses. Savings can be substantial for the very large fields typical of Australian farms. An added advantage is that these savings can be realized regardless of driving conditions. Benefits are provided directly and, perhaps most important, almost immediately, whereas benefits from yield map interpretation are longer-term but more difficult to acquire. Controlled traffic, which has been most intensely adopted on the heavy soils found in the wheat and cotton areas of southern Queensland and northern New South Wales, also applies guidance technology with a straightforward application—in this case to separate traffic zones from cropping zones, and thus reduces yield losses attributable to compaction from machinery wheels.

### *Use of PA to Manage Grain Protein*

Wheat and barley growers in Australia are paid according to the protein content of the grain. This premium can be substantial and in some years may boost income by 15 to 20 percent. The potential financial benefit of managing variable grain quality has been demonstrated by Stewart et al. (2002). Segregation of the grain at harvest would be a logical option that capitalizes on this premium. However, the feasibility of separating grain on the go appears limited due to the mechanical adjustments necessary to direct grain into two or more bins. If prior knowledge of the spatial distribution of grain protein is available, harvesting operations could target specific zones. Furthermore, maps of anticipated protein variation could be used to highlight areas more likely to provide a higher gross margin. A risk-averse approach would then be to harvest these areas first and minimize any losses due to rain and hail.

Before the recent introduction of grain protein sensors, capturing data on within-field variation in grain protein proved very difficult. Researchers in Australia initially used grab sampling but progressed to semiautomated stream sampling at the grain-up elevator (Jensen et al., 2001b) to measure protein variation. Both strategies require post-harvest protein assay. An on-the-go protein sensor is currently being tested in Australia to automate the entire process (Jensen et al., 2001a). A higher frequency of protein measurements will only improve the coincidental interrogation of data sets, and provide a greater certainty in delineating zones of similar protein within the field.

However, the practical implementation of improved grain quality streaming also depends on the capacity of the infrastructure to handle it. Farmers complain that the premiums for higher quality grain do not always justify the effort to acquire it, and (at least in Western Australia) that the infrastructure is not available to receive it even if they do.

Within-field measurement of grain protein has an agronomic management spin-off. Interpretations obtained from coincidental yield and protein maps can be used to improve overall agronomic management. Protein content in combination with yield can indicate, with a high degree of certainty, whether N or moisture supplies were yield-limiting (Holford et al., 1992). For example, wheat grain with a protein content over 11.5 percent but relatively poor yield is likely to have been limited by moisture supplies. Coincidental yield and protein data can provide useful, albeit retrospective, assessments of within-field variation in N removal, N supply, and antecedent soil water supplies. When used in conjunction with an expert system, the chances that additional N may have improved yields can be inferred (Kelly et al., 2001a,b). Where patterns in N supply are consistent across years, variable rate fertilization may ensure that yields can be maximized while inputs minimized.

A parallel effort in northern New South Wales is researching an empirical methodology to implement variable rate N fertilizer, based on known relationships between soil, crop, and weather variables (Cupitt and Whelan, 2001). Variable rate applications are being compared to uniform application with respect to yield, grain protein content, and economic return.

### *Weed Mapping*

A further practical application of GPS technology is weed mapping at the time of harvest. Although it is sometimes dangerous to rely on the presence of weeds at harvest to identify the problem of weeds during the growing season, this practice is deemed useful by some farmers simply to remind themselves in the following season where weed patches occurred. Given the uncertainties associated with our current capability to predict weed population dynamics within fields, the economics of precision weed management do not appear to support the use of patch spraying (Bennett and Pannell, 1998).

In addition to at-harvest mapping of weed populations, work has been performed in Australia on the most appropriate methods for sampling weed populations to create maps for site-specific management of weeds (Rew et al., 2001). The main conclusion of this work is that it is more cost effective to use aerial remote sensing methods in generating weed maps than it is to field sample. Telling the difference between broadleaf and grass weeds, which require different classes of chemicals for control, using aerial remote sensing methods was not discussed. Lamb (2000) echoes the same conclusions as Rew et al. (2001), but also does not deal with the economics of a full precision weed management package.

### *Development of Self-Help Groups*

An interesting feature of the Australian industry is that, as state governments have withdrawn partially from their traditional extension function, farmer groups throughout Australia have expanded in activity and scope. The most powerful aspect of this development is that groups develop solutions for themselves that are perceived as locally specific and relevant yet based on sound principles. The largest of these is currently the Birchip Cropping Group (http://www.bcg.org.au/), based in southeastern Australia. Several farm improvement groups are also becoming interested in developing precision agriculture solutions. The Yuna Farm Improvement Group on the northern fringes of the Western Australian Grain Belt and Conservation Farmers Incorporated (http://www.cfi.org.au) in the eastern states are two such groups.

#### *PA in the Australian Sugar Industry*

The potential gain of managing within-block variability of sugarcane was demonstrated by Bramley et al. (1997) and Bramley and Quabba (2001). Yield mapping was enabled with an experimental monitor developed by Cox et al. (1998). Following this early promise, however, the Australian sugarcane industry has not sought to develop this work, citing problems of low commodity prices as a reason for lack of support (Bramley and Quabba, 2001). At the time of writing (2001), the development of precision sugarcane production in Australia has been confined to issues of machine guidance (M. Smith, Bundaberg Sugar—pers. comm.)

#### *PA in the Australian Cotton Industry*

The Australian cotton industry was introduced early to PA with the support of the ACPA, operating through an industry-focused cooperative research center (McBratney and Whelan, 1995). The industry should be ripe for PA; it has high input costs, is subject to political pressure to manage within strict environmental limits, and competes for increasingly precious irrigation water. It is well organized and has good access to the Internet. However, perhaps only 30 cotton harvesters are fitted with yield mapping capability, which suggests that, as in the grains industry, farmers are waiting for more practical methods to be delivered before investing in the technology. Boydell et al. (2000) describe limitations to yield monitor accuracy in cotton and some options for managing N fertilizer.

The technology to apply different rates of fertilizer within a single cotton field already exists. However no method for determining fertilizer rates at this scale has been developed relevant to cotton. Consequently research at ACPA is aimed initially at quantifying the degree of spatial variability of soil nutrients within and between cotton fields to then be combined with other spatial data layers, such as yield maps and remotely sensed images.

A further project from ACPA is collecting evidence useful in evaluating the suitability of precision agricultural management techniques when applied to the Australian cotton industry. This research has three main objectives: The first is to characterize the variability of yield magnitude within a field. Here difference between low yield and high yield should give a measure of "opportunity" to improve current single-rate management using within-field management techniques. The second objective of the study is to characterize the spatial variability of yield. The relative size of similar yielding regions within an average field will give a measure of practicality of within-field management options. The third objective of research is to identify the predictability of yield outcomes at a subfield scale.

With an ability to predict yield potential differences comes the opportunity to apply "risk-management" techniques to input application and move PA from exploratory to implementation stages.

### *Precision Viticulture in the Australian Wine Industry*

The Australian wine industry was a latecomer to PA, producing its first yield map in 1999 (Bramley, 2001). Nevertheless, for the following reasons, this industry probably holds most promise within Australia for widespread adoption of technological change:

1. It has a strong export focus and is worth around three billion Australian dollars in annual export earnings.
2. It is highly organized. Approximately 80 percent of wine capacity is contained within just three companies.
3. It is well integrated vertically and has a thorough understanding of the benefits of product streaming. In many cases, the grape grower and winemaker are closely linked within one company.

4. It demonstrates very strong market differentiation on the basis of quality. Successful marketing ensures a wide price differential between good and poor quality wine.
5. It recognizes that rapid expansion and change is stretching conventional viticultural knowledge beyond traditional techniques.

Although little, if any, philosophical differences exist between precision viticulture (PV) and the application of PA to broadacre cropping (Bramley and Proffitt, 2000; Bramley, 2001; Cook and Bramley, 1998), the greater importance that grape growers and winemakers attach to crop quality than might occur in broadacre cereal cropping systems, for example, together with any yield/quality interactions, means that the development of targeted management strategies has to be done in such a way that productivity in terms of both yield and quality is optimized. The simple input-output system described by Figure 19.4 therefore becomes more complex in the case of viticulture and knowledge of within-vineyard variability in both yield and quality attributes and of the factors that might be driving such variability is essential to the successful development of targeted management strategies.

Current research on precision viticulture partially follows the lessons learned from PA in grain, but also looks more closely at the processing of grapes through to marketable wine. Bramley (2006) summarizes the current status of the practical application of precision viticulture. While researchers in the Commonwealth Scientific and Industrial Research Organization (CSIRO), ACPA, and their commercial partners examine viticultural relationships between crop growth, yield, and site conditions, emphasis is also placed on the capacity of the industry to use this information during the process of winemaking. The ability to separate the crop at harvest and to harvest according to quality specification should have a major impact on industry profitability (Bramley and Proffitt, 1999, 2000; Bramley et al., 2005b). This is because the potential value added by producing quality wine from premium fruit is substantial. Thus, winemakers aim to make their premium wines as high a quality as possible, and an ability to assure that the quality of premium wine is not reduced through the inclusion of suboptimal fruit has considerable potential value. Crop separation also affords possibilities with respect to tailoring harvesting according to market demand and expectations and winery storage capacity. The improved control over the production process foreshadowed through the adoption of PV should provide winemakers with an improved basis for paying growers—based on both yield and quality.

## IMPEDIMENTS TO ADOPTION

### Shortage of Advisors Skilled in the Interpretation of Yield Maps

The improvements PA technologies offer seem obvious to scientists working in this area. They include enhanced likelihood of profit, product of consistent quality, and reduced risk to the environment. Even without sophisticated variable rate applications it is not hard to demonstrate how simple improvements in practical decision making, which PA can enable, can benefit growers adequately to justify the investment in PA technology.

Yet the adoption of PA in Australia remains patchy. Partially, this repeats experience in North America. For example, some reports say that in some parts of the Midwest United States the *majority* of farmers use PA, whereas others say penetration of this technology is less than 5 percent (e.g., CropLife Purdue Survey 2001. In Australia less than 5 percent of broadacre cereal growers, along with some cotton growers, have adopted yield mapping, though many more appear to be "sitting on the fence."

More intensive activities, such as horticulture, viticulture, or sugarcane farming, which might be quicker to determine benefit, have been even slower to adopt (although precision viticulture appears to be accelerating well). Worse still, we hear through suppliers that, similar to experience in other countries, some early adopters have become sufficiently disillusioned by a lack of acquired benefit to disadopt.

Clearly a range of possible reasons for poor adoption exists, which needs to be appreciated in order to ensure that research is delivering real improvement. Cook et al. (2000) and Cook and Bramley (2001) discuss the range of problems. These include the following:

1. Cost of adoption
2. Lack of perceived benefit from adopting
3. Unwillingness to be early adopters
4. Lack of mechanism to deliver support

Of the reasons for slow adoption of PA, the last problem listed—poor delivery of support—appears to be the most taxing in Australia. Farmers are not generally irrational or unreasonably conservative, and in Australia they have frequently shown themselves very adept at change so long as they perceive sufficient benefit. However, to achieve benefit requires a set of specialist skills to analyze the complex data that few who currently advise farmers possess. A major obstacle to adoption of PA is the shortage of support services to enable farmers to *complete* the cycle of enhanced decision making represented in Figure 19.1. In part this is due to the skill base of the consultancy sector. It may also be due to lack of a good business case to provide the service or conservatism among consultants, who currently have less incentive to change than the clients they serve. In Australia, far from being a problem of the conservative farmer, the slow adoption of PA may well be a problem of the conservative advisor.

### *Capacity Building: Education and Training Facilities in Precision Agriculture Within Australia*

Although individual components of PA are touched on at most universities with an agricultural faculty, the University of Sydney is the only education institution in Australia teaching PA to degree level. The use of PA management tools is taught as a subject in the fourth year. A postgraduate course is also held a number of times a year for industry participants or those wishing to advise them. The subjects covered include:

- GPS technology
- Variable-rate application theory and technology
- Soil sampling principles
- Yield mapping
- Soil-map production
- Analysis of the spatial variability in soil attributes and crop yield
- Examining the effect of temporal variability on crop yield
- Within-field experimental design
- Options for determining within-field management units
- Interpreting yield and soil information for management decisions

In addition to formal training, several institutions within Australia have organized meetings, symposia, and workshops to try and keep abreast of developments in PA—at one stage these numbered several per year. Typically these invite a mixture of participants from research, government, and industry to ensure that outcomes are directed toward practical goals.

Probably the most successful of these has been the annual symposium organized by ACPA, which has been established to promote the center's activities, attract further collaboration, provide a forum for the exchange of ideas, and disseminate information to members of the farming and wider communities. The ACPA is also involved in presenting PA research and teaching through the *Leading Edge,* a supplement to the *Australian Grain* magazine for the rural industry.

## *CONCLUSIONS*

Precision agriculture presents exciting opportunities to improve management within a range of Australian agribusinesses. The potential gains in terms of profitability, product quality, environmental protection, and practical management are evident. They are already being explored actively by small groups in grains, wine, and cotton, mainly in response to the intense pressure for improvement that is perceived by these unprotected export industries operating in frequently challenging growing environments.

However, adoption remains limited, with less than 5 percent of growers mapping yields in 2001. Research effort remains small, accounting for far less than 1 percent of R&D funds spent over the past five years. Partially as a result of this, the pressure for change appears to exceed the capacity to respond. Some excellent research on PA has been carried out over the past ten years in Australia, but the number of researchers remains small. Similarly, the capacity to develop specific aspects of technology exists in Australia, but the scale of the domestic market may limit investments. Domestic suppliers of hardware and software are slowly increasing capacity to meet potential demand and now await the arrival of technical advisers to help delivery within the farming community.

Skeptics of PA abound in Australia. This must be expected, especially given the high profile that PA has acquired and its hi-tech image of "farming by satellite." However, few in Australia doubt that PA, *in some form,* will become standard practice by the next generation of growers. It is always hazardous to predict exactly how this will occur, but adoption is more likely to be based on improved management rather than complex technology. Such practices might include the following:

- Progressive emergence of management zones and retirement of unprofitable or environmentally sensitive land from production
- Diagnosing specific diseases or nutritional limitations
- Learning about site-specific performance of crops through on-farm experimentation, as individual farmers or networked communities of farmers
- More rapid adoption of new varieties through on-farm trials
- Tighter supply chain management

## NOTE

1. For an unconventional view of modern Australian history, see Knightley (2001).

## REFERENCES

Adams, M.L. and S.E. Cook. 1997. Methods of on-farm experimentation using precision agriculture technology. American Society for Agricultural Engineering Paper 97-3020. St Joseph, MI: ASAE.

Adams, M.L., S.E. Cook, P.A. Caccetta, and M.J. Pringle. 1998. Machine learning methods in site-specific management research: An Australian case study. In *3rd International Conference on Precision Agriculture,* eds. P.C. Robert, R.H. Rust, and W.E. Larsen. Madison, WI: ASA/CSSA/SSSA, pp. 1321-1333.

Adams, M.L., S.E. Cook, and J.W. Bowden. 2000a. Using yield maps and intensive soil sampling to improve nitrogen fertilizer recommendations from a deterministic model in the Western Australian wheat belt. *Australian Journal of Experimental Agriculture,* 40:959-968.

Adams, M.L., S.E. Cook, and R. Corner. 2000b. Managing uncertainty in site-specific management: What is the best model? *Precision Agriculture,* 2:39-54.

AFFA (Australian Fisheries and Forestry Australia). 2001. News. Available at: www.affa .gov.au/.

Antle, J. 1987. Econometric estimation of producers' risk attitudes. *American Journal of Agricultural Economics,* 69:509-522.

Bennett, A.L. and D.J. Pannell. 1998. Economic evaluation of a weed-activated sprayer for herbicide application to patchy weed populations. *Australian Journal of Agricultural and Resource Economics,* 42:389-408.

Boydell, B., and A.B. McBratney. 1999. Identifying potential within-field management zones from cotton yield estimates. In *Precision Farming 99: Proceedings of the 2nd European Conference on Precision Agriculture,* ed. J.V. Stafford. Sheffield, U.K.: Sheffield Scientific Press, pp. 331-341.

Boydell, B., C.M. Stewart, and A.B. McBratney. 2000. New challenges from precision agriculture. Presented at the 10th Australian Cotton Conference, August 16-18, 2000, Brisbane, Australia.

Bramley, R.G.V. 2001. Progress in the development of precision viticulture: Variation in yield, quality and soil properties in contrasting Australian vineyards. In *Precision Tools for Improving Land Management,* eds. L.D. Currie and P. Loganathan. Occasional report No. 14. Palmerston North: Fertilizer and Lime Research Centre, Massey University, pp. 25-43.

Bramley, R. 2003. Smarter thinking on soil survey. *Australian and New Zealand Wine Industry Journal,* 18(3):88-94.

Bramley, R.G.V. 2006. Acquiring an informed sense of place: Practical applications of Precision Viticulture. *Australian and New Zealand Wine Industry Journal,* 21(1):26-33.

Bramley, R.G.V., S.E. Cook, M.L. Adams, and R. Corner. 1999. *Designing your own on-farm experiments: How precision agriculture can help.* Canberra: Grains Research and Development Corporation.

Bramley, R.G.V., S.E. Cook, and G. McMahon., eds. 1997. *Precision agriculture—What can it offer the Australian sugar industry?* Proceedings of a workshop, July 1997, Townsville, Australia. Canberra: CSIRO Land and Water.

Bramley, R.G.V., T.E. Evans, and A.P.B. Proffitt. 2001. Digital elevation models and EM38 survey—Useful tools in the assessment of constraints to vineyard productivity. Presented at the 11th Australian Wine Industry Technical Conference, October 7-11, 2001, Adelaide, Australia. Adelaide: AWRI.

Bramley, R.G.V. and R.P. Hamilton. 2004. Understanding variability in winegrape production systems. 1. Within vineyard variation in yield over several vintages. *Australian Journal of Grape and Wine Research,* 10:32-45.

Bramley, R.G.V., D.M. Lanyon, and K. Panten. 2005. Whole-of-vineyard experimentation: An improved basis for knowledge generation and decision making. In *Proceedings of the Fifth European Conference on Precision Agriculture,* ed. J.V. Stafford. Wageningen, The Netherlands: Wageningen Academic Publishers, pp. 883-890.

Bramley, R.G.V. and A.P.B. Proffitt. 1999. Managing variability in viticultural production. *The Australian Grapegrower and Winemaker,* 427:11-16.

Bramley, R.G.V. and A.P.B. Proffitt. 2000. Managing variability in agricultural production: Opportunities for precision viticulture. Presented at the 5th International Symposium on Cool Climate Viticulture and Oenology (Workshop 12: Precision Management), January 16-20, 2000, Melbourne, Australia.

Bramley, R.G.V., A.P.B. Proffitt, R.J. Corner, and T.D. Evans. 2000. Variation in grape yield and soil depth in two contrasting Australian vineyards. In *Soil 2000: New Horizons for a New Century. Australian and New Zealand Second Joint Soils Conference.* Volume 2: *Oral Papers,* eds. J.A. Adams and A.K. Metherell. New Zealand: Lincoln University, New Zealand Society of Soil Science, pp. 29-30.

Bramley, R.G.V., Proffitt, A.P.B., Hinze, C.J., Pearse, B., and Hamilton, R.P. 2005. Generating benefits from precision viticulture through selective harvesting. In *Proceedings of the 5th European Conference on Precision Agriculture,* ed. J.V Stafford. Wageningen, The Netherlands: Wageningen Academic Publishers, pp. 891-898.

Bramley, R.G.V. and R.P. Quabba. 2001. Opportunities for improving the management of sugarcane production through the adoption of precision agriculture—An Australian perspective. *Proceedings of the 24th International Society of Sugar Cane Technologists Congress,* 1:38-45.

Carberry, P.S. 2001. Are science rigor and industry relevance both achievable in participatory action research? Presented at the 10th Australian Agronomy Conference, January 29-February 1, 2001, Hobart, Tasmania. Available at: www.regional.org.au/au/asa/2001/.

Carruthers, G. 1999. Environmental management systems and ISO 14000 in Australia—On and off-farm implementation and policy development. In *Adding Value through Environmental Marketing: Opportunities for Food Producers, Processors and Retailers.* Conference proceedings, December 6-7, 1999, Madison, WI. Madison, WI: Institute for Agriculture and Trade Policy. Available at: www.iatp.org/labels/envcommodities/print-part4.pdf.

Chisholm, A. 1992. Australian agriculture: A sustainability story. *Australian Journal of Agricultural Economics,* 36:1-30.

Connell, P. and S. Hooper. 2001. *Australian grains industry 2001.* Report of the Australian Agricultural and Grazing Industries Survey. Canberra: ABARE.

Cook, S.E., M.L. Adams, and R.J. Corner. 1998. On-farm experimentation to determine site-specific responses to variable inputs. In *Proceedings of the 4th International Conference on Precision Agriculture,* eds. P.C. Robert, R.H. Rust, and W.E. Larsen. Madison, WI: ASA/CSSA/SSSA, pp. 611-622.

Cook, S.E. and R.G.V. Bramley. 1998. Precision agriculture—Opportunities, benefits and pitfalls of site-specific crop management in Australia. *Australian Journal of Experimental Agriculture,* 38:753-763.

Cook, S.E. and R.G.V. Bramley. 2000. Coping with variability in agricultural production—Implications for soil testing and fertilizer management. *Communications in Soil Science and Plant Analysis,* 31:1531-1551.

Cook, S.E. and R.G.V. Bramley. 2001. Is agronomy being left behind by precision agriculture? Presented at the 10th Australian Agronomy Conference, January 29-February 1, 2001, Hobart, Tasmania. Available at: www.regional.org.au/au/asa/2001/.

Cook, S.E., R.G.V. Bramley, and M.L. Adams. 2000. What's obstructing the wider adoption of precision agriculture technology? In *Proceedings of the 5th International Conference on Precision Agriculture,* eds. P.C. Robert, R.H. Rust, and W.E. Larson. Madison, WI: ASA/CSSA/SSSA.

Cook, S.E., R.J. Corner, R.J. Groves, and G. Grealish. 1995. Application of airborne gamma radiometric data for soil mapping. *Australian Journal of Soil Research,* 34:183-194.

Cook, S.E., R.J. Corner, and G. Riethmuller. 1996. Precision agriculture and risk: An Australian example. In *Proceedings of the 3rd International Conference on Precision Agriculture*, eds. P.C. Robert, R.H. Rust, and W.E. Larsen. Madison, WI: ASA/CSSA/SSSA, pp. 1123-1132.

Cooke, P. 1999. Encouraging investment in new agricultural industries. In *Proceedings of the National Agricultural and Resources Outlook Conference,* Vol. 2: *Agriculture.* March 17-18, 1999, Canberra, Australia. Canberra: ABARE, pp. 183-189.

Corner, R.J. 2000. Achiever: A GIS-based achievable yield and fertilizer recommendation system. Presented at the Western Australia Department of Agriculture Crop Updates for Agribusiness Conference, February 16-17, 2000, Perth, Australia. Available at: www.agric.wa.gov.au/cropupdates/2000/.

Corner, R.J., S.E. Cook, G.A. Wheaton, and P.A. Caccetta. 1998. The effect of scale on the utility of remotely sensed grain yield estimates. In *Proceedings of the 9th Australasian Remote Sensing Conference.* CD-ROM. Sydney: The Remote Sensing and Photogrammetry Commission of the Spatial Sciences Institute.

Corner, R.J., and G.A. Moore. 1999. *The Expector soil attribute mapping method.* Canberra: CSIRO Land and Water.

Cox, G., H. Harris, and D. Cox. 1998. Application of precision agriculture to sugar cane. In *Proceedings of 4th International Conference on Precision Agriculture,* eds. P.C. Robert, R.H. Rust, and W.E. Larsen. Madison, WI: ASA/CSSA/SSSA, pp. 753-766.

Cupitt, J. and B.M. Whelan. 2001. Determining potential within-field crop management zones. In *Proceedings of the 3rd European Conference on Precision*

*Agriculture,* June 18-20, 2001, Montpellier, France, eds. G. Grenier and S. Blackmore. Montpellier, France: Agro-Montpellier, pp. 7-12.

Felton, F.R. and P.G. Nash. 1998. Role of reflectance techniques in precision weed management. In *Precision Weed Management in Crops and Pastures,* eds. R.W. Medd and J.E. Pratley. Waite, South Australia: CRC for Weed Management Systems, pp. 62-70.

French, R. and G. Schultz. 1984. Water use efficiency of wheat in a Mediterranean-type environment. *Australian Journal of Agricultural Research,* 35:743-64.

Hall, A., D.W. Lamb, B. Holzapfel, and J. Louis. 2002. Optical remote sensing applications in viticulture—a review. *Australian Journal of Grape and Wine Research* 8:36-47.

Hansen, J.W. and J.W. Jones. 2000. Scaling-up crop model applications for climate variability applications. *Agricultural Systems,* 65:43-72.

Harris, H.D. and G.J. Cox. 1997. Yield mapping—Current capabilities in yield mapping for sugar cane, are these limiting to the introduction of PA to the sugar industry? In *Precision Agriculture—What can it Offer the Australian Sugar Industry?* Proceedings of a workshop, eds. R.G.V. Bramley, S.E. Cook, and G.G. McMahon). Townsville, Australia. Canberra: CSIRO Land and Water, pp. 35-41.

Holford, I.C.R., A.D. Doyle, and C.C. Leckie. 1992. Nitrogen response characteristics of wheat protein in relation to yield responses and their interactions with phosphorus. *Australian Journal of Agricultural Research,* 43:969-86.

IPCC 2001. Intergovernmental Panel on Climate Change. In *Third Assessment Report: Climate Change 2001.* Cambridge, UK: Cambridge University Press. Available at: www.ipcc.ch.

Jensen, T., R. Kelly, W. Strong, and D. Butler. 2001a. Making better protein maps. In *Proceedings of the Geospatial Information and Agriculture,* July 17-19, 2001, Sydney, Australia. Sydney: NSW Agriculture, pp. 856-869.

Jensen, T., L. Zeller, and R. Kelly. 2001b. Mechanizing on-the-run grain sampling. Presented at the 10th Australian Agronomy Conference, January 29-February 1, 2001, Hobart, Tasmania. Available at: www.regional.org.au/au/asa/2001/.

Johnston, R.M. 1991. *Report of a workshop on the use of remote sensing in agriculture.* Bureau of Rural Resources Report No. R/3/91. Canberra: Australian Government Public Service.

Kaydos, W. 1991. *Measuring, managing and maximizing performance.* Cambridge, MA: Productivity Press.

Kelly, R., W. Strong, T. Jensen, and D. Butler. 2001a. Strategies to interpret the yield map: Defining yield-limiting factors. Presented at the 10th Australian Agronomy Conference, January 29-February 1, 2001, Hobart, Tasmania. Available at: www.regional.org.au/au/asa/2001/.

Kelly, R., W. Strong, T. Jensen, D. Butler, S. Norng, S.E. Cook, and B. Town. 2001b. Integrating yield and protein spatial information: A framework for analysis and interpretation. In *Proceedings of the Geospatial Information and Agriculture,* July 17-19, 2001, Sydney, Australia. Sydney: NSW Agriculture, pp. 528-533.

Kingwell, R.S. 1994. Risk attitude and dryland farm management. *Agricultural Systems* 45:191-202.

Knightley, P. 2001. *Australia: A biography of a nation.* Sydney: Random House.

Kolb, D.A. 1984. *Experiential learning: Experience as the source of learning and development.* Upper Saddle River, NJ: Prentice Hall.

Lamb, D.W. 2000. The use of qualitative airborne multispectral imaging for managing agricultural crops—A case study in south eastern Australia. *Australian Journal of Experimental Agriculture,* 40:725-738.

Lamb, D.W. and R.G.V. Bramley. 2001. Managing and monitoring spatial variability in vineyard variability. *Natural Resource Management,* 4:25-30.

Lamb, D.W., A. Mitchell, and G. Hyde. 2005. Vineyard trellising with steel posts distorts data from EM soil surveys. *Australian Journal of Grape and Wine Research,* 11:24-32.

Lamb, D.W., M.M. Weedon, and R.G.V. Bramley. 2004. Using remote sensing to map grape phenolics and colour in a cabernet sauvignon vineyard: The impact of image resolution and vine phenology. *Australian Journal of Grape and Wine Research,* 10:46-54.

Lawrence, D.N., S.T. Cawley, and P.T. Hayman. 2000. Developing answers and learning in extension for dryland nitrogen management. *Australian Journal of Experimental Agriculture,* 40:527-539.

Martin, W. and D. Mitra. 1999. *Productivity growth and convergence in agriculture and manufacturing.* Washington, DC: World Bank. Available at: wbln0018 .worldbank.org/research/workpapers.nsf/.

McBratney, A.B. and B.M. Whelan. 1995. *The potential for site-specific management of cotton farming systems.* Narrabri, Australia: CRC for Sustainable Cotton Production.

McBratney, A.B. and B.M. Whelan. 1999. The 'null hypothesis' of precision agriculture. In *Precision Farming 99: Proceedings of the 2nd European Conference on Precision Agriculture,* ed. J.V. Stafford. Sheffield, U.K.: Sheffield Scientific Press. pp. 947-957.

McBratney, A.B., B.M. Whelan, J.A. Taylor, and M.J. Pringle. 2000. A management opportunity index for precision agriculture. In *Proceedings of the 5th International Conference on Precision Agriculture and Resource Management,* eds. P.C. Robert, R.H. Rust, and W.E. Larsen. Madison, WI: ASA/CSSA/SSSA, CD-ROM.

McCown, R.L. 2001. Farming systems research and farming practice. Presented at the 10th Australian Agronomy Conference, January 29-February 1, 2001, Hobart, Tasmania. Available at: www.regional.org.au/au/asa/2001/.

Minasny, B., A.B. McBratney, and B.M. Whelan. 1999. VESPER Version 1.0. Sydney: Australian Centre for Precision Agriculture, University of Sydney. Available at: www.usyd.edu.au/su/agricacpa.

Moore, G.A. and J.P. Tyndale-Biscoe, 1999. Estimation of the importance of spatially variable nitrogen application and soil moisture holding capacity to wheat production. *Precision Agriculture,* 1:27-38.

Moran, M.S, Y. Inoue, and E.M. Barnes. 1997. Opportunities and limitations for image-based remote sensing in precision crop management. *Remote Sensing Environment,* 61:319-46.

OECD 2000. *Agricultural policies in OECD countries monitoring and evaluation 2000.* Paris: Organization for Economic Co-operation and Development. Available at: www.oecd.org/.

Passioura, J.B. 1996. Simulation models—Science, snake oil, education or engineering? *Agronomy Journal,* 88:690-694.

Pringle, M.J., A.B. McBratney, B.M. Whelan, and J.A. Taylor. 2003. A preliminary approach to assessing the opportunity for site-specific management in a field, using yield monitor data. *Agricultural Systems,* 77:273-292.

Rew, L.J., B. Whelan, and A.B. McBratney. 2001. Does kriging predict weed distributions accurately enough for site-specific weed control? *Weed Science,* 41:245-263.

Robertson, M.J., P.S. Carberry, and M. Lucy. 2000. Evaluation of a new cropping option using a participatory approach with on-farm monitoring and simulation: A case study of spring-sown mungbeans. *Australian Journal of Agricultural Research,* 51:1-12.

Stewart, C.M. and A.B. McBratney. 2001. Using bare soil imagery to determine management zones for the variable-rate applications of inputs for cotton. In *Proceedings of the 3rd European Conference on Precision Agriculture,* eds. G. Grenier, and S. Blackmore. Montpellier, France: Agro-Montpellier, pp. 319-324.

Stewart, C.M., A.B. McBratney, and J.H. Skerritt. 2002. Site-specific durum wheat quality and its relationship to soil properties in a single field in northern New South Wales. *Precision Agriculture,* 3:155-168.

Taguchi, G. and M.S. Phadke. 1984. Quality engineering through design optimization. In *Global Telecommunications Conference., Globecom '84.* Atlanta, GA: IEEE, pp. 1106-1113.

Viscarra Rossel, R.A. and A.B. McBratney. 1998. Soil chemical analytical accuracy and costs: Implications from precision agriculture. *Australian Journal of Experimental Agriculture,* 38:765-775.

Viscarra Rossel, R.A. and A.B. McBratney. 1999. Calibration of a lime requirement buffer for site-specific lime applications in south-eastern Australia. In *Precision Agriculture '99,* Vol. I, ed. J.V. Stafford. Sheffield, U.K: Sheffield Scientific Press, pp. 429-440.

Wallace, J.F., N.A. Campbell, G.A. Wheaton, and D.J. McFarlane. 1993. Spectral discrimination and mapping of waterlogged cereal crops in Western Australia. *International Journal of Remote Sensing,* 14:2731-2743.

Whelan, B.M. 2001. Moving to PA management in the grains industry. In *Proceedings of the Geospatial Information and Agriculture,* July 17-19, 2001, Sydney, Australia. Sydney: NSW Agriculture, pp. 406-416.

Williams, J., G.R. Walker, and T.J. Hatton. 2002. Dryland salinization: A challenge for land and water management in the Australian landscape. In *Agriculture, Hydrology and Water Quality,* eds. P.M. Haygarth and S.C. Jarvis. Wallingford, UK: CAB International, pp. 457-475.

Wong, M.T.F., R.J. Corner, and S.E. Cook. 2001. A decision support system for mapping the site-specific potassium requirement for wheat in the field. *Australian Journal of Experimental Agriculture,* 41:655-661.

# Chapter 20

# Europe

Simon Blackmore
Hans W. Griepentrog
Søren M. Pedersen
Spyros Fountas

## *INTRODUCTION*

Precision farming (PF) can be defined as the management of spatial and temporal variability to improve economic returns and reduce environmental impact. This can be achieved through using appropriate technologies within a coherent management structure. The technology gives the tools and the management decides how the tools should be used. PF uses these new technologies to reduce the unit area of treatment from, in some cases, the farm down to field or even subfield level. At the farm level management is relatively easy, as it is a process of managing by averages. Inputs are calculated to a certain level that can be applied across the whole farm. At the field level local conditions can be taken into account such as soil type and previous crop. These added data sources increase the decision-making complexity, but when used correctly they improve the efficiency. At the subfield level the complexity rises further when talking into account changes in soil type, shading from trees, areas of compaction, etc. At this level, the complexity can become so great that data overload occurs and the manager cannot use all of the available data. PF technology now has the ability to produce data about soils and crops at submeter levels across the whole field, but the capability to use this data is very limited until suitable information systems are developed.

Although this chapter does not set out to describe all PF activities in Europe, it does attempt to cover the significant trends in UK, Denmark, Germany, and Greece.

*Handbook of Precision Agriculture*

doi:10.1300/5627_20

## THE CURRENT SITUATION IN EUROPEAN FARMING

A general consistency exists in the conventional methods of arable farming in northern Europe, especially regarding the production of major crops such as cereals, oil seeds, pulses, and root crops. Most farmers tend to follow recommended practices regarding field operations and input levels, having to make relatively minor modifications to suit local conditions, especially regarding soils and climate. The conventional production system is a relatively high input: high output system whereby mechanized field operations, new varieties, and chemical controls are used in combinations which, subject to the uncertainty of weather conditions, try to achieve on average a reasonably consistent yield performance.

### *Conventional Practice*

Conventional farming practices have evolved from a combination of experience and scientific knowledge. They are modified according to local circumstances such as soils and climate, but for the most part they represent prescriptions for achieving the financial sustainability of the arable farm business. High yields and output are perceived to be the key to success. This is best achieved by relatively high variable inputs such as chemicals, which give relatively average returns per unit applied (even though in many years or in many parts of the farm marginal returns may be low or negligible). In this respect, an element of the expenditure in the conventional package is an insurance premium, which accommodates the lack of knowledge about the future or the inherent variability of the production system. For instance, in many years the capacity of the machinery inventory or the level of pesticide application may be greater than necessary. In any one year the variation in crop condition or soil type across the farm may not warrant uniform treatment. In an uncertain business environment, however, overcapacity or blanket prescriptions are often preferred to taking risks, even though over a period of years lower cost, risky strategies may give better average financial returns. Furthermore, the conventional package reduces the demands on farm management expertise. Risky systems are much more difficult to manage. Conventional methods economize on management skill.

Conventional crop management practices have evolved as a product of many years of personal experience by growers observing the consequences of management decisions and by the adoption over time of the conclusions of scientific research as interpreted by the individual grower, usually via some intermediate advice or extension agency. This formula has been successful over several decades in that it has generated large increases in crop performance, although not always in financial performance. Indeed, the current economic situation deems increases in crop performance essential to avoid decline in financial performance.

For the individual grower who is growing wheat as a commercial enterprise, the single factor in his or her mind above all others that affect profitability is yield. Historically the highest output crops have been the most profitable and growers, as astute business people, have geared their inputs so as to generate high yields. They have not been slow to appreciate the favorable ratios of economic response to the major inputs, with that of nitrogen fertilizer being around 5:1 (money output : money input), fungicides 2.5:1, and herbicides 2:1. Despite, for example, 90 percent of the yield of a wheat crop coming from 75 percent of the standard recommended application rate, a marginal economic response of more than 2:1 still occurs from using the additional 25 percent. Clearly an irresistible attraction exists for growers to be using this extra nitrogen, especially if, as usual, the overall margins are very tight.

It is because of financial pressure then that growers are driven toward high output practices. Because they have been acutely aware of the financial penalties of not achieving high outputs, and because the economic response to major inputs is so good, the tendency has been to err on the side of input excess to minimize the risk of failure to achieve that high output.

This gives rise to the notion of high-input:high-output management systems, which is actually an erroneous supposition frequently made on behalf of growers. Their true strategy is merely to achieve high outputs. In doing this their desire is to achieve it by using inputs in as financially efficient ways as possible. The problem is that, at present, most growers are not doing the latter very well, but they are willing to accept the cost risk of input excess or wastage as being preferable to the risk of output shortfalls.

### *Organic Farming*

Many people believe that organic farming is turning the clock backward to achieve the stated goals, but PF can help to turn the clock forward by achieving the same goals through the use of new appropriate clean technologies. Therefore the use of high technology in PF, achieved with the ideals of organic agriculture, is compatible and complimentary.

One of the main purposes of organic farming is that the farm should use organic fertilizers and eliminate sprays. Fertilizer is still applied in the form of farmyard manure and slurries, and herbicides are replaced by mechanical weeding. Both of these derivations can be better understood and controlled with PF technologies.

### *Economies of Scale*

One noticeable trend in European agriculture has been the increase in size of farms, fields, and equipment. As smaller farms come on the market they are often bought by neighboring farmers and incorporated into a larger enterprise. This not only increases the production capability but also allows investment costs to be spread over a wider area. Since the 1960s, a trend, especially in the United Kingdom, has been to remove fences and hedgerows to amalgamate smaller fields into larger ones.

Larger fields can accommodate larger tractors and equipment that have higher work rates and use less manpower. This increase in size is utilizing the principle of economies of scale in which larger production units and smaller investment in manpower have lead to a higher economic efficiency, especially in a time of high labor costs. One farmer in Denmark, who uses very large equipment, has reduced the time spent by his operator by 3.5 hours per ha in the production of winter wheat. In recent discussions with some of the main European tractor manufacturers, this trend of increasing equipment size is likely to continue into the foreseeable future.

Another advantage of bigger, faster equipment is that the periods of time when various operations can be carried out are relatively small, such as autumn cultivation, when all the tillage and seeding must be done within a few weeks. The larger equipment with high work rates can accomplish these tasks in the short windows of opportunity. Another example would be spraying when the air is still in the mornings and evenings.

The reasons for developing and using such large machines are to take advantage of the economies of scale. The unit cost of production falls when more units are produced (or more area worked) since the costs can be spread over more units. This is nothing new; industrial processes have been using these ideas since the industrial revolution. When taken to the extreme, fields can become very large with repetitive monocultures and high levels of chemical control, which can be detrimental to the biodiversity in the natural environment.

### *Environmental Problems*

Pollution and increased use of nonrenewable fuels are two major environmental effects of current farming practices. Modern agriculture is heavily dependent on nonrenewable energy sources such as farm machinery, fertilizers, and pesticides. N-fertilizer represents by far the largest single use of energy in modern arable farming and is a major source of pollutants such as nitrous oxide ($N_2O$) to atmosphere and nitrates to water. The use of toxic substances in the form of pesticides may pose a risk to human health because of residues in food, risk to people engaged in farming, and indirect exposure through environmental pollution. Fertilizers and pesticides are, however, major determinants of crop yields. High intensity agriculture usually has adverse effects on the farmland ecosystem. Examples include the buildup of pesticide resistance among target organisms, loss of predators that may be exerting a stabilizing effect on pest numbers, loss of desirable visible species, and possible damage to soil structure through physical and biological changes. Pollution of watercourses and underground aquifers is another serious issue, which represents a waste of applied inputs and causes problems for other organizations, such as water suppliers.

### *Agricultural Sustainability*

The concept of sustainable agriculture involves producing quality food and nonfood products now and into the future in an environmentally benign, socially acceptable, and economically efficient way. The improvement of social welfare and the preservation and enhancement of environmental quality are central to sustainability.

Public attitude has changed toward farmers in recent years. Farmers were seen as the providers of food for the nation but, as long as food supplies are adequate, they are seen more as custodians of the countryside. As the populace has become more aware of the rural environment it is demanding that intensive agriculture should have a less damaging effect on the environment. Any changes in agricultural practice that are likely to benefit the environment are likely to be encouraged rather than discouraged by the public.

## *PRECISION FARMING RESEARCH IN EUROPE*

A number of PF research programs have been instigated on a national basis throughout Europe. Although this chapter is not meant to be an exhaustive study, it should reflect a representative part of the work done in this area.

### *United Kingdom*

In April 1996 a 63-month project on precision farming (Project No. 1743) was initiated at Silsoe Research Institute of the University of Cranfield with funding support from the Home Grown Cereals Authority (HGCA) of the United Kingdom, Massey-Ferguson, and Hydro Agri (UK). At the end of this project, HGCA Report No. 264 was prepared (Godwin et al.,

2001). A series of papers was also published during and after completion of the report (Blackmore, 2000; Blackmore and Moore, 1999; Earl et al., 2003; Godwin et al., 2003; Godwin and Miller, 2003; Taylor et al., 2003; Welsh et al., 2003a,b; Wood et al., 2003a,b).

As per the final report, PF is the term given to a method of crop management by which areas of crop within a field may be managed with different levels of input. The benefits of so doing are threefold:

1. The economic margin from crop production may be increased by improvements in yield or a reduction in inputs.
2. The risk to environmental pollution from agrochemicals applied at greater levels than those required by the crop can be reduced.
3. Greater assurance from precise targeting and recording of field applications to improve traceability.

It is an excellent example of where both economic and environmental considerations are working together. This five-year study, principally involving five fields in southern and eastern England, covered a total of thirteen soil types that represented approximately 30 percent of the soils producing arable crops in England and Wales. The overall aim of the project was to determine guidelines to maximize profitability and minimize environmental impact of cereal production using PF. The objectives were:

1. To develop a methodology for identifying the causes of within-field variation in crop performance.
2. To develop practical guidelines required to implement PF technology to achieve best management practice.
3. To explore possibilities of using remote sensing methods to enable decisions to be made in "real time" during the growth of the crop.
4. To determine potential economic benefits of using PFtechnology for cereal production.
5. To collaborate with a range of farmers with interests in PF to ensure that research findings are appropriate for adoption.

The study concentrated on the interaction between soil/water variability and nitrogen applications. The harvest years 1995-97 (which included a harvest before the formal start of the program) concentrated on identifying the in-field variability and the development of the "real time" sensing techniques. Studies in harvest years 1998-2000 compared spatially controlled inputs with uniform agronomic practice. A number of techniques were used to decide upon the variable application strategy.

These included information on:

1. Yield variability from historic yield maps,
2. Variability in shoot density in the spring, and
3. Variability in the subsequent development of the canopy (green area index); the latter two enabling the development of the concept of "real-time" agronomic management.

The major outcomes of the project were as follows:

1. Yield maps are indispensable for targeting areas for investigation and treatment by PF practices and subsequent monitoring of results. They provide a valuable basis for estimating replenishment levels of P and K fertilizers, however, they do not provide a useful basis for determining a variable nitrogen application strategy to optimize management in a particular season.
2. The possible extent and potential causes of yield variability can be determined using low capital cost yield mapping systems together with electromagnetic induction techniques to assess variation in soil factors such as texture and water holding capacity. An objective methodology has been developed to use these techniques to determine within-field management zones. Both individually and together these systems provide a means for assessing the degree of variability within a field and provide a basis for targeting soil and crop sampling points, which is the only cost-effective method for commercial use.
3. The spatial variation in canopy development within a field can be estimated using an ADP technique developed by Cranfield University for this project for "real-time" agronomic management. This technique can be extended from field scale to farm scale for crops of similar varieties and planting dates. The processing of data from cameras mounted in light aircraft is sufficiently fast to enable application rate plans to be produced within a few hours of the aircraft landing. The technique can be used as a basis for determining the most appropriate application rate for nitrogen, and as a guide for herbicide and plant growth regulator application. It is feasible to adapt the system for use with tractor-based systems.
4. The application of nitrogen in a spatially variable manner can improve the efficiency of cereal production through managing variations in the crop canopy. Depending upon field and year, between 12 percent and 52 percent of the area of fields under investigation

responded positively to this approach. In 2000 seven out of eight treatment zones gave positive economic returns to spatially variable nitrogen with an average benefit of €36 (£22) $ha^{-1}$.

5. Simple nitrogen balance calculations have shown that in addition to a modest increase in yield, the spatially variable application of nitrogen can have an overall effect on reducing the nitrogen surplus by approximately one-third.
6. Common problems, such as waterlogging and fertilizer application errors, can result in significant crop yield penalties. PF can enable these problems to be identified, lost revenue to be calculated and resultant impact on cost/benefit to be determined. This provides a basis from which informed management decisions can be taken. It is critical that these problems are corrected prior to the spatial application of fertilizers and other inputs.
7. At current prices, benefits from spatially variable application of nitrogen outweigh costs of the investment in PF systems for cereal farms greater than 75 ha if basic systems costing €7380 (£4,500) are purchased, and greater than 200-300 ha for more sophisticated systems costing between €18860 (£11,500) and €26240 (£16,000).
8. Integrating the economic costs with the proportion of the farmed area that has benefit potential enables the break-even yield increase to be estimated. Typically a farmer with 250 ha of cereals where 20 percent of the farmed area could respond positively to spatially variable nitrogen would need to achieve a yield increase of 1.1 t $ha^{-1}$ on that 20 percent to break even.
9. The net effect of combining the benefits of spatially variable application of nitrogen (€36 or £22 $ha^{-1}$) with the benefits from both the spatial application of herbicides (up to €33 or £20 ha) and fungicides (up to €33 or £20 $ha^{-1}$), found from other studies, should provide valuable returns from the adoption of PF concepts. However, this should not be considered as a simple sum of maximum levels quoted.
10. These economic advantages linked to the environmental benefits should improve the longer-term sustainability of cereal production.

*Conclusions from the HGCA PF Report*

When this research project was proposed in 1996 one of the key driving forces was the concern that the technologies associated with PF were advancing faster than the understanding of the causes of yield variation within fields and the potential benefits of overcoming that variation.

At that time the techniques were available to accurately identify specific areas within a field and to precisely position equipment in specific areas of a field. However the only agronomic tools that were available were yield maps. Although they were very useful, they demonstrated "effect" rather the "cause" of variability. If the arable sector was going to capitalize on the potential benefits of the precision technology then it was vital that an understanding of the causes and extent of within-crop variability were obtained.

This project set out to explore the causes and extent of variability within fields, to develop techniques to measure variability during the course of the growing season and to develop methods of informing the arable grower of the potential benefits of PF. Two phases of the project were developed to explore the two primary issues:

- What is the extent of the variability that exists within crops and can any causes for this variability be identified? This phase lasted for two years and focused in very great detail on four fields that represented major soil types in the UK.
- Can within-field variability be monitored during the course of the growing season, rather than historically by the use of yield maps, to allow real-time agronomy to be implemented to optimize economic crop production? This phase lasted three years and monitored a total of 6 fields.

The main conclusions from the project are outlined in the next section and they have been grouped into the primary topic areas. One of the key objectives during the course of the five-year project was to release information and conclusions to enable the farming community to comment upon and benefit from the research findings of the project. This has led to the situation where some of the research findings have already found their way into regular usage and their initial origins in this project must not be overlooked.

### *The Importance of Variation in Soil Structure, Texture, and Moisture Holding Capacity*

- Long-term causes of yield variation can be attributed to soil and its associated water holding capacity.
- Historic yield maps, adapted to illustrate spatial trends, can be used to target specific areas of fields for investigation.
- Commercially viable soil survey techniques can provide valuable information on soil variability across fields and its potential impact upon yield.

- Electromagnetic induction techniques can further refine the delineation of soil types and water holding capacity variation within a field.
- Unless nutrient deficiencies are severe the current techniques of soil and tissue analysis did not provide useful information to assist in interpreting the causes of yield variation within a field.
- The combination of yield maps and accurate soil mapping further improves the ability to target areas for intensive soil and crop sampling.

*The Role of Yield Maps*

- Yield maps are valuable as a means of historically describing the yield variability existing in a field in any one season.
- Yield maps can identify specific areas within fields, particularly low yielding, that require attention.
- Yield maps, together with soil data, can be used to zone fields into potential management zones.
- Significant potential and actual errors in yield mapping systems have been identified and corrected using expert filter systems.
- By developing a spatial trend map it has been possible to identify variation within a field over a number of seasons. Variability within a field has been shown to be less predictable when using several seasons of yield mapping. Fields should therefore be managed according to the current year's conditions rather than by the use of historic yield maps.
- Yield maps can accurately determine the cost of localized damage such as that arising from pests (rabbits), poor drainage, and compaction or pH problems. They can also be used to measure the benefits of remedial actions.
- Yield maps verify the outcome of any spatially applied PF inputs to enable cost/benefit analysis to be undertaken.
- They can be used to determine the replenishment levels required of nutrients such as P and K by accurately measuring off-take levels.

*The Extent and Measurement of Within Field Crop Variation Using Remote Sensing Techniques*

- The need to understand and control within field variability has been highlighted by the discovery of variation within one field of yields ranging from 6 t·ha$^{-1}$ to over 14 t·ha$^{-1}$, although a more typical range is 5 t·ha$^{-1}$ t·to 10 t·ha$^{-1}$.

- An aerial digital photography (ADP) technique has been developed to accurately measure within field crop variability during the course of the growing season.
- The Cranfield University ADP technique, with the assistance of ground calibration, can determine within-field variations in both shoot densities and GAI (canopy development).
- Herbicide and plant growth regulator applications can be varied as a result of the information produced through the use of the techniques.
- The technique has been extended from field scale to farm scale for crops of similar varieties and planting dates.
- The technique could be adapted for tractor-based systems as well as the currently operated light aircraft system.

*The Management of Within-Crop Variation*

- Studies on both wheat and barley indicated that historic yield maps did not provide a valuable basis for determining a variable nitrogen application strategy to optimize yield.
- The extent of crop variability revealed by the Cranfield University ADP technique has allowed real-time agronomy input to be applied during the growing season.
- During the course of the project between 12% and 52% of the area of fields responded positively to differential nitrogen applications identified by the Cranfield University ADP technique.
- Variable nitrogen management has been demonstrated to be cost effective in the majority of circumstances where it was employed in this project.
- Variable nitrogen management presents the additional possibility of reducing total nitrogen usage in some circumstances.

*Economic Benefits*

- Interpreting the information from the Cranfield University ADP technique and using real-time agronomy (variable N management) produced an overall benefit of €36 (£22) $ha^{-1}$ compared to a standard N management policy.
- Using historic yield maps as the basis for N application in a subsequent crop did not improve financial returns over those generated from standard N management.
- Current estimates of capital and associated costs for equipping a farm to be capable of adopting precision farming techniques range from €8

(£5) $ha^{-1}$ to €29 (£18) $ha^{-1}$ depending upon system chosen and farm size.
- Data collection and interpretation to enable real time agronomy to be conducted could be available from €11 (£7) $ha^{-1}$ depending upon land area surveyed by aircraft or tractor mounted radiometry.
- The potential benefits available from the correction of localized agronomy/management problems should not be overlooked. Benefits of up to £195$ha^{-1}$ from correcting water logging and up to €104 (£65) $ha^{-1}$ from correcting uneven nitrogen applications have been identified.
- The total potential benefits from precision farming nitrogen (€36 or £22$ha^{-1}$), herbicide (€33 or £20 $ha^{-1}$), fungicides (€33 or £20 $ha^{-1}$) are financially very attractive.
- A typical farm, with 250 ha of cereal where 20 percent of the area could respond to variable spatial nitrogen application would only need a 1.1/t $ha^{-1}$ yield increase in that 20 percent of land to create a breakeven for a precision farming system costing €18400 (£11,500).

*The Environmental Benefits of Precision Farming*

- Spatial applications of nitrogen, herbicides and pesticides are likely to result in lower overall applications in the majority of circumstances but in some cases higher levels of application will be justified.
- Whilst this project did not address the environmental implications of spatial applications it was clear that nitrogen rates were generally lower than with standard application procedures. One fully documented example produced a 34 percent net reduction in N added to the soil compared to standard application.

This project has clearly identified a range of opportunities for growers to adopt precision farming techniques. However it is very clear that before PF is adopted routine agronomic and management practices must be optimized otherwise they can seriously undermine any financial benefits from PF.

Entering PF is a stepwise procedure and part of the remit of this project was to develop guidelines for precision farming to enable the decision-making process to be undertaken by the grower. The potential benefits of PF, up to €36 (£22)per ha from variable N application has been demonstrated in this project with the possible additive effect of variable application of other inputs, suggest that many growers should be investigating the concept.

## *Germany*

Since January 1999, site-specific arable crop production has been studied throughout Germany in a Federal Ministry of Education and Research (BMBF)-funded research project called Preagro (www.preagro.de). This research and development project has promoted the development of innovative management and technology for agriculture. Based on PF principles, it has created new tools and methods on a subfield level to enable agricultural management to be more precise according to agronomic, economic, and ecological criteria.

The goal of this collaborative project was to design a management system for site-specific cultivation based on agronomic rules and algorithms. These can be realized and integrated in software packages that will help the farmer achieve a more optimal management of the practices at a subfield level.

These and other issues are to be clarified in the interdisciplinary Preagro research project with 22 subprojects involving 17 research institutes and industrial companies. Scientific research and practical experience are closely linked between all participants. This ensures that research results will be successfully introduced to agricultural practice. The project's challenge lies in identifying the geological, geographical, climatic, and agronomical interactions and in applying guidelines for the development of cultivation methods. The attention is focused on the concept that the individual steps such as tillage, seeding, fertilization, and plant protection should not be undertaken separately but rather interlinked.

From the beginning, four farms, three contractors, and a machine cooperative have been integrated in the project in order to achieve practical application. The farms were selected to represent different climatic and geographical conditions of German landscape and agriculture. These farms are in Schleswig-Holstein, Bavaria, Westphalia, and Brandenburg. The developed PF guidelines together with the newest available equipment are tested on these farms. The selected farms are already experienced in using GPS. In the second project phase, at least three additional farms will be joining the project in order to validate the management system.

One focus area of the project was to develop software modules to link soil and crop information to decision-making processes. These software modules were built on results from first-year research activities. Modules for liming, P and K fertilization, determination of yield potentials, weather, and cultivation depending expected yields, seeding, and other factors were developed tested on the project farms.

Machine developments were also carried out during the project. Variable soil cultivation was realized in cooperation with a commercial company. The system consists of a software tool and new tillage machine design (VoBhenrich et al., 2001). Field experiments under practice conditions show that site-specific tillage can successfully be realized. Less primary tillage intensity at selected and mapped parts of the field provide equal plant development and yield compared with the conventionally tilled field.

Determining the optimum nitrogen fertilizer dose rate is still a big challenge in site-specific management. The development of software modules including simulation models lead to a substitution of costly soil-sampling operations (Kersebaum et al., 2001). It was shown during the project that soil-sample analysis for determination of spatially distributed mineralized nitrogen is unreliable because the variability within small subfield areas was as big as within the whole field.

During the project a management strategy adapted to the south German landscape called "transborder farming" was developed (Rothmund et al. 2002). Virtual land consolidation by transborder farming gives farmers in small structured regions the option of taking the short-term and long-term advantages of larger field sizes without implementing difficult real land consolidation measures. The reduction of labor input required, side and headlands, and lower overlapping benefits the environment, the soil, and farmers' pocketbooks at the same time. By using automatic data recording, all inputs and yields on the cooperatively operated land can be accurately calculated for all the participating farmers.

The objective of Preagro subproject "economy," is to extensively work out and analyze the economic aspects of precision agriculture. In principle, three levels, are distinguished in the economic evaluation: the process level, the operational level and the multifarm level (cooperatives, machinery rings, and contractors). Regarding the economic effects, the focus has been on machinery management and crop management. In operation planning and controlling bigger enterprises such as machinery contractors and machine rings organized by farmers have invested in new technologies. The benefits are better job documentation and billing, better fleet management, and keeping the customers (farmers) more satisfied with free extra information about, e.g., yield magnitude and variability by handing out yield maps of a combined field.

The PF management system proven in practice will support the user in the wide range of tasks. In the future, the system is intended to manage not only site-specific data but also the documentation of the whole crop-production process. Such documentation will make it easier for farmers to have their environmental compatibility recognized by the authorities and to demonstrate the production quality to their customers. These benefits will also be useful for private contractors as providers of different services for farmers.

### *Denmark*

The Danish Ministry of Food, Agriculture and Fisheries established a research program in 1998 dealing with development of techniques for application of fertilizers and pesticides according to spatial variability. The program "Precision Agriculture" was launched in 1998, running for five years with a total budget of €2 million (14 million DKK). It involves activities at Risø National Laboratory, The Danish Research Institute of Food Economics and the Danish Institute of Agricultural Sciences.

Since the Danish application of nitrogen is based on a quota system (N-norm) that is below the economic optimal application level, all commercial decision support systems (e.g., Kemira Loris and Agrosat from DataLogisk) are based on application strategies that utilize the whole amount of nitrogen. Consequently, the optimization is focused on the utilization of a given amount of nitrogen across the field instead of optimizing a fixed amount of nitrogen according to the site-specific soil and crop conditions. National environmental legislation and norms are therefore vital for a spatial N-application strategy. In Sweden, for instance, farmers face a tax on fertilizers without norms, which implies that N costs about 0.9 €/kg (7 DKK/kg) compared to about 0.5 €/kg (4 DKK/kg) in Denmark.

The N-sensor trials showed in 1999 that better utilization of nitrogen gives an average extra yield of about 200 kg per ha in cereals (Halmø, 2001). Other trials conducted by DLG (Dansk Landbrugs Grovvareselskab, Denmark's largest feedstuff company) with site-specific application have shown little effect on yield improvements but minor improvements on protein content (average + 0.3 percent). These trials are primarily based on yield maps and soil samples. The same trials also show a significant variation in yields among homogeneous and nonhomogeneous soils.

### *Greece*

Recently a number of projects have started in various crops with interest in Greece. One of these is being carried out from Aristotle University of Thessaloniki and the Agricultural University of Athens with the assistance of the Ministry of Agriculture. The basic aim of the project is the development of a precision agriculture system and evaluate its possible use under Greek agricultural conditions. The main objectives are tracing variability using remote sensing, identification of variability using field data, monitoring of variability using remote sensing and field data, and as a final stage providing recommendations to end users (Karydas and Silleos, 2000; Karydas et al., 2004).

One current project aims to construct a soil-properties GIS to be used for the PF purposes of a tobacco crop, providing detailed soil thematic maps by soil sampling and interpolation (Karydas and Silleos, 2000; Karydas, 2003).

Another research project, a first for its kind in Greece, concerned cotton. Cotton is the crop that gives the highest economical returns to farmers in Greece and Spain due to high yield and considerably good prices. Therefore, this crop is of great importance in both countries, which are the only cotton producers in Europe. During cotton harvesting of 2001, a system of yield mapping was installed on the cotton picker. The instrumentation consisted of two yield sensors installed in the two tubes conveying seed cotton from the picking unit to the store (basket). In addition, a GPS sensor was installed on the harvester cabin's roof, a magnetic sensor on the harvester wheel to measure the machine's speed, and a sensor for the position of the picking drum. The latter defined the time when measurements were taken and interrupted data storage when picking was not taking place. Signals from the sensors are stored in a unit in the machine cabin. The yield data are calibrated at the end of picking by weighing the actual yield. Sampling time interval was 2 seconds. Yield mapping was applied in two fields of 1.7 and 4.3 ha. The results in Figure 20.1, show that even in small fields the spatial variability is still very high (Markinos and Gemtos, 2002).

The very high spatial resolution images of the satellite Ikonos enable every feature larger than 1 to 2 m to be viewed and therefore appear to be suitable for the large heterogeneity of the Mediterranean. In order to exploit these new tools for precision agriculture, a technique that combined Ikonos image analysis with observations in the field and soil-plant sampling for analysis was developed. Figure 20.2 shows, for example, areas with increased soil moisture in an olive tree plantation in Crete. A GPS, together with visual guidance, was used to locate the features of interest in the field (see Figure 20.2). Once the variability was identified it was then recorded, and a database for the crop within the parcel was constructed (Gitas et al., 2004; Karydas et al., 2005a,b).

## VARIABILITY AND MANAGEMENT

### Measuring Variability

The first stage in the PF process is to measure important factors that indicate or affect the efficiency of the growing crop. The two main approaches

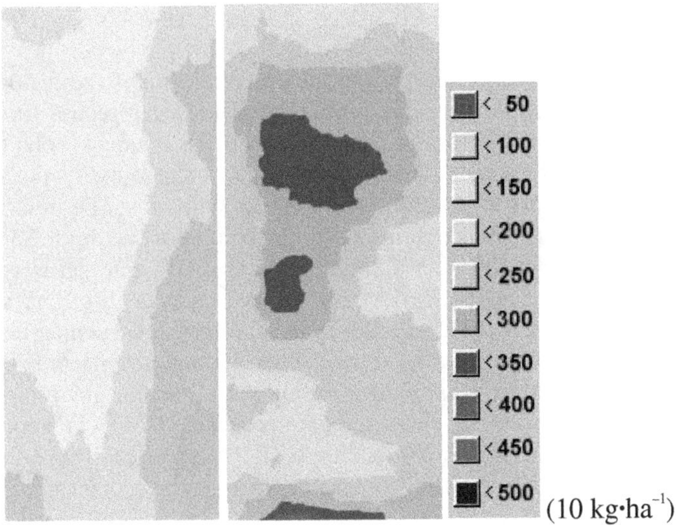


FIGURE 20.1. Yield maps of a small cotton field in Greece. (See corresponding PLATE 20.1 in the central gallery.)

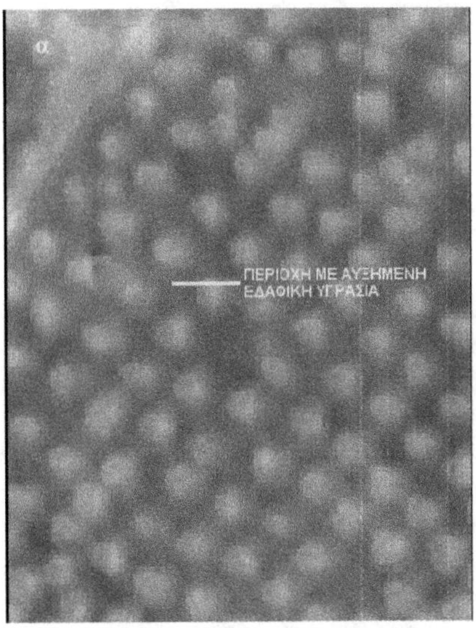


FIGURE 20.2. Area with increased soil moisture in olive-tree plantation in Crete, Greece.

are to create yield maps through instrumenting the harvesting system or assessing soil parameters by sampling. Both techniques give information about different parts of the cropping system. Yield maps are historic and cannot be used while the crop is growing, but they can record the actual yield during harvest. Soil sampling can be expensive, but many soil parameters such as texture and horizon depths do not change over time and so are good investments. Measuring soil nutrient status must be treated with care as repeatability, let alone accuracy, is difficult to achieve. Sampling strategies based on a simple grid tend to be expensive, and better-targeted sampling techniques are being developed (Thomas et al., 1999). Asset surveys can also be carried out to record physical features, such as field and crop boundaries, high trees that may cause shading, compaction in gateways, etc. Other high-density measurement techniques are rapidly becoming more important, such as remote sensing and ADP or noncontact sensing (e.g., electromagnetic induction). ADP can give real-time information of the crop canopy and allow management to be modified while the crop is growing.

PF is very information intensive. A large amount of base data are required to generate maps, and many techniques to gather this data are under development or have been developed during the past decade. The only real commercial and applied systems are yield mapping systems and soil conductivity mapping systems. Other techniques such as remote sensing and digital pictures from satellites and airplanes are available but not operational in a way that is suitable for the manager. Table 20.1 highlights some advantages and disadvantages of some different sensing systems that are commercial or nearly commercial on the European market.

### *Issues of Sampling Frequencies*

To understand the true amount of spatial and temporal variability, measurements must be taken that can give a representative sample. In spatial terms this means that the spatial sampling frequency must be close enough to be able to capture all of the desired features. Similarly, temporal sampling must occur enough times, otherwise events might be missed. Heightened sampling may often involve extra costs, so appropriate sampling frequencies must be chosen. The difficulty is determining the correct sampling frequency and the amount of variability that should be measured. This is a classical instrumentation problem. In engineering terms this is overcome by sampling at six times the highest frequency desired to be measured. In practical terms this does not make sense in the field, and many spatial parameters are measured on a 1-ha grid. If real-time or noncon-

TABLE 20.1. Advantages and disadvantages of various PF sensing systems.

| Data produced | Sampling frequency | Sensing system | What is it used for? | Advantages/ Disadvantages | Commercial? |
|---|---|---|---|---|---|
| Yield maps | 5m | Usually GPS fitted to harvester with yield monitor. | Map quantifying spatial variability of harvested crop. Gives an indication of the site-specific variation of the yield—and that something has to be done on certain spots | + Helps to guide attention to low (or high) yield spots.<br>+ Quantifies off take for replacement of P and K.<br>+ A useful synoptic measure of variability<br>–Difficult to interpret<br>–Expensive for small farms | Yes |
| Soil and crop maps | 100m | Samples taken from known locations on a grid, are sent to a laboratory for analysis | To analyze for: P, K, pH; Soil type: Cu, Mg, B, N-min. | + Gives answers to many different parameters and can be applied to control application of fertilizer and lime<br>–Difficult to control N based only on soil samples<br>–Expensive to get repeatable results | Yes |
| EMI maps | 5m | Noncontact electromagnetic induction device | Soil conductivity, which can be taken as a surrogate for soil moisture and the factors that affect soil moisture holding capacity. | + Rapid assessment method<br>+ A complement to conventional soil samples.<br>–Can be interpreted in many ways | Yes |
| Remote sensing—aerial digital photos | 2m | Photographing the fields from an airplane with multispectral cameras. | Images processed to give soil and vegetation indices.<br>Calibrated for plant density, tiller density, LAI, GAI. | + Aircraft can be flown when good weather and at specific growth stages.<br>–Expensive (cost of flight) | Yes |
| Remote sensing—satellite imagery | 20m | Same as ADP but from space with lower resolution | Images processed to give soil and vegetation indices.<br>Calibrated for plant density, tiller density, LAI, GAI. | + Used for rapid assessment over large areas<br>–Dependant on cloud cover<br>–Expensive over small areas<br>–Ikonos pictures available after 60 days. | Yes |

TABLE 20.1 *(continued)*

| Data produced | Sampling frequency | Sensing system | What is it used for? | Advantages/ Disadvantages | Commercial? |
|---|---|---|---|---|---|
| Nitrogen status of crop | 5m | Handheld radiometer or fitted on to tractor with fertilizer spreader behind | Real-time measurement of crop's nitrogen status based on amount of chlorophyll | +/– Proprietary automatic (closed) system<br>–Still to be properly validated | Yes |
| Weed mapping | 10m | Hand held GPS equipment and human effort | Manual detection and recording of weed patches | –Needs a lot of human effort | Yes |

LAI = leaf area index
GAI = green area index

tacting sensing can be used (such as the soil photo spectrometer the Hydro N-sensor system, or EMI [electromagnetic induction] scanning) then the sampling rates can be increased dramatically. A guide to practical soil sampling strategies is given in Oliver et al. (1997) and soil classification using EMI in James et al. (2000).

### *Managing Variability*

Three forms of arable variability have been identified: spatial, temporal, and predictive. Spatial variability is the variation seen across the field. It can be easily seen in any yield or soil map. Whether the spatial variability is significant is something the manager must decide, but often yield maps have a characteristic low yielding boundary around the edge of the field. Temporal variability can be seen when comparing yield maps from year to year, when again significance must be deduced, but trend maps can be developed to show underlying features. Predictive variability is where assumptions about the future are made but not realized. To be able to choose the appropriate levels of inputs, assumptions must be made as to the expected crop yield. If that yield is not reached the methods of prediction may need to be improved. New technologies allow new management strategies, but it is the management of the technology that has the greatest impact on the environmental well-being and economic viability.

The ultimate goal is a full understanding of the whole process and rational decision making following the management philosophy. Unfortunately this will never comopletely happen because too many unknowns exist in this very complex process, the weather being the most obvious. Nevertheless, IT (information technology) can still automate many of the repetitive tasks and help provide accurate information at the right time. Levels of automation are bound to rise, as is the use of computers and controllers. The trend is toward a better understanding and a smaller unit of treatment. Information sources will be on-farm as well as off-farm. Examples of on-farm sources are yield maps, soil maps, weather station data, quantities held in storage bins, weigh bridge, etc. Examples of off-farm sources are market prices, pest prediction models, expert advice, etc. Both of these sources could be integrated into an information system that allows the manager easy and timely access to the data precisely when the decision must be made.

Three levels of decision making can be taken on the farm: strategic, practice, and operational. Strategic decision making will occur only occasionally from year to year and will affect the whole purpose of the farm enterprise, and it is likely to be personal to an individual manager. Three approaches to strategic formulation have been identified:

- Yield protection with high physical inputs
- Reduced inputs with optimal return and low environmental concern
- Reduced inputs with high environmental concern

Many farmers will say that their main purpose is to make money, but that is too simplistic; sustainability and environmental issues are becoming more important. The practice level concerns the management options that apply to a particular sector. Operational elements describe the particular field operations.

*Management Zones and Treatment Maps*

Much of the research being carried out investigating PF has shown that the crop and soil homogenous areas are often much smaller than the existing fields and may even cross field boundaries. Significant variations in soil characteristics often have significant impact on the crop yield and a great amount of the yield variation found can be attributed to the variation in the available nitrogen and moisture capacity of the soil. These research findings indicate that many fields should be logically subdivided (not physically) into reasonably homogenous management areas, and each area should be treated according to its particular needs. Each management area may differ when considering different treatments, for example, a homogenous spray area will be different from a homogenous fertilizer treatment area. Equipment manufacturers have responded to this situation by producing more advanced controllers that can be connected to a differential global positioning system (DGPS).

The farm manager then must decide how to treat the field in terms of what areas should be covered, with what application rates, at what time, and under what conditions. All of the parameters are represented within the treatment map and downloaded to the controller. The driver guides the tractor through the appropriate areas and the controller adjusts the output to the predefined rate. This process combines the benefits from the economies of scale that the big equipment brings along with the improved control over the smaller management areas.

### *Treating Variability*

Spatially variable operations require equipment that can vary treatments according to their position within the field, defined by a treatment map or using sensors in real time. Equipment for spraying, cultivating, and seeding are all available in prototype form, although fertilizer application is currently available as equipment or as a service.

Treatment maps that show the segmented areas of the field and the treatment type or rate of application define all of the spatially variable operations. The field equipment controller can then implement the managerial decision based on the desired treatment map. A record should be kept by the controller of what actually happened in the field (as an actual treatment map), as this may be different from what was desired. This in itself can be a very useful piece of information. Figure 20.3 shows the variable cultivation equipment, developed at Silsoe Research Institute, that can change the amount of energy put into the soil according to the clod size (Scarlett et al., 1997).

## *TECHNOLOGY-LED OPPORTUNITIES*

PF is associated by many to be synonymous with high technologies such as GPS, yield mapping systems, GIS, variable rate controllers, and so on. In many ways PF has been technology led. Certainly the advent of what we now call precision farming was developed from discovering what could be done with GPS in the early 1990s. Usefulness in the agricultural sector was explored (and still is) by combining GPS positioning with relevant

FIGURE 20.3. Silsoe Research Institute's real-time, variable rate tillage system.

agricultural sensors and controllers, and the spatial variability of many factors were quantified.

Although a general need exists for better and more timely information from the agronomic management sector, most advances come from the technology sector, where a new device is developed or introduced from another sector. This is one reason why most PF projects are lead by agricultural engineers. Maps are produced and the agricultural relevance is then explored afterward. This may seem a strange sequence, but it seems that more opportunities exist in the technology sector than in agronomic management. It has developed to such a level that we have the capability to measure many significant climate, soil, and crop factors but currently have the inability to integrate them into a practical management system.

Examples of opportunities from new technology-led management requirements are yield mapping (GPS from U.S. military combined with existing yield monitors), soil conductivity (GPS combined with noncontact electromagnetic induction devices from geological surveying), and remote sensing (hyperspectral imaging from space, derived from military applications). Each of these technologies has been developed first and projects set up afterward to explore the relevance of the data.

One example of a new technology being developed from a previously defined agronomic need is the soil photo spectrometer developed by Shibusawa and colleagues (2000). Field trials looking at nitrogen sensitivity identified that the residual nitrogen left over in the soil from the previous year was a factor of vital importance if an efficient spatially variable nitrogen application map was to be produced. The soil photo spectrometer is a soil-engaging tine that can measure soil nitrates, moisture, organic matter, cation exchange capacity (CEC) and pH. (Further details of this device can be found in Chapter 3.)

### *Technology Adoption Levels*

In a report for the Scottish Natural Heritage organization, three technology adoption levels of PF were identified (Blackmore et al., 1994). *Technology level one* represents conventional practice with no IT. *Technology level two* has some IT investment and provides the farmer with an increased understanding of the enterprise but does not include the ability to vary application rates automatically. Farmers can however achieve patch application variation by manually influencing machinery settings. This technology level is seen as an interim to *technology level three,* which has fully supported variable application rate capability. The financial and environmental implications associated with each technology level and farming strategy were outlined. The important issue from this work was that although new technologies offer new opportunities, it is the management of the technologies and other resources that will have the greatest impact on the economic returns and environmental impact.

### *Yield Mapping*

Since the advent of differential global positioning systems into the agricultural sector, positional information and yield data from a combine harvester have been used to produce yield maps. These maps are often central to PF management. As the yield maps highly influence the decision-making process, it is important that the maps do actually represent the variation in the yield and not other systemic errors.

During harvest, recording the spot yield within a crop as it is harvested and the position of the combine harvester at that same time produce the data required for a yield map. The spot yield data is taken from a flow sensor, which is usually mounted in or around the clean grain elevator. Borgelt (1992) described various techniques of gaining spot yield data, ranging from volumetric measurement to mass flow to change in capacitance, each with their own advantages and disadvantages. The positional data comes from the GPS. These two data sets are combined together and stored as randomized triplets with the syntax: latitude, longitude, and yield. Some systems may include further data such as date, time, moisture, quality of GPS signal, etc. A detailed description of collecting, recording, processing, and displaying yield maps can be found in Blackmore (1998).

Many consider yield mapping to be the primary technology that kicked off PF in the early nineties. This may be so, but even ten years later we are still trying to understand what the causes of the spatial and temporal variability are. The yield maps quantify the variability, but the causes are often hard to understand.

### *Electromagnetic Induction*

An interesting lateral move of a technology from one application area to another is the EM38 from Geonics. (Figure 20.4) Originally designed for geophysical surveys, it has now been widely used in Europe as a means of rapidly assessing soil characteristics by measuring the soil conductivity through EMI. It is a noncontact sensor that gives readings in millisiemens per meter (conductivity) and is most sensitive to conditions at about 1.5 m deep, which is a reasonable approximation of the rooting zone for many crops. These devices were used to map salt concentrations in soil, but since many European soils do not have significant salt problems, conductivity is mostly affected by moisture and the factors that affect the soil moisture holding capacity. Since this is such a cheap, repeatable, noncontact way of assessing the soil, it would seem to have many potential uses.

Waine et al. (2000) described a methodology for using EMI to produce maps of available water content. EMI was used to target the location of neutron probe access tubes in soil textures representative of the range of soils

FIGURE 20.4. EM38 with wooden sledge.

found within a field. By examining moisture release curves for UK soils and plotting calibrated data on a texture-moisture graph, the available water content and soil moisture deficit can be calculated. This information is then used to calibrate EMI data and produce available water content maps. By sampling the conductivity and measuring moisture content at points of known texture, the spatial and temporal variation in soil moisture deficit can be calculated at key dates throughout the growing season. A practical technique for using EMI to estimate soil textural class, based on taking EMI conductivity measurements when the field is at field capacity, is also outlined.

### *Real-Time Sensing and Control*

Although nitrogen is the most sensitive nutrient for variable dose rate applications with the highest economic and environmental potential, other nutrient applications, such as P, K, and lime are better understood within PF research. The reason is that these nutrients are much more stable in the soil and can persist for years compared with nitrogen, which is very volatile and must be supplied every year. Because nitrogen changes the most (among all nutrients) both spatially and temporally, rapid data collection methods to determine variability in nitrogen demands of a plant population at acceptable prices and at the appropriate time are crucial for PF.

Real-time plant monitoring and control is seen as a good method to manage the crop's nitrogen demand. The reason for this is because soil analysis is costly and usually does not provide a sufficient resolution. In 1987 Hydro Agri in Germany started to develop real-time in-field methods to determine nitrogen fertilizer dose rates for fields. First, these methods determined the N fertilizer demand indirectly based on chlorophyll measurement by transmittance and reflectance, respectively, and they were calibrated in the field trials for cereals. The Hydro N-Sensor is now a commercialized system for tractor-mounted crop sensing and application of fertilizer in real time or on the go (Wollring et al., 1998).

As of 2005, nearly 300 Hydro N-sensor systems are in use worldwide, but most of them are used in Europe (Germany, Sweden, United Kingdom, Denmark, and the Czech Republic), and are mainly used for wheat and barley. Because of perceived environmental benefits, the Hydro N-Sensor systems have been subsidized in Europe. The N-sensor is mounted on the tractor and measures the multispectral response of the crop to sunlight. It is correlated with the crop needing more or less nitrogen, which is applied by the spreader behind the tractor. A rough estimate shows that about 20,000 ha in Sweden have been treated with the Hydro N-Sensor. In this respect the

Swedish approach to online sensing and canopy management differs from Danish practices. In Denmark, variable fertilizer application has primarily been based on historic data such as yield and soil maps.

### *Automatic Steering, Autonomous Vehicles*

To further improve the efficiency of developed agriculture, horticulture, and forestry found in northern Europe, a new concept is being developed that has identified that multiple small autonomous machines would be more efficient than traditional large tractors. In order to meet this hypothesis a small tractor with intelligent control is proposed. These vehicles will be able to work longer hours at a slower rate, giving the same or even greater overall output as conventional systems. Each vehicle would be capable of working 24 hours a day, year-round, in most weather conditions, and will have the intelligence embedded within it to behave sensibly in a seminatural environment such as horticulture, agriculture, parks, and forestry while carrying out a useful task. Moreover, it may have less environmental impact if it can replace the overapplication of chemicals and the high usage of energy, for example, diesel and fertilizer use, by control that is more intelligent. In addition, it will require smaller incremental investment and will have lower labor costs. Finally, it may have very low soil compaction that would lead to a more sustainable production system (Blackmore et al., 2002).

Agriculture, horticulture, and forestry has benefited in the past from a succession of technological developments that have brought greater productivity and economic efficiency to systems operated in many regions of the globe. Historically, the emphasis of these developments has been on the mechanization of field operations to increase work rates achievable by individual operators. Today, however, the general trend of increased efficiency through the use of larger, more powerful machines may be superseded by the adoption of newer information-based technologies that may ultimately enable reliable autonomous field operations to be viable (Earl et al., 2000). A review of possible platforms was carried out by (Ellenrieder, 1996) and a review of automatic steered tractors is given in Wilson (2000).

This scale-reduction process embodied in PF may lead to the possibility of individual plant care systems called *phytotechnology* (Shibusawa, 1996). Due to the increased data processing required to cover a complete field at the individual plant level, only certain operations are carried out with human intervention, but these processes lend themselves to different forms of automation, especially in high value crops. Two good examples of this process are mechanical weeding and field scouting.

Current known projects to develop autonomous "tractors" in Europe are

- GeoTec's GT 2000 automatic steering system
- Commercial tractors with automated steering at Renault, John Deere, and New Holland
- Autonomous horticultural toolbar at Silsoe Research Institute, United Kingdom
- Autonomous plant instrumentation system, DIAS (Danish Institute of Agricultural Sciences), AUC (Aalborg University Centre), KVL (Royal Veterinary and Agricultural University), Denmark
- Mechanical weeding platform, Technology University of Denmark (DTU), Denmark
- Mechanical weeding robot, Halmstad University, Sweden
- Automatic steered small tractor, Industrial Automation Institute (IAI), Spain

In Spain, (Garcia-Alegre et al., 2001) developed an automatic steered vehicle for agricultural tasks. The robot deals with the automation of a low-cost commercial lawnmower to perform unmanned precision agriculture tasks such as the spraying of pesticides. To this aim the vehicle has been fitted with a set of additional subsystems and sensors to perform remote motion control and to obtain relative and absolute location in the field. The subsystems are composed of pneumatic actuator systems, potentiometers, DGPS, and a dead-reckoning system. A fusion algorithm is proposed to take into consideration the lack of a DGPS signal, based upon the integration of DGPS and dead-reckoning systems. Both the robot remote control and the autonomous navigation have been tested in limited conditions.

In order to be economically attractive for the farmer, it is necessary to develop low-cost automatic sensing systems. Several research centers have focused on the development of autonomous detection systems and robotic weed control systems, but so far no commercial systems have been seen. One reason is the difficulties recognizing the various weed species during high speed. In northern Europe about 60 weed species exist, and about 5 to 10 species exist in a typical field with cereals (Christensen, 1995). The equipment is still relatively slow due to a slow image/digital analysis and need for more computer power. Moreover, accurate decision support systems still have to be refined. The timing of detecting weeds autonomously is also a serious obstacle. An automatically steered mechanical weeding platform has been developed at the Technology University of Denmark in 2001. The Danish Institute of Agricultural Sciences and Aalborg University in Denmark continued the development in 2002.

## ISSUES OF ADOPTION AND FARMER ATTITUDES

The development and dissemination of precision agriculture is driven and motivated by public and private interests. Although the promises of PF are attractive, the performance in practice remains largely unproven both in terms of economic returns and environmental impact. A key task will be to reveal whether this technology path is beneficial for the producers compared with other farming practices.

The highest adoption rate of PF practices has been in the United States and Europe. Some farmers in the United Kingdom as well as Denmark have used PF practices since the early 1990s. Precision farming has primarily been adopted in the cereal producing countries. A general trend is that large producers with more than 300 ha tend to be the first to invest in the new technology, whereas small producers are more reluctant to invest in GPS equipment. However, some divergence in the adoption speed has occurred between North European countries. To date, the adoption rate in France and Germany has been relatively slow compared with Denmark, UK, and Sweden. A relatively high adoption rate in Denmark is probably related to an early production of combine harvesters with GPS equipment. Massey Ferguson was a pioneer in the development of commercial yield monitors with GPS.

Similar breaks in the trends in other high-technology sectors have occurred. It took for instance about seven years before the mass market adopted the Internet and mobile phones. In this respect we might see a similar adoption pattern for PF technologies.

### *Precision Farming Surveys in 1998 and 2000*

In 1998 a PF adoption survey was carried out in the UK (Fountas, 1998). The postal interview indicated that 15 percent of the farmers used some PF techniques and 85 percent did not. Of the 85 percent nonusers, 22 percent planned to look at PF in the future. Five percent were likely to use PF in the next two or three years, 14 percent were likely to use slightly later, 31 percent considered it quite unlikely, and 15 percent considered it very unlikely. The sample was considered representative of the whole of Britain, which may be the current situation in PF in the United Kingdom today.

The majority of the users believed that PF was a combination of the following two definitions: "Use of GPS satellites, alongside positioning of the field characteristics, such as yield or soil mapping" and "targeted crop management, whereby inputs are used specifically according to soil and plant diagnosis." On the other hand, the nonusers related more closely to the first definition, which is more technically oriented, rather than the second one, which is more management oriented. This could be because users are convinced of the value of PF as a crop management tool, whereas the nonusers perceive PF as a new development and do not view it as a management tool.

All the respondents learned about PF mostly from farming magazines and agricultural events. This indicates that press as a very important source for promoting new ideas in agriculture. Additional sources of information are agronomists. However, regarding this source, a discrepancy between users and nonusers can be noted. Agronomists had a significantly higher influence on the knowledge of PF for the users, whereas fewer nonusers learned about PF from their agronomists. Implications of these findings are that agronomists who are familiar with PF can influence farmers regarding the use of this technology. The nonusers believe that a lower set-up cost, more convincing benefits, and advice from agronomists are the key factors to encourage them to use PF. On the other hand, endorsement from agricultural organizations or other farmers are not perceived to be very important.

Regarding the problems of adoption, predominantly the cost of investment was the major problem of PF. Eighty-one percent of the farmers viewed this to be the largest disadvantage. However, the increased financial pressure placed on British agriculture is also encouraging farmers to seek new technology and to reduce inputs. As (Lowenberg-DeBoer and Boehjle, 1996) stated, "history has shown that most of the benefits of any new agricultural technology go to the early adopter" (p. 937). Those who lag have often been forced out of farming. PF is expected to follow the same pattern. Those who begin to accumulate data and experience now will be ready to use improved precision technology as it matures.

In order to assess the farmer's experiences with different PF practices a farm survey was conducted in the autumn of 2000 in Denmark, Great Britain, and the state of Nebraska, United States (Pedersen et al., 2001a).

Findings from this survey shows that many producers are generally quite optimistic about increased profitability with precision agriculture, but figure it will take five to ten years to achieve their expected level of profitability. Variable rate application of fertilizer was the practice most commonly cited as the one likely to increase profits on their farm—either through increased yield or decreased total applications of phosphorous and potassium (Denmark and UK). Participants in all three countries felt that total nitrogen use on their farms would remain unchanged with the use of precision practices.

The farmers felt the greatest disincentive to adoption of precision agriculture was the cost of the equipment coupled with a current lack of evidence of increased profit. In spite of this, producers were generally quite optimistic about the potential for increased profit with PF in the future.

Although many admit an incentive to their adopting PF was a fascination with the technology, the second major disincentive was the time required to work on the data. A majority of producers also felt that the use of precision practices would help them use inputs more efficiently and reduce the environmental impact of crop production. More thorough and scientifically based advice about input application is also needed. Compatibility of hardware and software used in PF is also a significant concern of producers. Greater compatibility of hardware and software, and ease of use, should increase the potential for the "average" producer—those not so fascinated by the technology—to adopt one or more precision practices.

Figure 20.5 shows the relative proportions of respondents' attitude to specific PF issues. It is interesting to note the discrepancy between Danish farmers and American farmers when it comes to looking at profitability. Even though some major developments have occurred within PF lately, many farmers have not seen any specific economic and environmental benefits from the technology in practice.

### *General Economics of Precision Farming*

It is obvious that a precondition for variable rate treatment of nutrients and pesticides is that some spatial variation occurs on the field in order to

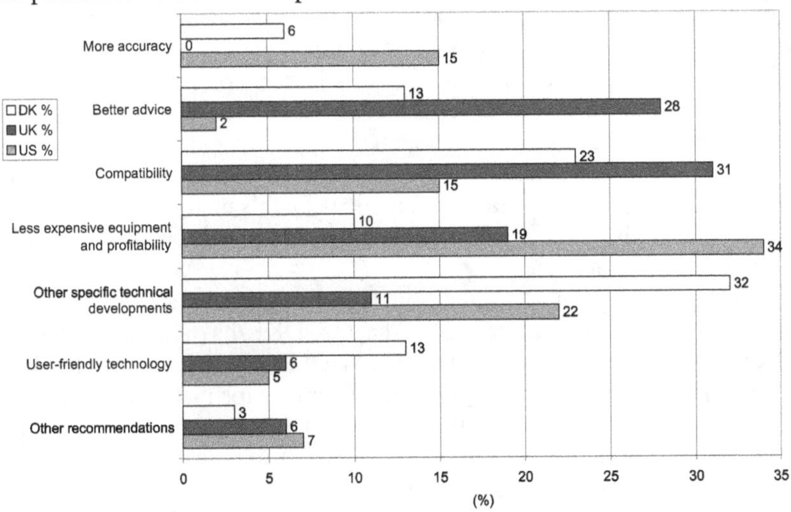


FIGURE 20.5. Farmer recommendations to improve the development of precision farming technology in Denmark (DK), United Kingdom (UK), and the United States (US).

create an additional economic benefit. Many farmers are usually aware of these conditions. It is however at least as important that the farm structure (size, crop rotation) and general economic conditions are suitable for PF. Even though the fields have a high spatial variation in soil conditions and yield potentials, it might very well be the case that it is not economically viable to invest in site-specific equipment simply because the arable farm area is too small.

The results in Table 20.2 present an example in which two cereal producers with 100 ha and 500 ha, respectively, implement variable rate nitrogen application. It is assumed that both producers invest in all current relevant equipment and software for conducting variable application of commercial fertilizers and slurry in practice. With these investments the producers should be able to gather site-specific information and use computer software for decision making and finally to conduct variable treatment of nitrogen.

This example is static in the sense that it does not consider the cost of implementing and establishing the equipment and extra time needed to conduct site-specific application with yield mapping and production of application maps. The calculation in Table 20.2 shows a substantial difference in the cost of implementing site-specific application of variable rate treatment between a medium-sized farm (100 ha) and a large-sized farm (500 ha).

The example shows the cost of full implementation, which includes soil sampling and analysis, yield mapping, automatic N-sensing equipment, and electronic devices and computers for variable rate treatment of commercial fertilizers and slurry. In this example it is hardly possible that a marginal benefit from documentation and higher yield from spatial application will cover the extra costs of implementation. A further spatial application of lime, seed, and pesticides will imply that the farmer has to invest in additional equipment for these farming practices. In this matter it is vital that the farmer match his investments to the actual needs on the farm and according to his management strategies.

The cost of implementing additional electronic equipment for spatial application of lime is between 2.0 €/ha (15 DKK/ha) and 10.0 €/ha (74 DKK/ha) for a medium-sized and large-sized farm, respectively (see Table 20.3). Similar electronic equipment on a sprayer for measuring the water/chemical flow is about 7,000 DKK (942 €/ha). Investments in injection systems for separating the various chemicals is far more expensive compared with a flow measuring system. The most expensive part of site-specific application of herbicides (which so far is the only relevant and practical feasible input of the various pesticides) is labor cost for weed and pest scouting. Calculations have shown that labor for scouting weeds on a conventional field costs about 16,1 €/ha (120 DKK/ha).

TABLE 20.2. Cost of site-specific application of nitrogen: An example.

| | Investment | | Yearly costs | |
|---|---|---|---|---|
| | **100 ha (€)** | **500 ha (€)** | **100 ha (€/ha)** | **500 ha (€/ha)** |
| Yield monitor | 3362.6 | 3362.6 | 6.7 | 1.3 |
| GPS receiver | 2690.1 | 5380.1 | 5.4 | 2.2 |
| Computer, combine harvester (portable) | 4035.1 | 4035.1 | 8.1 | 1.6 |
| Computer, tractor[a] | 672.5 | 4035.1 | 1.3 | 1.6 |
| DGPS reference signal[b] | 470.8 | 807.0 | 4.7 | 1.6 |
| Handheld GPS receiver | 134.5 | 134.5 | 0.3 | 0.1 |
| *Total* | | | 26.5 | 8.4 |
| Gathering information | | | | |
| Soil samples[c] | | | 6.7 | 6.7 |
| Electricity conductivity mapping (EM38)[d] | | | 1.3 | 1.3 |
| N-sensor[e] | | | 8.7 | 8.7 |
| *Total* | | | 16.8 | 16.8 |
| Variable treatment | | | | |
| N-spreader—computer | 2017.5 | 2017.5 | 4.0 | 0.8 |
| Slurry spreader—computer and flow measuring device | 4035.1 | 4035.1 | 8.1 | 1.6 |
| *Total* | | | 12.1 | 2.4 |
| Decision support | | | | |
| Software for precision farming[f] | | | 4.0 | 1.3 |
| Other software | | | 2.4 | 0.5 |
| *Total* | | | 6.5 | 1.8 |
| *Total costs* | | | 61.6 | 29.5 |

TABLE 20.2 *(continued)*

| | Investment | | Yearly costs | |
|---|---|---|---|---|
| | **100 ha (€)** | **500 ha (€)** | **100 ha (€/ha)** | **500 ha (€/ha)** |
| Capital costs | | | 4.9 | 2 |
| *Total costs* (including capital costs) | | | 66.5 | 31.5 |

*Source:* Calculation based on S.M. Pedersen et al., 2001a.

*Notes:* 1 € = 7,43478 DKK; depreciation of electronic equipment is 5 years, excluding farm personal computer, extra time spent on PF, and cost of capital; Cost of capital, interest rate is 6 percent and inflation is 2 percent (real interest rate is 3.9 percent.)

[a]Portable tractor computer can be used for various machines.

[b]Lower costs with several reference signals.

[c]200 DKK/ha (26.9 €/ha) conducted every 4th year.

[d]200 DKK/ha (26.9 €/ha) conducted every 20th year.

[e]Based on contracting marked price for a N-sensor is 140,000 DKK (18831 €). Depreciation is 5 years plus additional 12,000 DKK (1614 €) for yearly services. Own investments in a N-sensor system will cost between 80 (10.8 €/ha) and 400 DKK/ha (53.8 €/ha) with a farm size at 500 and 100 ha respectively.

[f]Software for conduction application maps, e.g. Kemira Loris, or AgroSat from DataLogisk ApS.

Table 20.4 shows the costs of various PF practices in 2001. It is expected that the costs would decrease further as the technology may provide higher yields and other potential savings of inputs in the future. The additional economic returns are still difficult to verify because of limited results from research trials.

The first two scenarios show the cost of conduction variable application of lime based on soil samples in grids and soil conductivity mapping, respectively. The third scenario describes where a contractor carries out the application. The next three scenarios show a combination of full implementation and investment in equipment for variable N, lime, and patch spraying of herbicides in different combinations. Four scenarios follow this with different information systems for conducting variable rate fertilizer application. Finally, two systems are presented to carry out patch spraying based on

TABLE 20.3. The cost of different precision farming systems, 2001.

| GPS receivers etc. | Cost (€) | €/ha (100 ha) | €/ha (500 ha) |
|---|---|---|---|
| DGPS receiver (tractor mounted) +/– 1 m[a] | 2690.1 | 8.1 | 1.6 |
| GPS Focus receiver[b] | 807.0 | 1.6 | 0.3 |
| Handheld GPS receiver | 134.5 | 0.3 | 0.1 |
| Board computer/monitor (applied for various instruments) | 4035.1 | 8.1 | 1.6 |
| Reference signal DGPS (yearly costs) | 470.8 | 4.7 | 0.9 |
| Personal computer | 1345.0 | 2.7 | 0.5 |
| *Gathering information* | | | |
| Yield monitor | 3362.6 | 6.7 | 1.3 |
| Soil sample and maps, P, K, pH (every four years)[c] | 27.0 | 6.7 | 6.7 |
| EM38 (every 20 years) | 27.0 | 1.3 | 1.3 |
| Satellite picture Ikonos 1 or 4 m = 1 pixel[d] | | 2.0 | 2.0 |
| Aerial photos 20 cm = 1 pixel[e] | | 4.7 | 1.3-2.6 |
| Hydro N-sensor[f] | | 53.8 | 10.8 |
| Chlorophyll meter (handheld) | 1345.0 | 2.7 | 0.5 |
| Manual weed detection[g] | | 16.1 | 16.1 |
| *Electronics and flow measuring devices* | | | |
| Fertilizer spreader | 2017.5 | 4.0 | 0.8 |
| Lime spreader | 4977.0 | 10.0 | 2.0 |
| Slurry spreader | 4035.1 | 8.1 | 1.6 |
| Sprayer | 942.0 | 1.9 | 0.4 |

TABLE 20.3 *(continued)*

| GPS receivers etc. | Cost (€) | €/ha (100 ha) | €/ha (500 ha) |
|---|---|---|---|
| *Decision support* | | | |
| Software, yield, and application maps (incl. yearly services) | | 4.0 | 1.3 |
| Other software | | 2.4 | 0.5 |

*Source:* Calculation based on Pedersen et al. (2001a).

*Note:* Period of depreciation of all electronic equipment is assumed to be five years.

[a]Racal Landstar (DGPS ref. signal is needed).

[b]GPS receiver, FM radiowave band (DGPS ref. signal is needed).

[c]200 DKK /ha (26.9 €/ha).

[d]Estimated cost based on the acquisition of 11*11 km from Space Imaging Europe, including additional costs for subdivision and distribution of imagery.

[e]Estimated cost based on calculations of the authors and of Dr. Bennedsen at KVL after acquisition of aerial photos "Ortofoto" from Blominfo AB company.

[f]Average cost for contracting: 65 DKK/ha (8.74 €/ha).

[g]Weed scouting costs were calculated to be about 120 DKK/ha (16.1 €/ha) per year. Scouting costs in the first year were estimated to be 440 DKK/ha (58.9 €/ha) based on assumption that 220 labor hours are required for weed detection in 100 ha at a cost of 200 DKK (26.9 €) per labor hour; Costs for updates from year 2 through year 5 were estimated at 40 DKK/ha (5.4 €/ha) per year.

manual weed detection. Electronic equipment for variable rate technology is based on a flow-measuring device without injection systems. The costs of these different systems varies between 7.9 €/ha and 95.6 €/ha according to farm area and level of implementation.

The economic studies described by Olesen (1997) show that the potential benefit from spatial application of nitrogen varies between 6.7 €/ha (50 DKK/ha) and 26.9 €/ha (200 DKK/ha) among 4 different Danish research fields with different cereal rotational systems.

Trials with Hydro N-sensors in Denmark have shown very small yield increases from variable application in 1999 and no increases of yield in trials conducted in 2000 and 2001 (Pedersen, 2001). It is also difficult to verify

TABLE 20.4. Additional costs of site-specific application of lime, N (nitrogen), and herbicides in wheat, 2001.

| Scenario | Variable input | Equipment used for site-specific application | 100 ha (€/ha) | 500 ha (€/ha) |
|---|---|---|---|---|
| 1 | Lime | With soil samples and electronic equipment on the lime spreader[a] | 41.9 | 16.6 |
| 2 | Lime | With (EM38) and electronic equipment on the lime spreader[a] | 36.5 | 11.2 |
| 3 | Lime | Contracting with additional costs for soil samples (without yield mapping) | 8.6 | 7.9 |
| 4 | N (commercial and slurry) | With soil samples, yield mapping, EMI (EM38), N-sensor plus electronic equipment on fertilizer spreader (commercial), and slurry spreader[a] | 66.5 | 31.5 |
| 5 | N (commercial and slurry) and lime | With soil samples, yield mapping, EMI (EM38), N-sensor plus electronic equipment on fertilizer spreader (commercial), slurry spreader, and lime spreader[a] | 77.4 | 33.7 |
| 6 | N (commercial and slurry), lime, and herbicides (patch spraying) | (With soil samples, yield mapping, EMI (EM38), N-sensor plus electronic equipment on fertilizer spreader (commercial), slurry spreader, lime spreader, and sprayer. Based on manual weed detection[a] | 95.6 | 50.2 |
| 7 | N (commercial) | With N-sensor (contracting), electronic equipment for fertilizer spreader, and yield mapping[a] | 44.0 | 19.7 |
| 8 | N (commercial) | Based on yield maps and electronic equipment for fertilizer spreader[a] | 39.7 | 11.8 |
| 9 | N (commercial) | Based on EMI (EM38) with yield mapping electronic equipment for fertilizer spreader[a] | 41.6 | 13.7 |
| 10 | N (commercial) | EMI (EM38) and electronic equipment for fertilizer spreader (without yield mapping)[a] | 32.8 | 9.3 |

TABLE 20.4 *(continued)*

| Scenario | Variable input | Equipment used for site-specific application | 100 ha (€/ha) | 500 ha (€/ha) |
|---|---|---|---|---|
| 11 | Herbicides, patch spraying | Patch spraying based on manual weed detection and electronic equipment on sprayer (with yield mapping)[a] | 53.5 | 27.5 |
| 12 | Herbicides, patch spraying | Patch spraying based on manual weed detection and electronic equipment on sprayer[a] | 44.6 | 24.2 |

*Source:* Based on internal unpublished economic calculations, Pedersen et al., 2001a.

*Note:* Depreciation is 5 years for electronic equipment and software; 20 years for soil conductivity mapping; 4 years for soil samples. Cost of capital, interest rate is 6 percent and inflation is 2 percent (real interest rate is 3.9 percent).

[a]Include investments in GPS receiver, tractor computer, DGPS-reference signal, and decision-support programs.

increases in protein content in cereals from variable applications. A similar trial from Risø laboratory shows that the increase in protein content was modest or even nonexistent. Yield increases from this N-sensor trial was about 170 kg/ha compared with a uniform application (Jørgensen and Jørgensen, 2001). Trials with variable N application based on EM38 soil conductivity mapping have at best shown yield improvements between 40 and 240 $kg \cdot ha^{-1}$ (Pedersen, 2001). The market price for cereals is about 106 €/ton (800 DKK/ton), which corresponds to an economic benefit between 4.3 and 25.4 €/ha.

In certain cases it might be possible to gain an extra economic benefit from higher protein contents in cereals. The Danish and European market price for bread wheat or for export depends on the protein content as well as other quality parameters. A reduction of the protein content from, e.g., 11.5 percent to 11.0 percent (export wheat) will reduce the intervention price with about 2.5 €/tons, which is similar to about 12 €/ha (yield 8 $tons \cdot ha^{-1}$). Similar economic penalties are used when producing malt barley on a contractual basis. In this respect, variable rate application of fertilizers might be an important tool to obtain a certain protein content and market price for cereals. The potential savings from application of lime is about 7-13 €/ha (50-100 DKK/ha) according to soil variability.

Trials with variable application of chemicals in cereals have shown that patch spraying can reduce the application of herbicides by 50 to 75 percent. A Danish trail from 1999 of spring barley shows that the application of herbicides could be reduced with 47 to 62 percent compared with uniform treatment. (Heisel et al., 1999). A 60 percent saving in chemicals is equivalent to 17 €/ha (126 DKK/ha). These benefits are well below the cost of patch spraying and manual weed detection. In this matter it is necessary to encounter other environmental or qualitative benefits in order to create an economic or sustainable application of patch spraying.

The results from this analysis conclude that PF is a technology for relatively big farmers. The various technologies should be used for several purposes at the same time in order to distribute the costs of basic equipment (GPS equipment and tractor computers) on many applications.

A combination of variable lime and nitrogen application might be economically viable on large farm areas with nonhomogeneous soils and spatial variation. For many farmers however, it might, be necessary to focus and direct the investments and implementations on specific technologies with which potential site-specific benefits and savings are possible for any given crop producer. Until now, most trials have shown only little yield improvements from variable fertilizer application or mixed results at best. In this matter it is necessary to improve the decision algorithms in order to manage spatial variation, including local climate and soil conditions. Variable lime application seems to be economically viable when conducted by a contractor or on large size areas. Patch spraying is still too costly primarily due to manual detection of weeds. New automatic weed detecting systems are necessary in order to make this management practice economic viable in the future.

Cereals cover most of the agricultural areas in southern Europe, but the yields are considerably less than in northern Europe (Denmark, France, Germany, and UK, whose yields are approximately 6 to 7 $t \cdot ha^{-1}$ due to the drier and warmer climate with less rain in most of the areas), such countries as Greece, Italy, Portugal, and Spain produce yields of 2 to 5 $t \cdot ha^{-1}$ (Commission of the European Communities, 2000). The lower yields may explain the low adoption rates and reluctance of southern European farmers to invest in new technologies due to lower economic returns. However, this could prove to be an opportunity for PF in the future to assist in quantifying these lower yields and offering ways to improve the efficiency based on a very limited water supply.

## FUTURE RESEARCH

Research has developed many new ways to measure spatial and temporal variability, but in the process has made PF too complex for most farmers to deal with. Some suggested solutions to this problem are the following:

- Develop simple methodology for farmers to adopt that can take into account individual farmer requirements (Identify priorities).
- Improve the links between farmers, extension, and research.
- Engineers should develop the knowledge of how to use new technologies in addition to the technologies themselves. Complete subsystems should be developed so we can have knowledge transfer as well as technology transfer.

Individual yield maps show great heterogeneity, but yield map trends over time become more homogenous. This leads to the need to manage each crop according to the current year's conditions in order to accomplish and determine the following:

- Optimize crop growth based on limiting resources (water)
- Soil moisture holding characteristics
- Topography (macro and micro)
- Climatic conditions (macro and micro)

Spatial and temporal variability is hard to analyze, so geostatistics (the branch of statistics that deals with spatial data) must be developed much further to provide the tools needed to scientifically analyze heterogeneity, trends over time, spatial correlation, etc.

The development of good reliable sensors, preferably accurate, reliable, cheap, and noncontacting, are always welcome. The following are some examples:

- EMI maps of soil (moisture maps, texture maps)
- Hyperspectral imaging of crop response (classify stresses, measure growth parameters)
- High accuracy positioning (RTK [real-time kinematic] GPS) (Mechanical weeding, navigation)
- Integration of sensors and models (Real-time modeling)

Many European farmers have shown an interest in PF because they can see that it should work in principle, but many also find it difficult to understand because of the complexity. It has been suggested that commercial demonstration farms be established to show how these ideas and principles can be put into practice. Such farms would be necessary to do the following:

- Develop clear demonstrable alternative management strategies, such as real-time canopy control strategies and precision organic farming
- Demonstrate economic and environmental benefits
- Develop added value systems including traceability
- Develop innovative methods for employing new technologies, rapid assessment techniques (e.g., EMI, ADP, SAR [synthetic aperture radar], GPR [ground-penetrating radar]), and approaches (e.g., speaking plant approach)

## CONCLUSIONS

PF is a management concept that is highly information intensive. Several new technological developments, which seem fully developed and commercialized, can be used for gathering information and taking action. Even so, it seems that neither of these systems are able to provide the whole story alone. It is still the farmer who is going to make the final decision about management strategy, and this approach seems to be even more important than ever before, simply as a result of an enormous pool of data and information.

It is likely that some farmers will benefit from PF, both economically and environmentally, especially the large producers who are able to cover the investments, and producers with less homogeneous soils will probably benefit from site-specific application. Manufacturers should strive to provide low-cost options for producers interested in precision agriculture but who are unable to afford large-scale investment.

Researchers and advisors need to make producers aware of precision practices that are most likely to be profitable for local conditions. Specifically, little research evidence exists to date that shows increased profitability with variable rate fertilizer application. Most studies evaluating the economic viability of site-specific application of fertilizers have shown mixed results at best (Schnitkey et al., 1996; Schmerler and Jürschik, 1997; Swinton and Lowenberg-DeBoer, 1998; Schmerler and Basten, 1999). It is likely that in some cases precision technologies have helped producers learn that they were applying fertilizers at rates above the economic optimum and have been able to reduce total fertilizer application without yield reduction, however, variable rate application was not necessarily critical to accomplish this. Even in countries (e.g., Denmark) where environmental regulation with N quotas limits the total N application to below the economic optimal level, PF has provided a "teachable moment" and helped producers use fertilizers more efficiently.

It would be profitable for researchers to study situations in which producers feel variable rate fertilization and other variable rate technologies are profitable farm practices. Some farmers might be able to gain an economic benefit due to better logistics during harvest and less fuel consumption due to fewer overlaps during tillage. It might also be easier to find drainage systems with GPS or to point out future set-aside areas. These benefits might be difficult to generalize but could be crucial for the individual farmer. Some research shows that variable rate lime application can be a profitable practice (Bongiovanni and Lowenberg-DeBoer, 2000), although the potential return is low due to low cost input.

Site-specific control of pesticides is a practice that many researchers believe has considerable economic potential in some cropping systems, but it is not currently viewed as attractive by farmers (Audsley, 1993; Leiva et al., 1997; Daberkow, 1997; Gerhards et al., 1999; Pedersen et al., 2001b). However, since autonomous and robust weed detection systems still need to be developed, manual weed detection is still too costly in most cropping systems because it needs a great amount of human effort.

In order to assess the wider technical and social impact on surrounding society as well as on the farming sector it could be relevant to apply technology assessment and impact analysis with an interactive and participatory approach. Manufacturers of PF technology have tried within the last decade to develop various methods for gathering information and conducting variable rate treatments, but many of them have developed equipment and decision support systems regardless of other developments and regardless of compatibility between different systems. In this respect, the manufacturers of PF technology should consider whether it is possible to more involve stakeholders in the development process of future precision farming technologies.

## REFERENCES

Audsley, E. 1993. Operational research analysis of patch spraying. *Crop Protection* 12: 111-119.

Blackmore, B. S. 1998. A yield map primer. In *Proceedings from the Conference on Precision Farming in Japan, USA and Europe,* October 1998, Sapporo, Japan (pp. 11-23). The Robotics and Mechatronics Division of the Japanese Society of Mechanical Engineers and the Hokkaido branch of Japanese Society of Agricultural Machinery.

Blackmore, B. S. 2000. The interpretation of trends from multiple yield maps. *Computers and Electronics in Agriculture* 26: 137-51.

Blackmore, B. S., H. Have, and S. Fountas. 2001. A specification of behavioural requirements for an autonomous tractor. In M. Zude, B. Herold, and M. Guyer (Eds.), *The Sixth International Symposium on Fruit, Nut and Vegetable Production Engineering* (pp. 25-36). Potsdam, Germany: Institute für Agrartechnik Bornim e.V. Germany.

Blackmore, B. S. and M. R. Moore. 1999. Remedial correction of yield map data. *Precision Agriculture* 1: 53-66.

Blackmore, B. S., P. N. Wheeler, R. M. Morris, J. Morris, and J. Graham. 1994. *Information Technology in Arable Farming.* Edinburgh, Scotland: Scottish Natural Heritage.

Bongiovanni, R. and J. Lowenberg-DeBoer. 1998. Economics of variable rate lime in Indiana. In P.C. Roberts, R.H. Rust, and W. E. Larson (Eds.), *Proceedings of the 4th International Conference on Precision Agriculture* (pp. 1653-1665). Madison, WI: ASA/CSSA, SSSA/ASAE.

Borgelt, S. C. 1992. Sensing and measurement technologies for site specific management. In Robert P.C., R.H. Rust, and W. E. Larson (Eds), *Soil specific crop management: A workshop on research and development issues* (pp. 141-157). Madison, WI: ASA/CSSA, SSSA/ASAE.

Christensen, S. 1995. Site specific weed control. In *Seminar on Site Specific Farming* (pp. 161-164). SP Report No. 26. Denmark, Danish Institute of Agricultural Sciences.

Commission of the European Communities. 2000. The agricultural situation in the European Union: 1999 report. Available online at: http://ienica.csl.gov.uk/policy/ec1999agreport.pdf.

Daberkow, S. G. 1997. Adoption rates for recommended crop management practices: Implications for Precision Farming. In J.V. Stafford (Ed.) *Precision Agriculture 1997: Proceedings of the 1st European Conference in Precision Agriculture* (pp. 941-948). Oxford, UK: BIOS Scientific Publishers.

Earl, R., J. C. Taylor, G. A. Wood, I. Bradley, I. T. James, T. Waine, J. P. Welsh, R. J. Godwin, and S. M. Knight. 2003. Soil factors and their influence on within-field crop variability part I: Field observation of soil variation. *Biosystems Engineering* (Special Issue on Precision Agriculture) 84(4): 425-440.

Earl, R., G. T. Thomas, and B. S. Blackmore. 2000. The potential role of GIS for autonomous field operations. *Computers and Electronics in Agriculture* 25(1-2):107-120.

Ellenrieder, K. 1996. Design of an autonomous agricultural vehicle. Unpublished master's of philosophy thesis. Cranfield University, Silsoe, UK.

Fountas, S. 1998. Market research on the views and perceptions of farmers about the role of crop management within precision farming. Unpublished master's thesis. Cranfield University, Silsoe, UK.

Garcia-Alegre, M. C., A. Ribeiro, L. Garcia-Perez, R. Martinez, D. Guinea, and A. Pozo-Ruz. 2001. Autonomous robot in agricultural tasks. In G. Grenier and B. S. Blackmore (Eds.) *Proceedings of the 3rd European Conference on Precision Agriculture* (pp. 25-30). Montpellier, France: Agro-Montpellier.

Gerhards, R., M. Sökefeld, C. Timmermann, S. Reichart, W. Kübauch, and M. M. Williams. 1999. Results of a four-year study on site-specific herbicide application. In J. V. Stafford (Ed.). *Proceedings of the 2nd European Conference in Precision Agriculture* (pp. 689-697). Sheffield, UK: Sheffield Academic Press.

Gitas, I. Z., C. G. Karydas, G. V. Kazakis, and D. R. Ghosn. 2004. The potential of high spatial resolution satellite imagery for studying complex Mediterranean landscapes: The case of Crete. In *Proceedings of the 3rd workshop of EARSeL: Special Interest Group on Remote Sensing for Developing Countries,* September, 2004, Cairo, Egypt, pp. 284-296.

Godwin, R. J., R. Earl, J. C. Taylor, G. A. Wood, R. I. Bradley, J. P. Welsh, T. Richards, B. S. Blackmore, M. C. Carver, S. M. Knight, and B. Welti. 2001. Precision farming of cereal crops: A five-year experiment to develop management guidelines. Project report 264e, London, UK: Home Grown Cereals Authority.

Godwin, R. J. and P.C.H. Miller. 2003. A review of the technologies for mapping within-field variability. *Biosystems Engineering (Special Issue on Precision Agriculture)* 84(4): 393-407.

Godwin, R. J., T. E. Richards, G. A. Wood, J. P. Welsh, and S. M. Knight, 2003. An economic analysis of the potential for precision farming in UK cereal production. *Biosystems Engineering (Special Issue on Precision Agriculture)* 84(4): 533-545.

Halmø, S. 2001. The Hydro N-Sensor: Spatial resolution and requirements of fertiliser applicators. Germany: PreAgro.

Heisel, T., S. Christensen, and A. M. Walter. 1999. Whole-field experiments with site-specific weed management. In J.V. Stafford (Ed.). *Proceedings of the 2nd European Conference in Precision Agriculture* (pp.759-768). Sheffield, UK: Sheffield Academic Press.

James, I.T., T. Waine, I. Bradley, R. J. Godwin, and J. C. Taylor. 2000. A comparison between traditional methods and EMI scanning for determining soil textural boundaries. EurAgEng, Warwick, UK. Silsoe Research Institute, Silsoe, UK. Paper No. 00-PA-014.

Jørgensen, J. R. and R. N. Jørgensen. 2001. Impact on grain quality parameters when nitrogen is sensor applied by the Hydro precise system. In G. Grenier and B. S. Blackmore (Eds.) *Proceedings of the 3rd European Conference on Precision Agriculture* (pp. 929-934). Montpellier, France: Agro-Montpellier.

Karydas, C.G. 2003. The Mediterranean precision farming potential. Mediterranean Agronomic Institute of Chania, Chania, Greece. Available online at: http://www.cpf.kvl.dk/Countries/MED_PF.pdf.

Karydas, C.G., T. Sekuloska, and I. Sarakiotis. 2005a. Fine scale mapping of agricultural landscape features to be used in environmental risk assessment in an olive cultivation area. *IASME Transactions (World Scientific and Engineering Academy and Society)* 2(4): 582-589.

Karydas, C.G., T. Sekuloska, and I.Sarakiotis. 2005b. Use of EO data to identify landscape features associated to assessment of environmental risk caused by olive farming and olive oil production. In *Proceedings of the 2005 IASME/WSEAS International Conference on Energy, Environment, Ecosystems and Sustainable Development,* July, 2005, Athens, Greece. CD-ROM.

Karydas, C. G. and N. G. Silleos. 2000. Precision agriculture: Description of the method: Current status and prospects. In *Proceedings of the 2nd Special Conference 'Information Systems in Agricultural Sector' of the Hellenic Operational Research Association,* October, 2000, Chania, Greece (pp. 136-146). Athens: Ekdoseis Neon Technologion.

Karydas, C.G., N.G. Silleos, and N.G. Karydas. 2004. Study of heterogeneity in a Mediterranean agricultural area, using high resolution multi-spectral imagery. In *Proceedings of the 2nd International Conference of HAICTA,* Volume I, March, 2004, Thessaloniki, Greece (pp. 57-66). Thessaloniki, Greece: Publishing Centre of Technological Educational Institute (TEI).

Kersebaum, K. C., K. Lorenz, O. Wendroth, H. I. Reuter, J. Schwarz, and P. Jürschik. 2001. Effects of site specific nitrogen fertilization on nitrogen leaching: Comparison of different strategies in arable fields based on observations and simulations. In G. Grenier and B. S. Blackmore (Eds.), *Proceedings of the 3rd European Conference on Precision Agriculture* (pp. 683-688). Montpellier, France: Agro-Montpellier.

Leiva, F. R., J. Morris, and B. S. Blackmore. 1997. Precision farming techniques for sustainable agriculture. In J.V. Stafford (Ed.), *Precision Agriculture 1997: Proceedings of the 1st European Conference in Precision Agriculture* (pp. 957-966). Oxford, UK: BIOS Scientific Publishers.

Lowenberg-DeBoer, J. and M. Boehjle. 1996. Revolution, evolution or dead-end: Economic perspectives on precision agriculture. In P. C. Robert, R. H. Rust, and W. E. Larson (Eds.), *Proceedings of the 3rd International Conference on Precision Agriculture* (pp. 923-944). Madison, WI: ASA/CSSA, SSSA/ASAE.

Markinos, A. and T. Gemtos. 2002. Yield mapping of cotton crop in Greece. Presentation at the 1st Conference of the Hellenic Association for Information and Communications Technology in Agriculture, October, 2002. Aristotle University of Thessaloniki, Greece. 11pp.

Olesen, S. E. 1997. Positionsbestemt dyrkning. Report No. 17. Copengagen, Denmark: Danmarks Jordbrugsforskning (DJF). 47pp.

Oliver, M. A., Z. L. Frogbrook, R. Webster, and C. J. Dawson. 1997. A rationale for determining the number of cores for bulked sampling of soil. In J.V. Stafford (Ed.), *Precision Agriculture 1997: Proceedings of the 1st European Conference in Precision Agriculture* (pp. 155-162). Oxford, UK: BIOS Scientific Publishers.

Pedersen, C. 2001. *Landsudvalget for Planteavl.* Skejby, Denmark: Landbrugets Rødgivningscenter, 246pp.

Pedersen, S. M., R. B. Ferguson, and R. M. Lark. 2001a. A comparison of producer adoption of precision agricultural practices in Denmark, the United Kingdom and the United States. *IPL 177-01 Technical Report No. 2.* Lyngby, Denmark: Institut for Produktion og Ledelse. 46pp.

Pedersen, S. M., R. B. Ferguson, and R. M. Lark. 2001b. A multinational survey of precision farming early adopters. *Farm Management* 11(3): 147-162.

Rothmund, M., H. Auernhammer, and M. Demmel. 2002. Eine Gemeinde umfassende Gewannebewirtschaftung in Zeilitzheim [Possibilities of Transborder Farming—the example Zeilitzheim]. *Landtechnik* 57(1):10-11.

Scarlett, A. J., J. C. Lowe, and D. A. Semple. 1997. Precision tillage: In-field, real-time control of seedbed quality. In J.V. Stafford (Ed.), *Precision Agriculture 1997: Proceedings of the 1st European Conference in Precision Agriculture* (pp. 503-510). Oxford, UK: BIOS Scientific Publishers.

Schmerler, J. and M. Basten. 1999. Cost/benefit analysis of introducing site-specific management on a commercial farm. In J.V. Stafford (Ed.), *Proceedings of the 2nd European Conference in Precision Agriculture* (pp. 959-967). Sheffield, UK. Sheffield Academic Press.

Schmerler, J. and P. Jürschik. 1997. Technological and economic results of precision farming from a 7200 hectares farm in East Germany. In J.V. Stafford (Ed.), *Precision Agriculture 1997. Proceedings of the 1st European Conference in Precision Agriculture* (pp. 991-997). Oxford, UK: BIOS Scientific Publishers.

Schnitkey, G. D., J. W. Hopkins, and L. G. Tweeten. 1996. An economic evaluation of precision fertilizer applications on corn-soybean fields. In P. C. Robert, R. H. Rust, and W. E. Larson (Eds.), *Proceedings of the 3rd International Conference on Precision Agriculture* (pp. 977-987). Madison, WI: ASA/CSSA, SSSA/ASAE.

Shibusawa, S. 1996. PhytoTechnology: An introduction to the concept and topic of a new project. Available online at: http://phytech.ishikawa-c.ac.jp/WhatIs.html.

Shibusawa, S., I. M. Anom, H. Sata, A. Sasao, S. Hirako, A. Otomo, and B. S. Blackmore. 2000. On-line real-time soil spectrophotometer. In P. C. Robert, R. H. Rust, and W. E. Larson (Eds.), *Proceedings of the 5th International Conference on Precision Agriculture*. CD-ROM. Madison, WI: ASA/CSSA, SSSA/ASAE.

Swinton, S. M. and J. Lowenberg-DeBoer. 1998. Evaluating the profitability of site-specific farming. *Journal of Production Agriculture* 11(4): 439-446.

Taylor, J. C., G. A. Wood, R. Earl, and R. J. Godwin. 2003. Soil factors and their influence on within-field crop variability part II: Spatial analysis and determination of management zones. *Biosystems Engineering (Special Issue on Precision Agriculture)* 84(4): 441-453.

Thomas, G., J. C. Taylor, T. Musthill, and J. C. Fuller. 1999. The development of remote sensing based products in support of precision farming. In J.V. Stafford (Ed.), *Proceedings of the 2nd European Conference in Precision Agriculture* (pp. 191-200). Sheffield, UK: Sheffield Academic Press.

Voßhenrich, H., J. Marquering, B. Gattermann, W. Täger-Farney, B. Ortmeier, and C. Sommer. 2001. Management system for DGPS-supported primary soil tillage. In G. Grenier and B. S. Blackmore (Eds.), *Proceedings of the 3rd European Conference on Precision Agriculture* (pp. 731-736). Montpellier, France: Agro-Montpellier.

Waine, T. W., B. S. Blackmore, and R. J. Godwin. 2000. Mapping available water content and estimating soil textural class using electro-magnetic induction. Paper No. 00-SW-044. EurAgEng, Warwick, UK. Silsoe Research Institute, Silsoe, UK. 28pp.

Welsh, J. P., G. A. Wood, R. J. Godwin, J. C. Taylor, R. Earl, B. S. Blackmore, and S. M. Knight. 2003a. Developing strategies for spatially variable nitrogen application in cereals part I: Winter barley. *Biosystems Engineering (Special Issue on Precision Agriculture)* 84(4): 481-494.

Welsh, J. P., G. A. Wood, R. J. Godwin, J. C. Taylor, R. Earl, B. S. Blackmore, and S. M. Knight. 2003b. Developing strategies for spatially variable nitrogen application in cereals part II: Wheat. *Biosystems Engineering (Special Issue on Precision Agriculture)* 84(4): 495-511.

Wilson, J. N. 2000. Guidance of agricultural vehicles: A historical perspective. *Computers and Electronics in Agriculture* 25: 3-9.

Wollring, J., S. Reusch, and C. Karlsson. 1998. Variable nitrogen application based on crop sensing. Proceedings No. 423. York, UK: The International Fertilizer Society, 28pp.

Wood, G. A., J. C. Taylor, and R. J. Godwin. 2003a. Calibration methodology for mapping within-field crop variability using remote sensing. *Biosystems Engineering (Special Issue on Precision Agriculture)* 84(4): 409-423.

Wood, G. A., J. P. Welsh, R. J. Godwin, J. C. Taylor, R. Earl, and S. M. Knight. 2003b. Real time measures of canopy size as a basis for spatially varying nitrogen at different seed rates in winter wheat. *Biosystems Engineering (Special Issue on Precision Agriculture)* 84(4): 513-531.

# Chapter 21

# Argentina

Rodolfo Bongiovanni
Jess Lowenberg-DeBoer

## *INTRODUCTION*

Precision agriculture does nothing to improve commodity prices, transportation infrastructure, or credit availability in Argentina, but some farm managers look for precision technology to help them maintain their position as low-cost producers of grain. Similar to producers in North America, they ask themselves what parts of this technology will prove practical and profitable for their conditions. These observations are based on information gathered from various sources and by several trips to the Argentine Pampas visiting farms and meeting farm managers, crop consultants, custom operators, and agribusiness people in the provinces of Buenos Aires, Santa Fe, and Córdoba. These visits were part of a long-term collaboration in Argentina of Purdue University, especially with the precision farming program at the National Institute for Agricultural Technology (INTA), Manfredi Experimental Station, Province of Córdoba; and with the Fundacíon Producir Conservando (FPC).

Overall, the potential of precision farming in Argentina's agriculture is to reduce costs in grain production, increase productivity, and make input use more efficient. A broader perspective on precision agriculture includes the use of whole farm information systems for input management over time and space, to optimize field operation logistics, to supervise employees in the field, to manage production risk, to market differentiated products, to provide traceability for food safety, and to document environmental com-

The authors would like to appreciate anonymous reviewers as well as Cristiano Casini, PhD, coordinator of the Department of Crop Production, INTA Manfredi, Argentina, for their contributions to this chapter.

*Handbook of Precision Agriculture*

doi:10.1300/5627_21

pliance. Yield monitors may also be used as a tool to diagnose weeds, pests, diseases, drainage, and fertility problems and differences in tillage systems, and therefore as a decision-making aid for choice of hybrids, varieties, or pesticides.

The objective of this article is to present Argentina as a case study in the adoption of precision agriculture because the country is growing the same crops, with the same technology, and in a similar climate as North America where precision agriculture has become widespread, although with very different economic and institutional situations.

The first section of this chapter outlines some background information on Argentine agriculture, the second section describes the current status of precision agriculture, the third section explains the research efforts currently underway, the fourth section details the factors that affect adoption, and the fifth and last section summarizes the prospects for the future.

## *ARGENTINE AGRICULTURE*

### *Characteristics of the Agricultural Sector*

Argentina is the second largest country in South America (after Brazil), with a total continental area of 2,791,810 km$^2$, about three-tenths the size of the United States. Nine percent of the area is arable land, located mainly in the fertile plains of the Pampas, in central Argentina. Grain production in Argentina is about 70 million tons per year. Agriculture and the agricultural food industry represent 12.8 percent of gross domestic product (GDP). Primary agricultural production (PP) represents 4.8 percent of GDP, and manufactures from agricultural origin (MOA) represents another 8 percent of GDP. As much as 56 percent of the country's total exports come from PP and MOA (CEI, International Economy Center, 2006). In the 2004/2005 crop season, total grain production from main crops was 84,066,000 metric tons. Main annual crops planted during the 2004/2005 crop season were soybeans (38,300,000 tons in 14,399,998 hectares), wheat (15,970,130 tons in 6,262,865 hectares), corn (20,482,572 tons in 3,403,837 hectares), sunflower (3,662,109 tons in 1,966,599 hectares), sorghum (2,894,250 tons in 617,452 hectares), and peanuts (444,800 tons in 211,090 hectares) (SAGPyA, Secretary of Agriculture, Livestock, Fisheries and Food of the Ministry of Economy and Production 2006). About 40 percent of the current planted area is under no-tillage systems (Kirschbaum, 2004).

### *Characteristics of Farms and Farm Structure*

Argentina is home to about 333,000 farms, and about two-thirds of them are in the Humid Pampas, which includes the Provinces of Buenos Aires, Santa Fe, and Córdoba. This is down from 421,000 farms in 1988. Average farm size in this region is 533 ha, up from 400 ha in 1988 (INDEC, 2002). About 35 percent of the farms are specialized in crop production, 20 percent are beef ranches, and 45 percent are mixed production, although these proportions vary with grains and beef price cycles. According to Kirschbaum (2004), about 30 percent of the farms produce 70 percent of the total agricultural production. In the last few years an increasing number of large, professionally managed operations have been established that plant more than 5,000 ha, paying rental rates ranging from 750 kg of soybean·ha$^{-1}$ per year to 1500 kg of soybean·ha$^{-1}$ per year of good cropland (i.e., approximately from 150 to 300 U.S. dollars/ha/year). Average purchase price for the best cropland ranges from 8,000 U.S. dollars/ha in the inner ring of the Pampas to 4,000 U.S. dollars/ha in the outer ring of the Pampas. The average rate of return for the average farm is 15 percent, although some large farms obtain average rates of returns as high as 90 percent due to economies of scale and to adoption of technology (no-till, biotechnology, irrigation, and precision agriculture). Some farm managers plant more than 50,000 ha of corn and soybeans per year in owned and rented land.

It is important to note that many crop decisions in Argentina are made by farm managers. Unlike much of the rest of the world where farms are typically operated by farmers who manage the farm business as well as doing some of the physical work (e.g., driving tractors and combines), many farms in the Argentine Pampas are owned by absentee landlords who hire a manager to run the farm business. The farm manager usually hires employees or custom operators that do the physical work of production.

In Argentina high taxation is placed on the farm sector, in contrast to subsidization elsewhere. Taxation is done directly through export taxes and indirectly through protection and other restrictions on trade. Currently, export taxes on grains are as high as 23.5 percent for soybeans. On the other hand, macropolicies in the 1990s that caused the real exchange rate to decline (overvaluing the peso) have also had a negative effect on agriculture, which is more tradable than nonagriculture. In addition, a large deficit financed by borrowing had a negative effect on agriculture because it is more capital intensive than nonagriculture (Mundlak et al., 1989). Economic policies implemented in Argentina caused agricultural growth to lag behind compared to other countries with similar potential. Strong changes took place during the 1990s in Argentina's agriculture: opening to free trade, strengthening of the South American Common Market (MERCOSUR), and internal reforms (deregulation, privatization, and tax reform) that caused Argentine agricultural exports to almost double (Casaburi et al., 2001). In 2002, the Argentine peso—which had been attached 1:1 to the U.S. dollar—was devalued to the current relationship of 3:1 following the 2001 socioeconomic and political crisis. This helped boost agricultural production and exports, despite high export taxes. According to *The Economist* ("The Big Mac index," 2006), the Argentine peso is 49 percent undervalued against the dollar.

Public agricultural research in Argentina is carried out mainly by INTA (www.inta.gov.ar), an autonomous organization of the Secretary of Agriculture, Livestock, Fisheries and Food that operates 12 research institutes in the capital city area, 15 regional centers for research and extension, and 42 experimental stations throughout the country. New research results from these stations are provided to farm managers through a network of 240 extension agencies. About 1,500 staff members are actively involved in the provision of research and extension services (INTA, 2006). Universities also do research in agriculture, although they are more focused on education. Other public research agencies also provide funding and grants for research, such as the Secretary of Science, Technology, and Innovation (SECyT), Agencia Córdoba Ciencia, etc. It is worth noting that Argentina invests only 0.35 percent of its GDP in science and technology, one-fifth as much as Brazil (Román, 2006).

## CURRENT STATUS

Adoption of precision agriculture in Argentina began in early 1996 with the launching of the precision agriculture project at the Manfredi Experiment Station of INTA. This program was extended to the national level in 1999, and it currently includes five experimental stations in five provinces (Buenos Aires, Córdoba, Santa Fe, Tucumán, and Entre Ríos), with headquarters at Manfredi. Current adoption and evolution of precision agriculture technologies is shown in Table 21.1.

Updating the statistics provided by Griffin et al. (2004), with 1,500 yield monitors, Argentina is the second country in number of yield monitors (after the United States), and with 64 yield monitors per million ha, it is the fifth country in number in yield monitor density (after the United States with 391, Denmark with 247, Sweden with 119, and Great Britain with 107).

The 1,200 combines with yield monitor and GPS are potentially able to harvest up to five percent of the total annual harvest of the major crops due to their large capacity and because they are owned by large farm managers (40 percent) or custom harvestors (60 percent). The major North American

TABLE 21.1. Current adoption and evolution of precision-agriculture technologies in Argentina.

| Equipment (units) | 1997 | 1998 | 1999 | 2000 | 2001 | 2002 | 2003 | 2004 | 2005 |
|---|---|---|---|---|---|---|---|---|---|
| Yield monitors with GPS | 25 | 75 | 155 | 270 | 400 | 420 | 600 | 850 | 1,300 |
| Yield monitors without GPS | 25 | 125 | 145 | 180 | 160 | 180 | 250 | 350 | 200 |
| Yield monitors TOTAL | 50 | 200 | 300 | 450 | 560 | 600 | 850 | 1,200 | 1,500 |
| VRT planters | 1 | 2 | 3 | 4 | 5 | 6 | 7 | 8 | 20 |
| VRT fertilizer applicators (Terra-Gator trucks) | 2 | 2 | 2 | 2 | 6 | 6 | 6 | 6 | 7 |
| VRT fertilizer applicators (urea, knifed) | 0 | 0 | 0 | 0 | 0 | 0 | 4 | 4 | 30 |
| VRT fertilizer applicators (urea, spreaders) | 0 | 0 | 0 | 0 | 0 | 0 | 0 | 10 | 20 |
| VRT fertilizer applicators (UAN,[a] knifed) | 0 | 0 | 0 | 0 | 1 | 2 | 3 | 3 | 3 |
| VRT self-propelled fertilizer applicators (UAN)[a] | 0 | 0 | 0 | 0 | 0 | 0 | 5 | 9 | 60 |
| VRT fertilization, *total* | 3 | 4 | 5 | 6 | 12 | 14 | 25 | 40 | 120 |
| GPS guidance systems in airplanes | 35 | 60 | 100 | 160 | 200 | 230 | 300 | 450 | 470 |
| GPS guidance systems in applicators | 0 | 10 | 70 | 200 | 400 | 500 | 2,000 | 2,600 | 4,000 |
| GPS auto guidance | 0 | 0 | 0 | 0 | 0 | 0 | 0 | 3 | 20 |
| GPS guidance systems, *total* | 35 | 70 | 170 | 360 | 600 | 730 | 2,300 | 3,053 | 4,490 |
| N sensor | 0 | 0 | 2 | 2 | 4 | 5 | 6 | 7 | 7 |

[a]UAN = urea ammonium nitrate solution (32-0-0)

combine manufacturers are marketing equipment in Argentina that either has a yield monitor as standard equipment or is prewired for yield monitor installation. The Argentine market leader combine manufacturer, Don Roque, also prewires combines for yield monitoring. AgLeader, a U.S. yield monitor manufacturer, has been willing to work with Argentinean dealers and farm managers to modify their yield monitor for Argentine conditions. Because no killing frost occurs before harvest in Argentina, the stalks and straws are usually more humid than in the United States, and this has created some problems for U.S.-built yield monitors (J. Amuchástegui, 2005, personal communication, TecnoCampo, Córdoba). Thus, technical adoption requires some adaptations.

Differential global positioning system (DGPS) is available from a private company, D&E, who provides amplitude-modulated (AM) signal for 300 U.S. dollars (plus tax) per year or for 900 U.S. dollars per lifetime. The company broadcasts from three antennas that cover twelve central and northern provinces of Argentina, located at San Carlos Sur (Province of Santa Fe), Bolívar (Province of Buenos Aires), and Las Lajitas (Province of Salta). D&E is also the dealer and representative in Argentina for several North American and European brands of precision-agriculture hardware and software. Satellite DGPS is also available from Omnistar and from Racal at a cost of 1,500 U.S. dollars (plus tax) per year (www.omnistar.com, 2006). DGPS signal from John Deere (StarFire) is expected to be provided soon.

Guidance systems for agricultural chemical applicators are the fastest growing precision agriculture technology because they are easy to use and because of their immediate benefits. Several brands are available in the Argentine market. Relatively few applicators use foam markers; most use flaggers in both ends of the fields, i.e., two employees that indicate to the machine or to the airplane the width of the application. Guidance systems allow applicators to reduce employee health risks and to decrease labor costs.

In Argentina, only 50 percent of cropland is tested periodically for nutrient levels, 25 percent is tested annually before planting, and about only one-percent of the total crop area undergoes intensive soil testing (García, 2001). Only a small number of variable rate technology (VRT) machines for planting or fertilizer spreading have been sold. Local manufacturers of agricultural machinery, such as Agrometal and VHB (planters and fertilizer spreaders) have become associated with local manufacturers of variable rate equipment (Verion [www.verion.com.ar] and DiRocco [www.viarural.com.ar/mapa/buenosaires/tresarroyos.htm]) for planters and for fertilizers, especially nitrogen. Some big agricultural input suppliers such as Agroservicios Pampeanos, a subsidiary of the Canadian company Agrium, offer VRT fertilizer custom services in the northeast of the province of Buenos Aires, using applicator trucks equipped with Falcon VRT technology (Amuchástegui, 2005). The chemical company PASA Petrosur also offers variable rate application (VRA) with high-volume spreader trucks in the southeast of the Province of Buenos Aires and VRA of urea-ammonium nitrate solution (UAN) in Córdoba Province (J. Bassi, 2004, personal communication, PASA Petrosur), but so far its use in commercial farms is limited due to lack of soils information. The Argentinean dealer of the Norwegian company

Norsk Hydro also offers the Hydro Precise fertilizer system, a scanning system for tractors that uses a variable-rate fertilizer equipped with Dickey-john controllers to vary the rate of nitrogen on the go. Seven of these greenness sensors are currently available in Argentina, and they are marketed through D&E. TecnoCampo, an agricultural input supplier of Córdoba, and A&T, a crop consultant and service provider company, also report that some farm managers of the provinces of Córdoba and Santa Fe are using manual VRT by topography (Amuchástegui, 2005 personal communication, Tecno-Campo, Córdoba).

More than 80 companies provide integral precision agriculture consultant services to farm managers. Service packages include soil survey with GPS, including digital elevation maps, soil compaction measurement, georeferenced soil sampling, soil testing in U.S. laboratories, georeferenced crop scouting, satellite images, creation of soil management zones, fertilization recommendations, soil mapping, etc. (AgroSat Argentina, 2005).

Remote sensing for agriculture in Argentina is also becoming popular, motivated by four key drivers, in descending order of importance: First, satellite images are required by law (in the province of Buenos Aires) to obtain emergency tax exemptions for flooding (Testa, 2001). Second, some banks also use satellite images before granting loans for production or for land purchase. Third, large farm managers use remote sensing to determine the quality of the land and the risk of flooding before buying or renting. Last, remote sensing is used by agronomists/crop consultants to determine aptitude areas, green indexes, topography, etc.

The reception of satellite information (Landsat 5 and 7) in Argentina has been done since early 1997 by the federal government through antennas of the National Commission on Space Activities (CONAE), located 30 km west of the city of Córdoba. Landsat 5 images have been relatively inexpensive compared to costs in the United States, which has allowed public institutions and private companies to profit from this technology. Currently, Landsat 7 images are offered at 150 U.S. dollars per full image with a resolution of 30 meters by 30 meters. Farm managers can obtain digitized images of their farms, ranging from 20 U.S. dollars for the basic image up to 100 U.S. dollars for images with a digitized map, area of the fields, and a written report with an interpretation from INTA Castelar (Mercuri, 2005).

The program "Harvest Forecast" of the Climate and Water Institute of INTA Castelar (province of Buenos Aires) has been using remote sensing and GIS since 1991 to determine the planted area for each crop (P. Mercuri, 2006, personal communication, INTA Castelar). This information, combined with crop simulation models (e.g., CERES-Wheat), has allowed INTA Castelar to predict grain production. The importance of this tool has been proven during the large-scale flooding that took place in the past few years. Currently it is possible to obtain Landsat satellite images for any given location every sixteen days, corresponding to the satellite recurrence cycle (Mercuri, 2005). Many farm managers and agronomists also use this information, in combination with the available soil maps, to determine the productive aptitude of soils before buying or renting land. Some of them have also used this information for restructuring field divisions.

It is important to highlight that the greatest advantage of the use of satellite images in Argentina is due to three main factors: The first is the lack of detailed cartographic information. The available soil maps are only at a scale of 1:50,000, and no digital elevation models, such as those used in the United States to analyze spatial variability in agricultural areas, are available. Second, the images from multispectral cameras are very expensive because they are offered by foreign companies. INTA Castelar has recently purchased the required instruments, and they expect to begin offering the service soon at a lower price. Other local private companies are also entering the market. The third factor that leads to an increased use of satellite images is that the panchromatic aerial photographs currently available are outdated because they are the same photos taken for the development of the soil maps in the late 1960s and early 1970s. Therefore, satellite images are an important resource when it comes to time series images and to up-to-date information.

Regarding GIS software, Farm Works is by far the most used among farm managers. D&E is the representative in Argentina. Using a push marketing strategy, they have been offering training courses for farm managers and agronomists in the use this software in order to provide incentives for the sale of GIS and other precision agriculture products. This software has the advantage of being available in Spanish and it has a relatively low price.

Most researchers use ArcView (from ESRI) GIS software, but its cost in Argentina is a disadvantage: 2,300 U.S. dollars for the basic software, as compared to 1,250 U.S. dollars in the United States. The company CentroGIS (http://www .cursogis.com.ar), which is related to the Department of Architecture of the University of Buenos Aires, offers training workshops for professionals of different disciplines on a regular basis, on ArcView, ENVI, remote sensing, and aerial photography. Some crop consultants use SSToolbox, a GIS software built on ArcView especially designed for agriculture by SST Development Group, Stillwater, Oklahoma. (http://www .sstdevgroup.com/). The representative of ESRI in Argentina is Aeroterra (http://www.aeroterra.com.ar/), a firm that also offers training courses and remote sensing images from different satellite companies, such as Ikonos and Radarsat. The price of the high-definition Ikonos images (6,000 U.S. dollars) in Argentina also reflects a big markup as compared to the prices in the U.S. (3,000 U.S. dollars).

The number of users of IDRISI, a GIS software developed by Clark University, is also increasing. Main advantages are its low price (600 U.S. dollars for the academic version), it is raster oriented, and it comes with extensive client support. IDRISI has expanded its popularity at many Argentine universities and most recently to other public and private institutions (INTA, Secretary of Agriculture, Livestock, Fisheries and Food of the Ministry of Economy and Production [SAGPyA], etc.). An IDRISI Resource Center (IRC) (http://crean.org.ar/Idrisi/) located at the University of Córdoba, Department of Agronomy (CREAN [Centro de Relevamiento y Evaluación de Recursos Agrícolas y Naturales]), provides training courses for agronomists, biologists, geologists, etc. CREAN also provides consultant services on remote sensing and GIS application in agriculture, forestry, and weather impact.

Insurance companies, such as the cooperative "La Dulce" in the southeast of the Province of Buenos Aires, have also pushed precision agriculture in Argentina. They organized the Agro Informática 2000 together with crop consultant company INTA, Potash and Phosphate Institute (PPI-INPOFOS) and Purdue University (http://www.expoinfoag.com.ar/), which took place December 8-10, 2000. Agro Informática is the Argentine version of the InfoAg Conference (http://farmresearch.com/infoag/) that is organized every other year by the PPI in the United States.

### *Research Efforts: Precision Agriculture Project*

The Manfredi experimental station of INTA, province of Córdoba, is the base for the precision agriculture project. Main tasks of this project are extension and research. The project also includes researchers in other experimental stations, such as Marcos Juarez (Province of Córdoba), Castelar (Province of Buenos Aires), Paraná (Province of Entre Ríos), Famaillá (Province of Tucumán), Oliveros (Province of Santa Fe), and Venado Tuerto (Province of Santa Fe).

Extension activities include extension workshops for farm managers and agronomists in the use of GIS software and fertilization recommendations as well as training courses for farm managers/agronomists and operators of agricultural machinery. The project is also working closely with other government research agencies from Chile (Instituto Nacional de Investigacíon Agropecuria [INIA]), Brazil (Empresa Brasileira de Pesquisa Agropecuária [Embrapa]), Uruguay (INIA), and Paraguay (Centro Regional de Investigacícola [CRIA]). All of these research institutions sponsor a precision agriculture avenue in the Cooperative Program for the Technological Development of Food and Agribusiness of the Southern Cone of America

(PROCISUR). As of March 2006, nine meetings and training workshops have been carried out so far in Argentina, Bolivia, Brazil, Chile, Paraguay, and Uruguay as part of this program.

In mid-2000 the project launched a Web site: http://www.agriculturadeprecision.org/, with updated information about precision agriculture technologies and the research being carried out in Argentina. This Web page currently has an average of 685 daily visits from Argentina and from other Latin American countries.

On-farm trials are currently being conducted in farms of the area of influence of INTA Manfredi as part of a precision agriculture project funded by SECyT to study the contributions of site-specific management to sustainable agriculture. The model involves the estimation of the site-specific crop response to nitrogen using spatial econometrics techniques. This study (Bongiovanni, 2002) is significant because it is the first application of spatial econometrics to yield monitor data. From the standpoint of the Argentine farm manager, this analysis is important because it offers a way to estimate the profitability of one type of precision agriculture technology.

Several other on-farm trials involving precision agriculture techniques are being conducted in the Manfredi Experiment Station area by multidisciplinary teams aiming to show the effects of different management practices on crop yields. On-farm trials are also being carried out in collaboration with AAPRESID and with the Potash and Phosphate Institute of Argentina (INPOFOS).

Efforts are underway at INTA Manfredi to improve techniques of analysis of satellite images in order to detect crop stress in peanut and to correlate it with the incidence of diseases, especially aflatoxins. Their goal is to differentiate areas of low and high disease risk, so they can be harvested and stored separately.

## *FACTORS THAT FAVOR ADOPTION*

Several factors suggest that yield mapping and other information systems technologies will be adopted more rapidly in Argentina than they have been in the United States, at least on larger farms.

### *Large Farm Operations with Relatively High Capital per Worker in the Pampas*

Norton and Swinton (2001) hypothesize that precision-agricultural technologies will be adopted first in regions with large farming area and high capital investment per ha. Using Food and Agriculture Organization of the

United Nations (FAO) statistics, they show that when it comes to countries as whole, only the United States, Canada, and Australia fit those conditions, but they hypothesize that although small, subsistence farms pull down national average farm area, some regions within developing countries may have the necessary preconditions for profitable use of precision technology in the near future (e.g., the Argentinean Humid Pampas). Norton and Swinton (2001) suggest that the early adopters of precision agriculture technologies might be the countries where agricultural land and capital are abundant. Based upon the relative abundance of land and capital relative to labor, their work proposes that Canada, the United States, and Australia would be the prime candidates for adoption, followed by Western Europe and Argentina. Argentina has a higher ratio of land to labor than does Eastern Europe, Brazil, Asia, and the rest of Latin America. Nevertheless, in Argentina the adoption patterns may differ substantially from those in the United States and Canada because of differences in agronomic and economic conditions.

### *Highly Educated and Professional Farm Management and Crop Consulting*

Since total Argentine literacy rate is 96.2 percent, almost all farm employees are literate. Nearly all farm managers have at least a high school degree, and some farm managers have a university degree. Approximately 90 percent of farm managers in the Humid Pampas use cellular phones, which may be a proxy for technology adoption given the long distances between farms and cities. About 10 percent of the agronomists/crop consultants speak English at a working level, although that percentage drops to about 1 percent for farm managers. Table 21.2 indicates the percentage of farm managers in the Humid Pampas who use computers and of farm managers who are connected to Internet. According to a recent survey, 18 percent of agribusiness companies have already developed a Web site, and 70 percent are planning to develop one in the near future (Campo21, 2001).

TABLE 21.2. Percentage of farm managers in the Humid Pampas who use computers and who are connected to Internet.

| Percentage of farm managers who | 2000 | 2001 | 2002 | 2003 |
|---|---|---|---|---|
| Use computers | 14 | 16 | 33 | 39 |
| Use computers and Internet | 28 | 32 | 39 | 50 |

### *Technology Available from North America*

Due to the similarities in the agricultural production of the U.S. Corn Belt and the Argentine Humid Pampas, most of the mechanical, chemical, and biotechnological technology available in the United States can be imported with little modification to Argentina, except for no-till planters, for which local developments or adaptations need to be made due to the particular conditions and to the double crop of wheat and soybeans. Approximately 1,000 Argentinean farm managers, agronomists, and manufacturers visit the Farm Progress Show in the United States every year to gather the latest developments in technology. As an example, INTA Manfredi has been organizing technical tours of the United States for the past fifteen years, visiting manufacturers, universities, and the Farm Progress Show. Nevertheless, machinery and equipment are more expensive in Argentina due to transportation costs and tariffs.

### *Farm Managers Are Hungry for Better Crop Data*

As farm size increases in Argentina, farm managers willing to expand their operations need more and better information to rent or to purchase land. At the same time, large farm managers need information about soils and yields to make fertilization and other management plans.

Yield maps, as-applied maps, and other precision data are more valuable in Argentina because the manager of the farm often does not observe crop conditions and yield from the tractor or combine seat. In the United States, where farmers usually operate their own equipment, yield maps often help to quantify variability that producers already know about. In Argentina the maps often reveal variability that managers had not observed directly. In addition, yield maps, as-applied maps, and other precision data provide an important quality control on work being done by employees or custom operators.

Use of yield monitor and other spatial data in Argentina will be facilitated because they will have less need to pool data across farms. In the United States, farmers will probably need to pool precision data to achieve the full benefit of the technology. Most Corn Belt farms are not big enough to provide the range of conditions needed to begin to identify patterns without designed experiments. On the many large farms in Argentina, the need to pool data is reduced. Many farms plant more than 5,000 ha. Depending on soil variability and management, 5,000 ha may provide enough data to begin to identify patterns without specifically designed trials.

If data pooling is necessary, it may be easier in Argentina than it has been in the United States. Argentine agriculture has a long history of data sharing. For example, the groups of the Argentine Association of Regional Experimental Consortia (AACREA) meet regularly to share agronomic and accounting information. In part, this occurs because Argentinean growers have only recently started to compete for rental land. In the past they usually focused on improving management on the acreage they already controlled. In the United States, one of the reasons for reluctance to share data is that producers do not want yield and other information in the hands of those who will be competing with them to rent the farm down the road.

Yield monitors in Argentina are usually used over larger areas than in the United States. Because of the long harvest season and custom harvesting, large combines often cover 3,500 to 5,000 ha per year, more than twice as much as the area covered in the United States. By spreading the cost of the yield monitor and GPS over more area, the per-ha cost is lowered.

## *FACTORS THAT DISCOURAGE ADOPTION*

For the adoption of precision agriculture in Argentina at least four crucial differences from the U.S. situation exist: higher investment cost, more risk, less management-induced soil variability, and the widespread use of custom operators.

### *High Investment Cost*

Currently, average real interest rates for production loans are in the range of 16 to 24 percent annually, and longer-term credit is not easily available. Argentina is having problems restructuring its external debt after the default of 2002, so interest rates are not expected to fall in the short term. Because most precision technology must be imported into Argentina, the cost is somewhat higher than the same technology in the United States due to import duties. For instance, a yield monitor with GPS costs twice as much in Argentina than in the United States.

### *High Risk*

Grain prices in Argentina fluctuate greatly, which affects farm income. The Argentine government has no programs to stabilize grain prices or farm incomes. High interest rates, difficulty in obtaining credit, higher technology costs, and risk discourage investment in all new technology, including precision agriculture. Access to insurance is limited: the Argentinean Secretary of Agriculture, Livestock, Fisheries and Food of the Ministry of Economy and Production (SAGPyA) indicates that only 9 percent of the agricultural production is insured as opposed to 42 percent in the United States, 52 percent in Canada, and 30 percent in Spain. In the crop season 2000-2001, insurance companies insured crops for a value of 52 million U.S. dollars, which is considered low, because it is expensive, and because only large operations can afford it. Nevertheless, hail insurance is increasing, and some farm managers are hedging risks with futures and options contracts. At the same time, low public investment in infrastructure—transport, drainage, and research—does not help to decrease production risks. It costs about twice as much to barge grain on Argentine rivers as what it does on the Mississippi (Tennessee, 1998). Most of the paved roads are two lanes, two ways, and access to farms usually includes a large portion of unpaved roads, making it difficult for trucks to transport grains to the local elevators, especially in the rainy season. Large areas of the Humid Pampas have serious drainage problems that cause seasonal flooding.

### *Low Management-Induced Soil Variability*

Much of the Pampas is characterized by large fields that have only recently been converted from pasture rotations to continuous cropping. In most places management-induced variability is less than it would be in Europe or North America. Few former fence rows, old lanes, or feedlots exist. Fertilizer use is relatively new in Argentina, and little opportunity is available to create soil fertility patterns through the application or misapplication of fertilizer. INTA officials and crop consultants involved in precision agriculture argue that Argentina has more natural soil variability than the United States. The natural soil variability in Argentina is substantial, even on the flat, seemingly homogenous Pampas, where characteristics such as salinity and small differences in topography can greatly influence yields.

Currently, many farms in Argentina use no fertilizer at all given the current relative prices. Even the well-managed, highly capitalized farms use rates that are low by U.S. standards, on corn 35 to 65 $kg \cdot ha^{-1}$ of nitrogen (N) and 20 to 45 $kg \cdot ha^{-1}$of $P_2O_5$ on soils of 1 to 3 percent organic matter, and Bray phosphorus (P) tests of 6 to 15 ppm. Normal values of Bray P tests should be 20 to 25 ppm (García, 2006). According to García (2006), farms are losing between 1.7 and 4 ppm of p every 3 years. This does not provide much opportunity to profit from cutbacks or reallocation of fertilizer. Fertilizer consumption in Argentina has notably increased in the past ten years from 300,000 metric tons in 1991 to more than 2,500,000 metric tons in 2004. However, nutrient balance is still negative for soils of the Pampas. Recent estimates show replenishment levels of 25 to 30 percent for N, 45 to 50 percent for P, 7 to 10 percent for Sulfur (S), and less than 1 percent for potassium (K), being almost nil for other essential nutrients (García, 2006).

### *Widespread Use of Custom Operators*

In Argentina farm management and equipment operation are often separate enterprises. Many farm operations, including planting, are done by custom operators. Many producers own almost no equipment. Harvesting is often done by crews similar to the custom cutters that operate in the U.S. Great Plains and paid with a percentage of the harvest, often 8 or 9 percent. For yield monitoring, custom harvesting can pose a problem. Combine operators are often employees of the custom harvesting business. They often have only a grade school education and have never used a computer. They are motivated to harvest as much grain as they can as quickly as possible. They are not necessarily motivated to collect good data. A few custom harvestors have seen yield mapping as an extra service that they can provide their clients. Even if no added fee is charged for yield mapping, it cements long-term relationships with clients. Custom operators also like having an estimate of harvest quantities from the yield monitor, which is independent of the farm scales or the grain dealer. Therefore, several incentives exist that are prompting its adoption. At some point in time, data management will become a challenge for both custom operators and farm managers when it comes to transforming information into knowledge, and then to transform that knowledge into management decisions. Effort is being made by INTA Manfredi and by local input suppliers to offer training courses for operators.

### *Training of Agronomists and Farm Managers to Gather Useful Site-Specific Information*

Site-specific agricultural technologies make possible the precision application of fertilizers and other inputs. However, a key information gap lies between the precision mapping of soil characteristics and the precision application of inputs. That gap is the fertilizer recommendation. Recommendations are typically based on large-scale regional research that has been synthesized across many sites. By design, state- or regional-level agronomic recommendations omit local site characteristics, which undermine their suitability for guiding site-specific input applications (Swinton et al., 2001).

Few resources for local research to adapt technologies to Argentine conditions exist: Both INTA and the universities do not have the necessary budget for research on precision agriculture. All research carried out so far is possible through the collaboration of commercial companies and farm managers who are willing to do on-farm trials.

## *PROSPECTS*

### *Rapid Adoption of Yield Monitoring*

The most successful application of precision agriculture has been yield monitors. The monitors provide information on something that farm managers are passionately interested in: crop yields. From field testing of a few units in 1996, yield monitor use has grown rapidly (Lowenberg-DeBoer, 2000). New combines already come with a yield monitor installed or optional, which provides large farm managers who own 40 percent of the large combines with a direct way to measure productivity of the fields they farm. Custom harvesters own 60 percent of the new large combines, and they see yield monitoring as a way to increase customer loyalty.

### *Rapid Adoption of GPS Guidance*

Guidance systems for agricultural chemical applicators are the fastest growing precision-agriculture technology because they are easy to use and because the benefits are immediate.

### *Lag in VRT*

We expect the use of VRT in Argentina to grow slowly because of the high cost of soil sampling, low management-induced variability, and current low rates of fertilizer use. Crop consultants and farm managers cite soil laboratory analysis fees ranging from 20 to 40 U.S. dollars per sample for a complete soil analysis, as compared to around 6 U.S. dollars per sample in the United States, which makes unfeasible the grid-sampling approach. The United States has a large volume of samples, and the labs can justify owning equipment for inductively coupled plasma-membrane analysis (ICP), which has a high initial cost (125,000 U.S. dollars in the United States) and high maintenance costs due to the price of consumables and the need for trained technicians. Laboratories using ICP enjoy economies of scale because they provide services nationwide (sometimes worldwide) and because, in addition to soil tests, they use it for water analysis, detection of pollutants in the air, etc. In Argentina (and Australia as well), the volume of samples has been low and so they use more expensive manual lab testing methods.

Those costs are prohibitive for grid sampling of the type used in the United States. Even for a guided sampling plan with larger management zones the cost would be quite high.

## *CHALLENGES*

Key challenges for precision agriculture in Argentina are discussed in the following sections.

### *Training Combine Operators to Collect High-Quality Data*

In response to this problem, INTA Manfredi offers courses for combine operators at the experiment station as well as upon request from several organizations. The agronomists at INTA hope to convey the message that combines are harvesting data as well as grain, and that the data are valuable. The challenge is to build in a price incentive for the system to really work.

### *Training Agricultural Economists and Agronomists to Carry Out Profitability Analysis of VRA*

Yield monitors have been in the market for almost ten years, but everywhere it has proven difficult to link yields to crop conditions (e.g., soil tests, input application, topography). Profitability analysis often indicates that fertilizer VRA is not profitable. One hypothesis is that the whole-field recommendations used to make VRA do not reflect site-specific response differences. Because the economics of precision agriculture are site specific, analyses are needed in Argentina.

### *Interpreting Data*

Interpreting data is a challenge everywhere precision agriculture is used. In Argentina a special challenge may exist in helping farm managers who do not spend hours in the tractor or combine seat to understand the variability of their fields. An additional challenge is adapting on-farm research designs. On-farm research methods developed in the United States may not transfer directly. For example, in the United States split-planter designs work well for corn because many producers have corn heads that are exactly half the width of their planters. This is not the case in Argentina. On-farm research design need to be developed to fit Argentine equipment and the questions that producers there ask.

### *Adapting VRT to Argentine Conditions*

Adapting VRT to Argentine conditions will probably require finding ways to make use of low-cost data instead of heavy reliance on grid-soil sampling. This low-cost data might include yield maps, topographical maps, satellite images, aerial photographs, and eventually remote sensing and soil sensors.

### *Developing a Local Precision Farming Research Base*

Developing such a base will probably be the most important challenge for Argentine agriculture. Precision-farming technology developed in North America or elsewhere will need to be adapted to Argentine conditions. In the past Argentine agriculture has relied heavily on borrowing North American technology. For precision agriculture, borrowing may not be entirely successful, because both the agronomics and economics of precision agriculture are site specific. As in North America, a combination of private-sector and public-sector research will probably be necessary to help precision technology fulfill its potential in Argentina.

## REFERENCES

AgroSat Argentina. 2005. Commercial brochure, 6 pp.

The Big Mac index. 2006. The Economist. January 12, 2006. Available at: http://www.economist.com/markets/bigmac/displayStory.cfm?story_id=5389856. (Accessed on March 1, 2006.)

Bongiovanni, R. 2002. A Spatial Econometric Approach to the Economics of Site-Specific Nitrogen Management in Corn Production. Unpublished doctoral thesis, Agricultural Economics, Purdue University, West Lafayette, IN, 306 pp.

Campo21. 2001. Soluciones de internet para los agronegocios. Available at: http://www.campo21.com/jsp/Principal.jsp. (Accessed September 1, 2005.)

Casaburi, G., E. Perona, and A. Reca. 2001. Preparados para alimentar al Mundo? El Agro Argentino Frente al Siglo 21. IERAL de Fundación Mediterránea, pp. 7-27.

CEI (International Economy Center). 2006. Statistics. Available at: http://www.cei.gov.ar/home.htm. (Accessed March 1, 2006.)

García, F. 2006. La nutrición de los cultivos y la nutrición de los suelos. Instituto de la Potasa y el Fósforo (INPOFOS). Informaciones Agronómicas 29: 13-16.

Griffin, T. W., D. M. Lambert, and J. M. Lowenberg-DeBoer. 2004. Testing appropriate on-farm trial designs and statistical methods for precision farming: A simulation approach. Presented at the Seventh International Conference on Precision Agriculture and Other Precision Resources Management, July 25-28, 2004, Minneapolis, MN, 15 pp.

INDEC (Instituto Nacional de Estadísticas y Censos). 2002. Statistics of the Argentinean agricultural sector. Available at: http://www.indec.mecon.gov.ar/. (Accessed December 30, 2004.)

INTA. 2006. National Institute for Agricultural Technology. Available at: http://www.inta.gov.ar/ins/en/englishvers.htm. (Accessed March 1, 2006.)

Kirschbaum, R. (Ed.). 2004. El Gran Libro de la Siembra Directa. Buenos Aires, Argentina: Clarín.

Lowenberg-DeBoer, J. 2000. Precision agriculture in the U.S.A. and potential in developing countries. Proceedings of the PROCISUR Meeting, June, 2000, Buenos Aires, Argentina.

Mundlak, Y., D. Cavallo, and R. Domenech. 1989. Agriculture and economic growth in Argentina. International Food Policy Research Institute. Research Report 76, November.

Norton, G.W. and S.M. Swinton. 2001. Precision agriculture: Global prospects and environmental implications. In G.H. Peters and P. Pingali (Eds.), Tomorrow's Agriculture: Incentives, Institutions, Infrastructure and Innovations: Proceedings of the 24th International Conference of Agricultural Economists (pp. 269-286). London: Ashgate.

Palermo, A. 2001. El tipo de cambio, los vaivenes del mercado, el clima y la tecnología exigen al hombre de campo mantener el paraguas abierto. La Nación. April 7, 2001. Available at: http://www.lanacion.com.ar/suples/campo/0114/p01.htm.

Román, V. 2006. Argentina fue elegida como uno de los referents. Clarín. February 15. Available at: http://www.clarin.com/diario/2006/02/15/sociedad/s-02601.htm.

SAGPyA (Secretaría de Agricultura, Ganadería, Pesca y Alimentación del Ministerio de Economía y Producción). 2006. Statistics. Available at: http://www.sagpya.mecon.gov.ar/, http://www.mecon.gov.ar/default_english.htm. (Accessed on March 1, 2006.)

Swinton, S., R. Bongiovanni, J. Lowenberg-DeBoer, and D. Bullock. 2001. Assessing the value of precision agriculture data: On-farm nitrogen response in Argentina. AAEA Spatial Analysis Learning Workshop, August 4, 2001. American Agricultural Economists Association Annual Meeting, August 4-8, 2001, Chicago, Illinois.

Tennessee, A. 1998. Global agriculture: Challenges for the future. Available at: http://www.cargill.com/today/speeches/tennessen.htm (Accessed on September 1, 2001)

Testa, A. 2001. Inundaciones: Utilizarán imágenes satelitales. Lo dispuso el gobierno bonaerense mediante un decreto. La Nación, April 21, 2001. Available at: http://www.lanacion.com.ar/suples/campo/0116/supl.asp?pag=P03.HTM&a=prem

# Chapter 22

# Brazil

Glaucio Roloff
Daniele Focht

## *INTRODUCTION*

In Brazil, the preferred term for site-specific crop management using information technology is precision agriculture (PA), probably because of the ease of translation into Portuguese and the image of modernity that it conveys. When progressive farmers are asked what they think of PA they commonly reply that they are aware of it, that it is an important technological development, and that sooner or later they will have to move toward adopting it. Their greatest concern related to adoption is the additional cost in relation to the perceived benefits. Second, they worry about the knowledge needed from their crop consultants and the training needs for their employees who handle the majority of field operations. When crop consultants are asked the same question, the common reply is that they are aware of PA but it is too complicated and too expensive, implying that it is still premature to recommend adoption. Their greatest concerns are centered around the cost:benefit ratio of PA within the current state of Brazilian agriculture as well as the need for training.

Even though first impressions are not generally favorable, when farmers and consultants take the time to analyze and realize that the soil attribute or yield maps of their fields often confirm their perceptions of yield zones within the field, their attitude often changes toward acceptance of the technology. They become enthused about new and improved ways to manage this variability. Interestingly, others have requested mapping services to prove that their fields are well-managed and uniform. These farmers often become somewhat frustrated when they are confronted by the mapped variability. The comprehension of this human element, made up also of managers and machinery operators, is important to direct extension activities toward long-term adoption of the PA system. It is also important to understand

*Handbook of Precision Agriculture*

doi:10.1300/5627_22

how the human element interacts with the conditions underlying crop production.

Brazil is a major producer of soybeans, sugar, cotton, orange juice, and coffee, which are important export commodities. Corn, dry beans, wheat, cassava, and potatoes are important as food sources and to reduce imports. These crops are mostly grown using modern, input-demanding technologies. The absence of subsidies for agriculture, the relatively high costs of agriculture chemicals and fertilizers that are set in U.S. dollars and their proportion in relation to total costs (Table 22.1), and the high cost of transportation require the Brazilian farmer be efficient to stay financially afloat. The promise of increased efficiency of the PA system is therefore alluring, but it must be made evident.

Even though the discussion on the actual and potential benefits of PA is ongoing worldwide, it certainly needs to be confronted at the country, region, field, and crop-specific level. Although recognizing the importance of off-site and environmental benefits of PA, this chapter will focus on PA's potential benefits within the field and the farm. We will also discuss the actual and perceived constraints for a wider adoption, current usage and strategies, and near-future developments in Brazil. Due to a lack of surveys and other sources of hard numbers and to few available scientific publications dealing with field size and commercial PA operations, the discussion has been confined largely to case studies, anecdotal evidence, and inferences from practical work and past experience. We have worked intensively with PA for more than seven years, both as researchers and consultants, testing and adapting the technology so that it can be adopted with small start-up costs and prioritizing actions that can potentially produce benefits on the

TABLE 22.1. Example of the relative costs of fertilizers and chemicals and area planted for some crops in 2005 in Brazil.

| Crop | Region | Fertilizers[a] | Chemicals[a] | Area ($10^6$ ha)[b] |
|---|---|---|---|---|
| Cotton | Savannas | 19 | 44 | 1.3 |
| Corn | South | 36 | 20 | 11.5 |
| | Savannas | 38 | 26 | |
| Spring wheat | South | 33 | 23 | 1.4 |
| Sugarcane | South | 13 | 5 | 5.9 |

[a]Percent of the variable costs from seeding to harvest, under no-tillage where possible (adapted from www.conab.gov.br/download/Safra/ CustodeProducao-SafradeVerao.pdf)

[b]Total area countrywide (www.agricultura.gov.br/spa/pagespa/index .html)

short term. Once farmers realize these benefits they may be more willing to move toward the PA system.

## *A BRIEF HISTORY OF PRECISION AGRICULTURE IN BRAZIL*

Studies dealing with components of the PA system, such as spatial variability, have been mentioned in the Brazilian literature since the 1980s (e.g., Vieira et al., 1983). A conceptual paper on variable rate technology was presented in the early 1990s (Balastreire, 1994). Launching PA as a practical system, though, is associated with two key happenings in the mid-1990s—the establishment of a consortium to provide services on PA, located in a major agricultural region on the central savannas in 1998 (www.plantiodireto.com.br/?body=cont_int&id=264), and the first Brazilian symposium on PA (I Simpósio Brasileiro Sobre Agricultura de Precisão) in 1996. The fates of these two pioneer studies are used to illustrate the evolution of PA in Brazil.

The original objective of the consortium was to provide a range of services from yield mapping to variable rate fertilization and seeding. Unfortunately, yield mapping was limited regionally with only four large-capacity combines available. Clients were sparse because the benefits regarding gains in return for the added cost of monitored yield and how this fits within a management system were not made clear. In addition, progressive farmers in Brazil tend not to rely on service providers for basic operations such as harvesting, fertilization, or spraying, and few if any less expensive variable-rate machinery or retro-fitting kits are available on the Brazilian market. The consortium suffered from the typical mistake of trying to do too much too soon and it folded in late 2000 after the parent companies stopped financing its operational deficit. This made potential service providers wary of entering the PA market, and left farmers wondering why they should venture into this new technology if fellow farmers that had the opportunity did not.

On the other hand, the first symposium, consisting of less than 50 participants (Balastreire, 2000), focused on conceptual and review papers. This humble beginning has grown to two conferences, held every alternating year, that bring together a few hundred researchers, practitioners, students, and a handful of farmers. In the most recent conference (III Simposio Internacional de Agricultura de Precisão, 2005), of the 64 papers, 17 were about crop monitoring, 13 about spatial variability of chemical and physical soil attributes, 8 about yield variability, 7 on remote sensing, and the remainder on a variety of issues. Since the first symposium the following has occurred:

1. A large increase in papers dealing with crop monitoring
2. A broadening of issues and crops
3. An increase in the relative participation of academia
4. A noticeable rise in the number of papers dealing with local technology

Notably absent in the symposia are papers dealing with the potential or actual economical and managerial benefits of PA within the various cropping systems in Brazil.

Other relevant activities include an academic seminar held in 2000 that had significant attendance and generated a book on PA (Borém et al., 2000), and included regional workshops, courses, and training sessions, held especially in southern and central Brazil, attended by crop consultants and farmers in numbers varying from tens to hundreds. A few academic dissertations have been produced on PA, with the bulk of them on spatial variability and remote sensing. On the other hand, PA seminars that once were a standard feature of yearly geoinformation meetings have been discontinued.

Even though degree of interest is fair among crop consultants, progressive farmers, and researchers, the current numbers of users of PA technology are dismally small (Table 22.2) compared with the size of Brazilian agriculture (Table 22.1). This is probably a result of several compounding factors, which are the following:

1. The relatively high cost of PA products in Brazil (Table 22.3), mainly due to high importation taxes and costs
2. The downturn on the farming economy beginning in 2004, especially because of depressed grain prices, droughts, and an increase in disease-control costs
3. The uncertainty of the economic benefits of PA
4. The steep learning curve required of crop consultants, farmers, and their employees.

On the other hand, factors favoring PA are the relatively small cost of soil-attribute mapping (Table 22.3) and the rising proportional costs of fertilizer and agricultural chemicals that follow the exchange rates. The challenge now is to demonstrate to farmers and crop consultants how they can adopt the PA system with reasonable start-up costs and whether its tangible and intangible benefits are worth the extra effort and cost.

## *PRECISION AGRICULTURE ON HIGHLY WEATHERED SOILS*

Crops and planted pastures in Brazil can be found on a large diversity of soils, climates, and field sizes. Yet, the bulk of grain, fiber, and sugarcane crops in Brazil are grown on rainfed and highly weathered Oxisols, with field sizes ranging from a few tens to several hundred ha within subtropical and tropical humid climates. Large amounts of land are also devoted to planted pastures and pasture-crop systems. No-tillage is increasingly becoming the preferred planting system for grain crops. This range of conditions is visualized in Figures 22.1 to 22.4 and detailed in Table 22.4. These fields are from widely separated locations and represent the range of the main agricultural soils. Particle size distribution varies widely among soils,

TABLE 22.2. Adoption of PA technology in Brazil.

| Description | Quantity[a] |
|---|---|
| Lightbar for surface guidance | 2,500 |
| Yield monitors (grain) | 1,200 |
| Self-propelled variable rate spreaders | 20 |
| Dedicated GIS software packages | 50 |
| Mapping of soil attributes (ha) | 2,000,000 |
| Corrective fertilization/liming using variable rate (ha) | 1,800,000 |
| Spreaders with variable rate kits | 100 |

[a]These are approximate rounded numbers as of March 2006, encompassing major brands and service providers and gathered using informal surveys.

TABLE 22.3. Comparison of costs[a] in U.S. dollars of PA products and services in the United States and in Brazil.

| | | United States | Brazil |
|---|---|---|---|
| Yield monitoring kit for grain combines | | 4,000 | 10,000 |
| Variable rate kit for planters, one drive | | 4,500 | 6,750 |
| Soil sampling and analysis (per ha): | 1 ha grid | 20-25 | 19-22 |
| | 2 ha grid | — | 13-15 |
| | 5 ha grid | — | 9-10 |

[a]Prices are approximate for major brand products and services in the United States–Midwest, and Brazil–South and Midwest, March 2006.

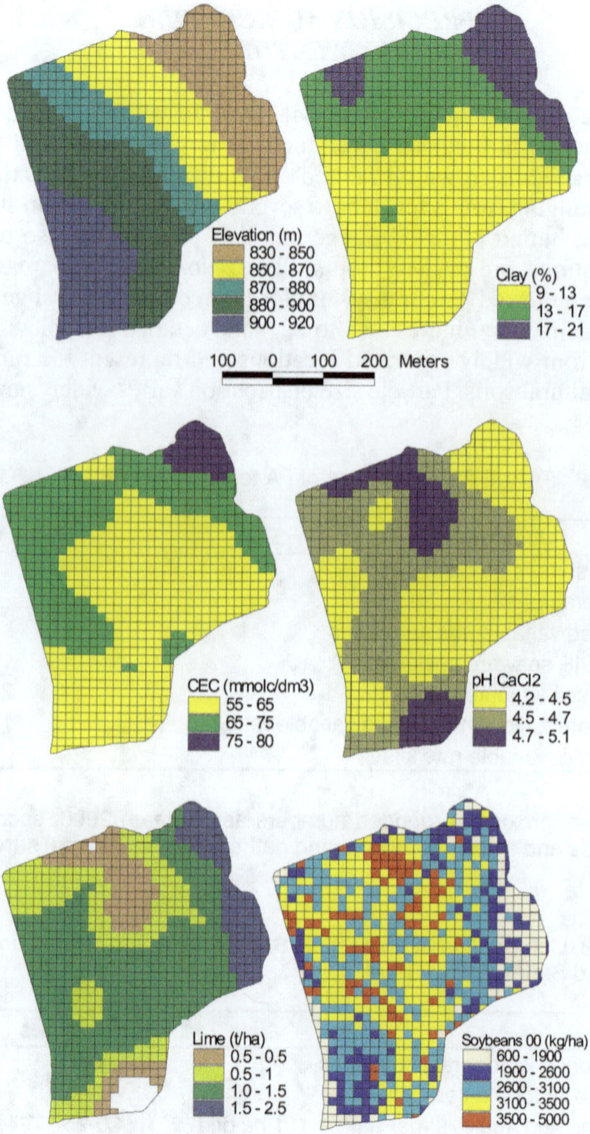


FIGURE 22.1. Soil topographic, physical, chemical, and productivity attributes of a commercial field with a rotation of grain crops under no-tillage; coordinates 24°54'S and 50°17'W, State of Parana, Brazil. (See corresponding PLATE 22.1 in the central gallery.)

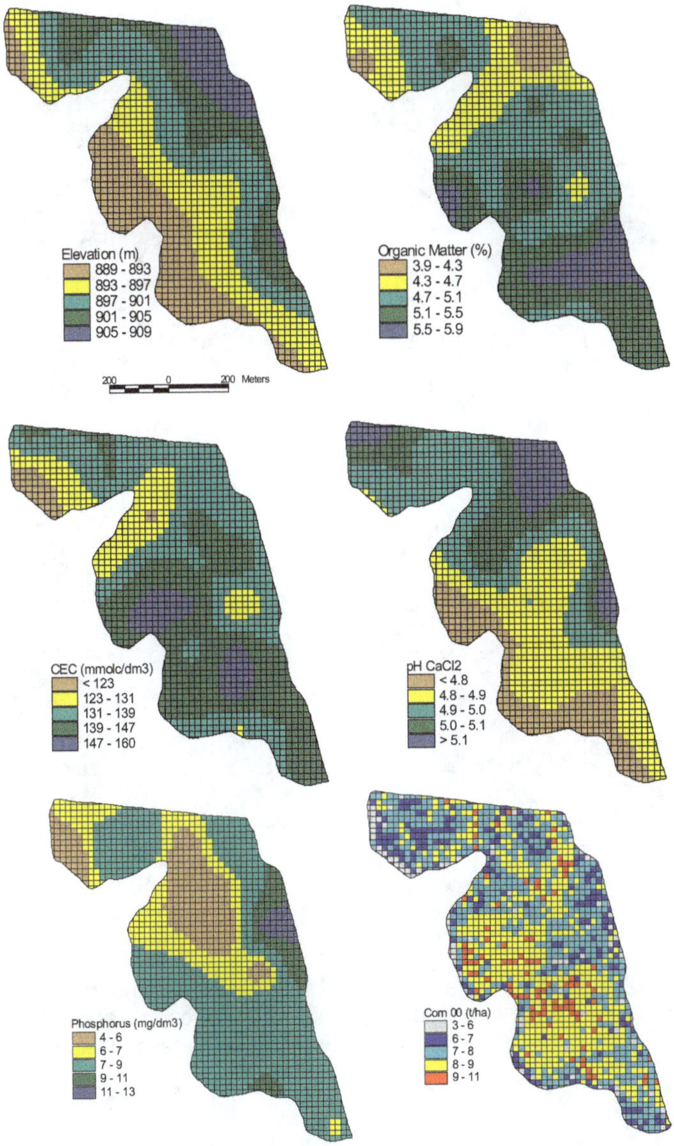


FIGURE 22.2. Soil topographic, physical, chemical, and productivity attributes of a commercial field with a rotation of grain crops under no-tillage; coordinates 25°44'S and 51°54'W, State of Parana, Brazil. (See corresponding PLATE 22.2 in the central gallery.)

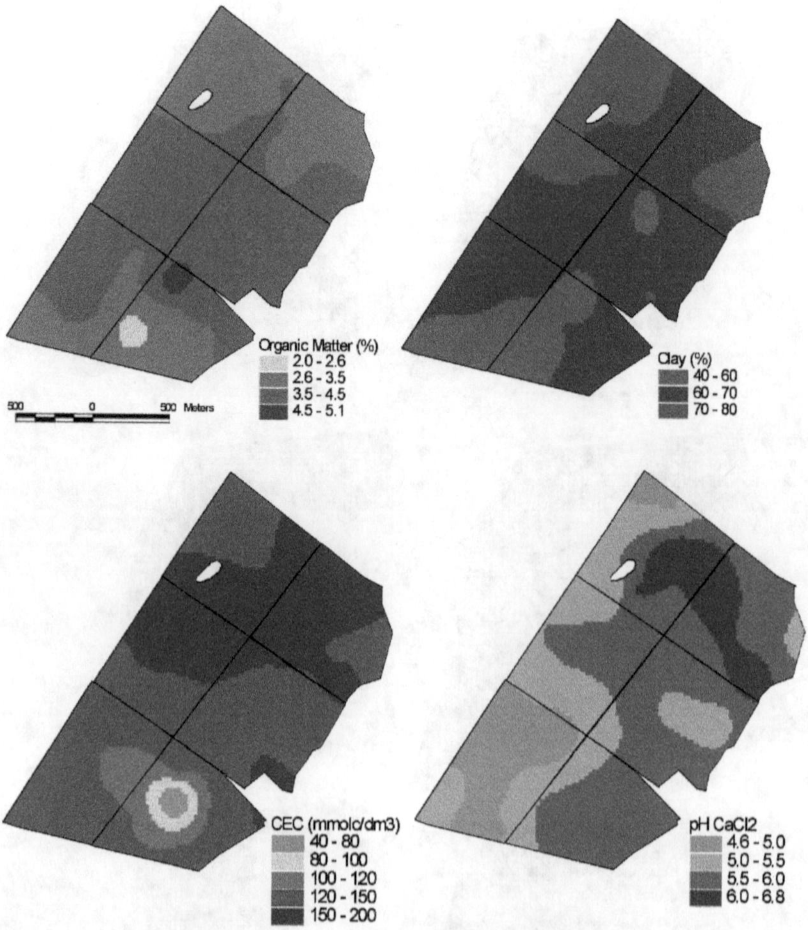


FIGURE 22.3. Soil physical and chemical attributes of a commercial field with a rotation of grain crops and pasture under no-tillage; coordinates 21°46'S and 54°24'W, State of Mato Grosso do Sul, Brazil. (See corresponding PLATE 22.3 in the central gallery.)

and often within a field. Cases such as that of Figure 22.2 (Table 22.4), though, have soils derived from basalt, with high clay content (about 65 to 70 percent) and little variation across the field, and usually are not analyzed for this attribute. Clay mineralogy is most often oxidic, but commonly includes noticeable amounts of kaolinite. Organic matter content is also

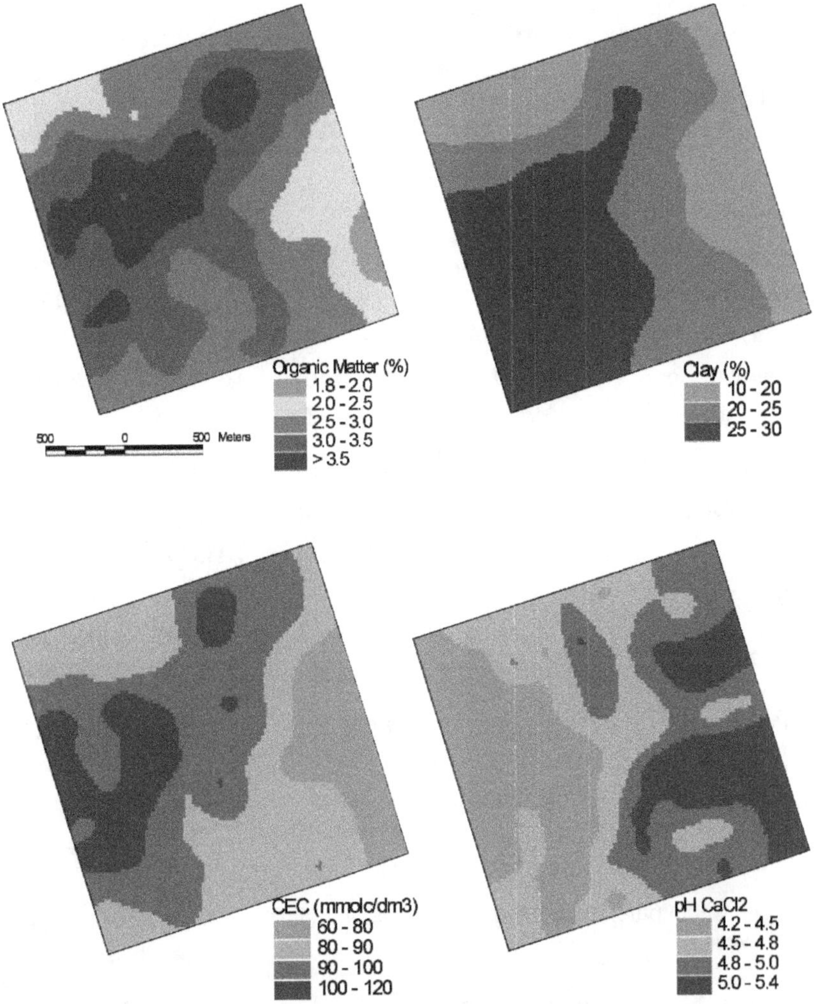


FIGURE 22.4. Soil physical and chemical attributes of a commercial field with a rotation of grain crops under no-tillage; coordinates 8°38'S and 46°10'W, State of Piaui, Brazil. (See corresponding PLATE 22.4 in the central gallery.)

highly variable across regions because of the effects of climate, particle size distribution, and past management. It is important to note that all fields used as examples here are currently under no tillage. Organic matter variability within fields is due to variations in particle-size distribution and effects of

TABLE 22.4. Means and coefficients of variation obtained from grid sampling for selected soil attributes from fields described in Figures 22.1 to 22.4.

| Characteristics (details) | Figure 22.1 | Figure 22.2 | Figure 22.3 | Figure 22.4 |
|---|---|---|---|---|
| Area (ha) | 33.3 | 56.8 | 291.8 | 402.8 |
| Latitude (S) | 24°54' | 25°44' | 21°46' | 8°38' |
| Longitude (W) | 50°17' | 51°54' | 54°24' | 46°10' |
| Clay (percent) | 14 (0.24)[a] | —[b] | 66 (0.11) | 21 (0.21) |
| Sand | 76 (0.06) | — | 18 (0.37) | 78 (0.06) |
| Organic matter | 1.8 (0.11) | 5.2 (0.10) | 3.6 (0.16) | 2.3 (0.23) |
| Base saturation | 36 (0.21) | 48 (0.11) | 71 (0.17) | 35 (0.22) |
| pH $CaCl_2$ (0.1 M) | 4.6 (0.04) | 4.9 (0.03) | 5.6 (0.09) | 4.6 (0.07) |
| K ($mmol_c \cdot dm^{-3}$) | 2.1 (0.19) | 2.2 (0.14) | 9.8 (0.41) | 1.3 (0.34) |
| Ca | 11 (0.30) | 47 (0.12) | 72 (0.37) | 21 (0.25) |
| Mg | 10 (0.23) | 17 (0.16) | 24 (0.31) | 6 (0.28) |
| CEC | 65 (0.09) | 138 (0.07) | 105 (0.31) | 83 (0.16) |
| Al | 4.6 (0.32) | 0.5 (1.46) | 0.0 (0.00) | 2.0 (0.90) |
| P (Mehlich) ($mg \cdot dm^{-3}$) | 16.2 (1.07) | 7.9 (0.24) | 8.1 (0.84) | 26.6 (0.45) |
| Corn ($t \cdot ha^{-1}$) | 8.3 (0.19) | 9.2 (0.18) | — | — |
| Soybeans | 3.0 (0.23) | — | — | — |

[a]Values within parentheses are coefficients of variation.
[b]Not available.

past erosion, especially for the fields in Figures 22.1 and 22.2, which have steeper slopes. Fields in Figures 22.3 and 22.4, on the other hand, are essentially flat, with slopes only up to 3 percent. Cation exchange capacity (CEC) is controlled mostly by organic matter content, and secondarily by clay content and mineralogy, as typical of highly weathered tropical soils.

Most chemical attributes, although partially controlled by other soil and landscape attributes, are relatively more influenced by past management. Their variability, therefore, is usually an expression of the compounded effects due to natural and management causes. For instance, K levels for the examples are mostly below ideal (3 $mmol_c k \cdot dm^{-3}$),[1] but their coefficients of variation are higher than most other attributes. Similarly, two of the fields have P levels below ideal (12 $mgp \cdot dm^{-3}$)[2] and high coefficients of variability. Part of this variability is due to decades of fixed-rate fertilizer applications, based on a composite soil sample, but variable extraction by the crops. Yield variability, though, is not as high. In our research and consulting work we have found coefficients of variation only as high as 0.30 (Roloff et al., 2000).

Similar to other countries, soil sampling strategy is a source of disagreement among PA practitioners and researchers in Brazil. The two main strategies discussed are zone and grid sampling. Mainly the machinery industry people, remote-sensing experts, and agricultural engineers, favor zone sampling (M. Moore, AGCO Corp., personal communication, 2001). Grid sampling is defended mainly by soil experts (Demattê, 2001), including ourselves, and service providers. We summarized the positive and negative aspects as we see them of both approaches in Table 22.5. We believe that major differences in Brazilian soils, sampling windows, and cost structure exist in relation to other countries, which justify the choice of grid sampling over zone sampling, especially when starting a PA system. The concept of

TABLE 22.5. Positive and negative aspects of zone versus grid soil sampling in Brazil.

| | Positive | Negative |
|---|---|---|
| Zone | • Lower sampling and analysis costs<br>• Better related to the economics (yield) | • Few and expensive yield monitors<br>• Eletroconductivity zoning not thoroughly tested and equipment expensive<br>• Inability to spatially separate multiple attributes related to one low yield zone<br>• No established criteria for zonation<br>• Several years of yield data for a reliable zonation |
| Grid | • Provides single-attribute variability across the field<br>• No preconceived cause-and-effect relations<br>• Allows short-term decision and action | • Higher sampling and analysis costs (about 20 to 30 percent)<br>• No established criteria for grid size and architecture |

zone sampling emanated from North America and Europe where (1) often grid spacing needs to be less than 1 ha to detect small-scale but important soil variability, associated to complex and relatively young landscapes, and (2) costs of sampling and analysis are relatively high (Table 22.3). Therefore, zone sampling is a much lower cost alternative to generate variable rate recommendation maps. In Brazil, on the other hand, landscapes are very old, commonly with long slope lengths, gentle slopes, and low drainage density. These result in soil and landscape variability on a larger scale, which can be appropriately described by grid spacing more than 1 ha. A more scientific determination of ideal grid spacing, though, is hindered by results that show little or no consistency among semi-variograms for different attributes within the same field (e.g., Couto et al., 1999; Marques et al., 2000), while even less among different fields. In our experience, we concluded that a 2 ha grid for Oxisols in southern Brazil is a good compromise between appropriate variability representation and costs. Other service providers set grid spacing according to the sampling and analysis costs in relation to crop value. For example, a 1 or 2 ha grid is used for sugarcane; grain and cotton a 5 ha grid; and pastures a 10 ha grid. Even though this may be construed as merely commercial pragmatism, this sampling strategy is certainly an improvement over the one composite sample per field of traditional agriculture, and its true representation of actual variability is aided by the usual large-scale variation of soil and landscape attributes. A further important argument for grid sampling is the immediate detection of pH and nutrient imbalances that can then be corrected in the short term as opposed to waiting several crops to develop stable sampling zones based on yield or vegetative index mapping. On the other hand, zones determined by inductive eletromagnetism were correlated with soil attributes, but their correlation with yield is yet to be established (Roloff et al., 2001).

Establishing cause-and-effect relationships between yield and field attributes in Brazil is difficult and complex. Yield monitors using grain mass plate sensors work relatively well for most grains but are influenced by operator ability, combine design and maintenance, and landscape. In addition, dry beans and soybeans often complicate mass and moisture sensors by clogging or contaminating them with soil and/or green stems. All these impart a degree of variability to the yield data in addition to field variability, therefore it is necessary to filter and aggregate yield data before studying trends. We usually filter the data by establishing reasonable minimum and maximum values and eliminating values that extrapolate these. Calculating average values for a circular zone centered at the soil-sampling point aggregates the yield data and eliminates short-distance noise. Our most common way of soil sampling is to take ten subsamples on a 10 m radius around the grid cell center to produce one composite sample per grid cell. Relations are then sought using parametric statistical methods, mainly correlation matrices, multiple linear regression, and regression trees. Such analysis is further complicated by the year-to-year variability, mostly associated with climate. Therefore, one must rely on scouting reports and farmer's observations to piece together the yield puzzle.

The analysis process usually starts with calculating a regression matrix. Correlation coefficients between yield, soil, and landscape attributes are usually low, suggesting that several attributes are controlling yield or the type of analysis is not appropriate. Several common findings from our experimental and commercial practice related to factors affecting yield are illustrated by the example on Table 22.6 and include the following:

1. One or more attributes affected by liming (pH, Ca) display the highest correlation coefficients
2. Nonmanageable attributes (clay, organic matter) display high coefficients, suggesting possible candidates for a zoning criterium
3. Bulk density is somewhat correlated with yield, but also and more strongly with sand or clay content
4. Negative correlation between phosphorus content and yield

This last item is not as frequent as the others but it has been noticed in several fields associated with liming-related attributes whose values are below minimum. We believe this is caused by phosphorus accumulation on lower yield potential zones and vice versa because of continuous fixed rate phosphorus fertilization but variable crop extraction. In addition to the examples listed on Table 22.5 we have investigated crop features, landscape parameters, micronutrients, weeds, pests, and diseases, but with little success in describing grain yield variability thus far. The lack of correlation between yield and weeds, pests, and diseases is justified by the farmers usually controlling them appropriately. The use of multiple linear regression of the aggregated yield data versus field attributes has allowed us to explain up to 78 percent in a study of wheat yield variability, but most often $R^2$ values are lower than 0.5.

The use of regression trees (Wilkinson, 1997) is another, perhaps more robust, way to search for relations between yield and attributes. Similar to the example on Figure 22.5, we commonly find proportional reduction in error (PRE)[3] values that explain a greater proportion of yield variability than multiple linear regression. In this example, liming-affected parameters appear on the highest split and phosphorus level on the second highest split for both crops and showing a direct relation. This suggests that liming and corrective phosphorus variable rate applications are necessary to maximize

TABLE 22.6. Correlation matrix of soil attributes and 1999 wheat yield for a 46.7 ha field in Southern Brazil (24°34' S, 50°00' W).

| | Yield | Elevation | Sand | Clay | pH | Al | Ca | K | CEC[a] | P | OM[a] | BS[a] | BD[a] |
|---|---|---|---|---|---|---|---|---|---|---|---|---|---|
| Yield | 1.00 | | | | | | | | | | | | |
| Elev | 0.55 | 1.00 | | | | | | | | | | | |
| Sand | −0.31 | −0.14 | 1.00 | | | | | | | | | | |
| Clay | 0.30 | 0.28 | −0.54 | 1.00 | | | | | | | | | |
| pH | 0.45 | 0.62 | −0.11 | 0.25 | 1.00 | | | | | | | | |
| Al | −0.32 | −0.42 | 0.11 | −0.31 | −0.79 | 1.00 | | | | | | | |
| Ca | 0.36 | 0.35 | −0.34 | 0.28 | 0.60 | −0.59 | 1.00 | | | | | | |
| K | 0.11 | −0.03 | −0.33 | −0.01 | −0.20 | 0.19 | 0.10 | 1.00 | | | | | |
| CEC | 0.09 | −0.20 | −0.47 | 0.14 | −0.32 | 0.17 | 0.46 | 0.47 | 1.00 | | | | |
| P | −0.22 | −0.41 | 0.05 | −0.56 | −0.47 | 0.52 | −0.24 | 0.34 | 0.23 | 1.00 | | | |
| OM | 0.44 | 0.37 | −0.59 | 0.50 | 0.08 | −0.09 | 0.41 | 0.55 | 0.56 | −0.20 | 1.00 | | |
| BS | 0.34 | 0.48 | −0.13 | 0.24 | 0.89 | −0.82 | 0.81 | −0.18 | −0.08 | −0.39 | 0.08 | 1.00 | |
| BD | −0.33 | −0.09 | 0.64 | −0.39 | 0.19 | −0.25 | 0.02 | −0.42 | −0.35 | −0.02 | −0.52 | 0.26 | 1.00 |

Multiple linear regression: Yield = $f$(pH, OM, Elev) $R^2$=0.36, $p < 0.001$

*Note:* Yield values used were averaged per hectare around soil sampling point ($n = 46$).
[a]CEC = cation exchange capacity; OM = organic matter; BS = base saturation; BD = bulk density.

yield potential in that field. The diagrams also suggest that a certain degree of variability exists that will always be present due to nonmanageable attributes, such as clay size distribution and CEC in this case. These attributes could be used as zoning criteria for future zone sampling.

Our experience in relation to attributes connected to yield variability in grain crops thus far suggests that

1. the most frequent soil attributes related to yield are affected by liming, therefore variable rate liming should be the first priority;
2. phosphorus or potassium deficiencies are common but less frequent causes;
3. elevation, particle size distribution, and CEC are the most frequently nonmanageable attributes;

4. Spring wheat and corn yields usually result in better correlations with soil and landscape attributes, whereas soybean yields usually require additional information from late-season scouting to account for lodging; and
5. only scant evidence is available to suggest that compaction may be affecting yields.

Based on these observations and the previous discussion, our suggestion to grain and cotton farmers is to start PA by grid soil sampling to detect below-ideal zones of lime-affected attributes, potassium, and phosphorus, but always including soil particle size distribution and elevation in the survey and noting eventual wet or shallow spots. The last three items would be relevant mostly for southern Brazil. After this base diagnosis is accomplished, corrective variable rate applications of lime and single nutrients should be performed to maximize yield potential across the diverse soil zones. Yield monitoring should then be implemented to allow (1) the generation of potassium and phosphorus replenishment maps based on grain or fiber removal and fertilizer efficiencies; and (2) the establishment of management zones based on relative or stable-unstable yield zones that should be soil sampled every three or four years. Nutrient replenishment, in this case, would be accomplished by variable rate application of a P and K formula at seeding of the next crop, requiring a planter or seeder with variable rate technology. If yield monitoring is not implemented, soil sampling should be done at the same interval, but zone sampling could be based on a nonmanageable soil attribute such as particle-size distribution or CEC.

## *MANAGING VARIABILITY*

The common Brazilian soils, with low pH and low base saturation, require liming about every three years to raise and maintain base saturation in the 50 to 70 percent range while being careful not raise pH $CaCl_2$ over about 6.0 to avoid lowering the availability of micronutrients, especially Zn. Therefore, correct liming is very important to maximize yield potential. On traditional, single-rate applications, lime is actually applied at lower-than-necessary rates on the more highly buffered soil zones because of higher CEC and vice versa. Variable rate liming, on the other hand, is usually calculated using base saturation and CEC, resulting in appropriate rates throughout the field. Extensive data from our research and commercial practice indicate that lime quantities required by variable rate application are often similar to or about 10 to 20 percent lower than single rate. These small savings are not enough to offset the additional sampling and application costs

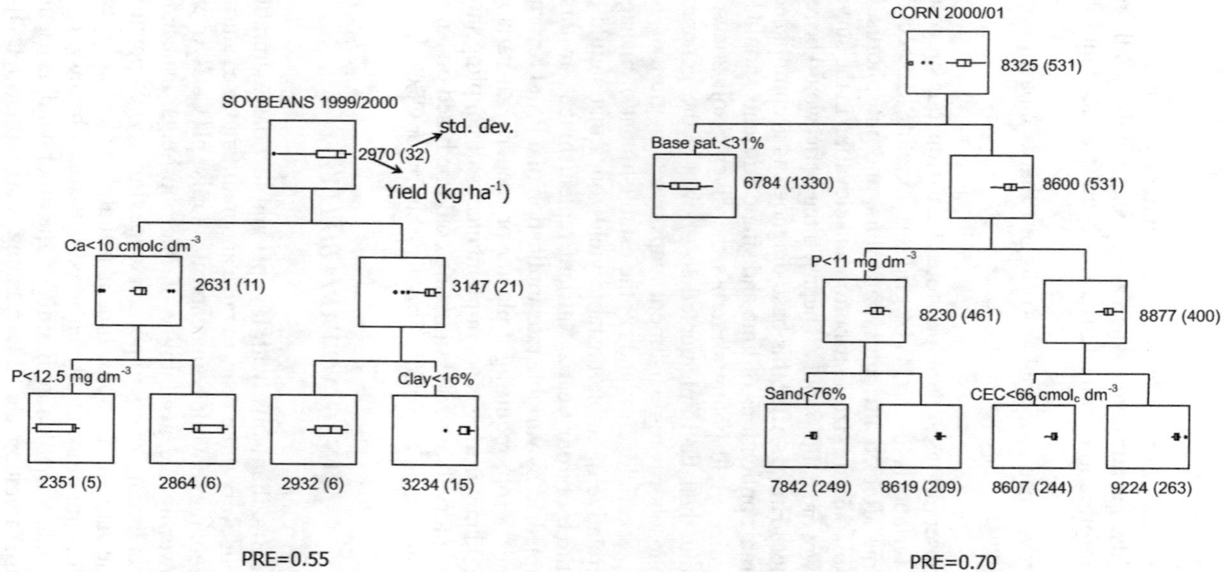


FIGURE 22.5. Tree diagrams of the relations between crop yield and significant soil attributes, for field shown in Figure 22.1. PRE is percent reduction in error. Units for P are mg·dm$^{-3}$, for Ca and CEC are mmol·dm$^{-3}$. (See corresponding PLATE 22.5 in the central gallery.)

associated with variable rate applications because lime is a relatively low-cost input due to plentiful regional supplies. Therefore, the benefits of variable rate liming should be evident only in the medium and long term by augmenting the yield potential of specific soil zones, provided proper variable rate fertilization is also accomplished.

We argue that the importance of proper liming has long been established, as evidenced by essentially all agricultural fields regularly receiving lime applications. Variable rate application then is just an improvement on the way lime is applied, not necessarily requiring extensive testing. Initially, a few large, self-propelled variable rate disk spreaders are available for custom liming on fields that are large with regularly shaped boundaries, such as those typical of the better sugarcane fields in the state of São Paulo or grain and cotton on the savannas. Per uniform distribution these rigs require lime that is of uniform particle size and dry, which is uncommon and thus more expensive than the standard lime. Currently, custom applications and farmers are favoring pulled spinning disks or drop spreaders, with after-market variable rate kits. The majority of farmers willing to try PA, though, are faced with the lack of variable rate spreaders. To overcome this we have developed a simple technique with the following three steps:

1. Lime application maps are produced with only three or four rates, rounded to the nearest metric ton per hectare
2. The application map is imprinted on the field with a low-end navigation GPS linked to a handheld computer with field-mapping software, both located on a tractor dragging a heavy object to scratch the residue on the soil surface, or with a rear mower when the surface has a cover crop, essentially reproducing the application map on a 1:1 scale
3. Lime is spread using a single-rate spreader that applies a given rate on all the same-rate zones, subsequently the spreader setup is changed to the next rate and so on

In southern Brazil, where no tillage with a winter cover crop is prevalent and the landscape is mostly rolling hills, these imprints are easily identifiable by the operators with the help of a printed map. On the flatter savannas some practioners use flags instead. Farmers are increasingly using this technique for their maintenance liming, which is done without incorporation, and applying KCl to bring potassium to ideal levels. Anectodal evidence suggests that this low-cost technique can have a positive impact on yields and profit. Farmers comment that their fields look more uniform after this application.

The liming question can also be used to illustrate one of the difficulties faced by PA in Brazil until recently. Equipment manufacturers claim that they do not believe that such a market for variable rate spreaders exists and that current spinning-disk spreaders do not offer enough commercial attraction to deserve improvement. Farmers and crop consultants who are conscious of the importance of liming claim they cannot work with variable rate because of the lack of appropriate spreaders. Even adapting existing spreaders is seen as dubious because of their poor distribution across the swath. This way, the low cost technique previously described can fulfill an important gap until an appropriate variable rate spreader is acquired after the decision to invest in PA is made.

Nutrients such as N, P, and K are commonly applied at seeding by deep banding. This is especially important for superphosphates on high clay soils to minimize P immobilization. The few variable-rate custom applicators in Brazil, though, use only surface spreaders. This limits them to variably spreading KCl on all soils or NPK formulas on low clay soils. An alternative for the clay-rich soils under no tillage prevalent in southern Brazil is the use of natural soluble phosphates variably spread before the winter crop but aimed at the summer crop. This has been used successfully with fixed rates. Another option involves a variable rate, deep banding equipment for preplant application. This is a "longer term" prospect, though, because the existing North American machine would be expensive, and the development of a local machine is not being currently contemplated.

Limited evidence available in Brazil suggests that variable rate application of fertilizers can result in either more or less quantities used as compared to fixed rates. For instance, Maraschin and Couto (2001) and Harger and Bodnar (2001) indicated reductions in phosphate needs, whereas França et al. (2000) found an increase in the need for potassium. We compared past fixed rate liming and KCl fertilization costs and past yields with real variable rate costs and yields during a 4-year rotation period (2001 to 2004) following adoption of variable rate for three fields in southern Brazil, focusing on corn and soybeans (unpublished data). Overall lime and KCl costs were lower for variable rate because of reduced quantities. Yield increases occurred as well, resulting in a net gain for the farmer of 56 U.S dollars $ha^{-1} \cdot yr^{-1}$ for corn and US\$33 $ha^{-1} \cdot yr^{-1}$ for soybeans. Menegatti et al. (2004) suggested overall fertilization cost reductions of about 40 percent with the use of variable rate fertilization in sugarcane. Therefore, variable rate application should be profitable in most cases for grains and sugarcane because it will likely reduce lime and fertilizer quantities and increase yields.

## PRECISION AGRICULTURE FOR SPECIFIC CROPS

PA in Brazil has been tried and researched mostly for grains, but more recently research and usage has been expanding into the important sugarcane and citrus crops. As we have discussed, PA can increase yield potential and fertilizer-use efficiency with positive effects on profits. This experience does not include variable rate application of agricultural chemicals. Even though agricultural chemicals account for a large proportion of the costs of some crops, such as soybeans, for example (Table 22.1), little work has been done toward reducing them with variable rate applications, probably because intensive crop scouting is not common for soybeans and it would be too costly to do implement. Such costs would only come down when near–real time, remotely sensed crop scouting or real-time sensors become commercially viable. Soybean and corn yields would probably benefit from variable rate seeding to minimize lodging in the case of soybeans and maximize yield potential in corn. Modifying existing seeding or planting equipment for variable rate application is a relatively low cost adaptation that could be easily accomplished with the technology available in Brazil, if the machinery industry invests the needed resources and support.

The relative cost of agriculture chemicals for the cotton crop is also high (Table 22.1). In this case, though, scouting for pests is a common practice and is adaptable to the use of GPS with handheld field computers. Variable rate application of agriculture chemicals could be as simple as *on* or *off*, or it could use sprayers adapted for variable rate. The cost of this adaptation is probably within a favorable cost:benefit ratio. Although no reports of such technique exist, its implementation should be simple. Cotton farmers would also benefit from variable rate liming and fertilization, with probably greater profits than grain farmers.

For the sugarcane crop, fertilization costs are low in relative terms (Table 22.1) because total costs are heavily influenced by mechanization and transportation costs. Nevertheless, we believe potential exists for noticeable profit augmentation through variable rate liming and fertilization. Hydraulic motors already run some modern planters and spreaders used in sugarcane production. Swapping them for electronically controlled motors to do variable rate application should be a relatively inexpensive proposition. However, a short-term study simulating sugarcane yield from variable rate phosphate concluded that no profit gain was made from it (Sparovek and Schnug, 2001). On the other hand, because of the high cost of mechanization and the increased efficiency of longer straight stretches, the same authors concluded that a 3 percent reduction in mechanization costs would result from site-specific erosion control. In their view, PA technology could be used to determine sites within a watershed where erosion levels were low enough to allow the elimination of terraces, resulting in longer straight rows. More recently, growers have become interested in precise guidance systems to avoid trampling of rows during harvest, which causes lower yield of the ratoon crop. Sugarcane yield monitoring is not a commercial reality yet, although a preliminary tests with an ultrasonic volume sensor and a load cell sensor were promising. Two methods have also been successfully tried for yield mapping of hand-cut sugarcane. In any case, the absence of sugarcane yield maps could be compensated by using a surrogate method such as (1) the variation of vegetative index obtained from satellite images or (2) the harvest data from the usual small sub-field sizes (3 to 5 ha) used to monitor harvest mass and quality. Because of the current worldwide interest in ethanol, sugarcane growers comprise the largest group of PA users.

High-volume and high-cost crops such as potato and cassava would certainly have increased profits with variable rate fertilization and liming because they demand relatively large quantities of these inputs. Yield monitoring is problematic because only a handful of mechanical harvesters exist, and they are mainly developed for potatoes. The cost of grid soil sampling, though, is probably small compared with the potential savings. Coffee and fruit tree groves of citrus and apples, for example, should benefit from variable rate liming and fertilization and absorb the added costs of the PA technology. A trial of a coffee yield monitor indicated that the technology is viable, although further development is required (Leal, 2002). Fruit yield monitoring would have to rely on georeferencing of the crates or bags holding the handpicked fruits. A trial system has been successfully tested for oranges (www.gpsglobal.com.br/Artigos/Agricola/Laranja.html).

## *CONCLUSIONS*

Although hard evidence is limited, qualitative analysis indicates several instances of beneficial and profitable uses in Brazil for technologies that are part of the PA system, yet the potential usage and benefits of the PA system in Brazil are far from being realized. Steady and sure adoption will require a concerted effort by industry, educators, and researchers to (1) clearly demonstrate the short- and long-term benefits of PA within regional cropping systems; (2) train the people involved in all levels, from crop consultants to machinery operators; and (3) reduce the cost of the required equipment, either by developing it locally or by reducing importation costs.

Crop yields in Brazil are more dependent on appropriate liming and fertilization because of the highly weathered soils. Therefore, we believe PA is an important tool at this moment to increase the efficiency in nutrient use and preserve or augment farm profitability, which is increasingly important in the face of global competition. In the future we envision PA as a technology to reduce environmental impacts of agricultural chemicals and as the standard field data management system. We also believe the ingenuity of Brazilian farmers and practitioners will help to develop novel and more efficient ways to practice PA.

## NOTES

1. 1.0 $mmol_c\ dm^{-3}$ is approximately 0.1 $cmol_c{\cdot}kg^{-1}$.
2. 1.0 mg $dm^{-3}$ is approximately 1.0 $mg{\cdot}kg^{-1}$.
3. Percent reduction in error: a parameter equivalent to $R^2$.

## REFERENCES


Balastreire, L.A. 1994. Aplicação localizada de insumos—ALI: Um velho conceito novo. In *Congresso Brasileiro de Engenharia Agricola 23.* Campinas, Brazil: UNICAMP.

Balastreire, L.A. 2000. Apresentação. In *O Estado-da-Arte da Agricultura de Precisão no Brasil,* ed. L.A. Balastreire. Piracicaba, Sao Paulo: Brazil: L.A. Balastreire.

Borém, A., M.P. del Giudice, D.M. Queiroz, L.R. Ferreira, F.X.R. do Valle, and R.L. Gomide. (eds.) 2000. *Agricultura de Precisão.* Viçosa, Brazil: A. Borém e outros.

Couto, E.G., A. Stein, and E. Klamt. 1999. Large area spatial variability of soil chemical properties in central Brazil. *Agriculture Ecosystems & Environment,* 66(2):139-152.

Demattê, J.A.M. 2001. O pedólogo e a agricultura de precisão. *Boletim Informativo* (Sociedade Brasileira de Ciência do Solo), 26(1):17-19.

França, G.E., A.C. Oliveira, E.C. Mantovani, C.L.T. Andrade, and I.E. Marriel. 2000. Análise preliminar de mapas de variabilidade espacial da fertilidade de solo. In *Agricultura de Precisão.* ed. A. Borém et al. Viçosa, Brazil: A. Borém e outros, pp. 337-356.

Harger, N. and A. Bodnar. 2001. Variabilidade horizontal e no perfil do solo nos índices de fertilidade em área de plantio direto no norte do Paraná. In *Annals of the XXVIII Congresso Brasileiro de Ciência do Solo,* July 1-6, 2001, Londrina, Brazil. Sociedade Brasileira de Ciência do Solo.

Maraschin, L. and E.G. Couto. 2001. A correção do solo realizada espacialmente: um caminho para uma agricultura de precisão. In *Annals of the XXVIII Congresso Brasileiro de Ciência do Solo,* July 1-6, 2001, Londrina, Brazil. Sociedade Brasileira de Ciência do Solo.

Marques Jr, J., R.B. Sanchez, G.T. Pereira, and J.E. Corá. 2000. Variabilidade espacial de propriedades químicas e físicas de latossolos em áreas de Cerrado sob cutivo de café, em Patrocíno, MG. In *O Estado-da-Arte da Agricultura de Precisão no Brasil,* ed. L.A. Balastreire. Piracicaba, Sao Paulo: Brazil: L.A. Balastreire, pp. 105-112.

Roloff, G., D. Focht, P.S. Marthaus, P.H. Cervi, M.A.D. Fernandes, and R.A. Rodrigues. 2000. Variability in spring wheat yields under no-tillage in Southern Brazil. In *Proceedings of the Fifth International Conference on Precision Agriculture,* Madison, WI: ASA-CSSA-SSSA, CD-ROM.

Roloff, G., J. Hammer, D. Focht, and U. Schmidhalter. 2001. Condutividade do terreno na determinação de zonas de amostragem de solos altamente intemperizados. In *Annals of the XXVIII Congresso Brasileiro de Ciência do Solo,* July 1-6, 2001, Londrina, Brazil. Sociedade Brasileira de Ciência do Solo.

Sparovek, G. and E. Schnug. 2001. Soil tillage and precision agriculture: A theoretical case study for soil erosion control in Brazilian sugar cane production. *Soil & Tillage Research,* 61: 47-54.

Vieira, S.R., I.C. De Maria, O.M. Castro, S.C.F Dechen, and F. Lombardi Neto. 1983. Geostatistical theory and application to variability of some agronomic properties. *Hilgardia,* 51:1-75.

Wilkinson, L. 1997. Classification and regression trees. In *SYSTAT 7.0: New statistics.* Chicago: SPSS Inc., pp. 13-30.

# Index

Page numbers followed by the letter "b" indicate boxed material; those followed by the letter "i" indicate illustrations; and those followed by the letter "t" indicate tables.





doi:10.1300/5627_23

www.ingramcontent.com/pod-product-compliance
Lightning Source LLC
LaVergne TN
LVHW021040011026
849160LV00007BA/214
* 9 7 8 1 5 6 0 2 2 9 5 5 1 *